OXFORD ENGINEERING SCIENCE SERIES

SERIES EDITORS

THE OXFORD ENGINEERING SCIENCE SERIES

---

1 D. R. RHODES: *Synthesis of planar antenna sources*
2 L. C. WOODS: *Thermodynamics of fluid systems*
3 R. N. FRANKLIN: *Plasma phenomena in gas discharges*
4 G. A. BIRD: *Molecular gas dynamics*
5 H.-G. UNGER: *Planar optical waveguides and fibres*
6 J. HEYMAN: *Equilibrium of shell structures*
7 K. H. HUNT: *Kinematic geometry of mechanisms*
8 D. S. JONES: *Methods in electromagnetic wave propagation (Two volumes)*
9 W. D. MARSH: *Economics of electric utility power generation*
10 P. HAGEDORN: *Non-linear oscillations (Second edition)*
11 R. HILL: *Mathematical theory of plasticity*
12 D. J. DAWE: *Matrix and finite element displacement analysis of structures*
13 N. W. MURRAY: *Introduction to the theory of thin-walled structures*
14 R. I. TANNER: *Engineering rheology*
15 M. F. KANNINEN and C. H. POPELAR: *Advanced fracture mechanics*
16 R. H.T. BATES and M. J. McDONNELL: *Image restoration and reconstruction*
18 K. HUSEYIN: *Multiple-parameter stability theory and its applications*
19 R. N. BRACEWELL: *The Hartley transform*
20 J. WESSON: *Tokamaks*
21 P. B. WHALLEY: *Boiling, condensation, and gas-liquid flow*
22 C. SAMSON, M. LeBORGNE, and B. ESPIAU: *Robot control: the task function approach*
23 H. J. RAMM: *Fluid dynamics for the study of transonic flow*
24 R. R. A. SYMS: *Practical volume holography*
25 W. D. McCOMB: *The physics of fluid turbulence*
26 Z. P. BAZANT and L. CEDOLIN: *Stability of structures: principles of elastic, inelastic, and damage theories*
27 J. D. THORNTON: *Science and practice of liquid–liquid extraction (Two volumes)*
28 J. VAN BLADEL: *Singular electromagnetic fields and sources*

Science and Practice of Liquid–Liquid Extraction

Volume 1

Phase Equilibria; Mass Transfer and Interfacial Phenomena; Extractor Hydrodynamics, Selection, and Design

JOHN D. THORNTON

Emeritus Professor of Chemical Engineering, University of Newcastle upon Tyne

CLARENDON PRESS • OXFORD

This book has been printed digitally and produced in a standard specification in order to ensure its continuing availability

OXFORD
UNIVERSITY PRESS

Great Clarendon Street, Oxford OX2 6DP

Oxford University Press is a department of the University of Oxford.
It furthers the University's objective of excellence in research, scholarship, and education by publishing worldwide in

Oxford New York

Auckland Cape Town Dar es Salaam Hong Kong Karachi
Kuala Lumpur Madrid Melbourne Mexico City Nairobi
New Delhi Shanghai Taipei Toronto

With offices in

Argentina Austria Brazil Chile Czech Republic France Greece
Guatemala Hungary Italy Japan South Korea Poland Portugal
Singapore Switzerland Thailand Turkey Ukraine Vietnam

Oxford is a registered trade mark of Oxford University Press in the UK and in certain other countries

Published in the United States
by Oxford University Press Inc., New York

Reprinted 2007

ISBN 978-0-19-856236-8

TO

THE LATE F. H. GARNER OBE

One time Professor of Chemical Engineering
and Dean of the Faculty of Science
University of Birmingham

PREFACE

VOLUMES 1 AND 2

Since the adoption of liquid extraction as a separative procedure towards the end of the last century, many industries have employed this technique as a means of separating the constituents of liquid mixtures. In the early days, the petroleum industry used extraction as a means of refining kerosenes and, from this beginning, developed a wide range of other extraction operations for hydrocarbon processing.

The use of liquid extraction for separating metal ions in solution is a later development which received great impetus from the need of the nuclear industry to extract uranium ores, to reprocess spent reactor fuels, and to purify other materials of nuclear importance. The reprocessing of nuclear fuels presented unprecedented problems not only in the field of extraction chemistry but also in the areas of contactor design and engineering. These problems were solved and, as a result, new advances were made in the technology of the process as a whole. Today, extraction provides viable procedures for separating and purifying many non-nuclear metal systems, and has, for example, been proposed for the production of high-purity materials for electronic applications.

Just as the advent of the nuclear age catalysed developments in both the theory and practice of extraction, so the emergence of the biologically based industries is generating new requirements for novel extractants as well as short residence time contactors capable of processing a wide range of fermentation liquors with a minimum of solute degradation.

The present volumes are an attempt to bring together some of the more important developments of recent years and to present them in a way that might be helpful to those engaged in research in the extraction field, as well as to engineers and chemists concerned with the industrial aspects of extraction processes.

The presentation assumes a working knowledge of the subject and is therefore not designed as an undergraduate text, although it is hoped that many post-graduate students will find much that is useful to them. The chapters have been contributed by writers with a first-hand knowledge of their respective fields and each is supplemented by references to the published literature.

Any work of this kind involves an element of selection if the text is not to become unduly long and expensive. I have therefore restricted the material to those topics which illustrate either the fundamental ideas behind the subject or new requirements and developments at the process level.

Wherever possible, the treatment of the various topics has been as fundamental as current knowledge allows with a minimum of empiricism.

For the convenience of readers, the work has been divided into two volumes. The first is primarily concerned with the fundamental aspects of extraction leading up to extractor selection and design while the second volume is devoted to process chemistry and extraction operations in a number of selected industries where new developments are rapidly emerging.

Volume 1 contains much that is likely to be of interest to research workers and design engineers and opens with a discussion of the measurement, correlation, and prediction of equilibrium data from a thermodynamic standpoint. Although vapour–liquid equilibria have long been interpreted in terms of activity coefficients, the same cannot be said of liquid–liquid systems and it is hoped that this presentation will help to stimulate renewed interest in a more fundamental approach to the interpretation of liquid–liquid equilibrium measurements rather than a continued reliance upon essentially empirical procedures.

The subject of mass transfer has been dealt with at some length and the fundamental ideas behind the diffusion of a solute across a liquid-liquid interface are first described in detail for 'simple' systems in which complicating factors such as interfacial turbulence are absent. It is not always appreciated that the transfer of certain solutes across a liquid–liquid interface can set up intense turbulence in the vicinity of the phase boundary with the result that mass transfer coefficients calculated purely on the basis of molecular diffusion can considerably underestimate the observed transfer rates.

It might be thought that such effects are a bonus in so far as actual transfer rates will be higher than those predicted by diffusion theory. Unfortunately this is not necessarily true since a concomitant effect is enhanced droplet coalescence rates which in turn result in a smaller interfacial area for solute transfer. Thus whilst the mass transfer coefficient may be enhanced, the interfacial area may be reduced to an extent that is not yet quantifiable so that the overall effect remains uncertain at the present time. The nature of these so-called Marangoni effects has therefore been dealt with in some detail and possible mechanisms by which they can influence the mass transfer rates are discussed. Although the prediction of interdroplet coalescence rates can be notoriously unreliable when traces of surface active materials are present, an account of the underlying theories of droplet coalescence in the absence of surfactants has been included with particular reference to coalescence mechanisms and their application to settler design.

The material dealing with contactor design has been divided into three separate chapters. Thus Chapter 6 deals with the analysis of contactor performance based upon plug flow considerations, whereas the more complex situation involving axial mixing is dealt with in Chapter 7. Here we are concerned with methods of modelling axial mixing together with appropriate methods of solving the equations involved. Finally, in Chapter 8, the selection, design, and scale-up of contactors is discussed with reference to a range of typical units.

Volume 2 is concerned with the process chemistry and extraction operations carried out in a number of key industries. These are industries which either demonstrate novel extraction technology or else have generated a need for unconventional extractants and/or high efficiency contactors with unusually short residence times.

The applications of extraction in the hydrometallurgical field are described in detail for a wide range of metal systems including the extraction of uranium from ores. The more circumscribed use of liquid extraction in a radioactive environment where special requirements prevail has been dealt with separately in the context of nuclear fuel reprocessing.

The extraction of complex labile molecules poses a number of problems to both the pharmaceutical and the foodstuffs industries where there is frequently a need for new types of extractants which are compatible with the solutes to be extracted coupled with high efficiency low residence time contactors in order to minimize solute degradation. Thus, for example, the possible use of two-phase aqueous-aqueous systems and supercritical extraction may well offer attractive process routes in the foreseeable future. The potential of such techniques has been examined, as have the future requirements for extractants and equipment in the biologically based industries.

The volume concludes with an account of some current developments in the extraction field including supercritical extraction, liquid membranes, biochemical processes, and electrically enhanced extraction.

When this work was first envisaged, I had the task of persuading the many contributors to give of their time and undertake the many hours of labour involved in the preparation of each chapter. This they all agreed to do and I should like to record my appreciation and thanks to each and every author for so generously agreeing to be part of this project.

For my part, I must accept the responsibility for the contents of the two volumes and if my readers feel that I have omitted anything of moment, I can only plead that I have had to bear production costs in mind continually and therefore set a limit to the ultimate number of pages if the volumes were not to become prohibitively expensive.

Finally, I should like to thank Oxford University Press for all their help in the production of this work and for so kindly producing the Index to each volume.

Worcestershire J.D.T.
August 1991

CONTENTS

VOLUME 1

CONTRIBUTORS TO VOLUME 1 xvii

CONTENTS OF VOLUME 2 xviii

1 LIQUID-LIQUID EQUILIBRIA 1
D. M. T. Newsham
1.1 Introduction 1
1.1.1 The phase rule 1
1.1.2 Phase diagrams 1
1.2 Thermodynamics of equilibrium 5
1.2.1 Conditions for equilibrium 5
1.2.2 Activities and activity coefficients 6
1.2.3 Temperature and pressure dependence of activity 6
1.3 Correlation and prediction of activity coefficients 7
1.3.1 The Gibbs–Duhem equation 7
1.3.2 Correlating equations 8
1.3.3 Predictive equations 10
1.3.4 Correlations for electrolyte solutions 13
1.3.5 Ion interactions and association equilibria 17
1.4 Prediction and computation of equilibria 18
1.4.1 Selection of activity coefficient correlation 18
1.4.2 Computation of equilibrium compositions 19
1.5 Measurement of liquid–liquid equilibria 20
1.5.1 Cell equilibration method 20
1.5.2 Continuous measurements 22
1.5.3 The titration method 22
1.5.4 Graphical representation of data 24
1.5.5 Correlation of tie-lines 27
1.6 Choice of solvent 27
1.6.1 General 27
1.6.2 Empirical methods 28
1.6.3 Methods based on the regular solution model 28
1.6.4 Methods based on the UNIFAC model 31
1.7 Applications 32
1.7.1 Organic systems 32
1.7.2 Ionic systems 32
1.8 Sources of data 34
Acknowledgement 35
Nomenclature 35
References 38

2 INTERPHASE MASS TRANSFER 40
A. H. P. Skelland
2.1 Diffusion in liquid systems 40
2.1.1 Flux relations and Fick's first law 40
2.1.2 Unsteady-state diffusion 46
2.1.3 Molecular diffusivities in liquids 49
2.2 Mass transfer coefficients and mechanisms of mass transfer 55
2.2.1 Individual and overall coefficients 55
2.2.2 The two-film theory 59
2.2.3 The penetration theory 60
2.2.4 The theory of penetration with varied surface renewal 62
2.2.5 The theory of interfacial boundary layers in immiscible, cocurrent, laminar streams 66
2.2.6 Other theories 72
2.3 Mass transfer associated with jet phenomena 75
2.3.1 Hydrodynamic outline 75
2.3.2 Mass transfer inside the jet 82
2.4 Mass transfer associated with drop phenomena 87
2.4.1 Hydrodynamic outline 87
2.4.2 Mass transfer during drop formation 95
2.4.3 Mass transfer in and around detached drops 99
2.4.4 Mass transfer during drop coalescence 124
2.5 Selection of the disperse phase in spray and plate columns 126
2.5.1 Formulation of a criterion for greater extraction 126
2.5.2 Evaluation of the criterion 127
2.5.3 Application 130
2.6 Stage efficiencies and transfer units 131
2.7 Mass transfer with chemical reaction 135
Nomenclature 140
References 147

3 MARANGONI PHENOMENA 157
E. S. Perez de Ortiz
3.1 Introduction 157
3.2 Mechanisms of interfacial convection 159
3.3 Types of Marangoni phenomena 162
3.4 Mathematical analysis of instability 163
3.4.1 General boundary conditions 166
3.5 Applications of the stability analysis to the derivation of stability criteria 167
3.5.1 Transfer of a single solute across a flat interface 167
3.5.2 Mass transfer with heat effects: binary liquid–liquid systems 172
3.5.3 Mass transfer with chemical reaction 179
3.5.4 Theoretical studies 180
3.5.5 Mass transfer in an electric field 182

3.6 Effect of Marangoni phenomena on specific interfacial area 183
3.6.1 Break-up of liquid jets 183
3.6.2 Drop size from jet break-up 190
3.6.3 Drop coalescence 193
3.6.4 Drop drag coefficients 194
3.7 Effect of Marangoni instabilities on mass transfer coefficients 195
3.8 Interactions between Marangoni phenomena and surface renewal in the rate of mass transfer to drops 198
Nomenclature 204
References 205

4 MASS TRANSFER COEFFICIENTS, INTERFACIAL EFFECTS, AND SURFACE RENEWAL 210
K. H. Javed
4.1 Introduction 210
4.2 Basic transport models 210
4.2.1 Two-film model 210
4.2.2 Penetration model 211
4.2.3 Surface renewal model 211
4.2.4 Film-penetration model 212
4.3 Modified surface renewal models 213
4.4 Turbulent diffusion models 214
4.5 Mass transfer to droplets 223
4.5.1 Droplet models 223
4.5.2 Partial mass transfer coefficients 227
4.6 Mass transfer and interfacial effects 230
Nomenclature 239
References 240

5 MIXING AND COALESCENCE PHENOMENA IN LIQUID–LIQUID SYSTEMS 244
G. A. Davies
5.1 Introduction 244
5.2 Liquid–liquid mixing correlations for drop size 247
5.3 The ambivalence range and phase inversion in liquid–liquid systems 252
5.4 Separation of liquid dispersions 255
5.4.1 Introduction 255
5.4.2 Coalescence of single drops 257
5.4.3 Analysis of film drainage 270
5.4.4 Coalescence and dispersions 274
5.4.5 Application to the design of industrial settlers 306
5.4.6 Phase entrainment in mixer–settlers 316
Nomenclature 334
References 337

INTRODUCTION TO CHAPTERS 6, 7, AND 8 343
H. R. C. Pratt and G. W. Stevens
Design procedures for countercurrent extractors 343
Types of operation and equipment 343
Basis of modelling procedures 344

6 PLUG FLOW ANALYSIS 345
G. W. Stevens and H. R. C. Pratt
6.1 Introduction 345
6.1.1 General 345
6.1.2 Concentration units 345
6.1.3 Graphical representation of equilibrium data 348
6.2 Single and multiple contact extraction 350
6.2.1 Single-stage extraction 350
6.2.2 Multistage crossflow extraction 352
6.3 Simple countercurrent extraction (I) ideal stages 354
6.3.1 Graphical methods 354
6.3.2 Analytical solutions 358
6.3.3 Allowance for operating line curvature 360
6.4 Stage efficiency 362
6.4.1 General 362
6.4.2 Overall efficiency 362
6.4.3 Murphree efficiencies 362
6.4.4 Hausen efficiency 364
6.5 Simple countercurrent extraction (II) 365
6.5.1 Non-ideal stages 365
6.5.2 Differential extractors 367
6.5.3 Nonlinear equilibria 379
6.6 Countercurrent extraction with reflux (single-solvent fractional extraction) 379
6.6.1 Process description 379
6.6.2 Graphical solution 379
6.7 Dual-solvent fractional extraction 381
6.7.1 Process description 381
6.7.2 Graphical solution 383
6.7.3 Material balances and limiting flows 389
6.7.4 Analytical solutions 392
6.7.5 Numerical methods 394
6.8 Effect of chemical reaction 395
6.8.1 Mass transfer with homogeneous reaction 395
6.8.2 Mass transfer with heterogeneous reaction 397
6.9 Multicomponent separations 398
6.9.1 Types of operation 398
6.9.2 Crossflow extraction 399
6.9.3 Countercurrent extraction 400
6.9.4 Dual-solvent systems 402

6.10	Batch processes	403
	6.10.1 Single and multiple-contact batch extraction	403
	6.10.2 Simulation of countercurrent extraction	403
	6.10.3 Countercurrent distribution	405
	6.10.4 Differential extraction	408
	Nomenclature	411
	References	414
7	AXIAL DISPERSION	416
	H. R. C. Pratt and G. W. Stevens	
7.1	Introduction	416
	7.1.1 Nature of axial dispersion	416
	7.1.2 Axial dispersion models	417
7.2	Development of models	418
	7.2.1 Diffusion model	418
	7.2.2 Backflow model	428
	7.2.3 Comparison of models	433
	7.2.4 Special solutions	435
	7.2.5 Generalized solutions	437
7.3	Graphical solution methods	448
	7.3.1 Differential columns	448
	7.3.2 Stagewise contactors	454
7.4	Numerical solution methods	457
	7.4.1 Stage-to-stage calculation	458
	7.4.2 Unsteady state method	459
	7.4.3 Matrix solution	460
	7.4.4 Dual-solvent extraction	468
7.5	Axial dispersion and polydispersivity	470
	7.5.1 General	470
	7.5.2 Polydispersivity and forward mixing	471
7.6	Measurements of axial dispersion and mass transfer parameters	474
	7.6.1 Determination from measured concentration profiles	474
	7.6.2 Tracer injection methods	477
	7.6.3 Discussion	484
	Nomenclature	485
	References	488
8	SELECTION, DESIGN, PILOT-TESTING, AND SCALE-UP OF EXTRACTION EQUIPMENT	492
	H. R. C. Pratt and G. W. Stevens	
8.1	Selection of extraction equipment	492
	8.1.1 Extractor types	492
	8.1.2 Design requirements	492
	8.1.3 Extractor selection chart	499

8.1.4 Examples of use of selection chart 500
8.2 Basis of extractor design 503
8.2.1 Introduction 503
8.2.2 Hydrodynamics of liquid extraction equipment 503
8.2.3 Mass transfer coefficients 512
8.2.4 Fundamental design procedures 513
8.3 Pilot testing and scale-up: columns and related types 514
8.3.1 General 514
8.3.2 Pilot test procedure 514
8.3.3 Scale-up procedures 516
8.3.4 Specific procedures: gravity-separated extractors 517
8.4 Pilot testing and scale-up: mixer–settlers (Class Ac3) 573
8.4.1 General 573
8.4.2 Pilot-scale testing 575
8.4.3 Scale-up procedure 578
8.5 Centrifugally-separated extractors 579
8.5.1 Continuous-contact units 579
8.5.2 Mixer–settler type 580
Nomenclature 580
References 583
Index 591

CONTRIBUTORS

VOLUME 1

G. A. DAVIES	Department of Chemical Engineering, University of Manchester Institute of Science and Technology, Manchester, UK
K. H. JAVED	Department of Chemical Engineering, University of Qatar, Doha, Qatar
D. M. T. NEWSHAM	Department of Chemical Engineering, University of Manchester Institute of Science and Technology, Manchester, UK
E. S. PEREZ DE ORTIZ	Department of Chemical Engineering and Chemical Technology, Imperial College of Science, Technology and Medicine, London
H. R. C. PRATT	Department of Chemical Engineering, University of Melbourne, Parkville, Victoria 3052, Australia
A. H. P. SKELLAND	School of Chemical Engineering, The Georgia Institute of Technology, Atlanta, Georgia, USA
G. W. STEVENS	Department of Chemical Engineering, University of Melbourne, Parkville, Victoria 3052, Australia

CONTENTS

VOLUME 2

1 LIQUID-LIQUID EXTRACTION IN HYDROMETALLURGY
Michael Cox

2 NUCLEAR FUEL REPROCESSING USING SOLVENT EXTRACTION
W. Batey

3 LIQUID-LIQUID PARTITION IN THE PHARMACEUTICAL INDUSTRY
Michael S. Verrall

4 LIQUID-LIQUID EXTRACTION IN FOOD PROCESSING
W. Hamm

5 SOME CURRENT DEVELOPMENTS AND EXTRACTION TECHNIQUES
L. R. Weatherley

1

LIQUID–LIQUID EQUILIBRIA

D.M.T. NEWSHAM

1.1 Introduction

The proper design of solvent extraction processes depends crucially upon an accurate knowledge of the relevant liquid–liquid equilibria. The latter must be either measured or predicted. This chapter is concerned with methods for the correlation of experimentally determined liquid–liquid equilibria and with methods for the prediction of unknown equilibria.

1.1.1 The phase rule

In an equilibrium system in which the only significant intensive variables are the temperature, pressure, and chemical potential, the phase rule provides a means of determining the number of intensive variables that can be independently varied (within limits) without changing the state of the system. This number of variables is called the number of degrees of freedom (f) and is related to the number of components (c) and number of phases (ϕ) as follows:

$$f = c + 2 - \phi \tag{1.1}$$

The important assumptions in the derivation of this relation are that no chemical reactions take place and that the interfaces between phases are plane.

1.1.2 Phase diagrams

The graphical representation of phase equilibria is facilitated by application of the phase rule. Thus, in a binary liquid–liquid system, $f = 2$ and each phase can be represented by a surface in a space diagram where, for example, the three co-ordinates could be the pressure, temperature, and composition. If, in addition, the pressure is fixed, as is usually the case in liquid–liquid extraction, then $f = 1$ and a plane representation of the system can be secured.

Figure 1.1. is a typical $T - x_i$ diagram for a binary system. The phase boundary separates the two-phase region ($L' + L''$) from the homogeneous regions (L' or L''). The maximum of this curve occurs at the critical solution temperature which is also known as the upper consolute point. Lower

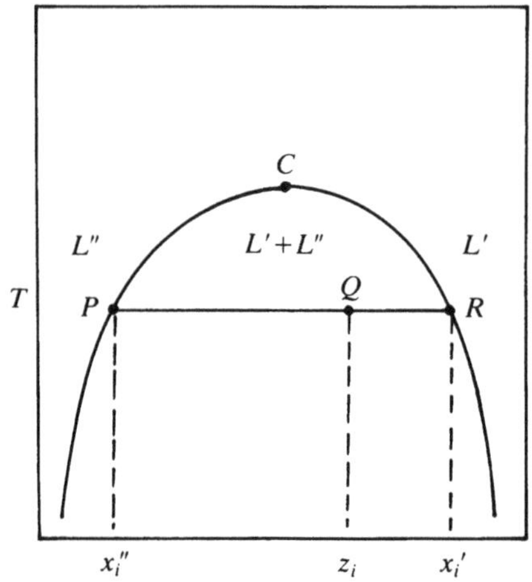

Fig. 1.1. Binary liquid–liquid system.

consolute points, representing a minimum on the phase boundary, also occur. A mixture of overall composition z_i that lies within the two-phase region at a given temperature separates into two liquid phases of compositions x_i' and x_i'' connected by a horizontal tie line. The relative amounts of the phases are given by the lever rule. Thus

$$n'/n'' = (z_i - x_i'')/(x_i' - z_i) = \overline{PQ}/\overline{QR} \tag{1.2}$$

where n' and n'' are the total number of moles in each phase.

In a ternary liquid–liquid system, $f = 3$ and a plane representation is only possible if two variables are fixed—usually pressure and temperature. Ternary phase diagrams are often plotted on a Gibbs equilateral triangle where the pure components are represented by the vertices, the binary mixtures by the edges, and ternary mixture by points within the triangle. A typical isothermal, isobaric phase diagram for a ternary mixture in which only one of the binary pairs is partially miscible is shown in Fig. 1.2a. This is known as a Type I system. The two branches of the phase boundary (the binodal curve) meet at the critical point C, which is also known as the plait point. Compositions of the liquid phases in equilibrium are connected by tie lines, that, in general, are not horizontal, although the lever rule still applies to them. Type II and Type III systems also exist, in which there are two and three partially miscible binary pairs, respectively, as shown in Fig. 1.2b,c. In the Type III system a three-phase region exists for which the system is invariant ($f = 0$). The compositions of the three equilibrium phases are represented by three isolated points connected by a tie triangle. The relative amounts of the phases can be determined from the overall composition of the

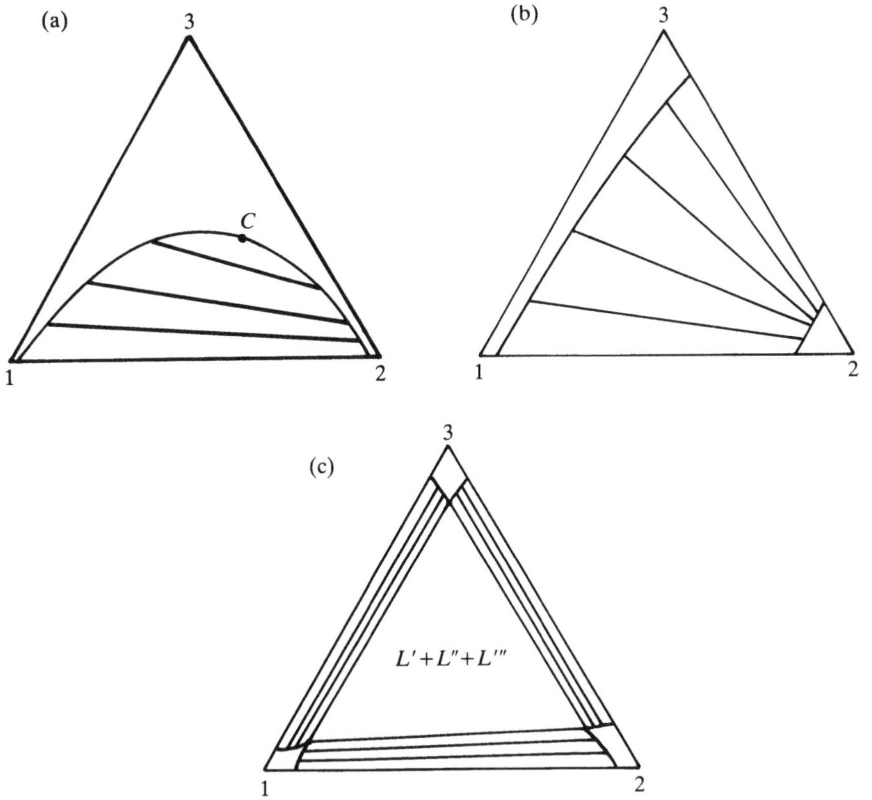

Fig. 1.2. Ternary liquid–liquid systems. (a) Type I, (b) Type II, (c) Type III.

mixture by a double application of the lever rule. Both the Type II and Type III systems can be considered to be formed by the merging of two or three Type I systems.

The temperature dependence of these equilibria can be represented by plotting the temperature on the vertical axis of a triangular prism. Such a diagram is shown in Fig. 1.3a for a Type I system. Also shown is the projection of an isothermal section and of the monotonic critical curve on the base plane of the prism. Systems having critical curves with maxima (or minima) also exist and these have closed phase boundaries in the ternary region over part of the temperature range. Fig. 1.3b shows how a Type I binodal curve can intersect a binary face of the prism to generate a Type II system, and Fig. 1.3c shows the merging of Type I binodal curves to give a Type II system. In the former case the critical curve is monotonic and in the latter it has a minimum corresponding to the merging of two plait points.

Ternary liquid–liquid regions can also be generated from a completely

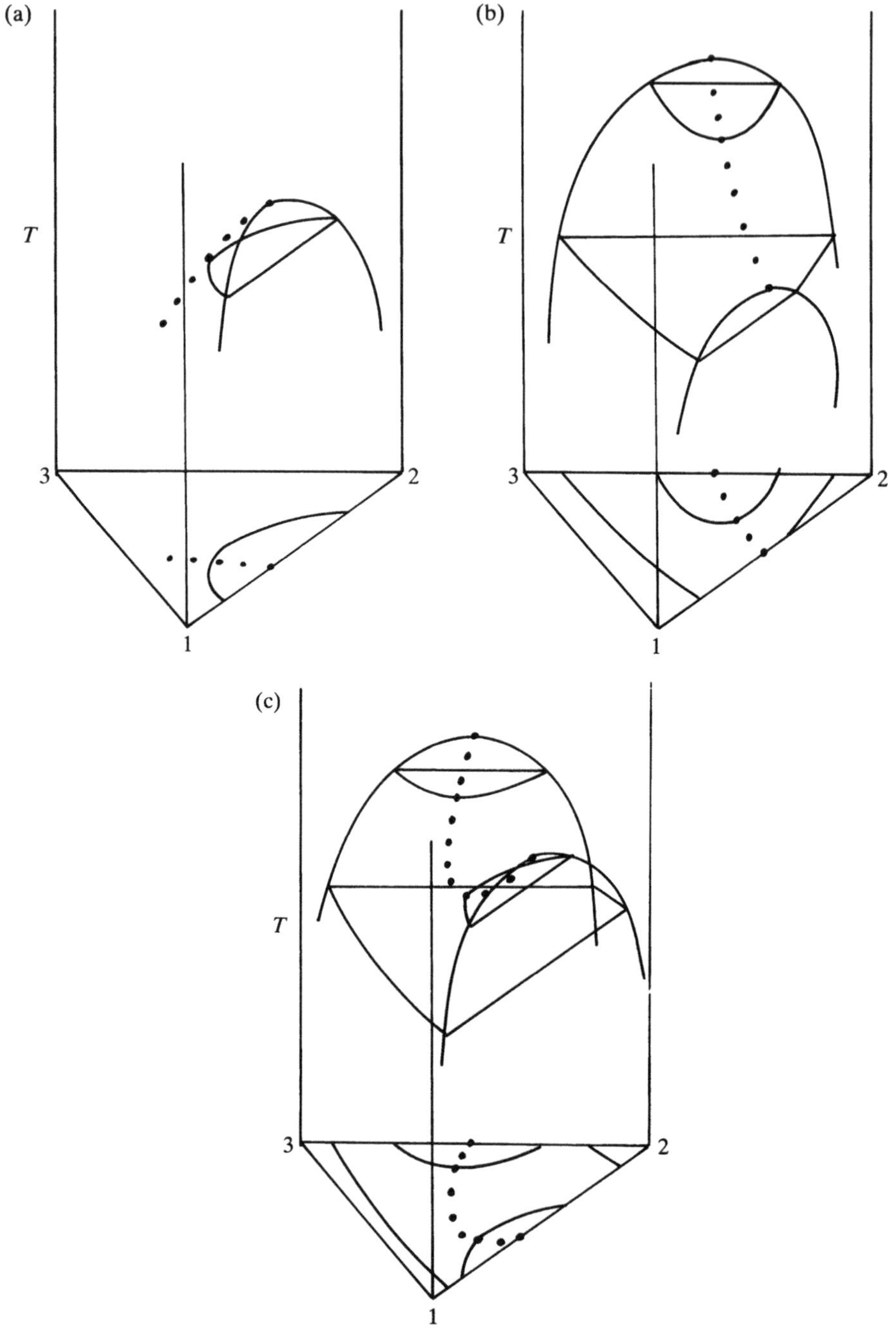

Fig. 1.3. Temperature dependence of ternary liquid–liquid equilibria, ••• critical curve.

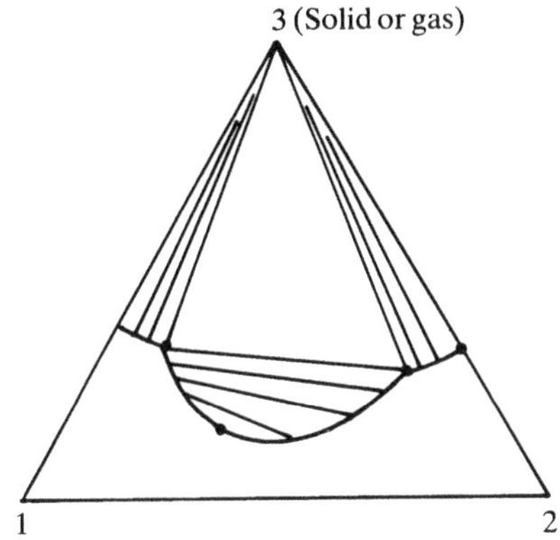

Fig. 1.4. Type IV system.

miscible binary system by contacting with a soluble solid (salting out) or a gas above its critical temperature (supercritical extraction). A typical phase diagram is shown in Fig. 1.4 which is sometimes known as a Type IV system.

1.2 Thermodynamics of equilibrium

1.2.1 Conditions of equilibrium

In a heterogeneous system comprising c components and ϕ phases the conditions for thermodynamic equilibrium may be written as follows:

$$\begin{aligned} T_{(1)} &= T_{(2)} = \ldots = T_{(\phi)}, \\ P_{(1)} &= P_{(2)} = \ldots = P_{(\phi)}, \qquad (1.3) \\ \mu_{i(1)} &= \mu_{i(2)} = \ldots = \mu_{i(\phi)},\ i = 1, \ldots, c \end{aligned}$$

Here μ_i is the chemical potential of component i which is related to the total Gibbs free energy $\mathbf{G}_{(\alpha)}$ of any phase α by eqn (1.4):

$$\mu_{i(\alpha)} = (\partial \mathbf{G}_{(\alpha)}/\partial n_{i(\alpha)})_{T,P,n_{j(\alpha)}} \qquad (1.4)$$

Thus, the criteria for equilibrium require that the temperature, pressure, and chemical potential of each component are uniform throughout the system.

An alternative, thermodynamically equivalent, and sometimes more useful criterion for equilibrium is that at constant temperature and pressure the total Gibbs free energy of the entire system is at a minimum.

$$\mathrm{d}\left(\sum_{\alpha=1}^{\phi} \mathbf{G}_{(\alpha)}\right)_{T,P} = 0 \qquad (1.5)$$

1.2.2 Activities and activity coefficients

If the conditions of eqn (1.3) are applied to a liquid–liquid system, the problem of relating the equilibrium compositions of the liquid phases is reduced to that of determining the composition dependence of the chemical potential. Thermodynamics alone cannot provide the solution to this problem and at this stage one must resort to a physical model. To this end it is useful to introduce the concept of an ideal solution for which the chemical potential of every component is related to its mole fraction as follows:

$$\mu_i = \mu_i^*(T, P) + RT \ln x_i \tag{1.6}$$

The quantity $\mu_i^*(T, P)$ is the standard chemical potential of pure i in the same state of aggregation and at the same temperature and pressure as the solution. The form of eqn (1.6) can be retained for a real solution by defining a quantity a_i known as the activity and which is identical to the product of the activity coefficient γ_i and the mole fraction x_i. Thus,

$$\mu_i = \mu_i^*(T, P) + RT \ln a_i = \mu_i^*(T, P) + RT \ln \gamma_i x_i \tag{1.7}$$

The definition of the activity is completed by using the following convention for liquid mixtures whose pure components are liquids at the pressure and temperature of the solution.

$$\lim_{x_i \to 1} (a_i/x_i) = \lim_{x_i \to 1} \gamma_i = 1 \tag{1.8}$$

Under this convention the quantity $\mu_i^*(T, P)$ has the same meaning as in eqn (1.6). In general, a_i and γ_i are functions of temperature, pressure, and composition.

Using eqns (1.3 and 1.7) it is easily shown that the condition for equilibrium between two liquid phases (L' and L'') is also that

$$a_i' = a_i'' \qquad i = 1 \ldots, c \tag{1.9}$$

Not only is the chemical potential of each component uniform throughout the system but so, too, is the activity of each component.

1.2.3 Temperature and pressure dependence of the activity

The temperature and pressure dependence of the activity and the activity coefficient can be obtained using the following standard thermodynamic relations.

$$\left[\frac{\partial(\mu_i/T)}{\partial(1/T)}\right]_{P, x_i} = \bar{H}_i \equiv (\partial \mathbf{H}/\partial n_i)_{P,T,n_j} \tag{1.10}$$

$$(\partial\mu_i/\partial P)_{T,x_i} = \bar{V}_i \equiv (\partial \mathbf{V}/\partial n_i)_{P,T,n_j} \tag{1.11}$$

In these equations $\bar{H}_i$ and $\mathbf{H}$ are the partial molar and total enthalpy and $\bar{V}_i$ and $\mathbf{V}$ are the partial molar and total volume, respectively. If these relations are applied to eqn (1.7) then it can be shown that

$$\left[\frac{\partial \ln a_i}{\partial(1/T)}\right]_{P,x_i} = \left[\frac{\partial \ln \gamma_i}{\partial(1/T)}\right]_{P,x_i} = (\bar{H}_i - H_i^0)/R \tag{1.12}$$

$$\left[\frac{\partial \ln a_i}{\partial P}\right]_{T,x_i} = \left[\frac{\partial \ln \gamma_i}{\partial P}\right]_{T,x_i} = (\bar{V}_i - V_i^0)/RT \tag{1.13}$$

Here, H_i^0 and V_i^0 are the pure component molar enthalpy and volume, respectively. These equations apply only for the convention adopted in eqn (1.8).

1.3 Correlation and prediction of activity coefficients

1.3.1 The Gibbs–Duhem equation

Although the methods of thermodynamics do not allow prediction of the composition dependence of the chemical potential, a relation between the activity coefficients of a given phase can be obtained from the Gibbs–Duhem equation. This equation relates the intensive variables that characterize the state of the phase, namely the temperature, pressure, and chemical potential.

$$S\,\mathrm{d}T - V\,\mathrm{d}P + \sum_{i=1}^{c} x_i\,\mathrm{d}\mu_i = 0 \tag{1.14}$$

If both P and T are constant then

$$\sum_{i=1}^{c} x_i\,\mathrm{d}\mu_i = 0 \tag{1.15}$$

If this relation is applied to eqn (1.7) it is easily shown that

$$\sum_{i=1}^{c} x_i\,\mathrm{d}(\ln \gamma_i) = 0 \tag{1.16}$$

It is therefore a necessary thermodynamic condition that the activity coefficients of a phase are related in this way. Furthermore, any empirical or theoretical relations representing the composition dependence of the activity coefficients must be solutions of this equation. It should be noted, however, that in a binary mixture it is not possible to vary the composition whilst keeping both T and P constant. This is a consequence of the phase rule, but usually the corrections required for variations in T or P are small and are often neglected.

The phase rule was first derived using the Gibbs–Duhem equation which must apply to each of the ϕ phases in a multi-phase system. If the temperature, pressure, and chemical potentials are uniform throughtout the system, there are $c + 2$ variables and therefore $c + 2 - \phi$ degrees of freedom as stated in eqn (1.1).

1.3.2 Correlating equations

A large number of semi-empirical equations have been proposed for the correlation of the activity coefficients of mixtures of non-electrolytes. All of these can provide useful correlations of experimental data for binary mixtures, but in many cases their extension to multicomponent mixtures is not successful unless additional parameters are included. The latter can only be obtained from measurements on the multicomponent mixture. Consequently, the most widely used correlations are those which can be accurately extended to multicomponent mixtures using only binary parameters. Such equations then become predictive when applied to a multicomponent system. For liquid–liquid systems the models of Renon and Prausnitz (1968) and Abrams and Prausnitz (1975) have given very satisfactory correlations.

1.3.2.1 *The NRTL equation*

The semi-empirical model of Renon and Prausnitz (1968) takes account of the non-random nature of liquid mixtures by using the concept of a local composition. When this is combined with the 'two-liquid' theory of Scott (1956) the following expressions result for the activity coefficients of a binary mixture:

$$\ln \gamma_1 = x_2^2 \left\{ \frac{\tau_{21} \exp(-2\alpha_{12}\tau_{21})}{[x_1 + x_2 \exp(-\alpha_{12}\tau_{21})]^2} + \frac{\tau_{12} \exp(-\alpha_{12}\tau_{12})}{[x_2 + x_1 \exp(-\alpha_{12}\tau_{12})]^2} \right\} \tag{1.17}$$

$$\ln \gamma_2 = x_1^2 \left\{ \frac{\tau_{12} \exp(-2\alpha_{12}\tau_{12})}{[x_2 + x_1 \exp(-\alpha_{12}\tau_{12})]^2} + \frac{\tau_{21} \exp(-\alpha_{12}\tau_{21})}{[x_1 + x_2 \exp(-\alpha_{12}\tau_{12})]^2} \right\} \tag{1.18}$$

where $\tau_{12} = (g_{12} - g_{22})/RT$, $\tau_{21} = (g_{21} - g_{11})/RT$, $g_{21} = g_{12}$. In these expressions α_{12} is an empirical 'non-randomness' parameter and the g_{ij} are residual Gibbs energies. The general expression for a multicomponent mixture is

$$\ln \gamma_i = \frac{\Sigma_j x_j \tau_{ji} G_{ji}}{\Sigma_k x_k G_{ki}} + \sum_j \left\{ \frac{x_j G_{ji}}{\Sigma_k G_{kj} x_k} \left[\tau_{ij} - \frac{\Sigma_m x_m \tau_{mj} G_{mj}}{\Sigma_k G_{kj} x_j} \right] \right\} \tag{1.19}$$

where $G_{ij} = \exp(-\alpha_{ij}\tau_{ij})$, $G_{ii} = 0$, $\tau_{ii} = 0$, $\alpha_{ij} = \alpha_{ji}$. The corresponding expression for the excess Gibbs energy (the difference between the Gibbs free energy of mixing of a real solution and an ideal solution) is

$$G^{E}/RT = \sum_i x_i \ln a_i - \sum_i x_i \ln x_i = \sum_i x_i \ln \gamma_i$$

$$= \sum_i \left[x_i \left(\sum_j \tau_{ji} G_{ji} x_j \right) \Big/ \left(\sum_k G_{ki} x_k \right) \right] \tag{1.20}$$

The non-randomness parameter α_{ij} can, in a crude way, be identified with the inverse of the number of nearest neighbours of a molecule in the mixture and would be expected to have a value of ~0.1–0.3. This is indeed found to be the case empirically. Larger values of α_{ij} are sometimes found but it should be noted that for a binary symmetrical system ($\tau_{ij} = \tau_{ji}$) values of $\alpha_{ij} > 0.426$ forbid representation of liquid–liquid equilibria.

The binary form of the NRTL equation contains three parameters (τ_{12}, τ_{21}, α_{12}), but it is effectively a two-parameter equation since the non-randomness parameter can be specified according to a set of rules devised by Renon and Prausnitz (1968) that depend on the chemical nature of the components in the mixture. In the representation of liquid–liquid equilibrium, α_{ij} is often assigned a value of 0.2 (Sorensen *et al.*, 1979).

The binary parameters of the NRTL equation can be determined by fitting to activity coefficients obtained from experimental vapour–liquid equilibrium data or by choosing values of the parameters that give the best fit to the mutual solubilities. In the latter case α_{ij} must be given a fixed value. In practice, it is found that the parameters $(g_{12} - g_{22})$ and $(g_{21} - g_{11})$ often vary linearly with temperature.

1.3.2.2 *The UNIQUAC model*

In order to develop a truly two-parameter equation Abrams and Prausnitz (1975) combined the local composition concept with the quasi-chemical lattice model of Guggenheim (1952). The resulting UNIQUAC (universal quasi-chemical) equation for the excess Gibbs energy of a binary mixture is

$$\begin{aligned} G^{E}/RT = {} & x_1 \ln(\phi_1/x_1) + x_2 \ln(\phi_2/x_2) \\ & - (z/2)\left[q_1 x_1 \ln(\phi_1/\theta_1) + q_2 x_2 \ln(\phi_1/\theta_1)\right] \\ & - q_1 x_2 \ln(\theta_1 + \theta_2 \tau_{21}) - q_2 x_2 \ln(\theta_2 + \theta_1 \tau_{12}) \end{aligned} \tag{1.21}$$

where $\tau_{21} = \exp[-(u_{21} - u_{11})/RT]$, $\tau_{12} = \exp[-(u_{12} - u_{22})/RT]$. In this equation ϕ_i and θ_i are the average segment and area fractions, respectively, and are related to the pure component structural parameters r_i and q_i as follows:

$$\phi_i = r_i x_i / \left(\sum_j r_j x_j\right) \tag{1.22}$$

$$\theta_j = q_i x_i / \left(\sum_j q_j x_j\right) \tag{1.23}$$

The parameters r_i and q_i can be determined from values of the van der Waals molecular volume and area given by Bondi (1968) using the following relations:

$$r_i = V_{wi}/15.17 \tag{1.24}$$

$$q_i = A_{wi}/(2.5 \times 10^9) \tag{1.25}$$

The lattice coordination number z is arbitrarily given a value of 10. The two adjustable parameters are $(u_{21} - u_{11})$ and $(u_{12} - u_{22})$ with $u_{21} = u_{12}$.

The corresponding expressions for the binary and multicomponent activity coefficients and the multicomponent excess Gibbs energy are given below.

$$\ln \gamma_1 = \ln(\phi_1/x_1) + 1 - \phi_1/x_1 - (z/2)q_1 [\ln(\phi_1/\theta_1) + 1 - \phi_1/\theta_1] - q_1 \ln(\theta_1 + \theta_2\tau_{21}) + \theta_2 q_1 [\tau_{21}/(\theta_1 + \theta_2\tau_{21}) - \tau_{12}/(\theta_2 + \theta_1\tau_{12})] \tag{1.26}$$

$$\ln \gamma_i = \ln(\phi_i/x_i) + 1 - \phi_i/x_i - (z/2)q_i[\ln(\phi_i/\theta_i) + 1 - \phi_i/\theta_i] - q_i \ln\left(\sum_j \theta_j\tau_{ji}\right) + q_i \sum_j \left(\theta_j\tau_{ij} \Big/ \sum_k \theta_k\tau_{kj}\right) \tag{1.27}$$

where

$$\tau_{ji} \equiv \exp - [(u_{ji} - u_{ii})/RT] \tag{1.28}$$

$$G^{\mathrm{E}}/RT = \sum_i x_i \ln(\phi_i/x_i) - (z/2)\sum_i q_i x_i \ln(\phi_i/\theta_i) - \sum_i q_i x_i \ln\left(\sum_j \theta_j\tau_{ji}\right) \tag{1.29}$$

1.3.3 Predictive equations

In the absence of experimental data with which to obtain the parameters of the NRTL or UNIQUAC equations, resort must be made to purely predictive models. Two of these are discussed below.

1.3.3.1 *The regular solution model*

Regular solution theory was developed by Scatchard (1931) and Hildebrand and Wood (1933). Detailed accounts of the theory are given by Hildebrand and Scott (1962) and Hildebrand *et al.* (1970). A regular solution is one for which the excess entropy of mixing is zero. For a binary mixture, the excess Gibbs energy is expressed in terms of the volume fraction ϕ_i and solubility parameter δ_i as follows:

$$G^E = V_m \phi_1 \phi_2 (\delta_1 - \delta_2)^2 \tag{1.30}$$

In this expression, $V_m = \Sigma_i x_i V_i^0$ where V_i^0 is the pure component liquid molar volume and $\phi_i = x_i V_i^0 / V_m$. The solubility parameter is related to the internal energy of vaporization ΔU_i^0.

$$\delta_i = (\Delta U_i^0 / V_i^0)^{1/2} \tag{1.31}$$

At normal pressures

$$\Delta U_i^0 \simeq \Delta H_i^0 - RT \tag{1.32}$$

where ΔH_i^0 is the heat of vaporization.

The activity coefficient in a binary mixture is given by

$$\ln \gamma_1 = V_1^0 \phi_2^2 (\delta_1 - \delta_2)^2 / RT \tag{1.33}$$

and for multicomponent mixtures

$$\ln \gamma_i = V_i^0 (\delta_i - \bar{\delta})^2 / RT \tag{1.34}$$

where

$$\bar{\delta} = \sum_i \phi_i \delta_i \tag{1.35}$$

The excess Gibbs energy of a multicomponent mixture is

$$G^E = \sum_i x_i V_i^0 (\delta_i - \sum_j \phi_j \delta_j)^2 \tag{1.36}$$

The regular solution model is suitable only for non-polar mixtures composed of molecules that are not too different in size, and the activity coefficients are always greater than unity and the excess enthalpy is always positive. Nevertheless, in the absence of any other information, it can give useful qualitative predictions in certain cases. A useful compilation of solubility parameters is given by Barton (1983).

1.3.3.2 *The UNIFAC model*

The UNIFAC model is a group contribution method in which the mixture is represented by a mixture of the constituent functional groups. Since the number of groups ($-CH_3$, $-OH$, $-CO_2H$, etc.) that are encountered is much smaller than the number of different mixtures that have to be modelled, the method is a powerful one. It was first developed by Fredenslund *et al.* (1975) who used the UNIQUAC equation to describe the group activity coefficients and hence the name UNIFAC (UNIQUAC Functional group Activity Coefficients).

The activity coefficients have a combinatorial contribution (γ_i^C) that depends upon the size and shape of the molecules and a residual contribution (γ_i^R) that characterizes the energetic interactions between molecules. Thus

$$\ln \gamma_i = \ln \gamma_i^C + \ln \gamma_i^R \tag{1.37}$$

$$\ln \gamma_i^C = \ln(\phi_i/x_i) + 1 - \phi_i/x_i - (z/2)\, q_i\, [\ln(\phi_i/\theta_i) + 1 - \phi_i/\theta_i] \tag{1.38}$$

where

$$\phi_i = x_i r_i / \sum_j x_j r_j \tag{1.39}$$

$$\theta_i = x_i q_i / \sum_j x_j q_j \tag{1.40}$$

The pure component parameters r_i and q_i are measures of the van der Waals molecular volumes and surface areas, respectively. They can be obtained from the group volume and surface area parameters, R_k and Q_k.

$$r_i = \sum_k \nu_{ki} R_k \tag{1.41}$$

$$q_i = \sum_k \nu_{ki} Q_k \tag{1.42}$$

where ν_{ki} is the number of groups of type k in molecule i. The values of R_k and Q_k can in turn be estimated from the van der Waals volumes and areas, V_k and A_k, given by Bondi (1968).

$$R_k = V_k / 15.17 \tag{1.43}$$

$$Q_k = A_k / (2.5 \times 10^9) \tag{1.44}$$

The coordination number z is usually assigned a value of 10. The residual contribution is given by the following expression:

$$\ln \gamma_i^{\mathrm{R}} = \sum_k \nu_{ki} (\ln \Gamma_k - \ln \Gamma_{ki}) \tag{1.45}$$

where Γ_k is the residual group activity coefficient and Γ_{ki} is the residual activity coefficient of group k in a reference mixture containing only molecules of type i. The latter definition is necessary so that $\gamma_i \to 1$ as $x_i \to 1$. The residual activity coefficients are given by

$$\ln \Gamma_k = Q_k \left[1 - \ln \left(\sum_m \theta_m \psi_{mk} \right) - \sum_m \left(\theta_m \psi_{km} \Big/ \sum_n \theta_n \psi_{nm} \right) \right] \tag{1.46}$$

where

$$\theta_m = Q_m X_m / \sum_n Q_n X_n \tag{1.47}$$

and the fraction of group m in the mixture X_m is given by

$$X_m = \sum_i \nu_{mi} x_i / \sum_i \sum_k \nu_{ki} x_i \tag{1.48}$$

The interaction parameters ψ_{nm} have a temperature dependent form

$$\psi_{nm} = \exp(-a_{nm}/T) \tag{1.49}$$

Each group-group interaction is characterized by two parameters (a_{nm}, a_{mn}). Also, $a_{nn} = 0$ and $\psi_{nn} = 1$. In eqns (1.37–1.49) the running variables i and j refer to molecular properties, and k, m, and n refer to group properties.

The UNIFAC method was originally devised for the prediction of vapour-liquid equilibria and the parameters a_{nm} were estimated from a large data base. Continual revisions and extensions of the parameter tables are being made and further details are given by Fredenslund *et al.* (1977), Fredenslund *et al.* (1977), Skjold-Jorgensen *et al.* (1979), and Gmehling (1986). Whilst the parameters given in the references can be used to predict liquid-liquid equilibria, it has been found that greater accuracy can be achieved if a separate parameter table is used with values of the parameters estimated mainly from binary and ternary liquid-liquid equilibria. Such a compilation is given by Magnussen *et al.* (1981).

1.3.4 Correlations for electrolyte solutions

In the description of electrolyte solutions it is common to express compositions in terms of the molality scale. The molality of a solute m_i is normally defined as the number of moles of the solute per kg of solvent. In the formulation of the chemical potential relations, the activity coefficient of the solvent is defined on the mole fraction scale but the solute activity coefficient is defined on the molality scale. Thus, for the solvent,

$$\mu_0 = \mu_0^* + RT \ln \gamma_0 x_0 \qquad \gamma_0 \to 1 \text{ as } x_0 \to 1 \tag{1.50}$$

and, for the solutes,

$$\mu_i = \mu_i^\ominus + RT \ln \gamma_i m_i/m_i^\ominus \qquad \gamma_i \to 1 \text{ as } x_0 \to 1 \tag{1.51}$$

where $m_i^\ominus$ is unit molality. The standard state is often referred to as a 'hypothetical ideal solution of unit molality' and it is so chosen that $\gamma_i \to 1$ as all $m_i \to 0$ (or $x_0 \to 1$). It is assumed that the linearity between a_i ($= \gamma_i m_i$) and m_i is maintained up to $m_i = 1$. For this reason, activity coefficients defined in this way are called practical rather than rational as they are when defined on the mole fraction scale.

It is not possible to obtain activity coefficients for individual ionic species. Thermodynamic measurements yield only the properties of the electrolyte as a whole. However, eqn (1.51) can still be applied formally to the individual ions. For electrolytes of general formula $M_{\nu_+}X_{\nu_-}$ we can write

$$\mu_+ = \mu_+^\ominus + RT \ln \gamma_+ m_+$$

$$\mu_- = \mu_-^\ominus + RT \ln \gamma_- m_- \tag{1.52}$$

The chemical potential of the electrolyte as a whole is given by

$$\mu = \nu_+\mu_+ + \nu_-\mu_- = \nu_+\mu_+^\ominus + \nu_-\mu_-^\ominus + \nu RT \ln \gamma_\pm m_\pm \tag{1.53}$$

where $\nu = \nu_+ + \nu_-$, $\gamma_\pm$ is the mean ionic activity coefficient, and $m_\pm$ is the mean ionic molality:

$$\gamma_\pm = (\gamma_+^{\nu_+}\gamma_-^{\nu_-})^{1/\nu} \tag{1.54}$$

$$m_\pm = (m_+^{\nu_+} m_-^{\nu_-})^{1/\nu} \tag{1.55}$$

In any electrolyte solution the electroneutrality condition must be observed. If z_+ and z_- are the ionic valences for cation and anion, respectively, then

$$\nu_+ z_+ = \nu_- z_- \tag{1.56}$$

1.3.4.1 *Ion interaction models*

In very dilute electrolyte solutions the classic description of the composition dependence of the mean ionic activity coefficient is given by the Debye–Hückel limiting law:

$$\ln \gamma_\pm = -A_\gamma z_+ z_- I^{1/2} \tag{1.57}$$

where I is the ionic strength and is defined as follows:

$$I = \frac{1}{2} \sum m_i z_i^2 \tag{1.58}$$

and the summation is taken over all ions. For water at 298.15 K, $A_\gamma = 1.172\,\mathrm{mol}^{-1/2}\,\mathrm{kg}^{1/2}$.

The Debye–Hückel limiting law has been thoroughly tested experimentally and gives an accurate description of the behaviour of the mean ionic activity coefficients at low concentrations. The activity coefficient of the solvent at low solute concentrations is, on the other hand, very close to unity and for this reason it is convenient to define an osmotic coefficient ϕ that exaggerates the deviation from ideality as follows:

$$\phi = -1000 \ln a_0 / \left(M_0 \sum m_i \right) \tag{1.59}$$

In this equation the solvent activity $a_0 = \gamma_0 x_0$, M_0 is the molar mass of the solvent, and the summation is taken over all ions. The corresponding Debye–Hückel expression for the osmotic coefficient in a form that is not restricted to valence type is given by Pitzer (1979):

$$\phi - 1 = -2A_\phi I^{3/2} \Big/ \sum m_i \tag{1.60}$$

where A_ϕ is the Debye–Hückel parameter for the osmotic coefficient and $A_\phi = A_\gamma/3$. For water at 298.15 K, $A_\phi = 0.391\,\mathrm{mol}^{-1/2}\,\mathrm{kg}^{1/2}$.

1.3.4.2 *Pitzer's ion interaction model*

The Debye–Hückel theory has been extended in various ways, but one of the most useful extensions is that developed by Pitzer (1979) who expressed the excess Gibbs energy of an electrolyte solution as a virial expansion.

$$G^E/RT = n_0 f(I) + (1/n_0) \sum_i \sum_j \lambda_{ij}(I) n_i n_j + (1/n_0^2) \sum_i \sum_j \sum_k \mu_{ijk} n_i n_j n_k \tag{1.61}$$

Here, G^E is the total excess Gibbs energy of a solution containing n_0 kg of solvent and n_i, n_j, etc. mol of solutes i, j, etc., $f(I)$ is a function of ionic strength (and of temperature and solvent properties), $\lambda_{ij}(I)$ is a function of ionic strength of such a form that it represents qualitatively the behaviour of real systems, and μ_{ijk} is a triple ion interaction term. The λ and μ matrices are symmetric and the corresponding expressions for the activity and osmotic coefficients are as follows:

$$RT \ln \gamma_i = (\partial G^E/\partial n_i) = (z_i^2/2) f' + 2 \sum_j \lambda_{ij} m_j + (z_i^2/2) \sum_j \sum_k \lambda'_{jk} m_j m_k + 3 \sum_j \sum_k \mu_{ijk} m_j m_k \tag{1.62}$$

$$\phi - 1 = -(\partial G^E/\partial n_0)/RT\sum m_i$$

$$= \left(\sum m_i\right)^{-1}\left[(If' - f) + \sum_i\sum_j (\lambda_{ij} + I\lambda'_{ij})\, m_i m_j\right.$$

$$\left. + 2\sum_i\sum_j\sum_k \mu_{ijk} m_i m_j m_k\right] \qquad (1.63)$$

In these expressions $f' = \mathrm{d}f/\mathrm{d}I$, $\lambda'_{ij} = \mathrm{d}\lambda_{ij}/\mathrm{d}I$, and $m_i = n_i/n_0$.

By expanding these equations and by choosing suitable functional forms for $f(I)$ and $\lambda_{ij}(I)$, Pitzer showed that a set of working equations could be developed that contained certain measurable combinations of second and third virial coefficients (λ and μ). Li *et al.* (1987) give a simplified form of Pitzer's original working equation for the activity coefficient of a single positive ion M.

$$\ln \gamma_M = (z_M^2/2)f' + 2\sum_a m_a\left[B_{Ma} + \left(\sum mz\right)C_{Ma}\right]$$

$$+ \sum_c\sum_a m_c m_a(z_M^2 B'_{ca} + z_M C_{ca}) \qquad (1.64)$$

where $B' = \mathrm{d}B/\mathrm{d}I$ and M and c refer to cations, and a refers to anions. The corresponding expression for an anion X is obtained by substituting M for X, and interchanging a and c. The chosen function for $f(I)$ is

$$f(I) = (-4A_\phi I/b)\ \ln\ (1 + bI^{1/2}) \qquad (1.65)$$

with $b = 1.2$. The form of the second virial term depends upon the type of electrolyte. For 1–1, 1–2, and 1–3 electrolytes,

$$B = \beta^{(0)} + (2\beta^{(1)}/\alpha^2 I)\ [1 - (1 + \alpha I^{1/2})\exp(-\alpha I^{1/2})] \qquad (1.66)$$

and for 2–2 electrolytes,

$$B = \beta^{(0)} + (2\beta^{(1)}/\alpha_1^2 I)\ [1 - (1 + \alpha_1 I^{1/2})\exp(-\alpha_1 I^{1/2})$$

$$+ (2\beta^{(2)}/\alpha_2^2 I\,[1 - (1 + \alpha_2 I^{1/2})\exp(-\alpha_2 I^{1/2})] \qquad (1.67)$$

where $\alpha = 2.0$, $\alpha_1 = 1.4$, $\alpha_2 = 12$. The parameters $\beta^{(0)}$, $\beta^{(1)}$, $\beta^{(2)}$ for specific ions are given by Pitzer (1979). The third virial contribution is given by

$$C = C_\phi/2(z_M z_X)^{1/2} \qquad (1.68)$$

and values for C_ϕ for specific solutes are also given by Pitzer (1979).

The simplified working expression for the osmotic coefficient given by Li *et al.* (1987) is

$$\phi - 1 = \left(\sum_i m_i\right)^{-1}\left\{f'I - f + 2\sum_c\sum_a m_c m_a\left[B_{ca} + IB'_{ca} + 2\left(\sum mz\right)C_{ca}\right]\right\} \tag{1.69}$$

and the terms in this equation are defined in the same way as eqn (1.64).

1.3.5 Ion interactions and association equilibria

Although the activity coefficients of strong electrolytes can be described by the ion interaction model detailed above, for weak electrolytes it is necessary to recognize the association of ionic species. Pitzer (1979) has shown how to combine the association (or dissociation) equilibria with the ion interaction model by introducing additional virial coefficients for the neutral species. The dissociation equilibrium itself is described by the usual thermodynamic equilibrium constant. For example, for a weak acid HA then

$$HA \rightleftharpoons H^+ + A^-$$

$$\exp(-\Delta G^0/RT) = K_a = \frac{m_{H^+}\gamma_{H^+}m_{A^-}\gamma_{A^-}}{m_{HA}\gamma_{HA}} \tag{1.70}$$

The introduction of the equilibrium constant eliminates the need for virial coefficients representing the H^+/A^- ion interaction. The virial coefficients (λ_{ij}) for the H^+/H^+ and A^-/A^- interactions cannot be determined separately and B_{H^+,A^-} is assumed to be zero. For interactions between neutral species the second and third virial coefficients ($\lambda_{HA,HA}$ and $\mu_{HA,HA,HA}$) can be used directly. For ion–neutral interactions only sums or differences of second virial coefficients that correspond to electrical neutrality are included, e.g. ($\lambda_{H^+,HA} + \lambda_{A^-,HA}$). For a solution containing the weak acid HA and a strong electrolyte MX there is a total of six sums and differences of which only three are independent. Third virial coefficients for ion–neutral interactions are usually neglected. The application of this method to phosphoric and sulphuric acids has been reported in detail by Pitzer and Silvester (1976) and by Pitzer *et al.* (1977).

Further correlations for aqueous electrolyte solutions and for solutions of electrolytes in mixed solvents can be found in the following references.

Edwards *et al.* (1978)
Maurer (1983)
Ball *et al.* (1985)
Renon (1986)
Cabezas and O'Connell (1986)
Mock *et al.* (1986)

Sander *et al.* (1986)
Zemaitis *et al.* (1986)

1.4 Prediction and computation of equilibria

1.4.1 Selection of activity coefficient correlation

In the correlation and prediction of liquid–liquid equilibria the selection of the model for the representation of the activity coefficients requires experience but a general guide can be devised as shown in Fig. 1.5. The figure, which excludes models for electrolyte solutions, requires some explanation.

1. UNIQUAC (SPECIFIC) refers to parameters that are fitted specifically to experimental data for the mixture (Sorensen and Arlt 1979, 1980).
2. UNIQUAC (COMMON) refers to parameters obtained either from binary liquid–liquid equilibria or from data for a number of ternary mixtures that include a common binary (Sorensen and Arlt 1979, 1980).
3. UNIFAC (LLE) parameters are obtained from liquid–liquid equilibria (Magnussen *et al.* 1981).

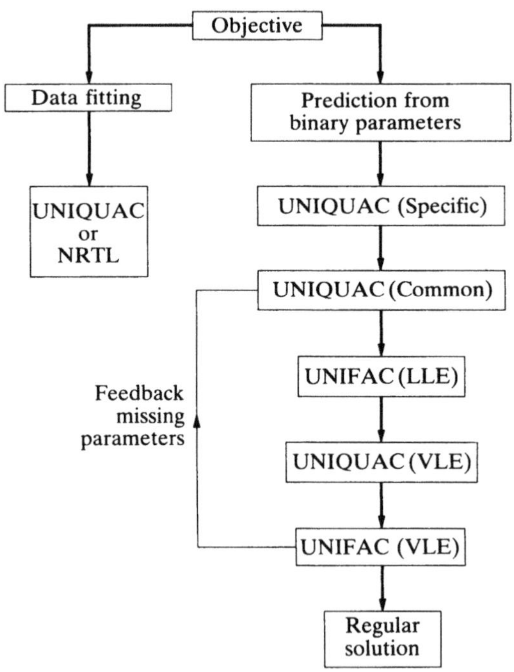

Fig. 1.5. Hierarchy of activity coefficient models.

4. UNIQUAC (VLE) parameters are obtained from vapour-liquid equilibrium data (Gmehling *et al.* 1977–1983).

5. UNIFAC (VLE) refers to parameters obtained from vapour-liquid equilibria (Gmehling 1986).

6. The possibility of feeding back missing parameters to the UNIQUAC model should not be overlooked. For example, suppose that for a given multicomponent mixture UNIQUAC parameters are available for all but one binary mixture. Clearly it is worthwhile to preserve the accuracy inherent in the known binary parameters. The missing binary parameters can then be obtained from, say, the UNIFAC model. This can be achieved by first calculating the infinite dilution activity coefficients from the UNIFAC model and then calculating the UNIQUAC parameters from these. Only one such feedback possibility is shown in Fig. 1.5.

The hierarchy of activity coefficient models given in Fig. 1.5 is not definitive or absolute but, at least, it allows the novice to avoid serious mistakes.

1.4.2 Computation of equilibrium compositions

Having selected a model for the activity coefficients, the liquid–liquid equilibrium compositions may be estimated in one of two ways.

1.4.2.1 *The iso-activity method*

In this method the relevant equilibrium conditions, at constant temperature and pressure are

$$a_i' = a_i'' \qquad i = 1, \ldots, c \tag{1.9}$$

and these relations must be satisfied subject to the mass balance constraints

$$n_i = n_i' + n_i'' \qquad i = 1, \ldots, c \tag{1.71}$$

where n_i' and n_i'' are the number of moles of component i in the two phases. There are, therefore, $2c$ equations and $2c$ unknowns (n_i', n_i''). Alternatively, $c - 2$ compositions can be specified and eqns (1.9) solved under the constraints

$$\sum_{i=1}^{c} x_i' = 1 \qquad \sum_{i=1}^{c} x_i'' = 1 \tag{1.72}$$

In this case there are $c + 2$ equations and $2c - (c - 2) = c + 2$ unknowns.

Since the iso-activity criterion does not ensure that the Gibbs free energy of the system is at a minimum, it is also necessary to check that the Gibbs free energy surface is convex at the composition of interest. This requires the matrix with elements $\mu_{ij} = \partial^2 G/\partial x_i \partial x_j\ (i, j = 1, \ldots, c-1)$ to be positive definite. For a binary mixture

$$\mu_{11} > 0 \tag{1.73}$$

and for a ternary mixture

$$\mu_{11} > 0,\ \mu_{22} > 0,\ \mu_{11}\mu_{22} - \mu_{12}^2 > 0 \tag{1.74}$$

Equations (1.9) are highly non-linear, but they may be solved using a variety of numerical methods. Sorensen and Arlt (1980) give a listing of a FORTRAN computer program for construction of binodal curves for ternary mixtures using the UNIQUAC or NRTL equations.

1.4.2.2 *Minimization of the Gibbs free energy*

Equation (1.5) gives the general condition for equilibrium in a multiphase system.

$$\mathrm{d}\left(\sum_{\alpha=1}^{\phi} \mathbf{G}_{(\alpha)}\right)_{T,P} = 0 \tag{1.5}$$

As it stands the equation does not exclude maxima or saddle points. The difficulties that are encountered in ensuring that the true minimum has been established are illustrated in Fig. 1.6 where the molar free energy of mixing is plotted for a binary mixture. This figure illustrates false solutions corresponding to maxima, false solutions arising from local minima and thermodynamically true but erroneous solutions also arising from local minima. Many of the difficulties arise from the algebraic complexity of the functions used to represent the activity coefficients. A full discussion is given by Sorensen *et al.* (1979).

The solution of eqn (1.5) under the constraints of eqns (1.71 or 1.72) can be achieved, for example, by the method of steepest descent as described by Dluzniewski and Adler (1972). The free energy minimization technique is especially useful for systems where chemical reactions occur in any or all of the phases. Since eqn (1.5) is general this problem can be tackled using the same mathematical framework as for non-reacting systems.

1.5 Measurement of liquid–liquid equilibria

1.5.1 Cell equilibration method

The determination of liquid–liquid equilibria is, in principle, straightforward and may be carried out, for example, using a simple cell of the type shown in Fig. 1.7 (Ashton *et al.* 1983). The cell takes a charge of about

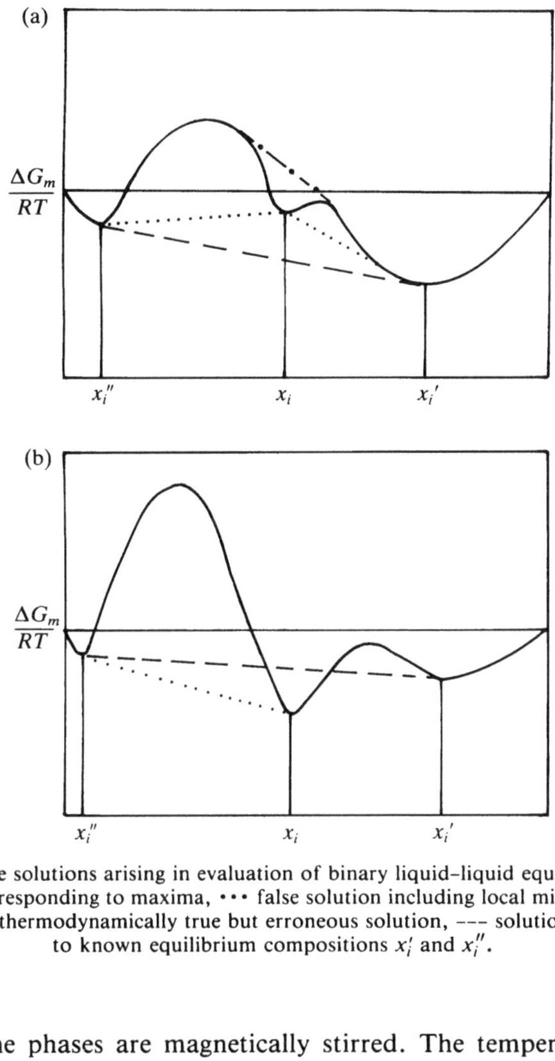

Fig. 1.6. Multiple solutions arising in evaluation of binary liquid–liquid equilibria. (a) -•-•- false solution corresponding to maxima, ••• false solution including local minimum, --- true solution. (b) ••• thermodynamically true but erroneous solution, --- solution corresponding to known equilibrium compositions x_i' and x_i''.

$20\,cm^3$ and the phases are magnetically stirred. The temperature is controlled by circulating fluid from a thermostat. After equilibration and settling, samples of each phase can be removed with a syringe for analysis, usually by gas chromatography. Several difficulties can arise in this type of experiment. First, equilibration times may be long, although this problem is minimized by stirring. Second, settling times may be very long, especially if secondary dispersions are formed or if the phases have similar densities. Third, phase separation in a sample of an individual phase may occur if the temperature of the experiment differs from ambient temperature. In this case the sample must be dissolved in a separate solvent before analysis. Fourth, if a sample is stored prior to analysis, the vapour space of the sample

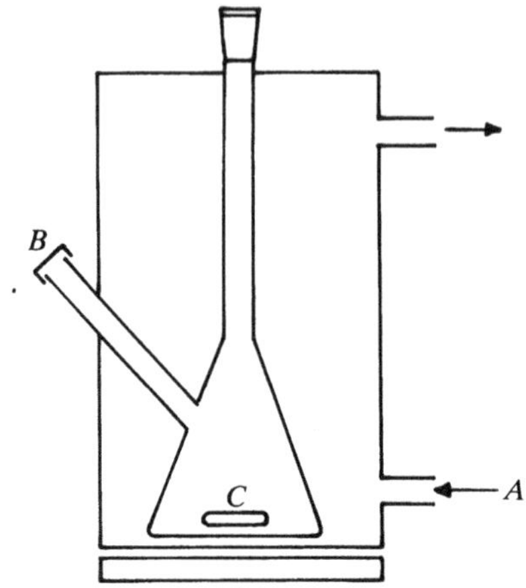

Fig. 1.7. Simple cell for determination of liquid–liquid equilibria. (a) temperature control fluid, (b) septum, (c) magnetic stirrer, (d) stirrer drive.

bottle must be kept to a minimum since the composition of the vapour phase can be very different from that of the liquid. Fifth, entrainment of one phase may occur if the samples are not carefully removed from the cell.

1.5.2 Continuous measurements

Many of the difficulties mentioned above can be avoided by use of the flow system described by Reinhardt and Rydberg (1969). As shown in Fig. 1.8, this apparatus consists of a mixing chamber to allow intense agitation of the phases which are then separated in a centrifuge. Provision is made for recirculation of the phases and for on-line determination of the compositions. The type of detector used depends upon the type of system being investigated. The method is fast, precise and allows several choices of methodology (Liljenzin 1979).

1.5.3 The titration method

A technique for ternary mixtures that avoids use of expensive analytical equipment and which depends only on simple titrations and material balances has been described by Newsham and Ng (1972). The method requires first the determination of the binodal curve. This can easily be done by titrating binary or ternary mixtures of known composition with one of the pure components. The end point is detected by the onset or disappearance of turbidity depending upon the starting composition.

The experimental procedure for determining the tie line compositions is

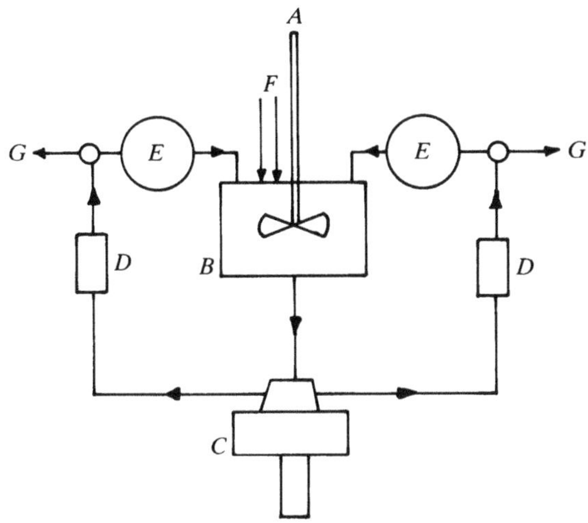

Fig. 1.8. Continuous apparatus for determination of liquid–liquid equilibria. (a) stirrer, (b) mixing chamber, (c) centrifuge, (d) analytical device, (e) heat exchanger, (f) feed supply, (g) sampling point.

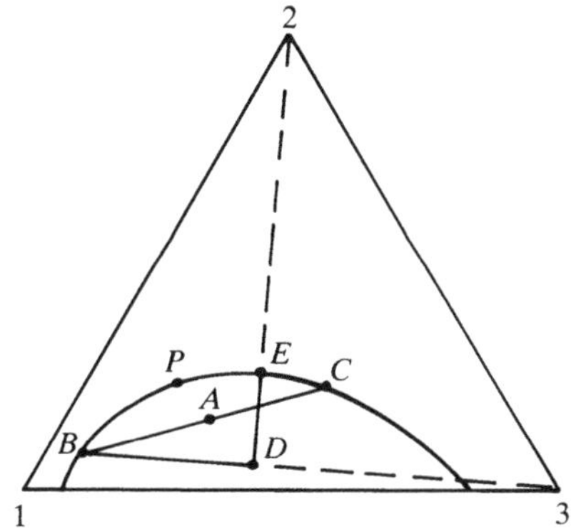

Fig. 1.9. Illustration of experimental procedure for determining composition of liquid layer B.

illustrated in Fig. 1.9. The original mixture A is separated into two layers, B and C. A weighed amount of B is taken and, to this, a known amount of pure 3 is added when the composition changes to that at D. The mixture is then titrated with pure 2 until the end point at E when, of course, the turbidity in the sample disappears. Care must be taken that E does not lie near the plain point P as this can result in the end point of the titration being obscured by critical opalescence. This can be done by addition of the

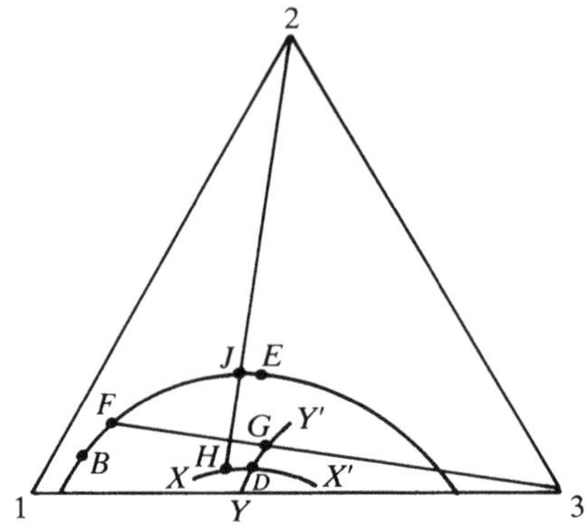

Fig. 1.10. Graphical procedure for estimating composition at point *B*.

appropriate amount of pure 3. The composition of *B* cannot be obtained directly from the known masses of the materials used in the titration since, initially, only the composition at *A* is known. A graphical procedure for determining the composition of *B* is illustrated in Fig. 1.10. The curve *YGY′* is the locus of points *G* which satisfy the material balance $FG/G3 = M_3/M_B$ where M_3 is the mass of pure 3 added to the sample *B* and M_B is the mass of sample *B*. The curve *XHX′* is the locus of points *H* which satisfy the material balance $JH/J2 = M_2/M_D = M_2/(M_3 + M_B)$. The intersection of *YGY′* and *XHX′* gives the composition at *D* from which the required composition of *B* can readily be obtained by calculation or by extrapolating the line *D*3 back to the binodal curve. A similar experimental and analytical procedure is used for determining the composition of layer *C*. The success of the method depends largely on an accurate knowledge of the binodal curve. The results can be checked, however, because the points *A*, *B*, and *C* must lie on a straight line.

1.5.4 Graphical representation of data
[see also Pratt and Stevens, Ch. 6]

The ternary phase diagrams presented so far in this chapter have been constructed on a Gibbs equilateral triangle. The advantage of this lies in the impartial treatment of all components, so that it is frequently most convenient for theoretical discussion. In practice, it is often of less value as it requires special graph paper and it is sometimes advantageous to give more importance to a particular part of the diagram.

An alternative is to use rectangular coordinates as shown in Fig. 1.11 which can be imagined as a deformed Gibbs triangle. The mole fractions

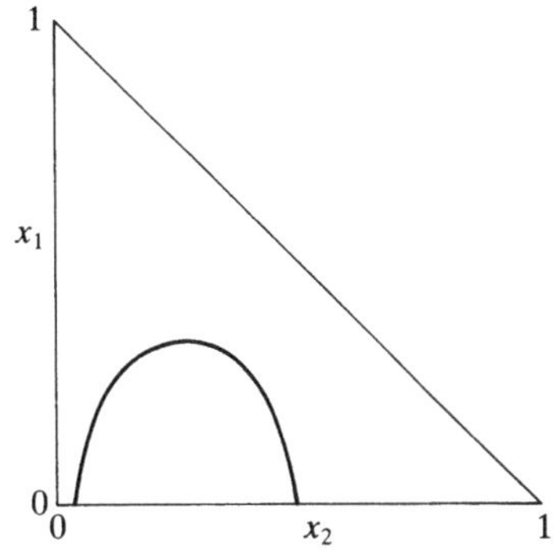

Fig. 1.11. Ternary phase diagram using rectangular coordinates.

of components 1 and 2 are plotted as ordinate and abscissa, respectively. Pure 3 lies at the origin and, of course, $x_3 = 1 - x_1 - x_2$.

Sometimes, however, a less symmetric choice of composition variables is useful. If one constituent (c) is the solvent we can use mole ratio variables on a 'solvent-free' basis. These are defined as

$$X_i = n_i / \sum_{i=1}^{c-1} n_i = x_i / \sum_{i=1}^{c-1} x_i = x_i/(1 - x_c) \tag{1.75}$$

so that

$$\sum_{i=1}^{c-1} X_i = 1, X_c = x_c/(1 - x_c) \text{ and } \sum_{i=1}^{c} X_i = 1/(1 - x_c) \tag{1.76}$$

A ternary system can then be represented by plotting X_3 as ordinate and X_2 as abscissa, as shown in Fig. 1.12. This diagram, a semi-infinite strip, expands the area of the Gibbs triangle around the solvent by allowing the solvent vertex to recede to infinity.

Another possibility is to express the concentrations of the solutes as their mole ratios with the solvent on a 'solutes-free' basis. Thus

$$X_i = n_i/n_c = x_i/x_c \tag{1.77}$$

so that

$$\sum_{i=1}^{c-1} X_i = (1 - x_c)/x_c \text{ and } X_c = 1 \tag{1.78}$$

As shown in Fig. 1.13, a ternary system can be represented using the first quadrant with pure 3 at the origin, pure 1 at infinity on the abscissa and pure 2 at infinity on the ordinate. This diagram has the feature of expanding the

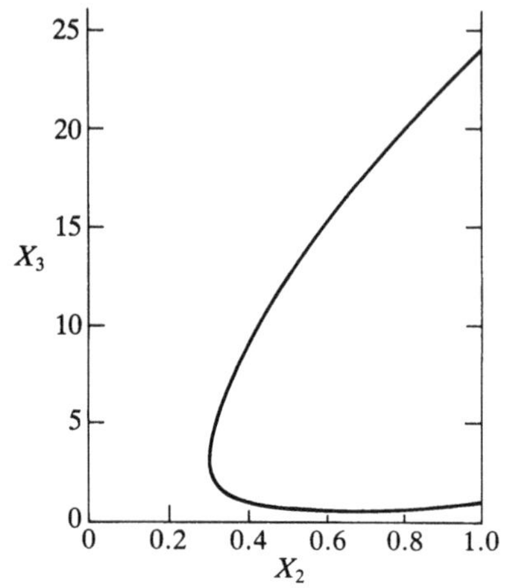

Fig. 1.12. Solvent-free composition diagram.

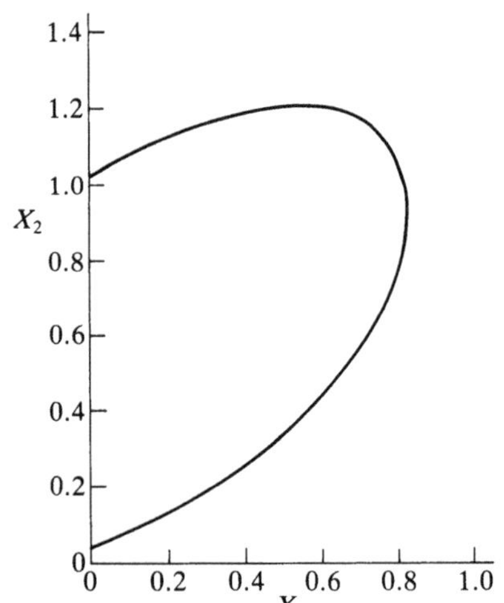

Fig. 1.13. Solutes-free composition diagram.

area of the Gibbs triangle around the 12 binary by allowing that side of the triangle to recede to infinity.

Similarly, three dimensional diagrams can be employed for quaternary systems (tetrahedra, slabs, and octants, cf. triangles, strips, and quadrants) but the difficulty of constructing such models usually precludes their quantitative use.

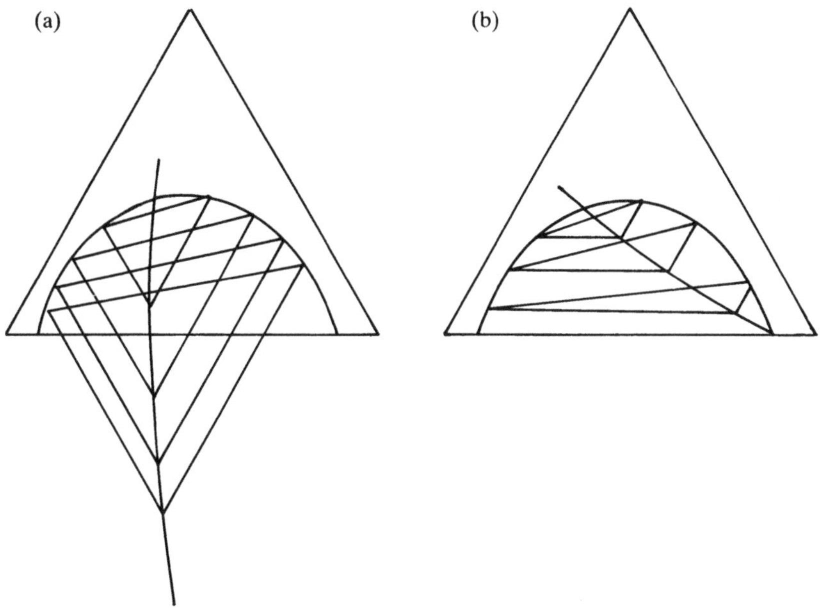

Fig. 1.14. Interpolation of tie lines: (a) external construction, (b) internal construction.

1.5.5 Correlation of tie lines

The quantitative use of ternary phase diagrams requires the construction of a sufficiently dense set of tie lines to permit interpolation. This may be achieved by use of auxiliary curves constructed by drawing lines through the ends of the available tie lines and parallel to two sides of the Gibbs triangle. Two possible choices are shown in Fig. 1.14. The auxiliary curves usually show only slight curvature and can be extrapolated to estimate the plait point composition. Construction of additional tie lines is simply a reversal of the above procedure.

Whilst the graphical methods described above are still used in process engineering calculations it is true to say that they have been largely superceded by computer-aided design methods.

1.6 Choice of solvent

1.6.1 General

The selection of an effective solvent for use in a solvent extraction process is controlled by several factors. The solvent must preferably be chemically stable, of low toxicity, non-corrosive, cheap, and easily recoverable from the extract. In addition, the solvent must have a favourable selectivity. The latter quantity is defined as follows:

$$\beta_{ij} \equiv k_i/k_j = \frac{x_i'}{x_i''} \cdot \frac{x_j''}{x_j'} = \frac{\gamma_i''}{\gamma_i'} \cdot \frac{\gamma_j'}{\gamma_j''} \quad (1.79)$$

where k_i and k_j are the partition coefficients for components i and j. The maximum value of the selectivity usually occurs at infinite dilution when

$$\beta_{ij}^{\infty} \simeq (\gamma_j^{\infty}/\gamma_i^{\infty})' \quad (1.80)$$

Not only is a high value of the selectivity required but it is also necessary to have a high solvent capacity or solvent power so as to reduce the amount of solvent required for the process. This property can be quantified, approximately, using the following relation

$$x_i'(\max) \simeq (1/\gamma_i^{\infty})' \quad (1.81)$$

Unfortunately, a high selectivity does not necessarily imply a high capacity and vice versa.

A further requirement is that the amount of solvent lost in the raffinate phase should be as small as possible. A measure of the solvent loss is given by the inverse of the activity coefficient of the solvent at infinite dilution in the raffinate phase (Brignole *et al.* 1986).

$$\text{Solvent loss} = (1/\gamma_S^{\infty})'' \quad (1.82)$$

1.6.2 Empirical methods

Many empirical techniques have been proposed to determine the selectivity of solvents. Only a few of these are outlined here.

Ewell *et al.* (1944) devised a system in which the deviations from ideal behaviour of liquids could be described by classifying liquids into five classes depending upon their hydrogen bonding characteristics. Francis (1944) extended this concept to define eight classes of liquid together with a scheme for predicting the immiscibility of any pair as shown in Fig. 1.15. An even more sophisticated diagram including 31 classes of liquid has been devised by Godfrey (1972) and this probably represents the best available empirical method.

1.6.3 Methods based on the regular solution model

Ashton *et al.* (1983) have described a method for selecting solvents based on a modified regular solution model. They used the following expression for the activity coefficient:

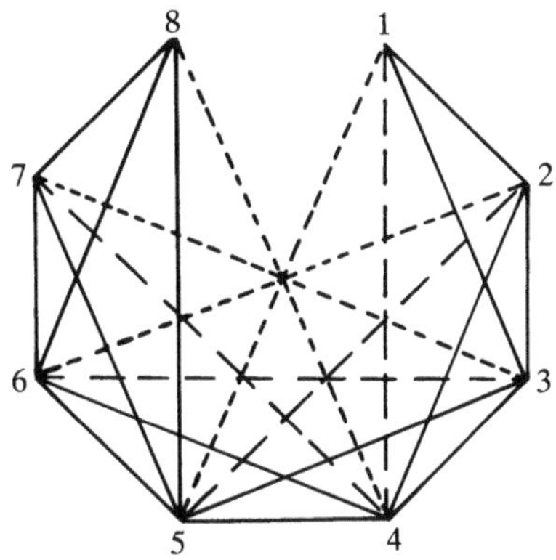

Fig. 1.15. Mutual solubilities of various liquids. —, complete miscibility; ---, high mutual solubility, ----, low mutual solubility; no line, very low mutual solubility. The following substances are typical of the eight classes of liquid numbered on the diagram: 1 = water, 2 = diethylene glycol, 3 = triethylene glycol, 4 = furfural, 5 = diethyl ether, 6 = benzene, 7 = cyclohexane, 8 = heptane (after Francis 1944).

$$\ln \gamma_i = \ln \phi_i/x_i + 1 - \phi_i/x_i + \frac{V_i}{RT}\{(\delta_{D,i} - \bar{\delta}_D)^2 + A_i[(\delta_{p,i} - \bar{\delta}_p)^2 + (\delta_{H,i} - \bar{\delta}_H)^2]\} \tag{1.83}$$

In this equation the solubility parameter has been separated into three contributions; δ_D is the solubility parameter for the dispersion forces, δ_p that for dipole interactions, δ_H that for hydrogen bonding interactions and A_i is a weighing factor. The overall solubility parameter can be estimated as follows:

$$\delta^2 = \delta_D^2 + \delta_p^2 + \delta_H^2 \tag{1.84}$$

Ashton *et al.* (1983) give a table from which values of δ_p and δ_H can be estimated using a group contribution method together with typical values of the weighting parameter A_i. For the extraction of component i from component j using a solvent S, these authors show that it is the dipole and hydrogen bonding solubility parameters that are most important in determining the selectivity and partition coefficient. They show that

$$k_i \propto r_{i,S}^{-2} \tag{1.85}$$

and

$$\beta_{ij} \propto (r_{j,S}/r_{i,S})^2 \tag{1.86}$$

where $r_{i,S}$ and $r_{j,S}$ are the distances between the points i and S and j and S

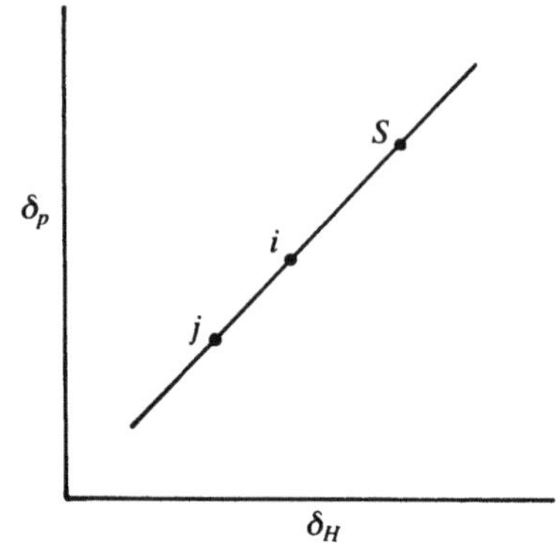

Fig. 1.16. Solvent selection on the δ_p–δ_H plane.

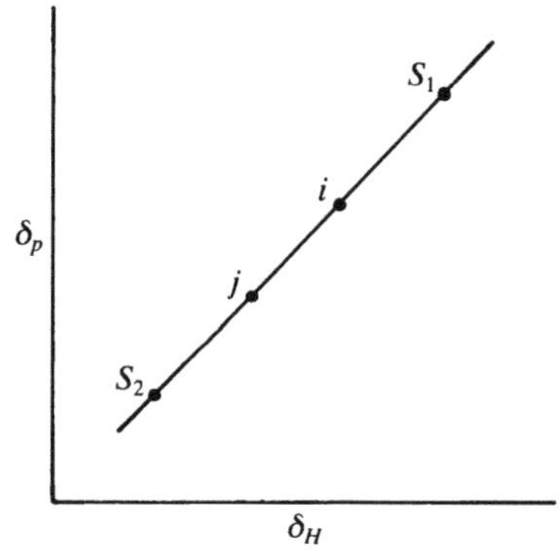

Fig. 1.17. Diagram for two-solvent systems.

plotted on a δ_p–δ_H plane as shown in Fig. 1.16. The maximum selectivity is obtained for the maximum separation of points j and S and the minimum separation of points i and S. The best solvent, therefore, lies on the line joining i and j, closer to i than j. However, as the distance between i and S is increased the partition coefficient decreases and therefore the solvent capacity. Point S must therefore be located so that the solvent capacity does not decrease to an unacceptably low value.

The method can also be applied to two-solvent systems and to mixed-solvent systems. The former case is illustrated in Fig. 1.17 and the latter in Fig. 1.18 where it should be noted that the solvents S_1 and S_2 should be positioned so that they do not tend to demix as would happen for solvents S_3 and S_4.

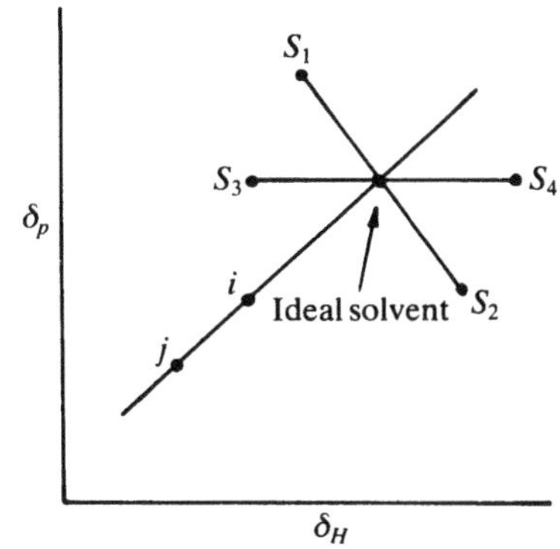

Fig. 1.18. Diagram for mixed solvents.

1.6.4 Methods based on the UNIFAC model

Since the UNIFAC model allows prediction of liquid–liquid equilibria from a knowledge only of group properties, it can be used to advantage for the selection of solvents. The required methodology has been devised by Gani and Brignole (1983) and Brignole *et al.* (1986). These authors identified four steps in the selection of a solvent.

1. Characterization of the groups in terms of their valency, geometry and chemical type.

2. Selection of groups for a particular problem using the interchange energy (w) defined by Magnussen *et al.* (1981) and which can be calculated from a summation of the group interaction parameters,

$$a_{mn} + a_{nm} = \left[u_{mn} - \frac{1}{2}(u_{mm} + u_{nn})\right]\frac{2}{R} = \frac{w \cdot 2}{R} \tag{1.87}$$

   where u_{mn} is the interaction energy between groups m and n. Large differences between the values of w for two groups interacting with the same solvent gives an indication of high selectivity.

3. Solution of the combinatorial problem. Here, the groups are combined according to rules dictated by the classification of step 1.

4. Evaluation of the solvent properties and final selection of the solvent. Here, the best solvent is selected by inspecting the value of the solvent loss in the raffinate and the product of the selectivity and solvent power.

1.7 Applications

1.7.1 Organic systems

The results of applying the NRTL and UNIQUAC equations to a typical Type I ternary system are shown in Fig. 1.19 where the experimental tie lines and fitted binodal curves have been plotted. Also shown is the binodal curve predicted from UNIQUAC parameters determined from experimental results for other systems. The average deviations between calculated and experimental mole fractions are given below.

NRTL	(specific parameters)	0.0069
UNIQUAC	(specific parameters)	0.0103
UNIQUAC	(common parameters)	0.0202

In this case the NRTL equation gives the best fit, although in general the UNIQUAC equation is marginally superior. The UNIQUAC common parameters do not give a satisfactory prediction in this case.

Whilst organic systems of this relatively simple type can usually be satisfactorily correlated as shown above, it should be noted that one of the major applications of solvent extraction is in the processing of complex biochemical products and the modelling of such systems is far more difficult.

1.7.2 Ionic systems

Another important application of solvent extraction is in the field of hydrometallurgy. Li *et al.* (1987) have modelled a process for the extraction

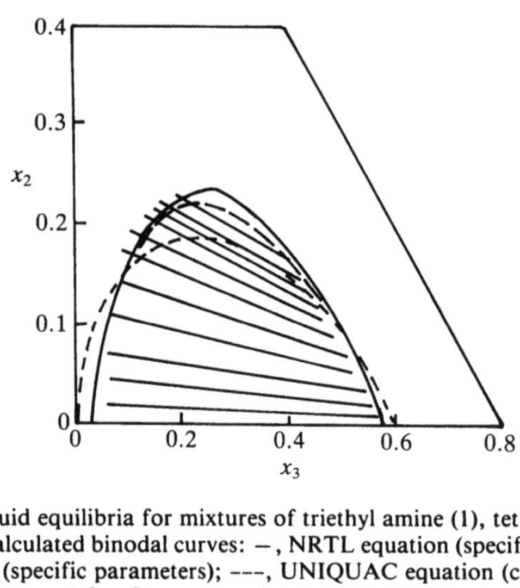

Fig. 1.19. Liquid–liquid equilibria for mixtures of triethyl amine (1), tetrachloromethane (2), and acetic acid (3). Calculated binodal curves: —, NRTL equation (specific parameters); — —, UNIQUAC equation (specific parameters); ---, UNIQUAC equation (common parameters); (after Sorensen and Arlt 1979, 1980).

of cobalt and copper from aqueous solutions using the complexing agent EHEHPA $[(CH_3(CH_2)_3CH(C_2H_5)CH_2)_2PO_3H]$.

The aqueous phase contains sulphuric acid, cobalt sulphate and copper sulphate. The sulphuric acid dissociates as follows:

$$H_2SO_4 \rightarrow H^+ + HSO_4^- \tag{1.88}$$

$$HSO_4^- \rightleftharpoons H^+ + SO_4^{2-} \tag{1.89}$$

and the ionic species in the solution are H^+, Co^{2+}, Cu^{2+}, HSO_4^- and SO_4^{2-}. It is assumed that the first dissociation is complete so that no molecular H_2SO_4 is present in the solution and OH^- is neglected because of the low pH. The concentrations of the ionic species are constrained by the following relations.

$$m(H_2SO_4) + m(CoSO_4) + m(CuSO_4) = m(HSO_4^-) + m(SO_4^{2-}) \tag{1.90}$$

$$m(H^+) + 2m(Co^{2+}) + 2m(Cu^{2+}) = m(HSO_4) + 2m(SO_4^{2-}) \tag{1.91}$$

$$K_a = \frac{m(H^+)\, m(SO_4^{2-})\, \gamma(H^+)\, \gamma(SO_4^{2-})}{m(HSO_4^-)\, \gamma(HSO_4^-)} \tag{1.92}$$

where K_a is the equilibrium constant for the second dissociation of sulphuric acid.

The activity coefficient of single ions are represented by the simplified Pitzer equation (eqn (1.64)). By combining eqns (1.91–1.93) and eqn (1.64), the equilibrium concentrations and activity coefficients of all ions can be evaluated from a set of experimental results. Typical values are given in Table 1.1 from which it can be seen that the experimental and calculated values of the pH are in close agreement.

There are five components in the organic phase. These are as follows: (1) H_2O, (2) $n-C_7H_{16}$—a diluent, (3) H_2R_2—the dimer of EHEHPA, (4) $Co(HR_2)_2$, (5) $Cu(HR_2)_2$. The extraction reactions and corresponding thermodynamic equilibrium constants are detailed below.

TABLE 1.1 *Comparison or experimental and calculated values of the pH of $H_2SO_4/CoSO_4/CuSO_4$ solutions*

$m(CO^{2+})$	$m(Cu^{2+})$	$m(HSO_4^-)$	$m(SO_4^{2-})$	$m(H^+)$	pH(expt)	pH(calc)
0.075	0,013	0.0022	0.007	0.0023	2.77	2.84
0.107	0.020	0.0024	0.127	0.0023	2.84	2.87
0.078	0.020	0.0012	0.098	0.0012	3.06	3.12
0.108	0.260	0.0012	0.134	0.0011	3.15	3.18

$$Co^{2+} + 2H_2R_2 \rightleftharpoons Co(HR_2)_2 + 2H^+ \quad (1.93)$$

$$Cu^{2+} + 2H_2R_2 \rightleftharpoons Cu(HR_2)_2 + 2H^+ \quad (1.94)$$

$$K_a(Co) = \frac{[m(H^+)\gamma(H^+)]^2\gamma_4 x_4}{m(Co^{2+})\gamma(Co^{2+})\gamma_3^2 x_3^2} = K(Co)\gamma_4/\gamma_3^2 \quad (1.95)$$

$$K_a(Cu) = \frac{[m(H^+)\gamma(H^+)]^2\gamma_5 x_5}{m(Cu^{2+})\gamma(Cu^{2+})\gamma_3^2 x_3^2} = K(Cu)\gamma_5/\gamma_3^2 \quad (1.96)$$

Further relationships between the activity coefficients can be obtained by differentiating eqns (1.95 and 1.96) and also from the Gibbs–Duhem equation.

$$d \ln K(Co) + d \ln \gamma_4 - 2d \ln \gamma_3 = 0 \quad (1.97)$$

$$d \ln K(Cu) + d \ln \gamma_5 - 2d \ln \gamma_3 = 0 \quad (1.98)$$

$$\sum_{i=1}^{5} x_i d \ln \gamma_i = 0 \quad (1.99)$$

Since the activity of water in the organic phase is equal to that in the aqueous phase, the activity coefficient of water can be obtained from eqns (1.59 and 1.69). The activity coefficient of heptane can be obtained from measurements of its partial pressure over the organic phase.

$$\gamma_2 = p_2/x_2 P_2^0 \quad (1.100)$$

The values of γ_3, γ_4 and γ_5 can be obtained by solving eqns (1.97–1.99) but it should be noted that for these components a pure liquid reference state is inconvenient and a working reference state must be defined equivalent to some arbitrary but fixed concentration at which the activity coefficients are unity. Thus the activity coefficients of all five components can be obtained from measured equilibrium concentrations. Li *et al.* (1987) fitted these to an improved regular solution model (Li *et al.* 1986) and obtained the results shown in Table 1.2. The good agreement between experimental and calculated activity coefficients and the constancy of the thermodynamic equilibrium constants confirms the general validity of the model.

1.8 Sources of Data

All published liquid–liquid equilibrium data have been collected and correlated with the NRTL and UNIQUAC equations to form a computerized data bank by Sorensen and Arlt (1979, 1980). In addition, a computer-aided bibliography has been published by Wisniak and Tamir (1980). All else failing, one can always make a measurement!

TABLE 1.2 *Comparison of experimental and calculated values of activity coefficients for the $H_2O/n-C_7H_{16}/H_2R_2/Co(HR_2)$, $Cu(HR_2)_2$ organic phase*

x_1	γ_1(expt)	γ_1(calc)	x_2	γ_2(exp)	γ_2(calc)	x_3	γ_3(calc)
0.056	17.73	17.39	0.824	1.046	1.049	0.113	2.943
0.067	14.83	17.49	0.822	1.049	1.048	0.095	2.720
0.073	13.50	13.69	0.822	1.049	1.048	0.085	2.916
0.086	11.64	11.89	0.817	1.055	1.052	0.069	3.336
0.093	10.76	10.50	0.816	1.059	1.055	0.055	3.717

x_4	γ_4(calc)	x_5	γ_5(calc)	K_a(Co) × 10^5	K_a(Cu) × 10^5
0.0010	6.052	0.0061	1.848	1.238	10.79
0.0025	5.129	0.013	1.820	1.239	10.79
0.0044	4.745	0.016	1.788	1.239	10.79
0.0064	4.056	0.022	1.686	1.239	10.79
0.0068	4.000	0.029	1.655	1.239	10.79

Acknowledgement

I am indebted to Mr R. Stevenson for preparation of the illustrations.

Nomenclature

A_i	weighting parameter for component i
A_k	van der Waals group surface area
A_{wi}	van der Waals molecular surface area
A_γ	Debye-Hückel parameter for the activity coefficient
A_ϕ	Debye-Hückel parameter for the osmotic coefficient
a_i	activity of component i
a_{nm}	group-group interaction parameter
B, B_{Ma}, B_{ca}	second virial coefficients
b	constant in Pitzer's equation
$C, C_\phi, C_{Ma}, C_{ca}$	third virial coefficients
c	number of components
f	number of degrees of freedom or function of ionic strength
G	molar Gibbs free energy
$\mathbf{G}$	total Gibbs free energy
G^E	molar excess Gibbs free energy
$\mathbf{G}^E$	total excess Gibbs free energy
ΔG^0	standard Gibbs free energy change
G_{ij}	parameter in NRTL equation
g_{ij}	parameter in NRTL equation
H	molar enthalpy

$\mathbf{H}$	total enthalpy
H_i^0	pure component enthalpy
$\bar{H}_i$	partial molar enthalpy
ΔH_i^0	heat of vaporization of pure component i
I	ionic strength
K	equilibrium constant
K_a	thermodynamic equilibrium constant
k_i	distribution ratio
M	mass
M_0	relative molecular mass
m_i	molality
$m_i^{\ominus}$	unit molality
$m_{\pm}$	mean ionic molality
n	number of moles
P	pressure
Q_k	group surface area parameter
q_i	surface area parameter in the UNIQUAC and UNIFAC equations
R	gas constant
R_k	group volume parameter
r_i	volume parameter in the UNIQUAC and UNIFAC equations
r_{ij}	distance between points i and j in Fig. 1.16
S	molar entropy
T	temperature
ΔU_i^0	internal energy change on vaporization of pure component i
u_{ij}	group interaction energy or parameter in the UNIQUAC equation
V	molar volume
$\mathbf{V}$	total volume
V_i^0	pure component molar volume
$\bar{V}_i$	partial molar volume
V_k	van der Waals group volume
V_{wi}	van der Waals molecular volume
V_{m}	weighted molar volume
w	interchange energy
X_i	mole ratio
X_m	group fraction
x_i	mole fraction
z	co-ordination number
z_i	overall mole fraction or ionic valence
z_+	cation valence
z_-	anion valence

Greek symbols

$\alpha, \alpha_1, \alpha_2$	parameters in Pitzer's equation
α_{ij}	non-randomness parameter in the NRTL equation
$\beta^{(0)}, \beta^{(1)}, \beta^{(2)}$	parameters in Pitzer's equation

β_{ij}	selectivity
Γ_k	residual group activity coefficient
Γ_{ki}	residual group activity coefficient in a mixture containing only molecules of type i
γ_i	activity coefficient of component i
γ_i^C	combinatorial contribution to the activity coefficient
γ_i^R	residual contribution to the activity coefficient
γ_0	solvent activity coefficient
$\gamma_\pm$	mean ionic activity coefficient
δ_i	solubility parameter
$\bar{\delta}$	weighted value of δ_i
θ_i	area fraction of the UNIQUAC and UNIFAC equations
λ_{ij}	second virial coefficient
μ_i	chemical potential
μ_{ij}	second derivative of Gibbs free energy
μ_{ijk}	third virial coefficient
ν_+	cation stoichiometric number
ν_-	anion stoichiometric number
ν	$\nu_+ + \nu_-$
ν_{ki}	number of groups of type k in molecule i
τ_{ij}	parameter in the NRTL and UNIQUAC equations
ϕ	osmotic coefficient or number of phases
ϕ_i	segment fraction in the UNIQUAC and UNIFAC equations and volume fraction in the regular solution model
ψ_{nm}	parameter in the UNIFAC equation

Subscripts

a	anion
c	cation or component
D	dispersion force contribution
H	hydrogen bonding contribution
i, j, k, m, n	component or group
M	cation
p	dipole contribution
S	solvent
X	anion
α	phase
0	solvent
+	cation
−	anion
±	mean ionic value

Superscripts

*	standard state
⊖	unit molality
′	liquid phase L'
″	liquid phase L''

0 pure component
∞ infinite dilution

References

ABRAMS, D.S. and PRAUSNITZ, J.M. (1975). *American Institute of Chemical Engineers Journal*, **21**, 62.

ASHTON, N.F., MCDERMOTT, C., and BRENCH, A. (1983). In *Handbook of solvent extraction* (ed. T.C. LO, M.H.I. BAIRD, and C. HANSON), Chap. 1, pp. 3–35. Wiley, New York.

BALL, F.-X., PLANCHE, H., FÜRST, W., and RENON, H. (1985). *American Institute of Chemical Engineers Journal*, **31**, 1233.

BARTON, A.F.M. (1983). *Handbook of solubility parameters and other cohesion parameters*. CRC Press, Boca Raton, Florida.

BONDI, A. (1968). *Physical properties of molecular crystals, liquids and gases*. Wiley, New York.

BRIGNOLE, E.A., BOTTINI, S., and GANI, R. (1986). *Fluid Phase Equilibria*, **29**, 125.

CABEZAS, H.J. and O'CONNELL, J.P. (1986). *Fluid Phase Equilibria*, **30**, 213.

DLUZNIEWSKI, H.J. and ALDER, S.B. (1972). *Institution of Chemical Engineers Symposium Series*, No. 35, 4 : 21.

EDWARDS, T.J., MAURER, G., NEWMAN, J., and PRAUSNITZ, J.M. (1978). *American Institute of Chemical Engineers Journal*, **24**, 966.

EWELL, R.H., HARRISON, J.M., and BERG, L., (1944). *Industrial and Engineering Chemistry*, **36**, 871.

FRANCIS, A.W. (1944). *Industrial and Engineering Chemistry*, **36**, 1096.

FREDENSLUND, A., GMEHLING, J., MICHELSON, M.L., RASMUSSEN, P., and PRAUSNITZ, J.M. (1977). *Industrial and Engineering Chemistry: Process Design and Development*, **16**, 450.

FREDENSLUND, A., JONES, R.L., and PRAUSNITZ, J.M. (1975). *American Institute of Chemical Engineers Journal*, **21**, 1086.

FREDENSLUND, A., GMEHLING, J., and RASMUSSEN, P. (1977). *Vapour–liquid equilibria using UNIFAC*. Elsevier, Amsterdam.

GANI, R. and BRIGNOLE, E.A. (1983). *Fluid Phase Equilibria*, **13**, 331.

GMEHLING, J. (1986). *Fluid Phase Equilibria*, **30**, 119.

GMEHLING, J., ONKEN, U., ARLT, W., GRENZHEUSER, P., and WEIDLICH, U. (1977–1983). *Vapour–liquid equilibrium data collection*, Chemistry data series, Vol. I, Parts 1, 1a, 2a–d, 3–5, 6a–c, 7,8. DECHEMA, Frankfurt, Germany.

GMEHLING, J., RASMUSSEN, P., and FREDENSLUND, A. (1982). *Industrial and Engineering Chemistry: Process Design and Development*, **21**, 118.

GODFREY, N.B. (1972). *Chemical Technology*, **2**, 359.

GUGGENHEIM, E.A. (1952). *Mixtures*, Oxford University Press.

HILDEBRAND, H.J. and SCOTT, R.L. (1962). *Regular solutions*. Prentice-Hall, Englewood Cliffs, New Jersey.

HILDEBRAND, J.H. and WOOD, S.E. (1933). *Journal of Chemical Physics*, **1**, 817.

HILDEBRAND, H.J., PRAUSNITZ, J.M., and SCOTT, R.L. (1970). *Regular and related solutions*. Van Nostrand Reinhold, New York.

LI, Y., TENG, T., LU, J., CHEN, G., and LI, J. (1986). *Fluid Phase Equilibria*, **30**, 297.

LI, Y., LU, J., LI, Z., BAO, T., LI, J., and TENG, T. (1987). In *Separation processes in hydrometallurgy* (ed. G.A. DAVIES), pp. 238-47. Ellis Horwood, Chichester.

LILJENZIN, J.L. (1979). In *Proceedings of the international solvent extraction conference*, ISEC 77, Vol. 1, CIM Special Vol. 21, 295. The Canadian Institute of Mining and Metallurgy, Montreal, Canada.

MAGNUSSEN, T., RASMUSSEN, P., and FREDENSLUND, A. (1981). *Industrial and Engineering Chemistry: Process Design and Development*, **20**, 331.

MAURER, G. (1983). *Fluid Phase Equilibria*, **13**, 169.

MOCK, B., EVANS, L.B., and CHEN, C.-C. (1986). *Fluid Phase Equilibria*, **30**, 213.

NEWSHAM, D.M.T. and NG, S.B. (1972). *Journal of Chemical and Engineering Data*, **17**, 205.

PITZER, K.S. (1979). In *Activity coefficients in electrolyte solutions* (ed. R.M. PYTKOWICZ), Vol. I, Chap. 7, pp. 157-208. CRC Press, Baton Rouge, Florida.

PITZER, K.S. and SILVESTER, L.F. (1976). *Journal of Solution Chemistry*, **5**, 269.

PITZER, K.S., ROY, R.N., and SILVESTER, L.F., (1977). *Journal of the American Chemical Society*, **99**, 4930.

REINHARDT, H. and RYDBERG, J. (1969). *Acta Chemical Scandinavica*, **23**, 2773.

RENON, H. (1986). *Fluid Phase Equilibria*, **30**, 181.

RENON, H. and PRAUSNITZ, J.M. (1968). *American Institute of Chemical Engineers Journal*, **14**, 135.

SANDER, B., FREDENSLUND, A., and RASMUSSEN, P. (1986). *Chemical Engineering Science*, **41**, 1171.

SCATCHARD, G. (1931). *Chemical Research*, **8**, 321.

SCOTT, R.L. (1956). *Journal of Chemical Physics*, **25**, 193.

SKJOLD-JORGENSEN, S., KOLBE, B., GMEHLING, J., and RASMUSSEN, P. (1979). *Industrial and Engineering Chemistry: Process Design and Development*, **18**, 714.

SORENSEN, J.M. and ARLT, W. (1979,1980). *Liquid-liquid equilibrium data collection*, Chemistry Data Series, Vol. V, Parts 1-3. DECHEMA, Frankfurt, Germany.

SORENSEN, J.M., MAGNUSSEN, T., RASMUSSEN, P., and FREDENSLUND, A. (1979). *Fluid Phase Equilibria*, **3**, 47.

WISNIAK, J. and TAMIR, A. (1980). *Liquid-liquid equilibrium and extraction*. Elsevier, Amsterdam.

ZEMAITIS, J.F., CLARK, D.M., RAFAL, M., and SCRIVNER, N.C. (1986). *Handbook of aqueous electrolyte thermodynamics*. American Institute of Chemical Engineers, New York.

2

INTERPHASE MASS TRANSFER

A.H.P. SKELLAND

2.1 Diffusion in liquid systems

1.1 Flux relations and Fick's first law

The applications of liquid–liquid extraction are increasing in diversity and this has led to a corresponding proliferation of extraction devices for industrial use. This chapter will accordingly focus on those fundamental aspects of diffusion and mass transfer that provide a unifying foundation for many of the extraction processes that may be encountered.

Two relatively immiscible liquid phases are brought into intimate contact to enable a component of one liquid phase to be transferred into the other. This is accomplished by the transport of the component in question—the solute—from the bulk of one phase to the interface between phases, then across the interface, and finally from the interface into the bulk of the adjacent phase. Depending upon the hydrodynamics of the situation, this may be brought about by molecular diffusion alone, or by a combination of molecular diffusion and convection. The former process will first be considered in some detail.

To quantify concentrations, the moles and mass of component A per unit volume of solution will be symbolized by c_A and ρ_A, respectively. If the total moles and total mass per unit volume are then written as c and ρ, the mole fraction of A becomes c_A/c or x_A, and the mass fraction of A is ρ_A/ρ or w_A.

A non-uniform fluid mixture of n components is undergoing bulk motion. Suppose that the statistical mean velocity of component i in the z-direction with respect to stationary coordinates is u_i. The definition of the molal average velocity of the mixture in the z-direction is then as follows:

$$U = \frac{1}{c}\sum_{i=1}^{n} c_i u_i \tag{2.1}$$

and the mass average velocity in the z-direction is

$$u = \frac{1}{\rho}\sum_{i=1}^{n} \rho_i u_i \tag{2.2}$$

The molal and mass fluxes of component i in the z-direction may then be defined in three different coordinate systems as shown below.

Molal flux relative to stationary coordinates: $N_{iz} = c_i u_i$ (2.3)

Mass flux relative to stationary coordinates: $n_{iz} = \rho_i u_i$ (2.4)

Molal flux relative to the mass average velocity: $I_{iz} = c_i (u_i - u)$ (2.5)

Mass flux relative to the mass average velocity: $i_{iz} = \rho_i (u_i - u)$ (2.6)

Molal flux relative to the molal average velocity: $J_{iz} = c_i (u_i - U)$ (2.7)

Mass flux relative to the molal average velocity: $j_{iz} = \rho_i (u_i - U)$ (2.8)

This permits the various fluxes to be interrelated as illustrated in the following examples.

The difference between the mass fluxes j_{iz} and i_{iz} will first be formulated. Consider eqns (2.1, 2.2, 2.3, 2.4, 2.6, and 2.8):

$$j_{iz} = \rho_i u_i - \rho_i U$$

$$i_{iz} = \rho_i u_i - \rho_i u$$

$$\begin{aligned} j_{iz} - i_{iz} &= \rho_i (u - U) = \rho_i u - \rho_i U \\ &= \frac{\rho_i}{\rho} \sum_{i=1}^{n} \rho_i u_i - \frac{\rho_i}{c} \sum_{i=1}^{n} c_i u_i \\ &= w_i \sum_{i=1}^{n} n_{iz} - \frac{c_i M_i}{c} \sum_{i=1}^{n} N_{iz} \end{aligned} \tag{2.9}$$

And for the binary system of A and B,

$$\begin{aligned} j_{Az} - i_{Az} &= w_A (n_{Az} + n_{Bz}) - \frac{c_A M_A}{c} (N_{Az} + N_{Bz}) \\ &= w_A (n_{Az} + n_{Bz}) - x_A \left(n_{Az} + \frac{M_A}{M_B} n_{Bz} \right) \end{aligned} \tag{2.10}$$

since $N_{Az} M_A = n_{Az}$; $N_{Bz} M_B = n_{Bz}$. When $M_A = M_B$, $w_A = x_A$, so $j_{Az} = i_{Az}$. In this case, therefore, $u = U$, a coincidence that occurs also in the case of very dilute A in B, and again in the bulk flow of uniform mixtures, regardless of unequal molecular weights.

The relation between the molal fluxes J_{iz} and N_{iz} will next be determined. Consider eqns (2.1, 2.3, and 2.7):

$$J_{iz} = c_i u_i - c_i U$$

$$= N_{iz} - \frac{c_i}{c}\sum_{i=1}^{n} c_i u_i = N_{iz} - x_i \sum_{i=1}^{n} N_{iz} \qquad (2.11)$$

and for the binary system of A and B,

$$J_{Az} = N_{Az} - x_A\,(N_{Az} + N_{Bz}) \qquad (2.12)$$

also

$$\sum_{i=1}^{n} J_{iz} = \sum_{i=1}^{n} c_i u_i - U\sum_{i=1}^{n} c_i = Uc - Uc = 0 \qquad (2.13)$$

In a non-uniform fluid mixture each component undergoes a net migration in the direction of decreasing concentration of that component. When the bulk of the fluid is either motionless or in laminar flow in a direction perpendicular to the concentration gradient, this mass transfer occurs as the result of random motion of the molecules and is called molecular diffusion. The process is described by Fick's first law of diffusion, and is written for non-reacting binary systems of A and B undergoing steady one-dimensional transfer as

$$J_{Az} = -D_{AB}\frac{\mathrm{d}c_A}{\mathrm{d}z}\,;\; J_{Bz} = -D_{BA}\frac{\mathrm{d}c_B}{\mathrm{d}z} \qquad (2.14)$$

$$i_{Az} = -D_{AB}\frac{\mathrm{d}\rho_A}{\mathrm{d}z}\,;\; i_{Bz} = -D_{BA}\frac{\mathrm{d}\rho_B}{\mathrm{d}z} \qquad (2.15)$$

where constant c and constant ρ are required along z for eqns (2.14 and 2.15), respectively. More generally,

$$J_{Az} = -cD_{AB}\frac{\mathrm{d}x_A}{\mathrm{d}z}\,;\; J_{Bz} = -cD_{BA}\frac{\mathrm{d}x_B}{\mathrm{d}z} \qquad (2.16)$$

$$i_{Az} = -\rho D_{AB}\frac{\mathrm{d}w_A}{\mathrm{d}z}\,;\; i_{Bz} = -\rho D_{BA}\frac{\mathrm{d}w_B}{\mathrm{d}z} \qquad (2.17)$$

Equations (2.16 and 2.17) do not require constancy of c and ρ, respectively, along z. However, their greater generality is often offset, in practical terms, by lack of detailed knowledge of the variation of c or ρ over the relevant path of transport, z, rendering integration difficult. Equation (2.14 or 2.15) is consequently more often used as an expediency. This important restriction of constant c along z for eqn (2.14), and its relaxation for eqn (2.16), will now be demonstrated by expanding eqn (2.1) for the binary system $A + B$ to obtain

$$Uc = Uc_A + Uc_B = u_A c_A + u_B c_B \qquad (2.18)$$

and from eqns (2.7 and 2.14),

$$u_A c_A = Uc_A + J_{Az} = Uc_A - D_{AB}\frac{dc_A}{dz} \tag{2.19}$$

$$u_B c_B = Uc_B + J_{Bz} = Uc_B - D_{BA}\frac{dc_B}{dz} \tag{2.20}$$

Combining eqns (2.18–2.20) and assuming $D_{AB} = D_{BA}$,

$$\frac{d}{dz}(c_A + c_B) = \frac{dc}{dz} = 0 \tag{2.21}$$

which specifies the restriction on eqn (2.14). Equation (2.16) may next be expanded as follows:

$$J_{Az} = -cD_{AB}\frac{dx_A}{dz} = -D_{AB}\frac{dc_A}{dz} + \frac{D_{AB}}{c}c_A\frac{dc}{dz} \tag{2.22}$$

and with eqns (2.19 and 2.20),

$$u_A c_A = Uc_A - D_{AB}\frac{dc_A}{dz} + \frac{D_{AB}}{c}c_A\frac{dc}{dz} \tag{2.23}$$

$$u_B c_B = Uc_B - D_{BA}\frac{dc_B}{dz} + \frac{D_{BA}}{c}c_B\frac{dc}{dz} \tag{2.24}$$

Substituting eqns (2.23 and 2.24) in eqn (2.18) shows that

$$\frac{d}{dz}(c_A + c_B) = \frac{dc}{dz} \tag{2.25}$$

The truism in eqn (2.25) confirms that eqn (2.16) does not require constant c along the diffusion path z. It was assumed below eqn (2.20) that $D_{AB} = D_{BA}$. This in fact follows from eqn (2.16), noting that $dx_A/dz = -dx_B/dz$ for this binary system, so that

$$J_{Az} + J_{Bz} = c\frac{dx_A}{dz}(-D_{AB} + D_{BA}) \tag{2.26}$$

Equation (2.13) shows that this equation equals zero and therefore $D_{AB} = D_{BA}$.

The relationships developed so far enable analysis of the total flux of component A contained in a non-uniform binary mixture of A and B that is undergoing steady bulk flow in the direction z in which A is diffusing. Combination of eqns (2.12 and 2.14) leads to

$$N_{Az} = (N_{Az} + N_{Bz})\frac{c_A}{c} - D\frac{dc_A}{dz} \tag{2.27}$$

where the first and second terms on the right side of the equation represent the convective (bulk flow) and diffusive fluxes, respectively.

Equation (2.27) embraces two special cases of molecular diffusion that are often encountered. In equimolal counterdiffusion components A and B diffuse at equal molal rates through each other, but in opposite directions. This mechanism is often characteristic of binary distillation and requires that $N_{Az} = -N_{Bz}$, so that the convective flux term in eqn (2.27) is zero. Unimolal unindirectional diffusion is that process in which only one molecular species, A, diffuses through stationary B, such that N_{Bz} equals zero. This is often typical of gas absorption, liquid–liquid extraction, and adsorption operations.

In preparing eqn (2.27) for integration, careful consideration must be given to deciding which quantities remain constant along the transfer path. In the particular case of gaseous systems the kinetic theory shows that D_{AB} is essentially independent of solute concentration for the binary system A plus B. For liquid systems, on the other hand, D_{AB} frequently varies with concentration. In addition, the geometry of the system will determine whether N_{Az} is a function of z at steady state. When N_{Az} varies with z one proceeds by noting that $N_{Az} = q_{Az}/A_z$, where q_{Az} is the rate of mass transfer (moles of A/time) and A_z is the area normal to transfer at z. From the geometry of the system A_z is a known function of z, enabling substitution in eqn (2.27), separation of variables (c_A and z), integration, and solution for q_{Az}. The following example should make this procedure clear.

2.1.1.1 *Steady-state diffusion from a sphere*

Component A is diffusing steadily in a radial direction from all points on the surface of a stationary spherical droplet into an infinite, stagnant, surrounding fluid B. It is desired to describe this process analytically, assuming D and c are constants.

Since component B is stagnant eqn (2.27) may be written in the radial direction with $N_{Br} = 0$. Then

$$N_{Ar}\left(1 - \frac{c_A}{c}\right) = -D\frac{\mathrm{d}c_A}{\mathrm{d}r}$$

But $N_{Ar} = f(r) = q_{Ar}/4\pi r^2$, so

$$\frac{q_{Ar}}{4\pi Dc}\int_{d_s/2}^{\infty} r^{-2}\,\mathrm{d}r = -\int_{c_{As}}^{c_{A\infty}} \frac{\mathrm{d}c_A}{c - c_A}$$

where d_s is the sphere diameter, at which point $c_A = c_{As}$, and q_{Ar} is the total radial mass transfer rate at any $r \geq d_s/2$. Integrating,

$$q_{Ar} = 2\pi Dcd_s \ln\frac{c - c_{A\infty}}{c - c_{As}} \qquad (2.28)$$

Now a mass transfer coefficient k_c is defined as

$$q_{Ar} = k_c \pi d_s^2 (c_{As} - c_{A\infty}) \tag{2.29}$$

Eliminating q_{Ar} from eqns (2.28 and 2.29) and recalling that $c - c_A = c_B$,

$$N_{\text{Sh}} = \frac{k_c d_s}{D}\left(\frac{c_{Blm}}{c}\right) = 2 \tag{2.30}$$

where N_{Sh} is the Sherwood number and

$$c_{Blm} = \frac{c_{B\infty} - c_{Bs}}{\ln (c_{B\infty}/c_{Bs})} \tag{2.31}$$

It is instructive to note that the convective flux radially outward is not zero in this case, but is of such magnitude as to exactly counter the diffusive flux of B radially inward [due to $(c_{B\infty} - c_{Bs}) > 0$], so that B is motionless relative to stationary coordinates.

A simpler situation for eqn (2.27) is one in which N_{Az} and N_{Bz} are not functions of z. In this case, for constant D,

$$D\int_{c_{A1}}^{c_{A2}} \frac{dc_A}{N_{Az} - [(N_{Az} + N_{Bz})/c]\, c_A} = -\int_{z_1}^{z_2} dz$$

Integrating for $(N_{Az}, N_{Bz}) \neq f(z)$, and with $\gamma = N_{Bz}/N_{Az}$,

$$N_{Az} = \frac{Dc}{z}\left(\frac{1}{1+\gamma}\right) \ln\left[\frac{1 - (1+\gamma)(c_{A2}/c)}{1 - (1+\gamma)(c_{A1}/c)}\right] \tag{2.32}$$

Equation (2.32) reduces to the two special cases of equimolal counterdiffusion and unimolal unidirectional diffusion described earlier, in the following manner.

2.1.1.2 *Steady-state equimolal counterdiffusion*

In this mechanism $N_{Az} = -N_{Bz}$ so that $\gamma = -1$. L'Hôpital's rule may be applied to eqn (2.32) for this case, to obtain

$$N_{Az} = \frac{D}{z}(c_{A1} - c_{A2}) = \frac{Dc}{z}(x_{A_1} - x_{A_2}) \tag{2.33}$$

where z is the length of the diffusion path, $z_2 - z_1$, and c_{A1} and c_{A2} are the concentrations of A at each end of this path. The concentration distribution is evidently linear at steady state for this mechanism.

2.1.1.3 *Steady-state unimolal unidirectional diffusion*

In this case the flux of component B in one direction, because of molecular diffusion, is equal to the flux of B in the opposite direction because of the bulk flow. This causes B to be motionless relative to stationary coordinates, so that $N_{Bz} = 0$. This reduces γ to zero in eqn (2.32), and since $c - c_A = c_B$, the result is

$$N_{Az} = \frac{Dc}{z} \ln \frac{c_{B2}}{c_{B1}} \tag{2.34}$$

which may be written as

$$N_{Az} = \frac{D}{z}\left(\frac{c}{c_{Blm}}\right)(c_{A1} - c_{A2}) = \frac{Dc}{z}\frac{(x_{A1} - x_{A2})}{x_{Blm}} \tag{2.35}$$

where

$$c_{Blm} = \frac{c_{B2} - c_{B1}}{\ln(c_{B2}/c_{B1})}, \; x_{Blm} = \frac{x_{B2} - x_{B1}}{\ln(x_{B2}/x_{B1})}$$

Equation (2.34) shows that the steady-state concentration distribution is non-linear in this case.

It is important to recall that eqns (2.32–2.35) can only be applied for constant values of D, N_{Az}, and N_{Bz} along the transfer path. When these requirements are not fulfilled one must return to the steady-flow eqn (2.27) and proceed as outlined earlier.

The constant diffusivity assumption in eqns (2.32–2.35) renders them more rigorous for gases than for liquids. However, these expressions are often pressed into service for liquids despite some variation in D and with $c = (c_1 + c_2)/2$ when changes in c are not excessive.

2.1.2 Unsteady-state diffusion

In unsteady-state diffusion processes, the concentration at a given point varies with time, and we will now examine continuity and Fick's second law for such cases. Consider the volume element $dx\,dy\,dz$ in an isotropic medium that is free from convection. The concentration of diffusing solute A at the center of the element (x,y,z) is c_A (see Fig. 2.1).

The rate at which solute A enters through the left face of the element by molecular diffusion is

$$dx\,dy\left[J_{Az} - \frac{\partial J_{Az}}{\partial z}\frac{dz}{2}\right]$$

where J_{Az} is the flux through the plane at (x, y, z) parallel to the left face. The rate at which solute A leaves by diffusion through the right face is

$$dx\,dy\left[J_{Az} + \frac{\partial J_{Az}}{\partial z}\frac{dz}{2}\right]$$

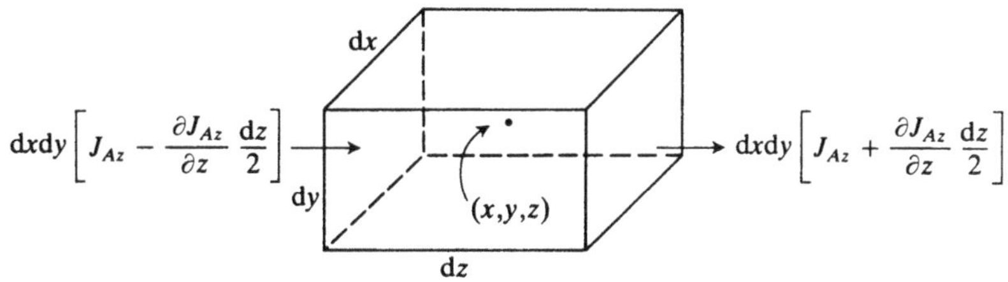

Fig. 2.1. Diffusion into and out of the volume element $dx\ dy\ dz$.

The rate of accumulation of A within the volume element from these two sources is therefore

$$-dx\ dy\ dz\frac{\partial J_{Az}}{\partial z}$$

Similarly, from the top and bottom faces the accumulation rate is

$$-dx\ dy\ dz\frac{\partial J_{Ay}}{\partial y}$$

and from the front and rear faces it is

$$-dx\ dy\ dz\frac{\partial J_{Ax}}{\partial x}$$

Another expression for the rate of accumulation of A in the volume element is

$$dx\ dy\ dz\frac{\partial c_A}{\partial t}$$

so

$$\frac{\partial c_A}{\partial t} = -\frac{\partial J_{Ax}}{\partial x} - \frac{\partial J_{Ay}}{\partial y} - \frac{\partial J_{Az}}{\partial z}$$

Combining with eqn (2.14) for component A, and with $D_{AB} = D_{BA} = D$ where D is constant,

$$\frac{\partial c_A}{\partial t} = D\left(\frac{\partial^2 c_A}{\partial x^2} + \frac{\partial^2 c_A}{\partial y^2} + \frac{\partial^2 c_A}{\partial z^2}\right) = D\nabla^2 c_A \qquad (2.36)$$

For one-dimensional diffusion, as in a direction normal to the two large surfaces of a slab, $\partial^2 c_A/\partial x^2$ and $\partial^2 c_A/\partial y^2$ are zero and eqn (2.36) reduces to

$$\frac{\partial c_A}{\partial t} = D\frac{\partial^2 c_A}{\partial z^2} \qquad (2.37)$$

which is Fick's second law of molecular diffusion. This will later be found useful in formulating the penetration theory, mass transfer in jets, and mass transfer with chemical reaction.

Equation (2.36) in the x, y, z coordinate system may be expressed in spherical polar coordinates, r, θ and ϕ, by use of the transformation equations and the assumption of radially symmetrical diffusion:

$$x = r \sin\theta \cos\phi$$

$$y = r \sin\theta \sin\phi$$

$$z = r \cos\theta$$

from which it follows that

$$r^2 = x^2 + y^2 + z^2 \tag{2.38}$$

$$\frac{\mathrm{d}(r^2)}{\mathrm{d}r}\frac{\partial r}{\partial x} = 2r\frac{\partial r}{\partial x} = 2x + 0 + 0, \text{ therefore } \frac{\partial r}{\partial x} = \frac{x}{r} \tag{2.39}$$

$$\frac{\partial c_A}{\partial x} = \frac{\partial c_A}{\partial r}\frac{\partial r}{\partial x} = \frac{\partial c_A}{\partial r}\frac{x}{r} \tag{2.40}$$

$$\frac{\partial^2 c_A}{\partial x^2} = \frac{\partial c_A}{\partial r}\frac{1}{r} - \frac{x^2}{r^3}\frac{\partial c_A}{\partial r} + \frac{x^2}{r^2}\frac{\partial^2 c_A}{\partial r^2} \tag{2.41}$$

Similarly,

$$\frac{\partial^2 c_A}{\partial y^2} = \frac{\partial c_A}{\partial r}\frac{1}{r} - \frac{y^2}{r^3}\frac{\partial c_A}{\partial r} + \frac{y^2}{r^2}\frac{\partial^2 c_A}{\partial r^2} \tag{2.42}$$

$$\frac{\partial^2 c_A}{\partial z^2} = \frac{\partial c_A}{\partial r}\frac{1}{r} - \frac{z^2}{r^3}\frac{\partial c_A}{\partial r} + \frac{z^2}{r^2}\frac{\partial^2 c_A}{\partial r^2} \tag{2.43}$$

Therefore

$$\frac{\partial^2 c_A}{\partial x^2} + \frac{\partial^2 c_A}{\partial y^2} + \frac{\partial^2 c_A}{\partial z^2} = \frac{3}{r}\frac{\partial c_A}{\partial r} - \frac{(x^2 + y^2 + z^2)}{r^3}\frac{\partial c_A}{\partial r} + \frac{(x^2 + y^2 + z^2)}{r^2}\frac{\partial^2 c_A}{\partial r^2}$$

and with eqn (2.38),

$$\frac{\partial^2 c_A}{\partial x^2} + \frac{\partial^2 c_A}{\partial y^2} + \frac{\partial^2 c_A}{\partial z^2} = \frac{\partial^2 c_A}{\mathrm{d}r^2} + \frac{2}{r}\frac{\partial c_A}{\partial r}$$

Substituting in eqn (2.36),

$$\frac{\partial c_A}{\partial t} = D\left(\frac{\partial^2 c_A}{\partial r^2} + \frac{2}{r}\frac{\partial c_A}{\partial r}\right) \tag{2.44}$$

Equation (2.44) is the differential equation describing unsteady-state diffusion in a stagnant sphere, subject to appropriate boundary conditions.

This will later provide the basis for analysis of unsteady-state mass transfer in liquid droplets which frequently may be approximated by stagnant spheres in liquid–liquid extraction.

The transformation of eqn (2.36) to cylindrical coordinates, r, θ, and z, is performed with the equations

$$x = r\cos\theta$$
$$y = r\sin\theta$$

For radially symmetrical diffusion in the cylinder with negligible end effects, $\partial c_A/\partial\theta = 0$, $\partial^2 c_A/\partial z^2 = 0$. From the transformation equations,

$$r^2 = x^2 + y^2$$

Equation (2.39) then follows once more. The development is now the same from eqns (2.40 to 2.42), eqn (2.43) being zero. It follows as before that

$$\frac{\partial^2 c_A}{\partial x^2} + \frac{\partial^2 c_A}{\partial y^2} = \frac{\partial^2 c_A}{\partial r^2} + \frac{1}{r}\frac{\partial c_A}{\partial r}$$

Substitution in eqn (2.36) gives

$$\frac{\partial c_A}{\partial t} = D\left[\frac{\partial^2 c_A}{\partial r^2} + \frac{1}{r}\frac{\partial c_A}{\partial r}\right] \tag{2.45}$$

which is the differential equation for unsteady-state diffusion in a cylinder, subject to prescribed boundary conditions. This will later be useful in analyzing unsteady-state mass transfer in jets.

Equations (2.37, 2.44, and 2.45) may all be regarded as forms of Fick's second law of diffusion in rectangular, spherical, and cylindrical coordinates, respectively, although the name is commonly applied to eqn (2.37) only.

2.1.3 Molecular diffusivities in liquids

Theoretical developments have been hindered by the fact that the kinetic theory of liquids is of course much less developed than that of gases. Several reviews are in the literature (Crank 1975; Bird *et al.* 1960; Jost 1952, 1960; Bird 1956; Hirschfelder *et al.* 1954; and Barrer 1941), and several correlations available for prediction of diffusivities in electrolytes and non-electrolytes under various conditions are critically compared by Reid *et al.* (1987). Skelland (1985) reviews expressions for estimating diffusivities for electrolytes and non-electrolytes at conditions of low and high solute concentration. Some of these relationships are presented below, but more detailed treatments are given in the references just cited.

On the theoretical side, a hydrodynamic approach by Einstein featured the application of Stokes' law to evaluate the drag on large spherical molecules of solute, A, moving in a continuum of small solvent molecules, B. The resulting expression is

$$\frac{D_{AB}\mu_B}{T} = \frac{k_B}{6\pi r_A} \tag{2.46}$$

where k_B is Boltzmann's constant and r_A is the radius of a molecule of A.

Eyring's theory of absolute reaction rates regards the molecules of the liquid as forming a quasi-crystalline lattice containing a random scattering of 'holes'. Solute molecules jump from hole to hole, depending upon their acquisition of the necessary activation energy. The theory leads to the equation (Jost 1952, p. 472)

$$\frac{D_{AB}\mu_B}{T} = f\,(\text{molal volume of mixture}) \tag{2.47}$$

Reid *et al.* (1987) review progress based on statistical mechanical and other theories.

Equations (2.46 and 2.47) showed generally poor agreement with experimental measurement, so that one of their major functions has been to prescribe a form for the correlation of laboratory data. This will be illustrated below for the cases of both non-electrolytes and electrolytes under dilute and concentrated conditions.

2.1.3.1 *Non-electrolytes in dilute solution*

Ten semiempirical relationships for diffusivity in very dilute binary solutions of non-electrolytes are listed by Skelland (1974, 1985, Ch. 3). Three of the more widely used of these are shown below.

Scheibel's (1954) expression is a modification of that presented by Wilke and Chang (1955) and is

$$\frac{D_{AB}\mu_{AB}}{T} = \frac{K}{V_{bA}^{1/3}} = \frac{8.2(10^{-8})}{V_{bA}^{1/3}}\left[1 + \left(\frac{3V_{bB}}{V_{bA}}\right)^{2/3}\right] \tag{2.48}$$

with the following special solvent cases: water: $V_{bA} < V_{bB}$, $K = 25.2\,(10^{-8})$, benzene: $V_{bA} < 2V_{bB}$, $K = 18.9\,(10^{-8})$, other solvents: $V_{bA} < 2.5V_{bB}$, $K = 17.5\,(10^{-8})$. D_{AB} is in cm^2/s; μ_{AB}, the solution viscosity, is in cP; T is in K, V_{bA} and V_{bB} are the molal volumes of the solute (A) and solvent (B) at their normal boiling temperatures, in cm^3/g-mol. Water is excluded as solute.

For diffusion in organic solvents, Lusis and Ratcliff (1968) give

$$\frac{D_{AB}\mu_B}{T} = \frac{8.52\,(10^{-8})}{V_{bB}^{1/3}}\left[1.40\left(\frac{V_{bB}}{V_{bA}}\right)^{1/3} + \frac{V_{bB}}{V_{bA}}\right] \tag{2.49}$$

where μ_B is viscosity of solvent B and units are as in eqn (2.48). Acceptable results were obtained in most cases with water as solute, provided it was assumed to diffuse as a tetramer—i.e. V_{bA} is calculated for $4H_2O$.

The correlation by King *et al.* (1965) also uses the term to accommodate the ratio of solvent size to solute size, (V_{bB}/V_{bA}), and additionally attempts to allow for intermolecular association forces with a function of the latent heats of vaporization, as follows:

$$\frac{D_{AB}\mu_B}{T} = 4.4(10^{-8})\left(\frac{V_{bB}}{V_{bA}}\right)^{1/6}\left(\frac{\Delta H_{mB}}{\Delta H_{mA}}\right)^{1/2} \tag{2.50}$$

where units are as in eqn (2.48) and ΔH_m is the molal latent heat of vaporization at the normal boiling temperature. The data included water as solute.

The relative accuracy of the above three equations in predicting diffusivities in organic solvents at infinite dilution was investigated by Reid *et al.* (1977). The comparison involved 83 experimental values from 65 binary systems over the temperature range 0–98.4°C. The absolute errors were as follows: eqn (2.48), 19.5 per cent; eqn (2.49), 28.6 per cent; eqn (2.50), 16.6 per cent.

A relationship developed particularly for dilute aqueous solutions is that of Hayduk and Laudie (1974):

$$D_{AB} = \frac{13.26\,(10^{-5})}{\mu_{wt}^{1.4}\,V_{bA}^{0.589}} \tag{2.51}$$

where units are as above and μ_{wt} is the viscosity of water at T(K), in cP.

Reid *et al.* (1977) compared the relative accuracy of eqns (2.48 and 2.51), using 54 experimental values from 31 highly dilute aqueous systems. The absolute average error in prediction for eqn (2.48) was 10.6 per cent and that for eqn (2.51) was 11.1 per cent.

It has been indicated above that anomalies may arise regarding the prediction of the diffusivity of water as solute in organic solvents. Thus eqns (2.48 and 2.49) may give predictions in error by 200–250 per cent when water is treated as a monomer. Olander (1961) gave a relationship to overcome this problem, based on a score of data points for nine systems, and assuming that polymerization of water to a tetramer in solution accounted for the discrepancy. The success he achieved on this basis did not always extend to more recent data on water diffusing in various organic solvents. Thus Lees and Sarram (1971) appeared to show that water sometimes diffuses as a monomer and sometimes as a tetramer.

Somewhat improved general performance in prediction of binary non-electrolyte diffusivities at infinite dilution was found by Reid *et al.* (1987)

from the correlations by Tyn and Calus (1975, a, b) and Hayduk and Minhas (1982). The former introduces the ratio of the parachors of the solute and solvent, in addition to the ratio of their molal volumes at their normal boiling points. However, the expression is subject to a variety of restrictions for which the original papers should be consulted. Hayduk and Minhas (1982) give several correlations, depending on the type of solute–solvent system.

2.1.3.2 *Non-electrolytes in concentrated solution*

Five relationships purporting to accommodate the influence of solute concentration in ideal and non-ideal liquid systems were tabulated by Skelland (1985). The expression provided by Leffler and Cullinan (1970) is

$$(D_A \mu_{AB})_{\text{conc}} = (D^0_{AB} \mu_B)^{x_B} (D^0_{BA} \mu_A)^{x_A} \left(1 + \frac{\text{d}\ln\gamma_A}{\text{d}\ln x_A}\right) \tag{2.52}$$

where D^0_{AB} and D^0_{BA} are diffusivities of very dilute A in B and B in A, respectively, in cm^2/s, μ_A, μ_B, and μ_{AB} are viscosities of A, B, and the mixture of A and B, in cP; x_A and x_B are mole fractions of solute A and solvent B, and γ_A is the activity coefficient of A.

2.1.3.3 *Electrolytes in dilute solution*

The electrical neutrality of a given solution of an electrolyte is preserved by the fact that the negatively and positively charged ions diffuse at the same rate, despite differences in their size. For strong (fully dissociated) electrolytes at infinite dilution, Nernst's (1888) equation is

$$D^0_A = 8.931\ (10^{-10})\, T \left(\frac{l^0_+ l^0_-}{l^0_+ + l^0_-}\right) \left(\frac{z_+ + z_-}{z_+ z_-}\right) \tag{2.53}$$

where D^0_A is diffusivity, in cm^2/s; and l^0_+ and l^0_- are the cationic and anionic conductances at infinite dilution, in mho/equivalent; $l^0_+ + l^0_-$ is the electrolyte conductance at infinite dilution; z_+ and z_- are absolute values of the cation and anion valences; and T is the absolute temperature, in K.

Ionic conductances at 25°C and infinite dilution in water are given in Perry *et al.* (1963, p. 14.24), which also shows how to convert these values to other temperatures.

2.1.3.4 *Electrolytes in concentrated solution*

For this complex situation some success has been obtained with Gordon's (1937) equation:

$$(D_A)_{\text{conc}} = D^0_A \left(1 + \frac{m\,\partial\ln\gamma\pm}{\partial m}\right) \frac{1}{c'_B \bar{V}_B} \left(\frac{\mu_B}{\mu_{AB}}\right) \tag{2.54}$$

where D_A^0 is calculated from Nernst's eqn (2.53); m is molality; c_B' is the number of moles of water per cm^3 of solution; $\bar{V}_B$ is the partial molal volume of water in solution, in cm^3/mol; μ_B and μ_{AB} are the viscosities of water and of the solution; and $\gamma\pm$ is the mean ionic activity coefficient based on molality.

Tabulations of $\gamma\pm$ as a function of m in several aqueous solutions are given by Harned and Owen (1950) and Glasstone (1947). Partial molal volumes $\bar{V}_B$ may be estimated as described by Lewis and Randall (1923).

Experimental values of D are of course to be preferred when available. An extensive listing of experimental diffusivities for non-electrolytes is given by Johnson and Babb (1956) and by Reid *et al.* (1987). Harned and Owen (1950) and Robinson and Stokes (1959) give similar data for electrolytes. Methods of experimental measurement of diffusivity in liquids are described by Tyrrell and Harris (1984) and Cussler (1984).

2.1.3.5 *Multicomponent diffusion in liquids*

Rather accurate predictions can be made of effective diffusivities for multicomponent mixtures of gases, in terms of known values of the relevant binary diffusion coefficients. Unfortunately the situation is more complicated in the case of multicomponent liquid mixtures, for which one must rely more heavily on experimental measurement.

In a multicomponent mixture of n components, the minimum value for n is of course three. Furthermore, the diffusion of dilute solute A in a mixture of two liquid solvents is of widespread industrial occurence. In an early study of the relatively ideal, low-viscosity system toluene–chlorobenzene–brombenzene, Burchard and Toor (1962) found that diffusion could be represented by a modification of Fick's law, eqn (2.14), written as

$$J_{Az} = -D_{Am}\frac{dc_A}{dz} \qquad (2.55)$$

where D_{Am} is a molal average diffusivity formed from the three binary diffusivities in dilute solution.

Two subsequent studies of molecular diffusion in ternary liquid systems were published within a year of each other, one by Perkins and Geankoplis (1969), and the other by Leffler and Cullinan (1970). The former authors present the following expression for the case of dilute A in a concentrated mixture of B and C:

$$D_{Am}^0\,\mu_{ABC}^{0.8} = x_B\,D_{AB}^0\mu_B^{0.8} + x_C\,D_{AC}^0\,\mu_C^{0.8} \qquad (2.56)$$

where D_{Am}^0 is the diffusivity of dilute A in a mixture containing x_B and x_C mole fractions of B and C, respectively. μ_{ABC}, μ_B, and μ_C are the viscosities of the ternary mixture, of B, and of C, and the dilute binary diffusivities of

A in B and of A in C are estimated from one of the binary correlations given earlier—unless, of course, experimental values are available. Eqn (2.56) was shown to represent the diffusivities of several ternary hydrocarbon systems within 0.5–8.2 percent; of CO_2 in ethanol–water within 22 percent; of KCl in ethylene glycol–water (μ_{ABC} up to 18 cP) within 14.6 percent; and of acetic acid in ethanol–water, which is highly non-ideal, with an average deviation of 8.5 percent. The flux is predicted by insertion of D^0_{Am} from eqn (2.56) into eqn (2.55).

A widely used expression by Leffler and Cullinan (1970) for the same ternary liquid situation is

$$D^0_{Am}\,\mu_{ABC} = (D^0_{AB}\,\mu_B)^{x_B}\,(D^0_{AC}\,\mu_C)^{x_C} \tag{2.57}$$

Presumably one chooses that relationship leading to the most conservative estimate for the case at hand, pending a more definitive selection between these alternative expressions.

Turning to more general multicomponent diffusion, current theory and data indicate that the mass flux of any component relative to the mass average velocity is a linear function of the concentration gradients of all components in the solution. Thus for the z direction,

$$i_{iz} = -\sum_{k=1}^{n-1} D_{ik}\frac{\mathrm{d}\rho_k}{\mathrm{d}z} \tag{2.58}$$

A procedure analogous to that leading to eqn (2.13) can be applied to eqn (2.6) to show that $\Sigma_1^n i_{iz} = 0$. Thus there are $n - 1$ independent fluxes and the same number of independent concentration gradients, indicating $(n - 1)^2$ diffusion coefficients. However, Tyrrell and Harris (1984, Ch. 3) and Cussler (1984, Ch. 8) note that the Onsager reciprocal relations can be used to show that there are actually only $n(n - 1)/2$ independent diffusivities in eqn (2.58).

Complex computational procedures, which in some cases allow multicomponent diffusion calculations to be reduced to equivalent binary calculations, were developed by Toor (1964a, 1964b) and by Stewart and Prober (1964). More recently, Krishna and Standart (1976) and Krishna (1976, 1977) have presented a general matrix method of solution to the Maxwell-Stefan equations for multicomponent systems. (Extension of the complex treatment expressly to the liquid phase is performed in the second of these papers). However, it may be well to note the comments by Cussler (1984, pp. 194, 211) on the general subject of multicomponent diffusion as follows: '. . . multicomponent effects are minor in dilute solutions, and most solutions are dilute.' Also: 'When (multicomponent) systems are dilute, the diffusion of each solute can be treated with a binary form of Fick's law. In concentrated solutions, the fluxes and concentration profiles deviate significantly from binary expectations only in exceptional cases. These exceptions include . . . mixed weak electrolytes, partially miscible species, and some alloys.'

These observations are reiterated by Kirwan (1987), who also notes that exceptions may arise in the case of systems that are highly non-ideal thermodynamically, or in the presence of strong molecular associations or electrostatic effects. A practical expedient that is often resorted to in multicomponent diffusion is to compute the flux of a given component in terms only of its *own* concentration gradient and an effective diffusivity in the mixture, D_{im}, thus:

$$i_{iz} = -D_{im}\,\rho\,\frac{dw_i}{dz} \tag{2.59}$$

where D_{im} is sometimes crudely evaluated as the binary diffusivity of i in the component identified as the solvent. Kirwan (1987) notes that this approach '. . . is often employed with reasonable success unless some highly specific interactions exist.' In fact, as he points out, D_{im} would actually depend on the concentrations and concentration gradients of the other components present, and would preferably be measured for the system under consideration.

For more details on the subject of multicomponent diffusion, the reader is referred to Cussler's (1976) book on the subject, and also to Cussler (1984), Tyrrell and Harris (1984), Toor (1964a, 1964b), Stewart and Prober (1964), Krishna and Standart (1976, 1979), and Krishna (1976, 1977).

2.2 Mass transfer coefficients and mechanisms of mass transfer

2.2.1 Individual and overall coefficients

The transfer of material across the interface between two liquid phases is a fundamental characteristic of liquid–liquid extraction. A variety of mechanisms have been proposed to describe this process, some of which will now be described. It will first be necessary to introduce the concept of mass transfer coefficients and to outline the idea of equilibrium.

Consider the distribution of a solute such as component A between two immiscible liquid phases in contact with each other. Under conditions of dynamic equilibrium the rate of transfer of A from the first to the second phase is equal to the rate at which A is transferred in the reverse direction. The relationship may be represented by a plot such as that labelled 'Equilibrium curve' in Fig 2.2, over a range of compositions of each phase, where Y_A and X_A refer to the light and heavy liquids, respectively.

A point located above the curve signifies transfer of A from the light to the heavy liquid, whereas a point below the curve denotes transfer in

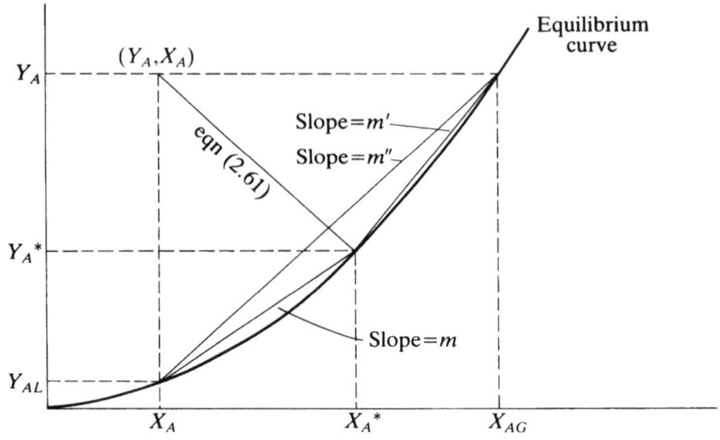

Fig. 2.2. Individual and overall concentration 'driving forces' for a two-phase system with distributed component A. Bulk concentrations of the two phases are, respectively, X_A and Y_A and interfacial equilibrium is assumed.

the other direction. The equilibrium curve is often located by experimental measurement, but can sometimes be predicted from thermodynamic considerations (see Ch. 1 of this volume, and also Abbott and Prausnitz 1987).

The compositions at the interface between two immiscible liquids will be written as Y_A^* and X_A^* and these are assumed to be local equilibrium values. Any convenient units are denoted by Y_A and X_A here. If transfer of A takes place from the lighter liquid to the heavier liquid, the individual coefficients k_Y and k_X for the light and heavy phases, respectively, are defined as follows:

$$N_A A = k_Y A(Y_A - Y_A^*) = k_X A(X_A^* - X_A) \tag{2.60}$$

A is the area of the interface, where the flux is N_A, and Y_A and X_A are concentrations of component A in the bulk of the light- and heavy-liquid phases. Equation (2.60) shows that

$$\frac{Y_A - Y_A^*}{X_A - X_A^*} = -\frac{k_X}{k_Y} \tag{2.61}$$

This relationship is plotted in Fig. 2.2 on the assumption of interfacial equilibrium.

Interfacial concentrations (X_A^*, Y_A^*) are often unknown at a given location within a two-phase system and it is then more convenient to use overall coefficients K_Y and K_X, defined in terms of overall concentration differences or 'driving forces', as shown below.

$$N_A A = K_Y A(Y_A - Y_{AL}) = K_X A(X_{AG} - X_A) \tag{2.62}$$

where Y_{AL} is the light-liquid concentration that would be in equilibrium with the existing heavy-liquid concentration; X_{AG} is the heavy-liquid concentration that would be in equilibrium with the existing light-liquid concentration. The location of these quantities is shown in Fig. 2.2. The relationships between the individual coefficients of eqn (2.60) and the overall coefficients of eqn (2.62) are developed in the following manner, using the definitions given below and taken from Fig. 2.2:

$$m = \frac{Y_A^* - Y_{AL}}{X_A^* - X_A} \quad m' = \frac{Y_A - Y_A^*}{X_{AG} - X_A^*} \quad m'' = \frac{Y_A - Y_{AL}}{X_{AG} - X_A}$$

$$N_A A = k_Y A(Y_A - Y_A^*) = K_Y A(Y_A - Y_{AL})$$

$$\frac{1}{K_Y} = \frac{1}{k_Y}\left(\frac{Y_A - Y_{AL}}{Y_A - Y_A^*}\right) = \frac{1}{k_Y}\left[\frac{(Y_A - Y_A^*) + (Y_A^* - Y_{AL})}{Y_A - Y_A^*}\right]$$

$$\frac{1}{K_Y} = \frac{1}{k_Y} - \frac{1}{k_Y}\left(\frac{Y_A^* - Y_{AL}}{Y_A^* - Y_A}\right) \tag{2.63}$$

$$Y_A^* - Y_{AL} = m(X_A^* - X_A) \tag{2.64}$$

and from eqn (2.61),

$$Y_A^* - Y_A = -\frac{k_X}{k_Y}(X_A^* - X_A) \tag{2.65}$$

Taking the ratio of eqn (2.64) to (2.65) and substituting the result in eqn (2.63) gives

$$\frac{1}{K_Y} = \frac{1}{k_Y} + \frac{m}{k_X} \tag{2.66}$$

Similarly,

$$\frac{1}{K_X} = \frac{1}{k_X} + \frac{1}{m' k_Y} \tag{2.67}$$

and

$$\frac{1}{K_X} = \frac{1}{m'' K_Y} \tag{2.68}$$

where the first two relationships express the overall resistance to transfer (in terms of the light and heavy liquids, respectively) as equaling the sum of the individual (or single-phase) resistances to mass transfer. The final expression relates the two forms of overall resistance to mass transfer, $1/K_X$ and $1/K_Y$ (note that constancy of K over a given concentration range requires constant k_X, k_Y, and m). The latter restriction becomes inconsequential, however, in systems which are either light-phase controlled ($1/k_Y \gg m/k_X$) or heavy-phase controlled ($1/k_X \gg 1/m'k_Y$)—a condition frequently determined by the magnitude of the distribution ratio, m.

When $m = m' = m'' = 1$ the equilibrium curve is linear and coincides with the 45° diagonal in Fig. 2.2. Equations (2.66 and 2.67) show that K_Y and K_X are then equal. This, of course, occurs rather rarely in mass transfer, in contrast to heat transfer where, at equilibrium, $T_{\text{phase 1}} = T_{\text{phase 2}}$, so that $m_{\text{heat transfer}}$ is always unity. (Here T denotes temperature and the coordinates of the heat transfer analog of Fig. 2.2 are the temperatures of the two phases between which heat transfer is taking place.) These considerations of course explain the occurrence of only one overall coefficient in two-phase heat transfer operations, in contrast to the usual appearance of two alternative overall coefficients in two-phase mass transfer processes.

The interfacial area A is often unknown in many types of mass-transfer equipment. In such cases, mass-transfer rates are frequently based on unit volume of the equipment instead of unit interfacial area. The rate eqns (2.60 and 2.62) are then written as

$$\begin{aligned} N_A A &= k_Y a(Y_A - Y_A^*)_m V_0 = k_X a(X_A^* - X_A)_m V_0 \\ &= K_Y a(Y_A - Y_{AL})_m V_0 = K_X a(X_{AG} - X_A)_m V_0 \end{aligned} \tag{2.69}$$

where subscript m denotes a suitable mean, V_0 is the contacting volume of the equipment, a is the interfacial area per unit volume, and the combined quantities $k_X a$, $k_Y a$, $K_X a$, and $K_Y a$ are called volumetric or capacity coefficients. Equations (2.66, 2.67, and 2.68) may be divided by a to yield

$$\frac{1}{K_Y a} = \frac{1}{k_Y a} + \frac{m}{k_X a} \tag{2.70}$$

$$\frac{1}{K_X a} = \frac{1}{k_X a} + \frac{1}{m' k_Y a} \tag{2.71}$$

$$\frac{1}{K_X a} = \frac{1}{m'' K_Y a} \tag{2.72}$$

Coefficients corresponding to low rates of mass transfer are usually denoted throughout this chapter by an asterisk as k^*.

Some of the mechanisms that have been proposed for predicting values of the individual coefficients k_X and k_Y have been classified (Sherwood *et al.* 1975, p. 151) as (1) *film theory*, (2) various *surface renewal theories*, and (3) forms of *boundary layer theory*. Prominent examples of these three categories of mechanism will now be developed.

2.2.2 The two-film theory

Lewis (1916) and Whitman (1923) are usually credited with developing this theory, which visualizes two non-turbulent fictitious films on either side of the interface between phases, and across which transfer occurs by purely molecular diffusion. The thickness of each film is such as to just contain the entire resistance to transfer offered by the appropriate phase, and local equilibrium is assumed at the interface. Steady-state diffusion in the films follows from the proposition that the time taken to set up the concentration distributions in the films is short compared to the total contact time.

Individual or single-phase mass transfer coefficients (k) are defined customarily as follows for a mechanism involving bulk molal flow:

Light phase $$N_A = k_y \Delta y_A = k_c \Delta c_A = k_\rho \Delta \rho_A / M_A \tag{2.73}$$

Heavy phase $$N_A = k_c \Delta c_A = k_x \Delta x_A = k_c (\rho/M)_{\mathrm{av}} \Delta x_A = k_\rho \Delta \rho_A / M_A \tag{2.74}$$

And from eqn (2.32), in the presence of bulk flow with finite N_{Bz}, for the light liquid phase:

$$N_{Az} = \frac{N_{Az}}{N_{Az} + N_{Bz}} \frac{D_G c}{z_{fG}} \ln \left(\frac{N_{Az}/(N_{Az} + N_{Bz}) - y_{A2}}{N_{Az}/(N_{Az} + N_{Bz}) - y_{A1}} \right) \tag{2.75}$$

so that from eqns (2.73 and 2.75),

$$k_y = \frac{D_G c}{z_{fG}} \left(\frac{N_{Az}}{N_{Az} + N_{Bz}} \right) \frac{1}{[N_{Az}/(N_{Az} + N_{Bz}) - y_A]_{\mathrm{lm}}} \tag{2.76}$$

The quantity $[N_{Az}/(N_{Az} + N_{Bz}) - y_A]_{lm}$ is the logarithmic mean of the values on the bulk-phase side and the interfacial side of the fictitious film. For the particular case of unimolal unidirectional diffusion N_{Bz} is zero, so that eqn (2.76) reduces to

$$k_y = \frac{D_G c}{z_{fG}(1 - y_A)_{lm}} = \frac{D_G c}{z_{fG} y_{Blm}} \tag{2.77}$$

where D_G is the solute diffusivity in the light phase, which has a fictitious film thickness z_{fG}; k_y is the individual light-phase coefficient of mass transfer for use with Δy_A.

Similarly, for the heavy liquid phase,

$$k_x = \frac{D_L c}{z_{fL}(1 - x_A)_{\mathrm{lm}}} = \frac{D_L c}{z_{fL} x_{B\mathrm{lm}}} \tag{2.78}$$

where D_L is the solute diffusivity in the heavy phase, which has a fictitious film thickness z_{fL}; k_x is the individual heavy-phase coefficient of mass transfer for use with Δx_A. Hydrodynamic complexities usually prohibit the evaluation of the fictitious film thickness, z_f, but it will surely decrease as the turbulence increases. A first-power dependence of k_y and k_x upon D will be noted in eqns (2.77 and 2.78).

2.2.3 The penetration theory

This theory, variously attributed to Boussinesq (1905) and Higbie (1935), among others, visualizes surface renewal by eddies which are carried to the interface from the bulk of the phase by turbulence. After a short but constant time each eddy is displaced back into the phase, where it mixes with the bulk fluid. While at the interface, solute 'penetrates' into each eddy by unsteady-state molecular diffusion as described by Fick's second law, eqn (2.37), written in terms of mass concentration, ρ_A, as

$$\frac{\partial \rho_A}{\partial t} = D\left(\frac{\partial^2 \rho_A}{\partial z^2}\right) \tag{2.79}$$

The boundary conditions are as follows, with negligible mass average velocity at and normal to the surface:

$$\rho_A(z, 0) = \rho_{A\infty};\ \rho_A(0, t) = \rho_A^*;\ \rho_A(\infty, t) = \rho_{A\infty}$$

The concentration at z effectively equal to infinity—measured from and normal to the interface—is constant at $\rho_{A\infty}$, as is the concentration ρ_A^* in the phase of interest at the interface. If the following abbreviations are adopted,

$$\rho_A' = \rho_A - \rho_{A\infty};\ \rho_A^{*\prime} = \rho_A^* - \rho_{A\infty} \tag{2.80}$$

then eqn (2.79) becomes

$$\frac{\partial \rho_A'}{\partial t} = D\frac{\partial^2 \rho_A'}{\partial z^2}$$

converting the boundary conditions to

$$\rho_A'(z, 0) = 0;\ \rho_A'(0, t) = \rho_A^{*\prime};\ \rho_A'(\infty, t) = 0$$

Taking the Laplace transform of both sides with respect to time t,

$$\int_0^\infty \exp\left[-pt\,(\partial^2 \rho_A'/\partial z^2)\right] \mathrm{d}t = \frac{1}{D}\int_0^\infty \exp\left[-pt(\partial \rho_A'/\partial t)\,\mathrm{d}t\right]$$

or

$$\frac{d^2F}{dz^2} = \frac{1}{D}pF - \rho_A'(z,0) = \frac{p}{D}F$$

where

$$F(z,p) = L_t[\rho_A'(z,t)]$$

so that

$$\frac{d^2F}{dz^2} - \frac{p}{D}F = 0$$

The second boundary condition transforms to $F(0,p) = \rho_A^{*\prime}/p$. The solution of the ordinary differential equation for F, subject to $F(0,p) = \rho_A^{*\prime}/p$ and F remains finite as $z \to \infty$, is

$$F(z,p) = \frac{\rho_A^{*\prime}}{p}\exp[-(pz^2/D)^{1/2}]$$

The function for which this is the Laplace transform may be found from tables, such as that given by Crank (1956, p. 327, item 8), as

$$\rho_A'(z,t) = \rho_A^{*\prime}\left[1 - \frac{2}{\pi^{1/2}}\int_0^{z/2(Dt)^{1/2}} e^{-\theta^2}d\theta\right] \tag{2.81}$$

and inserting the definitions of ρ_A' and $\rho_A^{*\prime}$ from eqn (2.80),

$$\frac{\rho_A^* - \rho_A}{\rho_A^* - \rho_{A\infty}} = \operatorname{erf}\frac{z}{2(Dt)^{1/2}} \tag{2.82}$$

where erf is the error function, defined generally as:

$$\operatorname{erf}\zeta = \frac{2}{\pi^{1/2}}\int_0^{\zeta} e^{-\theta^2}d\theta$$

in which θ is any dummy variable, used merely to describe the function to be integrated; θ is eliminated by the limits of the integration. Tabulation of the error function appears, for example, in Peirce and Foster (1956, pp. 96, 128–32). Equation (2.82) gives the concentration inside a given eddy at time t after its arrival at the interface and at location z. Since bulk flow is negligible in eqn (2.27) for the present case, the instantaneous rate of unsteady diffusion from unit interface is

$$n_{A0} = -D\left(\frac{\partial\rho_A}{\partial z}\right)_{z=0} \tag{2.83}$$

the concentration gradient at the interface is obtained from eqn (2.82), to give

$$n_{A0} = (\rho_A^* - \rho_{A\infty})\,(D/\pi t)^{1/2} \tag{2.84}$$

The total solute penetrating the eddy in an exposure time t_e is

$$\int_0^{t_e} n_{A0}\mathrm{d}t = (\rho_A^* - \rho_{A\infty})\,(\mathrm{D}/\pi)^{1/2} \int_0^{t_e} t^{-1/2}\,\mathrm{d}t$$

$$= 2(\rho_A^* - \rho_{A\infty})\,(Dt_e/\pi)^{1/2} \tag{2.85}$$

and the average rate of transfer per unit surface during exposure is obtained by dividing eqn (2.85) by t_e,

$$(n_{A0})_{\mathrm{av}} = 2(\rho_A^* - \rho_{A\infty})\,(D/\pi t_e)^{1/2} \tag{2.86}$$

Equation (2.86) shows that the mass transfer coefficient $k_\rho = 2(D/\pi t_e)^{1/2}$, and varies with the square root of the molecular diffusivity, instead of with D to the power unity, as predicted by the two-film theory. The exposure time per eddy, t_e, is reduced by increasing turbulence, but is usually not known.

It is of course important in applying the penetration theory that the time spent by each eddy at the interface, t_e, be so short that solute does not have time to 'penetrate' to the other side of the (stagnant) eddy; diffusion is then, in effect, taking place in a semi-infinite stagnant medium and the third boundary condition is fulfilled. If the depth of penetration, z_p, is defined as that depth at which the change in concentration reaches one percent of its final value, then eqn (2.82) can be applied to show that

$$z_p = 3.6\,(Dt_e)^{1/2} \tag{2.87}$$

Thus, for a typical liquid phase diffusivity of $1 \times 10^{-5}\,\mathrm{cm^2/s}$ and an exposure time as long as 5 s, the penetration depth is a mere 0.025 cm.

Consider the unsteady-state diffusion of a solute into a semi-infinite stagnant liquid that is initially free from the solute and in which the solute diffusivity is $2(10^{-5})\,\mathrm{cm^2/s}$. Equation (2.82) and a table of error functions can be used to prepare Fig. 2.3, showing the solute concentration distribution as a function of depth z into the liquid at various times t since the process began.

2.2.4 The theory of penetration with varied surface renewal

A major limitation in the application of the penetration theory developed above to a turbulent two-phase system has long been felt to lie in the assumption that each eddy spends the same time, t_e, at the interface, before being displaced back into the bulk of the phase from which it originated. The first attempt to overcome this objection was made by Danckwerts (1951). In his model, the interface is regarded as consisting of a mosaic of surface

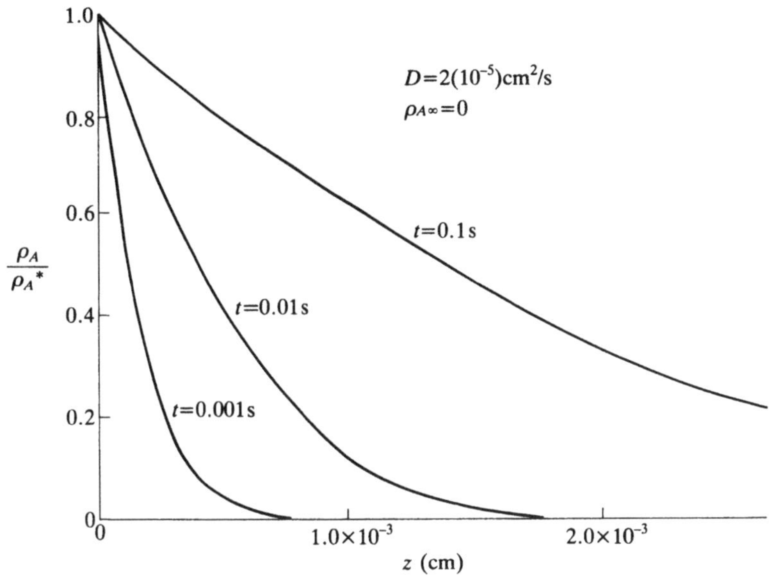

Fig. 2.3. Concentration profiles for unsteady-state diffusion into a stagnant semi-infinite liquid.

elements having a wide variety of 'ages'—i.e. a wide range of exposure times at the interface. Danckwerts (1951) proposed that 'The mean rate of production of fresh surface will be constant and equal to s, and the chance of an element of surface being replaced within a given time is assumed to be independent of its age; hence the fractional rate of replacement of the elements belonging to any age group is equal to s.'

In contrast, Kishinevsky (1954) contends that 'it is very difficult to agree with Danckwerts that the probability of replacement of an element of surface within a given time does not depend on its age . . . we cannot agree with . . . Danckwerts that a freshly formed element of the surface has the same chance of renewal as an old element, . . .'

Further shortcomings in Danckwerts' eddy residence time distribution have been noted by Danckwerts (1955) himself, and by Perlmutter (1961), who points out that it '. . . requires that the most probable eddy have a residence time of zero.' Perlmutter then proceeds to develop several new surface renewal models, based on some alternative eddy residence-time distributions.

In view of these controversial considerations, a new form of surface renewal model will be developed for illustration purposes here, based on the simple assumption that the fractional rate of surface renewal, s, is related to the age of the surface, t, through the constants ξ and n, as $s = \xi t^n$, where

ξ, $n + 1 > 0$. It will be found that the result reduces to that obtained by Danckwerts as a special case when $n = 0$.

Following Danckwerts, a surface age distribution function, $\phi(t)$, is defined, such that the fraction of surface with ages between t and $t + dt$ is $\phi(t)\,dt$. Then for unit surface,

$\phi(t)dt$	$= \phi(t - dt)dt$	$- \ [\phi(t - dt)dt]s\,dt$
differential area of surface at t of age t	differential area of surface at $t - dt$ of age $t - dt$	differential surface replaced during dt.

whence

$$\frac{\phi(t) - \phi(t - dt)}{dt} = -s\phi(t - dt)$$

or

$$\frac{d\phi(t)}{dt} = -s\phi(t) \tag{2.88}$$

Now assume that the fractional rate of surface renewal s is the following function of the age of the surface:

$$s = \xi t^n; \quad \xi, n + 1 > 0 \tag{2.89}$$

so

$$\frac{d\phi(t)}{dt} = -\xi t^n \phi(t)$$

Integrating and solving for the surface age distribution function,

$$\phi(t) = C\exp[-\xi t^{n+1}/(n + 1)] \tag{2.90}$$

The constant C may be obtained from the fact that

$$\int_0^\infty \phi(t)\,dt = 1 = \text{unit total surface}$$

so

$$C\int_0^\infty \exp[-\xi t^{n+1}/(n + 1)]\,dt = 1 \tag{2.91}$$

Let

$$\xi t^{n+1}/(n + 1) = \beta^{n+1}$$

then

$$t = \beta\left(\frac{n+1}{\xi}\right)^{1/(n+1)}, \quad dt = \left(\frac{n+1}{\xi}\right)^{1/(n+1)} d\beta$$

and substituting in eqn (2.91),

$$C\left(\frac{n+1}{\xi}\right)^{1/(n+1)} \int_0^{\infty} \exp(-\beta^{n+1})\, d\beta = 1$$

The integral can be shown to be the gamma function $\Gamma[(n+2)/(n+1)]$; then solving for C and inserting the result in eqn (2.90) gives

$$\phi(t) = \frac{[\xi/(n+1)]^{1/(n+1)}}{\Gamma[(n+2)/(n+1)]} \exp\left(\frac{-\xi t^{n+1}}{n+1}\right) \tag{2.92}$$

This is the surface age distribution function corresponding to eqn (2.89).

Next, to obtain the average mass flux of A at the surface and relative to stationary coordinates, consider eqn (2.84),

$$n_{A0} = (\rho_A^* - \rho_{A\infty})(D/\pi t)^{1/2} \tag{2.84}$$

This expression gives the instantaneous rate of transfer from unit surface for a given element. Thus for all elements with ages between t and $t + dt$ the instantaneous transfer rate per unit surface is given by

$$\phi(t)(\rho_A^* - \rho_{A\infty})(D/\pi t)^{1/2}\, dt$$

Therefore the flux of component A from a surface having elements with ages t distributed between zero and infinity is

$$(n_{A0})_{av} = (\rho_A^* - \rho_{A\infty}) \int_0^{\infty} \phi(t)(D/\pi t)^{1/2}\, dt \tag{2.93}$$

and with eqn (2.92),

$$(n_{A0})_{av} = (\rho_A^* - \rho_{A\infty})(D/\pi)^{1/2} \frac{[\xi/(n+1)]^{1/(n+1)}}{\Gamma[(n+2)/(n+1)]} \int_0^{\infty} t^{-1/2} \exp[-\xi t^{n+1}/(n+1)]\, dt \tag{2.94}$$

Let

$$\xi t^{n+1}/(n+1) = \beta^{2(n+1)}$$

then

$$t^{-1/2} = \frac{1}{\beta}\left(\frac{\xi}{n+1}\right)^{1/2(n+1)}, \quad dt = 2\beta\left(\frac{n+1}{\xi}\right)^{1/(n+1)} d\beta$$

Substituting into eqn (2.94) and condensing,

$$(n_{A0})_{av} = (\rho_A^* - \rho_{A\infty})\,(D/\pi)^{1/2}\,\frac{2[\xi/(n+1)]^{1/2(n+1)}}{\Gamma[(n+2)/(n+1)]}\int_0^\infty \exp\,(-\beta^{2(n+1)})\,d\beta$$

The integral can be shown to be the gamma function $\Gamma[(2n+3)/2(n+1)]$, so

$$(n_{A0})_{av} = (\rho_A^* - \rho_{A\infty})\,2(D/\pi)^{1/2}\left(\frac{\xi}{n+1}\right)^{1/2(n+1)}\frac{\Gamma[(2n+3)/2(n+1)]}{\Gamma[(n+2)/(n+1)]} \tag{2.95}$$

where the gamma functions may be evaluated from standard tables.

Consider the special case in which $n = 0$. Equation (2.89) shows that ξ then equals s. Substitution into eqn (2.92) gives

$$\phi(t) = \frac{s}{\Gamma(2)}\exp\,(-st)$$

From tables, $\Gamma(2) = 1.0$, so that

$$\phi(t) = se^{-st} \tag{2.96}$$

Also, in eqn (2.95),

$$(n_{A0})_{av} = (\rho_A^* - \rho_{A\infty})\,D^{1/2}\frac{2}{\pi^{1/2}}s^{1/2}\frac{\Gamma(3/2)}{\Gamma(2)}$$

and from tables, $\Gamma(3/2) = 0.88623 = \pi^{1/2}/2$, so

$$(n_{A0})_{av} = (\rho_A^* - \rho_{A\infty})\,(sD)^{1/2} \tag{2.97}$$

Equations (2.96 and 2.97) coincide with the results obtained by Danckwerts (1951), on the basis of the assumptions noted earlier. The last equation shows that the mass transfer coefficient $k_\rho = (sD)^{1/2}$. Evidently the coefficient is again found to vary with the square root of the molecular diffusivity, as in the case of the penetration theory, eqn (2.86). This square root dependence of k_ρ upon D is common to all forms of surface renewal models. The fractional rate of surface renewal, s, is usually unknown, but increases with increasing turbulence.

2.2.5 The theory of interfacial boundary layers in immiscible, cocurrent, laminar streams

The phenomenon of mass transfer between two immiscible liquids in cocurrent flow may arise, for example, in the case of one liquid jetting through another—instances of which are cited later, in the flow of a continuous phase around internally circulating droplets, in some forms of

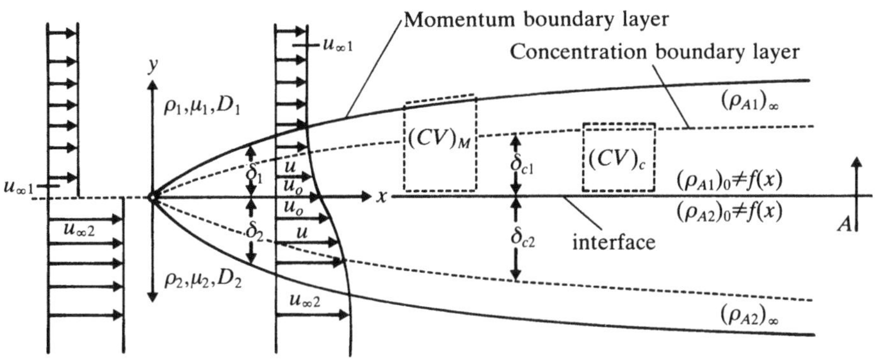

2.4. Momentum and concentration boundary layers at the interface between two immiscible cocurrent streams.

wetted wall column, in ejectors (Acharjee *et al.* 1978), and in short cocurrent packed columns (Leacock and Churchill 1961; Verma and Sharma 1975; Acharjee *et al.* 1977). The latter application has been proposed as an effective alternative to mixing vessels in mixer-settlers, because the 'mechanical agitators for such units are generally bulky, inefficient, and expensive' (Leacock and Churchill 1961). The velocity distributions in the two phases arising from their relative motion influence the rate of mass transfer from one phase to the other. Potter (1957) analysed this situation for the case of laminar boundary layers on each side of a stable interface, as represented in Fig. 2.4.

The upper and lower streams have uniform, free stream velocities of $u_{\infty 1}$ and $u_{\infty 2}$, respectively, and first come into contact with each other at 0; the coordinate system is as shown in Fig. 2.4, with the interface in the plane at $y = 0$. The respective boundary layer thicknesses at a given x, δ and δ_c, are arbitrarily defined as the normal distances from the interface at which the velocity u and solute concentration ρ_A have attained 99 percent of their free stream values.

Suppose the upper and lower immiscible liquids are B and C, with solute A transferring from C to B. Then, in the upper phase, $(\rho_{A1})_\infty < (\rho_{A1})_0$, so $(\rho_{B1})_\infty > (\rho_{B1})_0$, promoting diffusion of B towards the interface. However, if B cannot cross the interface (e.g. B and C are mutually saturated), there will be a convective bulk flow of velocity v_0 away from the interface to compensate for the diffusive inward flow of B. This is illustrated in Fig. 2.5, which shows the total mass flow rates into and out of the concentration boundary layer control volume, $(CV)_c$. When physical properties are constant and the flow is steady, Fig. 2.5 shows that

$$w \frac{\mathrm{d}}{\mathrm{d}x}\left[\int_0^{\delta_{c1}} \rho_1 u \,\mathrm{d}y\right] \mathrm{d}x = \rho_1 u' \, w \,\mathrm{d}x + \rho_1 v_0 w \,\mathrm{d}x$$

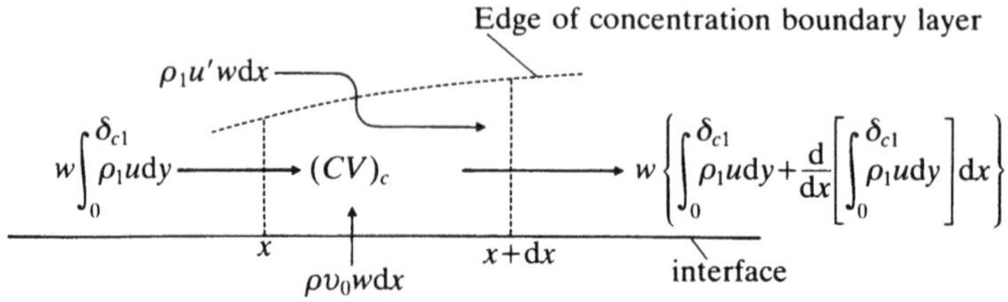

Fig. 2.5. Flow rates of total mass into and out of the concentration boundary layer control volume, $(CV)_c$, in Fig. 2.4.

so

$$u' = \frac{d}{dx}\left[\int_0^{\delta_{c1}} u \, dy\right] - v_0 \tag{2.98}$$

where u' is the velocity in the x direction at $y = \delta_c$, and w is width normal to flow.

Figure 2.6 is similar to Fig. 2.5, but written for component A instead of total mass. It shows that

$$n_{A0}\, w\, dx = w \frac{d}{dx}\left[\int_0^{\delta_{c1}} \rho_{A1}\, u\, dy\right] dx - (\rho_{A1})_\infty u'\, w\, dx$$

But

$$n_{A0} = (\rho_{A1})_0 v_0 - D_1 (\partial \rho_{A1}/\partial y)_{y=0+}$$

so that, combining these two expressions and with eqn (2.98) in place of u', we obtain

$$\frac{d}{dx}\int_0^{\delta_{c1}} [(\rho_{A1})_\infty - \rho_{A1}]\, u\, dy = D_1 (\partial \rho_{A1}/\partial y)_{y=0+} - v_0 [(\rho_{A1})_0 - (\rho_{A1})_\infty]$$

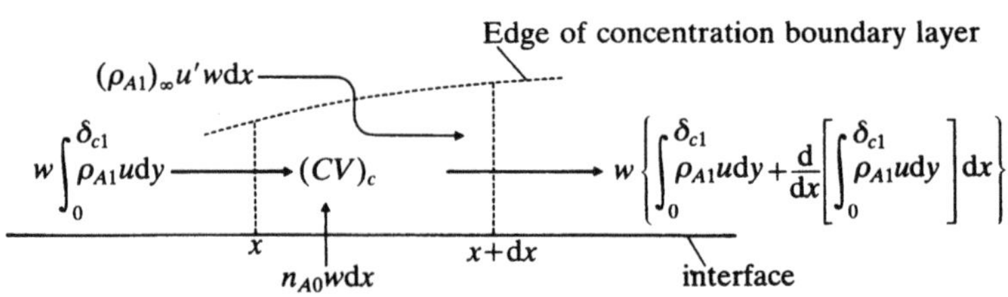

Fig. 2.6. Mass flow rates of component A into and out of the concentration boundary layer control volume, $(CV)_c$, in Fig. 2.4.

When mass transfer rates are low v_0 is negligible, so that this equation becomes

$$\frac{d}{dx}\int_0^{\delta_{c1}} [(\rho_{A1})_\infty - \rho_{A1}]\, u\, dy = D_1\, (\partial\rho_{A1}/\partial y)_{y=0+} \tag{2.99}$$

Similarly, for phase 2,

$$\frac{d}{dx}\int_0^{-\delta_{c2}} [(\rho_{A2})_\infty - \rho_{A2}]\, u\, dy = D_2\, (\partial\rho_{A2}/\partial y)_{y=0-} \tag{2.100}$$

Equations (2.99 and 2.100) are the integral mass or concentration boundary layer equations for the phases above and below the interface, respectively.

The momentum equation for the control volume $(CV)_M$ in Figs 2.4 and 2.7 equates the total efflux rate of x-directed momentum to the sum of all external forces in the x-direction acting on $(CV)_M$. Figure 2.7 shows that, for phase 1,

$$\Sigma\, g_c dF_x = g_c\left(\tau_0 w dx - \frac{dP}{dx} dx\, w\delta_1\right)$$
$$= w\frac{d}{dx}\left[\int_0^{\delta_1} \rho_1 u^2 dy\right] dx - w u_{\infty 1}\frac{d}{dx}\left[\int_0^{\delta_1} \rho_1 u\, dy\right] dx$$

When Bernoulli's equation is applied along a streamline in the frictionless flow just outside the momentum boundary layer it is found that $dP/dx = 0$. However, $dP/dy = 0$ for $0 \le y \le \delta_1$, so that $dP/dx = 0$ inside the boundary layer for the special case of a plane interface parallel to $u_{\infty 1}$ and $u_{\infty 2}$. If, in addition, the fluid is Newtonian, then

$$g_c\tau_0 = \mu_1\left(-\frac{\partial u}{\partial y}\right)_{y=0+}$$

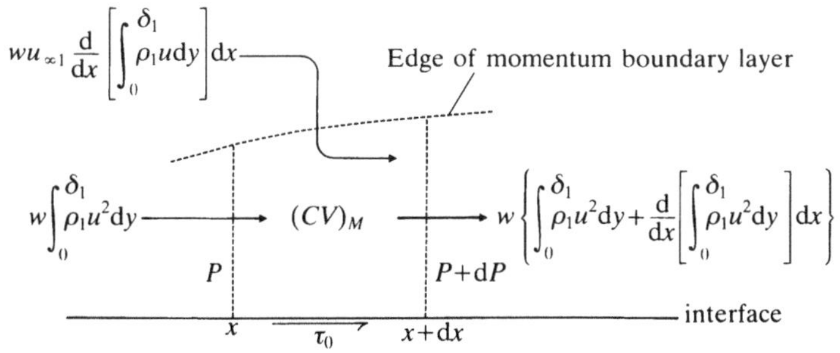

Fig. 2.7. Flow rates of x-directed momentum into and out of the momentum boundary layer control volume, $(CV)_M$, in Fig. 2.4.

and the equation for $\Sigma g_c \mathrm{d}F_x$ becomes

$$\frac{\mathrm{d}}{\mathrm{d}x}\int_0^{\delta_1} (u_{\infty 1} - u)u\,\mathrm{d}y = \frac{\mu_1}{\rho_1}\left(\frac{\partial u}{\partial y}\right)_{y=0+} \tag{2.101}$$

Similarly, for phase 2,

$$\frac{\mathrm{d}}{\mathrm{d}x}\int_0^{-\delta_2} (u_{\infty 2} - u)u\,\mathrm{d}y = \frac{\mu_2}{\rho_2}\left(\frac{\partial u}{\partial y}\right)_{y=0-} \tag{2.102}$$

Equations (2.101 and 2.102) are the integral momentum boundary layer equations for the phases above and below the interface, respectively.

Quartic polynomials were used by Potter (1957) to represent the velocity distributions in the two boundary layers. The expressions, which satisfy the usual boundary conditions, are

$$\frac{u}{u_{\infty 1}} = \frac{u_0}{u_{\infty 1}} + \left(1 - \frac{u_0}{u_{\infty 1}}\right)\left[2\left(\frac{y}{\delta_1}\right) - 2\left(\frac{y}{\delta_1}\right)^3 + \left(\frac{y}{\delta_1}\right)^4\right] \tag{2.103}$$

$$\frac{u}{u_{\infty 1}} = \frac{u_0}{u_{\infty 1}} + \left(\frac{u_{\infty 2}}{u_{\infty 1}} - \frac{u_0}{u_{\infty 1}}\right)\left[2\left(\frac{y}{-\delta_2}\right) - 2\left(\frac{y}{-\delta_2}\right)^3 + \left(\frac{y}{-\delta_2}\right)^4\right] \tag{2.104}$$

These equations are substituted for u on both sides of eqns (2.101 and 2.102), to obtain the local momentum boundary layer thicknesses

$$\delta_1^2 = \left(\frac{4\mu_1 x}{\rho_1 u_{\infty 1}}\right)\frac{630}{74 + 115(u_0/u_{\infty 1})} \tag{2.105}$$

$$\delta_2^2 = \left(\frac{4\mu_2 x}{\rho_2 u_{\infty 1}}\right)\frac{630}{74(u_{\infty 2}/u_{\infty 1}) + 115(u_0/u_{\infty 1})} \tag{2.106}$$

Combining eqns (2.103 and 2.104) with the requirement of continuity of shear stress at the interface gives u_0, the interfacial velocity, as

$$\frac{u_0}{u_{\infty 1}} = \frac{(\delta_2/\delta_1) + (u_{\infty 2}/u_{\infty 1})(\mu_2/\mu_1)}{(\delta_2/\delta_1) + (\mu_2/\mu_1)} \tag{2.107}$$

Appropriate manipulation then allows evaluation of δ_1, δ_2, and u_0 for any $u_{\infty 1}$, $u_{\infty 2}$, μ_1, μ_2, ρ_1, and ρ_2 from eqns (2.105, 2.106, and 2.107).

Concentration distributions in the boundary layers were represented by the quartic polynomials shown below, which satisfied the usual boundary conditions, including continuity of mass transfer across the interface.

$$\frac{\rho_{A1}}{(\rho_{A1})_\infty} = \frac{(\rho_{A1})_0}{(\rho_{A1})_\infty} + \left(1 - \frac{(\rho_{A1})_0}{(\rho_{A1})_\infty}\right)\left[2\left(\frac{y}{\delta_{c1}}\right) - 2\left(\frac{y}{\delta_{c1}}\right)^3 + \left(\frac{y}{\delta_{c1}}\right)^4\right] \tag{2.108}$$

$$\frac{\rho_{A2}}{(\rho_{A2})_\infty} = \frac{(\rho_{A2})_0}{(\rho_{A2})_\infty} + \left(1 - \frac{(\rho_{A2})_0}{(\rho_{A2})_\infty}\right)\left[2\left(\frac{y}{-\delta_{c2}}\right) - 2\left(\frac{y}{-\delta_{c2}}\right)^3 + \left(\frac{y}{-\delta_{c2}}\right)^4\right] \tag{2.109}$$

The velocity distributions from eqns (2.103 and 2.104) and the concentration distributions from eqns (2.108 and (2.109) are inserted throughout eqns (2.99 and 2.100), respectively. The following abbreviated result was shown to approximate Potter's (1957) more detailed solution in the usual range of liquid Schmidt numbers:

$$\frac{\delta_{c1}}{\delta_1} \doteq (N_{Sc1})^{-p_1};\ p_1 = f_{\text{known}}\left(\frac{u_0}{u_{\infty 1}}\right) \tag{2.110}$$

$$\frac{\delta_{c2}}{\delta_2} \doteq (N_{Sc2})^{-p_2};\ p_2 = f'_{\text{known}}\left(\frac{u_0}{u_{\infty 1}}, \frac{u_{\infty 2}}{u_{\infty 1}}\right) \tag{2.111}$$

The local mass transfer coefficients are then calculated, as shown below for the upper stream:

$$k^*_{\rho 1}(\rho_{A1})_\infty\left(1 - \frac{(\rho_{A1})_0}{(\rho_{A1})_\infty}\right) = D_1\left(\frac{\partial \rho_{A1}}{\partial y}\right)_{y=0+}$$

The derivative is evaluated from eqn (2.108), to obtain

$$\frac{k^*_{\rho 1}x}{D_1} = \frac{2x}{(\delta_{c1}/\delta_1)\delta_1}$$

Combining with eqns (2.105 and 2.110),

$$(N_{Sh1})^*_x = \frac{k^*_{\rho 1}x}{D_1} = \left(\frac{74 + 115(u_0/u_{\infty 1})}{630}\right)^{1/2}\left(\frac{xu_{\infty 1}\rho_1}{\mu_1}\right)^{1/2}(N_{Sc1})^{p_1} \tag{2.112}$$

Similarly for the lower stream,

$$(N_{Sh2})^*_x = \frac{k^*_{\rho 2}x}{D_2}\left(\frac{74(u_{\infty 2}/u_{\infty 1}) + 115(u_0/u_{\infty 1})}{630}\right)^{1/2}\left(\frac{xu_{\infty 1}\rho_2}{\mu_2}\right)^{1/2}(N_{Sc2})^{p_2} \tag{2.113}$$

When $u_0 = 0$, $p_1 = 1/3$, and eqn (2.112) reduces to the exact solution for mass transfer from a flat plate (Welty *et al.* 1984, pp. 595–601), except that the constant is 0.343 instead of 0.332.

When $u_0 = u_{\infty 1}$ the upper liquid is flowing without shear; p_1 becomes 1/2 and eqn (2.112) condenses to

$$k^*_{\rho 1} = 0.548\ (D_1 u_{\infty 1}/x)^{1/2} = 0.548\ (D_1/t)^{1/2} \tag{2.114}$$

where the contact time between phases at x is $t = x/u_{\infty 1}$. Equation (2.114)

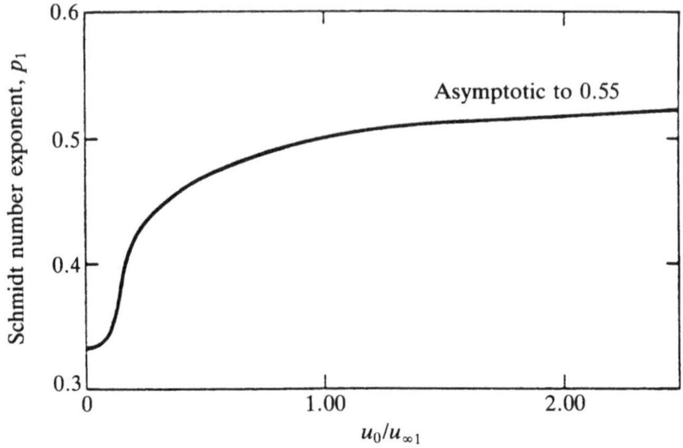

Fig. 2.8. Exponent on Schmidt number versus relative interfacial velocity, $u_0/u_{\infty 1}$ (Potter 1957).

is within 3 per cent of the value given earlier by the penetration theory in eqn (2.84).

Figure 2.8 shows Potter's evaluation of p_1, the exponent on the Schmidt group, for $N_{Sc1} = 100$, over a wide range of $u_0/u_{\infty 1}$. The plot was considered accurate to about 5 per cent for typical liquid N_{Sc} values.

The analysis is restricted to low transfer rates and stable interfaces; however, criteria for the latter have not been established.

2.2.6 Other theories

Many other theories have been developed to describe the mechanism of mass transfer either between two fluid phases or between a fixed surface or wall and an adjacent fluid. These may be approximately classified into the three categories of film, surface renewal, and boundary layer theories identified earlier, albeit with some overlap, as follows.

2.2.6.1 *Combined film and surface renewal theories*

A model usually referred to as the film-penetration theory was developed by Dobbins (1956) and by Toor and Marchello (1958). Those authors proposed the existence of a laminar surface film of thickness z_L, which contains the entire resistance to transfer offered by its own phase, and in which elements are continually replaced by fluid eddies drawn from the bulk of the phase. The concentration for all $z > z_L$ is constant at $\rho_{A\infty}$. The authors invoked Danckwerts' surface age distribution function, eqn (2.96), to express k_ρ in terms of z_L, D, and s. The result for short average element life is

$$k_\rho = \frac{(n_{A0})_{av}}{\rho_A^* - \rho_{A\infty}} = (sD)^{1/2}\{1 + 2\sum_{n=1}^{\infty} \exp[-2nz_L(s/D)^{1/2}]\} \quad (2.115)$$

and for long average element life,

$$k_\rho = \frac{(n_{A0})_{av}}{\rho_A^* - \rho_{A\infty}} = \frac{D}{z_L}[1 + 2\sum_{n=1}^{\infty} (1 + n^2\pi^2 D/sz_L^2)^{-1}] \quad (2.116)$$

At high surface renewal rates (sz_L^2/D is large) eqn (2.115) reduces to eqn (2.97), coinciding with Danckwerts' result. On the other hand, for low values of sz_L^2/D, eqn (2.116) reduces to the film theory result in eqn (2.78) at low concentrations of A, since $k_\rho = k_x/c$. The quantity z_L is usually unknown, but decreases with increasing turbulence.

In a later modification of the film-penetration model, Marchello and Toor (1963) postulated molecular diffusion between layers of fluid which are well mixed at random time intervals, but which are not necessarily at the surface.

A mechanism of consecutive film and surface renewal was proposed by Wasan and Ahluwalia (1969) to describe heat and mass transfer between a wall and an adjacent fluid stream. Predictions compared favorably with data in gas and liquid fluidized beds.

2.2.6.2 *Surface renewal theories*
[See also Javed, Ch. 4]

Kishinevsky and coworkers (1949–1954) disputed Danckwert's suggestion that the probability of replacement of a surface element is independent of its age. They further proposed that molecular diffusion into an eddy at the interface is negligible compared to transfer by convective mass flow. They therefore contend that k_ρ is proportional to D to the power zero.

Ruckenstein (1968) extended the penetration theory to allow for velocity distribution inside the eddies during penetration by the solute at the interface. Harriott (1962) presented a model that incorporates the penetration mechanism, but which includes thin, laminar, interfacial films that are stabilized by interfacial tension and are therefore excluded from intermittent mixture with the bulk fluid. The penetration theory has been extended by Angelo *et al.* (1966) to allow for 'stretching' surfaces, such as those that occur, for example, in large, oscillating droplets.

The modifications to surface renewal theory by Perlmutter (1961) included several alternative eddy residence time distributions, plus two ways of incorporating an interfacial resistance to mass transfer.

The eddy cell model of Lamont and Scott (1970) indicates the use of power input to the liquid phase as a factor in correlating the coefficient.

Hanratty (1956) extracted s values from corresponding experimental k_ρ

values in applying Danckwert's surface renewal model to mass transfer between a fixed solid surface and an adjacent fluid. Good agreement was obtained between measured and calculated concentration profiles near the solid surface.

A theory of penetration with periodically varying rate of surface renewal was developed by Skelland and Lee (1981) for two-phase dispersions in agitated vessels. This attempts to allow for the varying degrees of turbulence encountered by bubbles or drops as they circulate through all regions of the vessel.

2.2.6.3 *Boundary layer theories*

The turbulent boundary layer model, as outlined by Sherwood *et al.* (1975, pp. 156–7), refers to mass transfer between a fixed surface or wall and an adjacent turbulent fluid stream. Its formulation relies heavily on the analogies between momentum and mass transfer, as developed in detail by Skelland (1974, 1985).

King (1966) considered the case of turbulent liquid-phase mass transfer near the interface with a gas. He combines the penetration model with a variable eddy diffusivity, where the latter varies as a power function of the distance from the surface. As the contact time increases, King's development is characterized by a smooth transition from the penetration theory to the turbulent boundary layer model.

The model by Pinczewski and Sideman (1974) pictured the wall region as being a patchwork of part transient, part steady, developing boundary layers which are periodically replaced. Agreement with published data was good and the model has the advantage that only constants from fluid mechanics data are needed.

Comparisons between experimental results and the predictions of the film theory on the one hand and various forms of surface renewal and boundary layer theory on the other are briefly reviewed by Sherwood *et al.* (1975, Ch. 5) and by Skelland (1974, 1985). It is interesting that, although the evidence generally appears to favour the surface renewal or boundary layer mechanisms, the two-film theory contributes to the design of complex processes in a way that continues to be very useful. Examples of this include the formulation of transfer unit relationships for packed column design, as detailed by Skelland (1985), and the modelling of mass transfer with chemical reaction, as noted later.

Although application of some of these theories is hampered by lack of knowledge of some essential components (such as z_{fG}, z_{fL}, t_e, s, or z_L, as appropriate), they are nevertheless valuable in indicating limits on n between 0 and 1 in the expression $k_\rho \alpha D^n$. Empirical correlations of experimental data that show an exponent on D, the molecular diffusivity, outside this range, should probably be rejected.

2.3 Mass transfer associated with jet phenomena

The occurrence of a liquid jetting into another immiscible liquid is quite widespread industrially. As an example, consider the operation of liquid–liquid extraction in perforated plate columns. At disperse phase flow rates that exceed the jetting velocity, jets are formed from each perforation, breaking into drops at some distance from the plate. Advantages associated with this mode of operation were demonstrated by Mayfield and Church (1952), who found that plate efficiences are enhanced up to two and a half-fold, with accompanying increase in throughput rate. Several recent studies have examined the drop size formed from jet break-up in liquid–liquid spray columns (Perrut and Loutaty 1972; Horvath *et al.* 1978; Van Den Akker 1980; and Kumar and Hartland 1982). Such work may presumably be related to perforated plate columns eventually since, as Treybal (1980, p. 542) has noted, the use of spray columns is not usually recommended. This is because of the severity of axial mixing in such units, so that it is difficult to obtain much more extraction than that equivalent to a single stage with them. Jets are also encountered during the injection of the disperse phase into the mixing vessels in mixer–settler extraction units (Treybal 1963, p. 408); in the operation of injectors and jet mixers (Treybal 1963, pp. 432–5); in certain fibre spinning operations, and in liquid–liquid jet reactors (Coyle *et al.* 1981)—all in the presence of mass transfer.

2.3.1 Hydrodynamic outline

The relevant hydrodynamics can be outlined only briefly here, because the primary focus of this chapter is on mass transfer. Some quantitative data on jet breakup configurations are needed, however, for proper application of the mass transfer relationships.

When a liquid flows vertically with a mean linear velocity u_0 through an orifice or nozzle into another immiscible liquid, a variety of phenomena are observed as u_0 increases. At low flow rates drops are formed individually at the tip of the nozzle, and break off when they attain a particular size, determined by the nozzle diameter, d_n, flow rate, and physical properties. When the flow rate increases to the jetting velocity, u_{0j}, a small jet begins to form, the length of the jet, L_j, increasing as u_0 increases beyond u_{0j}. Eventually the jet reaches a maximum length when u_0 attains the value u_{0m}; as u_0 increases beyond u_{0m} the jet length diminishes until it becomes zero, with the production of small, non-uniform droplets at the nozzle itself. Drops detaching from the jet tip are rather uniform in size at a given u_0 for $u_{0j} < u_0 < u_{0m}$, and attain a minimum diameter (and hence maximum total interfacial area) when u_0 equals u_{0m}. The latter is therefore sometimes called the maximum area condition, with u_{0m} alternatively referred to as the critical velocity. [In fact, the maximum area condition was originally said to occur at 'a flow rate slightly below that for maximum jet length' (Christiansen and Hixson 1957). The distinction – if real – has been blurred, however in subsequent writings (Horvath *et al.* 1978; Kumar and Hartland 1982.)] The jet tapers from a diameter of d_n at the nozzle, to d_{jc} at the jet tip, where breakup into drops occurs. When u_0 reaches u_{0m} the corresponding value of d_{jc} will be denoted here by d_{jm}. In the range $u_{0j} < u_0 < u_{0m}$ the jet is said to be in the 'varicose' region, because of the appearance of occasional surface disturbances whose amplification results in breakup. For u_0 beyond u_{0m} the jet is described as 'sinuous' because of its irregular weaving motion. Drop sizes are much less uniform at a given u_0 in the sinuous region than in the varicose region.

The hydrodynamic aspects of jet behavior in liquid–liquid systems have been studied by many workers, but it has only recently been realized that the occurrence of mass transfer either into or out of the jet can have complex and substantial effects on both the jet length and the drop size resulting from breakup. Meister and Scheele (1969a) measured jet lengths during the transfer of acetone between benzene jets and water. They found that transfer into the jet produced significantly longer jets than transfer out, and that transfer in either direction had a stabilizing effect. They suggested that this stabilization due to mass transfer resulted from the reduction in interfacial tension induced by the solute. In contrast, Sawistowski (1973) noted that jet breakup in liquid–liquid systems is accelerated by mass transfer into a jet when $d\sigma/d\rho_A$ is negative, and decelerated by mass transfer out of the jet. These conflicting observations were somewhat reconciled by Burkholder and Berg (1974), who studied the effect of mass transfer on laminar jet breakup for liquid jets in liquids. They predicted, in a complex analysis, that mass transfer of a surface-tension-lowering solute either into or out of the jet may be either stabilizing (longer jets) or destabilizing (shorter jets) depending on system physical properties and the mass transfer rate. They contended that the mass transfer pheneomena are so complicated that it will often not be possible to predict whether mass transfer will be stabilizing or destabilizing in a given situation without numerical solution of a very complicated characteristic equation. Further developments of this analysis by Berg and coworkers include the paper by Tarr and Berg (1980) on the effect of non-linear concentration profiles on the break-up of jets undergoing mass transfer, and the article by Coyle *et al.* (1981). The latter work presents an elaborate linear hydrodynamic stability analysis of a liquid jet in a second immiscible liquid in the presence of mass transfer and predicts a significant but complicated dependence of the jet length vs velocity (L_j vs u_0) curves on the direction of mass transfer, on the mass transfer driving force, and on the extent of solute adsorption. Only qualitative agreement was obtained between most of the theoretical predictions and the data from experiments, in which acetone transferred between benzene jets and water. These complex influences of mass transfer rate and direction extend to the drop size and size distribution resulting from jet break up (Coyle *et al.* 1981). The important findings just outlined concerning the effects of the direction and rate of mass transfer must be kept in mind when consulting any of the following references on jetting hydrodynamics. Relevant aspects of *jetting velocity, jet length, jet contraction, and drop size from jet break up* will now be considered in turn.

The jetting velocity, u_{0j}, has been correlated by Shiffler (1965), Fujinawa *et al.* (1957), Scheele and Meister (1968), Ryan (1966), de Chazal and Ryan (1971), and Ruff *et al.* (1976). The last five correlations are presented in tabular form by Kumar and Hartland (1982), who successfully used the following expression due to Ruff *et al.* (1976) in their multi-orifice distributor:

$$u_{0j} = (2\sigma/\rho_{\mathrm{d}} d_n)^{1/2} \tag{2.117}$$

Horvath *et al.* (1978) were unable to obtain good results from the de Chazal and Ryan correlation for u_{0j} when applied to their multinozzle distributor. However, they found the Scheele and Meister (1968) correlation to work well, as follows:

$$u_{0j} = 1.73\left[\frac{\sigma}{\rho_d d_n}\left(1 - \frac{d_n}{d_p}\right)\right]^{1/2} \tag{2.118}$$

For a flat velocity distribution in the jet, as for flow through a perforation in a thin plate, instead of through a longer nozzle, the constant 1.73 is replaced by 2.0.

Jet length, L_j, has been the subject of many investigations; several studies have been performed for liquid jets either in a gas or in a partial vacuum, but here the effects of continuous phase density and viscosity are often slight. Attention will therefore be confined to the more pertinent case of jets in liquid–liquid systems, for which relevant references include Tyler and Watkin (1932), Tomotika (1935), Keith and Hixson (1955), Christiansen and Hixson (1957), Meister and Scheele (1967, 1969a), Sawistowski (1973), Duzubur and Sawistowski (1974), Burkholder and Berg (1974), Skelland and Johnson (1974), Skelland and Huang (1977, 1979), Tarr and Berg (1980), Coyle *et al.* (1981), and Skelland and Walker (1989).

Meister and Scheele (1969a) produced an analytical expression for jet length assuming a non-tapering jet and with forms of velocity distribution in both phases. Their analysis did not allow for the important effects of mass transfer. Correlations of jet length in liquid–liquid systems in the presence of mass transfer are very scarce; those obtained by Skelland and Huang (1977a, 1979) for transfer out of the jet are summarized in Table 2.1, where

TABLE 2.1 *Some hydrodynamics of liquid–liquid jetting with mass transfer out of the jet*

Mass transfer features								Validity with SAA*	
Rate	Direction‡	Controlling‡ phase	Equation No.	Equations for the varicose region (SAA absent)	σ Range (dynes/cm)	Mean abs error	Ref.**	SAA low conc. $\sigma > 15$ dyne/cm	All conc. of SAA
Jet length									
Zero	—	—	—	—	—	—	—	—	—
Very low	*d–c*	*c*	(2.119)	$L_j/d_n = 10.8341 (\Delta N_{We})^{0.5860} (\mu_c/\mu_d)^{0.5097}$		75.0%	SH, 79	(56.7%)	(57.6%)
Moderate	*d–c*	*d*	(2.120)	$L_j/d_n = 5.0767 (\Delta N_{We})^{0.5499} (\mu_c/\mu_d)^{0.5245}$	> 24 (high)	39.7%	SH, 77	(24.1%)†	(38.0%)
Jet diameter at tip									
Zero	—	—	(2.121)	$d_n/d_{jc} = 3.5037 (d_n/l_s) + 0.9491$	all	20.0%	SJ, 74	(30.2%)	(31.9%)
Very low	*d–c*	*c*	(2.122)	$d_n/d_{jc} = 2.6086 (d_n/l_s) + 0.8495$	all	21.6%	SH, 79	(14.9%)	(15.6%)
Moderate	*d–c*	*d*	(2.123)	$d_n/d_{jc} = 2.7350 (d_n/l_s) + 0.5718$	all	20.3%	SH, 77	(13.1%)†	(13.0%)†
Drop size									
Zero	—	—	(2.124)	$d_p/d_{jm} = 2.6051 - 0.7747 (u_o/u_{om}) + 0.3994 (u_o/u_{om})^2$	all	—	SH, 79	(14.6%)	(17.8%)
Very low	*d–c*	*c*	(2.125)	$d_p/d_{jm} = 2.7270 - 0.2028 (u_o/u_{om}) + 0.0731 (u_o/u_{om})^2$	all	14.6%	SH, 79	(10.4%)†	(11.0%)†
Moderate	*d–c*	*d*	(2.126)	$d_p/d_{jm} = 3.0704 - 0.1701 (u_o/u_{om}) + 0.0487 (u_o/u_{om})^2$	all	13.1%	SH, 77	(15.5%)	(15.7%)

* Parentheses show mean absolute error obtained when Skelland and Walker's (1989) data in the presence of surface active agents (SAA) are compared to values calculated using equations from column 5 with diminished value of interfacial tension due to presence of SAA. Initial solute concentration in Skelland and Walker's data was ca. 1 mass per cent throughout.

** SH, 77 = Skelland and Huang (1977a); SH, 79 = Skelland and Huang (1979); SJ, 74 = Skelland and Johnson (1974).

† This check mark denotes the recommended correlation for the data of Skelland and Walker (1989) with SAA.

‡ c and d are continuous and disperse phases.

'very low' and 'moderate' rates of mass transfer refer to rates resulting from the initial presence of about 1 mass per cent of benzoic acid or 2 mass per cent of acetic acid in the organic disperse phase, respectively (water was the continuous phase). Thus the 'moderate' rates of mass transfer corresponded to molal fluxes from the jet surface [$(N_{A,\text{jet}})_{\text{av}} = k_{dj}(\Delta c_{Aj})_{\text{av}}$] of the order of 0.4 (10^{-5}) g-mol/cm^{2}s. This exceeded the molal flux from the jet surface in the 'very low' range of mass transfer rates by one and sometimes two orders of magnitude.

In Table 2.1, column 5 contains empirical relationships in the varicose region for jet length, jet diameter at the tip, and resulting drop-size, determined in the absence of added surface active contamination by Skelland and Johnson (1974) and Skelland and Huang (1977a, 1979). The errors from comparison with experimental data appear in column 7, while the last two columns show the mean absolute errors resulting from the application of these empirical correlations to results from systems with moderate rates of mass transfer, to which various amounts of surface active agents (SAA) had been added (Skelland and Walker, 1989). In the latter application, the correlations were used with the diminished value of interfacial tension due to the presence of SAA. Equations (2.119 and 2.120) and the data of Skelland and Walker (1989) show an increase in jet length due to the presence of SAA (because of diminished σ) when mass transfer is also occurring. This is directionally in accordance with both the theoretical and experimental findings of Coyle *et al.* (1981). It is hoped that the relationships may provide some guidance in the absence of more definitive expressions, but they should not be extrapolated—for example, to mass transfer in the other direction or at higher rates. The ranges of variables pertaining to these correlations are given in the original papers. It should be noted that high accuracy in the correlation of instantaneously photographed jet lengths cannot be expected because of the increasing *oscillation* in jet length with decreasing flow velocity below u_{0m}, as described by Christiansen and Hixson (1957).

Jet contraction is observed in jets of Newtonian liquids with increasing distance from the nozzle; Meister and Scheele (1969a) observed decreases of 50% in jet diameter at the point of break-up. Christiansen and Hixson (1957) correlated jet contraction at the flow rate corresponding to minimum drop size after break-up in liquid–liquid systems by using the concept of a reduced nozzle diameter, defined as the ratio of nozzle diameter to a system length parameter, l_s, where $l_s = \pi(\sigma/\Delta\rho g)^{1/2}$. They found a relationship for jet contraction in a plot of d_n/d_{jc} vs d_n/l_s. Skelland and Johnson (1974) followed the concept of reduced diameter and constructed a similar plot to that of Christiansen and Hixson (1957) in the absence of mass transfer and for various jetting states not confined to the maximum area condition. They found reasonable correlation over a range of flow rates through the nozzle.

Photographic techniques were employed, so that an instaneous photograph records a d_{jc} somewhere between minimum and maximum values just before and just after drop detachment, respectively. This of course leads to considerable scatter in the data. The procedure just described was subsequently extended to systems undergoing 'very low' and 'moderate' rates of mass transfer out of the jet and into the continuous phase, leading to the correlations for d_{jc} in Table 2.1 (Skelland and Huang 1977a, 1979). As in the case of jet length, extrapolation cannot be recommended.

The drop size resulting from jet break-up has received considerable attention because of its relation to the interfacial area in liquid–liquid extraction columns. Since this section is devoted to jetting conditions, attention will now be confined to $u_0 > u_{0j}$; drop sizes for $u_0 < u_{0j}$ will be considered later.

Although reported studies may be divided into single nozzle and multinozzle distributors, Miller and Pilhofer (1976) and Kumar and Hartland (1982) showed that drop formation from multiple nozzles was essentially the same as that from single nozzles in the jetting region when typical nozzle spacings were used.

The drop size may be substantially dependent on both the direction and the rate of mass transfer (Perrut and Loutaty 1972; Coyle *et al.* 1981; Coyle and Berg 1984). In particular, Bayens and Laurence (1968) and Perrut and Loutaty (1972) both found that mass transfer from the continuous to the disperse phase had a negligible effect on drop size, whereas transfer in the opposite direction was accompanied by the formation of significantly larger drops.

Relationships for predicting drop size resulting from jet break-up are given by Perrut and Loutaty (1972, multiple nozzle, no mass transfer), Vedaiyan (1969, multiple nozzle, no mass transfer), Horvath *et al.* (1978, multiple nozzle, no mass transfer *or* transfer into the jet), Kumar and Hartland (1982, multiple and single nozzle, no mass transfer), Christiansen and Hixson (1957) and Treybal (1963, single nozzle, no mass transfer), Meister and Scheele (1969b, single nozzle, no mass transfer), Kumar (1983), Kumar and Hartland (1984), Dalingaros *et al.* (1986) (the last three being for multiple nozzle, no mass transfer), Skelland and Johnson (1974, single nozzle, no mass transfer), Skelland and Huang (1977a, 1979, single nozzle, mass transfer out of the jet), and Skelland and Walker (1989, single nozzle, mass transfer out of the jet, with added surface active agents).

The first five correlations listed above are tabulated by Kumar and Hartland (1982), and the last three, with particular emphasis on the presence of mass transfer out of the jet, with and without surface active contamination, will now be considered.

The drop diameter d_p resulting from jet break-up in uncontaminated systems is summarized for zero, very low, and moderate rates of mass transfer out of the jet by eqns (2.124, 2.125, and 2.126), respectively, in Table 2.1. These expressions contain d_{jm} and u_{0m} as normalizing factors, corresponding to the maximum interfacial area condition noted earlier. They are calculated from the equations given by Treybal (1963, pp. 467–8) as

$$u_{0m} = 2.69\left(\frac{d_{jm}}{d_n}\right)^2\left(\frac{\sigma/d_{jm}}{0.5137\rho_d + 0.4719\rho_c}\right)^{0.5} \tag{2.127}$$

where the ratio of nozzle-to-jet diameter is given empirically by

$$\frac{d_n}{d_{jm}} = \alpha\left(\frac{\pi d_n}{l_s}\right)^{\beta} + \gamma \tag{2.128}$$

when

$$\frac{\pi d_n}{l_s} < 0.785,\ \alpha = 0.485,\ \beta = 2,\ \gamma = 1$$

$$\frac{\pi d_n}{l_s} > 0.785,\ \alpha = 1.51,\ \beta = 1,\ \gamma = 0.12$$

An alternative expression for u_{0m} was deduced by Hughmark (1967) from an analysis of Christiansen and Hixson's (1957) data, for the particular case of a continuous phase with properties similar to water, as

$$u_{0m} = 2.94(\sigma/\rho_d d_n)^{1/2} \tag{2.129}$$

Comparison with eqn (2.117) for the jetting velocity, u_{0j}, shows that $u_{0m} \doteq 2u_{0j}$ as suggested by Ruff (1978)—at least for a water-like continuous phase.

The results of applying eqns (2.124–2.126) to data from moderate-transfer-rate systems in the presence of added surfactant appear in the last two columns of Table 2.1. In this application the correlations were used with the diminished values of σ caused by the SAA. Check marks show the application that best fitted Skelland and Walker's (1989) data with SAA in each case. Although their system exhibited 'moderate-rate' mass transfer out of the jet, the drop size was best predicted by the correlation for 'very low' rates of transfer. To explain this anomaly it was suggested that, by the time drop formation occurred, SAA had accumulated at the interface in sufficient amount to cut the mass transfer rate to 'very low' values (see e.g. Skelland and Caenepeel 1972).

As in the case of jet length and jet contraction, extrapolation of the drop size correlations beyond the ranges of variables for which they were obtained cannot be recommended.

2.3.2 Mass transfer inside the jet

Several analytical and experimental studies have been made of mass transfer in liquid jets, particularly with a gaseous continuous phase (Cullen and Davidson 1957; Scriven and Pigford 1958, 1959; Quinn and Jeannin 1961; Rehm *et al.* 1963; Kimura and Miyauchi 1966; Duda and Vrentas 1968; Caskey and Barlage 1972). Skelland and Huang (1977, 1979) attempted to apply a Graetz-type analysis, regarding the jet as a straight cylinder. They then tried to allow for the (usual) tapering, conical configuration, in an application of the penetration theory to the problem. Simple velocity distributions were assumed, with low rates of transfer, and the results compared rather successfully with experimental data in liquid–liquid extraction systems. Their approach will accordingly be given here.

Three alternative steady flow models are assumed in turn, on the basis of tube-plug flow, tube-parabolic flow, and the penetration theory. The latter will be applied to both the jet (disperse) phase and the continuous phase.

2.3.2.1 *Transfer in a jet with tube plug flow*

When the dominant resistance to mass transfer lies inside the jet, the solute concentration in the continuous phase can be considered as homogeneous with concentration $(\rho_{Aci} + \rho_{Acf})/2$. Then the dispersed-phase concentration at the jet surface can be regarded as constant at ρ^*_{Ad}. Solute transfers from the interior of the jet to its surface, and then out.

For plug flow the transport equation for solute A is obtained from eqn (2.45) by substituting z/V_j for t and ρ_A for c_A to obtain

$$\frac{\partial \rho_A}{\partial z} = \frac{D}{V_j}\left(\frac{\partial^2 \rho_A}{\partial r^2} + \frac{1}{r}\frac{\partial \rho_A}{\partial r}\right) \tag{2.130}$$

with boundary conditions

$$\rho_A = \rho^*_{Ad} \text{ at } r = r_t,\ 0 \le z \le L_j$$

$$\rho_A = \rho_{A0} \text{ at } z = 0,\ 0 \le r \le r_t$$

$$\frac{\partial \rho_A}{\partial r} = 0 \text{ at } r = 0,\ 0 \le z \le L_j$$

The solution to eqn (2.130) is given in detail by Skelland (1974, 1985, Ch. 5) as

$$\frac{\rho_A - \rho^*_{Ad}}{\rho_{A0} - \rho^*_{Ad}} = \sum_{i=1}^{\infty} \frac{2}{b_i r_t}\frac{J_0(b_i r)}{J_1(b_i r_t)} \exp\left(-\frac{D}{V_j} b_i^2 z\right) \tag{2.131}$$

The average mass transfer coefficient inside the jet, $(k^*_\rho)_{\text{lm}}$, is also shown to be

$$\frac{(k_\rho^*)_{\text{lm}} d_t}{D} = \frac{1}{4}\left(\frac{d_t}{L_j}\right) N_{\text{Re}} N_{\text{Sc}} \ln\left\{4\sum_{i=1}^{\infty} a_i^{-2} \exp\left[\frac{-2a_i^2(L_j/r_t)}{N_{\text{Re}} N_{\text{Sc}}}\right]\right\}^{-1} \tag{2.132}$$

where $a_i = b_i r_t$, and is the ith root of the equation $J_0(a_i) = 0$. The first thirty values of a_i are given by Kreyszig (1967). Also, $d_t = (d_n + d_{jc})/2$.

The $(k_\rho^*)_{\text{lm}}$ in eqn (2.132) is for use with the logarithmic mean concentration difference, as in

$$\frac{\pi d_t^2}{4} V_j (\rho_{AB0} - \rho_{A0}) = (k_\rho^*)_{\text{lm}} (\pi d_t L_j) \frac{(\rho_{Ad}^* - \rho_{A0}) - (\rho_{Ad}^* - \rho_{AB0})}{\ln[(\rho_{Ad}^* - \rho_{A0})/(\rho_{Ad}^* - \rho_{AB0})]} \tag{2.133}$$

where, rigorously, ρ_{AB0} is the bulk average or 'mixing cup' value of ρ_A at the tip of the jet.

2.3.2.2 *Transfer in a jet with tube parabolic flow*

For parabolic flow in a cylindrical jet the transport equation is

$$\frac{\partial \rho_A}{\partial z} = \frac{D}{2V_j[1 - (r/r_t)^2]}\left(\frac{\partial^2 \rho_A}{\partial r^2} + \frac{1}{r}\frac{\partial \rho_A}{\partial r}\right) \tag{2.134}$$

with boundary conditions

$$\rho_A = \rho_{Ad}^* \text{ at } r = r_t,\ 0 \le z \le L_j$$

$$\rho_A = \rho_{A0} \text{ at } z = 0,\ 0 \le r \le r_t$$

$$\frac{\partial \rho_A}{\partial r} = 0 \text{ at } r = 0,\ 0 \le z \le L_j$$

The solution is shown by Skelland (1974, 1985, Ch. 5) to be

$$\frac{\rho_A - \rho_{Ad}^*}{\rho_{A0} - \rho_{Ad}^*} = \sum_{i=1}^{\infty} \beta_i \psi_i(r_+) \exp(-\beta_i^2 z_+) \tag{2.135}$$

and the average mass transfer coefficient inside the jet, $(k_\rho^*)_{\text{lm}}$, is expressed as

$$\frac{(k_\rho^*)_{\text{lm}} d_t}{D} = \frac{1}{4}\left(\frac{d_t}{L_j}\right) N_{\text{Re}} N_{\text{Sc}} \ln\left\{\sum_{i=1}^{\infty} -\frac{4B_i}{\beta_i^2}\left(\frac{d\psi_i}{dr_+}\right)_{r_+=1} \exp\left[\frac{-\beta_i^2(L_j/r_t)}{N_{\text{Re}} N_{\text{Sc}}}\right]\right\}^{-1} \tag{2.136}$$

Hausen's (1943) convenient representation of eqn (2.136) is

$$\frac{(k_\rho^*)_{\text{lm}} d_t}{D} = 3.66 + \frac{0.0668 (d_t/L_j) N_{\text{Re}} N_{\text{Sc}}}{1 + 0.04[(d_t/L_j) N_{\text{Re}} N_{\text{Sc}}]^{2/3}} \tag{2.137}$$

2.3.2.3 *Transfer in and around a jet, as described by the penetration theory*

If the contact time and consequently the depth of solute penetration are small, then plug flow may be assumed in the vicinity of the interface in both the disperse and continuous phases. The individual coefficient of mass transfer on the disperse-phase side of the jet interface may then be estimated as follows from eqn (2.86):

$$k_{dj}^{*} = (k_{\rho}^{*})_{\text{av}} = 2(D/\pi t_e)^{1/2};\ t_e = L_j/\bar{V}_{ji} \qquad (2.138)$$

Similarly for the continuous phase,

$$k_{cj}^{*} = (k_{\rho}^{*})_{\text{av}} = 2(D_c/\pi t_e)^{1/2};\ t_e = L_j/\bar{V}_{ji} \qquad (2.139)$$

where the relevant area appears later in eqn (2.254) and D and D_c are the molecular diffusivities of the solute in the disperse and continuous phases, respectively. The time of exposure, t_e, relating to transfer between the two phases, requires knowledge of the mean interfacial velocity along the jet, $\bar{V}_{ji}$. This, in turn, necessitates that the velocity distribution be known—a question that is considerably more complex in liquid–liquid systems than in the case of liquid jets in gases. The steady-state velocity distribution of Garner *et al.* (1959) was derived by neglecting the inertial terms in the momentum equations. Unfortunately, jet break-up usually occurs before the steady-state distribution is reached. Vandegrift's (1963) analysis of velocity distribution in liquid–liquid jets neglected interfacial tension forces, thus often predicting jet expansion, instead of the observed contraction. The numerical procedure presented by Vrentas and Duda (1967) for calculating velocity distribution in liquid jets in gases could be extended to liquid–liquid systems by introducing an allowance for shear stress in the continuous phase, but a convenient mathematical relationship would not be forthcoming. Interfacial velocities predicted by the analysis of Meister and Scheele (1969a) relate to the idealized case of a non-tapering jet in the absence of mass transfer.

As an expediency, therefore, $\bar{V}_{ji}$ has been estimated as the average velocity along the jet, calculated on the basis of plug flow (Skelland and Huang 1977, 1979). Perhaps the best form of such a simplified approach is as follows:

The average linear velocity of flow in the tapering jet at distance z from the nozzle is

$$V_j = Q/A_z$$

where Q is the volumetric flow rate and A_z is the jet cross-section at z. Now approximating the jet as a truncated cone,

$$A_z = \frac{\pi d_j^2}{4} \doteq \frac{\pi}{4}\left[d_n - \left(\frac{d_n - d_{jc}}{L_j}\right) z\right]^2 \tag{2.140}$$

and

$$\bar{V}_j = \frac{\int_0^{L_j} V_j \, dz}{\int_0^{L_j} dz} = \frac{1}{L_j}\int_0^{L_j} V_j \, dz$$

$$= \frac{4Q}{\pi L_j}\int_0^{L_j} \frac{dz}{[d_n - ((d_n - d_{jc})/L_j)z]^2}$$

so

$$\bar{V}_j = \frac{4Q}{\pi d_{jc} d_n} = \frac{d_n}{d_{jc}} u_0 \tag{2.141}$$

This expression differs somewhat from an average $\bar{\bar{V}}_j$ calculated from the average of the diameters of the jet at the nozzle and at the tip $[\bar{\bar{V}}_j = 16Q/\pi(d_n + d_{jc})^2]$, and also from an average $\bar{\bar{\bar{V}}}_j$, computed as the average of the mean linear velocities at the nozzle and at the jet tip $\{\bar{\bar{\bar{V}}}_j = (2Q/\pi)[(d_{jc}^2 + d_n^2)/d_n^2 d_{jc}^2]\}$. That these variations are significant may be shown by considering differences in predicted k^*_{dj} using $\bar{V}_j$, $\bar{\bar{V}}_j$, and $\bar{\bar{\bar{V}}}_j$, for a jet contraction in which $d_n = 2d_{jc}$. Such drastic tapering is typical, as shown by Meister and Scheele (1969a) and Skelland and Huang (1977, 1979). Application of eqn (2.138), with $\bar{V}_{ji}$ replaced first by $\bar{V}_j$ and then by $\bar{\bar{V}}_j$, shows that the k^*_{dj} predicted with $\bar{\bar{V}}_j$ is about 6 per cent less than that predicted using $\bar{V}_j$. On the other hand, the k^*_{dj} using $\bar{\bar{\bar{V}}}_j$ exceeds that predicted with $\bar{V}_j$ by about 12 per cent.

Further sources of inaccuracy may arise from the assumption that penetration depths z_p are small enough to allow the jet to be regarded as a semi-infinite stagnant medium. This proposition is examined in Table 2.2, for the two experimental jetting runs described in some detail by Skelland and Huang (1979, col. 2, p. 83), for 'very low' and 'moderate' rates of mass transfer, respectively.

Although the jet length for the 'moderate' mass-transfer-rate case is only about half that for the other run, despite the same respective values of σ, ρ_d, μ_c, μ_d, d_n, and u_0, the diffusivity in the 'moderate' rate system is nearly twice that in the 'very low' rate case. The result, since $\bar{V}_j$ is nearly the same in the two systems, is that the individual coefficients k^*_{cj} and k^*_{dj} are of the same order of magnitude for the two runs. (The 'very low' and 'moderate' rate cases were continuous- and disperse-phase controlled, respectively, with solute content ≤ 2 mass per cent). However, the concentration driving

TABLE 2.2 *Mass transfer in and around jets at 25°C**

System**	Transfer rate	Surface*** flux $(N_{A,\,\mathrm{jet}})_{\mathrm{av}}$ g-mol/cm²s	L_j (cm)	d_n (cm)	d_{jc}† (cm)	u_o (cm/s)	$\bar{V}_j$ (cm/s)	z_p‡ (cm)	$\frac{100 z_p}{d_{jc}/2}$ (%)	k_j^{*}§ (cm/s)
C_6H_5Cl– C_6H_5COOH -H_2O	very low	$0.386(10^{-7})$	10.23	0.503	0.2964	13.13	22.282	0.0085	5.74	$k_{\mathrm{cj}}^{*} = 5.8(10^{-3})$
C_6H_5Cl– CH_3COOH -H_2O	moderate	$0.381(10^{-5})$	5.01	0.503	0.338	13.34	19.852	0.0082	4.85	$k_{\mathrm{dj}}^{*} = 10.2(10^{-3})$

* Skelland and Huang (1979, col. 2, p. 83).
** In the order disperse phase–solute–continuous phase. Transfer from jet.
*** $(N_{A,\,\mathrm{jet}})_{\mathrm{av.}} = k_j(\Delta c_A)_{\mathrm{av}}$.
† Calculated from eqns (2.122 and (2.123) in Table 2.1.
‡ Calculated from eqn (2.87).
§ Calculated from eqns (2.138 and 2.139).

forces were very different in the two systems, benzoic acid being only about 0.2 mass per cent soluble in water, whereas acetic acid is extremely soluble in chlorobenzene. The net result is that the average flux of solute across the jet interface is two orders of magnitude greater in the 'moderate' transfer rate case than in the 'very low' rate case. (See Table 2.2).

Skelland and Huang (1977, 1979) attempted to compare these three approaches to predicting mass transfer rates from jets in liquid-liquid systems with experimental data. However, only indirect comparison was possible, because their data consisted of the total or combined transfer occurring during jetting, during drop formation at the end of the jet, and during subsequent free fall of the drops through the continuous phase. In predicting total transfer for these three phenomena in combination, it was necessary to assume that transfer during drop formation at the jet tip was the same as that during drop formation on a nozzle for which $d_n = d_{jc}$. Visual experiments were also used to determine whether falling drops were internally stagnant, circulating, or oscillating, and due account of this was taken in the overall prediction procedure. On this slightly devious basis, it was found that the closest agreement between measured and predicted total transfer rate—with an average absolute deviation of 16.77 per cent—was obtained from application of the penetration theory to the jet. It should be noted, however, that they used $\bar{\bar{V}}_j$ instead of $\bar{V}_j$ from eqn (2.173) to estimate $\bar{V}_{ji}$ and hence k^*_{dj}. Results were considerably poorer in predicting total transfer (jet plus drop formation plus free fall) in the continuous-phase controlled systems (predictions underestimated measurements by 46 per cent). However, the penetration theory was the only method available for the estimation of k^*_{cj}.

More direct comparison with data confined to mass transfer across the jet interface in liquid-liquid systems is evidently needed, together with information on velocity distribution in both phases.

2.4 Mass transfer associated with drop phenomena

2.4.1 Hydrodynamic outline

Those aspects of drop hydrodynamics relevant to mass transfer considerations will first be reviewed briefly, with *drop size, terminal velocity, slip velocity and disperse phase hold-up, oscillation criteria, and drop shape* outlined in turn.

The drop size produced by orifices or nozzles operating above the jetting velocity ($u_0 > u_{0j}$) in liquid-liquid systems was discussed earlier. Many studies have been made of the drop sizes obtained in the non-jetting region, but only the briefest outline can be given here.

For drops formed very slowly, such that kinetic and drag forces can be neglected, Harkins and Brown (1919) developed an expression for predicting the so-called 'static drop volume'. For drops formed more rapidly, kinetic and drag forces become significant, and analyses or correlations under such conditions have been presented by Hayworth and Treybal (1950), Siemes (1956), Ueyama (1957), Null and Johnson (1958), Rao *et al.* (1966), Scheele and Meister (1968), Narayanan *et al.* (1970), Kumar (1971), Heertjes *et al.* (1971), de Chazal and Ryan (1971), Izard (1972), Grigar *et al.* (1972), Skelland and Raval (1972, for power-law non-Newtonian fluids), Kagan *et al.* (1973), Kumar and Hartland (1982, 1984), Skelland and Vasti (1985), Dalingaros *et al.* (1986), and Skelland and Slaymaker (1990).

It is unfortunate that small amounts of surface active contamination may significantly change the drop size, particularly since some workers actually added surfactants to modify σ. Hayworth and Treybal, for example, used various concentrations of Alkaterge C, while Rao *et al.* used Lissapol for this purpose. The possible presence of such trace amounts of unknown contamination may partially explain why the relationship presented by one investigator often fails to fit the experimental data of another. Skelland (1974, 1985, p. 389) details many such discrepancies and contradictions. Thus the popular equation of Scheele and Meister (1968) was found to overestimate the measurements of de Chazal and Ryan (1971) by more than 50 per cent in some regions, while providing a reasonable fit in others. Later, Horvath *et al.* (1978) found the correlation of de Chazal and Ryan (1971) to fit their multinozzle data on a single system better than did that of Scheele and Meister (1968) or of Kagan *et al.* (1973). However, as noted by Kumar and Hartland (1982), 'de Chazal and Ryan's (1971) model is complex because it requires the terminal velocity u_t to be calculated as a function of the drop diameter using Hu and Kintner's (1955) correlations.'

The question of which relationship to use is evidently somewhat vexed; in this regard, Clift *et al.* (1978, p. 333) reviewed the drop size correlations published up to 1978 and concluded that 'although appreciable discrepancies may exist (de Chazal and Ryan 1971; Kohli 1974) . . . Scheele and Meister's equation appears to be the best compromise between accuracy and ease of use for drop formation with $u_0 < u_{0j}$, especially for low-viscosity systems.'

This conclusion was restated independently as recently as 1985 and again in 1987 for single-nozzle systems by Koo *et al.* (1985) and Jeffreys (1987). The Scheele and Meister (1968) correlation for drop size in the non-jetting region is accordingly reproduced below:

$$v_p = F\left[\frac{\pi\sigma d_n}{g\Delta\rho} + \frac{20\mu_c Q d_n}{d_p^2 g\Delta\rho} - \frac{4\rho_d Q u_0}{3g\Delta\rho} + 4.5\left(\frac{Q^2 d_n^2 \rho_d \sigma}{(g\Delta\rho)^2}\right)^{1/3}\right] \qquad (2.142)$$

where Q is the volumetric flow rate through the nozzle and v_p is the drop volume. The drag term—containing μ_c—can be neglected when μ_c is less than 10 cP. Equation (2.142) was developed for a parabolic velocity distribution in the nozzle. For a flat velocity distribution, as in flow through an orifice in a thin plate, the constant 4/3 in the kinetic force term (third term on the right) is replaced by unity. The expression was shown to predict v_p successfully in the system benzene–acetone–water initially containing 5 mass per cent acetone in one phase and with transfer in either direction, using equilibrium σ values in the calculations. Equation (2.142) showed an average error of 11 per cent in correlating measurements from 15 liquid–liquid systems in which $1.8 \le \sigma \le 45.4$ dyne/cm. It appears, however, that its use should be confined to the range $0.0813 \le d_n \le 0.688$ cm (Skelland *et al.* 1987), which prescribes the range of d_n for which it was established. The Harkins and Brown (1919) correction factor F is obtainable from (Clift *et al.* 1978, p. 331).

$$F = \{0.92878 + 0.87638(d_n/v_p^{1/3}) - 0.261(d_n/v_p^{1/3})^2\}^{-1} \tag{2.143}$$

$$\text{for } (0.6 < d_n/v_p^{1/3} < 2.4)$$

and

$$F = 1.000 - 0.66023(d_n/v_p^{1/3}) + 0.33936(d_n/v_p^{1/3})^2 \tag{2.144}$$

$$\text{for } (0 < d_n/v_p^{1/3} < 0.6)$$

Alternatively, de Chazal and Ryan (1971) indicate that F is approximately 0.625 in most cases; a further estimating procedure is provided by Horvath *et al.* (1978) as

$$F = 0.6 + 0.4\exp(-2X) \tag{2.145}$$

where

$$X = d_n(\Delta\rho g/\pi\sigma d_n)^{1/3} \tag{2.146}$$

Skelland and coworkers (1987, 1990) have shown experimentally that, in contrast to the expectations of its authors, eqn (2.142) gives good prediction of drop size in the presence of anionic, cationic, and non-ionic surface active agents, when used with the diminished value of interfacial tension caused by the surfactant. This finding was made for drop formation times between 0.5 and 4 s.

For multinozzle distributors Jeffreys (1987) recommends use of the correlation provided by Kumar and Hartland (1982), as follows in the non-jetting region:

$$\frac{d_p}{d_n} = 1.591(\Delta\rho d_n u_0^2/\sigma)^{-0.068}(\Delta\rho d_n^2 g/\sigma)^{-0.278} \tag{2.147}$$

In contrast to the Scheele and Meister equation this expression is purely empirical, and was established in the absence of mass transfer. It is confined to 'low viscosity' fluids and surprisingly contains no ρ_d term, thereby ignoring the kinetic force or momentum effect tending to detach the drop from the nozzle. In this regard, Clift *et al.* (1978, p. 331) note that 'the momentum effect is particularly important, and causes drops to be smaller than those formed under near-static conditions ($Q \rightarrow 0$).' This point was also emphasized graphically by Scheele and Meister (1968). The success of eqn (2.147) despite the absence of ρ_d may therefore be due to the fact that ρ_d varied only about ± 12 per cent from its mean value in the data they used. (This contrasted with their ρ_c, which varied two-fold).

The multinozzle correlation by Dalingaros *et al.* (1986) is formulated by balancing the kinetic, buoyancy, and interfacial tension forces acting on the growing drop. This makes no specific allowance for the additional liquid that enters the drop during the 'necking' process that follows the balancing of the above-mentioned forces, and before detachment. In this connection Scheele and Meister (1968) state the following on the basis of observations of drop formation using high-speed movies: 'flow into the drop during the necking process can contribute significantly to the final drop volume and must be treated independently. Any empirical incorporation of this phenomenon into one of the force terms will cause errors in the dependence of drop volume on flow rate, nozzle diameter, and physical properties . . .'

Thus these multinozzle correlations may well be good only for data similar to those they were forced to fit.

The terminal velocity of drops in liquid–liquid systems has received considerable study, and an extensive review of work up to 1978 in this field is given by Clift *et al.* (1978). Fluid droplets differ from rigid spheres in three ways during rise or fall.

1. The frictional drag of the surrounding liquid may induce torroidal circulation of both the interface and the interior of the droplets.
2. The drops may depart from spherical shape, assuming the form of oblate spheroids.
3. Oscillation of the drops may occur, either between prolate and oblate forms or, more commonly, between more- and less-oblate forms.

The first of these features reduces the drag on each drop, thereby increasing its terminal velocity. The remaining two features have the opposite effect.

Perhaps the biggest single modifying influence on the terminal velocity of a drop of a given size in a particular liquid–liquid system is the frequent presence of trace amounts of unknown surface active contamination. Thus Garner and Skelland (1955) found that the terminal velocity of supposedly 'pure' nitrobenzene drops of about 0.5 cm in diameter falling in distilled water increased by about 25 per cent after redistillation of the nitrobenzene. The refractive index of the redistilled nitrobenzene at 20.2 °C was 1.5523, compared with the value in International Critical Tables (1930) at 20 °C of 1.5529. The refractive index of the nitrobenzene before and after distillation was the same. The residue from the distillation gave completely stagnant droplets, as revealed by the incorporation of grease-free aluminium powder inside the drops and viewing them closely, as they fell, with reflected light. Evidently the impurity inhibiting internal circulation was a relatively high-boiling substance, present in extremely small concentration. Additionally, a drop of the redistilled 'pure' nitrobenzene with a volume of 0.017 ccs in distilled water fell about 50 per cent faster than a drop of the same size in distilled water containing 0.003 mass per cent of sodium dodecyl sulphate 6. The former drop showed vigorous internal circulation but no oscillation, whereas the latter drop was completely stagnant.

Experimental values of terminal velocities of single drops falling or rising in liquid–liquid systems have been correlated by Hu and Kintner (1955), Klee and Treybal (1956), Johnson and Braida (1957), and Grace *et al.* (1976).

The latter authors utilized data from several sources and conducted many experiments of their own, enabling their correlation to cover a wide range of variables. They, too, noted that the purity of a system had a large effect on the terminal velocity of a drop. Because of this they separated the data in the literature in accordance with the degree of care taken by the respective authors to eliminate surfactants. They developed the following correlation for contaminated systems:

$$J = 0.94H^{0.757};\ (2 < H \leq 59.3) \tag{2.148}$$

$$J = 3.42H^{0.441};\ (H > 59.3) \tag{2.149}$$

where

$$H = (4/3)N_{\mathrm{E0}}M^{-0.149}(\mu_c/\mu_w)^{-0.14} \tag{2.150}$$

$$J = N_{\mathrm{Re}}M^{0.149} + 0.857 \tag{2.151}$$

and

$$M = \text{a modified Morton number} = g\mu_c^4\,\Delta\rho/\rho_c^2\sigma^3$$

$$N_{\mathrm{E0}} = \text{the Eotvos number} = g\Delta\rho d_p^2/\sigma$$

$$N_{\mathrm{Re}} = \text{the droplet Reynolds number} = d_p u_t \rho_c/\mu_c$$

The value of 59.3 for H corresponds to the transition from non-oscillating to oscillating drops. The terminal velocity is then given by

$$u_t = (\mu_c/d_p\rho_c)\,[(J - 0.857)/M^{0.149}] \tag{2.152}$$

For pure systems involving drops at low Reynolds numbers, the above equation was modified to

$$u_{t,\mathrm{pure}} = u_t[1 + 1/(2 + 3\mu_d/\mu_c)] \tag{2.153}$$

The terminal velocity relationships presented by Klee and Treybal (1956) appeared two decades earlier than those of Grace *et al.* (1976), but are simple to use and have found wide application. They consist of two separate equations, one for the non-oscillating, smaller drop region, and the other for the oscillating, larger drop region, thus:

$$u_t = 38.3\,\rho_c^{-0.45}\Delta\rho^{0.58}\mu_c^{-0.11}d_p^{0.70};\ d_p < d_{pc} \tag{2.154}$$

and

$$u_t = 17.6\,\rho_c^{-0.55}\Delta\rho^{0.28}\mu_c^{0.10}\sigma^{0.18};\ d_p > d_{pc} \tag{2.155}$$

where

$$d_{pc} = 0.33\,\rho_c^{-0.14}\Delta\rho^{-0.43}\mu_c^{0.30}\sigma^{0.24} \tag{2.156}$$

in c.g.s. units.

Equations (2.154–2.156) were developed by using 'some technical-grade materials without special purification.' They may thus embody a built-in accommodation for some degree of surface-active contamination commonly present in industrial systems. Indeed, Skelland and Vasti (1985), using presumably 'purer' systems, found their terminal velocities to be all higher than predicted by eqns (2.154 and 2.155), but (mostly) lower than calculated from the correlation for very pure systems given by Grace *et al.* (1976).

Equations (2.154–156) were shown by Skelland and Huang (1977b) and Skelland *et al.* (1987) to be roughly adequate for predicting u_t in systems containing various concentrations of different surface active agents (SAA), provided that they are applied using the diminished value of interfacial tension, σ, due to the presence of the SAA. This is of industrial significance both for eqn (2.142) and eqns (2.154–156), because the number of species, structure, and concentration of such contaminants are usually unknown. Apparently they may be characterized adequately by measurement of the single property σ for the purposes of applying eqns (2.142 and 2.154–2.156) to a given industrial system.

The effects of interaction between adjacent circulating or oscillating drops on drop formation, free fall, and mass transfer were studied experimentally by Skelland and Vasti (1985). They found effects of vertical interaction between adjacent drops, provided that $L/d_p < 20$, where L is the vertical distance between consecutive drops of diameter d_p. The most significant interaction occurred at $L/d_p < 5$, an enhancement in drop velocity of about 20 per cent being observed when L/d_p was reduced from 5 to 2. For $L/d_p > 20$, drops behaved as single, isolated drops. The presence of parallel drop streams did not appear to influence terminal velocity; however, the drops seldom fell side by side, so that this effect was difficult to study. The paper by Kumar *et al.* (1980) should also be consulted on this matter.

The slip velocity and disperse phase hold-up are closely related in spray and perforated-plate extraction columns. The slip velocity u_s is the net linear velocity between the two phases; however, it is not equal to the terminal velocity of a single drop in isolation because of the interactions between adjacent drops in a swarm, as just described. Weaver *et al.* (1959) and Beyaert *et al.* (1961) have shown that u_s/u_t is a unique function of the disperse phase hold-up, ϕ. The latter is defined as the fraction of the column volume occupied by drops of the disperse phase. This is exclusive of any coalesced layers and downcomers. According to Horvath *et al.* (1978), the 'best' relationship between u_s and ϕ is the rather complex expression—requiring an iterative procedure in application–presented by Pilhofer (1974). A simpler, but supposedly more approximate procedure uses an equation proposed by Thornton (1956):

$$u_s = u_t(1 - \phi) \tag{2.157}$$

In a countercurrent spray column we have

$$u_s = \frac{u_d}{\phi} + \frac{u_c}{1 - \phi} \tag{2.158}$$

where the superficial velocities u_c and u_d for the continuous and disperse phases are based on the empty-column cross section. In a perforated-plate column, u_c is considered zero because the horizontal flow of the continuous phase across the plate does not affect the holdup; u_d is then computed excluding the downcomer. Combining eqns (2.157 and 2.158) for a perforated-plate column,

$$u_t(1 - \phi) = u_d/\phi \tag{2.159}$$

Whether to use the simpler procedure or the Pilhofer (1974) relationship depends on the accuracy desired. Since both approaches require knowledge of u_t which, as noted above, is so sensitive to trace amounts of unknown surface active contamination, it is by no means clear which method would finally conform more closely to a given industrial situation. Horvath *et al.* (1978) describe sources of error in their attempts to predict hold-up in spray columns.

Droplet oscillation criteria are needed to predict the onset of droplet oscillation and to select the appropriate equations for use in estimating terminal velocity and mass transfer rates during free rise or fall. Some published criteria for the onset of oscillation are:

1. $N_{We} > 3.58$; $N_{We} = d_p u_t^2 \rho_c / \sigma$ (Hu and Kintner 1955).

2. $d_p > 0.33 \mu_c^{0.30} \sigma^{0.24} / \rho_c^{0.14} \Delta\rho^{0.43}$ cm (Klee and Treybal 1956).

3. $N_{Re,c} / N_{PG}{}^{0.15} > 20$; $N_{Re,c} = d_p u_t \rho_c / \mu_c$, $N_{PG} = \rho_c^2 \sigma^3 / g \mu_c^4 \Delta\rho$ (Johnson and Braida 1957).

4. $d_p > 0.162 (\rho_d / \Delta\rho)^{0.5}$ cm (Edge and Grant 1971).

5. $H > 59.3$ (Grace *et al.* 1976).

The influence of anionic, cationic, and non-ionic surfactants (simulating industrial contamination) on these criteria was examined experimentally by Skelland *et al.* (1987). They found the effect of surfactant type on the onset of oscillation to be small, if any. Johnson and Braida's criterion (using diminished σ due to the SAA) gave the least error but requires u_t, which may have to be estimated in practice. The criterion of Grace *et al.*, using the diminished value of σ due to the presence of the surfactant, otherwise seemed to be the best, with a mean absolute deviation of 20.2 per cent.

Drop shape determines the interfacial area available for mass transfer between the two phases for a drop of a given volume, v_p. Work on this subject has been reviewed by Clift *et al.* (1978), and the eccentricity for non-oscillating drops in liquid–liquid systems—which are typically oblate-spheroidal in form—is obtainable from the correlation of Wellek *et al.* (1966):

$$E = 1.0 + 0.093 (d_p u_s^2 \rho_c / \sigma)^{0.98} (\mu_c / \mu_d)^{0.07} \tag{2.160}$$

where E is the ratio of the major (horizontal) to the minor (vertical) axis of an oblate spheroid. The corresponding surface area of such a droplet is (Heertjes *et al.* 1954):

$$S_{os} = 2\pi \left(\frac{3 v_p E}{4\pi} \right)^{2/3} \left[1 + \frac{1}{E(E^2 - 1)^{1/2}} \ln [E + (E^2 - 1)^{1/2}] \right] \tag{2.161}$$

The ratio of the area of an oblate spheroid to that of a sphere of equal volume is (Garner and Tayeban 1960, p. 479).

$$\frac{S_{os}}{S_s} = \frac{1}{2} E^{2/3} + \frac{1}{2E^{1/3}(E^2 - 1)^{1/2}} \ln [E + (E^2 - 1)^{1/2}\} \tag{2.162}$$

For E of 2, 2.5, and 3, corresponding S_{os}/S_s values are 1.095, 1.17, and 1.26. Evidently realistic distortions in drop shape result in significant increases in interfacial area.

Equation (2.160) showed an average absolute deviation of 6 per cent from 198 measurements on 45 surfactant-free systems. The droplet Reynolds number range was 6–1354. However, Mekasut *et al.* (1979) were unable to adapt the expression in any simple way to the prediction of droplet eccentricities in the presence of surface active agents.

2.4.2 Mass transfer during drop formation

Several theoretical and experimental studies have been made of mass transfer in and around a drop during formation. Theoretical treatments have neglected mass transfer during drop detachment; the latter is negligible according to Popovich *et al.* (1964) in an experimental study. In contrast, Licht and Conway (1950) inferred from experiment that the effects of formation and detachment on mass transfer are about equal. The conflict remains unresolved at present.

The rate of mass transfer during droplet formation has received extensive theoretical and experimental attention. Mechanisms that have been proposed as relevant to the process include expansion of the surface of the growing drop by stretching of elements already at the surface; alternatively, extension of the surface has been said to occur by the addition of fresh elements to the surface. A further consideration of mechanism involves allowance for the effects of internal circulation currents in the drop during formation.

The surface-stretch mechanism has characterized analyses by Ilkovic (1934), Baird (1959), Michels (1960), Beek and Kramers (1962), Angelo, Lightfoot and Howard (1966), Heertjes and de Nie (1966), Angelo and Lightfoot (1968), and Ruckenstein and Constantinescu (1969).

The concept of surface extension by the addition of new or fresh elements was featured in the developments of Coulson and Skinner (1952), Heertjes *et al.* (1954), Groothuis and Kramers (1955), Beek and Kramers (1962), Sawistowski and Goltz (1963), Heertjes and de Nie (1966), Calderbank and Patra (1966), and Skelland and Hemler (1969).

The effects of internal circulation on mass transfer during drop formation were examined in studies by Rajan and Heideger (1971), Humphrey *et al.* (1974a, 1974b), Zimmerman *et al.* (1980), Heideger and Dubois (1985), and Heideger and Wright (1986).

Additional relationships appear in the papers by Licht and Pansing (1953), Zheleznyak (1967), and Walia and Vir (1976). The latter authors developed a complex double infinite series expansion for the fractional extraction during drop formation, which eliminates earlier simplifying assumptions of a planar drop surface and constant concentration within the drop. Summaries of many of the relationships referred to above are provided by Popovich *et al.* (1964) and by Walia and Vir (1976).

Popovich *et al.* (1964) found their measurements to be best fitted by Ilkovic's surface-stretch expression:

$$k_{df}^{*} = 1.31\,(D/\pi t_f)^{1/2} \tag{2.163}$$

In contrast, the data of Johnson and Hamielec (1960) were well represented by the following equation due to Heertjes *et al.* (1954), which assumes surface growth by the addition of fresh elements:

$$k_{df}^* = 3.429(D/\pi t_f)^{1/2} \tag{2.164}$$

In attempting to select a model for design use, it would be helpful if some indication could be found regarding whether the surface of a growing drop enlarges by stretching of the existing surface layer or by the addition of fresh elements. Recent experiments by Thornton and Anderson (1981), Thornton *et al.* (1985), Thornton (1987), and Javed *et al.* (1989) could bear on this question. In this work a colored spot was induced on the surface of a growing water drop in toluene in the absence of mass transfer. The drop contained a small amount of a photochromic tracer which yielded a blue spot of about 1 mm^2 on the growing surface in response to a finely focused UV laser pulse. The spot, which penetrated the surface by not more than 50 nm, was induced by triggering the laser early on during the droplet formation period. If, for example, the spot were induced when the drop had attained half its final volume, then the diameter of the spot should have increased by about 26 per cent at drop detachment, if the surface grew solely by uniform stretching of the surface layer. Inspection of Thornton's (1985, 1987) photographs appears to show no such enlargement of the spot. Some workers have assumed that the surface-stretch model implies uniform stretching all over the surface (Popovich *et al.* 1964; Walia and Vir 1976); others (Michels 1960; Walia and Vir 1976) have examined the consequences of uneven stretching. Further work along Thornton's (1987) lines may resolve this question more decisively. The technique could also be used to determine the interfacial velocity of a jet in liquid–liquid systems, the need for which was indicated earlier.

As an example of that category of analyses that assumes the surface of a growing drop to increase by the addition of fresh surface elements, the development by Skelland and Hemler (1969) may be considered. The assumptions are:

1. The drop grows spherically about a point source.
2. The concentration driving force is constant during drop formation at $\Delta\rho_{Af}$.
3. The fresh surface elements continually added to the growing surface are subject to mass transfer in accordance with the penetration theory of eqn. (2.86).

Consider the differential element of fresh surface, da, formed between t and $t + \mathrm{d}t$ during the formation of a single drop. This element da will be exposed for a time $t_f - t$, so that the solute transferred during the formation of one drop will be

$$k^*_{df}\pi d_p^2 \Delta\rho_{Af} t_f = \Delta\rho_{Af}\int_0^{\pi d_p^2} k^*_d (t_f - t)\mathrm{d}a \tag{2.165}$$

where t_f is the time of drop formation. The average mass transfer coefficient during drop formation is then

$$k^*_{df} = \frac{1}{\pi d_p^2 t_f}\int_0^{\pi d_p^2} k^*_d\,(t_f - t)\mathrm{d}a \tag{2.166}$$

for k^*_{df} based on the drop surface at detachment, πd_p^2. Invoking the penetration theory,

$$k^*_d = 2(D/\pi(t_f - t))^{1/2} \tag{2.167}$$

Let d_t and a be the diameter and surface area of the growing drop at time t. Then considerations of volumetric flow rate into the drop and the appropriate spherical geometry show that

$$d_t = d_p(t/t_f)^{1/3}, \quad a = \pi d_p^2 (t/t_f)^{2/3} \tag{2.168}$$

Equations (2.166, 2.167, and 2.168) may be combined to obtain

$$k^*_{df} = \frac{4}{3}\left(\frac{D}{\pi t_f}\right)^{1/2}\int_0^{t_f}\left(\frac{t_f - t}{t_f}\right)^{1/2}\left(\frac{t}{t_f}\right)^{-1/3}\frac{\mathrm{d}t}{t_f}$$

Let $\psi = t/t_f$, then

$$k^*_{df} = \frac{4}{3}\left(\frac{D}{\pi t_f}\right)^{1/2}\int_0^{1.0}(1 - \psi)^{1/2}\psi^{-1/3}\,\mathrm{d}\psi \tag{2.169}$$

The definite integral in eqn (2.169) is a standard form of the beta function; the latter is composed of gamma functions as follows (Jenson and Jeffreys 1963, pp. 154, 157):

$$\beta(p, q) = \frac{\Gamma(p)\Gamma(q)}{\Gamma(p + q)} = \int_0^{1.0}(1 - \psi)^{q-1}\psi^{p-1}\,\mathrm{d}\psi \tag{2.170}$$

where p and q are positive. Equations (2.169 and 2.170) reveal that $p = 2/3$ and $q = 3/2$, so that

$$\int_0^{1.0}(1 - \psi)^{1/2}\psi^{-1/3}\mathrm{d}\psi = \beta\left(\frac{2}{3}, \frac{3}{2}\right) = \frac{\Gamma(2/3)\,\Gamma(3/2)}{\Gamma(13/6)} \tag{2.171}$$

Tabulated values of the gamma function were used to evaluate the right-hand side of eqn (2.171), as described by Jenson and Jeffreys (1963, p. 152). The result is

$$k_{df}^{*} = 1.48(D/\pi t_f)^{1/2} \tag{2.172}$$

This expression would appear to be equally applicable to the continuous phase if D_c is substituted for D. The coefficient is based on the surface of the detached drop, does not include effects of any disturbances at detachment, and is for use in the left side of eqn (2.165) for predicting the solute mass transferred during formation of a single drop.

Skelland and Minhas (1971) measured mass transfer rates during drop formation in disperse-phase controlled systems. Their values, which include any transfer effects of droplet detachment, were higher than predicted by the models of Heertjes *et al.*, Coulson and Skinner, Groothuis and Kramers, Ilkovic, and Skelland and Hemler, and were empirically correlated by statistical methods as

$$k_{df}^{*} = 0.0432\frac{d_p}{t_f}\left(\frac{u_0^2}{d_p g}\right)^{0.089}\left(\frac{d_p^2}{t_f D}\right)^{-0.334}\left(\frac{\mu_d}{(\rho_d d_p \sigma)^{1/2}}\right)^{-0.601} \tag{2.173}$$

This equation correlated 23 measurements on three liquid–liquid systems with an average absolute deviation of about 26 percent. The coefficient is based on πd_p^2.

Michels and Pigford (cited in Perry, 1963, p. 14.64) developed a relationship for the continuous-phase mass transfer coefficient during drop formation; it has the same form as eqn (2.172), but with D replaced by D_c and the constant 1.48 converted to 4.6. However, the expression predicted values 3 times greater than the experimental values of Calderbank and Patra (1966) for gas bubbles, which were therefore in better agreement with the continuous-phase equivalent of eqn (2.172).

The continuous phase analog of eqn (2.173) was established by statistical correlation of experimental data by Skelland and Hemler (1969), as

$$k_{cf}^{*} = 0.386\left(\frac{D_c}{t_f}\right)^{0.5}\left(\frac{\rho_c \sigma}{\Delta\rho g t_f \mu_c}\right)^{0.407}\left(\frac{g t_f^2}{d_p}\right)^{0.148} \tag{2.174}$$

This coefficient is based on πd_p^2 and includes any effects of drop detachment from the nozzle. The equation correlated 20 data points on three liquid–liquid systems with an average absolute deviation of 11 percent, using statistical methods.

Skelland and Huang (1977a, 1979) obtained indirect evidence from experiment that eqns (2.173 and 2.174) are adequate for the prediction of the respective individual mass transfer coefficients during drop formation at the end of a jet in liquid–liquid systems (the correlations were of course originally established for drops forming at a nozzle tip).

2.4.3 Mass transfer in and around detached drops

Detached droplets, moving freely in a second immiscible liquid while undergoing solute exchange between the two phases, are a prime feature of most forms of liquid-liquid extraction processes. The drops may be internally stagnant, particularly when the interfacial tension is high, the drops are small, and trace amounts of surface active contamination are present. Alternatively, when the interfacial tension is low, the drops are larger, and the system is relatively free from surfactants, the drops may exhibit internal circulation currents resulting from the viscous drag of the surrounding liquid. The circulation currents typically assume a torroidal flow pattern. A further possibility, with still larger drops, is the occurrence of droplet oscillation, criteria for which were presented earlier between eqns (2.159 and 2.160). These three different droplet states—stagnancy, circulation, and oscillation—of course give rise to different corresponding analytical and empirical relationships for the rates of mass transfer on each side of the interface. The presence or absence of internal circulation currents in drops of a given system may be ascertained in preliminary glassware experiments in which aluminium particles within the drops are viewed under reflected light (Garner and Skelland 1956).

Further classifications of continuous-phase mass transfer may be made for the respective cases of non-agitated and agitated systems, and this spectrum of droplet phenomena is now outlined below.

2.4.3.1 *Transfer in stagnant drops, with and without external resistance*

The importance of stagnant drops, either with or without external resistance to mass transfer, is so great for many extraction systems that the derivation for these two cases will be shown in detail. The drops will be idealized to spherical form and the following assumptions made to describe the unsteady-state diffusion process:

1. The concentration of solute (component A) is uniform at ρ_{A0} throughout the sphere at the start of diffusion ($t = 0$).
2. The continuous phase mass transfer coefficient, k^*_{cr}, is constant and finite, as is the solute concentration at points remote from the sphere, $\rho'_{A\infty}$.
3. Diffusion is radial, there being no variation in concentration with angular position, physical properties are constant, and diffusion rates are low.

Fick's second law of diffusion in spherical coordinates is given for the present situation by eqn (2.44), expressed in the following alternative form, and with c_A replaced by ρ_A:

$$D\frac{\partial}{\partial r}\left(r^2\frac{\partial \rho_A}{\partial r}\right) = r^2\frac{\partial \rho_A}{\partial t} \tag{2.175}$$

The boundary conditions are

$$\frac{\partial \rho_A}{\partial r}(0, t) = 0$$

$$\rho_A(r, 0) = \rho_{A0}$$

$$-D\frac{\partial \rho_A}{\partial r}(r_s, t) = k^*_{\mathrm{cr}}(\rho'_{Ai} - \rho'_{A\infty})$$

The third boundary condition follows from equating the solute flux in the drop at the interface to the flux in the continuous phase, the latter expressed as the product of the continuous phase mass transfer coefficient and the concentration driving force in that phase, $\Delta\rho'_A$. The concentrations of solute A in the continuous phase, ρ'_A, must be converted to corresponding concentrations in the disperse phase, ρ_A, before the third boundary condition can be used. This is done via the distribution ratio, m, regarded here as constant at low solute concentrations (m is the ratio of disperse- to continuous-phase solute concentrations at equilibrium). Thus the third boundary condition becomes

$$-\frac{\partial \rho_A}{\partial r}(r_s, t) = \frac{k^*_{cr}}{mD}(\rho_{Ai} - \rho^*_{AB})$$

where ρ^*_{AB} is the disperse phase concentration in equilibrium with the bulk concentration in the continuous phase, $\rho^*_{AB} = m\rho'_{A\infty}$. Let

$$y' = \rho_A - \rho^*_{AB}$$

then eqn (2.175) is rewritten as

$$D\frac{\partial}{\partial r}\left(r^2\frac{\partial y'}{\partial r}\right) = r^2\frac{\partial y'}{\partial t} \tag{2.176}$$

and the boundary conditions are now

$$\frac{\partial y'}{\partial r}(0, t) = 0$$

$$y'(r, 0) = \rho_{A0} - \rho^*_{AB}$$

$$-\frac{\partial y'}{\partial r}(r_s, t) = \frac{k^*_{cr}}{mD}y'$$

Transformation from spherical to cartesian coordinates will be effected by substituting

$$\Omega(r,t) = ry'(r,t)$$

so that eqn (2.176) becomes

$$D\frac{\partial}{\partial r}\left[r^2\frac{\partial}{\partial r}\left(\frac{\Omega(r,t)}{r}\right)\right] = r^2\frac{\partial}{\partial t}\left[\frac{\Omega(r,t)}{r}\right]$$

from which

$$\frac{\partial^2\Omega}{\partial r^2} = \frac{1}{D}\frac{\partial\Omega}{\partial t} \tag{2.177}$$

and the boundary conditions transform to

$$\Omega(0,t) = 0$$

$$\Omega(r,0) = r(\rho_{A0} - \rho^*_{AB})$$

$$-\frac{\partial\Omega}{\partial r}(r_s,t) = \left(\frac{k^*_{\mathrm{cr}}}{mD} - \frac{1}{r_s}\right)\Omega$$

The transformed first and third boundary conditions follow from substituting Ω/r for y' in the forms stated below eqn (2.176) and performing the indicated differentiation at the r values shown.

In accordance with conventional procedure, separation of the variables will be achieved by assuming a solution of the following form:

$$\Omega(r,t) = R(r)T(t) \tag{2.178}$$

where R is some function of r only and T is some function of t only. Then evidently

$$\frac{\partial\Omega}{\partial t} = R\frac{\mathrm{d}T}{\mathrm{d}t};\ \frac{\partial\Omega}{\partial r} = T\frac{\mathrm{d}R}{\mathrm{d}r};\ \frac{\partial^2\Omega}{\partial r^2} = T\frac{\mathrm{d}^2R}{\mathrm{d}r^2};$$

and eqn (2.177) may be expressed as

$$T\frac{\mathrm{d}^2R}{\mathrm{d}r^2} = \frac{R}{D}\frac{\mathrm{d}T}{\mathrm{d}t}$$

or

$$\frac{1}{R}\frac{\mathrm{d}^2R}{\mathrm{d}r^2} = \frac{1}{DT}\frac{\mathrm{d}T}{\mathrm{d}t} = -\lambda^2 \tag{2.179}$$

The definitions of R and T show that the left side of eqn (2.179) is independent of t, while the right side is independent of r. Both sides are therefore equal to a constant which, if expressed as $-\lambda^2$, can be shown to give a non-trivial result. Solving the second differential equation,

$$T = C_1\exp(-\lambda^2 Dt) \tag{2.180}$$

The first differential equation is

$$\frac{d^2R}{dr^2} + \lambda^2 R = 0$$

This has the solution (Li 1960, p. 310)

$$R = C_2 \cos \lambda r + C_3 \sin \lambda r \tag{2.181}$$

so that, from eqn (2.178),

$$\Omega(r, t) = (C_4 \cos \lambda r + C_5 \sin \lambda r)\exp(-\lambda^2 Dt) \tag{2.182}$$

where C_1-C_5 are constants. The first boundary condition (at $r = 0$) shows that C_4 is zero. Also, from the third boundary condition,

$$\lambda \cot(\lambda r_s) = \frac{1}{r_s} - \frac{k_{cr}^*}{mD} \tag{2.183}$$

Next, define

$$\alpha = k_{cr}^* r_s / mD; \ \beta_n = \lambda r_s$$

so that eqn (2.183) may be written

$$\beta_n \cot \beta_n + \alpha - 1 = 0 \tag{2.184}$$

and the general form of eqn (2.182) is

$$\Omega(r, t) = \sum_{n=1}^{\infty} A_n \sin(\beta_n r/r_s) \exp(-\beta_n^2 Dt/r_s^2) \tag{2.185}$$

When $t = 0$, from the second boundary condition,

$$r(\rho_{A0} - \rho_{AB}^*) = \sum_{n=1}^{\infty} A_n \sin(\beta_n r/r_s)$$

This is a Fourier sine series, from which A_n may be evaluated by the Principle of Orthogonality (Li 1960, Ch. 5); i.e. multiply throughout by sin $(\beta_n r/r_s)$, integrate over $0 \leq r \leq r_s$, and extract A_n as

$$A_n = 2r_s(\rho_{A0} - \rho_{AB}^*) \frac{\alpha}{[\beta_n^2 + \alpha(\alpha - 1)] \sin \beta_n} \tag{2.186}$$

Substituting $\rho_A - \rho_{AB}^*$ for y' in ($\Omega = ry'$) and eqn (2.186) for A_n in eqn (2.185), then solving for ρ_A,

$$\rho_A = \rho_{AB}^* + \frac{2\alpha r_s}{r}(\rho_{A0} - \rho_{AB}^*) \sum_{n=1}^{\infty} \frac{\sin(\beta_n r/r_s)}{[\beta_n^2 + \alpha(\alpha - 1)] \sin \beta_n} \exp(-\beta_n^2 Dt/r_s^2) \tag{2.187}$$

The terms β_n are the roots of eqn (2.184), the first six values of which are tabulated by Crank (1956, p. 331) for α values between zero and infinity, where α is defined below eqn (2.183).

At any time t, the average concentration of solute A in the drop is $\bar{\rho}_A$, obtainable from the following solute balance:

$$(\bar{\rho}_A - \rho^*_{AB})\frac{4\pi r_s^3}{3} = \int_0^{r_s} 4\pi(\rho_A - \rho^*_{AB})\, r^2 \mathrm{d}r$$

Substituting for $\rho_A - \rho^*_{AB}$ from eqn (2.187) and defining the fractional extraction from the drop in time t as $E_f = (\rho_{A0} - \bar{\rho}_A)/(\rho_{A0} - \rho^*_{AB})$, we have

$$\frac{\bar{\rho}_A - \rho^*_{AB}}{\rho_{A0} - \rho^*_{AB}} = 1 - E_f = \frac{6}{r_s^2}\sum_{n=1}^{\infty}\frac{\alpha \exp\left(-\beta_n^2 Dt/r_s^2\right)}{[\beta_n^2 + \alpha(\alpha-1)]\sin\beta_n}\int_0^{r_s} \mathrm{r}\sin(\beta r/r_s)\mathrm{d}r$$

The integral equals (r_s^2/β_n^2) $(\sin\beta_n - \beta_n\cos\beta_n)$; dividing the contents of the second parentheses by $\sin\beta_n$ gives $1 - \beta_n \cot\beta_n$ which, according to eqn (2.184), equals α. The result is

$$E_f = \frac{\rho_{A0} - \bar{\rho}_A}{\rho_{A0} - \rho^*_{AB}} = 1 - \sum_{n=1}^{\infty}\frac{6\alpha^2}{\beta_n^2[\beta_n^2 + \alpha(\alpha-1)]}\exp\left(-\beta_n^2 Dt/r_s^2\right) \tag{2.188}$$

Curves of E_f vs $(Dt/r_s^2)^{1/2}$ for various values of α appear in Fig. 2.9, taken from Crank (1956, p. 92).

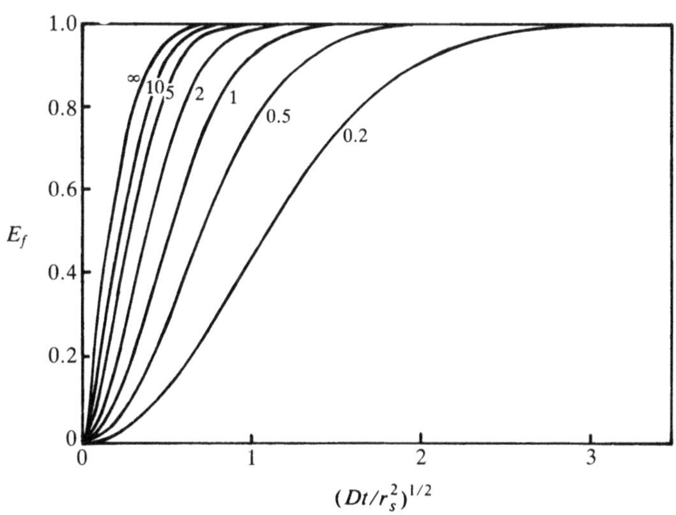

Fig. 2.9. Fractional extraction (E_f) by unsteady-state molecular diffusion in a sphere with external resistance to mass transfer. The curves are for various values of $\alpha = k^*_{cr}\, r_s/mD$. (Crank 1956.)

In the special case of zero resistance to mass transfer in the continuous phase, the coefficient k^*_{cr} and α become infinite and the solute concentration becomes uniform throughout the continuous phase. The latter is now in equilibrium with the solute concentration on the droplet side of the interface, which assumes the constant value of ρ^*_A. Equations (2.187 and 2.188) reduce to corresponding limiting forms as follows.

When α equals infinity, eqn (2.184) shows that β_n assumes values of $n\pi$, where n is any positive integer. For this condition, multiply eqn (2.187) by $(1/\alpha)/(1/\alpha)$, then replace α by $1 - \beta_n/\tan\beta_n$ from eqn (2.184) and use L'Hôpital's rule to overcome the resulting indeterminacy, to obtain

$$\rho_A = \rho^*_A + \frac{2r_s}{\pi r}(\rho_{A0} - \rho^*_A)\sum_{n=1}^{\infty}\frac{(-1)^{n+1}}{n}\sin\left(\frac{n\pi r}{r_s}\right)\exp\left(\frac{-Dn^2\pi^2 t}{r_s^2}\right) \tag{2.189}$$

Equation (2.188) may be multiplied by $(1/\alpha^2)/(1/\alpha^2)$ for $\alpha = \infty$ and $\beta_n = n\pi$ to find

$$E_f = \frac{\rho_{A0} - \bar{\rho}_A}{\rho_{A0} - \rho^*_A} = 1 - \frac{6}{\pi^2}\sum_{n=1}^{\infty}\frac{1}{n^2}\exp\left(\frac{-Dn^2\pi^2 t}{r_s^2}\right) \tag{2.190}$$

Equations (2.188 and 2.190) are commonly ascribed to Grober (1925) and to Newman (1931), respectively.

An individual coefficient of mass transfer for the disperse phase during free rise or fall may now be formulated. A balance on component A diffusing out of a rising, stagnant, spherical droplet during time dt is

$$k^*_{dr}\pi d_p^2(\bar{\rho}_A - \rho^*_A)\,\mathrm{d}t = -\frac{\pi d_p^3}{6}\,d\bar{\rho}_A \tag{2.191}$$

where, for the conditions of eqn (2.190), ρ^*_A is constant. Integrating between locations 1 and 2,

$$k^*_{dr} = -\frac{d_p}{6t}\ln\left[\frac{\bar{\rho}_{A1} - \rho^*_A}{\bar{\rho}_{A1} - \rho^*_A} - \frac{\bar{\rho}_{A1} - \bar{\rho}_{A2}}{\bar{\rho}_{A1} - \rho^*_A}\right] = -\frac{d_p}{6t}\ln(1 - E_f) \tag{2.192}$$

Combining eqns (2.190 and 2.192),

$$k^*_{dr} = -\frac{d_p}{6t}\ln\left\{\frac{6}{\pi^2}\sum_{n=1}^{\infty}\frac{1}{n^2}\exp\left[\frac{-Dn^2\pi^2 t}{(d_p/2)^2}\right]\right\} \tag{2.193}$$

The k^*_{dr} here is for use with a driving force expressed as $\Delta\rho_A$; to obtain a k^*_{dr} suitable for use with Δy_A the right hand side of eqn (2.193) must be multiplied by $(\rho/M)_{av}$ for the disperse phase between 1 and 2.

A useful empirical fit of eqn (2.193) when E_f is less than 0.5 is provided by the following expression (Vermeulen 1953; Johnson *et al.* 1958):

$$k_{dr}^* = -\frac{d_p}{6t}\ln\left[1 - \frac{\pi D^{1/2}t^{1/2}}{d_p/2}\right], \quad (E_f < 0.5) \tag{2.194}$$

If eqn (2.191) is rewritten with ρ_{AB}^* in place of ρ_A^*, the individual driving force on the left hand side has now been replaced with an overall driving force, $\bar{\rho}_A - \rho_{AB}^*$. This necessitates replacement of the individual coefficient k_{dr}^* by the overall coefficient based on the disperse phase, K_{dr}^*. The analytical form for K_{dr}^* then follows by substituting eqn (2.188) for E_f in the right hand term of eqn (2.192) giving, for the case of a finite resistance to mass transfer in the continuous phase,

$$K_{dr}^* = -\frac{d_p}{6t}\ln\left\{\sum_{n=1}^{\infty}\frac{6\alpha^2}{\beta_n^2[\beta_n^2 + \alpha(\alpha - 1)]}\exp\left(\frac{-\beta_n^2 Dt}{r_s^2}\right)\right\} \tag{2.195}$$

The contribution of the individual continuous phase mass transfer coefficient, k_{cr}^*, is of course accounted for in determining the value of α.

Earlier approximate applications of these relationships in describing extraction processes are well summarized by Korchinsky and Cruz-Pinto (1979).

2.4.3.2 *Transfer in circulating drops, with and without external resistance*

For freely moving drops in liquid–liquid systems in which the interfacial tension is low and surface active contamination is absent, the viscous drag of the surrounding phase induces torroidal circulation currents inside the drops. Kronig and Brink (1950) analysed this situation for the case of zero external resistance to mass transfer and drop Reynolds numbers below unity, assuming that the internal circulation pattern conformed to that developed by Hadamard (1911). Their result is

$$k_{dr}^* = -\frac{d_p}{6t}\ln\left\{\frac{3}{8}\sum_{n=1}^{\infty} B_n^2 \exp\left[-\lambda_n \frac{16\,Dt}{(d_p/2)^2}\right]\right\} \tag{2.196}$$

where the first seven values of B_n and λ_n are shown in Table 2.3, taken from Elzinga and Banchero (1960) and Heertjes *et al.* (1954).

The same expression was obtained by Elzinga and Banchero (1960) for the case of a finite continuous phase resistance. They provide the first three values of B_n and λ_n for various levels of continuous phase resistance to transfer.

Although these two developments were nominally confined to drop Reynolds numbers below one, Davies (1972, p. 304) notes that they should hold up to drop Reynolds numbers of about 50, due to persistence of the same flow pattern.

TABLE 2.3 *Values of B_n and λ_n in eqn (2.196)*

n	B_n	λ_n	Reference
1	1.29	1.656	Elzinga and Banchero (1960)
2	0.596	9.08	Elzinga and Banchero (1960)
3	0.386	22.2	Elzinga and Banchero (1960)
4	0.35	38.5	Heertjes *et al.* (1954)
5	0.28	63.0	Heertjes *et al.* (1954)
6	0.22	89.8	Heertjes *et al.* (1954)
7	0.16	123.8	Heertjes *et al.* (1954)

Equation (2.196) showed only moderate agreement with 115 measurements by Skelland and Wellek (1964) on four binary liquid–liquid systems of circulating, non-oscillating drops. In these runs the external resistance to transfer was virtually zero, the interfacial tension was between two and six dynes/cm, and the drop Reynolds number ranged from 37 to 546. Their data were correlated with an average absolute deviation of 34 percent by the expression

$$k_{dr}^* = 31.4\frac{D}{d_p}\left(\frac{4Dt}{d_p^2}\right)^{-0.34}\left(\frac{\mu_d}{\rho_d D}\right)^{-0.125}\left(\frac{d_p u_s^2 \rho_c}{\sigma}\right)^{0.37} \tag{2.197}$$

where t is the time of free rise or fall of a single drop, and subscripts d and c indicate disperse- and continuous-phase properties, respectively. The coefficient in eqn (2.197) is based on the area of a sphere having the same volume as the drop and is for use with a $\Delta\rho_A$ or Δc_A driving force.

The likelihood of occurrence of such internally circulating drops in industrial systems containing trace amounts of surface active contamination is questionable—particularly with small drops associated with agitated systems. Thus Davies (1972, p. 311) says 'Drops less than about 1 mm in diameter, as found in agitated systems, would be expected to obey ... equations (for) rigid or stagnant spheres.' And according to Korchinsky and Cruz-Pinto (1979): 'Viscous circulation is unlikely to appear in agitated extraction equipment, particularly when surface active contaminants are present.' Treybal (1980, p. 523) takes the opposite view in advocating Elzinga and Banchero's (1960) expression for k_{dr}^* in the design of agitated vessels in mixer- settler-type extractors.

2.4.3.3 *Transfer in oscillating drops, with and without external resistance*

The five criteria reviewed between eqns (2.159 and 2.160) may be applied to ascertain whether drops will be oscillating in a given system.

Handlos and Baron (1957) proposed a model which has been applied to oscillating droplets; they assumed fully developed internal circulation, in the form of circular tori. Radial vibrations—presumably originating from drop oscillation—are superimposed on the streamlines, promoting periodic mixing between streamlines. For zero continuous phase resistance to transfer their result leads to

$$k_{dr}^* = -\frac{d_p}{6t}\ln\left\{2\sum_{n=1}^{\infty} B_n \exp\left[\frac{-\lambda_n Dt\, N'_{Pe}}{128\, d_p^2}\right]\right\} \tag{2.198}$$

where $N'_{Pe} = (d_p u_t/D)/(1 + \mu_d/\mu_c)$.

Wellek and Skelland (1965) and Patel and Wellek (1967) extended this approach to include a finite resistance to mass transfer in the continuous phase. When contact times are large only the first term in the series summation was needed and, regarding $2B_1$ as unity, this led to

$$K_{dr}^* = \frac{\lambda_1 u_s}{768\,(1 + \mu_d/\mu_c)} \tag{2.199}$$

where λ_1 was given as a function of k_{cr}^*. When the continuous phase resistance is zero ($k_{cr}^* = \infty$), $\lambda_1 = 2.88$.

It must, however, be conceded that the Handlos and Baron model is based on an improbable mechanism; thus photographic evidence by Rose and Kintner (1966) showed very little of the circulating motion postulated within oscillating drops. Recently Jeffreys (1987) described the model as being 'based on an unconvincing picture . . .'; Korchinsky and Cruz-Pinto (1979), on the other hand, found it to be of some merit in correlating their data from a rotating disk column. However, the absence of a dependence on molecular diffusivity in eqns (2.198 and 2.199) does seem to be at variance with most experimental observation.

Alternative—and perhaps more appealing— models of mass transfer in oscillating drops have followed two lines. *The surface-stretch model* (Beek and Kramers 1962; Rose and Kintner 1966; Angelo *et al.* 1966) assumes no new elements enter the surface during an oscillation cycle; surface extension results from stretching and thinning the surface layer as the area increases. In contrast, *the new-surface-elements model* (Brunson and Wellek 1970) assumes that extension of the surface is accomplished by the addition of fresh elements as the surface expands, with the elimination of the oldest elements as the surface contracts. On the basis of certain additional assumptions (Clift *et al.* 1978, pp. 196–197), including the occurrence of spherical-oblate oscillation, *the surface-stretch model* predicts

$$\frac{\overline{k_{dr}^* A}}{A_p} = 2\left(\frac{Df_o}{\pi}\right)^{1/2}(1 + \epsilon + 3\epsilon^2/8)^{1/2} \tag{2.200}$$

where A_p is the surface of a sphere with the same volume as the drop; $\overline{k_{dr}^* A}$ is the time-average product of the drop surface and the disperse-phase mass transfer coefficient; $1 + \epsilon$ is the ratio of maximum to minimum surface area over the oscillation cycle (typically about 1.3 according to Clift *et al.* 1978, p. 197); and f_o, the oscillation frequency, is estimated from the relationship obtained experimentally by Schroeder and Kintner (1965), as

$$f_o = \left(\frac{48\sigma b}{\pi^2 d_p^3 (2\rho_c + 3\rho_d)} \right)^{1/2} \tag{2.201}$$

The factor b is a correction term to Lamb's original expression, accounting for the viscous damping at appreciable amplitudes of oscillation. Empirically,

$$b = 0.8 d_p^{0.225}$$

where d_p is in centimeters.

Equation (2.200) may be compared with the following prediction from *the new-surface-elements model*:

$$\frac{\overline{k_{dr}^* A}}{A_p} = 2(Df_o/\pi)^{1/2}(1 + 0.687\,\epsilon) \tag{2.202}$$

Whether the surface-stretch or the new-surface-elements model is closer to reality might be ascertained by observing any variation in the area of a photochromic spot, induced on the drop surface by a UV laser pulse, following the technique of Thornton *et al.* (1981, 1985, and 1987), referred to earlier. The question is not of great consequence, however, since an extreme value of ϵ of 0.5 results in a prediction from eqn (2.202) in excess of that from eqn (2.200) by only about 6 percent (Clift *et al.* 1978, p. 197).

Equations (2.200 and 2.202) should apply equally well to the continuous phase, with substitution of the continuous phase diffusivity for D.

A modified Reynolds number may be defined as

$$N'_{Re} = \frac{d_p^2 f_o \rho_d}{\mu_d} \tag{2.203}$$

and eqn (2.200) is then written in the following dimensionless from:

$$N_{Sh} = \frac{k_{dr}^* d_p}{D} = \left[\frac{4(1 + \epsilon + 3\epsilon^2/8)}{\pi} \right]^{1/2} (N'_{Re})^{1/2} N_{Sc,d}^{1/2} \tag{2.204}$$

where $N_{Sc,d} = \mu_d/\rho_d D$. Using $\epsilon = 0.3$ as the typical value for many systems cited by Clift *et al.* (1978, p. 197), this becomes

$$N_{Sh} = 1.30(N'_{Re})^{1/2} N_{Sc,d}^{1/2} \tag{2.205}$$

Yamaguchi *et al.* (1975a) obtained the following correlation of their experimental data

$$N_{\mathrm{Sh}} = 1.14\,(N'_{\mathrm{Re}})^{0.56} N_{\mathrm{Sc},d}^{1/2} \tag{2.206}$$

which agrees quite substantially with eqn (2.205). Other k^*_{dr} relationships for oscillating drops are reviewed by Brunson and Wellek (1970).

2.4.3.4 *Transfer in breaking drops in agitated systems*

Rather few studies have been published on mass transfer in agitated liquid–liquid systems in baffled vessels, and almost all have been confined to the continuous phase. The considerations below therefore represent a provisional attempt to remedy the dearth of published work on disperse-phase-controlled systems in such equipment. It is based on investigations by Skelland and Hu Xien (1990).

The merits of batch tests, as used in the paper by Skelland and Lee (1981), have not perhaps been fully appreciated. Thus droplet breakdown into successively smaller drops persists for several minutes after injection of the disperse phase (Skelland and Lee 1981; Bapat *et al.* 1983; Bapat and Tavlarides 1985; Hong and Lee 1985), whereas mass transfer is usually complete after a minute or two (Rushton *et al.* 1964; Treybal 1980, Skelland and Lee 1981). Batch experimentation enables this behavior to be quantitatively followed with time by measurement of k, drop size, and surface area from moment to moment for a single injected element of disperse phase. In contrast, experiments with continuous-flow-through operation give only average transfer conditions corresponding to many different elements of disperse phase which simultaneously occupy the vessel but which entered it at different times and which are therefore in different stages of mass transfer and drop break-up at any given instant. The time history of a given element of disperse phase in the vessel cannot then be followed in the manner necessary to properly model the operation.

The particular process to be considered consists of a cylindrical, flat-bottomed vessel with four radial baffles, agitated by a centrally mounted, 6-flat-bladed turbine. The vessel is filled to a height equal to its diameter, and the disperse phase (up to 20 percent of the total volume) is poured rapidly into the vicinity of the rotating impeller shaft at time t_0. Solute transfers between the phases, the transfer rate being monitored, for example, by recording the instantaneous electrical conductivity of the continuous phase. The equipment and procedure are shown in greater detail by Skelland and Lee (1981).

It should first be noted that the minimum impeller speed needed to ensure complete dispersion of one liquid in the other in such equipment was correlated by Skelland and Ramsay (1987) for five different types of impeller—including turbines—in different locations and sizes, for a variety of systems from three different published sources. Data from 481 runs were correlated with an average absolute deviation of 12.7 per cent by the expression

$$\frac{N_{\min}^2 \rho_m d_i}{g \Delta \rho} = C^2 \left(\frac{T}{d_i}\right)^{2\alpha} \phi^{0.106} \left(\frac{\mu_m^2 \sigma}{d_i^5 \rho_m g^2 \Delta \rho^2}\right)^{0.084} \tag{2.207}$$

where C and α are tabulated according to impeller type and location in the original paper, and

$$\rho_m = \phi \rho_d + (1 - \phi) \rho_c \tag{2.208}$$

$$\mu_m = \frac{\mu_c}{1 - \phi}\left(1 + \frac{1.5\, \mu_d \phi}{\mu_d + \mu_c}\right) \tag{2.209}$$

Skelland and Moeti (1989) found that eqn (2.207) correlated data in systems containing surface active agents (SAA) when the diminished value of σ due to SAA was used to compute $N_{\min}$. In addition, Skelland and Lee (1978) showed that N values exceeding $N_{\min}$ by an average of about 8 per cent were needed to obtain grossly uniform liquid–liquid dispersions. Equation (2.207) has been extended to non-Newtonian systems by Skelland and Kanel (1990).

The mass transfer process will be analysed between time t_1 at which the initial disturbance due to disperse phase addition has ceased (2–5 s after t_0), and time t_2 when a specified fractional approach to concentration equilibrium, E_f, is attained (0.26–1.31 min after t_0 in this experimental study, for $0.661 \le E_f \le 0.987$).

For low volume fractions of disperse phase, ϕ, such that droplet coalescence is slight, suppose that the total interfacial area from a given element of injected disperse phase increases 'continuously' for the duration of mass transfer, due to drop breakage. Each fresh element of interface da thus formed between time t and $t + \mathrm{d}t$ is assumed to be exposed for the remainder of the mass transfer period, namely for $t_2 - t$. Then

$$\mathrm{d}Q_T = -\, V_d \mathrm{d}\rho_A = k_d^*(t_2 - t)\, \delta \rho_A\, V_T \mathrm{d}a \tag{2.210}$$

where

$\delta \rho_A$ = a suitable driving force

k_d^* = a suitable coefficient

a = the interfacial area per unit volume of the system.

A property of the Sauter mean drop diameter, d_{32}, leads to

$$a = \frac{6\phi}{d_{32}} \tag{2.211}$$

Suppose that

$$d_{32} = \alpha t^{-\beta} + \gamma \tag{2.212}$$

where, from Hong and Lee (1985),

$$\alpha = 29.7 d_{32}^* (d_I/T)^{-2.015} (N_{\mathrm{We},c}/N_{\mathrm{Re},c})^{0.5508} N^{-0.7} \tag{2.213}$$

$$\beta = 0.7 \tag{2.214}$$

$$\gamma = d_{32}^* \tag{2.215}$$

$$d_{32}^*/d_I = 0.05\,(1 + 2.316\phi)\,(d_I/T)^{-0.75} N_{Fr,c}^{-0.13} N_{\mathrm{We},c}^{-0.6} \tag{2.216}$$

From eqns (2.211 and 2.212),

$$\mathrm{d}a = \frac{6\alpha\,\beta\,\phi t^{-(\beta+1)}}{(\alpha t^{-\beta} + \gamma)^2}\,\mathrm{d}t \tag{2.217}$$

If k_d^* is approximated by the penetration theory, then

$$k_d^* = 2\left(\frac{D}{\pi\,(t_2 - t)}\right)^{1/2} \tag{2.218}$$

Combining eqns (2.210, 2.217, and 2.218) and integrating over the mass transfer period,

$$\frac{V_d}{V_T}\int_{\rho_{A1}}^{\rho_{A2}} \frac{-\,\mathrm{d}\rho_A}{\rho_A - \rho_A{}^*} = 12\alpha\,\beta\,\phi\left(\frac{D}{\pi}\right)^{1/2}\int_{t_1}^{t_2} \frac{(t_2 - t)^{1/2} t^{-(\beta+1)}}{(\alpha t^{-\beta} + \gamma)^2}\,\mathrm{d}t \tag{2.219}$$

If the equilibrium curve is linear and the slope of the operating line $(-V_c/V_d)$ is constant, as for low total transfer, then the driving force $\rho_A - \rho_A^*$ is also linear with ρ_A. In this case eqn (2.219) is integrated to give

$$\frac{Q_T}{V_T} = \phi\,\frac{Q_T}{V_d} = 12\,\alpha\,\beta\,\phi\,\Delta\rho_{A\mathrm{lm}}\left(\frac{D}{\pi}\right)^{1/2}\int_{t_1}^{t_2} \frac{(t_2 - t)^{1/2} t^{-(\beta+1)}}{(\alpha t^{-\beta} + \gamma)^2}\,\mathrm{d}t \tag{2.220}$$

where

$$\Delta\rho_{A\mathrm{lm}} = \frac{(\rho_{A1} - \rho_{A1}^*) - (\rho_{A2} - \rho_{A2}^*)}{\ln\left[(\rho_{A1} - \rho_{A1}^*)/(\rho_{A2} - \rho_{A2}^*)\right]} \tag{2.221}$$

The fractional extraction E_f predicted by this model in time $t_2 - t_1$ is formulated as shown below where, for batch operation,

$$E_f = \frac{\rho_{A1} - \rho_{A2}}{\rho_{A_1} - \frac{1}{2}(\rho_{A_1}^* + \rho_{A_2}^*)} = \frac{Q_T}{V_T}\cdot\frac{V_T}{V_d}\Big/\left[\rho_{A1} - \frac{1}{2}(\rho_{A1}^* + \rho_{A2}^*)\right] \tag{2.222}$$

Combining eqns (2.220 and 2.222) yields

$$E_f = \frac{12\alpha\beta\Delta\rho_{ALM}(D/\pi)^{1/2}}{\rho_{A_1} - \frac{1}{2}(\rho^*_{A_1} + \rho^*_{A_2})} \int_{t_1}^{t_2} \frac{(t_2 - t)^{1/2} t^{-(\beta+1)}}{(\alpha t^{-\beta} + \gamma)^2} dt \qquad (2.223)$$

Skelland and Hu Xien (1990) measured values of Q_T/V_T in eighty-six dispersed-phase-controlled runs, using equipment and experimental procedure closely similar to the continuous-phase-controlled program described by Skelland and Lee (1981). Three different systems were used, covering 4.8- and 8.6-fold variations in diffusivity and viscosity, respectively, in the disperse phase. The experimental values of Q_T/V_T are compared with those predicted from eqn (2.220) in Fig. 2.10, where the solute transferred is expressed in moles. The mean absolute deviation between predicted and experimental values was 9.13 per cent.

The model is based on the penetration theory, which has the advantage of being independent of drop size in the estimation of k_d. In addition

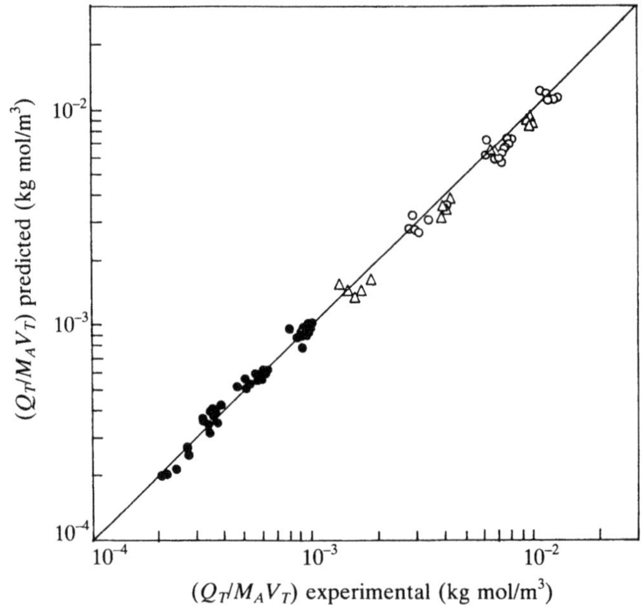

Fig. 2.10. Comparison between experimental values of Q_T/M_AV_T and those predicted by eqn (2.220). Total runs: 86; symbols •, ○, and △ denote systems 1, 2, and 3 respectively; $0.0141 \leq \phi \leq 0.0909$; $3.167 \leq N \leq 12.0$; $0.3134 \leq d_i/T \leq 0.4776$; Mean absolute deviation; 9.13%. Systems — in the order disperse - solute - continuous phase: (1) di-isobutyl ketone - succinic acid - deionized water; (2) 40 (vol) % chlorobenzene + 60% oil* - acetic acid - deionized water; (3) 30% chlorobenzene + 70% oil* - acetic acid - deionized water. *Water white, pharmaceutical grade, mineral oil. Transfer was from the disperse to the continuous phase throughout.

the interfacial area is known as a function of time through the use of eqns (2.211–2.216). The development involves the assumption that the mass transfer characteristics near a given element of interface are not influenced by disturbances due to subsequent breakage of the drop to which that element of surface belongs. This is in fact consistent with surprising recent findings by Thornton *et al.* (1981, 1985, 1987), to the effect that points on the surface of a newly formed drop remain undisturbed by the process of detachment or break-off from the nozzle in liquid–liquid systems free from Marangoni effects. Their observations prevailed at nozzle Reynolds numbers as high as 300 and actual break-off times of less than 50 ms. However, rigorous fulfillment of each of the boundary conditions upon which eqn (2.218) is based (shown below eqn (2.79)) seems unlikely—and of course the applicability of the planar relationship (eqn (2.218)) to a given element of spherical interface should decrease with increasing $t_2 - t$. Such deviations should be less severe for transfer near the surface of the drop, and it may be that most of the transfer occurs under conditions not far removed from a penetration theory framework, i.e. during the early stages of exposure of each new surface element arising from repeated drop breakages. Whether such considerations explain the degree of success achieved in Fig. 2.10 must remain the subject of further study.

Mention should be made of two alternative mechanisms of mass transfer in an agitated vessel, proposed by Howarth (1963). The first, based on unsteady-state diffusion in drops according to eqn (2.188), assumed that neither coalescence nor breakage occurs while drops are in the agitated vessel. The second assumed rapid coalescence and redispersion of the drops, with application of the penetration theory. It was concluded from comparison with a limited amount of data that 'the experimental points are not consistent with either mechanism,' but that they may represent rough limits for small and large nominal residence times, respectively.

2.4.3.5 *Transfer around drops in non-agitated systems*

Transfer in the continuous phase surrounding droplets in liquid–liquid systems is determined by a combination of molecular diffusion and convection in the continuous phase. In the absence of agitation, the latter contribution depends on gravitational effects arising from the density difference between the two phases, which induces relative motion between each drop and its surrounding liquid.

The continuous-phase mass transfer coefficient around a given drop, k_{cr}^*, depends upon whether the drop is internally stagnant, circulating, or oscillating. Creeping flow around a spherical drop occurs for drop Reynolds numbers $(d_p u_t \rho_c / \mu_c)$ less than unity; analytical procedures in this region are characterized by neglect of inertial forces in the Navier–Stokes equations. At Reynolds numbers around 20 a torroidal vortex begins to form in the wake behind the drop (Nisi and Porter 1923; Garner and Skelland 1955), with separation of the boundary layer from the spherical surface as $N_{\mathrm{Re},c}$ increases. Periodic shedding of the wake begins around Reynolds numbers of 200 for rigid spheres, but this may be delayed up to $N_{Re,c}$ of 800 for mobile surfaces of drops free from surface active contamination (Clift *et al.* 1978, pp. 184–85). Consider *flow around an internally stagnant sphere* at high Peclet number ($N_{\mathrm{Pe},c} = N_{\mathrm{Re},c} N_{\mathrm{Sc},c}$), such that the concentration boundary layer in the continuous phase adjacent to the surface of the sphere is both very thin and much thinner than the momentum boundary layer (so $N_{\mathrm{Sc},c} \gg 1$). Friedlander (1961) considers these conditions to correspond to $N_{\mathrm{Pe},c} > 100$. Lochiel and Calderbank (1964) derive the following expression in some detail for the continuous-phase mass transfer coefficient in these circumstances:

$$N_{\mathrm{Sh},c} = \frac{k_{cr}^* d_p}{D_c} = 0.641 \left\{ \int_0^{\pi} \left[\frac{(\partial u_\theta / \partial y)_{y=0} d_p}{2u_t} \sin\theta \right]^{1/2} \sin\theta \, d\theta \right\}^{2/3} N_{\mathrm{Pe},c}^{1/3} \tag{2.224}$$

where $y = r - r_s$ for $r \geq r_s$; θ is the angle with the direction of u_t, measured at the center of the sphere from the front stagnation point; u_θ is the θ-direction velocity in the continuous phase boundary layer around the spherical surface. The integration limits 0–π neglect any separation of the boundary layer from the surface, described earlier. Evidently knowledge of the radial gradient of the continuous phase velocity along the surface of the sphere, evaluated at the surface, will allow prediction of k_{cr}^*. For droplet Reynolds numbers above unity, Lochiel and Calderbank (1964) assumed a quartic velocity profile in the momentum boundary layer, and evaluated the thickness of this layer from Boltze's (1908) application of the Navier–Stokes equation to the boundary layer around the sphere. This approach was considerably simpler than that used originally by Froessling (1938) and resulted in

$$N_{\mathrm{Sh},c} = 0.62 N_{\mathrm{Re},c}^{1/2} N_{\mathrm{Sc},c}^{1/3} \tag{2.225}$$

However, boundary layer separation occurs at $\theta \doteq 108°$, so that substitution of this angle for π as the upper limit of integration in eqn (2.224) gives

$$N_{\mathrm{Sh},c} = 0.56 N_{\mathrm{Re},c}^{1/2} N_{\mathrm{Sc},c}^{1/3} \tag{2.226}$$

[Froessling's (1938) constant was 0.59 in plase of 0.56]. Transfer from the wake region of the sphere's surface has been estimated to be about 20 percent of that over the frontal region prior to boundary layer separation (Lochiel and Calderbank 1964), which would change the constant in eqn (2.225) to 0.67.

Many investigators—including Froessling (1938, 1940)—have sought to allow for the limiting case of zero Reynolds number in eqns (2.225 and 2.226) by adding a term to the right hand side which would account for radial diffusion from the sphere into a stagnant surrounding fluid. This term was derived here earlier as eqn (2.30). For dilute solutions the result, allowing for transfer in the wake region, is

$$N_{\mathrm{Sh},c} = 2 + 0.67 N_{\mathrm{Re},c}^{1/2} N_{\mathrm{Sc},c}^{1/3} \tag{2.227}$$

Skelland (1985, pp. 276–7) lists six published correlations of experimental data in which replacements of the constant 0.67 range from 0.544 to 0.95. He also lists an additional six correlations which omit the term 2 (one of which appears in Table 2.4), plus two additional expressions which introduce allowance for natural convection. The same reference discusses the differences between these correlations in terms of variations in free stream turbulence intensity, surface roughness of the sphere, the manner in which it was supported, and other factors.

Turning to *flow around an internally circulating sphere*, again for the case of high Peclet number, the analog of eqn (2.224) is (Baird and Hamielec 1962; Lochiel and Calderbank 1964; Redfield and Houghton 1965; Winnikow 1967)

$$N_{\mathrm{Sh},c} = \frac{k_{cr}^{*} d_p}{D_c} = \left(\frac{2}{\pi}\right)^{1/2} \left[\int_0^{\pi} \frac{(u_\theta)_{r=r_s}}{u_t} \sin^2\theta \, \mathrm{d}\theta\right]^{1/2} N_{\mathrm{Pe},c}^{1/2} \tag{2.228}$$

where $(u_\theta)_{r=r_s}$ is the velocity of the interface in the θ-direction. The derivation of eqn (2.228) is shown in detail by Lochiel and Calderbank (1964).

Several authors attempted to solve eqn (2.228) for finite drop Reynolds numbers, but with erroneous expressions for $(u_\theta)_{r=r_s}$. Weber (1975) identified the correct expression for $(u_\theta)_{r=r_s}$ as that given by Harper and Moore (1968), namely

$$\frac{(u_\theta)_{r=r_s}}{u_t} = \frac{3}{2}\sin\theta - \frac{2}{(\pi N_{\mathrm{Re},c})^{1/2}} \left[\frac{2 + 3\mu_d/\mu_c}{1 + (\mu_d\rho_d/\mu_c\rho_c)^{1/2}}\right] \frac{X^{1/2}}{\sin\theta} +$$

$$4\left(\frac{2}{\pi N_{\mathrm{Re},c}}\right)^{1/2} \left[\frac{\mu_d/\mu_c}{1 + (\mu_d\rho_d/\mu_c\rho_c)^{1/2}}\right] \frac{1}{X^{1/2}\sin\theta}$$

$$\int_0^{\infty} g(z) \exp\left(-4z^2/X\right) \mathrm{d}z \tag{2.229}$$

in which $X = 2 - 3\cos\theta + \cos^3\theta$; $g(z)$ is a function given by Harper and Moore (1968), and z is an integration variable.

Weber (1975) inserted eqn (2.229) in eqn (2.228) and performed the integration. He was able to correlate his numerical results by the following approximate form, provided that $\mu_d/\mu_c \leq 2$, $0 < \rho_d/\rho_c \leq 4$:

TABLE 2.4 *Other correlations for k^*_{cr} for single streams of drops***

State of drop	Correlation	Reference
Stagnant[†]	$k^*_{cr} = 0.74(D_c/d_p)(N_{Re})^{1/2}(N_{Sc})^{1/3}$	Skelland and Cornish (1963)
Circulating‡	$k^*_{cr} = (2/\pi^{1/2})(D_c u_t/d_p)^{1/2}(8\phi'_{os\pi}/d_p^3)^{1/2}$	Shirotsuka and Kawase (1975, 1976)
Circulating[§]	$k^*_{cr} = (D_c/d_p)[2 + 1.13(N_{Re}N_{Sc})^{1/2}k_v^{1/2}]$	Griffiths (1960)
Circulating	$k^*_{cr} = (D_c/d_p)(-178 + 3.62N_{Re}^{1/2}N_{Sc}^{1/3})$	Thorsen and Terjesen (1962)
Circulating	$k^*_{cr} = (D_c/d_p)(-126 + 1.8N_{Re}^{1/2}N_{Sc}^{0.42})$	Garner *et al.* (1959)
Circulating	$k^*_{cr} = 0.6(D_c/d_p)N_{Re}^{1/2}N_{Sc}^{1/2}$	Garner and Tayeban (1960)
Oscillating	$k^*_{cr} = 1.2D_c^{1/2}[48\sigma/\pi^2 d_p^3(2\rho_c + 3\rho_d)]^{1/4}$	Clift *et al.* (1978), p. 197
Oscillating	$k^*_{cr} = (D_c/d_p)(50 + 0.0085N_{Re}^{1.0}N_{Sc}^{0.7})$	Garner and Tayeban (1960)

**N_{Re} and N_{Sc} refer to the continuous phase properties throughout.
[†]For 13 other expressions, see Skelland (1974, 1985), pp. 276–7.
‡$8\phi'_{os\pi}/d_p^3$ is given graphically as $f(L/d_p)$ in Shirotsuka and Kawase (1975).
[§]Selected from Griffiths' (1960) Table 2 for $(1 - k_v)(\xi/\xi_0)(k_v N_{Sc}^{1/3})^{-1} < 1/3$.

$$N_{\mathrm{Sh},c} = \frac{k_{cr}^* d_p}{D_c} = \frac{2}{\pi^{1/2}}\left\{1 - \frac{1}{N_{\mathrm{Re},c}^{1/2}}\left[2.89 + 2.15\left(\frac{\mu_d}{\mu_c}\right)^{0.64}\right]\right\}^{1/2} N_{\mathrm{Pe},c}^{1/2} \tag{2.230}$$

This approximation fitted Weber's numerical values of $N_{\mathrm{Sh},c}/N_{\mathrm{Pe},c}^{1/2}$ within 5 per cent. Equation (2.230) was shown to be in good agreement with Griffiths' (1960) data for ethyl acetate drops in water—a system with a low interfacial tension of about 5 dyne/cm. However, poorer agreement is observed between eqn (2.230) and five other (mostly empirical) correlations for k_{cr}^* for circulating drops, shown in Table 2.4 and plotted in Fig. 2.11 for comparison with data from single and triple streams of falling drops. These data were obtained by Skelland and Vasti (1985) on the system chlorobenzene + 76.3 vol per cent toluene (dispersed)–benzoic acid (solute)–distilled water (continuous), with transfer from the organic to the aqueous phase. The interfacial tension was substantially higher (27 dynes/cm) and the drops were consequently more susceptible to reduction of internal circulation by trace amounts of unknown surface active contamination.

These considerations probably account for the differences between predictions from the various correlations for a given set of conditions. Nonreproducible and differing types of trace contamination in the various systems upon which the equations are based probably led to differing degrees of internal circulation and the corresponding differences in prediction seen in Fig. 2.11.

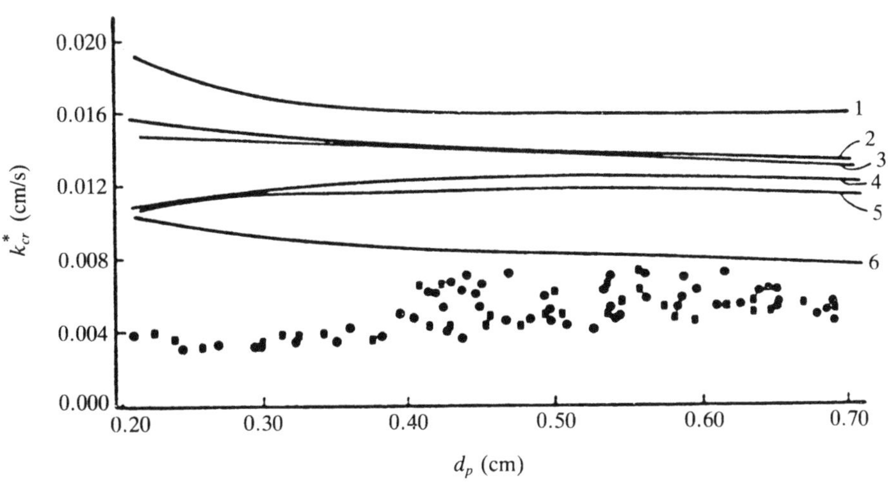

Fig. 2.11. k_{cr}^* vs drop diameter (circulating drops). ●, single nozzle; ■, triple nozzle; curves 1–6 are the correlations of Shirotsuka and Kawase (1976), Griffiths (1960), Weber (1975), Thorsen and Terjesen (1962), Garner *et al.* (1959), and Garner and Tayeban (1960), respectively, from Table 2.4.

Two limiting cases of eqn (2.228) are of interest; first, when the drop Reynolds number $N_{\mathrm{Re},c} \rightarrow \infty$, potential flow prevails around the drop and eqn (2.229) shows that

$$(u_\theta)_{r=r_s}/u_t = \frac{3}{2}\sin\theta \tag{2.231}$$

Insertion in eqn (2.228) gives

$$N_{\mathrm{Sh},c} = \frac{2}{\pi^{1/2}} N_{\mathrm{Pe},c}^{1/2} \tag{2.232}$$

This equation was first proposed by Boussinesq (1905); it is in fact identical with k_{cr}^* defined by the penetration theory in eqn (2.86) when t_e is equated to d_p/u_t. Thus eqn (2.232) is obtainable from $k_{cr}^* = 2\ (D_c/\pi t_e)^{1/2}$ in eqn (2.86) by setting $t_e = d_p/u_t$ and multiplying throughout by (d_p/D_c) $(\rho_c \mu_c/\rho_c \mu_c)^{1/2}$. If a typical value of 1000 is taken for the Schmidt group and eqn (2.232) divided by eqn (2.225), we find that the continuous phase mass transfer coefficient in the case of the fully circulating sphere is 5.76 times greater than that for the stagnant sphere.

For finite Reynolds numbers and a very viscous continuous phase, such that $\mu_d/\mu_c \rightarrow 0$, eqn (2.229) shows that only the first two terms remain to be considered. In this case, eqn (2.230) reduces to

$$N_{\mathrm{Sh},c} = \frac{2}{\pi^{1/2}} \left\{1 - \frac{2.89}{N_{\mathrm{Re},c}^{1/2}}\right\}^{1/2} N_{\mathrm{Pe},c}^{1/2} \tag{2.233}$$

The analysis and equation of Shirotsuka and Kawase (1975, 1976) in Table 2.4 allowed for vertical interaction between adjacent drops separated by distance L, but indicated that this was significant only for $L/d_p \leq 4$. For the system in Fig. 2.11 Skelland and Vasti (1985) found a reduction of about 45 per cent in k_{cr}^* when L/d_p was reduced from 10 to 2. This interaction was believed to explain the slight decrease in k_{cr}^* with decreasing d_p in Fig. 2.11, in contrast to most of the plotted correlations (L/d_p was smaller with smaller d_p values).

The only other circulating-drop correlation that attempts to allow for droplet interaction in drop swarms is (Treybal 1963, from the data of Ruby and Elgin 1955)

$$k_{cr}^* = 0.725\ (d_p u_s \rho_c/\mu_c)^{-0.43}\ (\mu_c/\rho_c D_c)^{-0.58} u_s\,(1-\phi) \tag{2.234}$$

for low interfacial tension systems. An interesting exercise is as follows; suppose, for a continuous-phase controlled system, that eqn (2.234) is combined with eqn (2.157) for u_s and applied to a spray column similar to that for which it was established. For fixed physical properties and drop size we have

$$k_{cr}^{*} = C_1 (1 - \phi)^{1.57}$$

The total interfacial area of the freely rising droplets is

$$A_r = \frac{(\text{column volume})\,\phi}{\text{volume/drop}} \begin{pmatrix} \text{surface area} \\ \text{per drop} \end{pmatrix} = C_2 \phi$$

And the mass transfer rate during free rise of the drops is

$$q_{Ar} = k_{cr}^{*} A_r \Delta \rho_A$$

or

$$\frac{q_{Ar}}{C_1 C_2 \Delta \rho_A} = (1 - \phi)^{1.57} \phi$$

Differentiating with respect to ϕ and equating to zero for maximum rate of mass transfer gives $\phi_{\text{optimum}} = 0.389$.

Considering *flow around an oscillating drop*, it was observed earlier that eqn (2.200) for *the surface-stretch model* and eqn (2.202) for *the new surface-elements model* should be equally applicable to the continuous phase, if the continuous phase diffusivity D_c is used in place of D.

Table 2.4 shows two other correlations for k_{cr}^{*} in oscillating drop systems, which are compared with single and triple drop stream data in Fig. 2.12. These data were obtained by Skelland and Vasti (1985) on the system distilled water (dispersed)–acetic acid (solute)–toluene (continuous),

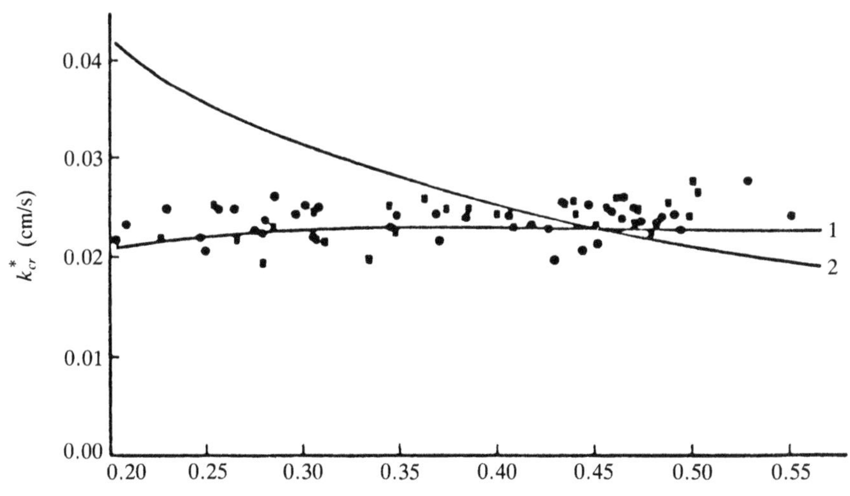

Fig. 2.12. k_{cr}^{*} vs drop diameter (incipiently oscillating drops). ●, single nozzle; ■, triple nozzle; 1 = correlation of Garner and Tayeban (1960); 2 = correlation of Clift *et al.* (1978), from Table 2.4.

with transfer from the organic to the aqueous phase. No drop-to-drop interaction on mass transfer was observed for oscillating drops in this study. One further correlation of data for k^*_{cr} around oscillating drops is given by Yamaguchi *et al.* (1975b)—it agrees quite closely with the surface stretch model.

2.4.3.6 *Transfer around drops in agitated systems*

The convective transport of solute in non-agitated systems depends on flow fields generated by gravity due to a density difference between the two phases. In contrast, for agitated systems, formulation of the convective contribution requires appropriate characterization of the transport features—such as the fluctuating velocity—of the turbulence adjacent to a given droplet. This turbulence arises not from gravitational effects but from the energy imparted to the system by the agitation.

The proper weighting of these turbulence transport features necessitates some understanding of the particular mass transport mechanism prevailing under a given set of operating conditions. This may be partially inferred from the way in which the continuous phase mass transfer coefficient depends on the solute molecular diffusivity in that phase. Thus, in the expression

$$k^*_{cr} \propto D_c^n \qquad (2.235)$$

the exponent n has a value of unity according to the film theory, whereas values between 1/2 and one are consistent with the film-penetration theory of Toor and Marchello (1958) and Dobbins (1956, 1964). Other theoretical mechanisms, including some that permit n to approach zero, are reviewed by Kozinski and King (1966). When the interface is fully mobile an n of 1/2 is indicated by the eddy cell model (Lamont and Scott 1970), by boundary layer theory (Calderbank 1967), by the large-eddy model (Fortesque and Pearson 1967), by turbulent boundary layer and slip velocity theory (Keey and Glen 1969), and by potential flow theory (Calderbank 1967). An n value of 1/2 also follows from Skelland and Lee's (1981) theory of periodically varying rate of surface renewal. On the other hand, for immobile interfaces, n equals 2/3 according to the eddy cell model (Lamont and Scott 1970), to boundary layer theory (Calderbank 1967), and to slip velocity theory for 'large' particles (Harriott 1962).

When attention is confined to mass transfer in two-phase suspensions in agitated vessels, the experimental findings are conflicting and often at variance with theory. Thus it was contended that n equals zero by Barker and Treybal (1960). This was reiterated in two subsequent publications by Treybal (1961, 1963, p. 417).

For mass transfer from solid particles freely suspended in agitated liquids (i.e. immobile interfaces), an n of 2/3 was obtained by Calderbank and Moo-Young (1961). This contrasts with n values of 1/2 found by Hixson and Baum (1941), Bieber (1962), Barker and Treybal (1960), and Sykes and Gomezplata (1967). In the work of Barker and Treybal, a 17-fold range of D_c was used to establish a value of 1/2 for n by graphical means. Sykes and Gomezplata specifically showed that an n of 1/2 satisfactorily represented their results, whereas 2/3 did not. They observed that 'the over correction for the effect of system Schmidt number (and therefore of D_c) . . . by the fixed particle correlations (i.e. by use of $n = 2/3$) has been noted by Miller (1964).' Unfortunately their conclusions are undermined by their invalid 'finding' that k_{cr}^* increases with increasing continuous phase viscosity, μ_c.

Both Harriott (1962) and Calderbank (1967) state that small drops and bubbles behave as rigid spheres when suspended in agitated liquids. The resulting immobile interface would suggest that n equals 2/3, according to some of the above theories. Thus Calderbank's (1967) correlation of measurements for small bubbles in aqueous electrolytes uses $n = 2/3$. Both Treybal, (1963, p. 416; 1980, p. 523) and Glen (1965), on the other hand, say that the surface of such liquid drops is mobile. This is consistent with Prasher and Wills' (1973) use of $n = 1/2$ to correlate mass transfer from gas bubbles in agitated liquids.

Humphrey and Van Ness (1957) and Miller (1971) arbitrarily assigned values of 1/2 and 2/3, respectively, to n, even though they each studied only one system of solid particles dissolving in an agitated liquid. Clearly D_c and $N_{Sc,c}$ were not variables in their work.

The only study known to the author that is specifically designed to establish the value of n in agitated liquid–liquid systems is that of Skelland and Moeti (1990). They used batch operation in a baffled, cylindrical vessel, agitated with a centrally mounted 6-flat-blade turbine. The organic disperse phase ($\phi = 0.01$) was poured rapidly into the vicinity of the rotating impeller shaft, and the progress of the acid transfer out of the drops recorded as the changing electrical conductivity of the aqueous continuous phase. Drop sizes were obtained photographically, enabling evaluation of an area-free continuous phase coefficient. Several continuous-phase-controlled systems at various temperatures were used to obtain a 7.8-fold range of D_c, with interfacial tensions between 11 and 35 dynes/cm, and impeller speeds, N, between 3 and 9 r.p.s.

After preliminary establishment of the dependency of k_{cr}^* first on N and then on μ_c, its dependency on D_c was evaluated. A typical result, with chlorobenzene as dispersed phase solvent, appears in Fig. 2.13. The regression line for this plot has a slope of 0.70 with 95 per cent confidence intervals of ±0.037. The line of slope 1/2 clearly shows that n was not 1/2 in these experiments. An overall value of 2/3 for n was indicated by this work, and it was also found that

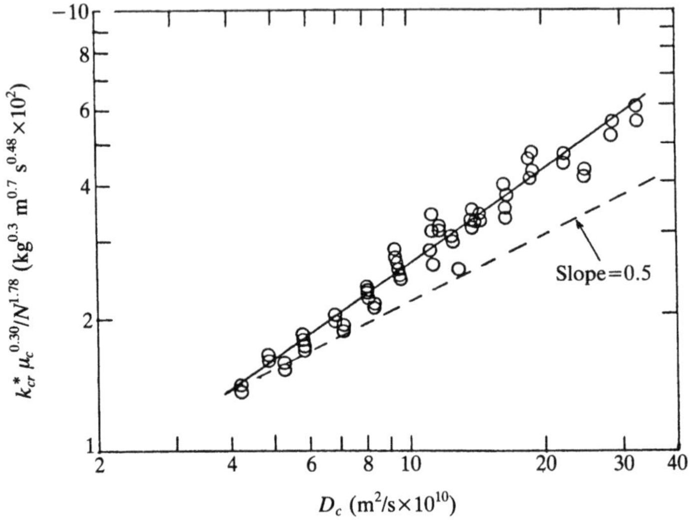

Fig. 2.13. $k^*_{cr}\mu_c^{0.30}/N^{1.78}$ vs D_c for chlorobenzene as disperse phase solvent, water as continuous phase, and nonanoic, heptanoic, and benzoic acids as solutes, between 5° and 70 °C, in an agitated vessel.

$$k^*_{cr} \propto D_c^{2/3}\mu_c^{-1/3}N^{3/2} \tag{2.236}$$

A literature survey revealed that coefficients for solid-particle dissolution in agitated systems usually showed a much higher apparent dependence on μ_c when sugar had been used to enhance the viscosity of the continuous phase. Apparently some interaction between the solute and sugar in solution was masking the true effect of μ_c. Additionally, for solid-particle dissolution the exponent on N is about half the value shown in liquid–liquid systems by eqn (2.236). This effect was previously known and predicted (Davies 1972, p. 239–240), the exponent on N in eqn (2.236) being about the same as that found earlier in liquid–liquid studies by Glen (1965), Schindler (1967), and Lee (1978).

Correlation of Skelland and Moeti's data was effected by invoking the customary exponential relationship between Sherwood, Reynolds, and Schmidt groups. However, the velocity term in the drop Reynolds number, $N'_{\mathrm{Re},c}$, was the turbulence fluctuating velocity in the continuous phase, scaled over a distance comparable to the drop size, d_p. The latter was shown to lie in the inertial subrange of eddy sizes and so, in accordance with the theory of local isotropic turbulence, Batchelor (1951) gives this fluctuating velocity as

$$(\overline{u^2})^{1/2} \propto \left(\frac{P}{V_s}\right)^{1/3}\left(\frac{d_p}{\rho_c}\right)^{1/3} \tag{2.237}$$

The agitator power input per unit volume of a baffled vessel is

$$\frac{P}{V_s} = \frac{4K_T}{\pi}\frac{d_i^5 N^3 \rho_c}{T^3} \tag{2.238}$$

with K_T equal to 2.5, the value of the Power number for turbines in the turbulent region (Bates *et al.* 1963). The Reynolds number was accordingly formulated by substituting eqns (2.237 and 2.238) into ($N'_{\mathrm{Re},c} = (\overline{u^2})^{1/2} d_p\rho_c/\mu_c$). The exponent on D_c in eqn (2.236) relates predominantly to that transfer mechanism associated with free travel of discrete drops around the vessel. This follows from Boyadzhiev and Elenkov's (1966) finding in high interfacial tension systems for $0 \le \phi \le 0.06$ that the break-up-coalescence process does not significantly affect the transfer mechanism in a turbulent isotropic flow field when the system is continuous-phase controlled, as here. The value of the exponent on D_c—namely 2/3—indicates that these small drops were internally stagnant, resembling solid spheres in this respect. But for solid spheres in this range of (d_p–$\Delta\rho$–μ_c), k_c is independent of d_p according to many studies (Hixson and Baum 1941; Humphrey and Van Ness 1957; Barker and Treybal 1960; Calderbank and Moo-Young 1961; Calderbank 1967; Davies 1972, p. 149; Sherwood *et al.* 1975, p. 221–4, etc.). Thus the adoption of $k^*_{cr} \propto d_p^o$, with additional manipulation to satisfy eqn (2.236), resulted in

$$\frac{k^*_{cr}d_p}{D_c} = 1.237\,(10^{-5})\left(\frac{\mu_c}{\rho_c D_c}\right)^{1/3}\left(\frac{d_i^2 N\rho_c}{\mu_c}\right)^{2/3}\left(\frac{d_i N^2}{g}\right)^{5/12}\left(\frac{d_i}{d_p}\right)^{2}$$
$$\left(\frac{d_p}{T}\right)^{1/2}\left(\frac{\rho_d d_p^2 g}{\sigma}\right)^{5/4}\phi^{-1/2} \tag{2.239}$$

in which d_p, d_i, and T are drop, impeller, and vessel diameters, respectively. The mean absolute deviation between eqn (2.239) and 180 data points from nine systems was 19.71 percent. The net exponent on d_i of 3.75 in eqn (2.239) is of the general order indicated for agitated liquid–liquid systems by Davies (1972, p. 240), in the form $k_c \propto (d_i^2N\rho_c/\mu_c)^{1.2\pm0.3}$. Schindler (1967, pp. 179–183) measured transfer rates from ethyl acetate drops to water in a turbine-agitated vessel. When d_i/T and (baffle-width)/T were similar to those in Skelland and Moeti's work he found that k_c varied with $(d_i^2N\rho_c/\mu_c)$ raised to exponents between 1.5 and 1.75, depending on ϕ, thus implying that $k_c \propto d_i^{3\ to\ 3.5}$. The exponent of -0.5 on T in eqn (2.239) coincides with that found by Keey and Glen (1969) in agitated liquid–liquid systems. The sixth group on the right in eqn (2.239) is a variant on the droplet Eotvos number, which has been used (Clift *et al.* 1978, p. 40) to characterize the reduced mobility of a droplet surface due to surfactant contamination.

2.4.4 Mass transfer during drop coalescence

Droplet coalescence at a plane liquid–liquid interface occurs in spray columns, and repeatedly in perforated-plate columns and settlers in mixer–settler cascades. Although the relevant hydrodynamics have been studied extensively, the mass transfer aspects of the process have received less attention. A preliminary approach by Johnson and Hamielec (1960) applied the penetration theory to the fluid layer resulting from the proposed spreading of each arriving drop over the entire coalescence surface. For such a situation eqn (2.84) shows that the instantaneous flux of solute A into the layer at time t since the arrival of the last drop is

$$q_{Ac}/A_c = n_{Ac} = \Delta \rho_{Ac} (D/\pi t)^{1/2}$$

where A_c is the total coalescence interface and q_{Ac} is the rate of solute transfer during coalescence, (mass/time). The average solute flux into the layer between the arrival of two consecutive drops—i.e. during drop formation time t_f in spray or perforated plate columns—is then

$$\frac{1}{t_f}\int_0^{t_f} n_{Ac}\,\mathrm{d}t = \frac{\Delta \rho_{Ac}}{t_f}\left(\frac{D}{\pi}\right)^{1/2}\int_0^{t_f} t^{-1/2}\,\mathrm{d}t = 2\Delta \rho_{Ac}\left(\frac{D}{\pi t_f}\right)^{1/2}$$

so that

$$k^*_{dc} = \frac{(n_{Ac})_{av}}{\Delta \rho_{Ac}} = 2\left(\frac{D}{\pi t_f}\right)^{1/2} \tag{2.240}$$

The mechanism is one of transient transfer between the arrival of consecutive drops at the coalescence plane.

Skelland and Minhas (1971) found that eqn (2.240) agreed well with their measured values for drops that coalesced instantaneously (namely chlorobenzene drops coalescing at a water–chlorobenzene interface with acetic acid transferring from the organic phase). Some overestimation was found for the case of water drops coalescing at an ethyl acetate-water interface, with ethyl acetate transferring into the water, although some delay in drop coalescence on reaching the interface was observed here. Ten-fold over-predictions were observed, however, for the case of drops formed from a 50 (vol) per cent mixture of CCl_4 and mineral oil, coalescing at their interface with water, and with acetic acid again diffusing out of the drops. The latter system showed marked delays in coalescence of drops resting at the interface; it had the same interfacial tension, but a disperse-phase viscocity five times greater than the chlorobenzene system.

An empirical relationship, also for the disperse phase coefficient during coalescence, was developed by Skelland and Minhas (1971) from statistical correlation of their measurements on the same three systems used to generate eq (2.173) for forming drops. Three parallel streams of drops were used, such that the coalescence area per stream was about the same as that in a typical perforated plate column. The equation is

$$k_{dc}^{*} = 0.173 \frac{d_p}{t_f} \left(\frac{\mu_d}{\rho_d D} \right)^{-1.115} \left(\frac{\Delta \rho g d_p^2}{\sigma} \right)^{1.302} \left(\frac{u_s^2 t_f}{D} \right)^{0.146} \tag{2.241}$$

The average absolute deviation between this expression and 23 data points was 25 percent. The coefficient k_{dc}^{*} is based on the plane area of interface per droplet stream.

The continuous phase analog of eq (2.241) was obtained by Skelland and Hemler (1969), using the same triple-nozzle equipment and the same three systems as those used to develop eq (2.174) for forming drops. Their statistical correlation is

$$k_{cc}^{*} = 5.959\,(10^{-4}) \left(\frac{D_c}{t_f} \right)^{0.5} \left(\frac{\rho_c u_s^3}{g \mu_c} \right)^{0.332} \left(\frac{d_p^2 \rho_c \rho_d u_s^3}{\mu_d \sigma} \right)^{0.525} \tag{2.242}$$

The average absolute deviation between eq (2.242) and data from 20 measurements was 22 percent.

2.5. Selection of the disperse phase in spray or plate extraction columns

Criteria were established by Skelland and Chadha (1981) for deciding whether the controlling phase should be dispersed or continuous in perforated plate or spray extraction columns. Their analysis was intended for those cases 'where comparable amounts of each phase are present,' with similar corresponding physical properties in the two streams, to accomodate the fixed dimensions in a given column. However, only minor modifications are needed to extend the treatment to cases where the flow rates of the two phases differ significantly from each other, perhaps with substantially different densities and viscosities of the two liquids, in designing a *new* column. These extensions will accordingly be included here. Relevant factors were shown to include the plate spacing or height of free rise or fall of the drops before coalescence, and the presence of surface active impurities – the findings are qualitatively similar for both high and low interfacial tension systems.

The topic will be treated in some detail, not only because of its intrinsic importance, but also because it will serve as a vehicle for illustrating how many of the separate phenomena dealt with earlier may be brought together in a unified approach to a particular problem.

2.5.1 Formulation of a criterion for greater extraction

A perforated-plate liquid–liquid extraction column is to be designed for a solute transferring from phase 1 to phase 2 and where phase 1 offers the controlling resistance to mass transfer. Attention is confined to the column section between consecutive plates *a* and *b*. Phase 1 is to be processed continuously at a fixed volume rate of V_1, regardless of whether it is dispersed or continuous. Its solute content will be reduced from ρ_{Aa} to ρ_{Ab} (phase 1 dispersed) or to ρ'_{Ab} (phase 1 continuous). The column active volume between plates *a* and *b* is V_c when phase 1 is dispersed, and V'_c when phase 1 is continuous. (For a given plate spacing, V_c will differ from V'_c when V_1 differs significantly from V_2 and/or when relevant physical properties of the two phases have different respective values). The extract phase flows countercurrently at the volume rate of V_2, with an average solute concentration in equilibrium with ρ^*_A between plates *a* and *b*. $V_{1,d}$ and $V_{1,c}$ denote phase 1 as disperse and continuous, respectively, and $V_{1,d} = V_{1,c} \doteq$ constant between *a* and *b* for dilute systems or low rates of transfer. When phase 1 is dispersed

$$q = V_{1,d}(\rho_{Aa} - \rho_{Ab}) = k_{do}a(\Delta\rho_A)_{0,lm}V_c \tag{2.243}$$

where $k_{do}a$ is the overall capacity coefficient (disperse phase controlling, so $K_d \doteq k_d$ from eqn (2.66 and below 2.68)); it includes transfer during drop formation at plate *a*, during free rise of the drops through the continuous phase, and during coalescence of the drops beneath plate *b*. Also

$$(\Delta\rho_A)_{0,lm} = (\rho_{Aa} - \rho_{Ab})/\ln\left[(\rho_{Aa} - \rho^*_A)/(\rho_{Ab} - \rho^*_A)\right] \tag{2.244}$$

Rearranging

$$\ln(1 - E_f) = \frac{-k_{do}aV_c}{V_{1,d}};\ E_f = \frac{\rho_{Aa} - \rho_{Ab}}{\rho_{Aa} - \rho^*_A} \tag{2.245}$$

When phase 1 is continuous the development is repeated with $k_{co}a$, $V_{1,c}$, V'_c, and ρ'_{Ab} replacing $k_{do}a$, $V_{1,d}$, V_c, and ρ_{Ab}, respectively. Taking ratios, and noting that E'_f, the fractional extraction between plates *a* and *b* when phase 1 is continuous, is $(\rho_{Aa} - \rho'_{Ab})/(\rho_{Aa} - \rho^*_A)$

$$\frac{\ln(1 - E_f)}{\ln(1 - E'_f)} = \frac{k_{do}a}{k_{co}a}\frac{V_c}{V'_c} \tag{2.246}$$

This shows that greater extraction will be obtained by dispersing the controlling phase when $(k_{do}a/k_{co}a)(V_c/V'_c)$ is greater than 1.0. This, then, is our criterion for greater extraction, and one should therefore check whether this k_0aV_c ratio is greater or less than unity when selecting the phase to be dispersed. Procedures for doing this are described in the next section.

2.5.2 Evaluation of the criterion

Equation (2.243) may alternatively be written as

$$q = k_{df} A_f (\Delta\rho_{Af})_{\text{lm}} + k_{dr} A_r (\Delta\rho_{Ar})_{\text{lm}} + k_{dc} A_c (\Delta\rho_{Ac})_{\text{lm}} \quad (2.247)$$

which is the sum of the mass transfer rates during drop formation on plate a, free rise, and coalescence beneath plate b. If transfer rates are low enough to regard $\Delta\rho_A$ as constant at its log mean value between plates a and b, then from eq (2.243 and 2.247)

$$\frac{q}{(\Delta\rho_A)_{0,\text{lm}}} = k_{do} a V_c = k_{df} A_f + k_{dr} A_r + k_{dc} A_c \quad (2.248)$$

where A_f is the surface of n_o drops at detachment from n_o holes per plate, A_r is the total surface of all rising drops between plates a and b at any instant, and A_c is the plane coalescence area (column cross section minus downcomer cross section).

Similarly, when the controlling phase is made continuous

$$k_{co} a V_c' = k_{cf} A_f' + k_{cr} A_r' + k_{cc} A_c' \quad (2.249)$$

where A_f', A_r', and A_c' are not necessarily the same as A_f, A_r, and A_c, respectively. Thus

$$\frac{k_{do} a}{k_{co} a} \frac{V_c}{V_c'} = \frac{k_{df} A_f + k_{dr} A_r + k_{dc} A_c}{k_{cf} A_f' + k_{cr} A_r' + k_{cc} A_c'} \quad (2.250)$$

In general, the interfacial areas and individual coefficients (which approximate the overall coefficients based on the controlling phase) during drop formation, rise, and coalescence should be estimated as in Tables I and II, respectively, in the paper by Skelland and Conger (1973). For illustration purposes here the interfacial areas will be obtained from the following expressions

$$A_f = n_o \pi d_p^2; \; A_f' = n_o' \pi (d_p')^2 \quad (2.251)$$

$$n_o = 4V_1/\pi d_n^2 u_o; \; n_o' = 4V_2/\pi d_n^2 u_o$$

$$A_r = \frac{H}{u_t t_f} n_o \pi d_p^2 = \frac{t}{t_f} n_o \pi d_p^2; \; A_r' = \frac{H}{u_t' t_f'} n_o' \pi (d_p')^2 = \frac{t'}{t_f'} n_o' \pi (d_p')^2 \quad (2.252)$$

$$A_c = \pi (r_c^2 - r_d^2); \; A_c' = \pi[(r_c')^2 - (r_d')^2] \quad (2.253)$$

in which H is the vertical height of free rise or fall at droplet terminal velocity u_t or u_t' through the continuous phase. (More accurately, u_s replaces u_t.)

It should be evident that eq (2.250) is readily extended to include operation under jetting conditions, by the addition of the terms $k_{dj}A_j$ and $k_{cj}A_j'$ to the numerator and denominator, respectively. The coefficients k_{dj} and k_{cj} might then be estimated from eq (2.138 and 2.139) with eq (2.141), and

$$A_j = \frac{n_o \pi}{2} (d_n + d_{jc}) [\tfrac{1}{4} (d_n - d_{jc})^2 + L_j^2]^{1/2}$$

$$A_j' = \frac{n_o' \pi}{2} (d_n + d_{jc}') [\tfrac{1}{4} (d_n - d_{jc}')^2 + (L_j')^2]^{1/2} \qquad (2.254)$$

Expressions used to estimate the coefficients in the absence of surface active agents (SAA) or impurities are listed in column 2 of Table 2.5; for brevity they are cited as equation numbers in this chapter. It will be noted that the coefficient during free rise depends upon whether the drops are internally stagnant, internally circulating due to the drag of the continuous phase, or oscillating.

Surface-active impurities are usually present to an unknown degree on an industrial scale. These accumulate at the interface between the two phases and inhibit circulation within the drops, modify the pattern of droplet oscillation, offer mechanical and adsorptive barriers to transfer across the interface, and change the shape and terminal velocity of the drops. The formulation of relationships to allow for these effects is prevented by the specific way in which they depend on the structure and concentration of the surface active contaminant. In the absence of any more definitive guidance, therefore, the findings of Skelland and Caenepeel (1972) will be used. These authors measured the effects of anionic and cationic surfactants on k_{df}, k_{dr}, k_{dc}, k_{cf}, k_{cr}, and k_{cc} over a wide range of surfactant concentration. The

TABLE 2.5 *Relationships used to estimate coefficients in eqn (2.250)*

Coefficient	No SAA added. Equation No.	Crude estimates, SAA present.* Equation No.
k_{df} (based on A_f)	(2.173)	(2.164)
$k_{dr}(E_f > 0.5)$ stagnant	(2.193)	(2.193)
$k_{dr}(E_f < 0.5)$ stagnant	(2.194)	(2.194)
k_{dr}, circulating	(2.197)	–
k_{dr}, oscillating	(2.206)	–
k_{dc}	(2.241)	(2.240)
k_{cf}(based on A_f)	(2.174)	(2.164) (D_c replaces D)
k_{cr}, stagnant	1st. eqn, Table 2.4	1st. eqn, Table 2.4
k_{cr}, circulating	6th. eqn, Table 2.4	–
k_{cr}, oscillating	8th. eqn, Table 2.4	–
k_{cc}	(2.242)	(2.240) (D_c replaces D)

* See Skelland and Caenepeel (1972).

relations listed in column 3 of Table 2.5 did not actually correlate their measured coefficients—they are merely those that came closest to their data in most instances. In fact, the highest average absolute deviations between experiment and the relations in column 3 of Table 2.5 were +59 percent and −44 percent for systems that, in the absence of SAA, had interfacial tensions of about 23 dyne/cm. Clearly only the crudest estimates can be expected from these expressions in the presence of SAA. (Average absolute percent deviation = $(100/n)\Sigma_1^n$ [|predicted-experimental|/predicted]; (+) or (−) means that the relation generally over- or underestimated the n measurements, respectively.)

Visualize now two different runs on the same liquid–liquid system, with the controlling phase first dispersed and then continuous. The first run requires a column with n_o, r_c, and r_d; the second needs a column with n_o', r_c', and r_d'. This gives rise to V_c and V_c', respectively, for a given plate spacing, as is evident from Table I in Skelland and Conger (1973). The value of $(k_{do}a/k_{co}a)\ (V_c/V_c')$ for these two runs can be formulated either in the absence or the presence of SAA, via eqn (2.250) and Table 2.5. As an example, in the presence of SAA, when the drops are stagnant, $E_f < 0.5$, and the interfacial areas are given by eqns (2.251–2.253):

$$\frac{k_{do}aV_c}{k_{co}aV_c'} = \frac{\text{Overall capacity coefficient when the controlling phase is dispersed}}{\text{Overall capacity coefficient when the controlling phase is continuous}}\left(\frac{V_c}{V_c'}\right)$$

$$= \frac{\dfrac{24}{7}\left(\dfrac{D}{\pi t_f}\right)^{1/2} n_o\pi d_p^2 - \dfrac{n_o\pi d_p^3}{6t_f}\ln\left[1-\dfrac{\pi D^{1/2}t^{1/2}}{d_p/2}\right] + 2\left(\dfrac{D}{\pi t_f}\right)^{1/2}\pi(r_c^2 - r_d^2)}{\dfrac{24}{7}\left(\dfrac{D}{\pi t_f'}\right)^{1/2} n_o'\pi(d_p')^2 + \dfrac{0.74n_o'\pi t' d_p' D}{t_f'}\left(\dfrac{d_p' u_s\rho}{\mu}\right)^{1/2}\left(\dfrac{\mu}{\rho D}\right)^{1/3} + 2\left(\dfrac{D}{\pi t_f'}\right)^{1/2}\pi[(r_c')^2 - (r_d')^2]} \tag{2.255}$$

Physical properties corresponding to either the dispersed or the continuous phase are substituted in the individual expressions for k_{df}, k_{dr}, k_{dc}, k_{cf}, k_{cr}, and k_{cc} as appropriate. [They will, of course, correspond to the *same* controlling-phase liquid, whether dispersed (numerator) or continuous (denominator). This is why no subscripts appear on D, ρ, and μ in eqn (2.255).]

The dependence of the middle terms in both the numerator and the denominator of eqn (2.255) upon the time of free rise of each drop, t (or t'), is such that the value of $(k_{do}a/k_{co}a)(V_c/V_c')$ decreases with increasing t. Evidently a critical rise (or fall) time t_c exists at which the $k_o aV_c$ ratio becomes unity. Thus, for $t < t_c$, the fractional extraction will be greater if the phase offering the greater resistance to transfer is dispersed. For $t > t_c$ the reverse is true.

The corresponding height of rise (or fall) is given by $H_c = u_s t_c$ or, for small ϕ, by $H_c = u_t t_c$. The selection of which phase should be dispersed, then, depends on H, the plate spacing chosen. For $H < H_c$, the fractional extraction will be greater if the phase offering the greater resistance to transfer is dispersed. For $H > H_c$ the opposite conclusion holds.

The question of interest is therefore whether H_c, the plate spacing corresponding to t_c, lies within the normal range of plate spacing for commercial systems, subject to surface-active contamination.

2.5.3 Application

Consider the transfer of acetic acid from chlorobenzene to water. This was studied without SAA by Skelland and Minhas (1971) and with SAA by Skelland and Caenepeel (1972). Internal droplet conditions without SAA for this system are known from the work of Skelland and Huang (1977a). In another paper, Skelland and Huang (1977b) give the effects of SAA on droplet fall velocities in this system.

Minhas' (1969) experiments were disperse-phase controlled, because he used chlorobenzene drops falling in water. For purposes of illustration we also conceive of a fictitious continuous-phase controlled experiment, consisting of water drops rising in chlorobenzene, with acetic acid again transferring from the chlorobenzene. When calculating both $k_{do}a$ and $k_{co}a$ we use Minhas' values for his smallest drop diameter and corresponding drop formation time and terminal velocity. These are all taken to be respectively unchanged by reversing the phase dispersed; this simplification seems to be justified by the respective closeness of ρ and μ in the two phases. The relevant physical properties are of course those of the controlling chlorobenzene phase throughout. The data appear in the caption of Fig. 2.14.

Application of the criterion $[(N_{\mathrm{We}})_{\mathrm{oscill.\,drops}} \geq 3.58]$, given by Hu and Kintner (1955), showed that these drops were non-oscillating. The internal stagnancy of drops in this system without SAA—at least for the case of chlorobenzene dispersed—was also shown visually, using aluminium particles inside the drops. The results of these calculations appear in Fig. 2.14, where corresponding primed and unprimed quantities (t and t', n_o and n_o', V_c and V_c', etc.) were taken to have the same respective values for this example (Skelland and Chadha 1981). Evidently, in the absence of SAA, $H_c = u_s t_c \doteq u_t t_c = 11.027(8.467) = 93.37$ cm (or 36.76 in). When SAA is present, $H_c \doteq 8.625(0.983) = 8.48$ cm (or 3.34 in).

For this case it follows that, in the absence of surface-active contamination, the fractional extraction in a perforated plate column will be greater if the chlorobenzene is *dispersed*, provided that the plate spacing is less than 36.76 in. In contrast, when surface active contamination similar to that in Caenepeel's work is present, the fractional extraction will be greater if the controlling chlorobenzene phase is *continuous*, provided that the plate spacing is more than 3.34 in.

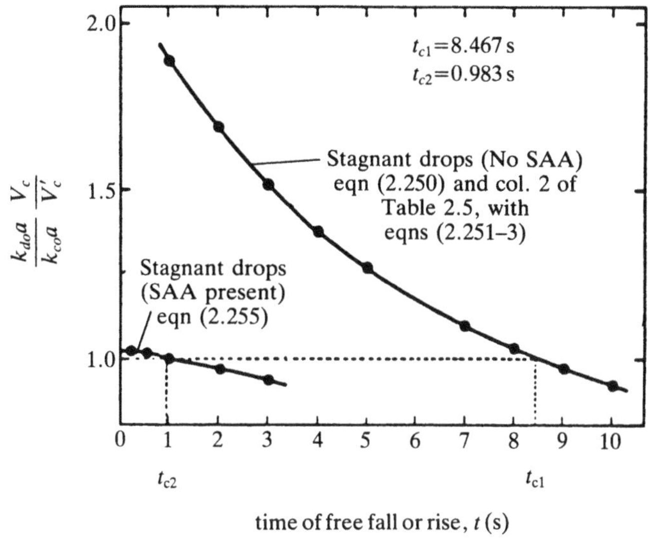

Fig. 2.14. Evaluation of t_c in the system acetic acid–chlorobenzene–water, with and without surface active agent (sodium dodecyl sulphate 1) for conditions given below at 23.5 °C, from Skelland and Minhas (1971), with $V_c = V_c'$.

$d_p = 0.563$ cm	$\rho C_6H_5Cl = 1.1005$ g/ml
$t_f = 0.505$ s	$\mu C_6H_5Cl = 0.755$ cP
$u_t^* = 11.027$ cm/s	D_{HAc} in $C_6H_5Cl = 2.072 \times 10^{-5}$ cm^2/s
$u_t^{**} = 8.625$ cm/s	$\sigma = 24.28$ dyne/cm

* No SAA added. ** A small amount of sodium dodecyl sulphate 1 present (Skelland and Huang 1977b).

These conclusions are significant because the calculated values of plate spacing relate to the range of industrial interest. (Plate spacings of 6–24 in are common, according to Treybal 1963, p. 503).

Skelland and Caenepeel (1972) found that very small amounts of SAA reduced the dispersed-phase mass transfer coefficients drastically for all three processes of drop formation, fall, and coalescence, as shown by Figs 2, 4, and 6 of their paper. The effect on the three continuous-phase mass transfer coefficients, on the other hand, was much less pronounced. This explains, at least in a general way, why $(k_{do}a/k_{co}a)(V_c/V_c')$ was reduced by small amounts of SAA at a given t, as demonstrated by Fig. 2.14.

2.6. Stage efficiencies and transfer units
[See also Stevens and Pratt, Ch. 6]

The *overall stage efficiency* E_o of an extraction cascade is the ratio of the number of ideal to actual stages needed to achieve a particular separation, thus

$$E_o = \frac{n_s, \text{ ideal}}{n_s, \text{ actual}} \tag{2.256}$$

The convenience of this definition is offset by the fact that E_o is a function of so many variables that its prediction is difficult.

The *Murphree efficiencies* (1925) express the extent to which one outgoing stream approaches equilibrium with the other outgoing stream from a given stage. Thus, for a stage in which the inlet and outlet disperse phase concentrations are ρ_{Ai} and ρ_{Ao}, and the inlet and outlet continuous phase concentrations are ρ'_{Ai} and ρ'_{Ao}, we have for the disperse phase,

$$E_{MD} = \frac{\rho_{Ai} - \rho_{Ao}}{\rho_{Ai} - \rho^*_{Ao}} \tag{2.257}$$

and for the continuous phase,

$$E_{MC} = \frac{\rho'_{Ao} - \rho'_{Ai}}{(\rho'_{Ao})^* - \rho'_{Ai}} \tag{2.258}$$

where ρ^*_{Ao} and $(\rho'_{Ao})^*$ would be in equilibrium with ρ'_{Ao} and ρ_{Ao}, respectively.

The *Hausen efficiency* E_H (1953) is defined in terms of the operating line unique to the stage in question, and is independent of the phase on which it is based. Its advantages in binary distillation have been described in detail by Ho and Prince (1970), while its disadvantages when applied to a countercurrent cascade have been well demonstrated by Pratt (1983). The latter author regards it as 'somewhat unrealistic when applied to countercurrent multistage processes', and it will therefore not be considered further here.

Formulation of the Murphree efficiency based on the disperse phase for a mixing vessel in terms of mass transfer coefficients and transfer units may be performed as follows. Suppose the disperse phase flows through the vessel at a volumetric rate of V_D, with inlet and outlet solute concentrations of ρ_{Ai} and ρ_{Ao}; the corresponding quantities for the continuous phase are V_C, ρ'_{Ai}, and ρ'_{Ao}. Solute transfers from the disperse to the continuous phase; the latter is thoroughly 'backmixed,' so that its concentration is everywhere uniform at its outlet concentration, ρ'_{Ao}. This condition has been demonstrated experimentally (Schindler and Treybal 1968; Keey and Glen 1969). For the disperse phase to similarly be of uniform concentration at its outlet value would require very rapid and extensive coalesence and redispersion of the drops in the vessel. This is not supported by the limited available evidence; for a three-component system in an agitated vessel, Keey and Glen (1969) showed that the assumption of a completely mixed continuous phase and an unmixed disperse phase gave the best correlation of their measurements. On this basis, the disperse phase concentration ρ_A^* in equilibrium with ρ'_{Ao} is everywhere given by $m\rho'_{Ao}$.

A unit volume of disperse phase is now defined as equal to the disperse-phase hold-up at any instant—i.e. $V_s\phi$, where V_s is the volume of the vessel. Although different portions of this unit volume enter and leave the vessel at different times, each portion will undergo the same average experience in the vessel. Thus each portion will enter with solute concentration ρ_{Ai}, exhibit concentration ρ_A at time t after entering, remain for an average residence time t_r, and leave with concentration ρ_{Ao}. A balance on solute transferred from any such unit volume of disperse phase between t and $t + \mathrm{d}t$ is then

$$-V_s\phi \mathrm{d}\rho_A = K_d a(\rho_A - \rho_A^*) V_s \mathrm{d}t \tag{2.259}$$

Equations (2.77 and 2.78) show that the relevant mass transfer coefficients vary with solute concentration for this mechanism. However, if attention is confined to low concentrations then $(1 - y_A)_{\mathrm{lm}} \to (1 - x_A)_{\mathrm{lm}} \to 1$; the coefficient may therefore be regarded as constant, as may the phase flow rates and hence ϕ. Rearranging,

$$(\mathrm{NTU})_{od} = -\int_{\rho_{Ai}}^{\rho_{Ao}} \frac{\mathrm{d}\rho_A}{\rho_A - \rho_A^*} = \frac{K_d a}{\phi}\int_0^{t_r} \mathrm{d}t \tag{2.260}$$

where t_r is the average residence time of the disperse phase in the vessel,

$$t_r = \frac{\phi V_s}{V_D} \tag{2.261}$$

Integrating,

$$-\ln\left(\frac{\rho_{Ai} - \rho_A^*}{\rho_{Ai} - \rho_A^*} - \frac{\rho_{Ai} - \rho_{Ao}}{\rho_{Ai} - \rho_A^*}\right) = \frac{K_d a t_r}{\phi}$$

and with eqns (2.257) and 2.261),

$$\ln(1 - E_{MD}) = \frac{-K_d a V_s}{V_D} \tag{2.262}$$

$$E_{MD} = 1 - \exp(-K_d a V_s / V_D) \tag{2.263}$$

The integral in eqn (2.260) is called the number of overall disperse-phase transfer units. More comprehensive expressions allowing for variation in coefficients and flow rates as transfer proceeds are available (Skelland 1985, pp. 346–349). However, as the left-side integral shows, the $(\mathrm{NTU})_{od}$ is given essentially by dividing the average overall driving force in terms of the disperse phase into the concentration change achieved in that phase. As a result, equation (2.263) may alternatively be written as

$$E_{MD} = 1 - \exp[-(\text{NTU})_{od}] \tag{2.264}$$

By analogy with large-diameter plate distillation columns, an effect may occur in relatively large perforated-plate extraction columns arising from a horizontal concentration gradient due to solute gain in the continuous phase as it flows across the plate. Thus, disperse phase with a given fractional approach to equilibrium after contact with continuous phase leaving the plate, would be richer in solute than disperse phase with the same fractional equilibrium after contact in the middle of the plate. Such a case would require evaluation of *local or point Murphree efficiencies*, thus

$$E_{MDP} = \frac{\rho_{Ai} - \rho_{AP}}{\rho_{Ai} - \rho^*_{AP}} \tag{2.265}$$

$$E_{MCP} = \frac{\rho'_{AP} - \rho'_{Ai}}{(\rho'_{AP})^* - \rho'_{Ai}} \tag{2.266}$$

where ρ_{AP} and ρ'_{AP} denote local concentrations of the disperse and continuous phases at point P along the plate and ρ^*_{AP} and $(\rho'_{AP})^*$ are corresponding equilibrium values at the same location. Equations (2.263 and 2.264) can then be written for E_{MDP}. Analogous expressions can of course be developed for E_{MC} and E_{MCP}. The relation between point and plate efficiencies depends on the degree of lateral mixing in each separate phase along the plate. Only two models for this are available in liquid–liquid systems, in contrast to distillation.

In the first model, the continuous phase is assumed to move along the plate in plug flow, without any mixing, while the coalesced layer of disperse phase beneath the next plate (for rising drops) is of uniform composition. (The spreading mechanism proposed by Johnson and Hamielec (1960) to develop eqn (2.240) could be invoked to generate these conditions). Lewis (1936) then gives

$$E_{MD} = \frac{\exp(mV_D E_{MDP}/V_C) - 1}{mV_D/V_C} \tag{2.267}$$

where E_{MDP} is assumed to be constant across the plate.

In the second model the continuous phase is fully mixed by the rising drops and has uniform composition at all points on the plate. In this case, local and average corresponding concentrations become the same, and so

$$E_{MD} = E_{MDP} \tag{2.268}$$

Treybal (1980, p. 538) considers that the second model is more likely to prevail in perforated plate extractors.

When the slopes of the equilibrium and operating lines are constant and

E_{MD} is constant for all stages, the following expression was derived by Lewis (1936) for the overall stage efficiency of the entire cascade, E_o:

$$E_o = \frac{\ln\{1 + E_{MD}[(mV_D/V_C) - 1]\}}{\ln(mV_D/V_C)} \tag{2.269}$$

A proliferation of algebraic relationships has been generated between E_{MDP}, E_{MCP}, E_{MD}, E_{MC}, E_H, E_o, and NTU for batch and continuous extraction in individual stages and in crossflow and countercurrent extractors with and without entrainment. Relations are available between transfer units, exit compositions, and stage efficiencies when the two phases are in plug flow, when either phase is completely mixed, or when both phases are completely mixed. The reader should consult Lewis (1936), Treybal (1963, pp. 398–403; pp. 450–452), Ho and Prince (1970), and Pratt (1983, pp. 182–187) for details.

2.7 Mass transfer with chemical reaction
[See also Pratt and Stevens, Ch. 6]

The rate of interphase mass transfer may be dramatically enhanced by chemical reaction of the solute with some component in the phase into which the solute transfers. It is, however, necessary to distinguish between heterogeneous and homogeneous reactions when analysing such processes. In the former case the relevant transport equation does not contain a term describing the rate of chemical reaction, which is instead treated as a boundary condition, consistent with the occurrence of the reaction at the phase boundary. On the other hand, when the reaction occurs within the phase containing the solute, the process is called a homogeneous reaction and the transport equation describing the situation will contain a reaction-rate term.

Entire books have been written on the subject of mass transfer with chemical reaction (Astarita 1967; Danckwerts 1970; Astarita *et al.* 1983), so that only the briefest introduction can be presented in this small portion of a single chapter. The reader is referred for more information to the three books noted above and to the treatments by Sherwood *et al.* (1975, Ch. 8), Sharma (1983, pp. 37–52), and Cussler (1984, Chs 13 and 14).

From a mathematical point of view the irreversible first-order reaction is the easiest to handle. Although most reactions of interest are not truly first order, it often happens that all reacting non-solute components, B, C, . . ., are present in concentrations substantially exceeding those stoichiometrically required for reaction with the solute, A. When the reaction is first-order with respect to the solute we then have what is called a pseudo-first order reaction, the rate of which is written as

$$R_A = -\frac{dc_A}{dt} = k'c_A c_B^m c_C^n \ldots = k_1 c_A \tag{2.270}$$

where reactant concentrations c_B, c_C, ... are essentially constant and k_1 is the pseudo-first order rate constant, with dimensions of $(\text{time})^{-1}$.

Quantitative considerations will begin with the case of quasi-steady-state molecular diffusion with irreversible first-order reaction in a semi-infinite stagnant liquid. Next, to illustrate use of a different boundary condition, this case will be modified by the insertion of an impermeable boundary oriented normal to transfer and at a finite distance from the solute source. Finally, transfer rates with and without irreversible first order reaction will be compared in an agitated liquid. This case has been modelled in terms of the film, penetration, and Danckwerts-surface-renewal theories. All three approaches generally give about the same results (Danckwerts, 1970, p. 109; Sherwood *et al.*, 1975, pp. 321–2; Cussler, 1984, p. 350). Accordingly, attention will be confined to the simplest approach, which is based on the film theory. The latter is of course also founded on considerations of transfer in a stagnant fluid, and so provides some continuity with the first two topics.

Suppose that, in the development of Fick's second law, eqns (2.36 and 2.37), component A is destroyed by chemical reaction in the volume element $dx\, dy\, dz$ of Fig. 1 at the local rate of R_A moles of A per unit volume in unit time. It follows that eqn (2.36) then becomes

$$R_A + \frac{\partial c_A}{\partial t} = D\left(\frac{\partial^2 c_A}{\partial x^2} + \frac{\partial^2 c_A}{\partial y^2} + \frac{\partial^2 c_A}{\partial z^2}\right) \tag{2.271}$$

and for one-dimensional diffusion, eqn (2.37) may be written with chemical reaction as

$$R_A + \frac{\partial c_A}{\partial t} = D\frac{\partial^2 c_A}{\partial z^2} \tag{2.272}$$

then for an irreversible first-order reaction,

$$k_1 c_A + \frac{\partial c_A}{\partial t} = D\frac{\partial^2 c_A}{\partial z^2} \tag{2.273}$$

Danckwerts (1950) provides the solution to eqn (2.273) for the rate of transfer of A and its concentration distribution for diffusion into a semi-infinite medium. Consider, however, the simplification introduced by steady-state transfer ($\partial c_A/\partial t = 0$) of solute A from one liquid to another, across the horizontal interface between the two immiscible, stagnant liquids. The solute is sparingly soluble and reacts with a large excess of another component in the lower liquid in a manner that is irreversible and essentially first order with respect to A. The concentration of A in the lower liquid at the interface is constant at c_A^*, and the lower liquid extends effectively to infinity in the z direction, where z is zero at the interface and normal to it. An expression is required for the concentration distribution of A in the lower liquid and for the molal flux of A at the interface.

For these conditions eqn (2.273) becomes

$$D\frac{d^2c_A}{dz^2} = k_1 c_A \qquad (2.274)$$

with boundary conditions

$$c_A = c_A^* \text{ at } z = 0 \qquad (2.275)$$

$$c_A = 0 \text{ at } z = \infty \qquad (2.276)$$

The solution to eqn (2.274) is shown by Li (1960, p. 277) to be

$$c_A = C_1 \exp[(k_1/D)^{1/2}z] + C_2 \exp[-(k_1/D)^{1/2}z] \qquad (2.277)$$

or

$$c_A = A \cosh[(k_1/D)^{1/2}z] + B \sinh[(k_1/D)^{1/2}z] \qquad (2.278)$$

The boundary condition (2.275) shows that $A = c_A^*$. Dividing eqn (2.278) by $\cosh[(k_1/D)^{1/2}z]$ and setting $c_A = 0$ at $z = \infty$, in accordance with boundary condition (2.276), shows that $B = -c_A^*$, so that

$$c_A = c_A^* \cosh[(k_1/D)^{1/2}z] - c_A^* \sinh[(k_1/D)^{1/2}z] \qquad (2.279)$$

which is the required concentration distribution.

For low transfer rates and hence negligible bulk flow, eqn (2.27) shows that the molal flux at the interface is essentially given by

$$(N_{Az})_{z=0} = -D(dc_A/dz)_{z=0} \qquad (2.280)$$

The concentration gradient at $z = 0$ is evaluated from eqn (2.279) to obtain

$$(N_{Az})_{z=0} = c_A^*(k_1 D)^{1/2} \qquad (2.281)$$

which is the flux of A into the lower liquid at the interface.

Next, suppose that the situation just considered is modified by the introduction of a plane impermeable boundary in the lower liquid at $z = H$. This problem has been discussed by Bird *et al.* (1960), Geankoplis (1972), and Hershey (1973).

Equation (2.274) and boundary condition (2.275) still hold, but the new boundary condition below replaces eqn (2.276):

$$(N_{Az})_{z=H} = -D(dc_A/dz)_{z=H} = 0 \qquad (2.282)$$

This expresses the fact that no solute transfers across the impermeable boundary at $z = H$, and so the solute concentration gradient is zero at that location.

The first boundary condition (2.275) shows that A is again c_A^* in eqn (2.278). Differentiating the latter equation with respect to z, equating to zero at $z = H$, and solving for B gives

$$B = -\frac{c_A^* \sinh\left[(k_1/D)^{1/2}H\right]}{\cosh\left[(k_1/D)^{1/2}H\right]} \tag{2.283}$$

Substituting in eqn (2.278) and invoking the general relation for $-\sinh u(\sinh v)$ leads to

$$c_A = \frac{c_A^* \cosh\left[(k_1/D)^{1/2}H(1 - z/H)\right]}{\cosh\left[(k_1/D)^{1/2}H\right]} \tag{2.284}$$

This defines the new concentration distribution, from c_A^* at $z = 0$ to $c_A^*/\cosh\left[(k_1/D)^{1/2}H\right]$ at $z = H$.

The flux of A at the interface is again given by eqn (2.280), where the concentration gradient at $z = 0$ is found from eqn (2.284), giving

$$(N_{Az})_{z=0} = c_A^*(k_1 D)^{1/2} \tanh\left[(k_1/D)^{1/2}H\right] \tag{2.285}$$

Comparison between eqns (2.281 and 2.285) shows that the presence of the impermeable boundary at $z = H$ reduces the flux of A at the interface by the factor $\tanh[(k_1/D)^{1/2}H]$, compared to the interfacial flux into a stagnant semi-infinite liquid.

Attention will now turn to the effect of an irreversible first order chemical reaction on the rate of solute transfer into an agitated liquid, as indicated by the film theory. The mechanics of such postulated films on each side of the interface between two immiscible agitated fluids were described earlier under the heading of 'The Two-Film Theory'. First consider transfer in the absence of reaction, for which eqn (2.274) becomes

$$\frac{d^2 c_A}{dz^2} = 0 \tag{2.286}$$

with boundary conditions

$$c_A = c_A^* \text{ at } z = 0 \tag{2.287}$$

$$c_A = c_{A\delta} \text{ at } z = \delta \tag{2.288}$$

where δ is the film thickness in the phase under consideration. Integrating,

$$-\frac{dc_A}{dz} = \text{constant} = \frac{c_A^* - c_{A\delta}}{\delta} = \frac{c_A - c_{A\delta}}{\delta - z} \tag{2.289}$$

from which

$$\frac{c_A}{c_A^*} = 1 - \frac{z}{\delta} + \left(\frac{c_{A\delta}}{c_A^*}\right)\frac{z}{\delta} \tag{2.290}$$

The solute flux at the interface (and throughout the film) is

$$(N_{Az})_{z=0} = (N_{Az})_{z=z} = -D\frac{dc_A}{dz} = \frac{D}{\delta}(c_A^* - c_{A\delta}) = k_0(c_A^* - c_{A\delta}) \tag{2.291}$$

where k_0 is the mass transfer coefficient without reaction. When an irreversible first order reaction occurs, the solution to eqn (2.274) is given by the alternative eqns (2.277 or 2.278). The first boundary condition, eqn (2.287), shows that $A = c_A^*$, and the second boundary condition, eqn (2.288), gives

$$B = \frac{c_{A\delta} - c_A^* \cosh[(k_1/D)^{1/2}\delta]}{\sinh[(k_1/D)^{1/2}\delta]} \tag{2.292}$$

Substitution in eqn (2.278), rearrangement, and use of the general relation for sinh $(u - v)$, leads to

$$\frac{c_A}{c_A^*} = \frac{\sinh[(k_1/D)^{1/2}(\delta - z)] + (c_{A\delta}/c_A^*)\sinh[(k_1/D)^{1/2}z]}{\sinh[(k_1/D)^{1/2}\delta]} \tag{2.293}$$

The solute flux at the interface $(z = 0)$ is given by eqn (2.280), where the concentration gradient at $z = 0$ is obtained from eqn (2.293), giving

$$(N_{Az})_{z=0} = \frac{Dc_A^*}{\delta}\left[\frac{(k_1/D)^{1/2}\delta\cosh[(k_1/D)^{1/2}\delta] - (c_{A\delta}/c_A^*)(k_1/D)^{1/2}\delta}{\sinh[(k_1/D)^{1/2}\delta]}\right] \tag{2.294}$$

If we define a mass transfer coefficient k_r with reaction in terms of the same concentration driving force as that in eqn (2.291) without reaction, then

$$\frac{(N_{Az})_{z=0}\,(\text{with reaction})}{(N_{Az})_{z=0}\,(\text{without reaction})} = \frac{k_r}{k_0} = \frac{\text{eqn (2.294)}}{\text{eqn (2.291)}}$$

And for the common case in which $c_{A\delta} = 0$, with δ replaced by D/k_0 from eqn (2.291), we find

$$\frac{k_r}{k_0} = (k_1 D/k_0^2)^{1/2}\coth(k_1 D/k_0^2)^{1/2} \tag{2.295}$$

This is the *enhancement factor* by which the mass transfer coefficient k_0 is increased by an irreversible first-order reaction, according to film theory. The mass transfer coefficient without reaction, k_0, is estimated by the appropriate relationship from those developed earlier, corresponding to the situation at hand. As noted between eqns (2.270 and 2.271), corrections for the effect of such chemical reaction on the mass transfer rate based on penetration—and Danckwerts—surface—renewal models are usually closely similar to those obtained from film theory.

Nomenclature

Any consistent system of force (F)—mass (M)—length (L)—time (T) units in which g_c (ML/T^2F) is unity may be used. Exceptions are as indicated; for example in the prediction of molecular diffusivities.

A_j, A_f, A_r, A_c, A_j', A_f', A_r', A_c'	total interfacial area between two consecutive plates for the respective stages of jetting, drop formation, free rise (or fall), and coalescence—primes show controlling phase is continuous.
A_n	coefficient in the nth term of an infinite series
A_p	surface of a sphere with the same volume as a droplet
A_z	area normal to transfer at z, *or* a jet cross-section at distance z from the nozzle
a	interfacial area per unit volume, *or* surface area of a growing drop at time t
a_i	the ith root of the equation $J_0\ (a_i) = 0$
B_i, B_n	coefficients in the ith and nth terms of an infinite series
b_i	a_i/r_t
c, c_A, c_{A1}, c_{A2} c_B, c_{B1}, c_{B2}, c_i	total moles per unit volume; of component A; at $z = z_1$; at $z = z_2$; of B; at z_1 and z_2; of component i, mol/(length)^3
$c_A^*, c_{As}, c_{A\delta}, c_{A\infty}$	molar concentration of A at the interface; at the surface of a sphere; at $z = \delta$; at infinity
c_{blm}	defined by eqn (2.31)
c_B'	number of g-moles of water per cm^3 of solution
$(\Delta c_{Aj})_{av}$	average concentration difference for component A between the bulk of a jet and its surface, mol/(length)^3
D, D	(volumetric) molecular diffusivity; in the disperse phase, (length)^2/time
D_{AB}, D_{BA}, D_{AB}^0 D_{BA}^0, D_{AC}^0	molecular diffusivity of A in B; of B in A; superscript zero denotes dilute solution, (length)^2/time
D_A^0	as in eqn (2.53)
D_c	molecular diffusivity in the continuous phase, (length)^2/time
D_{Am}	molal average diffusivity formed from the three binary diffusivities in dilute solution, eqn (2.55). (length)^2/time
D_{Am}^0	diffusivity of dilute A in a mixture of B and C, (length)^2/time
D_G, D_L	diffusivity in phase G; in phase L (length)^2/time

D_{im}	effective diffusivity of component i in a mixture, eqn (2.59), $(\text{length})^2/\text{time}$
D_1, D_2	diffusivity in phases 1 and 2, $(\text{length})^2/\text{time}$
d_i, d_j, d_{jc}, d_{jm}, d_n, d_p, d_{pc}, d_s, d'_{jc}, d'_p	diameter of an impeller; of a jet at z; of a jet tip; of a jet tip at maximum jet length; of a nozzle or orifice; of a drop; of the critical-size drop at which oscillation begins; of a sphere. Primes show controlling phase is continuous
d_t	tube diameter *or* diameter of a growing drop at time t
d_{32}, d^*_{32}	Sauter mean drop diameter; at drop-size equilibrium
E	droplet eccentricity, major axis/minor axis
E_f	fractional extraction
E_{MC}, E_{MCP}, E_{MD}, E_{MDP}	Murphree efficiency for the continuous phase; at a point; for the disperse phase; at a point
E_H	Hausen stage efficiency
E_o	overall stage efficiency, eqn (2.256)
F	Harkins and Brown correction factor
F_x	external force in the x direction
f_o	droplet oscillation frequency
g	acceleration due to gravity
g_c	conversion factor, (mass × acceleration)/force
H	dimensionless group defined by eqn (2.150) *or* plate spacing *or* location of a plane impermeable boundary at $z = H$
H_c	critical plate spacing at which $(k_{do}a/k_{co}a)(V_c/V'_c) = 1$
h	height of liquid in a vessel
$\Delta H_{mA}, \Delta H_{mB}$	latent heats of vaporization of A and B at their normal boiling temperatures, cal/g-mol
I_{iz}	molal flux of component i relative to the mass average velocity, both in the z direction, $\text{mol}/(\text{length})^2$ (time)
i_{Az}, i_{Bz}, i_{iz}	mass flux of component A relative to the mass average velocity, both in the z direction; same for component B; and for component i, $\text{mass}/(\text{length})^2$ time
J	dimensionless group defined by eqn (2.151)
J_{Az}, J_{Bz}, J_{iz}	molal flux of component A relative to the molal average velocity, both in the z direction; same for B; and for i, mol/$(\text{length})^2$ time
j_{Az}, j_{iz}	mass fluxes of components A and i relative to the molal average velocity, all in the z direction, $\text{mass}/(\text{length})^2$ time
J_0, J_1	Bessel functions of first kind of order zero and one
K_d	overall mass transfer coefficient based on the disperse phase, eqn (2.259), length/time
K_{dr}	overall mass transfer coefficient based on the disperse phase during free rise (or fall) of a drop, length/time
K_X, K_Y	overall mass transfer coefficients based on ΔX and ΔY, mol/(time) $(\text{length})^2$ (unit of ΔX or ΔY)
K^*	* denotes low rates of mass transfer
k_B	Boltzmann's constant
k_c, k_x, k_y, k_ρ	individual mass transfer coefficients defined in eqs (2.73 and 2.74)

$k_{cj}, k_{cf}, k_{cr}, k_{cc}$	individual continuous—phase mass transfer coefficients based on $\Delta\rho_A$ during jetting, droplet formation, free rise or fall, and coalescence, length/time
$k_{co}a, k_{do}a$	overall capacity coefficient (continuous phase controlling) including transfer during drop formation, rise, and coalescence; as for $k_{co}a$, but disperse phase controlling, 1/time
$k_{dj}, k_{df}, k_{dr}, k_{dc}$	individual disperse-phase mass transfer coefficients based on $\Delta\rho_A$ during jetting, droplet formation, free rise or fall, and coalescence, length/time
$\overline{k^*_{dr}A}$	the time—average product of the oscillating drop surface and the disperse—phase mass transfer coefficient, $(\text{length})^3$/time
k_0, k_r	mass transfer coefficients without and with chemical reaction, eqns (2.291 and 2.295), length/time
k_v	ratio of actual interfacial speed to the potential flow interfacial speed
k_X, k_Y	individual mass transfer coefficients based on ΔX and ΔY, mol/(time)$(\text{length})^2$ (unit of ΔX or ΔY)
k_1	first order or pseudo-first order chemical reaction rate constant, 1/time
$(k_\rho)_{\text{lm}}$	defined by eqn (2.133)
k^*	* denotes low rates of mass transfer
L	vertical distance between consecutive drops
L_c	characteristic length
L_j, L_j'	jet length; when controlling phase is continuous
l_s	system length parameter, $\pi(\sigma/\Delta\rho g)^{1/2}$
l^0_+, l^0_-	cationic and anionic conductances at infinite dilution, mho/equivalent
M	a modified Morton number, $g\mu_c^4\Delta\rho/\rho_c^2\sigma^3$
M_A, M_B, M_i	molecular weights of components A, B, and i
m	molality
m, m', m''	defined in Fig. 2 and above eqn (2.63); also m is the distribution ratio of disperse-to-continuous-phase solute concentrations at equilibrium
N, N_{min}	impeller speed; minimum N for complete dispersion of one immiscible liquid in another, without regard to uniformity of the mixture, revs/time
N_{Az}, N_{Bz}, N_{iz}	molal flux of components A, B, and i relative to stationary coordinates, all in the z direction, mol/$(\text{length})^2$ time
N_{Ar}, N_{Br}	molal flux of A and B in the r direction relative to stationary coordinates, mol/$(\text{length})^2$ time
N_{Eo}	the Eotvos number, $g\Delta\rho d_p^2/\sigma$
$N_{\text{Fr},c}$	Impeller Froude number for the continuous phase, $d_i^2N^2\rho_c/\Delta\rho hg$
$N_{\text{Pe},c}, N'_{\text{Pe}}$	Peclet number for the continuous phase, $d_\text{p}u_\text{t}/D_\text{c}$; $(d_\text{p}u_\text{t}/D)(1+\mu_\text{d}/\mu_\text{c})$.

N_{PG}	$\rho_c^2\sigma^3/g\mu_c^4\Delta\rho = 1/M$
$N_{\text{Re}}, N_{\text{Re},c}, N'_{\text{Re}}, N'_{\text{Re},c}$	forms of Reynolds number, i.e. $2r_t V\rho/\mu$; $d_p u_t \rho_c/\mu_c$; $d_p^2 f_o \rho_d/\mu_d$; $d_p\left(\overline{u^2}\right)^{1/2}\rho_c/\mu_c$
$N_{\text{Sc}}, N_{\text{Sc},c}, N_{\text{Sc},d}, N_{\text{Sc1 or 2}}$	forms of Schmidt number, i.e. $\mu/\rho D$; $\mu_c/\rho_c D_c$; $\mu_d/\rho_d D$; $(\mu/\rho D)_{\text{phase 1 or 2}}$
$N_{\text{Sh}}, N_{\text{Sh},c}, (N_{\text{Sh1 or 2}})_x^*$	forms of Sherwood number, i.e. (k or k^*) L_c/D; $k_\rho d_p/D_c$; local $k_{\rho 1}x/D_1$ or local $k_{\rho 2}x/D_2$
$(\text{NTU})_{od}$	number of overall disperse phase transfer units
$N_{\text{We}}, N_{\text{We},c}, \Delta N_{\text{We}}$	forms of Weber number, i.e. $d_p u_t^2 \rho_c/\sigma$; $d_i^3 N^2 \rho_c/\sigma$; $(u_0^2 - u_{0j}^2)\, d_n \rho_d/\sigma$
n_{Az}, n_{Bz}, n_{iz}	mass flux of components A,B, and i in the z direction relative to stationary coordinates, mass/(length)^2 time
n_{Ac}	instantaneous mass flux of component A into a coalesced layer at time t since the arrival of the last drop, mass/(length)^2 time.
$n_{A0}, (n_{A0})_{\text{av}}$	instantaneous mass flux of component A at the interface; average mass flux of A at the interface during exposure time t_e, mass/(length)^2 time
n_o, n'_o, n_s	number of orifices, perforations, or nozzles on a plate; as for n_o when continuous-phase controlled; number of stages in an extraction cascade.
P	pressure, force/(length)^2 *or* power input to a fluid by an agitator, length (force)/time
Q	volumetric flow rate through a nozzle or orifice, (length)^3/time
Q_T	total solute mass transferred between phases in a batch operation, mass
q	rate of mass transfer, mass/time
q_{Ac}	instantaneous mass transfer rate for component A into a coalesced layer at time t since the arrival of the last drop, mass/time
q_{Ar}	total radial mass transfer rate from a sphere at any $r \geq d_s/2$, *or* the mass transfer rate of A during free rise of drops, mass/time
q_{Az}	rate of mass transfer of A in the z direction, (mass or moles)/time
$R(r) = R$	some function of radial coordinate r only
R_A	rate of a first-order or pseudo-first order chemical reaction, (moles of A)/(length)^3 time
$r, r_A, r_c, r_d, r'_c, r'_d, r_s, r_t$	radial coordinate; radius of a molecule of A; of a column; of a downcomer; as for r_c and r_d when continuous-phase controlled; of a sphere; of a tube.
r_+	r/r_t
SAA	surface active agent
S_{os}	surface area of an oblate spheroid

S_s	surface area of a sphere of volume equal to the oblate spheroid with S_{os}
s	fractional rate of surface renewal, 1/time
T	absolute temperature, °K, *or* some function of time t only, *or* vessel diameter
$t, t', t_c, t_e, t_f, t_f', t_0, t_r$	time, and H/u_t; H/u_t'; critical time of rise or fall of drops at which $(k_{do}a/k_{co}a)(V_c/V_c')$ becomes unity; time of exposure; time of formation of a single drop; as for t_f when continuous-phase controlled; time at which disperse phase enters an agitated batch of continuous phase; average residence time of disperse phase in an agitated vessel
t_1, t_2	terminals of a time interval for analysis of two-phase mass transfer in a batch agitated vessel.
U,u	molal and mass average velocities in the z direction, length/time; *or* u = linear flow velocity at a point in a stream
u_A, u_B, u_i	statistical mean velocities of components A,B, and i in the z direction relative to stationary coordinates, length/time
u_c, u_d	superficial velocities of continuous and disperse phases, based on the empty column cross section, length/time
$u_o, u_{oj}, u_{om}, u_s, u_t$ u_t'	interface velocity *or* mean linear velocity through an orifice or nozzle; orifice velocity when jetting begins; orifice velocity at maximum jet length; slip velocity; droplet terminal velocity; as for u_t when continuous-phase controlled
u_θ, $(u_\theta)_{r=r_s}$	velocity in the continuous phase boundary layer in the θ-direction around a spherical surface; velocity of the drop interface in the θ-direction, length/time
$u_{\infty 1}, u_{\infty 2}$	velocities outside the momentum boundary layers in streams 1 and 2 (Fig. 2.4)
u'	velocity in the x-direction at $y = \delta_c$ (Fig. 2.5)
$(\overline{u^2})^{1/2}$	turbulence fluctuating velocity in the continuous phase, scaled over d_p
V	average linear velocity
V_{bA}, V_{bB}	molal volumes of pure liquid components A and B at their normal boiling temperatures, cm^3/g-mol
$\bar{V}_B$	partial molal volume of water in solution, cm^3/g-mol
V_C, V_D	volumetric flow rates of continuous and disperse phases through an agitated vessel, $(length)^3$/time
V_c, V_c'	column active volume between consecutive plates a and b, excluding downcomer; when continuous-phase controlled
V_c, V_d, V_T	volumes of continuous and disperse phases and their sum in a batch agitated vessel
$V_j, \bar{V}_{ji}, \underline{V}_j, \bar{V}_j, \underline{\underline{V}}_j$	plug-flow linear velocity in a non-tapering jet, *or* mean linear velocity at z along a tapering jet; mean interfacial velocity along a tapering jet; estimates of the mean plug-flow velocity along a tapering jet, given by—and below—eqn (2.141)
V_0	contacting volume of the equipment
V_s	vessel volume

v_0	linear velocity of the convective bulk flow in the y direction (Fig. 2.5)
v_p	volume of a single drop
w	width normal to flow
w_A, w_B, w_i	mass fractions of components A,B, and i
X_A	composition of heavy-liquid phase, any units
X_A^*	local equilibrium concentration in the heavy liquid at the interface, any units
X_{AG}	heavy-liquid concentration that would be in equilibrium with the existing light-liquid concentration, any units
x	coordinate direction
x_A, x_B, x_C, x_i	mole fractions of components A,B,C, and i
x_{A1}, x_{A2}	mole fractions of A at z_1 and z_2
$x_{b\mathrm{lm}}$	defined below eqn (2.35)
Y_A	composition of light-liquid phase, any units
Y_A^*	local equilibrium concentration in the light liquid at the interface, any units
Y_{AL}	light-liquid concentration that would be in equilibrium with existing heavy-liquid concentration, any units
y	coordinate direction
$y_{b\mathrm{lm}}$	$(y_{B2} - y_{B1})/\ln(y_{B2}/y_{B1})$, 1 at z_1, 2 at z_2
y'	$\rho_A - \rho_{AB}^*$
z, z_1, z_2	coordinate direction; locations along z
z_{fG}, z_{fL}	fictitious film thicknesses in the G and L phases
z_L	laminar surface film thickness
z_p	depth of penetration, eqn (2.87)
z_+, z_-	absolute values of the cation and anion valencies
z_+	dimensionless distance in the direction of flow, $(z/r_t)/N_{\mathrm{Re}}N_{\mathrm{Sc}}$
α	defined below eqn (2.183)
β_i	eigenvalues
β_n	defined below eqn (2.183)
γ	N_{Bz}/N_{Az}
γ_A	activity coefficient of A
$\gamma\pm$	mean ionic activity coefficient based on molality
$\delta, \delta_c, \delta_1, \delta_2, \delta_{c1}, \delta_{c2}$	momentum boundary layer thickness *or* film thickness in transfer with chemical reaction; concentration boundary layer thickness; δ in streams (phases) 1 and 2; δ_c in streams 1 and 2
ϵ	$1 + \epsilon$ is the ratio of maximum to minimum surface area over one cycle for an oscillating drop
θ,θ	a spherical polar coordinate [between eqns (2.37 and 2.38)]; angle with the direction of u_t, measured at the center of the sphere and from the front stagnation point (eqn 2.224)
λ, λ_n	constant; eigenvalue for the nth term in a series
$\mu, \mu_A, \mu_{AB}, \mu_{ABC}, \mu_B$	viscosity; of A; of the mixture A and B; of the mixture of A, B, and C; of B, cP
μ_c, μ_d, μ_m	viscosities of the continuous and disperse phases; of the mixture, defined by eqn (2.209)

μ_w, μ_{wt}	viscosity of water taken as 0.9 cP; viscosity of water at *T*K,cP
μ_1, μ_2	viscosity of immiscible streams 1 and 2
ξ	constant in eqn (2.89)
ξ/ξ_0	ratio of the concentration boundary layer thickness with interfacial flow to that without interfacial flow
ρ, ρ_c, ρ_d, ρ_1, ρ_2, ρ_m	density; of continuous and disperse phases; of immiscible streams 1 and 2; defined by eq (2.208), mass/(length)^3
ρ_A, ρ_B, ρ_i, ρ_{A1}, ρ_{A2}	mass of component *A* per unit volume of solution, also specifically in the disperse phase; similar mass concentrations of components *B* and *i*; for component *A* in streams 1 and 2, mass/(length)^3
ρ_{AB0}, ρ^*_{AB}	bulk average or 'mixing cup' value of ρ_A at the tip of a jet; disperse phase concentration in equilibrium with the bulk concentration in the continuous phase, mass/(length)^3
ρ_{Aci}, ρ_{Acf}, ρ^*_{Ad}	initial and final homogeneous concentrations of *A* in the continuous phase; dispersed phase concentration of *A* at the jet surface, mass/(length)^3
ρ_{Ai}, ρ'_{Ai}	interfacial concentration in the disperse phase, *or* inlet concentration of disperse phase in a stage; interfacial concentration in the continuous phase, *or* inlet concentration of continuous phase in a stage, mass/(length)^3
ρ_{Ao}, ρ'_{Ao}, $(\rho'_{Ao})^*$, ρ^*_{Ao}	initial solute concentration in a jet (at $z = 0$), *or* uniform concentration in a sphere at the start of diffusion, *or* outlet concentration of disperse phase in a stage; outlet concentration of continuous phase in a stage; in equilibrium with ρ_{Ao} in a stage; in equilibrium with ρ'_{Ao} in a stage, mass/(length)^3
ρ_{AP}, ρ'_{AP}, $(\rho'_{AP})^*$, ρ^*_{AP}	*local* solute concentrations of the disperse and continuous phases at point *P* along the plate; equilibrium value in the continuous phase at *P*; equilibrium value in the disperse phase at *P*, mass/(length)^3
$(\rho_{A1})_0$, $(\rho_{A1})_\infty$, $(\rho_{A2})_0$, $(\rho_{A2})_\infty$, $(\rho_{B1})_0$, $(\rho_{B1})_\infty$	mass concentrations of *A* in phase 1 at the interface and outside the boundary layer; same for *A* in phase 2; same for *B* in phase 1, mass/(length)^3
$\rho_{A\infty}$, $\rho'_{A\infty}$	mass concentration of *A* effectively at infinity; continuous-phase concentration of *A* at points remote from a sphere, mass/(length)^3
ρ^*_A, ρ'_A, $\rho^{*\prime}_A$, $\bar{\rho}_A$	phase concentration at the interface and in equilibrium with the adjacent phase *or* disperse phase concentration in equilibrium with ρ'_{Ao}; solute concentration in the continuous phase *or* $\rho_A - \rho_{A\infty}$; $\rho^*_A - \rho_{A\infty}$; average solute concentration in a drop at time *t*, mass/(length)^3
$\Delta\rho$, $\Delta\rho_{Ac}$, $\Delta\rho_{Af}$, $\Delta\rho'_A$	$\lvert\rho_d - \rho_c\rvert$; concentration difference in the coalesced layer of disperse phase, |bulk-interfacial concentration|; concentration difference in the disperse phase during drop formation, |bulk-interfacial concentration|; concentration difference in the continuous phase, $\rho'_{Ai} - \rho'_{A\infty}$, mass/(length)^3
σ	interfacial tension, mass/(time)^2

τ_0	shear stress at the interface, mass/(length)(time)^2
$\phi, \phi(t)$	a spherical polar coordinate [between eqns (2.37 and 2.38] *or* volume fraction occupied by the disperse phase in a two-phase mixture; a surface age distribution function
$\phi'_{0s\pi}$	approximate velocity potential at stagnation point, (length)^3
$\psi, \psi_i(r_+)$	t/t_f; eigenfunctions
Ω	transformation function defined between eqns (2.176 and 2.177)

References

ABBOTT, M.M. and PRAUSNITZ, J.M. (1987) *Handbook of separation process technology* (ed. R.W. ROUSSEAU), Ch. 1. Wiley-Interscience, New York.

ACHARJEE, D.K., MITRA, A.K., and ROY, A.N. (1977). *Indian Chemical Engineer*, **XIX** (No. 4), 38–41.

ACHARJEE, D.K., MITRA, A.K., and ROY, A.N. (1978). *Canadian Journal of Chemical Engineering*, **56** 37–42.

ANGELO, J.B. and LIGHTFOOT, E. N. (1968). *American Institute of Chemical Engineers Journal*, **14**, 531.

ANGELO, J.B., LIGHTFOOT, E.N., and HOWARD, D.W. (1966). *American Institute of Chemical Engineers Journal*, **12**, 751–60.

ASTARITA, G. (1967). *Mass transfer with chemical reaction*. Elsevier, Amsterdam.

ASTARITA, G., SAVAGE, D. W., and BISIO, A. (1983). *Gas treating with chemical solvents*. Wiley, New York.

BAIRD, M.H.I. (1959). *Chemical Engineering Science*, **9**, 267.

BAIRD, M.H.I. and HAMIELEC, A.E. (1962). *Canadian Journal of Chemical Engineering*, **40**, 119.

BAPAT, P.M. and TAVLARIDES, L.L. (1985). *American Institute of Chemical Engineers Journal*, **31**, 659.

BAPAT, P.M., TAVLARIDES, L.L., and SMITH, G.W. (1983). *Chemical Engineering Science*, **38**, 2003.

BARKER, J.J. and TREYBAL, R.E. (1960). *American Institute of Chemical Engineers Journal*, **6**, 289.

BARRER, R.M. (1941). *Diffusion in and through solids*. CUP, Cambridge.

BATCHELOR, G.K. (1951). *Proceedings of the Cambridge Philosophical Society*, **47**, 359.

BATES, R.L., FONDY, P.L., and CORPSTEIN, R.R. (1963). *Industrial and Engineering Chemistry: Process Design and Development*, **2**, 310.

BAYENS, C.A. and LAURENCE, R.L. (1968). *Industrial and Engineering Chemistry: Fundamentals*, **7**, 521.

BEEK, W.J. and KRAMERS, H. (1962). *Chemical Engineering Science*, **17**, 909–21.

BEYAERT, B.O., LAPIDUS, L., and ELGIN, J.C. (1961). *American Institute of Chemical Engineers Journal*, **7**, 46.

BIEBER, H. (1962). Sc. D. thesis. Columbia University, New York.

BIRD, R.B. (1956). In *Advances in chemical engineering* (eds T.B. Drew and J.W. Hoopes), Vol. 1, pp. 155–239. Academic Press, New York.

BIRD, R.B., STEWART, W.E., and LIGHTFOOT, E.N. (1960). *Transport phenomena*. Wiley, New York.

BOLTZE, E. (1908). Thesis. Gottingen University, Germany.
BOUSSINESQ, J. (1905). *J. Math. Pures. Appl.* **11**, 285.
BOYADZHIEV, L. and ELENKOV, D. (1966). *Chemical Engineering Science* **21**, 955.
BRUNSON, R.J. and WELLEK, R.M. (1970). *Canadian Journal of Chemical Engineering*, **48**, 267–74.
BURCHARD, J.K. and TOOR, H.L. (1962). *Journal of Physical Chemistry*, **66**, 2015.
BURKHOLDER, H.C. and BERG, J.C. (1974). *American Institute of Chemical Engineers Journal*, **20**, 872.
CALDERBANK, P.H. (1967). In *Mixing—theory and practice* (eds V.W. UHL and J.B. GRAY), Vol. II, Ch. 6. Academic Press, New York.
CALDERBANK, P.H. and MOO-YOUNG, M. (1961). *Chemical Engineering Science*, **16**, 39.
CALDERBANK, P.H. and PATRA, R.P. (1966). *Chemical Engineering Science* **21**, 719–21.
CASKEY, J.A. and BARLAGE, W.B. (1972). *Journal of Colloid and Interface Science*, **41**, 52.
CHRISTIANSEN, R.M. and Hixson, A.N. (1957). *Industrial and Engineering Chemistry*, **49**, 1017.
CLIFT, R., GRACE, J.R., and WEBER, M.E. (1978). *Bubbles, drops, and particles.* Academic Press, New York.
COULALOGLOU, C.A. and TAVLARIDES, L.L. (1977). *Chemical Engineering Science*, **32**, 1289.
COULSON, J.M. and SKINNER, S.J. (1952). *Chemical Engineering Science*, **1**, 5.
COYLE, R.W. and BERG, J.C. (1984). *Chemical Engineering Science*, **39**, 168–70.
COYLE, R.W., BERG, J.C., and NIWA, J.C. (1981). *Chemical Engineering Science*, **36**, 19–28.
CRANK, J. (1956). *The mathematics of diffusion*. Clarendon Press, Oxford.
CRANK, J. (1975). *The Mathematics of Diffusion* (2nd edn). Clarendon Press, Oxford.
CULLEN, E.J. and DAVIDSON, J.F. (1957). *Transactions of the Faraday Society*, **53**, 113.
CUSSLER, E.L. (1976). *Multicomponent diffusion*. Elsevier, Amsterdam.
CUSSLER, E.L. (1984). *Diffusion—mass transfer in fluid systems.* CUP, Cambridge.
DALINGAROS, W., KUMAR, A., and HARTLAND, S. (1986). *Chemical Engineering Processing*, **20**, 95.
DANCKWERTS, P.V. (1950). *Transactions of the Faraday Society*, **46**, 300.
DANCKWERTS, P.V. (1951). *Industrial and Engineering Chemistry* **43**, 1460.
DANCKWERTS, P.V. (1955). *American Institute of Chemical Engineers Journal*, **1**, 456.
DANCKWERTS, P.V. (1970). *Gas–liquid reactions*. McGraw-Hill, New York.
DAVIES, J.T. (1972). *Turbulence phenomena*. Academic Press, New York.
DE CHAZAL, L.E.M. and RYAN, J.T. (1971). *American Institute of Chemical Engineers Journal*, **17**, 1226.
DOBBINS, W.E. (1956). In *Biological treatment of sewage and industrial wastes* (eds M.L. MCCABE and W.W. ECKENFELDER) Part 2–1. Reinhold, New York.
DOBBINS, W.E. (1964). *Int. Conf. on Water Pollut. Res.*, Lond., Sept., 1962, p. 61. Pergamon Press, New York.

DUDA, J.L. and VRENTAS, J.S. (1968). *American Institute of Chemical Engineers Journal*, **14**, 286–94.
DUZUBUR, I. and SAWISTOWSKI, H. (1974). *Proceedings of the International Solvent Extraction Conference*, **1**, 379.
EDGE, R.M. and GRANT, C.D. (1971). *Chemical Engineering Science* **26**, 1001–12.
ELZINGA, E.R. and BANCHERO, J.T. (1960). *Chemical Engineering Progress Symposium Series*, **55** (No. 29), 149.
FORTESCUE, G.E. and PEARSON, J.R.A. (1967). *Chemical Engineering Science*, **22**, 1163.
FRIEDLANDER, S.K. (1961). *American Institute of Chemical Engineers Journal*, **7**, 347.
FROESSLING, N. (1938). *Gerlands Beitraege zurgeophysik*, **52**, 170.
FROESSLING, N. (1940). *Lund Universitet Arsskr. N.F. A VD.* **36**, No. 4.
FUJINAWA, K., MARUYAMA, T., and NAKAIKE, Y. (1957). *Kagaku Kikai*, **21**, 194.
GARNER, F.H., FOORD, A., and TAYEBAN, M. (1959). *Journal of Applied Chemistry*, **9**, 315–23.
GARNER, F.H., MINA, P., and JENSEN, V.G. (1959). *Transactions of the Faraday Society*, **48**, 1627.
GARNER, F.H. and SKELLAND, A.H.P. (1955). *Chemical Engineering Science*, **4**, 149–58.
GARNER, F.H. and SKELLAND, A.H.P. (1956). *Industrial and Engineering Chemistry*, **48**, 51–8.
GARNER, F.H. and TAYEBAN, M. (1960). *An. Real Soc. Esp. Fis. Quim. Ser. B-Quim.* LVI (B), 479.
GEANKOPLIS, C.J. (1972). *Mass transport phenomena*. Holt, Rinehart and Winston, New York.
GLASSTONE, S. (1947). *Thermodynamics for chemists*, p. 402. Van Nostrand, Princeton, New Jersey.
GLEN, J.B. (1965). Mass transfer in disperse systems. Ph. D. dissertation, Univ. of Canterbury, Christchurch, New Zealand.
GORDON, A.R. (1937). *Journal of Chemical Physics*, **5**, 522.
GRACE, J.R., WAIREGI, T., and NGUYEN, T.H. (1976). *Transactions of the Institution of Chemical Engineers*, **54**, 167–73.
GRIFFITHS, R.M. (1960). *Chemical Engineering Science*, **12**, 198–213.
GRIGAR, K., PROCHAZKA, J., and LANDAU (1972). *Collection of Czechoslovak Chemical Communications* **37**, 1943–51.
GROBER, H. (1925). *Z. Ver. dtsch. Ing.*, **69**, 705.
GROOTHUIS, H. and KRAMERS, H. (1955). *Chemical Engineering Science*, **4**, 17–25.
HADAMARD, J.S. (1911). *C.R. Acad. Sci.* **152**, 1735–1738.
HANDLOS, A.E. and BARON, T. (1957). *American Institute of Chemical Engineers Journal*, **3**, 127.
HANRATTY, T.J. (1956). *American Institute of Chemical Engineers Journal*, **2**, 359.
HARKINS, W.D. and BROWN, F.E., (1919). *Journal of the American Chemical Society*, **41**, 499.
HARNED, H.S. and OWEN, B.B. (1950). *The physical chemistry of electrolytic solutions*, American Chemical Society Monograph 95.
HARPER, J.F. and MOORE, D.W. (1968). *Journal of Fluid Mechanics*, **32**, 367.
HARRIOTT, P. (1962). *Chemical Engineering Science*, **17**, 149.
HAUSEN, H. (1943). *Verfahrenstechnik Beihette Z. Ver. dtsch, Ing.* **4**, 91.

HAUSEN, H. (1953). *Chemie-Ingenieur-Technik*, **25**, 595.
HAYDUK, W. and LAUDIE, H. (1974). *American Institute of Chemical Engineers Journal*, **20**, 611.
HAYDUK, W. and MINHAS, B.S. (1982). *Canadian Journal of Chemical Engineering*, **60**, 295.
HAYWORTH, C.B. and TREYBAL, R.E. (1950). *Industrial and Engineering Chemistry*, **42**, 1174.
HEERTJES, P.M. and DE NIE, L.H. (1966). *Chemical Engineering Science*, **21**, 755-68.
HEERTJES, P.M., DE NIE, L.H., and DE VRIES, H.J. (1971). *Chemical Engineering Science*, **26**, 441.
HEERTJES, P.M., HOLVE, W.A., and TALSMA, H. (1954). *Chemical Engineering Science*, **3**, 122.
HEIDEGER, W.J. and DUBOIS, S.E. (1985). *American Institute of Chemical Engineers Journal*, **31**, 483-4.
HEIDEGER, W.J. and WRIGHT, M.W. (1986). *American Institute of Chemical Engineers Journal*, **32**, 1372-6.
HERSHEY, D. (1973). *Transport analysis*. Plenum Press, New York.
HIGBIE, R. (1935). *Transactions of the American Institute of Chemical Engineers*, **31**, 365.
HIRSCHFELDER, J.O., CURTISS, C.F., and BIRD, R.B. (1954). *Molecular theory of gases and liquids*. Wiley, New York.
HIXSON, A.W. and BAUM, S.J. (1941). *Industrial and Engineering Chemistry*, **33**, 478.
HO, G.E. and PRINCE, R.G.H. (1970). *Transactions of the Institution of Chemical Engineers*, **48**, T101-T106.
HONG, P.O. and LEE, J.M. (1985). *Industrial and Engineering Chemistry: Process Design and Development*, **24**, 868.
HORVATH, M., STEINER, L., and HARTLAND, S. (1978). *Canadian Journal of Chemical Engineering*, **56**, 9-18.
HOWARTH, W.J. (1963). *Chemical Engineering Science*, **18**, 47-9.
HU, S. and KINTNER, R.C. (1955). *American Institute of Chemical Engineers Journal*, **1**, 42.
HUGHMARK, G.A. (1967). *Industrial and Engineering Chemistry: Fundamentals*, **6**, 408.
HUMPHREY, J.A.C., HUMMEL, R.L., and SMITH, J.W. (1974a). *Chemical Engineering Science*, **29**, 1496.
HUMPHREY, J.A.C., HUMMEL, R.L. and SMITH, J.W. (1974b). *Canadian Journal of Chemical Engineering*, **52**, 449.
HUMPHREY, D.W. and VAN NESS, H.C. (1957). *American Institute of Chemical Engineers Journal*, **3**, 283-6.
ILKOVIC, D. (1934). *Collection of Czechoslovak Chemical Communications*, **6**, 498.
IZARD, J.A. (1972). *American Institute of Chemical Engineers Journal*, **18**, 634-638.
JAVED, K.H., THORNTON, J.D. and ANDERSON T.J. (1989), *American Institute of Chemical Engineers Journal*, **35(7)**, 1125
JEFFREYS, G.V. (1987). *Chemistry and Industry*, 16th March, 181-5.
JENSEN, V.G. and JEFFREYS, G.V. (1963). *Mathematical methods in chemical engineering*. Academic Press, New York.

JOHNSON, A. I. and BRAIDA, L. (1957). *Canadian Journal of Chemical Engineering*, **35**, 165.
JOHNSON, A. I. and HAMIELEC, A. G. (1960). *American Institute of Chemical Engineers Journal*, **6**, 145.
JOHNSON, A. I., HAMIELEC, A. E., WARD, D., and GOLDING, A. (1958). *Canadian Journal of Chemical Engineering*, **36**, 221.
JOHNSON, P. A. and BABB, A. L. (1956). *Chemical Reviews*, **56**, 387.
JOST, W. (1952). *Diffusion in solids, liquids, and gases*. Academic Press, New York.
JOST, W. (1960). *Diffusion in solids, liquids, and gases* (Revised edn). Academic Press, New York.
KAGAN, S. Z., KOVALEV, YU N., and Zakharychev, A. P. (1973). *Theoretical Foundations of Chemical Engineering*, **7**, 514.
KEEY, R. B. and GLEN, J. B. (1969). *American Institute of Chemical Engineers Journal*, **15**, 942.
KEITH, F. W. and HIXSON, A. N. (1955). *Industrial and Engineering Chemistry*, **47**, 258.
KIMURA, M. S. and MIYAUCHI, T. (1966). *Chemical Engineering Science*, **21**, 1057.
KING, C. J. (1966). *Industrial and Engineering Chemistry: Fundamentals*, **5**, 1–8.
KING, C. J., HSUEH, L., and MAO, K. W. (1965). *Journal of Chemical Engineering Data*, **10**, 348–50.
KIRWAN, D. J. (1987). In *Handbook of separation process technology* (ed. R. W. Rousseau), Ch. 2. Wiley-Interscience, New York.
KISHINEVSKY, M. KH. (1951). *Journal of Applied Chemistry U.S.S.R.* (Engl. Trans.), **24**, 542.
KISHINEVSKY, M. KH. (1954). *Journal of Applied Chemistry U.S.S.R.* (Engl. Trans.), **27**, 359.
KISHINEVSKY, M. KH. and KERAIVARENKO, M. A. (1951). *Journal of Applied Chemistry* U.S.S.R. (Engl. Trans.), **24**, 413.
KISHINEVSKY, M. Kh. and Keraivarenko, M. A. (1953). *Novik*, **26**, 673.
KISHINEVSKY, M. KH. and PAMFILOV, A. V. (1949). *Journal of Applied Chemistry U.S.S.R.* (Engl. Trans.), **22**, 118.
KLEE, A. J. and TREYBAL, R. E. (1956). *American Institute of Chemical Engineers Journal*, **2**, 444.
KOHLI, A. (1974). Ph.D. thesis. University of Maryland.
KOO, K. K., JEON, Y. M., and LEE, W. K. (1985). *Journal of Chemical Engineering of Japan*, **18**, 142–147.
KORCHINSKY, W. J. and CRUZ-PINTO, J. J. C. (1979). *Chemical Engineering Science*, **34**, 551–61.
KOZINSKY, A. A. and KING, C. J. (1966). *American Institute of Chemical Engineers Journal*, **12**, 109.
KREYSZIG, E. (1967). *Advanced engineering mathematics* (2nd edn), p. 524. Wiley, New York.
KRISHNA, R. (1976). *Letters in Heat and Mass Transfer*, **3**, 153.
KRISHNA, R. (1977). *Chemical Engineering Science*, **32**, 1197.
KRISHNA, R. and STANDART, G. L. (1976). *Letters in Heat and Mass Transfer*, **3**, 173.
KRISHNA, R. and STANDART, G. L. (1979). *Chemical Engineering Communications*, **3**, 201–75.
KRONIG, R. and BRINK, J. C. (1950). *Applied Science Research*, **A-2**, 142.

KUMAR, R. (1971). *Chemical Engineering Science*, **26**, 177.
KUMAR, A. (1983). D. Tech. Sci. thesis. ETH Zurich.
KUMAR, A. and HARTLAND, S. (1982). *Transactions of the Institution of Chemical Engineers*, **60**, 35–9.
KUMAR, A. and HARTLAND, S. (1984). *Chemical Engineering Communications*, **31**, 193–207.
KUMAR, A., VOHRA, D.K., and HARTLAND, S. (1980). *Canadian Journal of Chemical Engineering*, **58**, 154.
LAMONT, J.C. and SCOTT, D.S. (1970). *American Institute of Chemical Engineers Journal*, **16**, 513.
LEACOCK, J.A. and CHURCHILL, S.W. (1961). *American Institute of Chemical Engineers Journal*, **7**, 196.
LEE, J.M. (1978). Mass transfer and liquid–liquid dispersion in agitated vessels. Ph.D. dissertation. University of Kentucky, Lexington.
LEES, F.P. and SARRAM, P. (1971). *Journal of Chemical Engineering Data*, **16**, 41.
LEFFLER, J. and CULLINAN, H.T. (1970). *Industrial and Engineering Chemistry: Fundamentals*, **9**, 84–93.
LEWIS, W.K. (1916). *Industrial Engineering Chemistry*, **8**, 825.
LEWIS, W.K. (1936). *Industrial Engineering Chemistry* **28**, 399–402.
LEWIS, G.N. and RANDALL, M. (1923). *Thermodynamics and the free energy of chemical substances*, p. 39. McGraw-Hill, New York.
LI, WEN HSIUNG (1960). *Engineering analysis*. Prentice-Hall, Englewood Cliffs, N.J.
LICHT, W. and CONWAY, J.B. (1950). *Industrial and Engineering Chemistry* **42**, 1151.
LICHT, W. and PANSING, W.F. (1953). *Industrial and Engineering Chemistry* **45**, 1885.
LOCHIEL, A.C. and CALDERBANK, P.H. (1964). *Chemical Engineering Science*, **19**, 471.
LUSIS, M.A. and RATCLIFF, G.A. (1968). *Canadian Journal of Chemical Engineering*, **46**, 385–7.
MARCHELLO, J.M. and TOOR, H.L. (1963). *Industrial and Engineering Chemistry: Fundamentals* **2**, 8.
MAYFIELD, F.W. and CHURCH, W.L. (1952). *Industrial and Engineering Chemistry* **44**, 2253.
MEISTER, B.J. and SCHEELE, G.F. (1967). *American Institute of Chemical Engineers Journal* **13**, 682–8.
MEISTER, B.J. and SCHEELE, G.F. (1969a). *American Institute of Chemical Engineers Journal* **15**, 689–99.
MEISTER, B.J. and SCHEELE, G.F. (1969b). *American Institute of Chemical Engineers Journal* **15**, 700–6.
MEKASUT, L., MOLINIER, J., and ANGELINO, H. (1979). *Canadian Journal of Chemical Engineering* **57**, 668–671.
MICHELS, H.H. (1960). Ph.D. thesis. University of Delaware.
MICHELS, H.H. and PIGFORD, R.L. (1963). Cited in *The chemical engineers' handbook* (4th edn) (eds R.H. Perry, C.H. Chilton, and S.D. Kirkpatrick, p. 14.64. McGraw-Hill, New York.
MILLER, D.N. (1964). *Industrial and Engineering Chemistry*, **56** (No. 10), 18.
MILLER, D.N. (1971). *Industrial and Engineering Chemistry: Process Design and Development*, **10**, 365.
MILLER, H.D. and PILHOFER, T. (1976). *Chemie-Ingenieur-Technik*, **48**, 1069.

MINHAS, S.S. (1969). Ph.D. dissertation. Univ. Notre Dame, Indiana.
MURPHREE, E. (1925). *Industrial and Engineering Chemistry* **17**, 747–50, 960–4.
NARAYANAN, S., BASU, A., and ROY, N.K. (1970). *Chemical Engineering Science*, **25**, 1950.
NERNST, W. (1888). *Zeitschrift fuer Physikalishe Chemie*, **2**, 613.
NEWMAN, A.B. (1931). *Transactions of the American Institute of Chemical Engineers*, **27**, 310–333.
NISI, H. and PORTER, A.W. (1923). *Philosophical Magazine*, **46**, 755.
NULL, H.R. and JOHNSON, H.F. (1958). *American Institute of Chemical Engineers Journal*, **4**, 273.
OLANDER, D.R. (1961). *American Institute of Chemical Engineers Journal* **7**, 175–176.
PATEL, J.M. and WELLEK, R.M. (1967). *American Institute of Chemical Engineers Journal* **13**, 384–386.
PEIRCE, B.O. and FOSTER, R.M. (1956). *A short table of integrals* (4th edn). Ginn – Blaisdell, Waltham, Mass.
PERKINS, L.R. and GEANKOPLIS, C.J. (1969). *Chemical Engineering Science*, **24**, 1035.
PERLMUTTER, D.D. (1961). *Chemical Engineering Science*, **16**, 287–296.
PERRUT, M. and LOUTATY, R. (1972). *Chemical Engineering Journal*, **3**, 286–293.
PERRY, R.H., CHILTON, C.H., and KIRKPATRICK, S.D. (1963). *Chemical engineers' handbook* (4th edn). McGraw-Hill, New York.
PILHOFER, T. (1974). *Chemie-Ingenieur-Technik*, **46**, 783.
PINCZEWSKI, W.V. and SIDEMAN, S. (1974). *Chemical Engineering Science*, **29**, 1969.
POPOVICH, A.T., JERVIS, R.E., and TRASS, O. (1964). *Chemical Engineering Science*, **19**, 357–65.
POTTER, O.E. (1957). *Chemical Engineering Science* **6**, 170–82.
PRASHER, B.D. and WILLS, G.B. (1973). *Industrial and Engineering Chemistry: Process Design and Development*, **12**, 351.
PRATT, H.R.C. (1983). In *Handbook of solvent extraction* (eds T.C. Lo, M.H.I. Baird, and C. Hanson) Ch. 5. Wiley-Interscience, New York.
QUINN, J.A. and JEANNIN, P.G. (1961). *Chemical Engineering Science*, **15**, 243.
RAJAN, S.M. and HEIDEGER, W.J. (1971). *American Institute of Chemical Engineers Journal*, **17**, 202.
RAO, E.V.L., KUMAR, R., and KULOOR, N.R. (1966). *Chemical Engineering Science*, **21**, 867–80.
REDFIELD, J.A. and HOUGHTON, G. (1965). *Chemical Engineering Science*, **20**, 131.
REHM, T.R., MOLL, A.J., and BABB, A.L. (1963). *American Institute of Chemical Engineers Journal* **9**, 760.
REID, R.C., PRAUSNITZ, J.M., and POLING, B.E. (1987). *The properties of gases and liquids* (4th edn). McGraw-Hill, New York.
REID, R.C., PRAUSNITZ, J.M., and SHERWOOD, T.K. (1977). *The properties of gases and liquids* (3rd edn). McGraw-Hill, New York.
ROBINSON, R.A. and STOKES, R.H. (1959). *Electrolyte solutions* (2nd edn). Academic Press, New York.
ROSE, P.M. and KINTNER, R.C. (1966). *American Institute of Chemical Engineers Journal*, **12**, 530–4.

RUBY, C.L. and ELGIN, J.C. (1955). *Chemical Engineering Progress Symposium Series*, **51**, No. 16, 17–29.
RUCKENSTEIN, E. (1968). *Chemical Engineering Science*, **23**, 363–71.
RUCKENSTEIN, E. and CONSTANTINESCU, D. (1969). *International Journal of Heat and Mass Transfer*, **12**, 1249.
RUFF, K. (1978). *Chemie-Ingenieur-Technik*, **50**, 441.
RUFF, K., PILHOFER, T., and MERSMANN, A. (1976). *Chemie-Ingenieur-Technik*, **48**, 759.
RUSHTON, J.H., NAGATA, S., and ROONEY, T.B. (1964). *American Institute of Chemical Engineers Journal*. **10**, 298.
RYAN, J.T. (1966). Ph. D. thesis. University of Missouri.
SAWISTOWSKI, H. (1973). *Chemie-Ingenieur-Technik*, **45**, 1114.
SAWISTOWSKI, H. and GOLTZ, G.E. (1963). *Transactions of the Institution of Chemical Engineers*, **41**, 174.
SCHEELE, G.F. and MEISTER, B.J. (1968). *American Institute of Chemical Engineers Journal* **14**, 9.
SCHEIBEL, E.G. (1954). *Industrial and Engineering Chemistry* **46**, 2007.
SCHINDLER, H.D. (1967). Ph.D. dissertation. New York University, NY.
SCHINDLER, H.D. and TREYBAL, R.E. (1968). *American Institute of Chemical Engineers Journal* **14**, 790.
SCHROEDER, R.R. and KINTNER, R.C. (1965). *American Institute of Chemical Engineers Journal* **11**, 5–8.
SCRIVEN, L.E. and PIGFORD, R.L. (1958). *American Institute of Chemical Engineers Journal* **4**, 439.
SCRIVEN, L.E. and PIGFORD, R.L. (1959). *American Institute of Chemical Engineers Journal* **5**, 397.
SHARMA, M.M. (1983). In *Handbook of solvent extraction* (eds T.C. Lo, M.H.I. Baird, and C. Hanson) Ch. 2.1. Wiley–Interscience, New York.
SHERWOOD, T.K., PIGFORD, R.L., and WILKE, C.R. (1975). *Mass transfer*. McGraw-Hill, New York.
SCHIFFLER, D.A. (1965). Ph.D. thesis. Cornell University, Ithaca, NY.
SHIROTSUKA, T. and KAWASE, Y. (1975). *Kagaku Kogaku Ronbunshu*, **1**, 191.
SHIROTSUKA, T. and KAWASE, Y. (1976). *Journal of Chemical Engineering of Japan*, **9**, 71.
SIEMES, W. (1956). *Chemie-Ingenieur-Technik* **28**, 727–31.
SKELLAND, A.H.P. (1974, 1985). *Diffusional mass transfer*. Wiley–Interscience, New York (1974); Reprint edn with revisions, Krieger, Malabar, Florida (1985).
SKELLAND, A.H.P. and CAENEPEEL, C.L. (1972). *American Institute of Chemical Engineers Journal* **18**, 1154–63.
SKELLAND, A.H.P. and CHADHA, N. (1981). *Industrial and Engineering Chemistry: Process Design and Development*, **20**, 232–239.
SKELLAND, A.H.P. and CONGER, W.L. (1973). *Industrial and Engineering Chemistry: Process Design and Development*, **12**, 448–54.
SKELLAND, A.H.P. and CORNISH, A.R.H. (1963). *American Institute of Chemical Engineers Journal* **9**, 73–76.
SKELLAND, A.H.P. and HEMLER, C.L. (1969). Unpublished work at the University of Notre Dame.
SKELLAND, A.H.P. and HUANG, Y.-F. (1977a). *American Institute of Chemical Engineers Journal* **23**, 701–14.

SKELLAND, A.H.P. and HUANG, Y.-F. (1977b). *Canadian Journal of Chemical Engineering*, **55**, 240–45.
SKELLAND, A.H.P. and HUANG, Y.-F. (1979). *American Institute of Chemical Engineers Journal* **25**, 80–7.
SKELLAND, A.H.P. and HU XIEN (1990). *Industrial and Engineering Chemistry Research*, **29**, 415–20.
SKELLAND, A.H.P. and JOHNSON, K.R. (1974). *Canadian Journal of Chemical Engineering* **52**, 732–8.
SKELLAND, A.H.P. and KANEL, J.S. (1990). *Industrial and Engineering Chemistry Research* **29**, 1300–6.
SKELLAND, A.H.P. and LEE, J.M. (1978). *Industrial and Engineering Chemistry Process Design and Development* **17**, 473–8.
SKELLAND, A.H.P. and LEE, J.M. (1981). *American Institute of Chemical Engineers Journal* **27**, 99–111; *also* **29**, 174 (1983).
SKELLAND, A.H.P. and MINHAS, S.S. (1971). *American Institute of Chemical Engineers Journal* **17**, 1316–24.
SKELLAND, A.H.P. and MOETI, L.T. (1989). *Industrial and Engineering Chemistry Research* **28**, 122–7.
SKELLAND, A.H.P. and MOETI, L.T. (1990). *Industrial and Engineering Chemistry Research* **29**, 2258–67.
SKELLAND, A.H.P. and RAMSAY, G.G. (1987). *Industrial and Engineering Chemistry Research*, **26**, 77–81.
SKELLAND, A.H.P. and RAVAL, V.K. (1972). *Canadian Journal of Chemical Engineering* **50**, 41–4.
SKELLAND, A.H.P. and SLAYMAKER, E.A. (1990). *Industrial and Engineering Chemistry and Research* **29**, 494–9.
SKELLAND, A.H.P. and VASTI, N.C. (1985). *Canadian Journal of Chemical Engineering* **63**, 390–8.
SKELLAND, A.H.P. and WALKER, P.G. (1989). *Canadian Journal of Chemical Engineering* **67**, 762–70.
SKELLAND, A.H.P. and WELLEK, R.M. (1964). *American Institute of Chemical Engineers Journal* **10**, 491–6, 789.
SKELLAND, A.H.P., WOO, S., and RAMSAY, G.G. (1987). *Industrial and Engineering Chemistry: Research* **26**, 907–11.
STEWART, W.E. and PROBER, R. (1964). *Industrial and Engineering Chemistry: Fundamentals* **3**, 224–35.
SYKES, P. and GOMEZPLATA, A. (1967). *Canadian Journal of Chemical Engineering* **45**, 189.
TARR, L.E. and BERG, J.C. (1980). *Chemical Engineering Science* **35**, 1465–67.
THORNTON, J.D. (1956). *Chemical Engineering Science* **5**, 201.
THORNTON, J.D. (1987). *Chemistry and Industry*, 16 March, 193–6.
THORNTON, J.D. and ANDERSON, T.J. (1981). *International Journal of Heat and Mass Transfer*, **24**, 1847–8.
THORNTON, J.D., ANDERSON, T.J., JAVED, K.H., and ACHWAL, S.K. (1985). *American Institute of Chemical Engineers Journal* **31**, 1069–76.
THORSEN, G. and TERJESEN, S.G. (1962). *Chemical Engineering Science* **17**, 137–48.
TOMOTIKA, S. (1935). *Proceedings of the Royal Society*, **A150**, 322.
TOOR, H.L. (1964a). *American Institute of Chemical Engineers Journal* **10**, 448–55.
TOOR, H.L. (1964b). *American Institute of Chemical Engineers Journal* **10**, 460–5.

TOOR, H.L. and MARCHELLO, J.M. (1958). *American Institute of Chemical Engineers Journal* **4**, 97–101.
TREYBAL, R.E. (1961). *Industrial and Engineering Chemistry* **53**, 597.
TREYBAL, R.E. (1963). *Liquid extraction* (2nd edn). McGraw-Hill, New York.
TREYBAL, R.E. (1980). *Mass transfer operations* (3rd edn). McGraw-Hill, New York.
TYLER, E. and WATKIN, F. (1932). *Philosophical Magazine*, **14**, 849.
TYN, M.T. and CALUS, W.F. (1975a). *Journal of Chemical Engineering Data*, **20**, 106.
TYN, M.T. and CALUS, W.F. (1975b). *Journal of Chemical Engineering Data*, **20**, 310.
TYRRELL, H.J.V. and HARRIS, K.R. (1984). *Diffusion in liquids—a theoretical and experimental study*. Butterworths, London.
UEYAMA, K. (1957). *Kagaku Kogaku*, **21**, 766.
VAN DEN AKKER, H.E.A. (1980). *Chemical Engineering Journal*, **19**, 255–9.
VANDEGRIFT, A.E. (1963). Ph.D. dissertation. University California Berkeley.
VEDAIYAN, S. (1969). Ph.D. thesis. University of Madras.
VERMA, R.P. and SHARMA, M.M. (1975). *Chemical Engineering Science*, **30**, 279.
VERMEULEN, T. (1953). *Industrial and Engineering Chemistry*, **45**, 1664.
VRENTAS, J.S. and DUDA, J.L. (1967). *Chemical Engineering Science* **22**, 855.
WALIA, D.S. and VIR, D. (1976). *Chemical Engineering Science* **31**, 525–33.
WASAN, D.T. and AHLUWALIA, M.S. (1969). *Chemical Engineering Science* **24**, 1535–42.
WEAVER, R.E.C., LAPIDUS, L., and ELGIN, J.C. (1959). *American Institute of Chemical Engineers Journal* **5**, 533.
WEBER, M.E. (1975). *Industrial and Engineering Chemistry: Fundamentals*, **14**, 365–6.
WELLEK, R.M., AGRAWAL, A.K., and SKELLAND, A.H.P. (1966). *American Institute of Chemical Engineers Journal* **12**, 854–62.
WELLEK, R.M. and SKELLAND, A.H.P. (1965). *American Institute of Chemical Engineers Journal* **11**, 557.
WELTY, J.R., WICKS, C.E., and WILSON, R.E. (1984). *Fundamentals of momentum, heat, and mass transfer* (3rd edn). Wiley, New York.
WHITMAN, W.G. (1923). *Chemical and Metallurgical Engineering* **29**, 147.
WILKE, C.R. and CHANG, P. (1955). *American Institute of Chemical Engineers Journal* **1**, 264.
WINNIKOW, S. (1967). *Chemical Engineering Science*, **22**, 477.
YAMAGUCHI, M., FUJIMOTO, T., and KATAYAMA, T. (1975a). *Journal of Chemical Engineering of Japan*, **8**, 361.
YAMAGUCHI, M., WATANABE, S., and KATAYAMA, T. (1975b). *Journal of Chemical Engineering of Japan*, **8**, 415.
ZHELEZNYAK, A.S. (1967). *Journal of Applied Chemistry U.S.S.R.*, **40**, 834.
ZIMMERMAN, V., HALWACHS, W., and SCHUGERL, K. (1980). *Chemical Engineering Communications*, **7**, 95.

3

MARANGONI PHENOMENA

E.S. PEREZ DE ORTIZ

3.1 Introduction

The term Marangoni phenomena is given to spontaneous interfacial flows driven by interfacial tension gradients. In solvent extraction, these gradients may be originated by local changes of those variables that affect the value of the interfacial tension such as solute concentration, temperature and interfacial electrical potential. Irrespective of their cause the interfacial convection generated by these movements provides an additional component to the interfacial flux that is not included in the general theories of mass transfer. Experimental evidence of increases in interfacial fluxes of up to one order of magnitude over predicted values indicates the importance of the Marangoni phenomena contribution to the rate of solute transfer. In addition, these interfacial movements modify the hydrodynamics of the fluid layers adjacent to the interface with consequent effects on drop coalescence, jet break-up and drop drag coefficients, so that the performance of contacting equipment is not only affected through the interfacial flux but also by changes in hydrodynamic behaviour.

Although spontaneous interfacial convection accompanying the mass transfer of a solute across a flat interface was first reported in 1888 (Quincke 1888), systematic research into its nature and effects on mass transfer processes only started expanding in the 1950's. An extensive and comprehensive review of interfacial phenomena in liquid–liquid extraction has been written by Sawistowski (1971) in which he discusses the main experimental contributions in the field, and outlines the linearized stability analysis of a flat interface under conditions of mass trasfer. Berg (1972) reviewed the subject including effects on absorption and distillation, focussing the discussion on thermodynamic fundamentals and hydrodynamic aspects of these phenomena. In the last few years Marangoni flows have been associated with phenomena apparently as wide appart as the formation of void structures in the casting of polymeric membranes (Hancock *et al.* 1980) and the deformation of biological cell membranes leading to cell motility (Sørensen 1980).

It is not intended to present here an exhaustive review of the literature. Instead an attempt is made to cover the aspects of Maragoni phenomena related to solvent extraction in a systematic and comprehensive way including some of the most recent developments. In the pursuit of this aim some overlapping with the previously mentioned reviews is unavoidable.

Following a description of the mechanisms of interfacial convection and types of Marangoni phenomena relevant to solvent extraction, the contents of this chapter deals with three main interrelated areas: mathematical analysis of instabilities, applications of the theoretical analysis to the the development of stability criteria for mass transfer under different conditions and the effect of Marangoni instabilities on specific interfacial area, drop drag coefficients and mass transfer coefficients.

3.2 Mechanisms of interfacial convection

Marangoni phenomena accompanying the transfer of propionic acid from a stationary benzene drop into water are illustrated in the photograph shown in Fig. 3.1(a). No such interfacial convection can be observed in Fig. 3.1(b) which is typical of diffusional transfer. Extensive investigations of the interfacial behaviour of different systems during transfer of one or more solutes (Sawistowski 1971) have shown that while in some systems spontaneous interfacial convection occur even for very low mass transfer driving forces, in others the regime of transfer is diffusional over a wide range of mass transfer rates. The question then arises on which are the conditions that lead to interfacial convection.

The mechanism of onset of interfacial turbulence is based on the response of the system to local variations of its interfacial tension. In a multicomponent system the interfacial tension is mainly a function of solute concentration, temperature and, to a lesser extent, interfacial electrical potential. A perturbation in the interfacial tension, σ_0, thus may be produced by variations of solute concentration, temperature and/or electrical charge density. Let us consider a system where a solute is transferring in steady state between two equilibrated liquid phases as shown in Fig. 3.2(a). The temperature profiles sketched in the figure are due to the heats of solution. At the initial state the interfacial tension is constant throughout the interface. Mechanical vibrations in the system may bring elements of fluid from the bulk to the interface causing local changes of interfacial concentration and temperature. The consequent local changes in interfacial tension will generate radial spreading flow from the points of lower interfacial tension, Fig. 3.2(b). Depending on the physical properties of the system and the magnitude of the disturbances these movements may restore the interfacial tension of the unperturbed state thus eliminating the destabilizing force. If, despite spreading, the interfacial tension gradient remains, interfacial flows will continue and, by dragging the adjacent liquid layers, will generate circulation patterns of the type shown in Fig. 3.2(c). In this case the system is interfacially unstable.

Thus the onset of interfacial flows depends on the possibility of having

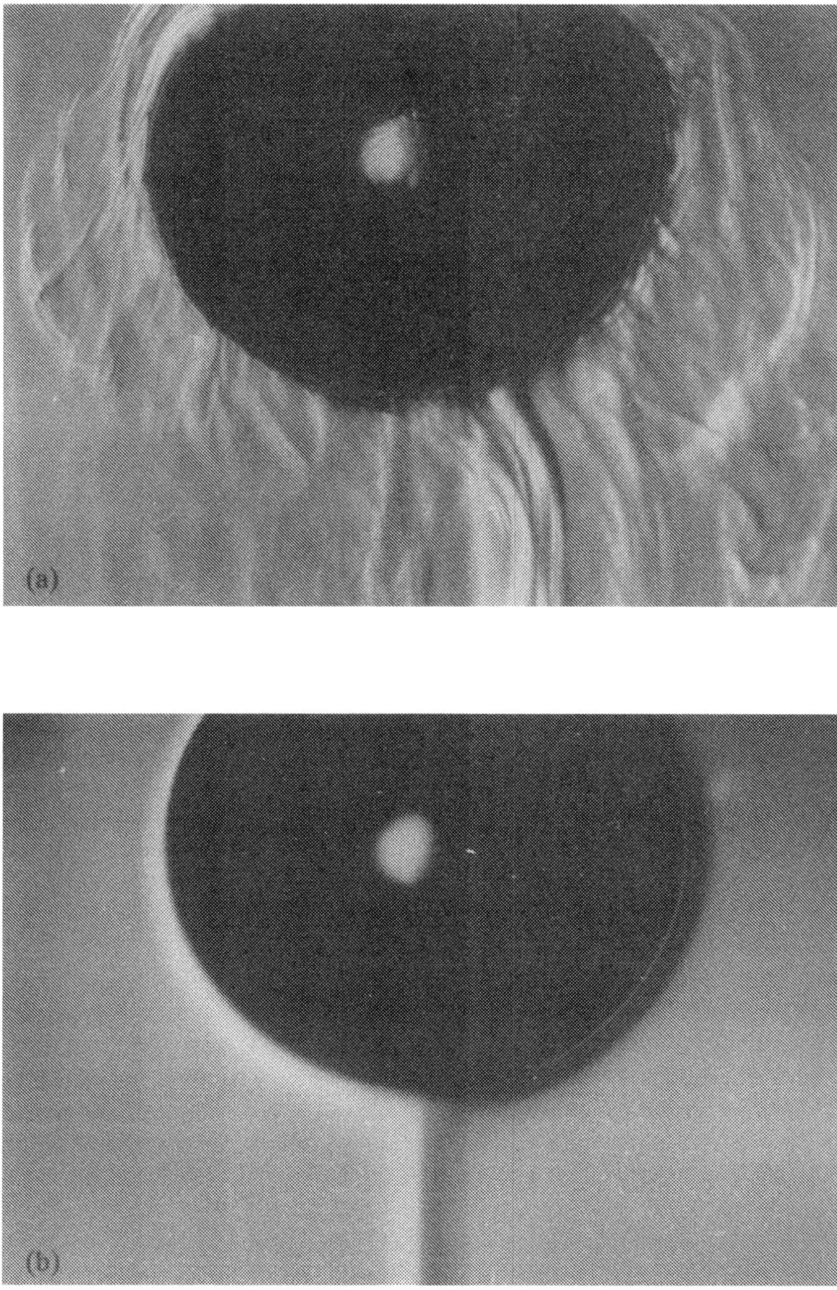

Fig. 3.1. (a) Interfacial turbulence in the transfer of propionic acid from a stationary benzene drop into water. (b) Diffusional transfer. (After Austin 1966.)

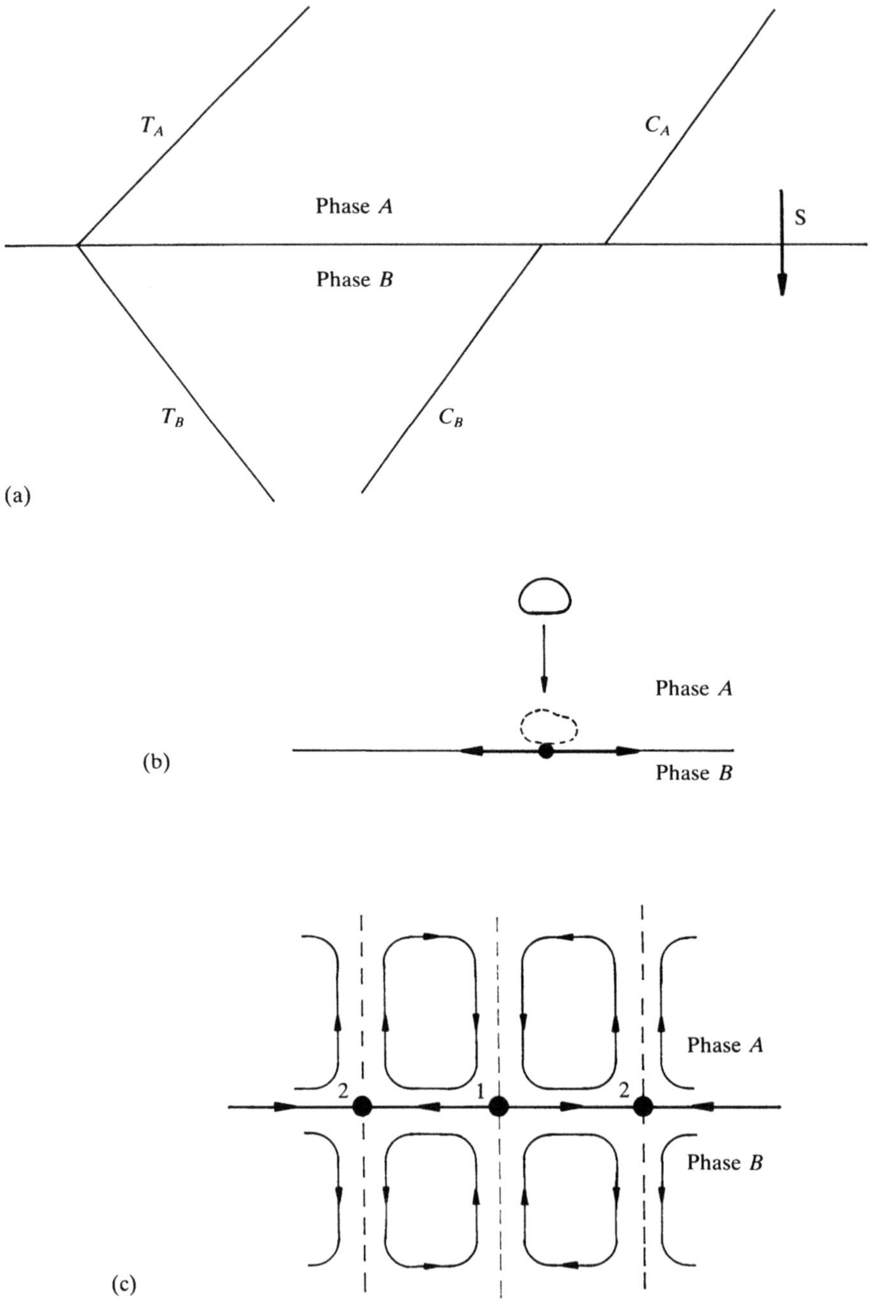

Fig. 3.2. (a) Concentration and temperature profiles in the steady state transfer of a solute between two equilibrated phases. (b) Mechanism of interfacial perturbation. (c) Convection cells.

local variations of interfacial tension while their development into sustained interfacial convection requires additional conditions. Two physical properties play an important role in interfacial convection: solute diffusivity, $\mathscr{D}$, and kinematic viscosity, ν. Rather than their absolute value, it is their relative value for the two phases that affects the behaviour of the system. The influence of relative phase viscosity is more easily seen in the case where the solute diffusivity is the same in both phases, i.e. $\mathscr{D}_A = \mathscr{D}_B$. With reference to Fig. 3.2, where the solute is transferring from phase A into phase B, it will be first assumed that $\nu_A > \nu_B$ and that the variation of interfacial tension with solute concentration is negative, i.e. $\partial\sigma_0/\partial C < 0$. A disturbance that increases the interfacial concentration, such as fluid from the bulk of phase A reaching the interface, will produce a local decrease in interfacial tension leading to the spreading movements shown in Fig. 3.2(b). Penetration of these movements into the bulk is deeper in the phase of higher viscosity, thus fluid from phase A replacing the displaced interfacial elements will come from farther away than fluid from phase B. Transfer of solute from phase A to phase B during flow along the interface will cause a decrease in the interfacial concentration maintaining the interfacial tension gradient. As a result convection will be sustained, as schematically shown in Fig. 3.2(c), and the system will be unstable.

In the case where $\nu_A < \nu_B$, fluid from phase A is drawn from a region closer to the interface than in the previous case, therefore its concentration will be lower. So will be the concentration of the fluid approaching the interface from phase B. During motion from points 1 to 2 the solute will be transferred to phase B but to an extent that will decrease its concentration to a value below that at point 2 thus providing the conditions for back–spreading from point 2. Consequently the perturbation will die out and the system will be stable. It can then be concluded that in a system with positive adsorption and similar values of solute diffusivities in both phases, interfacial convection is likely to be sustained when mass transfer takes place from the phase of higher viscosity. The reverse should be true for systems in which $\partial\sigma_0/\partial C > 0$.

A similar qualitative analysis can be conducted for a system in which the two phases have the same kinematic viscosity but different diffusivities (Sternling and Scriven 1959). In this case, for $\partial\sigma_0/\partial C < 0$, the system is predicted unstable when the solute is transferred from the phase in which solute diffusivity is lower.

In systems where chemical reactions take place close to the interfacial region it is not easy to predict the interfacial stability qualitatively.

The intensity of interfacial turbulence strongly depends on the magnitude of the gradient of interfacial tension and the relative rate of transfer of mass and momentum in the two phases. This partly explains the stabilizing effect of surfactants which affect the phenomena in several ways. By preferentially adsorbing at the interface, surfactants reduce the interfacial tension below the level at which it can be changed by local perturbations to the solute concentration. In this way the interfacial forces that may trigger interfacial instabilities are supressed. In addition surfactants alter the structure of the interface causing changes in rheological behaviour and in the rate of mass transfer.

Surface tension driven flows are not the only cause of interfacial turbulence. Although not included within the Marangoni phenomena, density gradients in the vicinity of the interface may produce interfacial convection and have a similar effect on the rate of interfacial transfer. The mechanism of this type of instability is the same as that of the Bénard cells appearing in evaporating liquid layers (Bénard 1900a,b, 1901). If a solute substantially decreases the density of both phases, its transfer from the lower phase will produce unstable density gradients (Berg and Morig 1969; Sawistowski 1971). Convective movements may then develop into sustained interfacial convection in systems that are stable to Marangoni flows. This gravitational convection can also be produced by volume contractions due to mixing and density gradients produced by heat of solution or reaction.

3.3 Types of Marangoni phenomena

According to its origin and relative depth of penetration into the liquid bulk phases, interfacial convection can be classified into

(i) Marangoni instabilities;

(ii) Marangoni disturbances;

(iii) thin-film phenomena.

In the first two cases the depth of penetration of the interfacial convection into the two phases is small with respect to the thickness of the phases. The difference between Marangoni instabilities and Marangoni disturbances is the magnitude of the mass transfer driving force required to produce interfacial perturbations. While Marangoni instabilities occur spontaneously under initial conditions of molecular diffusion, Marangoni disturbances take place in the presence of surface renewal caused by forced convection in systems that would be otherwise stable.

In very thin films, such as those trapped between drops before coalescence, the depth of the interfacial flow caused by a local change in interfacial tension is large enough to change the thickness of the film substantially. This phenomenon is illustrated in Fig. 3.3 for the transfer of a lowering tension solute into the film. In the section of the film where there is a minimum thickness the solute concentration increases faster than in the neighbouring

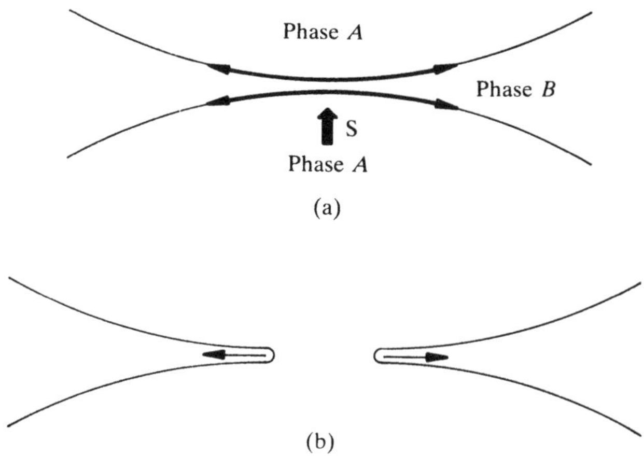

Fig. 3.3. Marangoni thin-film phenomena: (a) Transfer of an interfacial tension lowering solute into a thin film; (b) film breakage.

sections leading to a local drop in interfacial tension (Fig. 3.3(a)). The interfacial tension driven flow may be strong enough to break up the film (Fig. 3.3(b)). If the same solute were transferred out of the film, interfacial convection would be in the opposite direction thus thickening the film instead of breaking it. These phenomena may affect both drop coalescence and break up in dispersed phase contactors, as well as the stability of supported films, such as those occurring in packed columns.

Thus while both Marangoni instabilities and Marangoni disturbances are likely to increase the interfacial flux by introducing a convective component into the mass transfer mechanism, thin-film phenomena may affect contactor performance by changing the specific interfacial area.

3.4 Mathematical analysis of instability

The mechanism of Marangoni instability discussed previously suggests that the stability of the interface depends on the response of the system to interfacial flows due to local changes of interfacial tension. If the interfacial perturbation dies down the system is stable, if it grows in time the system is unstable. The mathematical treatment of stability problems generally proceeds along the following lines. Initially the system is in stationary state. The phases may be either stagnant or flowing as in the case of a liquid jet. The equations governing the perturbed system are obtained by introducing infinitesimal increments to the physical variables that describe the flow. These equations are linearized and solved in order to obtain the evolution of the perturbations in time. The theory derived on the basis of these equations is called the linear stability theory and it has been widely used in problems of hydrodynamic stability. A detailed description of the linear hydrodynamic stability analysis is given by Chandrasekhar (1961). This theoretical analysis was first applied to the diffusional transfer of a solute between two liquid phases by Sternling and Scriven (1959). Subsequent stability studies involving mass transfer have usually followed their approach.

The general procedure will be discussed briefly followed by its application to some problems relevant to solvent extraction. The model to be analysed consists of two liquid phases in contact along a planar or cylindrical interface. A solute is transferring from one of the phases into the other in steady state. This transfer may be accompanied by heat of solution. It is assumed that the liquids are Newtonian and incompressible and that neither concentration nor temperature gradients affect their physical properties. Thus in the undisturbed state the concentration and temperature profiles, as shown in Fig. 3.2(a) for a planar interface, are functions only of the coordinate x, and can be expressed as

$$C_A^0 = C_{iA}^0 + \mathscr{L}_A x \quad \text{at } x > 0 \tag{3.1}$$

$$C_B^0 = C_{iB}^0 + \mathscr{L}_B x \quad \text{at } x < 0 \tag{3.2}$$

$$T_A^0 = T_i^0 + \mathscr{G}_A x \quad \text{at } x > 0 \tag{3.3}$$

$$T_B^0 = T_i^0 + \mathscr{G}_B x \quad \text{at } x < 0 \tag{3.4}$$

where the superscript 0 indicates undisturbed values, the subscript i interfacial values, $\mathscr{L} = \mathrm{d}C/\mathrm{d}x$, $\mathscr{G} = \mathrm{d}T/\mathrm{d}x$, and subscripts A and B denote the phase.

A small disturbance is imposed upon the system that causes perturbations in the solute concentration, temperature and velocity components in both phases. Neglecting terms of the second and higher order the equations for the concentration, temperature and velocity in the perturbed state can be written as:

$$C = C^0 + C'\ (x, y, t) \tag{3.5}$$

$$T = T^0 + T'\ (x, y, t) \tag{3.6}$$

$$\mathbf{u} = \mathbf{u}^0 + \mathbf{u}'\ (x, y, t) \tag{3.7}$$

where $\mathbf{u}$ is the velocity vector and C', T', and $\mathbf{u}'$ are the corresponding perturbations. There are two sets of eqns (3.5–3.7), one for phase A and the other for phase B. In two-dimensional models the velocity vector has two components, one normal and the other tangential to the interface. If in the initial state the phases are stagnant then $\mathbf{u}^0 = 0$. In the perturbed state C, T and $\mathbf{u}$ are functions of x, y, and time and must satisfy the mass, energy, and momentum balances. These are given by the continuity, diffusion, energy and Navier–Stokes equations respectively:

$$\nabla \cdot \mathbf{u} = 0 \tag{3.8}$$

$$\partial C/\partial t + \mathbf{u}\cdot\nabla C = \mathscr{D}\nabla^2 C \tag{3.9}$$

$$\partial T/\partial t + \mathbf{u}\cdot\nabla T = \kappa\nabla^2 T \tag{3.10}$$

$$\partial \mathbf{u}/\partial t + \mathbf{u}\nabla\cdot\mathbf{u} = \nu\nabla^2\mathbf{u} - \nabla p/\rho \tag{3.11}$$

where p is the pressure, ρ the density, and κ the thermal diffusivity. Equations (3.11) are the linearized Navier–Stokes equations.

The cell patterns observed at the interface of unstable systems suggest that the disturbances can be analysed in terms of periodic waves. The disturbance wavelength represents the dimension of the convective cells in the interfacial plane at the onset of instability and its amplitude the penetration depth into the phases. Usually both the disturbance amplitude and wavelength vary with time. Thus solutions to eqns (3.8–3.11) are sought for perturbations of the form:

$$C' = A(x)\exp(i\alpha y + \beta t) \tag{3.12}$$

$$T' = B(x)\exp(i\alpha y + \beta t) \tag{3.13}$$

$$\varphi' = \varphi(x)\exp(i\alpha y + \beta t) \tag{3.14}$$

where φ is the stream function of the velocity field, $\alpha = 2\pi/\lambda$ is the wave number, λ is the wave length, and β is the disturbance growth constant to be determined for each value of α.

In general the growth constant β is a complex number

$$\beta = \hat{\beta} + \tilde{\beta}\mathrm{i} \tag{3.15}$$

The real part, $\hat{\beta}$ gives the rate of growth of the perturbation with time. Positive values of $\hat{\beta}$ indicate that the disturbance grows in time and the system is unstable. When $\hat{\beta} < 0$ the disturbance is damped and the interface remains stable. The stability analysis thus consists in determining the value of $\hat{\beta}$ for all possible wavelengths using the solutions of the perturbation equations.

In most unstable systems a minimum disturbance size is required for instabilities to set in, i.e. below this critical size $\hat{\beta} < 0$, and above it $\hat{\beta} > 0$. Of special interest is then the value of α for which $\hat{\beta} = 0$ since it separates the stable from the unstable regime. This is the state of marginal or neutral stability in which the perturbation neither grows nor decays.

The imaginary component $\tilde{\beta}$ gives the perturbation frequency. If $\tilde{\beta} = 0$ the perturbation either grows or decays in place. For $\tilde{\beta} \neq 0$ the perturbation oscillates with a frequency given by $\tilde{\beta}$. In the neutral regime of stability, i.e.

when $\hat{\beta} = 0$, the disturbance may remain in place or oscillate depending on whether or not $\tilde{\beta} = 0$. These two states are called neutral stationary regime and neutral oscillatory regime respectively. A system is considered stable only when $\hat{\beta} < 0$ for all values of α.

The solutions to eqns (3.8–3.11) must satisfy a set of boundary conditions. Conditions that are common to the problems of interfacial stability considered in this chapter are discussed in the following section.

3.4.1 General boundary conditions

The perturbation equations [eqns (3.8–3.11)] have been derived on the assumption that the disturbances are infinitesimal. It therefore follows that far from the interface **u**, C and T have the values of the undisturbed state. The Marangoni perturbations occur at the interface due to an interfacial tension gradient, $\partial\sigma/\partial y$, that produces interfacial flow. In turn these flows generate interfacial stresses that act as stabilizing forces. The balance of these forces must satisfy the continuity of shear stresses. Thus at $x = 0$

$$\tau_{xy}|_B - \tau_{xy}|_A = \partial\sigma/\partial y \tag{3.16}$$

where τ_{xy} is the shear stress in the direction y and σ is the dynamic interfacial tension. In accordance to the Boussinesq approximation the dynamic interfacial tension can be written as

$$\sigma = \sigma_0 + \xi, \quad \text{at } x = 0 \tag{3.17}$$

where σ_0 is the equilibrium interfacial tension and ξ contains terms that describe the rate of interfacial deformation. The equilibrium interfacial tension is generally a function of solute concentration, temperature and surface density of electrostatic charges.

Forces in the direction normal to the interface must satisfy the continuity of normal stresses at $x = 0$:

$$p_A - \tau_{xx}|_A - p_B + \tau_{xx}|_B = \sigma\left[\frac{1}{R_1} + \frac{1}{R_2}\right] + \partial\sigma/\partial x \tag{3.18}$$

where τ_{xx} is the normal stress and R_1 and R_2 are the principal radii of curvature of the interface. If the interface is planar and does not deform under the effect of the perturbations, the right hand side of eqn (3.18) vanishes.

Finally, the tangential and normal velocities at the interface are the same, i.e.:

$$\mathbf{u}_A = \mathbf{u}_B, \quad \text{at } x = 0 \tag{3.19}$$

Boundary conditions related to the conservation of solute and energy at the interface will be discussed for each particular problem.

3.5 Applications of the stability analysis to the derivation of stability criteria

3.5.1 Transfer of a single solute across a flat interface

In their pioneering study Sternling and Scriven (1959) applied a linearized stability analysis to systems undergoing mass transfer and derived the stability criteria for the transfer of a solute between two equilibrated liquid phases. Their model has the general characteristics described earlier. In addition it was assumed that in the perturbed state

(i) the two phases are in thermodynamic equilibrium at all points in the interface;

(ii) there is no accumulation of solute at the interface;

(iii) perturbations are so small that they do not deform the interface in the direction x;

(iv) there are no heat effects.

Conditions (i)–(iii) lead to the following relations:

$$mC^0_{Ai} = C^0_{Bi} \tag{3.20}$$

$$\mathscr{D}_A \partial C_A/\partial x = \mathscr{D}_B \partial C_B/\partial x \tag{3.21}$$

$$u_{xA} = u_{xB} = 0 \tag{3.22}$$

where m is the solute distribution coefficient and u_x the normal component of the velocity vector.

Since there are no heat effects the variation of interfacial tension along the y-direction is only a function of the interfacial concentration gradient. Thus boundary condition (3.16) becomes:

$$\tau_{xy}|_B - \tau_{xy}|_A = (\partial\sigma_0/\partial C)(\partial C/\partial y) + \mu_s(\partial^2 u_y/\partial y^2) \tag{3.23}$$

where μ_s is the composite interfacial viscosity and u_y is the tangential component of the velocity vector.

In this model the heat of solution is neglected so the equations of energy vanish. Solution of the equations of motion and the diffusion equations yields the characteristic equation relating the disturbance wave number α with the growth constant β. Expressed in terms of dimensionless numbers the characteristic equation is a functional relation of the form:

$$\bar{\beta}_A = f(Ma_D, \bar{\beta}_A, r^2, e^2, Sc, m, v_A) \tag{3.24}$$

where $\bar{\beta}$ is the dimensionless growth constant and Ma_D the Marangoni

TABLE 3.1 *Dimensionless groups*

$Ma_D \equiv$	$\dfrac{(\partial\sigma_0/\partial C)(\partial C/\partial x)}{\mu\mathscr{D}\alpha^2}$	Marangoni number for mass transfer
$\bar{\beta} \equiv$	$\dfrac{\mu\beta}{(\partial\sigma_0/\partial C)(\partial C/\partial x)}$	dimensionless growth constant
$Sc \equiv$	$\nu/\mathscr{D}$	Schmidt number
$e^2 \equiv$	ν_A/ν_B	
$r^2 \equiv$	$\mathscr{D}_A/\mathscr{D}_B$	
$v_A \equiv$	μ_s/μ_A	

number for mass transfer. The expressions for these dimensionless numbers are given in Table 3.1. The Marangoni number is the ratio between the destabilizing interfacial forces and the stabilizing viscous and diffusional forces. Because in this model the depth of the phases do not affect the stability of the interface, there is no characteristic length other than the disturbance wave length, λ, represented by α. Therefore for this case the Marangoni number becomes

$$Ma_D \equiv \frac{(\partial\sigma_0/\partial C)(\partial C/\partial x)}{\mu\,\mathscr{D}\alpha^2} \tag{3.25}$$

The dimensionless growth constant is defined as

$$\bar{\beta} \equiv \frac{\mu\beta}{(\partial\sigma_0/\partial C)(\partial C/\partial x)} \tag{3.26}$$

Once all the physical properties and concentration gradients are assigned, eqn (3.26) determines the growth constant β for each wave number α. Therefore stability criteria can be derived by analysing the behaviour of eqn (3.24) for different values of the dimensionless groups.

As previously discussed the condition for stability is that $\hat{\beta}$ be negative for all values of α. Unstable systems are usually stable with respect to disturbances of wave lengths lower than that leading to neutral stability. The wave number of neutral stability is obtained from eqn (3.24) for $\hat{\beta} = 0$. In the neutral stationary regime $\tilde{\beta}$ is also equal to zero while in the neutral oscillatory regime $\tilde{\beta}$ has a positive value.

Since the wave number of neutral stability separates the disturbances leading to instability from those that do not affect the interfacial behaviour, any system in which neutral stability is possible is considered unstable. Therefore stability criteria can be derived from the conditions leading to neutral stability. This is achieved by studying the limiting behaviour of

TABLE 3.2 *Prediction of stability behaviour in ternary systems (Sternling and Scriven 1959)*

Ma_D	r^2	e^2	f	predicted instability
		$= 1$	–	
> 0	> 1			stationary
		< 1	≥ 0	
		$= 1$	–	
< 0	< 1			stationary
		> 1	≥ 0	
> 0	$= 1$	< 1	–	oscillatory
< 0	$= 1$	> 1	–	oscillatory
> 0	> 1	< 1	< 0	stationary and oscillatory
< 0	< 1	> 1	< 0	stationary and oscillatory
> 0	> 1	$> r^2$	–	
or	< 1	$< r^2$	–	stationary and oscillatory
> 0	$\geq e^2$	> 1	–	
	$\leq e^2$	< 1	–	

eqn (3.24) for $\hat{\beta} = 0$ and $\tilde{\beta} \geq 0$ as discussed by Sternling and Scriven (1959).

The conditions for interfacial instability derived from eqn (3.24) are summarized in Table 3.2 where

$$f = \frac{r^2 + 1}{2} + \frac{m + 1}{2(1 + m/r^2)} + \frac{(1 + e^2\mu_B/\mu_A)/Sc_A}{4(1 + \mu_B/\mu_A + \alpha_{NS}v_A/2)} + \frac{(r^2e^2 - 1)}{4(r^2 - 1)Sc_A} \tag{3.27}$$

As seen in this table conditions for instability depend on the sign of the Marangoni number. Because the wave number is real, only solutions for positive values of α^2 have physical meaning. Thus the sign of the Marangoni number depends on the sign of the product $(\partial\sigma_0/\partial C)(\partial C/\partial x)$. While the sign of $\partial\sigma_0/\partial C$ is a characteristic of the system, that of $\partial C/\partial x$ depends on the direction of solute transfer. Therefore every system has both a positive and a negative Marangoni number according to the direction of transfer. On the other hand the values of r^2, e^2, and f depend only on the physical properties of the system. As an example let us consider a system in which $\nu_A = \nu_B$ (i.e. $e^2 = 1$), $\mathscr{D}_A > \mathscr{D}_B$ (i.e. $r^2 > 1$) and $\partial\sigma_0/\partial C < 0$. The stability criteria predict stationary instabilities for $Ma_D > 0$, i.e. for $\partial C/\partial x < 0$, which corresponds to solute transfer out of the phase where the solute diffusivity is lower. In the opposite direction of transfer the system is predicted stable. In a system with $r^2 = 1$ and $e^2 > 1$ the unstable direction of transfer is out of the phase of higher viscosity. In this case the instabilities are oscillatory. These theoretical results are in agreement with the qualitative analysis of instability discussed previously. When the two destabilizing conditions act in the same direction of transfer the type of instability depends on the sign of f.

A qualitative analysis cannot predict the interfacial stability when a destabilizing condition opposes a stabilizing one, as in the case in which $e^2 > 1$ and $r^2 > 1$. In this situation the Sternling and Scriven theoretical results indicate that the destabilizing effect overrules the other and the system can be unstable in both directions of transfer as shown in Table 3.2.

The response of unstable systems to disturbances of different wave length is illustrated in Fig. 3.6 where the growth constant $\hat{\beta}$ is plotted vs the wave number. Although the systems in the figure are binary their $\hat{\beta}$ vs α behaviour curves are similar to those of ternary systems. The curves show that the value of α_{NS} can vary substantially with the system. Thus disturbances in the interval $45 < \alpha < 118$ will destabilize systems (a) and (b) but will not be large enough to affect system (c).

This explains why some unstable systems do not show interfacial turbulence in the absence of bulk motion. The maximum value of $\hat{\beta}$ in each curve corresponds to the wave number of the faster growing disturbance. Since this will ultimately dominate the system the corresponding value of λ is called the dominant disturbance, λ_D. Disturbances of wave number $\alpha > \alpha_{NS}$ (i.e. $\lambda < \lambda_{NS}$) will be too small to overcome the viscous shear forces opposing them and will be damped.

The stability criteria derived by Sterling and Scriven are listed in Table 3.3.

The applicability of these criteria to the prediction of interfacial stability in ternary systems has been widely investigated (Berg 1972; Sawistowski 1971). Although some disagreements have been found between stability criteria and experimental observations, the criteria provide valuable guidelines for the prediction of interfacial behaviour. It should be noted that phenomena such as volume contraction on mixing, interactions between solvent and solute and surface deformation are not included in this analysis. Therefore any of these phenomena may lead to disagreement with the predicted behaviour.

Some of the restrictions of the Sternling and Scriven model have been

TABLE 3.3 *Conditions that promote interfacial instabilities in ternary systems (Sternling and Scriven 1959)*

(i)	Solute transfer out of the phase of higher viscosity
(ii)	Solute transfer out of the phase where its diffusivity is lower
(iii)	Large differences in kinematic viscosity and solute diffusivity between the two phases
(iv)	Steep concentration gradients near the interface
(v)	Interfacial tension highly sensitive to solute concentration
(vi)	Low viscosities and diffusivities in both phases

removed by subsequent studies. Some of these models and the main conclusions derived from their analysis will be briefly discussed here.

Marsh *et al.* (1965) extended the Sternling and Scriven analysis to solute transfer in unsteady state. The only contradictory criterion derived from the transient model is that transfer out of the phase where the solute diffusivity is higher is also predicted unstable. Although the other criteria are not changed the transient model predicts instability with respect to smaller disturbances and larger growth constants than the steady-state model. Gross and Hixon (1969a,b) investigated the effect of non-linear concentration profiles in the presence of an interfacial resistance to solute transfer. Their numerical results showed that the range of unstable disturbances in unsteady state was wider than in steady state. Increasing the interfacial resistance from low values had first a destabilizing effect until a point was reached beyond which further increases in interfacial resistance increased the system stability. Another effect of considering the interfacial resistance was that certain systems were predicted unstable in either direction of transfer. The authors also considered the influence of interfacial viscosity, which they found to be a stabilizing factor.

Solutions of the stability equation for deformable interfaces, transfer of surface active solutes, systems with density stratification, mass transfer with interfacial chemical reactions and electrical constraints have been obtained by Hennenberg, Sørensen and co-workers and by Sanfeld and co-workers. These two groups, who have often collaborated, derived their interest in interfacial stability problems from their possible relevance in the deformation of biological cell membranes. A comprehensive set of review papers by members of the two groups covering different problems of interfacial stability has been edited by Sørensen (1979). The stability of a flat deformable interface undergoing the transfer of a non-ionic surface active solute was studied by Sørensen (1978). They also considered the effect of non-linear concentration profiles. Contrary to expectations differences between their numerical results and those obtained using Sternling and Scriven's model were negligible for systems with interfacial tensions greater than 1 mN m^{-1}. However at lower interfacial tensions, oscillatory instabilities appeared in regions predicted stable by the Sternling and Scriven model. Their numerical results were in good agreement with the convection cell diameters measured by Orell and Westwater (1957) during the transfer of acetic acid between ethylene glycol and ethyl acetate.

Sørensen (1979) also investigated the effect of the coupling of Marangoni and gravitational instabilities. The latter, which refers to the instabilities of the plane interface when the heavier liquid is placed above the lighter, is called the Rayleigh–Taylor instability (Chandrasekhar 1961). This problem is relevant to liquid–liquid extraction in cases where there is volume contraction on mixing or when the density of the solute is such that the concentration gradients lead to unstable density gradients. The theoretical analysis of this problem showed that interference between the two types of instabilities is negligible except in cases in which the wavelengths of both instabilities are very close.

3.5.2 Mass transfer with heat effects: binary liquid–liquid systems

It was generally expected that the order of magnitude of the temperature driven destabilizing forces would be too small to modify the stability criteria derived for ideal ternary systems. This was confirmed by Ortiz and Sawistowski (1975) by adding heat effects to the Sternling and Scriven analysis and solving the resulting characteristic equation for typical values of physical properties and heats of solution.

However a different response was obtained in the case of heat effects in partially miscible liquid pairs. In these systems the solute in one phase is the solvent in the other, as for example in the acetylacetate–water pair. Thus the solute is not the same in the two directions of transfer.

Assuming that interfacial equilibrium is reached instantaneously and in isothermal conditions, the phase rule precludes local changes of interfacial concentration in binary systems. This seems to eliminate the possibility of Marangoni instabilities thus making these systems an obvious choice for fundamental studies in mass transfer. However this prediction was not confirmed experimentally (Austin *et al.* 1966; Ying and Sawistowski 1971; Davies and Thornton 1977). In Fig. 3.4 schlieren photographs of the systems acetylacetone–water and ethylacetate–water show that the type of phenomena encountered in binary systems are similar to those observed in ternary systems.

Ying and Sawistowski (1971) investigated 46 partially miscible liquid pairs using a schlieren system (Sawistowski 1971). Some of them were found to be stable while others showed varying degrees of interfacial activity. Directional effects, i.e. interfacial convection in only one direction of transfer, were also observed. Were these phenomena the result of interfacial tension gradients, there could be two situations that could make local variations of interfacial tension possible:

(i) the presence of thermal effects accompanying mass transfer, which would introduce the required degree of freedom in the phase rule, and/or;

(ii) dynamic interfacial effects, which would preclude the application of the phase rule to interfacial elements of short contact times.

Heat effects are known to produce interfacial instabilities in thin films. Bénard (1900a,b) observed regular cell patterns on the surface of liquid layers

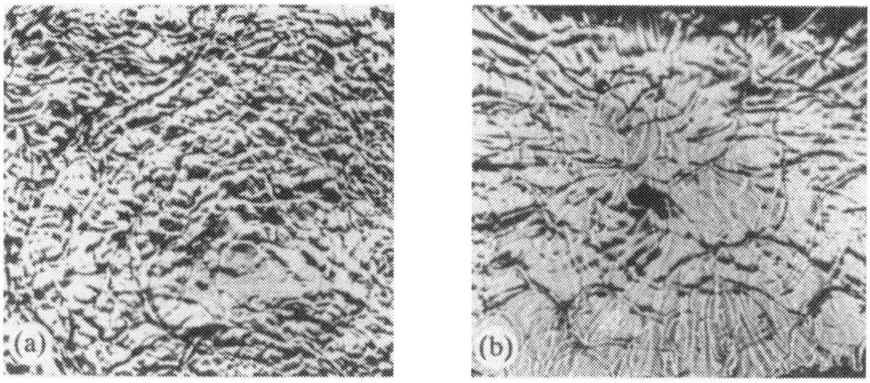

Fig. 3.4. Unstable binary systems: (a) acetylacetone–water; (b) ethylacetate–water. (After Austin *et al.* 1966.)

heated from below. Because of thermal expansion, the fluid at the bottom is lighter than the fluid at the top thus producing a potentially unstable density gradient. The ever present vibration in the system may trigger convection currents that will be resisted by the fluid viscosity. The viscous stabilizing forces may be overcome if the temperature gradient is increased above a critical value. This mechanism of instability, studied mathematically by Lord Rayleigh (1916), is gravitational.

However the same type of convective patterns were observed in thin films of paint in which the free surface was the lower boundary, i.e. when evaporation of the solvent cooled the film from below thus producing a stable density gradient. These observations led Pearson (1958) to propose an instability mechanism based on temperature driven Marangoni perturbations. His stability analysis shows that thin films can become unstable when a critical value of the thermal Marangoni number is exceeded. He also established that there is a critical film thickness over which the destabilizing mechanism changes from surface tension driven to buoyancy driven.

The possibility of similar phenomena taking place at the interface of binary systems partly depends on the steepness of the temperature gradients produced by heat of solution. Ortiz and Sawistowski (1973a) measured temperature differences of the order of 0.2–0.6°C between the interface and points 2 mm deep into one of the phases in four unstable binary systems. Experimental results were in good agreement with the transient temperature profiles obtained for the same four systems by solving the diffusion equation for simultaneous heat and mass transfer. Figure 3.5 shows the theoretical concentration and temperature profiles in the system methyl ethyl ketone–water for transfer of the ketone into water. An analysis of the stability of

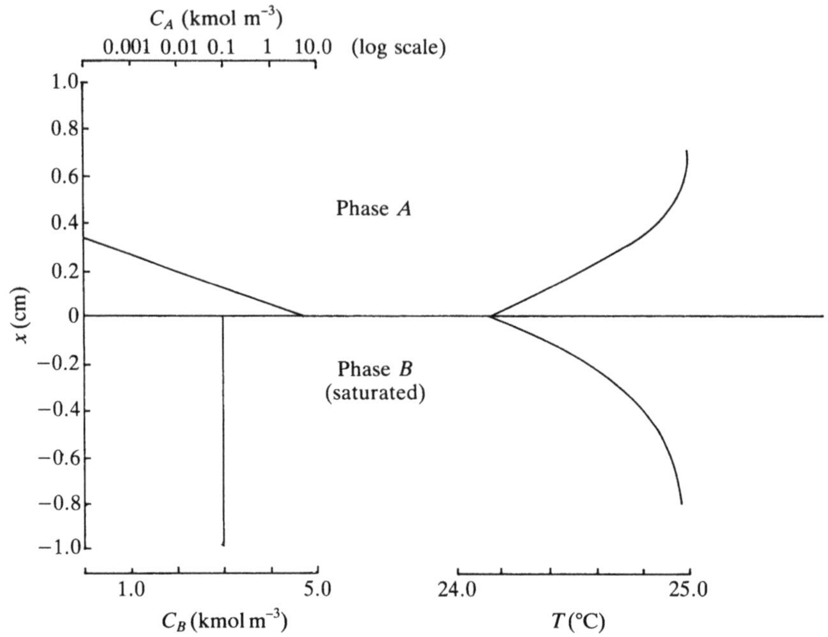

Fig. 3.5. Theoretical concentration and temperature profiles for the transfer of methyl-ethyl-ketone into initially solute-free water after 30 s contact time. (After Ortiz and Sawistowski 1973b.)

systems to local perturbations of the interfacial temperature is discussed in the next section.

3.5.2.1 *Stability analysis of binary systems*

The procedure (Ortiz and Sawistowski 1973a) follows closely Sternling and Scriven's analysis. The model is also similar to that for ternary systems. It is specific to binary systems that at the instantaneously equilibrated interface, concentration and temperature are not independent variables. Therefore local variations of interfacial concentration can only occur where there are changes in temperature. The variation of interfacial tension with y is given by:

$$\partial\sigma/\partial y = \partial\sigma_0/\partial T + \mu_s\,(\partial^2 u_y/\partial y^2) \qquad (3.28)$$

The heat generation at the interface results from the heat of solution produced by the interfacial molar flux. Assuming none of the phases is saturated with the other, i.e. there is simultaneous transfer of both solutes, the energy balance at the interface is given by

$$\mathscr{D}_A \frac{\partial C_A}{\partial x} \Delta H_A - \mathscr{D}_B \frac{\partial C_B}{\partial x} \Delta H_B = -k_A \frac{\partial T_A}{\partial x} + k_B \frac{\partial T_B}{\partial x} \tag{3.29}$$

where $-\Delta H_A$ and $-\Delta H_B$ are the heats of solution of component B in A and component A in B respectively (subscripts A and B refer to the solvent phase), C_A and C_B are the concentrations of solute in phases A and B, as previously, and k_A and k_B are the thermal conductivities in phases A and B. If one of the phases is saturated there is mass transfer in only one direction and the corresponding term on the left hand side will disappear. Simultaneous solution of the equations of motion, diffusion and energy gives a characteristic equation of the form:

$$Ma_T = f\,(Ma_T, \bar{\beta}_A, D_j) \tag{3.30}$$

where

$$Ma_T \equiv \frac{(\partial \sigma_0/\partial T)\,(\partial T/\partial x)_A}{\mu_A\, \kappa_A\, \alpha^2} \tag{3.31}$$

is the Marangoni number for heat transfer and D_j are dimensionless groups involving physical properties and temperature gradients. An analysis using a range of values of physical properties and heats of solution most commonly encountered in binary systems predicts the stability criteria summarized in Table 3.4. The derived stability criteria were applied to six selected binary systems that covered a wide range of values of physical parameters and represented different degrees of intensity of interfacial activity (Ortiz and Sawistowski 1973b).

3.5.2.2 *Comparison between predicted and experimental results*

Predicted behaviour and experimental observations obtained under quiescent conditions are given in Table 3.5. It should be noticed that a system is predicted stable only if it is stable to disturbances of all sizes. It is therefore feasible that the minimum disturbance wave length required to destabilize the system may be outside the range that can be expected under experimental conditions. This can be seen in Fig. 3.6 where the growth constant is plotted against wave number for three of the systems listed in Table 3.5. The curves show that, of the three systems, water–isobutanol is the most stable since it

TABLE 3.4 *Conditions that promote interfacial instabilities in binary systems*

(i)	Heat of solution and variation of interfacial tension with temperature are of opposite sign
(ii)	Solute transfer out of the phase of higher viscosity
(iii)	Large absolute values of the thermal Marangoni number

TABLE 3.5 *Comparison of predicted and experimental behaviour of binary systems (Ortiz and Sawistowski 1973b)*

Liquid A	Liquid B	Direction of transfer	Predicted response	Experimental behaviour*
ethyl acetate	water	$A \rightarrow B$	unstable	unstable
		$B \rightarrow A$	unstable	unstable
acetylacetone	water	$A \rightarrow B$	unstable	unstable
		$B \rightarrow A$	stable	stable
isobutanol	water	$A \rightarrow B$	unstable	stable, unstable**
		$B \rightarrow A$	unstable	stable, unstable†
MEK	water	$A \rightarrow B$	stable	stable
		$B \rightarrow A$	unstable	stable
cyclohexanol	water	$A \rightarrow B$	unstable	unstable
		$B \rightarrow A$	stable	unstable
hexane	aniline	$A \rightarrow B$	stable	unstable
		$B \rightarrow A$	unstable	unstable

* Ying and Sawistowski (1971)
** Aguirre *et al.* (1985)
† Merson and Quinn (1964)

requires the largest disturbance wavelength to be destabilized and has the lowest growth constant. This may explain why Ying and Sawistowski (1971) did not observe any interfacial activity under static conditions while Merson and Quinn (1964) reported instabilities for the transfer of water into isobutanol in a flow system in which perturbations of larger magnitude may have occurred. In the opposite direction of transfer, Aguirre *et al.* (1985) determined experimentally the concentration profile of isobutanol in water. On comparing their results with the theoretical profiles obtained by solving a diffusion model for a stationary system, they observed that experimental concentrations were consistently higher than their predicted values. Deviations between theoretical and experimental results were even higher for the system ethylacetate–water. They attributed this enhancement in the rate of solute transfer to convection and interfacial activity.

Table 3.5 shows disagreement between predicted and observed behaviour for transfer of hexane into aniline and water into cyclohexanol, both predicted stable contrary to experimental observations. In the former, conditions exist for gravitational instability to be possible. This is also the case in the transfer of acetylacetone into water.

In order to decouple the possible interaction between surface tension,

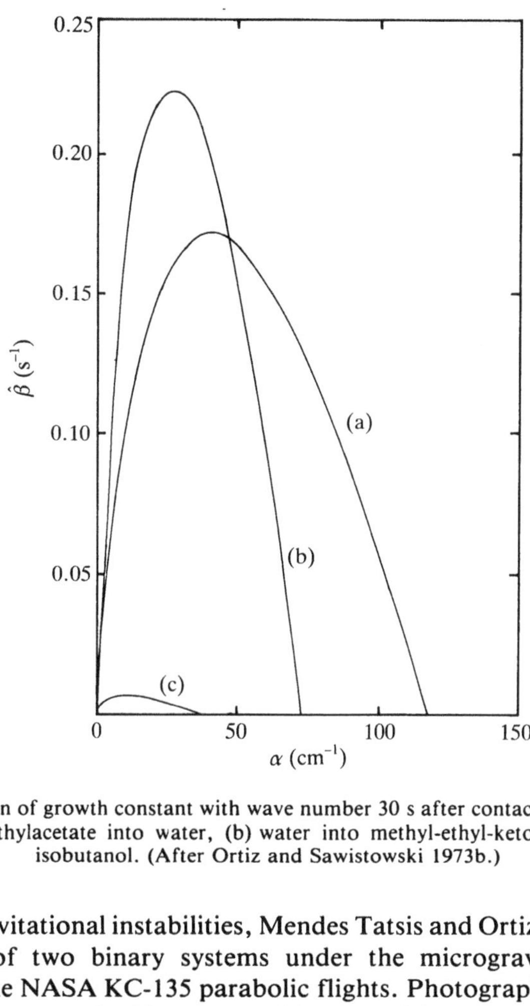

Fig. 3.6. Variation of growth constant with wave number 30 s after contacting the phases for transfer of: (a) ethylacetate into water, (b) water into methyl-ethyl-ketone, (c) water into isobutanol. (After Ortiz and Sawistowski 1973b.)

driven and gravitational instabilities, Mendes Tatsis and Ortiz (1989) studied the stability of two binary systems under the microgravity conditions provided by the NASA KC-135 parabolic flights. Photographs of a drop of acetylacetone saturated with water and immersed in pure water are shown in Fig. 3.7. Under normal gravity conditions Marangoni instabilities in the form of eruptions develop into streams due to density gradients. In a drop formed under microgravity conditions (Fig. 3.7(b)) intense turbulence is observed but confined to the drop surroundings. These experiments seem to confirm that thermal Marangoni disturbances can be strong enough to destabilize the interface.

The stability criteria for binary liquid systems have been used by Hancock *et al.* (1980) to explain the formation of void structures in the coagulation of membranes and spun fibres. These systems consist of a polymer solution brought into contact with a coagulant solvent. Hancock *et al.* observed that the void structures originated in a fingering phenomena observed at the

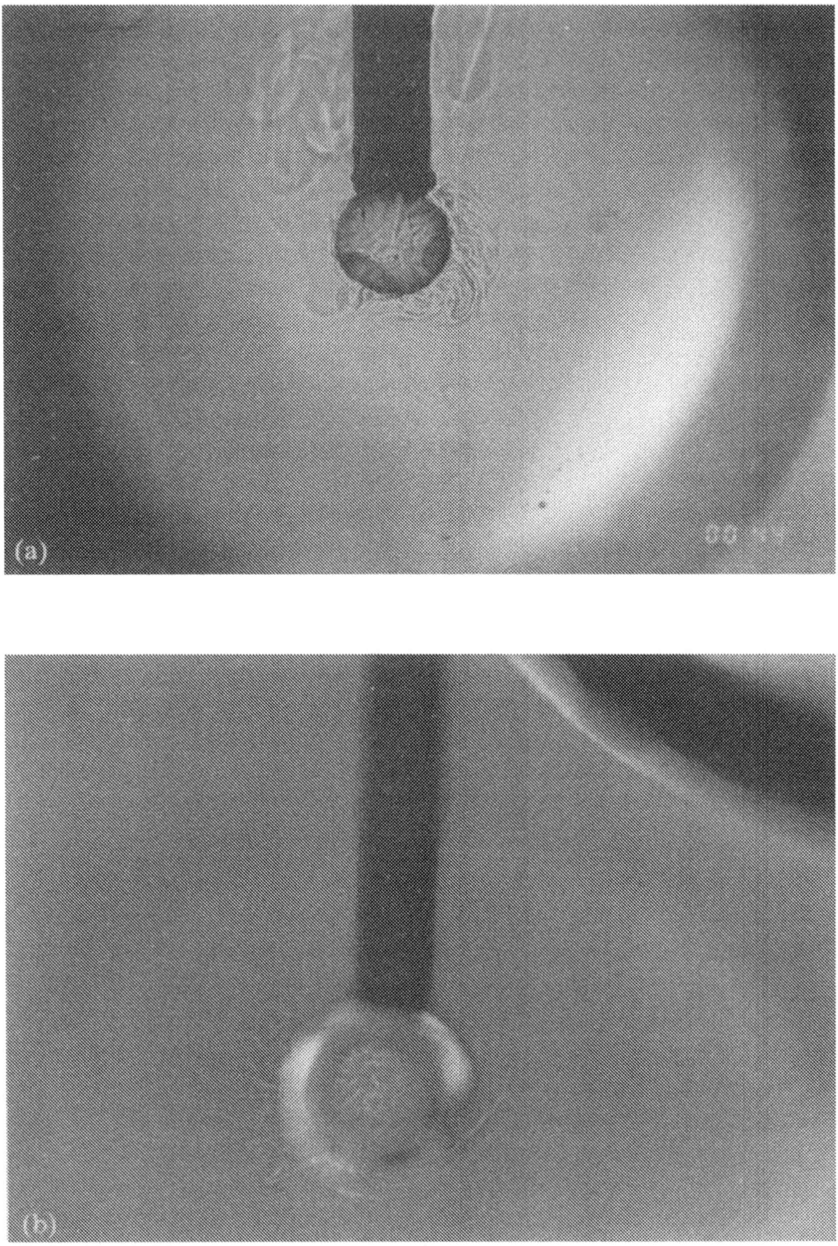

Fig. 3.7. Effect of gravity on interfacial turbulence during the transfer of acetylacetone from an acetylacetone drop saturated with water into pure water (a) normal gravity; (b) at microgravity conditions obtained in a NASA KC-135 parabolic flight. (After Mendes Tatsis and Ortiz 1989.)

polymer solution–coagulant interface. In the systems in which these deformations occurred they were also observed in the absence of the polymer, i.e. when the systems were just binary. The same mechanism has been suggested for the formation of finger-like cavities in coagulated membranes (Frommer and Messalem 1973).

3.5.3 Mass transfer with chemical reaction

Experimental evidence of interfacial turbulence in extraction accompanied by chemical reaction is abundant. Probably the first description of such phenomena is that of Quincke's (1888). He observed spontaneous emulsification on contacting a solution of lauric acid in oil with an aqueous solution of NaOH.

More recently Sherwood and Wei (1957) observed spontaneous emulsification in the extraction of acetic acid from an organic solvent into an alkaline solution. The phenomenon was attributed to heat effects due to the strongly exothermal reaction. Dupeyrat and Nakache (1977, 1978) observed interfacial turbulence accompanying the reaction of an alkyl ammonium ion with KI, KBr and picric acid at an oil–aqueous interface. They suggested a mechanism of instability based on the possible presence of oscillating reactions. The nature of these reactions would create the necessary conditions for local changes of interfacial tension. Subsequently Nakache *et al.* (1983, 1984) described the different types of interfacial instabilities appearing at the interface between an aqueous solution of a long chain alkylammonium chloride ($C_{16}Cl$) and an organic solution of picric acid (HPi). The observed phenomena included emulsification, interfacial movements and surface deformation. In a systematic study of the conditions leading to the different types of instabilities they measured the variation of interfacial tension and pH with contact time. Oscillations of interfacial tension in experiments where interfacial movements were evident indicated a correlation between surface convection and adsorption density. It was also observed that the system was always stable when the reactants were dissolved in the phases where they were less soluble, i.e. HPi in water and $C_{16}Cl$ in the organic solvent. The organic solvent also affected the interfacial stability. Replacing nitrobenzene by hexane eliminated interfacial convection. This was interpreted in terms of the effect of the dielectric constant of the solvent on the ionization of the reactants. Substitution of KPi for HPi also eliminated interfacial movements but emulsification was still observed. The proposed mechanism of instability included adsorption–desorption steps as well as chemical, diffusional and convective steps. The mechanism also explained the effect of reactant concentrations.

Similar interfacial movements accompanied by emulsification were reported by Nakache *et al.* (1986) in the extraction of copper by a mixture of di(2-ethylhexyl) phosphoric acid (DEPAH) and its sodium salt (DEPANa) in xylene. An even higher degree of interfacial activity and interfacial deformation was observed during extraction with DEPANa alone while extraction with DEPAH did not destabilize the interface. The behaviour of these systems was also explained in terms of interfacial phenomena.

Another example of spontaneous interfacial convection in systems with solute–extractant interactions was found in the area of nuclear fuel reprocessing. Thompson *et al.* (1984) observed interfacial instabilities during the extraction of uranyl nitrate, plutonium (IV) nitrate and nitric acid from aqueous solutions into an organic phase containing tri-butyl phosphate (TBP). These extraction systems involve solvation steps. Using both schlieren and shadowgraph techniques they studied the effect on spontaneous convection of concentrations, organic solvent and direction of transfer. Fig. 3.8(a) and (b) show two photographs of an aqueous drop containing uranyl nitrate in 3M nitric acid suspended in a TBP organic solution. They observed that interfacial turbulence was enhanced by increasing the mass transfer driving force, as is the case for physical transfer of a single solute. However the variation of the ratio of kinematic viscosities in the range 0.5–2.5, which was achieved by changing the organic diluent, did not change the interfacial behaviour. Thus the relative value of kinematic viscosities does not seem to affect the stability of this system.

3.5.4 Theoretical studies

The first linearized analysis of a liquid–liquid reactive system was conducted by Ruckenstein and Berbente (1964) for a bulk reaction of first order. Their numerical results showed that even for small values of the reaction rate constant the stability criteria of a non-reactive system cease to apply. The complexity of the resulting equations did not allow for the derivation of general stability criteria.

The effect of different types of interfacial reactions was investigated by Hennenberg *et al.* (1975 and Sørensen *et al.* (1976) for plane and spherical interfaces. Sanfeld *et al.* (1979) have presented a comprehensive review of the extensive work conducted mainly by Sanfeld and his collaborators. These studies include the effect on interfacial stability of disturbances to the electrical double layer produced by the adsorption of ionized and dipolar molecules. Results showed that for only one fluctuating species the necessary, but not sufficient, condition for instability was that the reaction itself be unstable. For more than one fluctuating species instabilities could be produced even by stable reactions. In all cases the stabilizing effect of interfacial tension and interfacial and bulk viscosities had to be overcome for instabilities to set in. Application of the characteristic equation to the

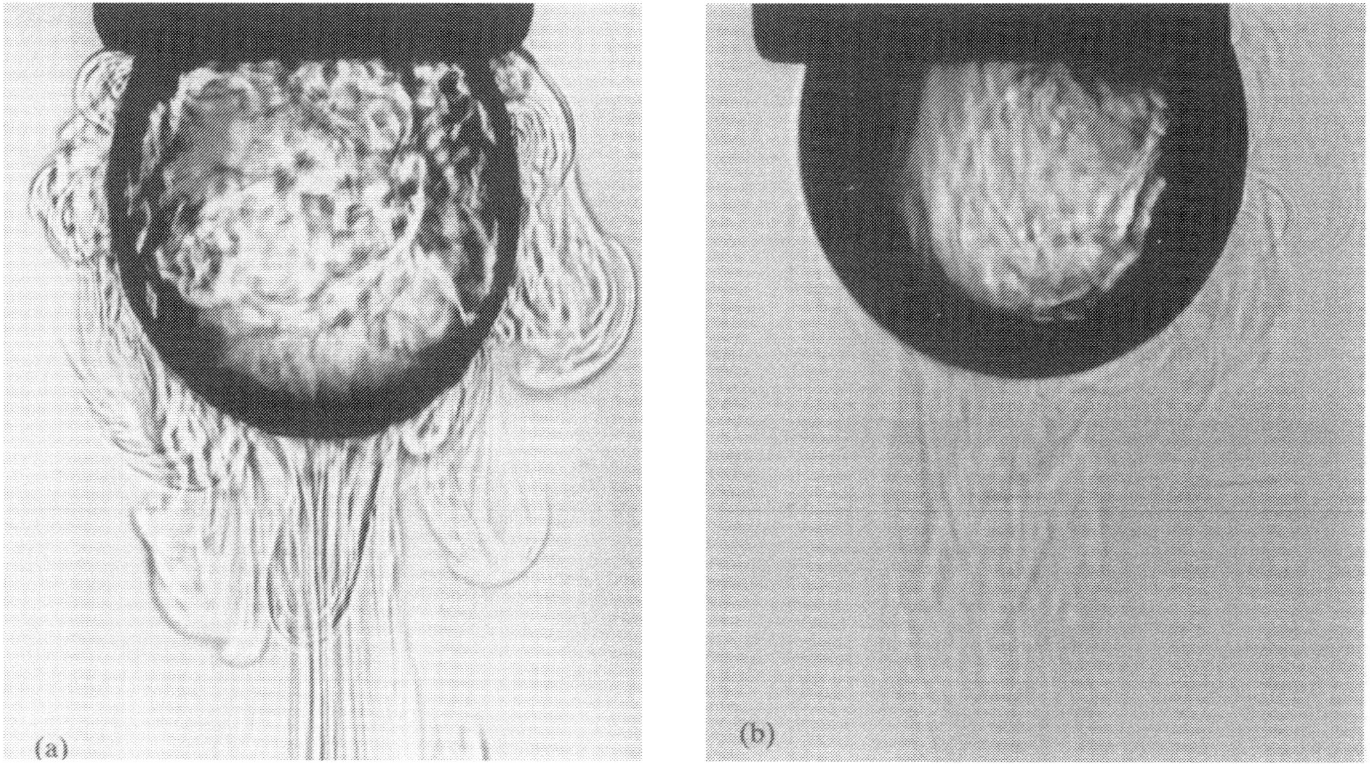

Fig. 3.8. Transfer of uranyl nitrate from an aqueous drop into an uranium free organic phase containing tri-butyl phosphate (TBP). (a) 200 g U/l - pure TBP; (b) 100 g U/l - 20% TBP in odourless kerosene. (Courtesy of Dounreay, UKAEA.)

reaction between alkyltrimethylammonium chloride and picric acid studied experimentally by Dupeyrat and Nakache (1977) did not predict the observed instabilities for the linear chemical kinetics suggested by these authors. However, oscillatory instabilities were predicted in this system when perturbations to the electrical double layer generated by the adsorption of ionized molecules were taken into consideration.

3.5.5 Mass transfer in an electric field
[See also Weatherley, Vol. 2, Ch. 5]

The enhancement of the rate of extraction on application of electric field was thoroughly investigated by Stewart and Thornton (1967a,b) and Thornton (1976). Using dispersed systems they studied the effect of electric forces on drop sizes, velocity and hydrodynamics. Austin *et al.* (1971) intended to complement these studies by investigating the effect of electric field on the mass flux across the plane interface of four partially miscible binary systems. Visual observations in a schlieren field at different voltages are shown in

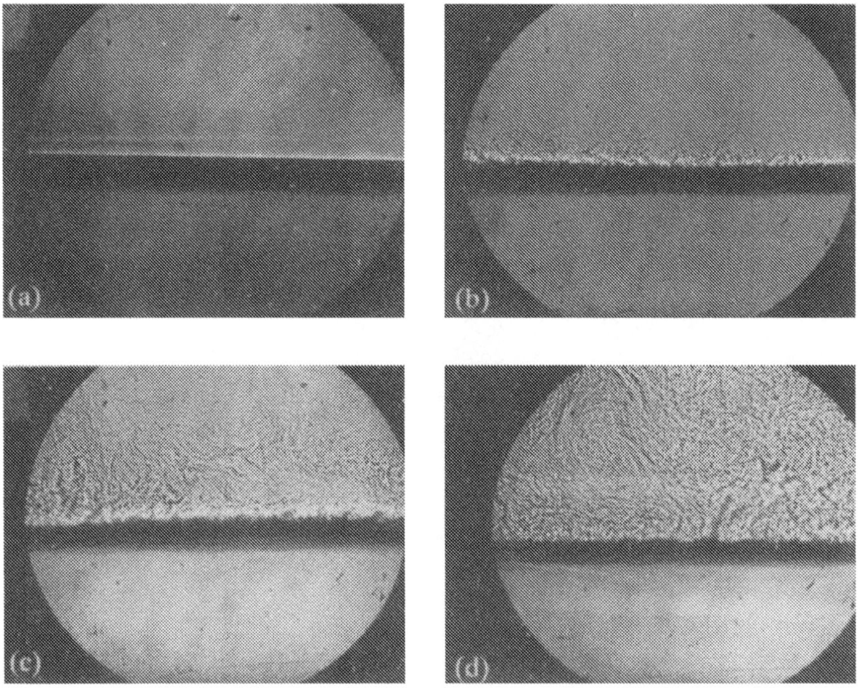

Fig. 3.9. Effect of voltage on interfacial behaviour during the transfer of isobutanol into water: (a) 0 V; (b) 450 V; (c) 850 V; (d) 1980 V. (After Austin *et al.* 1971.)

Fig. 3.9 for the system isobutanol–water. It can be seen that in the absence of electric field there is no interfacial convection. At low voltages interfacial convection is observed as well as turbulence in the organic phase in a region close to the interface. As the voltage was increased, turbulence propagated into the organic phase. The onset of interfacial instability was attributed to Marangoni instabilities driven by local variations of interfacial density changes. It should be noticed that the interfacial tension of these systems is lowered on application of electric field. These experiments were conducted in the absence of electrolytes. Watanabe *et al.* (1978) studied the effect of electric forces on an equilibrated liquid–liquid system containing KCl and an ionic surfactant. They observed that a stable emulsion was formed when the voltage reached a critical value. This value depended on electrolyte concentration.

A stability analysis of fluid–fluid interfaces with electrical double layer in the absence of mass transfer was performed by Bisch *et al.* (1983). The derived stability criteria, which involve mechanical and electrochemical properties of the system, were in good agreement with the observations of Watanabe *et al.* (1978).

3.6 Effect of Marangoni phenomena on specific interfacial area

3.6.1 Break-up of liquid jets

The stability of liquid jets is relevant to the design and operation of liquid–liquid contactors, particularly spray columns, in which the dispersion is produced by injection of one of the phases through nozzles and by redispersion of the coalesced layer by jetting through sieve plate holes. Disregarding coalescence in the bulk of the dispersion, the specific interfacial area depends on the drop sizes obtained by jet break-up. The length of the jet as well as the size of the drops resulting from jet break-up depends on the stability of the jet to disturbances that deform its surface. Figure 3.10 shows schematically how a disturbance of a certain wave length causes regular varicose deformations of amplitude growing with jet length that ultimately

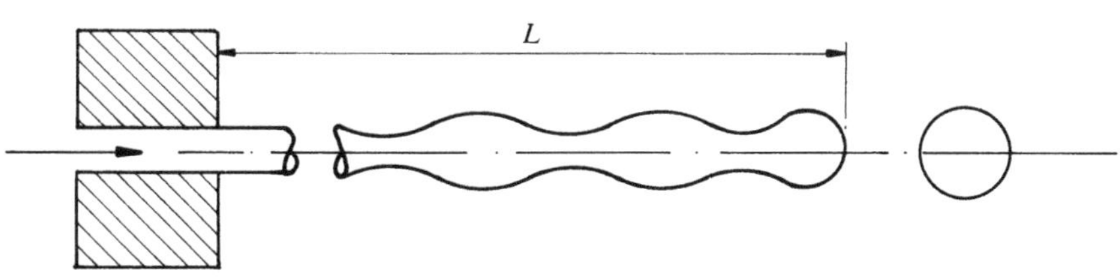

Fig. 3.10. Mechanism of jet break-up.

breaks the jet into drops. Out of a spectrum of disturbance sizes present in the system there is one that is preferentially amplified and will determine the drop size. The wave length of this dominant disturbance and its growth constant depend on the physical properties of the system and the experimental conditions.

The drop size obtained from jet break-up is proportional to the dominant wavelength. As for the jet length, the higher the perturbation growth constant the shorter the jet. Essentially this is a problem of hydrodynamic stability whose solution should provide equations suitable for the prediction of drop size and jet length.

3.6.1.1 *Experimental studies*

Meister and Scheele (1969) measured the effect on jet length of the transfer of acetone between a benzene jet and water in both directions of transfer and compared these results with those obtained with the pure liquids. They observed that mass transfer into the jet produced longer jets that transfer out and that in both cases the jet lengths were longer than in the absence of acetone. They interpreted this stabilizing effect of solute transfer as a consequence of the reduction in interfacial tension due to the presence of acetone. They did not elaborate on the effect of the direction of solute transfer.

Similar results were obtained by Dzubur and Sawistowski (1974) for the transfer of propionic acid between a benzene jet and water and a chlorobenzene jet and water. In their study they included experiments with the pure phases as well as with the equilibrated ternary systems so that the effect of interfacial tension could be examined in the absence of mass transfer. Figure 3.11 shows their results for the chlorobenzene jet plotted as jet length against jet Reynolds number.

Jet length is expressed in dimensionless form as L/d_N, where L is the jet length and d_N the nozzle diameter, and the jet Reynolds number has d_N as characteristic length. All curves pass through a maximum. In the systems containing the solute the maximum was reached at substantially lower Reynolds number than in the solute free system. Below the Reynolds number for maximum length, the equilibrated system produced the longest jets, i.e. the most stable, solute transfer into the jet yielded longer jets than transfer out and the pure system was the most unstable. Similar results were obtained with the benzene jet.

These experimental results, as well as those reported by Meister and Scheele (1969), contradicted the predictions of a mechanism of jet break-up based on thin-film Marangoni phenomena previously proposed by Sawistowski (1973). He had suggested that given the axial variation of volumes in the deformed jet, the rate of change of solute concentration in

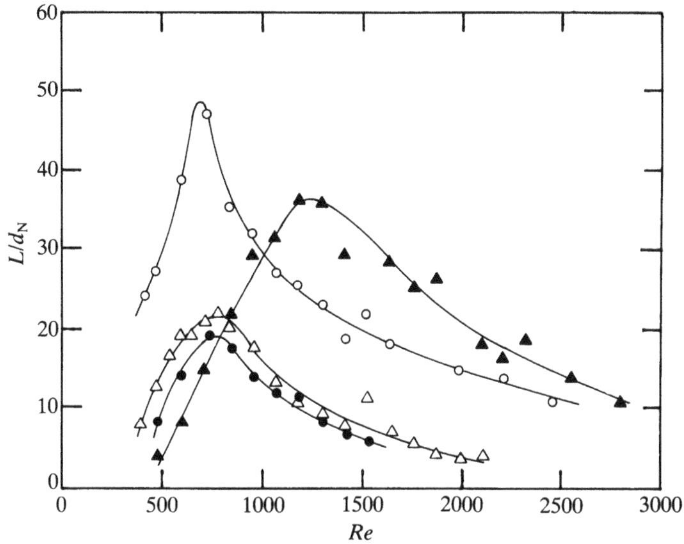

Fig. 3.11. Variation of jet length, with Reynolds number for a chlorobenzene jet in water. ▲ Solute free system. In the presence of propionic acid: • transfer out of the jet; Δ transfer into the jet; ○ phase equilibrium.

the nodes would be faster than in the bulges thus creating the conditions for axial variations of interfacial tension. For a lowering tension solute transferring into the jet, the higher solute concentration at the necks would generate surface flows from the necks to the bulges that would lead to premature jet break-up. Transfer out of the jet would produce flows in the opposite direction so increasing the jet resistance to break-up. Therefore the jet length should be longer for transfer out of the jet, not shorter as observed. In view of the fact that this mechanism did not explain the jet behaviour, Dzubur and Sawistowski (1974) put forward the hypothesis that directional effects were due to a change of interfacial tension with the direction of transfer. This is feasible if the mass transfer coefficients in the two phases are not the same. By assuming different values of the ratio between the continuous and jet mass transfer coefficients they estimated the interfacial concentrations of propionic acid for transfer in and out of the jet. The difference in interfacial tension between transfer in and out varied between −2.5 and −12 mN/m for mass transfer coefficient ratios between 1 and 4. Thus, assuming that the jet stability increases with decreasing interfacial tension, transfer of a tension lowering solute into the jet should yield longer jets than transfer out when the mass transfer coefficient in the continuous phase is greater than that in the jet.

Skelland and Huang (1977) considered the effect of mass transfer on jet length using a chlorobenzene jet in water and acetic acid as a solute. They compared the results for transfer out of the jet with those for the pure system at different jet flow rates. They found that the curves intersected before either curve reached a maximum. At flow rates above the intersection point the jet undergoing mass transfer was shorter than the one in the pure system, while below that point it tended to be longer.

Evidently the effect of mass transfer on jet stability is complicated and its dependence on interfacial tension and flow rate cannot be easily foretold on the basis of qualitative arguments. In the case of equilibrated systems the equations derived from stability theories and their semi-empirical simplifications predict well the jet lengths and drop sizes resulting from jet break-up. The theoretical analysis of jet behaviour in the presence of mass transfer discussed in the next section provides an insight into the interplay of different physical parameters in their effect on jet stability, as well as some useful stability criteria.

3.6.1.2 *Stability analysis of a cylindrical jet in the presence of mass transfer*

Lord Rayleigh (1878, 1879, and 1892) was the first to study theoretically the stability of jets. Although his model was highly idealized (an inviscid jet in a vacuum) he established a technique that was susbsequently applied to the study of more realistic models. The equation derived by Tomotika (1935) for a zero density viscous jet surrounded by a zero density viscous fluid predicted well the experimental results obtained previously by Taylor (1934). Meister and Scheele (1967, 1969) improved Tomotika's model by removing the non-zero density condition of the liquids. Their theoretical and experimental jet lengths are in good agreement. In all these models the jet is at equilibrium with the surrounding liquid so their applicability to the case in which there is interfacial mass transfer is limited to cases in which this process does not affect jet stability.

The stability analysis of a jet in the presence of mass transfer is due to Burkholder and Berg (1974). Although the resulting equations reflect the intrinsic difficulties of the problem they clarify certain aspects of the mechanism of jet break-up that were not foreseen in the qualitative treatment.

In Burkholder and Berg's model the only restrictions imposed upon the two liquids are that they are non-volatile, incompressible and Newtonian. The jet is in steady laminar plug flow, so slow that aerodynamic effects are negligible. The latter restriction was removed in a more recent analysis (Coyle *et al.* 1981). In the undisturbed state a solute is transferring in steady state and the phases are in local equilibrium at the interface.

It is central to the analysis of jet stability that the interface is deformed by small disturbances leading to perturbations in the radial and axial velocities, the pressure, and the solute concentration in both phases. These changes may be followed by axial variations of interfacial tension along the jet, which must satisfy eqns (3.16 and 3.18), now in cylindrical coordinates. It should be noticed that in this problem the surface deflection gives rise to the capillary pressure appearing on the right hand side of eqn (3.18).

Following the usual procedure Burkholder and Berg obtained the characteristic equation relating the growth constant and the perturbation wave number. This equation is not amenable to the type of analysis that provided general stability criteria in the cases of flat interfaces discussed previously. A numerical analysis of the characteristic equation revealed that the stability of the jet depends not only on the physical properties of the system but also on the rate of solute transfer. These conditions determine the effect of the direction of solute transfer as illustrated in Fig. 3.12, where the curves show the variation of the dominant dimensionless growth constant with the Marangoni number in the jet with the viscosity ratio μ_A/μ_B as a parameter (where A is the surrounding phase and B the jet phase). The Marangoni number in the jet is given by: $Ma_B \equiv (\partial\sigma_0/\partial C)(\partial C/\partial r)R^2/\mu_B\mathscr{D}_B$, where r is the radial distance and R the jet

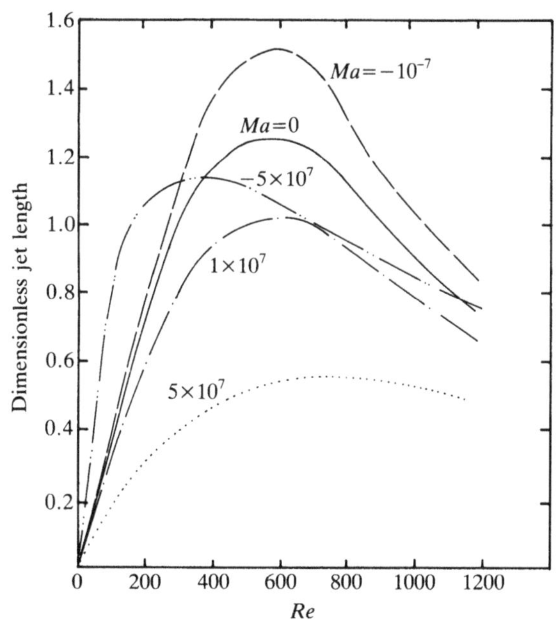

Fig. 3.12. Predicted variation of dimensionless jet length, $\bar{L}$, with Reynolds number, mass transfer driving force and direction of transfer of an interfacial tension lowering solute. $Ma>0$: transfer out of the jet; $Ma<0$: transfer into the jet. (After Coyle *et al.* 1981.)

radius. Since it includes the concentration gradient, the absolute value of the Marangoni number is proportional to the transfer rate and its sign indicates the direction of transfer. For a tension lowering solute, Ma_B is positive for transfer out of the jet. The graph shows that for $|Ma_B| < \sim 10^6$ the dimensionless growth constant is the same for both directions of transfer, i.e. there is no directional effect. Also the stability of the jet decreases with decreasing μ_A/μ_B. At higher transfer rates the curves for transfer in and out of the jet diverge and the effect of the direction of transfer depends on the ratio μ_A/μ_B. At $\mu_A/\mu_B = 0.1$, transfer out of the jet is stabilizing, as indicated by a decrease in the growth constant with increasing Ma_B. Transfer in the opposite direction leads to higher values of $\bar{\beta}$. However at $\mu_A/\mu_B = 10$ the effect of the direction of transfer is reversed and transfer out of the jet becomes destabilizing. Similar changes in the effect of the direction of transfer were observed when varying the density and diffusivity ratios. In general the stabilizing effect of mass transfer gradually disappears as the transfer rate is increased until the system eventually becomes unstable in both directions of transfer.

The conclusions drawn from the numerical analysis of the characteristic equation can be summarized as

(1) mass transfer of a surface tension lowering solute either into or out of the jet may be either stabilizing or destabilizing depending on system physical properties and mass transfer rate;

(2) surface adsorption may strongly counteract either the stabilizing or destabilizing effects of the mass transfer on the jet;

(3) surface adsorption, including that of strong surface active agents, has only a very small effect on jet stability in the absence of mass transfer and;

(4) the wavelength of the most unstable disturbance, hence the ultimate drop size, may either increase or decrease for transfer either into or out of the jet.

The theoretical study of Burkholder and Berg throws some light on the mechanism of jet stability in the presence of mass transfer. This can be better described with the help of Fig. 3.13. Figure 3.13(a) shows the jet in the undisturbed state with parallel lines indicating levels of equal solute concentration. When the jet boundary is deformed, fluid is brought to the peaks from the bulk phases and the concentration lines are shifted as shown in Fig. 3.13(b). For transfer out of the jet, the fluid reaching the interface from the jet is richer in solute than the fluid arriving from the surroundings. The new local interfacial concentration will depend on the physical

(a)

(b)

(c)

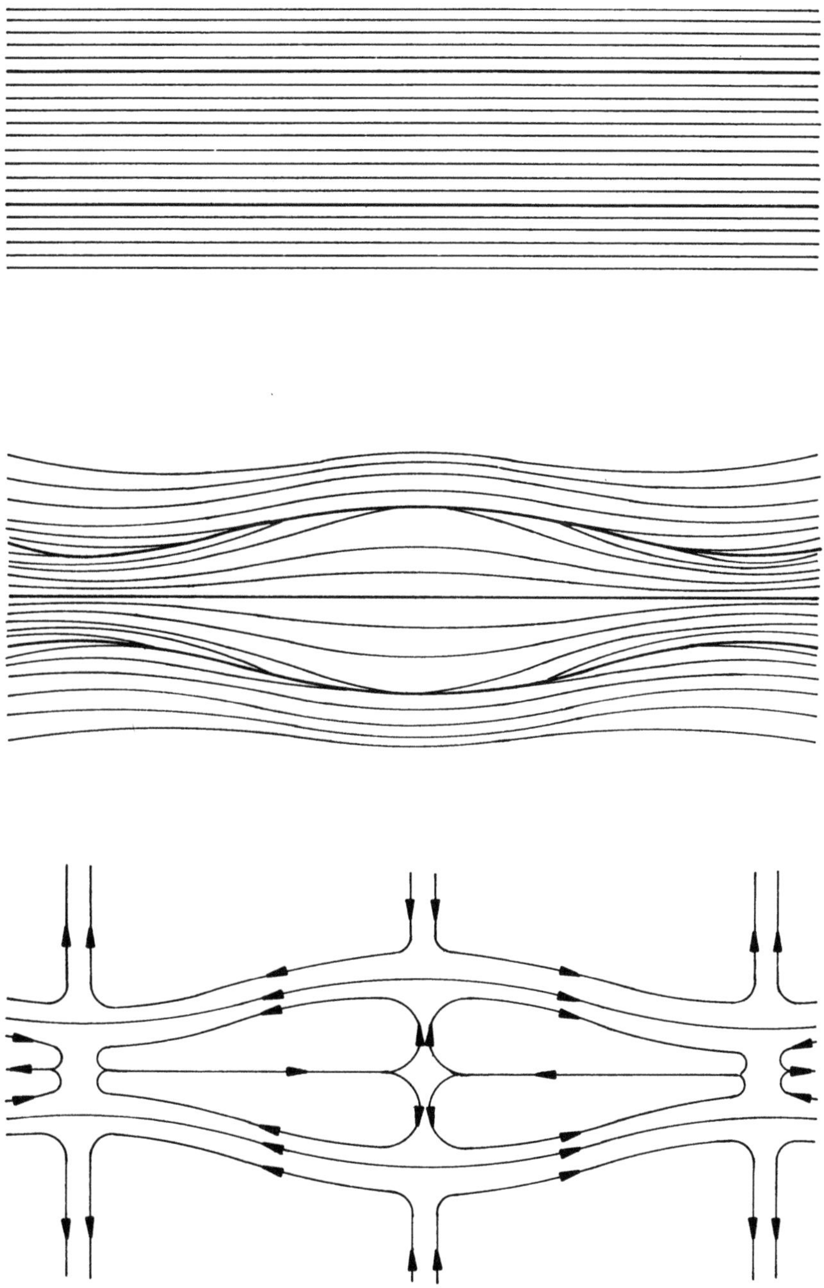

Fig. 3.13. Influence of mass transfer on jet stability. (a) Undeformed jet with lines indicating constant solute concentration; (b) deformed jet showing distortion of concentration lines; (c) internal flows caused by interfacial tension variation. (After Burkholder and Berg 1974.)

properties of the system and the relative convective diffusivities of the phases. If the net result is an increase in interfacial concentration at the peaks, Marangoni flows for a lowering tension solute will occur from the peaks to the valleys, as shown in Fig. 3.13(c). These flows will stabilize the jet. However the concentration at the peaks may be lower than at the valleys if the effective convection in the surrounding liquid, which is leaner in solute, is greater than in the jet. This is possible due to the difference in pressure between the two phases at the interface brought about by the jet deformation, as given by eqn (3.18). The decrease in pressure at the disturbance peaks may reinforce convection in the surrounding phase leading to a decrease in interfacial concentration at the peaks which in turn will generate destabilizing Marangoni flows. For example, in a system with $\nu_A/\nu_B = 1$ and $\mathscr{D}_A/\mathscr{D}_B = 1$, which would be in the state of marginal stability according to the Sternling and Scriven criteria for mass transfer across a planar interface, the deformed jet could be either stabilized or destabilized by mass transfer in a given direction depending on the ratio of convective diffusivities. These hydrodynamic parameters vary with the experimental conditions and cannot be easily predicted. This partly explains the difficulties encountered in establishing general stability criteria for liquid jets.

Burkholder and Berg solved their equation for the conditions investigated by Meister and Scheele (1969) in the presence of mass transfer and observed good agreement between predicted and experimental results.

More recently Coyle *et al.* (1981) extended the stability analysis of jets undergoing mass transfer to jets in relative motion through the surrounding liquid and in the presence of adsorbable solutes. Results of this analysis allow the prediction of the effect of mass transfer rate and solute adsorption on the variation of jet length with jet velocity. Numerical results are in qualitative agreement with the authors' experimental measurements for the transfer of acetone between benzene jets and water as well as with experimental results obtained by Meister and Scheele (1969) and Dzubur and Sawistowski (1974).

3.6.2 Drop size from jet break-up

The size of the drops formed by the break-up of jets depends on the wave length of the dominant disturbance, λ_D. It is usually assumed that the drops contain the same volume as a section of the undisturbed jet of length λ_D. The values λ_D for equilibrated jets calculated from the corresponding characteristic equations in general fit well drop sizes obtained experimentally. Dzubur and Sawistowski (1974) found very good agreement between the drop sizes predicted with Tomotika's equation (Tomotika 1935) and experimental values for the solute free system chlorobenzene jet–water. In

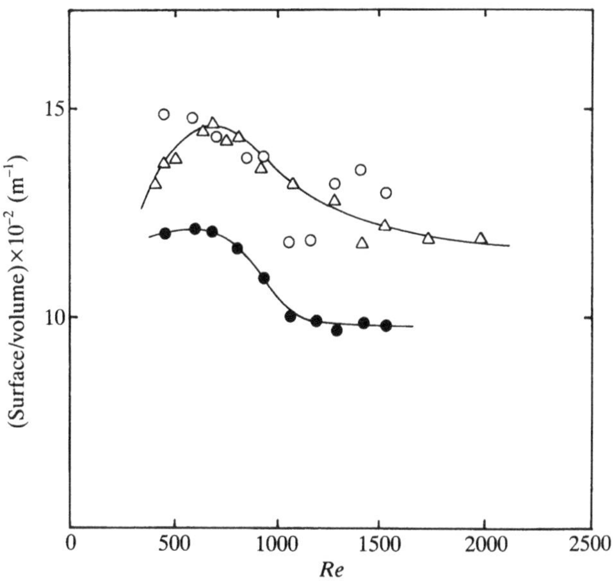

Fig. 3.14. Variation of drop surface to volume ratio with Reynolds number for transfer of propionic acid between a chlorobenzene jet and water. ○ Solute in phase equilibrium; • Transfer out of jet; Δ Transfer into jet. (After Dzubur and Sawistowski 1974.)

the presence of mass transfer, drop sizes were affected as much as jet lengths. Figure 3.14 shows the variation of the ratio of drop surface to drop volume with Reynolds number measured by Dzubur and Sawistowski (1974) for the transfer of propionic acid into and out of a chlorobenzene jet. In the unstable direction of transfer, i.e. transfer out of the jet, the figure shows that the dispersed phase specific interfacial area is about 20 per cent smaller than for transfer in the stable direction. These findings are consistent with their hypothesis that longer jets would be obtained in the direction of transfer in which the interfacial tension is lower, this leading to smaller drops than in the opposite direction.

Drop sizes measured by Skelland and Huang (1977, 1979) for transfer out of the jet in several systems confirm the trend that at more stable jet conditions drops resulting from break-up are smaller. In their study on the hydrodynamics of drop formation under jetting conditions at both moderate and slow transfer rates, they noted that stability decreased with increasing transfer rates, thus smaller drops were obtained at low rates of solute transfer. Since correlations for drop size and jet length could not cover the whole range of mass transfer rate, they proposed different correlations for slow and moderate rates of transfer. Theoretical studies (Burkholder and Berg 1974; Coyle and Berg 1984) confirm that correlations between jet length, drop size and transfer rate cannot be generalized.

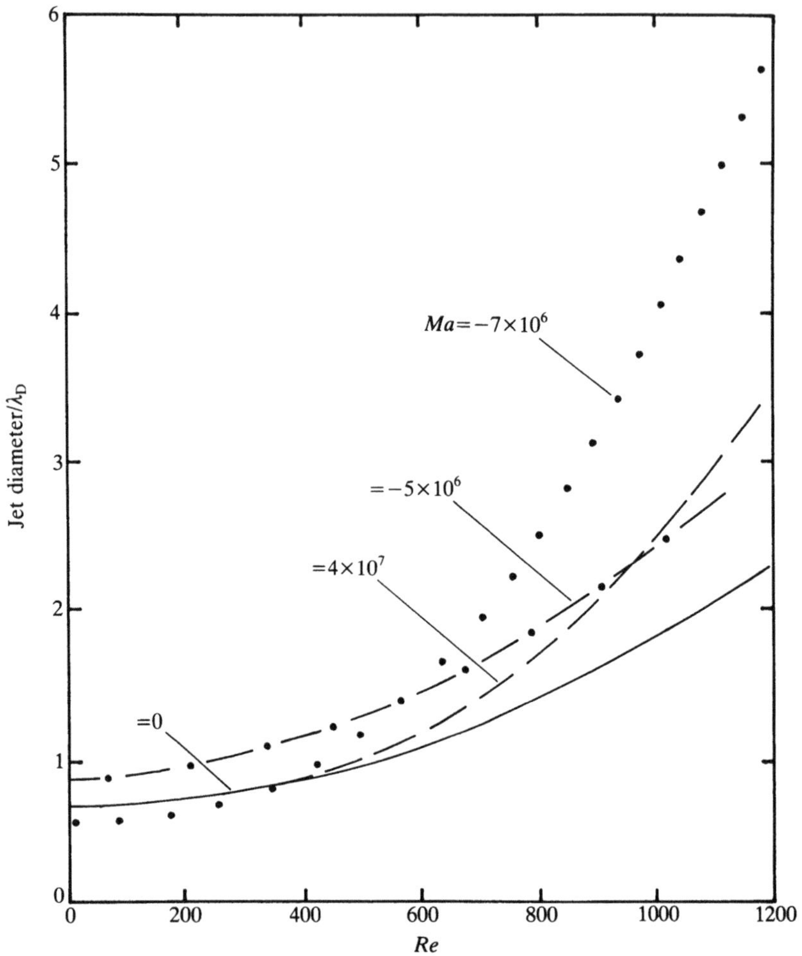

Fig. 3.15. Predicted dependence of drop size (represented by λ_D) on Reynolds number at different values of the Marangoni number for acetone transfer into and out of hexamethyl di-siloxane jets. $Ma > 0$: transfer out of the jet; $Ma < 0$: transfer into the jet. (After Coyle and Berg 1984.)

The predicted dependence of drop size represented by the dominant wave length, λ_D, on Reynolds number at different Marangoni numbers is shown in Fig. 3.15 for the transfer of acetone into and out of hexamethyl di-siloxane jets (Coyle and Berg 1984). It must be recalled that positive and negative Marangoni numbers correspond to transfer out and into the jet respectively and that the Marangoni number is proportional to the mass transfer rate. Transfer in either direction is predicted to decrease the drop size, particularly in the range of higher Reynolds numbers. At lower Reynolds numbers only transfer out of the jet has an effect on drop size. Intersection between curves for transfer in and out indicates that the relative effect of the direction of transfer depends on both the rate of mass transfer and the Reynolds number. Thus no general criteria on the correlation between drop size, direction of transfer and operating conditions can be derived from the theoretical analysis. Coyle and Berg's experimental results are in good qualitative agreement with the predicted trends.

An important consequence of the stability analysis of jets is the realization that correlations for jet length and drop diameter may not be applicable to conditions that are outside the experimental range used to develop them.

3.6.3 Drop coalescence

The effect of mass transfer direction on drop coalescence can be interpreted in terms of thin-film phenomena. The thickness of the film separating an approaching drop from a flat interface, or two colliding drops, is not uniform. For mass transfer out of the drops the rate of solute accumulation in the thinner sections of the film will be greater than in their surroundings thus creating the conditions for the appearance of interfacial tension gradients. For a lowering tension solute Marangoni driven flows would help film rupture as shown in Fig. 3.3. For transfer into the drops the reverse phenomena would take place thus preventing, or delaying, coalescence. Figure 3.16 shows the coalescence times of two drops of anisole in water for the solute free case and for transfer of acetone in either direction of transfer.

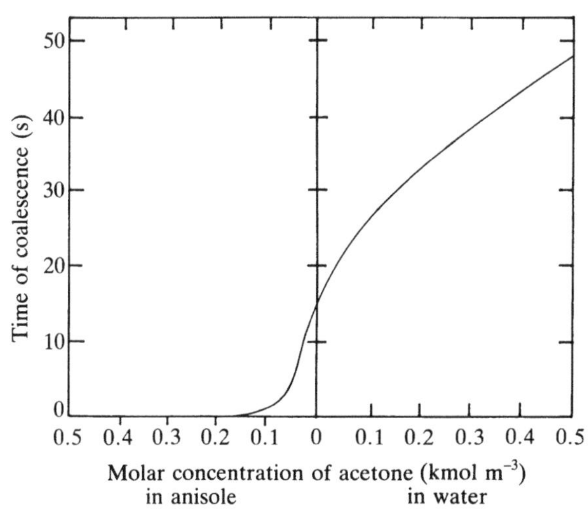

Fig. 3.16. Effect of direction of mass transfer on coalescence of two anisole drops in water. (After Sawistowski 1981.)

In this case $\partial\sigma/\partial C < 0$. Coalescence is almost instantaneous for transfer of acetone from the drops, i.e. under the conditions in which Marangoni thin film phenomena contribute to film break-up. In the opposite direction of transfer the coalescence time is shown to increase up to three times the value for the solute free system. Similar results were obtained by Sawistowski and James (1963) in a spray column for the transfer of propionic acid between benzene drops and water. For transfer out of the drops they observed a 40 per cent decrease in the number of drops that arrived at the top of the column with respect to the number of drops formed at the bottom. On the other hand, transfer into drops almost eliminated coalescence. The effect of a tension increasing solute is illustrated by the extraction of uranium by tributyl phosphate in odourless kerosene. Batey *et al.* (1984) conducted extraction and backwash experiments in a pulsed column in the aqueous phase continuous mode of operation. In the extraction experiments, i.e. for transfer into the drops, they observed large irregular drops in the top section of the column which made the determination of the average drop size very difficult. This type of drop was not observed in the backwash experiments. This contrasting drop behaviour could be due to Marangoni thin-film phenomena. In this system interfacial tension increases with uranium concentration. Thus transfer into the drops may create the conditions for faster film rupture. In addition this system shows strong interfacial instabilities in the water to organic direction of transfer as discussed previously. It should be noticed that although both the thin-film effect and Marangoni instabilities are caused by surface tension driven flows, they are independent phenomena. The effect of the latter in either drop coalescence or jet break-up has not yet been systematically investigated.

3.6.4 Drop drag coefficients

The presence of interfacial turbulence in rising and falling drops affects the flow patterns otherwise observed in the vicinity of stable interfaces. These changes in the velocity profiles lead to changes in drop drag coefficients. This was first reported by Sehrt and Linde (1967) who investigated the effect of the direction of transfer of acetic acid on the drag coefficients of benzaldehyde and benzene drops moving through water at different Reynolds numbers. They observed substantially higher drag coefficients in the unstable direction of transfer. This increase in the drag coefficient is due to the supression of drop circulation caused by interfacial turbulence. Results obtained by Sawistowski for the transfer of propionic acid between nitrobenzene drops and water are plotted in Fig. 3.17 for both directions of transfer as well as for the equilibrated and the solute free systems. The highest drag coefficients were measured for transfer of propionic acid out

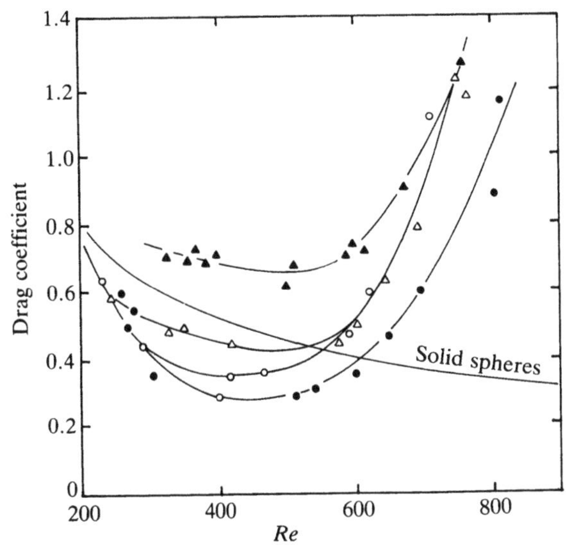

Fig. 3.17. Drag coefficient of nitrobenzene drops falling through water: ○ pure phase, • phases in equilibrium, Δ transfer of propionic acid into drops, ▲ transfer of acid out of drops. (After Sawistowski 1981.)

of the drops, i.e. in the unstable direction of transfer. The curve for solid spheres is shown for reference purposes.

3.7 Effect of Marangoni instabilities on mass transfer coefficients

[See also Javed, Ch. 4]

The effect of spontaneous interfacial turbulence on the rate of interfacial mass transfer can be measured in a relatively easy way. Experimental results reported in the literature for different types of contactors cover a range of mass transfer enhancement factors ranging between 1.5 and 10 (Berg 1972; Sawistowski 1971). However the complexity of the phenomena and the diversity of interfacial flows in which they result make a formal analysis of mass transfer enhancement difficult. There are few equations proposed to predict the mass transfer coefficients in unstable systems and most of them contain parameters that require experimental measurement. Since the intensity of interfacial turbulence is in general related to the steepness of the concentration profile and the physical properties of the system, the enhancement factor usually depends on these parameters.

An example of macroscale turbulence due to mass transfer is the 'kicking drop' observed by Lewis and Pratt (1953) during the transfer of a solute between the drop and the continuous phase. Haydon (1958) studied the conditions under which these violent and erratic oscillations occurred and suggested a mechanism based on the dependence of interfacial tensions on concentrations which was supported by experimental results obtained with twelve tertiary systems. He proposed the following equation for the minimum total concentration of solute required to produce a kick:

$$C_2 = \Delta\pi(1 + m)/\partial\sigma_0/\partial C \qquad (3.32)$$

where C_2 is the addition of the solute concentrations at both sides of the interface, m is the equilibrium distribution coefficient of the solute relative to the phase where it is being transferred, and $\Delta\pi$ is the interfacial pressure. In the systems studied by Haydon, $\Delta\pi$ varies between 0.4×10^{-3} and 4×10^{-3} N/m. Equation (3.32) gives a critical value of C_2 over which the rate of mass transfer is enhanced by droplet oscillations. Sawistowski and co-workers (Sawistowski 1971) studied the effect of mass transfer on interfacial behaviour during drop formation, and in different types of contactors. They investigated the single transfer of phenol, acetic acid, chlorobenzene and acetone between water and benzene. In all cases the stable direction of transfer is from benzene to water. In their experiments on mass transfer between benzene drops and water (Sawistowski and Göltz 1963; Sawistowski and James 1963, 1967) they observed that even in the stable direction of transfer three regimes of transfer could be distinguished. At low driving forces the mass transfer coefficients were independent of the driving force in both directions of transfer. This region corresponds to the diffusional regime. Increases in driving force led to the appearance of eruptions on the drop surface (Marangoni disturbances) and a consequent increase in mass transfer coefficient. Further increases in driving force gave rise to sustained interfacial turbulence. Sawistowski and James (1963, 1967) adopting a model similar to that of Haydon's correlated their results by the relation

$$K^* \equiv K/K_{\text{diff}} = 1 + b(\Delta\pi - \Delta\pi_{\text{crit}}) \text{ for } \Delta\pi \geq \Delta\pi_{\text{crit}} \qquad (3.33)$$

where K_{diff} is the mass transfer coefficient in the absence of spontaneous interfacial convection, b is a constant that depends on interfacial and bulk properties and $\Delta\pi_{\text{crit}}$ is the increment in surface pressure required to produce Marangoni perturbations in the system. The values of $\Delta\pi_{\text{crit}}$ varied between 1×10^{-3} and 4×10^{-3} N/m, i.e. the same range as those obtained by Haydon (1958).

Bakker *et al.* (1966) studied the effect of microscale convection at a flat interface between two unstirred phases on mass transfer coefficients at short contact times. They investigated the transfer of acetic acid in three systems: ethylene glycol–ethyl acetate, water–isobutanol and water–ethyl acetate. In all cases the bulk phases were quiescent. Plan viewing of the interface in a schlieren system revealed polygonal structures of pseudostationary convection cells, the sizes of which grew with time. The linear dimensions of the cells varied with the square root of time and were of the same order of magnitude as the penetration depth of the solute as given by the penetration theory of mass transfer.

These authors found that in the presence of cellular convection the mass transfer rate is increased by a factor

$$K^* = \pi d/2\delta \tag{3.34}$$

where $\delta = (\pi \mathscr{D} t)^{1/2}$ is the penetration of the concentration profile given by the penetration theory of mass transfer, and d is the time dependent mean linear size of the convection cell. Independent measurements of the ratio d/δ and K^* were in good agreement with eqn (3.34). The values of K^* for the studied systems varied between 1.57 and 3.14. It should be noticed that in this investigation the bulk phases were quiescent, thus no velocity components other than those caused by interfacial instabilities were present in the system. Therefore eqn (3.34) may not be directly applicable to stirred tanks or continuous contactors.

Different authors have studied the effect of the hydrodynamics of the bulk phases on the unstable rate of interfacial transfer. Davies (1972) proposed a treatment of the rate of mass transfer in turbulent flow that includes the contribution due to interfacial instability, by considering an additional component of the fluctuation velocity, tangential to the interface. This results in a decrease of the thickness of the diffusional layers with respect to those in the stable regime of transfer. For systems where there is surface renewal, Davies extended Danckwerts' treatment of the penetration theory (Danckwerts 1951, 1955) to allow for the spontaneous turbulence contribution to the overall rate of fractional surface renewal. He assumed that the rate of surface renewal s could be approximated by

$$s = s_E + s_\sigma \tag{3.35}$$

where s_E is the rate of renewal by eddies produced by bulk turbulence and s_σ is the additional rate of surface renewal due to spontaneous turbulence. Equation (3.35) leads to the following ratio of mass transfer coefficients:

$$K^* = (1 + s_\sigma/s_E)^{1/2} \tag{3.36}$$

Sethy and Cullinan (1975a,b) conducted experiments in a non-dispersive stirred cell employing partially miscible binary systems and ternary systems in which more than one solute was transferring. Schlieren observations indicated that transfer in the selected binary systems was interfacially stable. Ternary systems were obtained by adding a third component to the binary systems and coupled transfer of the two solutes was investigated in the unstable direction of transfer. Following a development similar to that of

Davies they derived the following expression for the mass transfer coefficient in the unstable direction of transfer:

$$K^* = (1 + \gamma \Delta C K^*/K_{\text{diff}})^{1/2} \tag{3.37}$$

where K_{diff} is the mass transfer coefficient in the binary system, which is taken as reference value for K^*, γ is a function of the total molar concentration, the matrix of diffusion coefficients and the direction of transfer, and ΔC is the concentration driving force for the second solute. Comparison of eqns (3.36 and 3.37) shows the complex way in which s_σ is affected by S_E and the driving force. The range of values of K^* between 1.5 and 3 is the same as that obtained by Bakker *et al.* (1966) in quiescent systems. This could indicate that the mechanism of transfer during the life of an interface element is that proposed by Bakker *et al.* (1966) for a stagnant film.

The effect of bulk turbulence on K^* was investigated by Peker *et al.* (1980) in one ternary system and four binary systems in the unstable direction of transfer. Using a non-dispersive stirred cell they measured the dependence of the mass transfer coefficients on stirring speed. Their results were well fitted by the equation

$$\ln K = N/101 - A \tag{3.38}$$

where N is the stirring speed and A is a constant that changes with the system. This correlation is of the same form as that proposed by Davies and Khan (1965) and used by Austin and Sawistowski (1967) to correlate mass transfer coefficients in stable systems. The enhancement factor K^* observed by Peker *et al.* (1980) varies between 1.82 and 3.97 throughout the investigated concentration range.

3.8 Interactions between Marangoni phenomena and rate of surface renewal in the rate of mass transfer to drops

In a series of papers Thornton and co-workers (Javed and Thornton 1984; Javed *et al.* 1989; Thornton and Anderson 1981; Thornton *et al.* 1985) investigated the interfacial dynamics of a pendant water drop suspended in toluene in the absence of mass transfer and during the transfer of three single solutes: propionic acid, acetone and diacetone alcohol. Using a novel technique based on the laser activation of a photochromic tracer they labeled an element on the drop surface and recorded its behaviour by using high speed cinephotography. They observed that in the absence of mass transfer the coloured spot remained undisturbed both during formation and detachment of the drop from the nozzle (Thornton *et al.* 1985). In the range of nozzle Reynolds number investigated ($42 \leq Re_N \leq 300$), circulation towards the rear of the droplet was visually evident by the displacement of

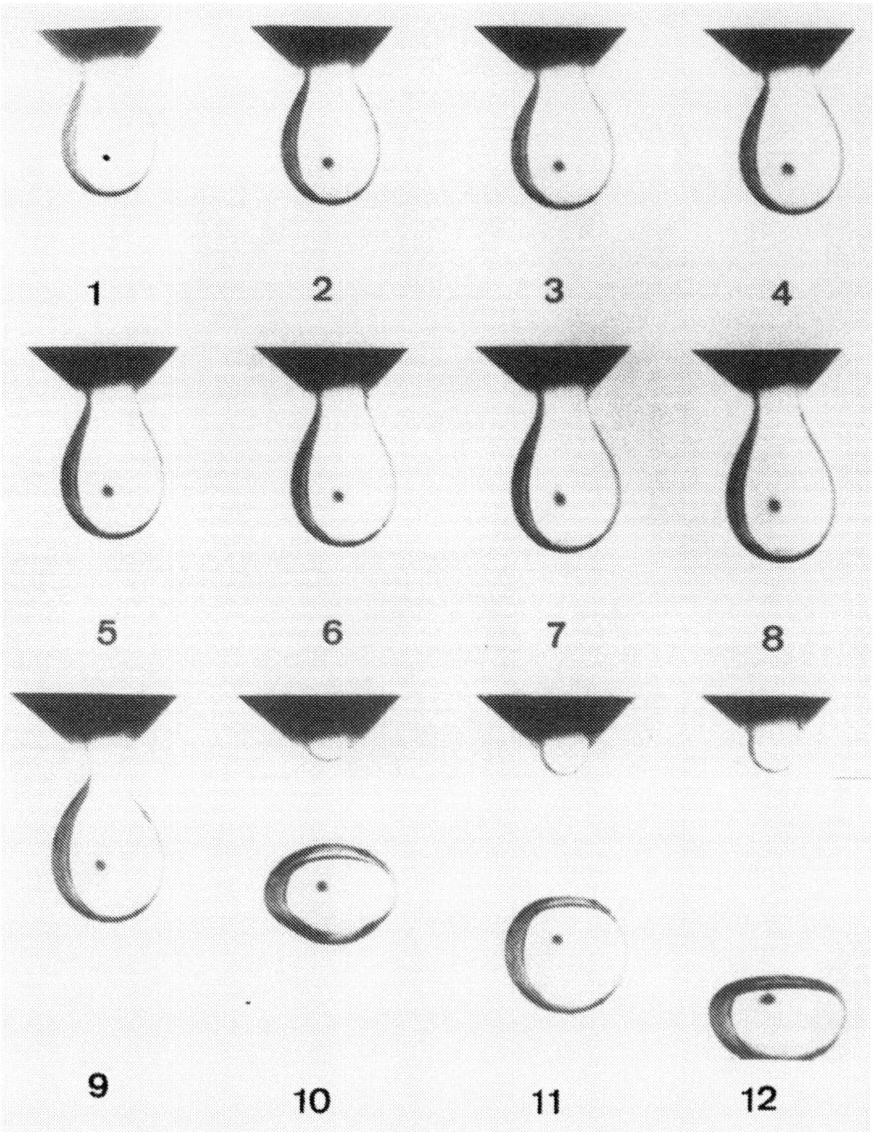

Fig. 3.18. Formation and detachment of a water droplet in toluene without mass transfer. $Re_N = 169$; nozzle diameter: 0.331 cm; nozzle velocity: 4.85 cm/s. Frame interval: 50 ms. (After Thornton *et al.* 1985.)

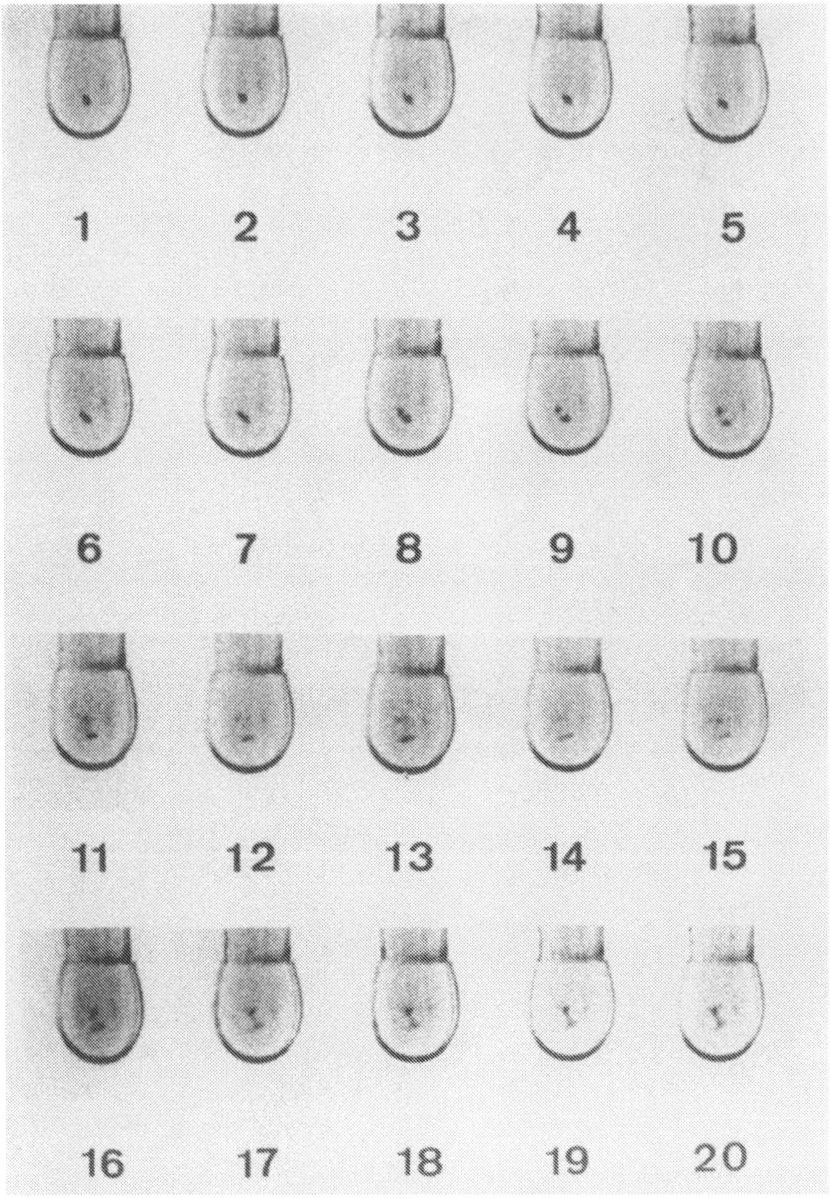

Fig. 3.19. Growing pendant droplet in toluene containing 3 v/v% diacetone alcohol, $Re_N = 42$; nozzle diameter = 0.417 cm; nozzle velocity = 0.920 cm/s. Frame interval: 4 ms. (After Thornton *et al.* 1985.)

the labelled spot. However there was no indication of surface renewal even at the highest values of Re_N. A typical sequence of photographs is shown in Fig. 3.18 for a nozzle Reynolds number, Re_N, equal to 169.

In the presence of mass transfer the labelled spot disintegrated as a result of violent, sporadic disturbances. This can be seen in the photographic sequence in Fig. 3.19. The intensity of the disturbances was never high enough to distort the droplet profile, therefore it was assumed that disturbances were due to the onset of Marangoni induced roll cells, the lifetime of which corresponded to the duration of the disturbance.

Two sets of measurements were obtained from a frame by frame analysis of the film:

(i) the time elapsed between the spot formation by laser activation and the onset of the first disturbance, and;

(ii) the time intervals between commencement of successive disturbances. This data was then interpreted in terms of the residence time distribution function of the drop surface elements, $\phi(\theta)$.

The interactions between rates of surface renewal and mass transfer rates were investigated using a hemi-spherical pendant water droplet immersed in toluene (Javed *et al.* 1989). In these experiments the droplet was formed at the tip of a concentric nozzle by a continuous feed of aqueous phase while by simultaneous withdrawal of aqueous liquor the droplet volume was kept constant. This set-up offered the advantages that experiments were run under conditions of forced circulation and that the variation of mass transfer rate with time could be estimated from the concentration of acetone in the aqueous outlet stream.

Instantaneous values of the overall mass transfer coefficient based on the dispersed phase, K_{od}, vs contact time are plotted in Fig. 3.20 for the three concentrations of acetone in toulene investigated. As is characteristic of mass transfer in the presence of interfacial turbulence, rate enhancement increases substantially with driving force, in this case with acetone concentration. However the enhancement factor rapidly decays with time until it reaches a steady state value. The high speed cine films showed a strong correlation between mass transfer rates and interfacial turbulence. As the concentration of acetone in toluene increased so did the intensity and frequency of the disturbances. However as the interface aged, both the intensity and frequency of the disturbances decayed following the same trend as the mass transfer coefficients. Fig. 3.21 shows the dispersion of the tracer spot on a drop suspended in toluene containing 5 v/v per cent acetone. As in the case of growing drops, no turbulence was observed in the absence of mass transfer.

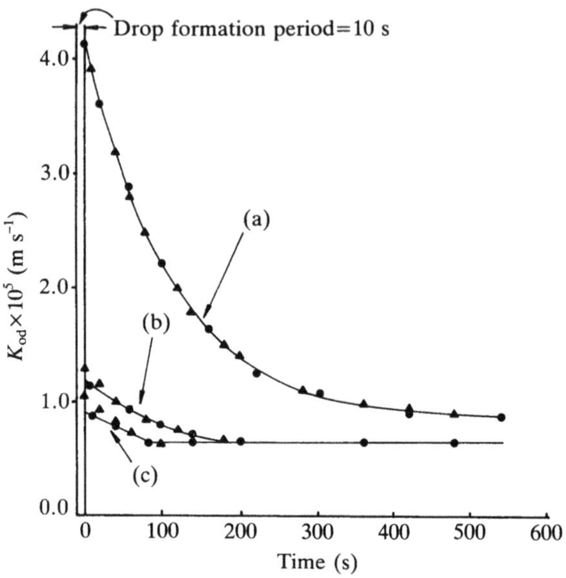

Fig. 3.20. Overall dispersed phase mass transfer coefficients vs actual time for the transfer of acetone from toluene into a water suspended drop. Acetone volume concentration: (a) 5%, (b) 1%, and (c) 0.5%. ● Experimental values; ▲ calculated values. (After Javed *et al.* 1989.)

Decreases in the mass transfer rate with interface age in unstable systems have been reported by several authors (Maroudas and Sawistowski 1964; Naikaike and Tadaki 1979; Naikaike *et al.* 1971; Orell and Westwater 1961, 1962). Javed *et al.* (1989) discussed different possible causes for the time dependence of the mass transfer coefficient, including surface contamination which seemed unlikely under the conditions of their experiments. They concluded that without further work no definite explanation for the coefficient time-dependence could be provided.

A link between mass transfer rates and the rate of surface renewal was established using the surface age distribution functions obtained from the photographic films. The average solute flux taken over the total drop surface, $\bar{N}$, is given by:

$$\bar{N} = \int_0^\infty N\phi(\theta)\ \mathrm{d}\theta \tag{3.39}$$

where N is the instantaneous flux. Two time intervals were considered in which the rate of surface renewal was different:

(i) $0 \leq \theta \leq A$

(ii) $A < \theta \leq \infty$

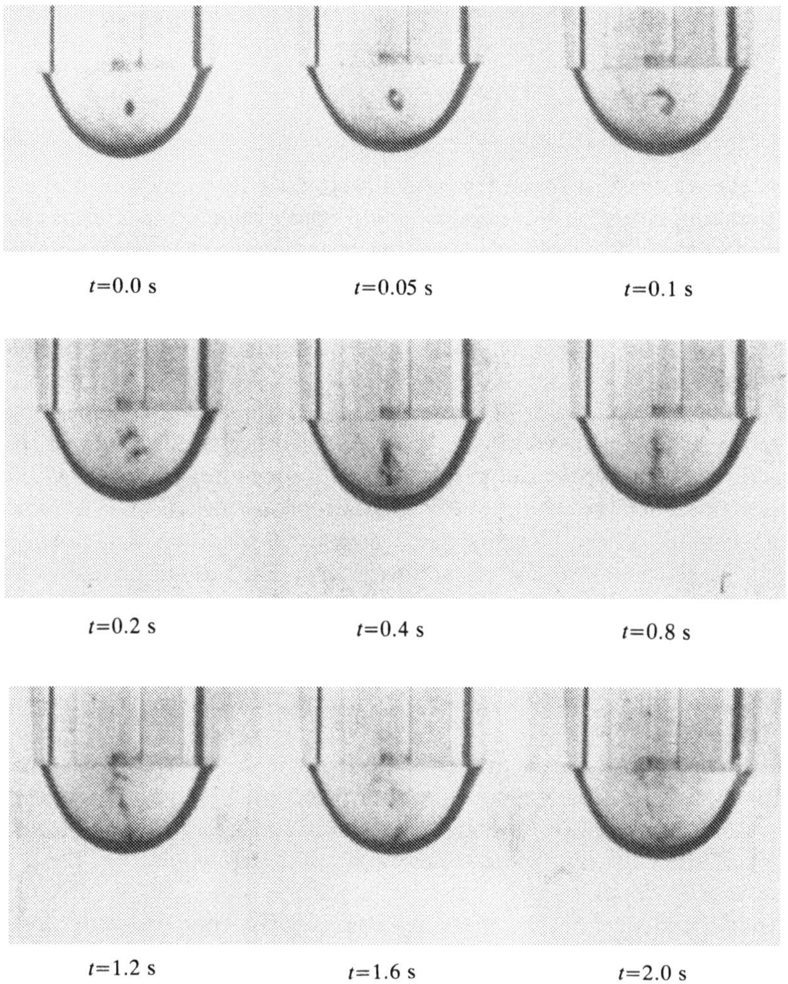

Fig. 3.21. Quasi-steady water drop suspended in toluene containing 5 v/v% acetone. (After Javed *et al.* 1989.)

where A is the minimum duration of a roll cell. Case (i) corresponds to the duration of a disturbance and can be approximated by the assumption that a surface element is created at $\theta = 0$ and it remains stagnant during a minimum time A. Thus, as in Higbie's theory (Higbie 1935), the rate of surface renewal during this period is equal to zero. Beyond this time, i.e. case (ii), the probability that the element be removed from the interface is independent of time, as in Danckwerts surface renewal model (Danckwerts 1951, 1955). Combining eqn (3.39) with the expressions for $\phi(\theta)$ in the time intervals (i) and (ii)(see Javed *et al.* 1989), the following expression was obtained for the single film mass transfer coefficient:

$$K = (\mathscr{D}A)^{1/2}\,[s/(1 + As)]\,\{3/\pi^{1/2} + 1/\eta\,[1 - \operatorname{erf}\eta]\,(3/2 - \eta^2)\exp\eta^2\} \tag{3.40}$$

where $\eta = (2As)^{1/2}$, s is the fractional rate of surface renewal, and erf η is the error function of η. Parameters A and s are estimated from film analysis (Thornton *et al.* 1985). For a water drop growing in toluene containing 3 vol per cent diacetone alcohol, A varied between 0.012 and 0.040 s and corresponding average values of s ranged from 13.8 to 4.5/s. Values of K predicted with eqn (3.40) were in good agreement with experimental determinations.

A similar correlation between the intensity and frequency of the disturbances and mass transfer coefficients has been observed in the extraction of uranyl nitrate with tri-butyl phosphate (Rogers *et al.* 1987). Using an aqueous drop with forced circulation (Javed *et al.* 1989) submerged in the organic phase, Rogers *et al.* measured the time dependent mass transfer coefficients for the extraction of uranyl nitrate at different driving forces. They found that the mass transfer enhancement factor increased with driving force and that mass transfer coefficients rapidly decreased with time down to an asymptotic value of about 10 per cent of the original one. In most of their experiments there was a 30 per cent decrease in transfer rate during the first 50 s.

However in the reverse direction of transfer, i.e. backwash from the organic phase, no time dependence was observed. It should be noticed that this system exhibits Marangoni instabilities for transfer into the organic phase, as shown in Fig. 3.8(a) and 3.8(b), while transfer from the organic phase is diffusional.

The dependence of the mass transfer coefficients on both time and concentration has important implications in the performance of solvent extraction contactors (Batey and Thorton 1989; Thornton 1987). In particular the time-scale over which drop ageing affects the coefficients makes the rate of generation of fresh interface, i.e. the frequency of coalescence and redispersion in the contactor, a major design parameter.

Nomenclature

C	concentration
$\mathscr{D}$	solute diffusivity
$\mathscr{G}$	dT/dx
e	$(\nu_A/\nu_B)^{1/2}$
f	dimensionless function defined by eqn (3.27)
k	thermal conductivity

H molar enthalpy
K mass transfer coefficient
$\mathscr{L}$ dC/dx
N stirring speed
p pressure
r $(\mathscr{D}_A/\mathscr{D}_B)^{1/2}$
s fractional rate of surface renewal
t time
$\mathbf{u}$ velocity vector
v_A μ_s/μ_A
x,y spatial coordinates

Greek letters
α wave number
β growth constant
ϕ surface age frequency distribution function
κ thermal diffusivity
λ wavelength
μ dynamic viscosity
μ_s surface viscosity
ν kinematic viscosity
π interfacial pressure
ρ density
σ_0 equilibrium interfacial tension
τ_{xy} y-component of the fluid shear stress
ψ x-part of the stream function

Subscripts
A phase A ($x > 0$)
B phase B ($x < 0$)
i interfacial value ($x = 0$)
N neutrally stable disturbance
S stationary disturbance

Superscripts
0 value in the undisturbed state
$'$ perturbation value
$\hat{}$ real part of complex variable
$\tilde{}$ imaginary part of complex variable

References

AGUIRRE, F. J., KLINZING, G. E., and CHIANG, S. H. (1985). *International Journal of Heat and Mass Transfer*, **18**, 1891.
AUSTIN, L. J. (1966). PhD Thesis. University of London.

AUSTIN, L.J. and SAWISTOWSKI, H. (1967). *Institution of Chemical Engineers Symposium Series*, No. 26, 3.

AUSTIN, L.J., BANCYK, L., and SAWISTOWSKI, H. (1971). *Chemical Engineering Science*, **26**, 2120.

AUSTIN, L.J., YING, W.E., and SAWISTOWSKI, H. (1966). *Chemical Engineering Science*, **21**, 1109.

BAKKER, C.A.P., VAN BUYTENEN, P.M., and BEEK, W.J. (1966). *Chemical Engineering Science*, **21**, 1039.

BAKKER, C.A.P., VAN VLISSINGEN, F.H.F., and BEEK, W.J. (1967). *Chemical Engineering Science*, **22**, 1349.

BATEY, W. and THORNTON, J.D. (1989). *Industrial and Engineering Chemistry Research*, **28**, 1101.

BATEY, W., LONIE, S.J., THOMPSON, P.J., and THORNTON, J.D. (1984). *Institution of Chemical Engineering Symposium Series*, No. 88, 57.

BÉNARD, H. (1900a). *Revue Générale des Sciences Pures et Appliquées*, **11**, 1261–71.

BÉNARD, H. (1900b). *Revue Générale des Sciences Pures et Appliquées*, **11**, 1309.

BÉNARD, H. (1901). *Annales de Chimie et de Physique*, **23**, 62–144.

BERG, J.C. (1972). In *Recent developments in separation science*, Vol. II. C.R.C. Press, Cleveland, Ohio.

BERG, J.C. and MORIG, C.R. (1969). *Chemical Engineering Science*, **24**, 937.

BISCH, P.M., STEINCHEN, A., and SANFELD, A. (1983). *Journal of Colloid and Interface Science*, **95**, 561.

BURKHOLDER, H.C. and BERG, J.C. (1974). *American Institute of Chemical Engineers*, **20**, 872.

CHANDRASEKHAR, S. (1961). *Hydrodynamic and hydromagnetic stability.* Clarendon Press, Oxford.

COYLE, R.W. and BERG, J.C. (1984). *Chemical Engineering Science*, **39**, 168.

COYLE, R.W., BERG, J.C., and NIWA, J.C. (1981). *Chemical Engineering Science*, **36**, 19.

DANCKWERTS, P.V. (1951). *Industrial and Engineering Chemistry, Process Development*, **43**, 1460.

DANCKWERTS, P.V. (1955). *Journal of the American Institute of Chemical Engineers*, **1**, 456.

DAVIES, J.T. (1972). *Turbulence phenomena.* Academic Press, New York.

DAVIES, J.T. and KHAN, W. (1965). *Chemical Engineering Science*, **20**, 713.

DAVIES, G.A. and THORNTON, J.D. (1977). *Letters in Heat and Mass Transfer*, **4**, 287.

DUPEYRAT, M. and NAKACHE, E. (1977) In *Physicochemical hydrodynamics* (ed. D.B, Spalding), Vol. 2, pp. 251. Advance Publications, Guernsey.

DUPEYRAT, M. and NAKACHE, E. (1978). *Bioelectrochemistry and Bioenergetics*, **5**, 134.

DZUBUR, I. and SAWISTOWSKI, H. (1974). *Proceedings of the international conference on solvent extraction* 1974, p. 379. Society of Chemical Industry, London.

FROMMER, M.A. and MESSALEM, R.M. (1973). *Industrial and Engineering Chemistry, Product and Research Development*, **12**, 328.

GROSS, B. and HIXON, A.N. (1969a). *Industrial and Engineering Chemistry Fundamentals*, **8**, 288.

GROSS, B. and HIXON, A.N. (1969b). *Industrial and Engineering Chemistry Fundamentals*, **8**, 296.

HAYDON, D.A. (1958). *Proceedings of the Royal Society*, **A243**, 483.
HANCOCK, T.A., WHITE, J.L., and SPRUIELL, J.E. (1980). *Polymer Engineering and Science*, **20**, 1126.
HENNENBERG, M., SØRENSEN, T.S., STEINCHEN, A., and SANFELD, A. (1975). *Journal de Chimie Physique*, **72**, 1202.
HIGBIE, R. (1935). *Transactions of the American Institute of Chemical Engineers*, **31**, 365.
JAVED, K.H. and THORNTON, J.D. (1984). *Institution of Chemical Engineers Symposium Series*, No. 88, 203.
JAVED, K.H., THORNTON, J.D., and ANDERSON, T.J. (1989). *Journal of the American Institute of Chemical Engineers* (in press).
LEWIS, J.B. and PRATT, H.R.C. (1953). *Nature*, London, **171**, 1155.
MAROUDAS, N.G. and SAWISTOWSKI, H. (1964). *Chemical Engineering Science*, **19**, 919.
MARSH, B.D., SLEICHER, C.A., and HEIDEGER, W.J., (1965). Paper presented at 57th Annual Meeting of the American Institute of Chemical Engineers, Philadelphia.
MEISTER, B.J. and SCHEELE, G.F. (1967). *Journal of the American Institute of Chemical engineers*, **13**, 682.
MEISTER, B.J. and SCHEELE, G.F. (1969). *Journal of the American Institute of Chemical Engineers*, **15**, 689.
MENDES TATSIS, M.S. and ORTIZ, E.S.P. de, (1989). In *Microgravity experiments during parabolic flights of KC-135 aircraft*, pp. 19 European Space Research and Technology Centre, Noorwijk, The Netherlands.
MERSON, R.L. and QUINN, J.A. (1964). *Journal of the American Institute of Chemical Engineers*, **10**, 804.
NAIKAIKE, Y. and TADAKI, T. (1979). *International Chemical Engineering*, **19**, 328.
NAIKAIKE, Y., TADENUMA, Y., SATO, T., and FUJINAWA, K. (1971). *International Journal of Heat and Mass Transfer*, **14**, 1951.
NAKACHE, E., DUPEYRAT, M., and VIGNES-ADLER, M. (1980). *Journal of Colloid and Interface Science*, **74**, 495.
NAKACHE, E., DUPEYRAT, M., and VIGNES-ADLER, M. (1983). *Journal of Colloid and Interface Science*, **94**, 187.
NAKACHE, E., DUPEYRAT, M., and VIGNES-ADLER, M. (1984). *Faraday Discussions of the Chemical Society*, **77**, 189.
NAKACHE, E., DUPEYRAT, M., and LEMAIRE, J. (1986). *Journal de Chimie Physique et de Physico-Chemie Biologique*, **83**, 339.
OLANDER, D.R. and REDDY, L.B. (1964). *Chemical Engineering Science*, **19**, 67.
ORELL, A. and WESTWATER, J.W. (1957). *Industrial and Engineering Chemistry*, **49**, 1030.
ORELL, A. and WESTWATER, J.W. (1961). *Chemical Engineering Science*, **16**, 127.
ORELL, A. and WESTWATER, J.W. (1962). *Journal of the American Institute of Chemical Engineers*, **8**, 351.
ORTIZ, E.S.P. de, and SAWISTOWSKI, H. (1973a). *Chemical Engineering Science*, **28**, 2051.
ORTIZ, E.S.P. de, and SAWISTOWSKI, H. (1973b). *Chemical Engineering Science*, **28**, 2063.
ORTIZ, E.S.P. de, and SAWISTOWSKI, H. (1975). *Chemical Engineering Science*, **30**, 1527.

PEARSON, J.R.A. (1958). *Journal of Fluid Mechanics*, **4**, 489.
PEKER, S., ÇÖMDEN, M., and ARAGÜNDÜZ, G. (1980). *Chemical Engineering Science*, **35**, 1679.
QUINCKE, G. (1888). *Annalen der Physik, Leipzig*, **35**, 593.
RAYLEIGH, Lord (1878). *Proceedings of the London Mathematical Society*, **10**, 4.
RAYLEIGH, Lord (1879). *Proceedings of the London Mathematical Society*, **29**, 71.
RAYLEIGH, Lord (1892). *Pilosophical Magazine*, **34**, 145.
RAYLEIGH, Lord (1900). Scientific papers, ii, pp. 200. Cambridge, England.
RAYLEIGH, Lord (1916). *Pilosophical Magazine*, **32**, 529.
ROGERS, D., THOMPSON, P.J., and THORNTON, J.D. (1987). *Institution of Chemical Engineers Symposium Series*, No. 103, 15.
RUCKENSTEIN, E. and BERBENTE, C. (1964). *Chemical Engineering Science*, **19**, 329.
SANFELD, A., LIN, M., BOIS, A., PANAIATOV, I., and BARET, J.F. (1984). *Advances in Colloid and Interface Science*, **20**, 101.
SANFELD, A., STEINCHEN, A., HENNENBERG, M., BISH, P.M. VAN LAMSWEERDE-GALLEZ, D., and DALLE-VEDOVE, W. (1979). In *Lecture notes in physics*, (ed. T.S. Sorensen), Vol. 105, pp. 108. Springer-Verlag, Berlin.
SAWISTOWSKI, H. (1971). Interfacial phenomena. In *Recent advances in liquid–iquid extraction* (ed. C. Hanson). Pergamon Press, Oxford.
SAWISTOWSKI, H. (1973). *Chemic-Ingenieur-Technik*, **45**, 1114.
SAWISTOWSKI, H. (1981). *Latin American Journal of Chemical Engineering and Applied Chemistry*, **11**, 169.
SAWISTOWSKI, H. and GÖLTZ, G.E. (1963). *Transactions of the Institution of Chemical Engineers*, **41**, 174.
SAWISTOWSKI, H. and JAMES, B.R. (1963). *Chemie-Ingenieur-Technik*, **35**, 175.
SAWISTOWSKI, H. and JAMES, B.R. (1967). Proceedings of the third international conference on surface active materials. Akademie Verlag, Berlin.
SEHRT, B. and LINDE, H. (1967). Proceedings of the third international conference on surface active materials. Akademie Verlag, Berlin.
SETHY, A. and CULLINAN, H.T. Jr. (1975a). *Journal of the American Institute of Chemical Engineers*, **21**, 571.
SETHY, A. and CULLINAN, H.T. Jr. (1975b). *Journal of the American Institute of Chemical Engineers*, **21**, 575.
SKELLAND, A.H.P. and HUANG, Y.F. (1977). *Journal of the American Institute of Chemical Engineers*, **23**, 701.
SKELLAND, A.H.P. and HUANG, Y.F. (1979). *Journal of the American Institute of Chemical Engineers*, **25**, 80.
SHERWOOD, T.K. and WEI, J.C. (1957). *Industrial and Engineering Chemistry*, **49**, 1030.
SØRENSEN, T.S. (1979). In *Lecture notes in physics*, (ed. T.S. Sorensen), Vol. 105, pp. 1–74. Springer-Verlag, Berlin Heidelberg.
SØRENSEN, T.S. (1980). *Journal of the Chemical Society, Faraday Transactions Section II*, **76**, 1170.
SØRENSEN, T.S., HENNENBERG, M., STEINCHEN, A., and SANFELD, A. (1976). *Journal of Colloid and Polymer Science*, **56**, 191.
SØRENSEN, T.S., HENNENBERG, M., and HANSEN, F.Y. (1978). *Journal of the Chemical Society, Faraday Transactions Section II*, **74**, 1005.
STERNLING, C.V. and SCRIVEN, L.E. (1959). *Journal of the American Institute of Chemical Engineers*, **5**, 514.

STEWART, G. and THORNTON, J.D. (1967a). *Institution of Chemical Engineers Symposium Series*, No. 26, 29.
STEWART, G. and THORNTON, J.D. (1967b). *Institution of Chemical Engineers Symposium Series*, No. 26, 37.
TAYLOR, G.I. (1934). *Proceedings of The Royal Society (London)*, **A 146**, 501.
THOMPSON, P.J., BATEY, W., and WATSON, R.J. (1984). *Institution of Chemical Engineers Symposium Series*, No. 88, 231.
THORNTON, J.D. (1976). *Birmingham University Chemical Engineer*, **27(1)**, 6.
THORNTON, J.D. (1987). *Chemistry and Industry*, **6** (March 16), 193.
THORNTON, J.D. and ANDERSON, T.J. (1981). *International Journal of Heat and Mass Transfer*, 1847.
THORNTON, J.D., ANDERSON, T.J., JAVED, K.H., and ACHWAL, S.K. (1985). *Journal of the American Institute of Chemical Engineers*, **31**, 1069.
TOMOTIKA, S. (1935). *Proceedings of the Royal Society, London*, **A150**, 322.
WATANABE, A., HIGASHITSUJI, K., and NISHIZANA, K. (1978). *Journal of Colloid and Interface Science*, **64**, 278.
YING, W.E. and SAWISTOWSKI, H. (1971). *Proceedings of the international solvent extraction conference* 1971, pp. 840. Society of Chemical Industry, London.

4

MASS TRANSFER COEFFICIENTS, INTERFACIAL EFFECTS, AND SURFACE RENEWAL

K.H. JAVED

4.1 Introduction

Mathematical models describing the mechanism of mass transfer across liquid–liquid interfaces are of basic importance in the prediction of mass transfer coefficients for use in the design of industrial contacting equipment.

Unfortunately, mass transfer across an interface between two fluid phases is a complex phenomenon and our understanding of the interactions between the transfer rate and the hydrodynamic environment near the interface is as yet incomplete; as a result, a purely theoretical analysis of the transport process is not possible. On the other hand, some progress can be made by idealizing the hydrodynamic situation in and around the interface so that the effects of turbulence on the transfer mechanism can be taken into account by suitable model parameters. The resulting models are simple but the relevant parameters do need to be determined experimentally.

4.2 Basic transport models

The underlying theories of mass transfer have been discussed at length in Chapter 2 where a number of transport models were developed. The resultant expressions for the mass transfer coefficients are now summarized and compared for the more important cases.

4.2.1 Two-film model (Lewis and Whitman 1924)

Whilst the postulate of resistances in the form of static films at each side of the interface is still of practical utility in analysing experimental data, it is no longer thought to be appropriate to many of the hydrodynamic situations encountered in liquid–liquid systems. In this model, the mass transfer flux and coefficient are defined by:

$$N = (D/L)\,\Delta c \tag{4.1}$$

$$k = (D/L) \tag{4.2}$$

Equation (4.2) states that the mass transfer coefficient is directly proportional to the molecular diffusivity of the solute in the phase in question and is inversely proportional to the film thickness. The latter cannot be measured directly and is therefore usually assumed to be an inverse function of the turbulence level within the system.

4.2.2 Penetration model (Higbie 1935)

The so-called penetration theory envisages mass transfer as an unsteady state process in which fluid eddies from the bulk phase travel by convection to the interface where they remain for an equal but limited period of time. During this exposure period, solute transfer proceeds by molecular diffusion until the eddies are replaced by fresh elements of fluid from the bulk phase. The corresponding expressions for the mass transfer flux and the mass transfer coefficient are now:

$$N_{av} = 2(c^* - c_0)(D/\pi t_e)^{1/2} \tag{4.3}$$

$$k = 2(D/\pi t_e)^{1/2} \tag{4.4}$$

By contrast with the two-film model, eqn (4.4) predicts that the transfer coefficient is dependent upon the square root of the molecular diffusivity and inversely proportional to the exposure time. Since values of t_e are generally unknown, the model is only of use in those instances in which the exposure time can be deduced from flow considerations.

4.2.3 Surface renewal model (Danckwerts 1951)

Unlike Higbie, Danckwerts proposed that bulk phase turbulence extended to the interface which was thereby renewed continuously by fresh fluid eddies of uniform bulk phase composition. The turbulent eddies were thought to remain at the interface for varying time intervals extending from zero to infinity before being swept back into the bulk phase by convection currents. Thus at any instant, the interface was assumed to consist of a population of eddies with different ages represented by a probability function, $\phi(t)$. It was further assumed that the probability of an eddy being replaced was independent of its age so that all the eddies had an equal chance of replacement. The probability function was defined by:

$$\phi(t) = s\mathrm{e}^{-st} \tag{4.5}$$

where s is the rate of surface renewal and is constant for any given set of hydrodynamic conditions. This concept leads to the following expressions for the mass transfer flux and the mass transfer coefficient:

$$N_{av} = (c^* - c_0)(Ds)^{1/2} \tag{4.6}$$

$$k = (Ds)^{1/2} \tag{4.7}$$

Once again, the mass transfer coefficient is seen to be proportional to the square root of the molecular diffusivity and, in common with the previous model, an unknown parameter, s, is present. As a result, the use of eqn (4.7) for the prediction of mass transfer coefficients is limited to those situations in which the rate of surface renewal has been measured.

4.2.4 Film-penetration model (Toor and Marchello 1958)

The above models have, in essence, been combined in the so-called film-penetration model. According to this theory, the resistance to mass transfer is confined within a surface film but, unlike the two film theory in which the film is assumed to be quiescent, the film is now regarded as composed of fluid elements which are of a finite thickness and which experience continual renewal as suggested by the penetration and surface renewal theories.

The solute transfer through the elements of the film is accomplished by steady state as well as unsteady state molecular diffusion, the particular mechanism being determined by the relative ages of the elements at the surface. Thus those fluid elements which have been freshly created follow an unsteady state mechanism of solute penetration, whilst those elements which are not replaced during their penetration stage permit transfer by a steady state mechanism since further solute accumulation is precluded.

The model equations describing these conditions give rise to the following expressions for the mass transfer coefficient for the two different surface age distribution functions used by these authors:

(a) Uniform surface age distribution (Higbie).

(i) For short exposure times:

$$k = 2\,(D/\pi t_e)^{1/2}\{1 + 2\pi^{1/2}\sum_{n=1}^{\infty} \text{ierfc}\,[nL/(Dt_e)^{1/2}]\} \tag{4.8}$$

where *ierfc* is the complementary error function.

(ii) For long exposure times:

$$k = (D/L)\{1 + (2L^2/\pi^2 Dt_e)[(\pi^2/6) - \sum_{n=1}^{\infty} \exp(-n^2\pi^2 Dt_e/L^2)]\} \tag{4.9}$$

Equations (4.8 and 4.9) reduce to eqns (4.4 and 4.2) of the penetration and two film theory for small and large values of t_e, respectively.

(b) Random surface age distribution (Danckwerts).

(i) For short exposure times:

$$k = (Ds)^{1/2}\{1 + 2\sum_{n=1}^{\infty} \exp[-2nL(s/D)^{1/2}]\} \qquad (4.10)$$

(ii) For long exposure times:

$$k = (D/L)\{1 + 2\sum_{n=1}^{\infty} [1 + n^2\pi^2(D/sL^2)]^{-1}\} \qquad (4.11)$$

Equation (4.10) reduces to Danckwerts' model eqn (4.7) when sL^2/D is large and eqn (4.11) reduces to the two film model eqn (4.2) for small values of sL^2/D.

Again the application of these expressions involves a knowledge of model parameters such as the eddy size, exposure times and the rate of surface renewal, none of which are readily calculable at the present time.

4.3 Modified surface renewal models

[See also Skelland, Ch. 2]

These models maintain the surface renewal concept coupled with unsteady state molecular diffusion as the operating mechanism for mass transfer but differ mainly from the original penetration and surface renewal models of Higbie and Danckwerts in adopting different age distributions for the surface elements.

Perlmutter (1961) criticized Higbie's notion of equal life times for all the eddies as statistically unrealistic and disagreed with Danckwerts' random surface age distribution since it predicted that the most likely residence time of an eddy was zero. As alternatives, Perlmutter suggested two forms of the age distribution. The first was based on the two well-mixed vessels in series model, which was taken to represent the interfacial region in which eddies exchange mass by unsteady state molecular diffusion. The second distribution was based upon a single well-mixed vessel with a dead time effect, implying that each eddy had a minimum residence time. This dead time reflected the likely presence of stagnant zones in the interfacial boundary which was otherwise equivalent to a completely mixed vessel in Danckwerts' model.

Other residence time concepts have been proposed by Koppel *et al.* (1966) who used a gamma distribution for the residence times of the eddies and a delayed exponential residence time distribution to account for stagnant fluid elements as suggested by Perlmutter.

Benjamin *et al.* (1971) pointed out that surface renewal models mainly focussed on steady state transfer processes and were therefore unrepresentative of unsteady state processes. They presented unsteady state forms of the age distributions for the Higbie, Danckwerts and multiple capacity models of Perlmutter while Rambabu and Narsimhan (1980) proposed a transient form of the gamma residence time distribution used by Koppel *et al.* There is thus a spectrum of models, each differing in the properties of their respective eddy residence time distributions.

A new concept was introduced by Harriott (1962) when he criticized the penetration and the surface renewal theories for assuming that mass transfer is due entirely to the turbulent eddies which travel right up to the interface thereby causing surface renewal. He argued that those eddies which approach but do not reach the interface can also influence the transfer rate and suggested a two parameter model (the random eddy model) which assumed that the eddies arriving at random times come to within random distances of the interface. This concept was formulated as a stochastic process by Bullin and Dukler (1972) by replacing the probability distribution used by Harriott with the gamma distribution for generating sequences of random eddy distances and lifetimes.

The age distribution functions used in these models are summarized in Table 4.1. It must be remembered however that although the more sophisticated distribution functions no doubt represent a closer approximation to reality, the practical use of the models is restricted by lack of experimental verification of the distributions coupled with an inability to predict them from first principles.

4.4 Turbulent diffusion models

Several models have been proposed which attempt to relate the mechanism of mass transfer to turbulence through eddy diffusivity and the parameters which influence the size of eddies and the characteristic turbulence velocity.

Kishinevskii with his co-workers (Kishinevskii and Pamfilov 1949; Kishinevskii 1955; Kishinevskii and Serebryansky 1956) proposed the idea of continuous surface renewal in which the solute transfer to surface eddies is predominantly by eddy diffusion, rather than by molecular diffusion as suggested in the penetration and surface renewal theories of Higbie and Danckwerts. They also disagreed with the suggestion that the probability of replacement of a surface element was independent of its age.

According to the present theory, the eddy diffusivity varies in magnitude with the normal distance from the interface where it persists at a sufficient level to dominate the mode of solute transfer. The mass transfer coefficient is thus independent of the molecular diffusivity, an argument which is in conflict with all other theories of mass transfer.

In the eddy diffusion model suggested by Levich (1962), the turbulent eddies arriving at the interface are retarded in their motion by the interfacial tension forces which oppose surface deformation and there exists a primary zone of damped turbulence near the interface. The solute transfer through this zone occurs by steady state molecular diffusion. The fluid velocity

TABLE 4.1 *Age/residence time distribution for surface eddies in surface renewal models*

Author	Age/residence time distribution function	Steady or unsteady form	Mathematical definition	Model parameter/s
Higbie (1935)	uniform age	steady state	$\phi(t) = \begin{bmatrix} \frac{1}{t_e} & 0 < t < t_e \\ 0 & t \geq t_e \end{bmatrix}$	t_e
Danckwerts (1951)	random age	"	$\phi(t) = s\exp(-st)$	s
Perlmutter (1961)	(i) two capacitance	"	$\phi(t) = \frac{1}{\tau}\left(1 + \frac{2t}{\tau}\right)\exp(-2t/\tau)$	τ
	(ii) single well-mixed vessel with dead-time	"	$\phi(t) = \begin{bmatrix} \frac{1}{\tau + A} & 0 \leq \theta < A \\ \frac{1}{\tau + A}\exp[-(t - A)/\tau] & \theta \geq A \end{bmatrix}$	τ, A
Harriott (1962)	random age and random approach distance	"	$P(z) = \frac{1}{\Gamma(\alpha)\beta^{\alpha z}} z^{\alpha - 1}\exp\left(-\frac{z}{\beta}\right)$ $\alpha = 1$ for $f(\theta)$ $\alpha = 1$ and 2 for $f(H)$ $\beta = \left(\frac{\bar{z}}{\alpha + 1}\right)$	$\bar{H},\tau,\alpha$

TABLE 4.1 *continued*

Author	Age/residence time distribution function	Steady or unsteady form	Mathematical definition	Model parameter/s
Koppel *et al.* (1964)	(i) gamma distribution	"	$\phi(t) = \frac{s\exp(-st)}{n+1}\sum_{i=0}^{n}(st)^i/i!$	s, n
	(ii) dead time model	"	$\phi(t) = \frac{1}{\tau}\left[u(t-A)\exp\left[\frac{A\tau - t(\tau)^{1/2}}{1-A}\right] + u(t) - u(t-\tau A)\right]$ $u(t) = \begin{pmatrix} 0 & t<0 \\ 1 & t\geq 0 \end{pmatrix}$	τ, A
Benjamin *et al.* (1971)	(i) uniform age	unsteady state	$\phi(t,\tau) = \begin{bmatrix} \frac{1}{\tau} + (1 - t'/\tau)\delta(t-t') & 0 \leq t \leq t' \leq \tau \\ 0 & t' < t \leq \tau \\ 0 & \tau < t \end{bmatrix}$	
	(ii) random age	"	$\phi(t,\tau) = \begin{bmatrix} s\exp(-st) + \exp(-st')\cdot\delta(t-t') & 0 \leq t \leq t' \\ 0 & t' < t \end{bmatrix}$	s

	(iii) multiple capacitance	"	$\phi(t,\tau) = s\exp(-Ns)\left[\sum_{i=0}^{N-1}\frac{(Nst)^i}{i!}\right] + \delta(t-t')\left[\frac{\exp(-Nst')}{N}\sum_{i=0}^{N-1}\sum_{n=0}^{i}\frac{(Nst')^n}{n!}\right]$	s, N
Rambabu and Narsimhan (1980)	gamma distribution	"	$\phi(t,\tau) = 0 \quad \text{for } 0 < t \le t'$; for $t' < t$: $\phi(t,\tau) = \beta(t')\left[1 - \frac{1 - e^{-nst}\sum_{k=0}^{n-1}[(nst)^k/k!]}{1 - e^{-nst}\sum_{k=0}^{n-1}[(nst')^k/ki]}\right]$ $\beta(t') = s\left[1 - e^{-nst'}\sum_{k=0}^{n-1}\frac{(nst')^k}{k}\right] \cdot \left[1 - \sum_{k=0}^{n-1} e^{-nst'}\sum_{i=0}^{k}\frac{(nst')^i}{i!} - e^{-nst'}\sum_{k=0}^{n-1}\frac{(nst')^{k+1}}{k!}\right]^{-1}$	s, n

normal to the interface is taken to be a linear function of the distance from the interface with the eddy diffusivity depending upon the square of this distance. On this basis, the expression for the mass transfer coefficient is given by:

$$k = (v_0)^{3/2} (\rho D/\sigma)^{1/2} \tag{4.12}$$

It will be noticed that eqn (4.12) shows a square root dependence of k upon D.

Levich's model was modified by Davies (1972) in order to take into account the additional effect of gravitational forces on the damping of eddy diffusivity. This was done by defining an equivalent or modified interfacial tension force in eqn (4.12).

King (1966) elaborated this theme in his proposal of a generalized model which combined the concepts of penetration/surface renewal with an eddy diffusivity which was damped near the interface due to the effects of interfacial tension forces. Thus it was proposed that a fresh liquid element formed as a result of surface renewal underwent unsteady state solute penetration. But at large values of the surface age, solute transfer through the eddy took place by the steady state mechanisms of molecular and eddy diffusion while, at the interface, the eddy diffusivity was zero and transfer was by molecular diffusion alone.

The eddy diffusivity, D_E, was assumed to follow a continuous profile with distance, x, normal to the interface according to the expression:

$$D_E = ax^m \tag{4.13}$$

where a and m are constants.

At low surface ages, the model reduces to the penetration model as the term D_E becomes unimportant. At large values of the surface age, the solution of the steady state form of the model equation gives the following expression for the mass transfer coefficient:

$$k = (m/\pi) \sin (\pi/m) \, a^{1/m} D^{(1-1/m)} \tag{4.14}$$

The constants m and a are unknown quantities in this equation but King suggested that the value of m should vary between 2 and 4 so that the exponent on D agreed with experimental observations. The constant a was related to the following variables by

$$a = C(\epsilon\rho/\mu^2) \tag{4.15}$$

where the constant C remains to be determined experimentally.

Steady state eddy diffusion mechanisms were also proposed by Lamourelle and Sandall (1972), Prasher and Fricke (1974) and Henstock and Hanratty (1979). The resulting expressions for the mass transfer coefficients again follow the form of eqn (4.14) but it is argued that m must be equal to 2 so that k is proportional to $D^{1/2}$ which is generally observed in practice. In each model however, a different relationship is suggested between the constant a and the physical parameters of the system. These are summarized in Table 4.2.

Models based on idealized turbulent structures for eddy renewal of the interface were proposed by Fortescue and Pearson (1967) and Lamont and Scott (1970). Fortescue and Pearson proposed a square roll cell structure for the eddies and attributed mass transfer to the large eddies of integral length scale and of intensity corresponding to the bulk phase turbulence. Lamont and Scott in their eddy cell model proposed that mass transfer is controlled by small scale eddies and therefore depends on parameters such as the rate of energy dissipation and kinematic viscosity which control the size of eddies.

The models described in this section have generally been used in interpreting gas–liquid mass transfer in turbulent liquid films and across flat interfaces, with the bulk liquid phase agitated. The articles by Davies and Lozano (1979), Bin (1983) and Sonin *et al.* (1986) provide a comprehensive overview of these models and of the experimental work concerning the measurements of the model parameters.

Table 4.2 lists the major models of mass transfer. Among these models the two film model completely ignores the flow conditions in the vicinity of the interface by assuming quiescent surface layers. The effect of turbulence is confined to its influence on the effective thickness of the layers which are unlikely to be physically present under turbulent flow conditions (Danckwerts 1951). The model thus provides an over-simplified and inadequate representation of the real mechanism of mass transfer. Nevertheless, the concepts of equilibrium conditions at the interface and the two film resistances acting in series have been very useful contributions as they are invariably used in the fundamental treatment of interphase transport processes and in the design of gas absorbers (Sherwood *et al.* 1975) and liquid–liquid extractors (Skelland 1974; Tung and Luecke 1986).

By contrast the surface renewal models assume hydrodynamic fluid exchange between the interface and the bulk phase. The influence of the turbulence is viewed in terms of refreshing the interface but no insight is offered concerning the hydrodynamics of the renewing eddies. The resulting effect of the renewal mechanism appears in the time scale for which diffusion occurs between the surface elements and the bulk phase and each renewal model attaches a different significance to this parameter.

Application of these models to mass transfer in liquid–liquid systems has been prevented by the difficulty of measuring the age distribution of the liquid eddies. The only work which has reported surface renewal measurements for liquid interfaces is due to Thornton *et al.* (1985). They

TABLE 4.2 *Major mathematical models of mass transfer*

Authors	Transfer mechanism	Model equation	Initial and boundary conditions	Mass transfer coefficient, k	Empirical model parameter/s	Value of n in $k \alpha D^n$
Lewis and Whitman (1924)	Stagnant surface films with steady state molecular diffusion	–	–	$k = D/L$	L	1.0
Higbie (1935)	Regular surface renewal with unsteady state molecular diffusion	$\frac{\partial c}{\partial t} = D\frac{\partial^2 c}{\partial x^2}$	$c(x,0) = c_0$ $c(0,t) = c^*$ $c(\infty,t) = c_0$	$k = 2(D/\pi t_e)^{1/2}$	t_e	0.5
Danckwerts (1951)	Random surface renewal	$\frac{\partial c}{\partial t} = D\frac{\partial^2 c}{\partial x^2}$	$c(x,0) = c_0$ $c(0,t) = c^*$ $c(\infty,t) = c_0$	$k = (Ds)^{1/2}$	s	0.5
Toor and Marchello (1958)	Surface film and surface renewal (Regular or random)	$\frac{\partial c}{\partial t} = D\frac{\partial^2 c}{\partial x^2}$	$c(x,0) = c_0$ $c(0,t) = c^*$ $c(L,t) = c_0$	eqns (4.8)–(4.11)	L, t_e or s	0.5–1.0
Perlmutter (1961)	Surface renewal (Two capacitances or a single well-mixed vessel with dead time	$\frac{\partial c}{\partial t} = D\frac{\partial^2 c}{\partial x^2}$	$c(x,0) = c_0$ $c(0,t) = k'(c^* - c)$ $c(\infty,t) = c_0$	eqns (21), (29), and (30), in Perlmutter (1961)	τ or τ and A	0.5

Koppel *et al.* (1966)	Surface renewal (Gamma distribution or a single vessel with dead time	$\frac{\partial c}{\partial t} = D\frac{\partial^2 c}{\partial x^2}$	$c(x,0) = c_0$ $c(0,t) = c^*$ $c(\infty,t) = c_0$	eqns (19) and (21) in Koppel *et al.* (1966)	τ or τ and A	0.5
Levich (1962)	Steady state molecular and eddy diffusion	–	–	$k = (v_0)^{3/2}(D\rho/\sigma)^{1/2}$	v_0	0.5
King (1966)	Surface renewal or steady state molecular and eddy diffusion	$\frac{\partial c}{\partial t} = (D + D_E)\frac{\partial^2 c}{\partial x^2}$ $D_E = ax^m$	$c(x,0) = c_0$ $c(0,t) = c^*$ $c(\rightarrow\infty,t) = c_0$	eqns (4.13), (4.14), and (4.15)	t_e, a, m $\left[a \propto \left(\frac{\epsilon\rho}{\mu^2}\right)\right]$	0.5–1.0
Lamourelle and Sandall (1972)	Steady state molecular and eddy diffusion	$0 = (D + D_E)\frac{\partial^2 c}{\partial x^2}$ $D_E = ax^2$	"	$k = \frac{2}{\pi}\sin\left(\frac{\pi}{2}\right)a^{1/2}D^{1/2}$	a $[a \propto Re^{1.678}]$	0.5
Prasher and Fricke (1974)	"	"	"	"	a $\left[a \propto \frac{\rho}{\sigma}\left(\frac{\delta^3\epsilon^3}{g\gamma}\right)^{1/2}\right]$	0.5
Henstock and Hanratty (1979)	"	"	"	"	a $\left[a \propto \left(\frac{v_0^3\delta}{\gamma^2}\right)\right]$	0.5

developed a novel photochromic dye trace technique (see Section 4.6; p. 230) which allows both the qualitative and quantitative analysis of the surface behaviour to be made. The technique has been shown to be useful for measurements of the age distribution of the surface elements during mass transfer augmented by the Marangoni effect in liquid–liquid droplet systems. Their results for the droplet systems under the conditions employed in the study indicated that the observed surface renewal was induced by interfacial turbulence accompanying mass transfer and not by the hydrodynamic flow as assumed in the renewal models.

From these models, Higbie's theory has been widely used to predict the transfer coefficients in liquid–liquid mass transfer. The most common application is to liquid drops moving in the continuous liquid phase, in which case the eddy exposure time is assumed to be the time required for the drop to move through a distance equal to its diameter (Laddha and Degaleesan 1976). The theory has also been extended to forming drops (Angelo *et al.* 1966) and the various models for mass transfer during drop formation are based on this theory (Popovich 1964; Walia and Vir 1976). Maroudas and Sawistowski (1964) and Bakker *et al.* (1967) used this model for mass transfer in wetted-wall columns. The systems studied showed higher transfer coefficients which were attributed to interfacial turbulence.

The turbulent diffusion models by comparison with the surface renewal models are founded on a more fundamental basis inasmuch as they view the effect of turbulence in terms of the characteristics of turbulent flow. The difficulty which, however, remains is that almost no direct information is available on the turbulent flow phenomena near the interface. The turbulent diffusion models are, therefore, equally limited for effective predictions of mass transfer coefficients and so are not more promising than the renewal models. Both types of models have a similar dependence of k on D.

The observed dependence of the mass transfer coefficient on the molecular diffusivity has often been used as a criterion for the validity of a transfer model and for indirect comparisons between the transfer models. The functional relationship between k and D as shown by various models can in general be represented by an equation of the type, $k \infty D^n$ where n assumes a value of unity in the two-film model and is equal to 0.5 in the penetration/surface renewal models as well as in the turbulent diffusion models. An exception is the model proposed by Kishinevskii (1955) in which n is equal to zero.

Based on the experimental data of a number of authors including their own, Kozinski and King (1966) concluded that, in general, for gas absorption the exponent on diffusivity lies between the extremes of 0.50 and 0.75. This was also shown in a review on gas absorption by Mangers and Ponter (1980). On the contrary, the work of Kishinevskii and Serebryansky (1956) in gas absorption indicated that there is no dependence of k on D.

Similar findings were reported by Lewis (1954) who studied mass transfer across liquid–liquid interfaces in a stirred vessel. Other investigators (McManamey 1961; Bulicka and Prochazka 1976), however, contradicted the results of Lewis and showed that the exponent n is equal to 0.5.

The direct quantitative evidence of the validity of one or the other model is, however, very scarce because of a lack of suitable techniques to measure directly either the surface eddies' exposure time distribution for use in the renewal models, or information about the eddy size and the velocity fluctuations in the interfacial regions to test the validity of the turbulent diffusion models.

Only recently some experimental techniques have been reported for such measurements. The laser induced tracer technique of Thornton and Anderson (1981) has shown reliability for surface renewal measurements during mass transfer in liquid–liquid pendant droplet systems.

The other experimental technique which has been used by various research workers is laser Doppler anemometry (LDA). Komori *et al.* (1982) used this technique to study the turbulence structure and transport mechanism at the free surface in open channel flow. Their findings favour surface renewal by turbulent eddies with a size close to the integral length scale. Luke and Lee (1986) using LDA, have measured the velocity fluctuations in a region extremely close, 90 μm, to an air–water interface in an unbaffled stirred tank. They found that the eddy velocities are generally lower in the diffusion boundary layer zone compared with the bulk region and that the local mass transfer coefficients are high in regions where small eddies are dominant. Experimental support for random surface renewal is also forthcoming from the work of Van Maanen and Fortuin (1983) who used LDA techniques to measure the time-interval distributions of fluid lumps at pipe walls in the presence of water flow.

With the advent of these modern experimental methods, direct measurement of rates of surface renewal and the characteristics of turbulent flow in interfacial regions may be just emerging. The situation however is far from satisfactory and it must be remembered that it will not be possible to formulate a unified theory which properly represents the real mechanisms of transfer until a lot more information is available concerning the hydrodynamic behaviour at, and very near to, the interface.

4.5 Mass transfer to droplets

4.5.1 Droplet models

So far the discussion has centred on transfer across a liquid–liquid interface without any reference to the nature of that interface. Droplet models are concerned with the specific cases of liquid drops which are rising or falling in a second substantially immiscible liquid phase. The mechanism of mass transfer from the core of the liquid drop to the surface or vice versa is therefore very much dependent upon the fluid dynamics of the combined system.

The theory behind the more important models has already been discussed in Chapter 2 and it will suffice here merely to summarize the characteristics of the principal models currently in use. These differ in the underlying assumptions regarding the nature of the liquid circulation which may prevail in the interior of a drop. In all the models however it is commonly assumed that the continuous phase concentration is uniform and that the drop is spherical. A further consideration in the solution of the model equations is whether the resistance to transfer lies entirely within the the droplet or is shared between the two phases.

4.5.1.1 *Stagnant drops*

Newman (1931) analysed this situation and derived the following expression for unsteady state mass transfer on the basis of no continuous phase resistance:

$$k_d = -(d/6t)\ln\left[(6/\pi^2)\sum_{i=1}^{\infty}(1/i^2)\exp\left(-4D\pi^2 i^2 t/d^2\right)\right] \quad (4.16)$$

If the concentration difference driving force is linear, the asymptotic (i.e. steady state) solution to this equation becomes:

$$k_d = 2\pi^2 D/3d \quad (4.17)$$

The solution for unsteady state diffusion into a rigid drop, taking into account a finite continuous phase resistance, was presented by Groeber (1925):

$$k_d = -(d/6t)\left[\sum_{i=1}^{\infty} B_i \exp\left(-\lambda_i^2 4Dt/d^2\right)\right] \quad (4.18)$$

where B_i and λ_i are functions of the continuous phase mass transfer coefficient, k_c, and are given in Table 4.3.

4.5.1.2 *Circulating drops*

Hadamard (1911) and Rybezinski (1911) derived, from the equations of motion, the stream functions describing the flow patterns inside and outside drops for the viscous flow regime i.e. for *Re* numbers less than 1. Using the internal circulation patterns suggested by these authors, Kronig and Brink (1950) developed a model for mass transfer in circulating drops when the

TABLE 4.3 *Values of B_i and λ_i for a stagnant drop with finite continuous phase resistance*

$\frac{k_c d}{D}$	B_1	B_2	B_3	B_4	λ_1	λ_2	λ_3	λ_4
4	0.159	0.00634	–	–	2.029	4.913	7.979	11.086
6	0.153	0.01090	–	–	2.289	5.087	8.096	11.173
8	–	–	–	–	2.456	5.233	8.205	11.256
10	0.142	0.01790	0.00408	–	2.570	5.354	8.303	11.335
12	–	–	–	–	2.654	5.454	8.391	11.409
14	0.134	0.02200	0.00600	–	2.717	5.358	8.470	11.477
16	–	–	–	–	2.765	5.608	8.541	11.541
18	0.129	0.02420	0.0119	0.00291	2.804	5.667	8.603	11.599
20	–	–	–	–	2.836	5.717	8.659	11.653
22	0.125	0.02550	0.00858	0.00359	2.863	5.761	8.708	11.703
32	0.118	–	–	–	2.948	–	–	–
42	–	–	–	–	2.993	–	–	–
62	0.111	–	–	–	3.041	–	–	–
82	–	–	–	–	3.080	–	–	–
102	0.107	–	–	–	3.080	–	–	–
∞	0.101	–	–	–	3.142	–	–	–

* Taken from Sideman (1966).

continuous phase resistance was zero. The mass transfer coefficient for the dispersed phase was given by:

$$k_d = (d/6t)\left[(3/8)\sum_{i=1}^{\infty} B_i^2 \exp\left(-64\lambda_i Dt/d^2\right)\right] \tag{4.19}$$

and the steady state solution for k_d is approximated by:

$$k_d \doteq 17.9D/d \tag{4.20}$$

Treybal (1963) has pointed out that 'comparison with the case for rigid spheres shows that one might consider such a drop as a rigid sphere for which the effective diffusivity equals 2.25–2.7 times the true molecular diffusivity.'

Kronig and Brink calculated the first two values of B_i and λ_i in eqn (4.19). Further values of these constants were computed by Heertjes *et al.* (1954) to provide up to seven terms in the solution of eqn (4.19); these values are shown in Table 4.4. Heertjes *et al.* also investigated experimentally the transfer of water into isobutanol drops and vice-versa. When transfer occurred into the organic drop, their measurements showed agreement with Kronig and Brink's model but anomalous results were obtained for transfer in the opposite direction.

TABLE 4.4 *Values of B_i and λ_i for Kronig and Brink's model for zero continuous phase resistance*

i	B_i	λ_i
1	1.290	1.658
2	0.596	9.080
3	0.386	22.200
4	0.350	38.500
5	0.280	63.000
6	0.220	89.800
7	0.160	123.800

TABLE 4.5 *Values of B_i and λ_i for Kronig and Brink's model for a finite continuous phase resistance*

$\frac{k_c d}{D}$	B_1	B_2	B_3	λ_1	λ_2	λ_3
3.20	1.49	0.107	–	0.262	4.24	–
5.33	–	–	–	0.386	–	–
8.00	–	–	–	0.534	–	–
10.70	1.49	0.300	–	0.680	4.92	–
16.00	1.48	0.382	–	0.860	5.26	–
21.30	1.47	0.428	–	0.982	5.63	–
26.70	1.49	0.495	0.205	1.082	5.90	15.7
53.30	1.43	0.603	0.298	1.324	7.04	17.5
107.00	1.39	0.603	0.384	1.484	7.88	19.5
213.00	1.31	0.588	0.396	1.560	8.50	20.8
320.00	1.31	0.583	0.391	1.600	8.62	21.3
∞	1.29	0.596	0.386	1.656	9.08	22.2

Kronig and Brink's model was extended by Elzinga and Banchero (1959) to account for the effect of a continuous phase resistance. The resulting expression for the mass transfer coefficient, k_d, is still given by eqn (4.19) but the values of the constants B_i and λ_i depend on the continuous phase resistance and the appropriate values are listed in Table 4.5.

The circulation patterns described by Hadamard (1911) and Rybezinski (1911) have also been used by many other workers (Johns and Beckmann 1966; Brauer 1978a, 1978b; Abramzon and Borde 1980; Oliver and Chung 1985) to obtain numerical solutions of the unsteady state differential equations for the concentration and temperature fields within a drop and in the surrounding fluid phase.

Handlos and Baron (1957) developed a model for vigorously circulating drops. They envisaged that, at high Reynolds numbers (about 1000), a drop is highly turbulent and/or oscillating. Under these conditions, the fluid elements experience both a tangential motion caused by internal circulation and a random radial motion resulting from drop vibrations. They proposed a system of circulation tori to allow for random displacement of eddies in the radial direction. The analysis was based on a negligible continuous phase resistance and the resulting expression for the mass transfer coefficient was given by:

$$k_d = 0.00375\, U/(1 + \mu_d/\mu_c) \tag{4.21}$$

These workers reported that their experimental data agreed with model predictions within 20 per cent. The experimental study of Johnson and Hamielec (1960) for transfer from an organic continuous phase to aqueous drops using the two component systems water–cyclohexanol, water–butanol and water–ethylacetate showed that at low Reynolds numbers, Kronig and Brink's model was appropriate whereas Handlos and Baron's model showed good agreement for oscillating drops. The work of Skelland and Wellek (1964) for low interfacial tension, two component, systems however showed appreciable differences (54 per cent and 38 per cent) between their measurements and the models of Kronig and Brink and Handlos and Baron for circulating and oscillating drops respectively.

Handlos and Baron's model was further developed by Wellek and Skelland (1965) to include a finite continuous phase resistance. This model was only valid for long contact times and they observed a deviation of 25 per cent between their model predictions and the experimental values of the overall mass transfer coefficients reported by Handlos and Baron (1957) and Garner and Skelland (1954) for droplet Reynolds numbers greater than 500.

Patel *et al.* (1967) extended the Handlos–Baron model for short contact times. They also compared the model with the experimental data for mass transfer from high Reynolds number water drops into ethyl acetate and with the data of Handlos and Baron (1957) for different droplet systems. The experimental and calculated coefficients showed significant deviations.

Thus at the present time, it has not been possible to formulate a generalized model of the mass transfer process capable of taking into account the physical chemistry as well as the fluid dynamics of the extraction system.

4.5.2 Partial mass transfer coefficients

The discussion so far has centred on mass transfer models in which the drop environment was considered to be constant. During the time of solute transfer, the fluid dynamics of the system did not change and could be used as a basis for modelling the transport process.

In practice, this is rarely true except perhaps in the case of spray towers where the drops move progressively through a continuous phase under the influence of constant gravitational forces. Even here however, the solute transfer rate during drop formation is likely to be different from that obtaining during steady state motion through the column. In most other types of equipment, the drops experience different hydrodynamic conditions, and hence different transfer rates, during their passage through the contactor.

A good example of this is the conventional packed column in which a range of events may be experiencd by each drop as it passes through the packing. Thus the drop may (Thornton 1964):

1. Move freely within the packing voids.
2. Collide with elements of packing and suffer subsequent distortion.
3. Collide with the packing and break down into two daughter drops.
4. Suffer distortion in passing narrow packing voids.
5. Coalesce with a second stationary drop in the interstices of the packing prior to redispersion.

In practice, the drops will experience most of these events during their passage through the column. Each situation will be characterized by a different mass transfer rate, and a knowledge of the respective mass transfer coefficients is essential if the behaviour of the packing is to be properly understood and the overall coefficient computed.

The term 'partial mass transfer coefficient' has been proposed by Batey and Thornton (1989) to represent the mass transfer characteristics of each of the above micro-events in the life of a drop passing through a packed column. The practical utility of this approach is two-fold; not only could packings be designed which promoted those events associated with high transfer rates but a knowledge of the partial coefficients coupled with the time of each event would enable the overall mass transfer coefficient to be synthesized provided that back-mixing effects were small.

This approach has not been possible until recently because it has not been feasible to measure transfer rates over the very small time intervals associated with each of the above micro-events. A solution to this problem was however developed by Thornton and his co-workers at the University of Newcastle upon Tyne in 1969 and was first described (by Thornton *et al.*) in 1976.

In essence, the technique involves placing a ruled graticule behind a drop undergoing mass transfer and recording its history on high speed cine film. The drop functions as a lens and produces an image of the graticule which is recorded on the film. The optical principle is illustrated in Fig. 4.1 and

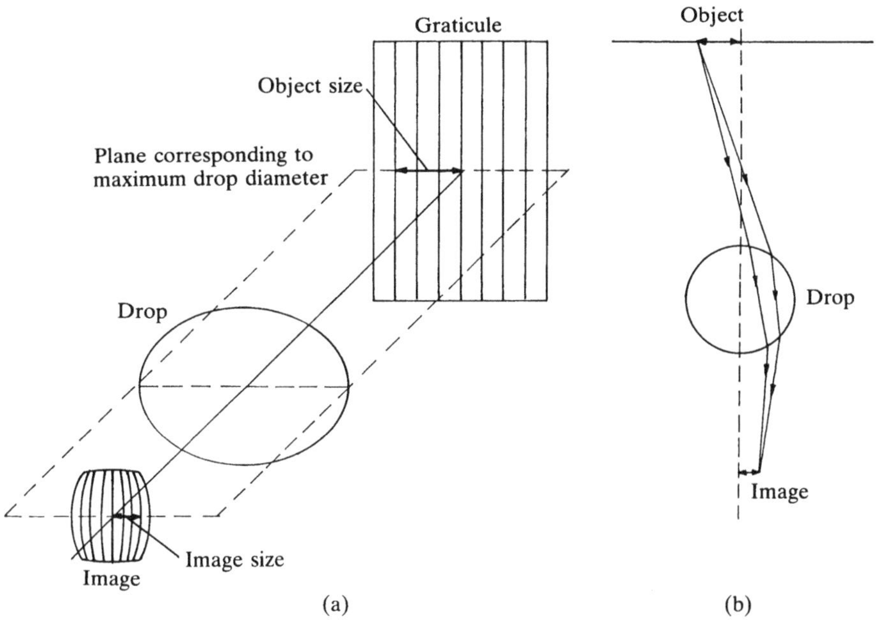

Fig. 4.1. Scale-displacement technique for monitoring solute concentrations in drops: (a) optical arrangement, (b) ray trace (Batey and Thornton 1989.)

typical images of a graticule produced by a moving drop and an impacting drop are shown in Fig. 4.2. From a knowledge of the geometries of the drop, the graticule, and the image, the refractive index of the drop may be computed. Refractive index may, by calibration, be expressed in terms of solute concentration inside the drop so that if filming is carried out at 250 frames/s, the solute concentrations are known at 1/250 s intervals without the need to remove drops for analysis. By this means, the rates of mass transfer during drop splitting, breakdown, distortion or coalescence may be measured over the very small time intervals involved.

Values of partial coefficients have been reported (Batey and Thornton 1989) for some of the events outlined above. One of the significant findings to emerge from this work is the important role played in the mass transfer process by drop distortion and oscillation, following drop collision with a stationary packing element. Values of the partial mass transfer coefficients can be up to five times greater than the corresponding coefficients for unimpeded free drop motion.

The conventional use of a single coefficient to represent the several regimes of mass transfer in a contactor has serious limitations since it gives no information about those individual events which result in high mass transfer rates. If the partial coefficients are known, contactor design can

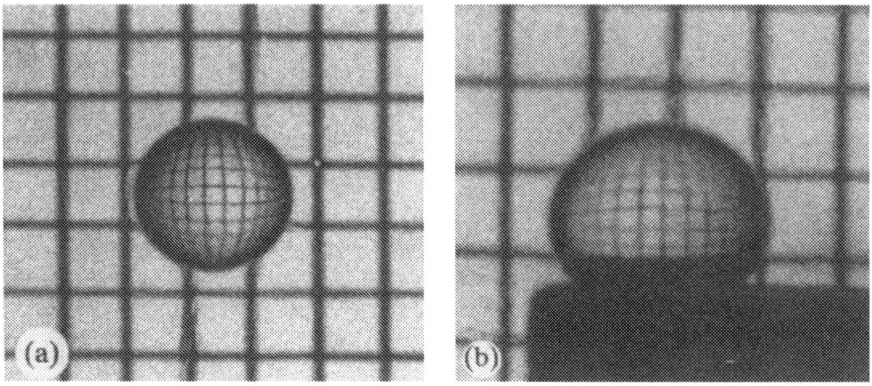

Fig. 4.2. Graticule images formed by aqueous drops in toluene: (a) freely falling drop, (b) impact with a fixed baffle (Fig. 2a from Batey and Thornton 1989).

proceed on a more rational basis and equipment can be developed which promotes the most favourable environment for high efficiency extraction. Unfortunately this is not yet possible and further progress will be dependent upon much more work in this area.

4.6 Mass transfer and interfacial effects [See also Perez de Ortiz, Ch. 3]

All the mass transfer models discussed so far have been based on the assumption that interfacial behaviour is governed entirely by hydrodynamic factors arising from the droplet motion or the actual phase flow rates themselves. This is only true of those systems in which interfacial turbulence or Marangoni effects are absent. The nature of such effects has been discussed in Chapter 3 so that, at this point, we shall only be concerned with their role in the mass transfer process.

In general, interfacial turbulence of this nature results in mass transfer coefficients which are greater than those predicted by conventional models. In many cases, transfer of solute across the interface is accompanied by gross disturbances which manifest themselves in the form of spontaneous agitation, localized eruptions, and surface rippling. In other instances, such disturbances may occur on a micro-scale but will still result in deviations from model predictions. They have been observed during transfer across planar as well as droplet interfaces and mostly for solute transfer from organic to aqueous phases (Sawistowski 1971).

Although mass transfer coefficients may be enhanced by interfacial phenomena, it should be noted that such effects also promote interdroplet coalescence and a consequent reduction in interfacial area. Whether or not the overall contactor efficiency is increased then becomes a matter of balancing an enhanced transfer coefficient against a reduced interfacial area, and the outcome can vary from system to system.

A considerable amount of work has been carried out to study the various aspects of the phenomenon both theoretically and experimentally. Sternling and Scriven (1959) presented a theoretical explanation for the occurrence of interfacial turbulence, according to which the spontaneously arising variations in the solute concentrations during mass transfer produce interfacial tension gradients which in turn induce flow in the interface. Theoretical studies, similar to Sternling and Scriven, have also been performed by Gross and Hixon (1969), Ortiz and Sawistowski (1973a, 1973b, 1975) and Imaishi and Fujinawa (1974) to examine hydrodynamic instability during mass transfer in liquid–liquid systems under different conditions.

Numerous experimental studies (Sawistowski 1971; Takeuchi and Numta 1977; Javed and Thornton 1984) have shown that when mass transfer is accompanied by interfacial disturbances, the rates of mass transfer are markedly increased. Also, that in the presence of this phenomenon, the mass transfer coefficients increase with increasing initial solute concentration and driving force. This effect is thought possibly to arise from the concentration dependence of the interfacial tension and the consequent increase in the intensity of interfacial turbulence.

The phenomena have also been observed qualitatively using schlieren photographic techniques (Sawistowski 1971; Naikaike and Tadaki 1979). These studies indicate that interfacial turbulence induces surface renewal and at higher driving forces the spontaneous interfacial disturbances increase the rate of surface renewal beyond that due to hydrodynamic flow alone; moreover the intensity of interfacial turbulence progressively decays with time.

The surface renewal characteristics and time dependent nature of the mass transfer coefficients in the presence of interfacial turbulence have both been investigated quantitatively for aqueous drops in an organic solvent by Thornton and Anderson (1981), Javed and Thornton (1984), Thornton *et al.* (1985) and Javed *et al.* (1989). The experimental procedure that they employed permitted direct visual observation of the interface as well as quantitative measurements of the rates of surface renewal associated with the interfacial disturbances.

The procedure relies upon a finely focussed ultraviolet laser pulse to induce a local colour change in a photochromic tracer dissolved in the dispersed phase, thereby labelling a small element of the interface. The subsequent behaviour of this spot is then recorded by high speed cine-photography. Care must be taken to ensure that neither the tracer nor its

TABLE 4.6 *Tracer Components (mixed in 2:2:1 ratio)*

Solution	Composition
A	18.4 g potassium ferric oxalate $K_3Fe(C_2O_4)_3.3H_2O$ 25 ml 1N H_2SO_4 Distilled water to make up volume to 250 ml
B	13.4 g potassium ferricyanide $K_3Fe(CN)_6$ Distilled water to make up volume to 250 ml
C	12.3 g sodium acetate, CH_3COONa 60 ml 1N H_2SO_4 Distilled water to make up volume to 250 ml

coloured complex is surface-active and the composition of the three solutions comprising the tracer are given in Table 4.6.

The freshly prepared pale yellow–green tracer solution on irradiation by light of the appropriate wavelength results in the simultaneous reduction of ferric ions and oxidation of oxalic acid. The ferrous ions so produced are detected by the formation of an intense blue suspension of ferrous ferricyanide and the sodium acetate solution is provided as a buffer. Subdued incandescent light does not materially affect the solution colour since the reaction is primarily sensitive to blue and ultraviolet wavelengths. The colour change on irradiation corresponds to strong absorption beyond 600 nm and can be accentuated by the use of a suitable optical filter. The tracer solution had a specific gravity of 1.059 and a viscosity approximately 4.6 per cent greater than water over the range 20–25°C.

A schematic diagram of their experimental arrangement is shown in Fig. 4.3. The cylindrical borosilicate glass test cell (A) of 8 cm diameter and 15 cm height, was fitted with a planar window to provide an undistorted view into the cell as well as a point of entry for the ultraviolet laser (wavelength 347 nm). The test cell was viewed by a 16 nm Bolex cine camera (B) through the planar mirror (C), mounted at 45° to the laser beam and equipped with a small hole to allow transmission of the light pulse. The pulsed ruby laser (D) provided output at two wavelengths (200 mJ at 694 nm and 50 mJ at 347 nm in 30 ns) and included a colinear 2 mW continuous He–Ne laser that irradiated the photocell (E) via two quartz focusing lenses (F). A droplet forming at the glass nozzle (G) obscured the beam to the photocell and activated the control box (H). This device has a number of functions. It immediately started camera (B) and switched the filter (I) into the beam path. This filter was chosen to exclude the He–Ne wavelength (633 nm) and, more importantly, to exclude the ruby laser output at 694 nm while allowing transmission of the ultraviolet energy at 347 nm. The control box also

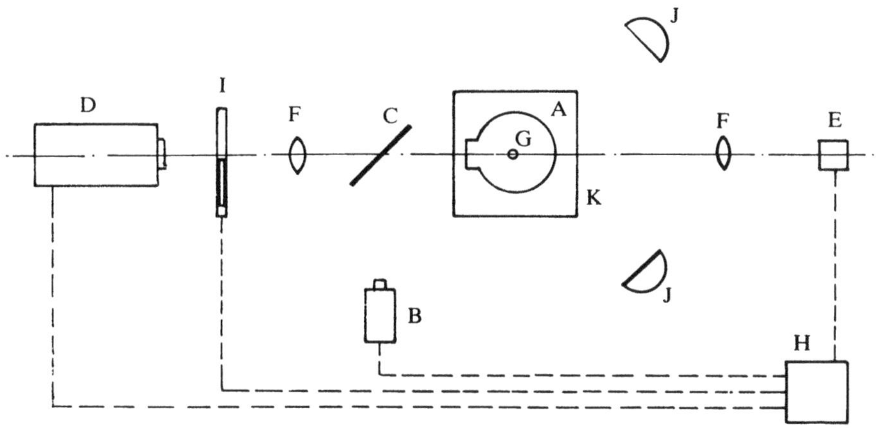

A = test cell
B = cine camera
C = mirror
D = pulsed ruby laser
E = photocell
F = quartz lenses
G = glass nozzle
H = control box
I = optical filter
J = photoflood lamps
K = thermostatted tank

Fig. 4.3. Experimental arrangement for determining rates of renewal of drop surface (Thornton *et al.* 1985).

triggered the ruby laser after a variable delay, suitably chosen to allow the camera to reach full running speed and the filter to be switched into the light path. Finally, the control box stopped the camera after a suitable running period. Illumination was provided by photoflood lamps (J) mounted to allow viewing of the pendant droplets transmitted light, and the camera lens was fitted with an optical filter to accentuate the green to blue color transition in the droplet surface. The test cell was enclosed in a constant temperature bath in order to eliminate all thermal convection currents.

This arrangement allowed a small element of surface ($\approx 1\ mm^2$) of a forming droplet to be labelled in a wide range of positions by varying the location of nozzle G relative to the fixed axis of the He–Ne laser beam and by varying the control box delay before the pulsed laser was activated.

Their first study (Thornton *et al.* 1985) was concerned with the surface dynamic behaviour of pendant aqueous droplets formed at a submerged nozzle in toluene with and without mass transfer.

The most significant result of their experiments conducted in the absence of mass transfer is that the droplet surface shows a complete absence of surface renewal. A typical sequence of photographs for an equilibrated

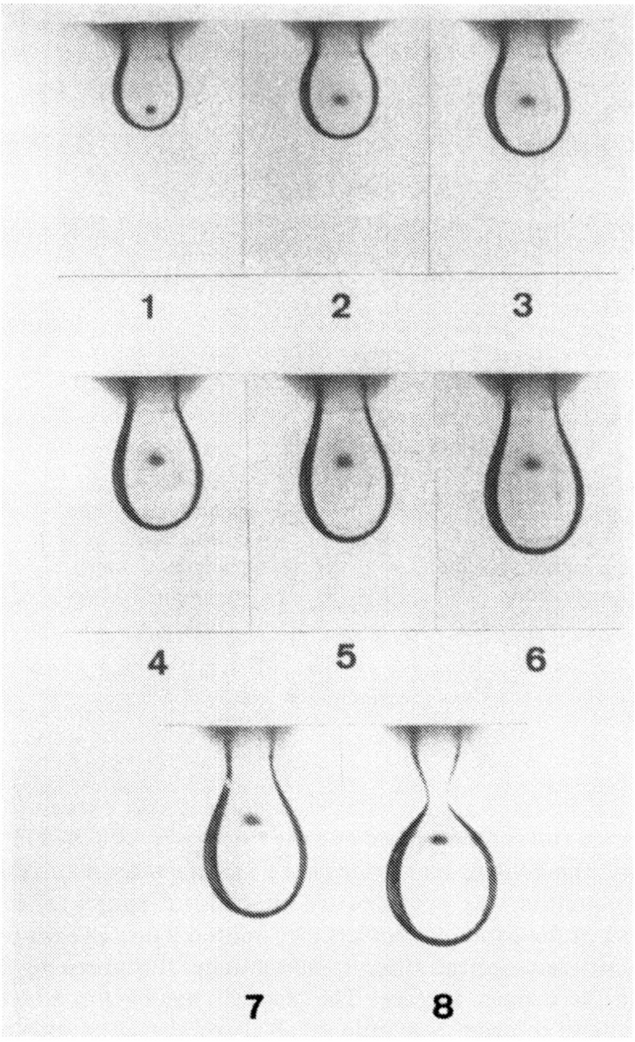

Fig. 4.4. Growing pendant droplets in toluene without mass transfer. (Nozzle Reynolds number 169; nozzle diameter 0.331 cm; nozzle velocity 4.85 cm s^{-1}; frame time interval 50 ms) (Thornton *et al.* 1985).

aqueous droplet growing in toluene is shown in Fig. 4.4. There is no indication of any disturbance to the tracer spot either during formation or after detachment (See also the photographic sequence in Fig. 3.18, p. 199, Chapter 3) They verified this behaviour for nozzle Reynolds numbers as high as 300 and they formed the same conclusions from a limited number of observations for freely falling droplets, some of which oscillated markedly and presumably were turbulent. They claimed that surface renewal due to hydrodynamic turbulence as proposed by Danckwerts (1951) does not occur in equilibrated droplet systems.

By contrast, in the mass transfer runs, the laser induced tracer spot experienced a series of violent, sporadic disturbances that gave rise to local renewal of the droplet surface. Figure 3.19 (p. 200, Chapter 3) presents a typical sequence of photographs showing the life history of the tracer spot during mass transfer. This behaviour was verified for a number of solutes including diacetone alcohol, propionic acid and acetone which were transferred from the continuous toluene phase to the aqueous droplet and also for the two-component system, water/ethyl acetate. This effect was qualitatively independent of nozzle Reynolds number and was apparent even at zero nozzle flow rates until the droplet had reached equilibrium.

The quantitative measurements of surface renewal rates were made only with diacetone alcohol solute labelling the droplet surface at three positions i.e. near the apex, at the major diameter and near the nozzle. In each case, from the cine film record, they measured the time intervals between the consecutive disturbances from which they evaluated the age frequency function for the surface elements using a model based on two first order processes with a dead time effect. The technique has thus been shown to be useful for quantifying Marangoni disturbances.

In a second study, Javed *et al.* (1989) applied the same technique to measure the age and residence time distributions of the surface elements for a fixed volume hemispherical quasi-steady state aqueous drop immersed in toluene containing acetone. The solute was transferred from toluene to the aqueous droplet phase. The drop was formed at the tip of a concentric glass nozzle and was maintained at a constant size over an extended period of time by injecting the dispersed phase liquid into the drop continuously through the inner tube and withdrawing it at the same rate through the annulus of the nozzle. This quasi-steady state droplet technique was used to fulfill two purposes. First, to make simultaneous measurements of mass transfer and of the rates of surface renewal for comparison. Second, to measure instantaneous values of the mass transfer coefficients during the decay of interfacial turbulence while maintaining constant hydrodynamic conditions in the bulk phases in order to quantify the time dependent nature of the phenomenon.

In this study, once again no surface renewal was observed due to bulk flow but when mass transfer induced interfacial turbulence, the tracer spot suffered a succession of spontaneous disturbances which eventually resulted in complete dispersion of the spot. This dispersion is shown in Fig. 3.21 (p. 203, Chapter 3) which is a photographic record of the disturbances in a typical run. The surface renewal measurements showed that the frequencies of the interfacial disturbances decayed with the age of the interface and that the intensity of surface renewal was higher at higher solute concentrations and at low ages of the interface.

This behaviour was also reflected in parallel mass transfer measurements which indicated that, for a given concentration, the overall dispersed phase mass transfer coefficient decreased progressively with the age of the interface, ultimately approaching a constant value. Furthermore, the initial value of the coefficient increased with increasing solute concentration as shown in Fig. 3.20 (p. 202, Chapter 3).

The overall coefficients estimated from the surface renewal measurements showed good agreement with those determined experimentally and the correspondence between the mass transfer measurements and measurements of surface renewal rates suggests that turbulence at the surface of pendant drops does not occur as a result of bulk flow but is generated only by Marangoni effects accompanying solute transfer. This in turn results in a greater rate of transfer and indeed the two effects, the frequency of surface perturbations (and hence rate of surface renewal) and the mass transfer flux are interactive.

The quasi-steady state drop arrangement used in Javed's study was also employed in later work by Rogers *et al.* (1987) to investigate the time dependence of the mass transfer of an inorganic solute, uranyl nitrate, between aqueous nitric acid and tri-butyl phosphate in odourless kerosene (TBP/OK). Their findings were consistent with those of Javed and Thornton and once again both the level of interfacial turbulence and the rate of mass transfer were linked and decreased with the 'age' of the drop interface. These phenomena are therefore present in both organic and inorganic systems. Further work is necessary to establish the generality of these effects but it seems likely that they are not restricted to the systems studied.

A convenient way to detect the presence of interfacial effects at a qualitative level is to photograph the drop against a back-ground graticule. If such effects are absent, the drop will form an image of the graticule, albeit distorted, which will be clearly visible provided that the refractive indices are of the correct order (Batey and Thornton 1989). Any interfacial turbulence will cause multiple refractions at the drop interface and no clear image of the graticule will be observed. This technique is well illustrated in Fig. 4.5 which depicts the time dependence of interfacial effects as uranyl nitrate transfers from the aqueous drop into the 20 per cent TBP/OK phase. The level of interfacial turbulence is seen to decrease with time and the instantaneous mass transfer coefficients follow a similar trend as shown in Fig. 4.6. When transfer occurs in the reverse direction, i.e. from the organic to the aqueous phase, no interfacial effects are apparent and the mass transfer coefficient is constant. This fact is demonstrated by the constancy of the solute concentration leaving the drop at any instant as shown by Fig. 4.7 (Rogers *et al.* 1987).

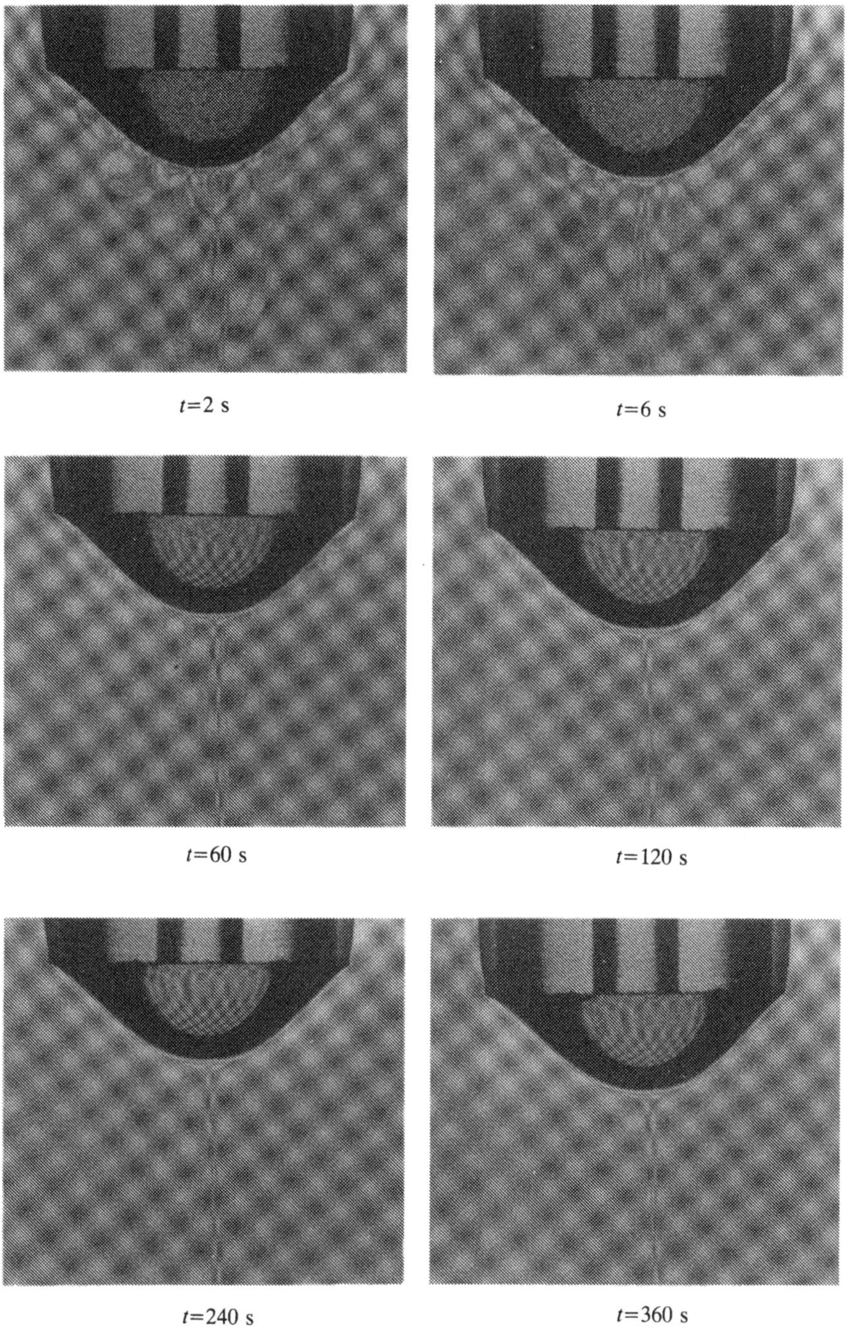

Fig. 4.5. Photographic sequence showing qualitatively the decrease with time of interfacial disturbances accompanying the extraction of 50 g/l Uranium from a quasi steady-state aqueous drop into 20 per cent TBP/OK (Courtesy: Dounreay Nuclear Power Development Establishment, UKAEA).

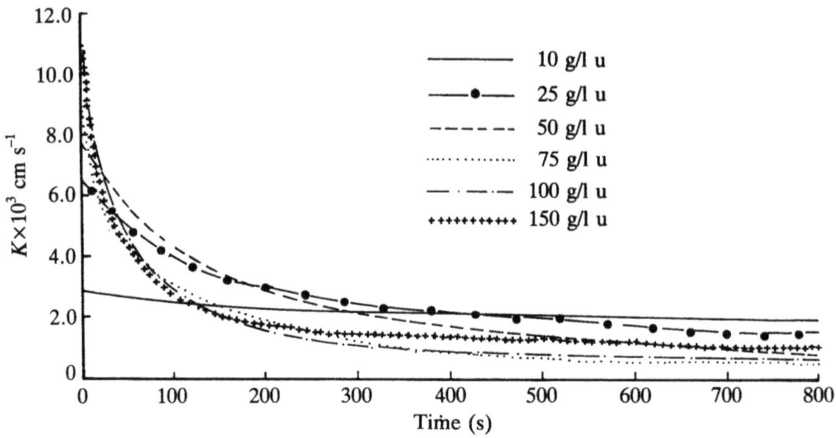

Fig. 4.6. Overall mass transfer coefficient as a function of time for the extraction of aqueous drops containing uranyl nitrate into 20% TBP/OK. Interfacial turbulence present (Rogers *et al.* 1987).

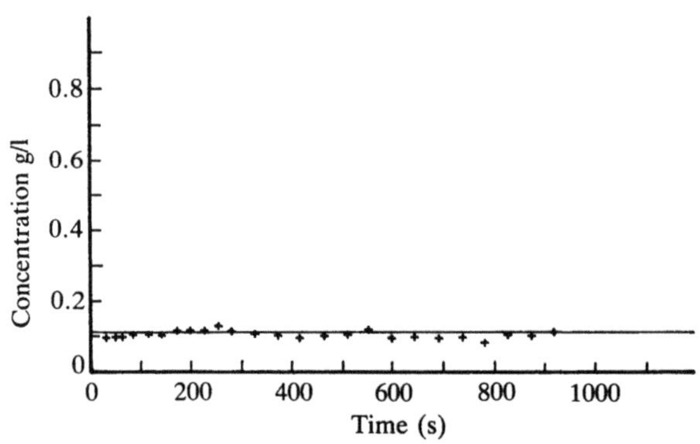

Fig. 4.7. Drop concentration vs time under 'backwash' conditions for the extraction of 18 g/l Uranium in 20 per cent TBP/OK into an aqueous drop containing 0.01 M HNO_3. Interfacial turbulence absent (Rogers *et al.* 1987).

The progressive decay of both the frequency of interfacial perturbations and the overall mass transfer coefficient has important implications for contactor design. If advantage is to be taken of the high transfer rates associated with interfacial turbulence, systems which exhibit this phenomenon should be processed in a contactor in which the drop surface is continually renewed through a rapid coalescence/redispersion cycle or other mechanism. It is important to ensure that the drop interface does not age and therefore mechanical contactors such as mixer–settlers in which the coalescence/redispersion cycle is rapid would be expected to take full advantage of the initially high mass transfer coefficients.

Nomenclature

A	dead time, s
a	constant
c	concentration, kmol/m^3
c^*	equilibrium concentration, kmol/m^3
c_0	bulk phase solute concentration, kmol/m^3
D	molecular diffusivity, m^2/s
D_E	eddy diffusivity, m^2/s
D_N	nozzle diameter, m
d	drop diameter, m
$f(H)$	approach distance distribution
$f(\theta)$	residence time distribution
g	gravitational acceleration, m/s^2
H	eddy's approach distance, m
$\bar{H}$	mean approach distance, m
k	mass transfer coefficient, m/s
k'	reciprocal of interfacial resistance, m/s
L	film thickness or eddy size, m
m	constant
N	mass flux, kmol/sm^2
N	number of vessels in series (Table 4.1)
n	shape factor of distribution function (Table 4.1)
$P(z)$	Probability density function
Re	Reynolds number
Re_N	nozzle Reynolds number defined as $D_N v_N \rho_d / \mu_d$
s	rate of surface renewal, s^{-1}
t	time or age, s
t'	process time, s
t_e	eddy's exposure time, s
U	droplet terminal velocity, m/s
$u(t)$	unit step function (defined in Table 4.1)
v_N	nozzle velocity, m/s
v_0	characteristic velocity, m/s
x	distance normal to interface, m
z	dummy variable (= θ or H)
$\bar{z}$	mean value of z

Greek symbols

α	distribution parameter
β	as defined in Table 4.1
$\beta(t')$	function as defined in Table 4.1

$\Gamma(\alpha)$ gamma function
ρ density, kg/m^3
μ viscosity, kg/ms
γ kinematic viscosity, m^2/s
ϵ rate of energy dissipation per unit liquid mass, m^2/s^3
σ surface tension, N/m
δ Dirac function (Table 4.1)
δ film thickness, m (Table 4.2)
θ residence time, s
τ mean residence, time, s
$\phi(t)$ age distribution function
$\phi(t,\tau)$ unsteady state age distribution function

Subscripts
c continuous phase
d dispersed phase

References

ABRAMZON, B. and BORDE, I. (1980). *American Institute of Chemical Engineers Journal*, **26**, 536–44.

ANGELO, J.B., LIGHTFOOT, E.N., and HOWARD, D.W. (1966). American Institute of Chemical Engineers Journal **12**, 751–9

BAKKER, C.A.P., FENTENER VAN VLISSINGEN, F.H., and BEEK, W.J. (1967). *Chemical Engineering Science*, **22**, 1349–55.

BATEY, W. and THORNTON, J.D. (1989). *Industrial Engineering Chemistry, Research*, **28**, 1096–1101

BENJAMIN, T.F.C., FAN, L.T., and HWANG, C.L. (1971). *American Institute of Chemical Engineers Journal*, **17**, 154–60.

BIN, A.K. (1983). *International Journal of Heat and Mass Transfer*, **26**, 981–91.

BRAUER, H. (1978a). *International Journal of Heat and Mass Transfer*, **21**, 445–53.

BRAUER, H. (1978b). *International Journal of Heat and Mass Transfer*, **21**, 445–65.

BULICKA, J. and PROCHAZKA, J. (1976). *Chemical Engineering Science*, **31**, 137–46.

BULLIN, J.A. and DUKLER, A.E. (1972). *Chemical Engineering Science*, **27**, 439–42.

DANCKWERTS, P.V. (1951). *Industrial and Engineering Chemistry, Process Development*, **43**, 1460–7.

DAVIES, J.T. (1972). *Turbulence phenomena*. Academic Press, New York.

DAVIES, J.T. and LOZANO, F.J. (1979). *American Institute of Chemical Engineers Journal*, **25**, 405–15.

ELZINGA, E.R., Jr. and BANCHERO, J.T. (1959). *Chemical Engineering Progress Symposium Series*, **55**, 149–61.

FORTESCUE, G.E. and PEARSON, J.R.A. (1967). *Chemical Engineering Science*, **22**, 1163–76.

GARNER, F.H. and SKELLAND, A.H.P. (1954). *Industrial and Engineering Chemistry*, **46**, 1255–64.

GROEBER, H. (1925). Z. Ver. Deutsch Ing. **69**, 705.

GROSS, B. and HIXON, A.N. (1969). *Industrial and Engineering Chemistry Fundamentals*, **8**, 288–96.

HADAMARD, J. (1911). C.R. Hebd. Seanc. Acad. Sci. Paris. **152**, 1745–1817.
HANDLOS, A.E. and BARON, T. (1957). *American Institute of Chemical Engineers Journal*, **3**, 127–35.
HARRIOTT, P. (1962). *Chemical Engineering Science*, **17**, 149–54.
HEERTJES, P.M., HOLVE, W.A., and Talsma, H. (1954). *Chemical Engineering Science*, **3**, 122–42.
HENSTOCK, W.H. and HANRATTY, T.J. (1979). *American Institute of Chemical Engineers Journal*, **25**, 122–31.
HIGBIE, R. (1935). *Transactions of the American Institute of Chemical Engineers*, **31**, 365–89.
IMAISHI, N. and FUJINAWA, K. (1974). *Journal of Chemical Engineering of Japan*, **7**, 81–7.
JAVED, K.H. and THORNTON, J.D. (1984). *Institution of Chemical Engineers Symposium Series*, **88**, 203–16.
JAVED, K.H., THORNTON, J.D., and ANDERSON, T.J. (1989), *American Institute of Chemical Engineers Journal*, **35**(7), 1125.
JOHNS, L.E. and BECKMANN, R.B. (1966). *American Institute of Chemical Engineers Journal*, **12**, 10–16.
JOHNSON, A.I. and HAMIELEC, A.E. (1960). *American Institute of Chemical Engineers Journal*, **6**, 145–149.
KING, C.J. (1966). *Industrial and Engineering Chemistry Fundamentals*, **5**, 1–8.
KISHINEVSKII, M.Kh. (1955). *Journal of Applied Chemistry USSR*, **28**, 881–6.
KISHINEVSKII, M.Kh. and PAMFILOV, A.V. (1949). *Journal of Applied Chemistry USSR*, **22**, 1173.
KISHINEVSKII, M.Kh. and SEREBRYANSKY, V.T. (1956). *Journal of Applied Chemistry USSR*, **29**, 29–33.
KOMORI, S., UEDA, H., OGINO, F., and MIZUSHINA. (1982). *International Journal of Heat and Mass Transfer*, **25**, 513–21.
KOPPEL, L.B., PATEL, R.D., and HOLMES, J.T. (1966). *American Institute of Chemical Engineers Journal*, **12**, 941–6.
KOZINSKI, A.A. and KING, C.J. (1966). *American Institute of Chemical Engineers Journal*, **12**, 109–16.
KRONIG, R. and BRINK, J.C. (1950). *Applied Science Research*, **A2**, 142–55.
LADDHA, G.S. and DEGALEESAN, T.E. (1976). *Transport phenomena in liquid extraction*, pp. 197–8. Tata McGraw-Hill, New Delhi.
LAMONT, J.C. and SCOTT, D.S. (1970). *American Institute of Chemical Engineers Journal*, **16**, 513–19.
LAMOURELLE, A.P. and SANDALL, O.C. (1972). *Chemical Engineering Science*, **27**, 1035–43.
LEVICH, V.G. (1962). *Physiochemical hydrodynamics*. Prentice-Hall, Englewood cliffs, N.J.
LEWIS, J.B. (1954). *Chemical Engineering Science*, **3**, 248–59.
LEWIS, W.K. and WHITMAN, W.G. (1924). *Industrial and Engineering Chemistry*, **16**, 1215–20.
LUKE, S. and LEE, Y.H. (1986). *American Institute of Chemical Engineering Journal*, **32**, 1546–54.
MANGERS, R.J. and PONTER, A.B. (1980). *Industrial and Engineering Chemistry, Process Design and Development*, **19**, 530–37.

MAROUDAS, N.G. and SAWISTOWSKI, H. (1964). *Chemical Engineering Science*, **19**, 919–31.
MCMANAMEY, W.J. (1961). *Chemical Engineering Science*, **15**, 251–4.
NAIKAIKE, Y. and TADAKI, T. (1979). *International Chemical Engineering*, **19**, 328–35.
NEWMAN, A.B. (1931). *Transactions of the American Institute of Chemical Engineers*, **27**, 203–11.
OLIVER, D.L.R. and CHUN, J.N. (1985). *International Journal of Heat and Mass Transfer*, **29**, 879–887.
ORTIZ, E.S.P. and SAWISTOWSKI, H. (1973a). *Chemical Engineering Science*, **28**, 2051–61.
ORTIZ, E.S.P. and SAWISTOWSKI, H. (1973b). *Chemical Engineering Science*, **28**, 2063–9.
ORTIZ, E.S.P. and SAWISTOWSKI, H. (1975). *Chemical Engineering Science*, **30**, 1527–8.
PATEL, J.M. and WELLEK, R.M. (1967). *American Institute of Chemical Engineers Journal*, **13**, 384–6.
PERLMUTTER, D.D. (1961). *Chemical Engineering Science*, **16**, 287–96.
POPOVICH, A.T., JERVIS, R.E., and TRASS, O. (1964). *Chemical Engineering Science*, **19**, 357–65.
PRASHER, B.D. and FRICKE, A.L. (1974). *Industrial and Engineering Chemistry, Process Design and Development*, **13**, 336–40.
RAMBABU, D. and NARSIMHAN, G. (1980). *The Chemical Engineering Journal*, **20**, 169–75.
ROGERS, D., THOMPSON, P.J., and THORNTON, J.D. (1987). *Institution of Chemical Engineers Symposium Series*, **103**, 15–27.
RYBEZINSKI, W. (1911). Bull. Int. Acad. Sci. Lett. Cracovie. A40.
SAWISTOWSKI, H. (1971). *Interfacial phenomena. Recent advances in liquid–liquid extraction*, Ch. 9. Pergamon Press, New York.
SHERWOOD, T.K., PIGFORD, R.L., and WILKE, C.R. (1975). *Mass transfer*. McGraw-Hill, New York.
SIDEMAN, S. (1966). Direct contact heat transfer between immiscible liquids. *Advances in Chemical Engineering*, Vol. 6, Ch. 4. Academic Press, New York.
SKELLAND, A.H.P. (1974). *Diffusional mass transfer*. John Wiley.
SKELLAND, A.H.P. and WELLEK, R.M. (1964). *American Institute of Chemical Engineers Journal*, **10**, 491–6.
SONIN, A.A., SHIMKO, M.A., and CHUN, J.H. (1986). *International Journal of Heat and Mass Transfer*, **29**, 1319–32.
STERNLING, C.V. and SCRIVEN, L.E. (1959). *American Institute of Chemical Engineers Journal*, **5**, 514–23.
TAKEUCHI, H and NUMTA, Y. (1977). *International Chemical Engineering*, **17**, 468–74.
THORNTON. J.D. (1964). *Industrial Chemist*, **40**(1), 13.
THORNTON, J.D. and ANDERSON, T.J. (1981). *International Journal of Heat and Mass Transfer*, **24**(11), 1847.
THORNTON, J.D., EGBUNA, D.O., and RAHMAN, M. (1976). Solvent extraction Symposium, 7th–9th September, University of Newcastle upon Tyne. Society of Chemical Industry and Institution of Chemical Engineers. (London).

THORNTON, J.D., ANDERSON, T.J., JAVED, K.H., and ACHWAL, S.K. (1985). *American Institute of Chemical Engineers Journal*, **31**, 1069–76.
TOOR, H.L. and MARCHELLO, J.M. (1958). *American Institute of Chemical Engineers Journal*, **4**, 97–101.
TREYBAL, R.E. (1963) *Liquid extraction* (2nd edn), p. 186, McGraw Hill, New York.
TUNG, L.S. and LUECKE, R.H. (1986). *Industrial and Engineering Chemistry, Process Design and Development*, **25**, 664–73.
VAN MAANEN, H.R.E. and FORTUIN, J.M.H. (1983). *Chemical Engineering Science*, **38**, 390–423.
WALIA, D.S. and VIR, D. (1976). *Chemical Engineering Science*, **31**, 525–33.
WELLEK, R.M. and SKELLAND, A.H.P. (1965). *American Institute of Chemical Engineers Journal*, **11**, 557–60.

5

MIXING AND COALESCENCE PHENOMENA IN LIQUID–LIQUID SYSTEMS

G.A. DAVIES

5.1 Introduction

The process of liquid–liquid extraction involves the contact of two partially miscible liquid solutions. These are not at chemical equilibrium so that one more component, solute(s) will transfer from one phase, the raffinate to the second phase, the extract. The maximum extraction of the solute will be achieved at a particular temperature if the two phases are allowed to come to chemical equilibrium. Then:

$$(\mu_i)_R = (\mu_i)_{E\,T,\pi} \tag{5.1}$$

where μ_i is the chemical potential of component i, usually specifically the solute and the subscripts R and E relate to the two liquid phases raffinate and extract respectively. If the two liquid phases were quiescent and left as two layers one above the other in the order of the respective phase densities then transfer would take place only by molecular diffusion. The transfer flux would be low and the time taken to achieve chemical equilibrium long. In order to accelerate this process the two solutions are mixed together to form a dispersion of droplets of one phase in the other. The molar transfer rate of component i from phase R is proportional to the interfacial area between the two phases. This area, a, is computed on the basis of a unit volume of mixture so that $N_{iR} \propto a$. In order to reduce the time to achieve chemical equilibrium one way would be to increase a. This can be achieved by increasing the level of turbulence in the mixing stage. The interfacial area, a, is related to the volume fraction of dispersed phase Φ and the Sauter mean drop size $\bar{d}_{VS}$ thus:

$$a = \frac{6\Phi}{\bar{d}_{VS}} \tag{5.2}$$

increasing a must result in a decrease in $\bar{d}_{VS}$. For the extraction process to be completed, the two phases must be separated. If the extraction were carried out as a batch process this would then allow the lower phase to be decanted off or if a continuous flow process was used two separate streams would be removed.

Any equilibrium stage process, mass transfer, gas absorption, stripping, distillation, liquid extraction involves two discrete steps,

(1) mixing to increase the interfacial area enhance mass transfer and minimise the contact time required to achieve (or approach) chemical equilibrium; and

(2) phase separation so that the two phases can be removed.

In these mass transfer processes the dispersions formed during the contact mass transfer steps are unstable so that when the contacting process is stopped the phases separate spontaneously, usually as a result of the gravitational force acting on the droplets or bubbles which causes them to coalesce and thereby bring about a reduction in contact area. In gas–liquid systems where the differences in density between the phases are usually high, typically from 500–1000 kg/m^3 and where the surface tension which has a major influence on coalescence is high ($20–70 \times 10^{-3}$ mN/m), the rate of phase separation is rapid. In gas–liquid systems the mass transfer process may be the rate limiting step. In liquid–liquid systems, as in liquid extraction, the rate of phase separation is much lower. This is due to the marked differences in the relevant physical properties. The differences in density are typically between 10–300 kg/m^3, the interfacial tension is lower at between $1–30 \times 10^{-3}$ mN/m and the viscosities of the phases may be higher. All three factors contribute to reducing the rates of phase separation. In liquid extraction it is not unusual for this to be the rate limiting step. Nowhere is this better exemplified than in the liquid extraction processes applied to the extraction of metal ions from dilute leach-ore liquors. These processes have been developed and applied in the last thirty-five years on a scale greater than any other application of the technology. Mixer–settlers are used, an example is shown in Fig. 5.1. In this photograph the relative sizes of the mixer boxes in which the extractive transfer is carried out to the settlers can be seen. The volume of the settlers is approximately ten times that of the mixer. Such as example serves to demonstrate an important principle for design. The overall process comprises two basic operations, phase contact, mass transfer and phase separation. In the first, increasing interfacial area increases the transfer flux and thereby will promote a decrease in equipment volume. In the second, increasing interfacial area will decrease the separation flux and will thereby cause an increase in equipment volume. The important first principle must be that the overall design must recognize that phase contacting must be

Fig. 5.1. Photograph of the mixer settlers of a copper solvent extraction plant. Courtesy of Davy McKee Ltd.

considered simultaneously with phase separation and that only by doing this can an efficient, economic optimum be achieved.

In this chapter aspects of both process steps important to design are considered. In Section 5.2 the contacting of liquid–liquid solutions is considered with special reference to the prediction of the drop size and interfacial area. When this process is considered an apparently simple question has to be addressed. That is, 'when two immiscible liquid phases are mixed together in an agitated vessel, which phase forms the drops or dispersed phase?' Clearly this is important, as from (5.1) the interfacial area depends on the volume fraction of the dispersed phase. Additionally it is easily shown that the rates of phase separation in a given system are not equal if the phase continuity, an expression to define the dispersion, is reversed. Thus the answer to the question will influence the design of both the phase contactor and separator. In most of the early work on liquid extraction it was assumed that droplets in a dispersion could be represented as spheres and then, if a monodispersion was assumed, phase inversion would occur once the volume representing the optimum packing efficiency of spheres has been exceeded. This equates to the packing efficiency of the face centred cubic structure and is 74.06 per cent v/v. In early work it was assumed that this upper limit represented the phase inversion point and could be used to predict the phase continuity if the volumes or volume flows of each phase were determined. This is not true and the problem will be considered in Section 5.3 (p. 252). In Section 5.4 (p. 255), the coalescence of drops, separation of dispersion leading to aspects of the design of separators, and settlers will be considered.

5.2 Liquid–Liquid mixing and correlations for drop size

In the liquid extraction process a range of equipment is used for contact, mixing of the phases. In mixer–settlers mechanical energy is transmitted to the mixture by an impeller or stirrer thereby inducing turbulence and causing break-up of one phase into droplets. A wide range of proprietary impeller designs are available, some including a pumping action and producing a pressure head which is used to overcome flow resistance in the settler. These are called 'pump-mix' contactors and are widely used in large-scale hydrometallurgical liquid extraction processes. Recently there has been interest in so-called pressure mixers. In these potential (pressure) energy is used to produce mixing of the two liquids in a structured packing enclosed in a tube. The main advantage claimed for these devices is the ability to control the drop size produced and to produce (approximately) monodispersions.

The most common equipment used is still the mechanical mixer and in the design of the *total* plant an important factor is the drop size distribution produced. In a mixer, drops are formed by a 'break-up' of dispersed phase in the shear field, in other parts of the flow drops are coalescing. If a characteristic size or size distribution is to be computed for given conditions these are only relevant if a dynamic equilibrium between break-up and coalescence is achieved in the contactor. An extensive literature exists on work related to drop break-up in turbulent fields and to coalescence. The theoretical work underpinning the description of drop break-up is Kolmogoroff's theory of isotropic turbulence (Kolmogoroff 1949). The essential point relating to drop break-up is that turbulent flow produces large-scale primary eddies. These are unstable and decay dissipating energy by disintegrating into smaller and smaller eddies until at the smallest scale energy is dissipated by viscous flow. Eddies in the primary flow are usually anisotropic, the components of velocity fluctuations resolved for any right set of coordinates will not be equal. The assumptions made are that the smaller eddies produced by the dissipative process are statistically independent, in size, from the primary eddies and that they are isotropic. The application of this theory to drop break-up was considered by Batchelor (1951). In this work he postulated that two different mechanisms were responsible for break-up, dependent on the dimension of the drops. If the drops are much larger than the micro-scale of turbulence, smallest eddies decaying by viscous dissipation of energy, then dynamic pressure forces rather than viscous shear forces control the breakage process. When the drop dimensions, diameter, are of the order of or less than the eddy length then viscous forces promote break-up. The velocity fluctuations over the surface of a drop causes deformations in relation to the scale of the small eddies. These shear deformations are counteracted by interfacial tension forces and viscous forces in the drop. Deformation and break-up is controlled by these three forces, one induced by the dissipative turbulent process and two restoring forces. Break-up will occur when the dissipative force exceeds the restoring forces. The restoring forces can be represented by dimensionless groups, the Weber number, $We = \tau d/\sigma$, and the viscosity group $\mu_d/(\rho_d \sigma d)^{1/2}$. The Weber number represents the ratio of the inertia to surface forces and is used to describe the conditions for break-up, the so-called critical Weber number (Hinze 1955). In a liquid–liquid contactor the critical Weber number defines the diameter of the largest stable drop in the dispersion from which

$$(We_d)_{crit} = \frac{\rho_c \overline{u^2(d)}\, d_{\max}}{\sigma} \tag{5.3}$$

The shear stress τ is related to the mean square velocity fluctuations over a length d, $\overline{u^2(d)}$ If d is large with respect to the smallest eddies in the flow, Kolmogoroff's first case, then from Batchelor (1951)

$$\overline{u^2(d)} = C_1 \epsilon^{2/3} d^{2/3} \tag{5.4}$$

where ϵ is the energy dissipation per unit mass and C_1 is a constant. The critical Weber number then becomes

$$(We_d)_{\text{crit}} = K_1 \left(\frac{\rho_c}{\sigma}\right)^{3/5} \epsilon^{2/5} d_{\max} \tag{5.5}$$

For a fully baffled turbine mixer operating at high Reynolds number

$$\epsilon \propto N^3 D^2 \tag{5.6}$$

where N and D are the speed and diameter of the impeller respectively. Then applying the theory to a practical case, noting that $(We_d)_{\text{crit}}$ is constant, and assuming that the energy dissipation near the impeller is described by Kolmogoroff's theory

$$d_{\max} = K_2 \left(\frac{\sigma}{\rho_c}\right)^{3/5} N^{-6/5} D^{-4/5} \tag{5.7}$$

In experimental studies most research workers have computed from the results the Sauter mean drop diameter $\bar{d}_{VS}$. This, via eqn (5.2), is a more useful statistic since it gives information on the interfacial area necessary for mass transfer calculations. Shinnar and Church (1960), Sprow (1967),

Chen and Middleman (1967), and Brown and Pitt (1970) have all shown that the maximum stable drop diameter d_{max} is linearly related to the Sauter mean drop diameter so that d_{VS} can be predicted from eqn (5.7):

$$\bar{d}_{VS} = K_3 \left(\frac{\sigma}{\rho_c}\right)^{3/5} N^{-6/5} D^{-4/5} \tag{5.8}$$

When the power number, N_p, and impeller Weber number We_I are introduced into eqn (5.8),

$$N_p = \frac{Pg}{\rho N^3 D^5} \tag{5.9i}$$

$$We_I = \frac{N^2 D^3 \rho_c}{\sigma} \tag{5.9ii}$$

P is the power input, then:

$$\frac{\bar{d}_{VS}}{D} \propto We_I^{-0.6} N_p^{-0.4} \left(\frac{D^3}{V}\right)^{-0.4} \tag{5.10}$$

For a particular contactor design the power number varies with the impeller Reynolds number $Re_I = (D^2 N \rho_c / \mu_c)$. At higher Reynolds numbers N_p asymptotes to a constant value. It is in this region that most contactors operate. Thus:

$$\frac{\bar{d}_{VS}}{D} = K_4 We_I^{-0.6} \tag{5.11}$$

Equation (5.11) has been the basis for correlation of a wide range of experimental results for mixing in turbine contactors at low dispersed phase hold up (Vermeulen *et al.* 1955; Chen and Middleman 1967; Sprow 1967). The equations together with others discussed below are given in Table 5.1. The basic derivation of eqn (5.11) depending on the assumption of local isotropic turbulence and spatial uniformity in the energy dissipation per unit mass, ϵ, within the vessel has been modified by some workers (Cutter 1966; Lee and Tasakorn 1979). Cutter demonstrated that ϵ or the turbulence intensity was much higher in the discharge stream of the impeller than in other parts of the vessel. Interpreting this Lee and his co-workers suggested a relationship between the maximum energy dissipated, which controls the maximum drop size, and the average energy dissipated $\bar{\epsilon}$ which is related to the power input and power number

$$\epsilon = \bar{\epsilon}\Theta \tag{5.12}$$

where Θ is termed the turbulence intensity factor. Θ enters into the constant term in eqn (5.12).

TABLE 5.1 *Correlation to predict the Sauter mean drop diameter* $\bar{d}_{vs}$

Investigators	Correlation	Range of physical properties					
		ρ_d (kg/m^3)	ρ_c (kg/m^3)	μ_d (Ns/m^2)	μ_c (Ns/m^2)	$\sigma \times 10^3$ (N/m)	X_V
Brown and Pit	$\bar{d}_{vs} = K\left(\frac{\sigma}{\rho\epsilon - t}\right)^{0.6}$	780–840	970–998	0.59–3.30	1.0–1.28	1.9–50	0.05
Shinnar	$\bar{d}_{vs} = K(\sigma D)^{-3/8} We^{-3/8}$			22.5	0.4		0.05
Vermeulen *et al.*	$\bar{d}_{vs} = K\, We^{-0.6}$	683–1595	693–1595	0.52–18.4	1.81–65.4	3.1–58.1	0.1–0.4
Rodger	$\bar{d}_{vs} = K\, We^{-0.36}\left(\frac{D_1}{DT}\right)^{-7}$	761–1101	1000	0.578–3.91	1.0	2.1–49.0	0.50
Sprow	$\frac{\bar{d}_{vs}}{D_1} = K(\rho_c\mu_c)^{-0.5}\left(\frac{\rho_c}{\sigma D_1}\right)^{-0.25} We^{-0.78}$	692	1005	0.51	1.14–3.37	1.5–1.0	0.50
Chen and Middleman	$\frac{\bar{d}_{vs}}{D_1} = 0.045\, We^{-0.57}$	703–1101	997–1001	0.52–25.8	0.89–1.27	4.75–48.3	4.75–48.3
Mlyneck and Resnick	$\frac{\bar{d}_{vs}}{D_1} = 0.058(1 + 5.4X_v)\, We^{-0.6}$	1055	1000	1.0	1.0	41	0.0245–0.34
Brown and Pitt	$\frac{\bar{d}_{vs}}{D_1} = 0.0571(1 + 3.14X_v)\, We^{-0.6}$	783–838	972–998	0.59–3.30	1.00–1.28	1.9–5.0	0.05–0.30
Van Heuven and Beck	$\frac{\bar{d}_{vs}}{D_1} = 0.047(1 + 2.5X_v)\, We$	-	998	-	8.5–49.5		0.04–0.35
Godfrey and Grilic	$\frac{\bar{d}_{vs}}{D_1} = 0.058(1 + 3.6X_v)\, We^{-0.6}$	783–829	1000	2.05–8.60	1.0	1.9–34.5	0.05–0.5

An important parameter in correlations for drop size is the dispersed phase hold-up Φ. As Φ increases eqn (5.11) is found to deviate from experiments. It is found that as Φ increases $\bar{d}_{vs}$ also increases. There are two underlying mechanisms responsible for this. As Φ increases, the number density of drops increases and the probability of collision leading to drop–drop coalescence increases. In addition as Φ increases, the increase in drop number density tends to damp the turbulent eddies which again promotes coalescence. To account for these factors eqn (5.11) is modified. From experiments it has been shown that the Sauter mean drop diameter is a linear function of Φ and so that the equation reduces to eqn (5.11) at low hold up, the factor used is $(1 + K_6\Phi)$. Thus,

$$\frac{\bar{d}_{VS}}{D} = K_5(1 + K_6\Phi)\, We_I^{-0.6} \tag{5.13}$$

This equation has been used to correlate data by numerous workers (Calderbank 1958; Mlyneck and Resnick 1972; Van Heuven and Beek 1971; Brown and Pitt 1972; Coulalogou and Tavlarides 1976; Godfrey and Grilic 1977; Lee and Tasakorn 1979). The equations giving the constants and ranges over which measurements were made are given in Table 5.1.

A similar dependence of $\bar{d}_{vs}$ on hold up was reported by Bouyatiotis and Thornton (1966). They correlated their data by an equation

$$\bar{d}_{vs} = d^0_{VS} + m\Phi \tag{5.14}$$

where d^0_{VS} was defined as the Sauter mean drop diameter at zero hold up ($\Phi \to 0$). This they correlated in terms of a fluid properties group:

$$\frac{d^{0\,3}_{vs}\rho_c^2 g}{\mu_c^2} = 29\left(\frac{P^3 g_c^3}{V^3\rho_c^2\mu_c g^4}\right)^{-0.32}\left(\frac{\rho_c\sigma^3}{\mu_c^4 g}\right)^{-0.14} \tag{5.15}$$

and m, the constant premultiplying Φ, was given as

$$m = 1.18\left(\frac{\sigma^2 g_c^2}{\mu_c^2 g}\right)\left(\frac{\Delta\rho\sigma^3 g_c^3}{\mu_c^4 g}\right)^{-0.62}\left(\frac{\Delta\rho}{\rho_c}\right)^{0.05} \tag{5.16}$$

In technical applications it is important to be able to predict the interfacial area of a dispersion and, because droplet velocities within extraction columns and settling and coalescence all depend on drop size, it is important to be able to predict the distribution of drop sizes. None of the correlations in Table 5.1 address this problem. The difficulties lie in being able to obtain a body of statistically significant data on which to work. Some workers have considered the problem; Bouytiotis and Thornton (1966) measured drop sizes at the bottom of a mixing vessel and found that a normal distribution fitted the data.

$$f_n(d_{vs}) = \frac{1}{s(\pi)^{1/2}} \exp - \frac{(d - \bar{d})^2}{2s^2} \tag{5.17}$$

where $f_n(d_{vs})$ is the probability density and s is the standard deviation and $\bar{d}$ the arithmetic mean drop size.

Brown and Pitt (1972) obtained similar results which when normalized with respect to $\bar{d}_{vs}$ were presented as

$$f\left(\frac{d}{d_{VS}}\right) = \frac{1}{0.2(2\pi)^{1/2}} \exp\left[-12.5\left(\frac{d}{\bar{d}_{vs}} - 1.07\right)^2\right] \tag{5.18}$$

The standard deviation of 0.2 reported implies that 68 per cent of the volume of the dispersed phase consisted of droplets with diameters within ± 20 per cent of $\bar{d}_{vs}$. More recently Godfrey *et al.* (1989) have reported on data measured in a pump-mix mixer. Again a normal distribution was used to fit the data:

$$f\left(\frac{d}{\bar{d}_{vs}}\right) = \frac{1}{0.3(2\pi)^{1/2}} \exp\left[-5.52\left(\frac{d}{\bar{d}_{vs}} - 0.8\right)^2\right] \tag{5.19}$$

The corresponding correlation they proposed for computing $\bar{d}_{vs}$ for a centrally located pump-mix mixer was

$$\frac{\bar{d}_{vs}}{D} = 0.187(1 + 1.18\Phi^{0.6})\, We^{-0.8}\left[\frac{\mu_c}{\mu_d}\right]^{0.37} \tag{5.20}$$

The difficulty in obtaining accurate data on drop size distributions is shown by the sparsity in the literature. Many workers have recommended equations to relate the mean drop size to the maximum drop size and the mode of the distribution but there is relatively little reliable data on drop size distributions. This data will undoubtedly become available with the improvements in measurement techniques and instrumentation.

5.3 The ambivalence range and phase inversion in liquid–liquid systems

The question of defining the dispersed phase in a given liquid–liquid mixture has already been posed. The answer is not trivial and indeed it may not be possible to define a priori. If the volume fraction of one phase, say the organic phase, is defined by Φ then at low values of $\Phi < 0.1$ the organic phase is likely to be dispersed. The contrary is true at high values of Φ, $\Phi \rightarrow 1.0$. In these extreme ranges the probability of correctly defining the phase continuity is high. Most liquid extraction systems, however, do not operate in these extreme ranges but rather where $0.3 < \Phi < 0.8$. In this region it is not at all certain which phase is dispersed, the conditions will

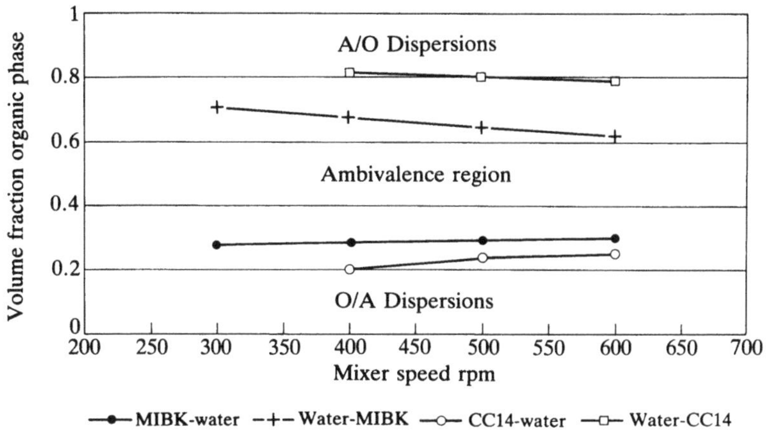

Fig. 5.2. Phase map of a liquid–liquid mixture showing the ambivalence region.

depend on the initial state. It is possible to define the limits of stable behaviour with respect to phase continuity. An example is shown in Fig. 5.2. In this graph Φ the volume fraction of organic phase is plotted as a function of the impeller speed for conditions in a fully baffled turbine mixer. The limits of stability are defined by two curves on this graph which divide the phase space into three regions. Region 1 is the space between the lower curve and the abscissa. In this region only organic in aqueous phases are stable, *O/A*. It corresponds to low values of Φ. Region 2 is the space between the upper curve and the graph limit $\Phi = 1$. In this region only aqueous in organic dispersions, *A/O*, are stable. In between these two curves, region 3, it is possible to produce conditions where either *O/A* or *A/O* dispersions are formed. This is called the ambivalence region. The simple theories based on the Oswald fraction giving the critical stability fraction at phase conversion of $\Phi = 0.7406$ is not correct. The ambivalence limits vary between systems. The physical phenomena defining these limits are most complex and involve dynamic equilibrium between turbulent shear forces which break up droplets and viscous dissipative forces which influence drop coalescence. Many workers have considered the problem of measuring the limits of phase inversion, correlating the data and finally trying to produce a robust physical model to interpret the data. Some progress has been made on measurements and correlations. Selker and Sleicher (1965), Fakhr-Din (1970), Luhring and Sawistowski (1971), Maclarcy and Mansoori (1975), but virtually no progress has been made on producing a satisfactory physical model. The problem is highly non-linear, within the ambivalence range the phase trajectory appears to depend critically on the initial conditions. This might well be a problem which may be approached using chaos theory.

The experimental results taken in agitated turbine contactors all show that the boundary curves depend on the impeller speed and tend to converge as the speed increases asymptoting to constant values of Φ. This corresponds to decreasing drop size and increasing number density of droplets which favours coalescence. The curves do not extend or extrapolate to high speeds since a point will be reached at which a significant proportion of the dispersion will be made up of a fine haze, secondary dispersion. This would represent a range where operation of *practical* extraction systems would not be possible.

The initial conditions, start-up, have been shown by numerous workers to influence the phase continuity in the ambivalence region, Rogers *et al.* (1956), Selker and Sleicher (1965). These workers were amongst the first to show that the desired dispersion could be obtained if the impeller, prior to start-up, was located in the phase to be continuous.

Quinn and Sigloh (1963) were amongst the first to try to correlate the inversion curves. They proposed that the inversion concentration in a given system is inversely proportional to the power input:

$$\Phi = \Phi_0 + \frac{a}{\bar{P}} \tag{5.21}$$

where Φ_0 is the asymptotic value at high impeller speeds. Luhring and Sawistowski (1971) carried out a more extensive and thorough study and proposed that the asymptotic values (at high N), for the upper and lower inversion curves could be related to the impeller Weber number. Thus:

(i) Inversion to A/O dispersions, at Φ_0',

$$\Phi_0' = 0.160 + 6.0 \times 10^{-5}(We_I) \tag{5.22}$$

(ii) Inversion to O/A dispersions, at Φ_0'',

$$\Phi_0'' = 0.470 + 2.0 \times 10^{-5}(We_I) \tag{5.23}$$

Selker and Sleicher (1965) proposed that the limits of ambivalence were primarily dependent on the kinematic viscosity ratio of the two phases. A conclusion supported by studies on isopyknic systems by Maclarcy and Mansoori (1975). Yeh (1964) also concluded that the phase inversion, promoted by rapid coalescence of droplets within the mixer, was controlled by viscous rather than inertial forces and proposed that:

$$\Phi' \propto \left(\frac{\nu_0}{\nu_{aq}}\right)^{1/2} \tag{5.24}$$

where ν_0 and ν_{aq} are the kinematic viscosities of the organic and aqueous phases respectively. The experimental difficulties in determining phase inversion are quite severe and perhaps account for differing interpretations of results. For example Selker and Sleicher concluded that the interfacial tension would have little influence on phase inversion whereas Luhrung and Sawistowski considered this to be of paramount importance and correlated the 'width' of the ambivalence range, $\Phi' - \Phi''$:

$$(\Phi' - \Phi'') = (0.09N - 64.0)\sigma^{(0.82 + 0.66 \times 10^{-3}N)} \tag{5.25}$$

Fakhri-Din studied over 20 liquid-liquid systems in a fully baffled turbine mixer and proposed two equations to correlate the inversion points Φ' and Φ''

$$\Phi' = 1.32 \times 10^{6}\left(\frac{\mu_d}{\mu_c}\right)^{0.32}\left(\frac{\Delta\rho}{\rho_c}\right)^{-0.11}(Fr)^{0.71}(Re)^{-1.06}(We_l)^{-0.25} \tag{5.26}$$

$$\Phi'' = 12.2\left(\frac{\mu_d}{\mu_c}\right)^{0.31}\left(\frac{\Delta\rho}{\rho_c}\right)^{0.04}(Fr)^{0.13}(Re)^{0.22}(We_l)^{-0.03} \tag{5.27}$$

Fr is the Froude number and *Re* the impeller Reynolds number.

There is considerable variation in the reported results and no really convincing correlations on which to base designs. The results discussed above have all been measured in equilibriated systems whereas the practice is that mass transfer by necessity must be accounted for. There are only one or two studies reported on the effect of mass transfer on the ambivalence limits ϕ' and ϕ''. Rowden *et al.* (1975) carried out a very detailed study on mixing and phase inversion in a copper extraction system. These workers demonstrated that mass transfer did in fact influence both ϕ' and ϕ''. The variations were dependent on the direction of solute transfer with respect to the actual phase continuity. These variations could be explained qualitatively in terms of interfacial phenomena, the variation of interfacial tension with adsorbed solute concentration, the resultant influence on coalescence and the guidelines set out by Sawistowski (1971). The quantitative predictions were however quite another matter and virtually no progress has been made to date.

This area of research is one where work and interpretation is urgently required. It is really quite remarkable that it is still almost impossible to define in certain ranges what will be the phase continuity in commercial extractors and therefore to compute all the parameters which depend on the properties of the dispersion.

5.4 Separation of liquid dispersions

5.4.1 Introduction

Separation of dispersions into two liquid layers takes place by droplets coalescing both together and at the phase boundary between the dispersion

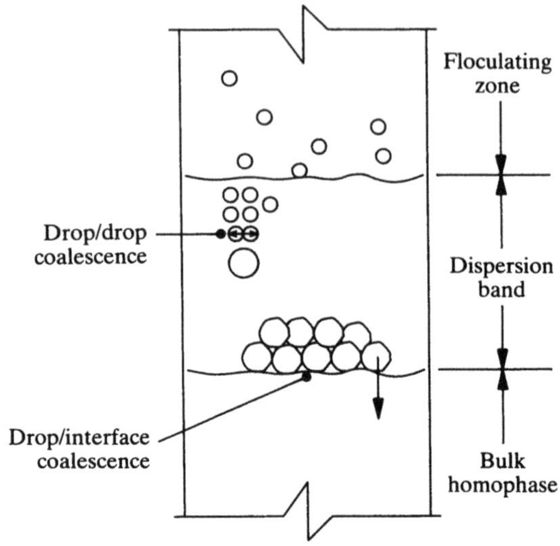

Fig. 5.3. Coalescence and separation of dispersions at a phase boundary.

and bulk phase. The process is shown diagrammatically in Fig. 5.3. If we consider the separation zone in a commercial extractor or settler the droplets must first setle to collect at the phase boundary. Here, depending on the flow rate, a bed of droplets will form, often called a dispersion band. For this band to form any turbulence in the flow leaving the contactor section must be damped. In the band, droplets coalesce together giving an increase in the mean drop size. Ultimately the droplets must coalesce at the phase boundary and the contents join the bulk phase. At steady state the volumetric flow rate of dispersed phase into the settling zone must be equal to the volume coalescing from the dispersion band. A simplified expression for this volume balance can be formulated if the time taken for a droplet, diameter d, to coalesce is assumed to be $\tau(d)$ and if the mean diameter of droplets entering the dispersion and coalescing at the phase boundary are d_0 and d_i respectively. Then

$$V = \frac{n_0 \pi d_0^3}{6} = \frac{\pi d_i^3}{6\tau_i} \tag{5.28}$$

n_0 is the number of drops entering the dispersion band per unit time. The problem to determine the rate of separation might appear simple if d_i can be determined and if the relationship between the coalescence time and drop size is known. Binary droplet coalescence within the band will increase the local drop size and one might assume that the droplet diameter, defined as the mean, would increase with distance from the inlet. The problem is far from simple, the coalescence process is only qualitatively understood for single drops and the extrapolation of results to dispersions is a problem of significant complexity. In the next section we will review our understanding of the coalescence of a single drop and then extend this to a raft of drops and finally to a dispersion band to lead into the engineering science which must underpin separator design.

5.4.2 Coalscence of single drops

The scientific study of coalescence and separation of dispersions was initiated with research into the behaviour of single drops resting at a plane interface. It was recognized that coalescence did not generally take place instantaneously and that the rest time of a drop at any interface varied between one physical system and another and therefore must be related to the physico-chemical properties of the system. Additionally the rest time was found to vary with the size of the drop. Most experimental measurements were made by forming a droplet at a nozzle and then allowing the drop to settle towards a liquid–liquid interface, maintained in a suitable cell, so that the rest time could be measured and the sequence of events leading to coalescence observed. The first experimental study was carried out by Reynolds (1881) who observed the coalescence of rain droplets into a pool of water. This was followed by Smoluckowki (1917) and Worthington and Cole (1895) who, using flash photography were able to demonstrate the complex phenomena which lead to the formation of small 'daughter' droplets, a subject which will be discussed later. The rest time of a droplet at any interface is defined as the time from arrival to the point when the contents of the drop are transferred into the homophase. A thin film of continuous phase fluid is trapped between the drop and interface. This film must drain before it ruptures then transfer into the homophase takes place. The drainage of this thin film controls the coalescence process. In liquid–liquid systems the time for drainage up to film rupture is of the order of seconds whereas transfer into the homophase after rupture is of the order of milliseconds. Over the last thirty years much research has been carried out into this basic process. This has been driven by a desire to understand the underlying mechanism by which drainage and rupture of the film takes place and a desire to be able to predict the rest or coalescence time. The use of the latter was seen in relation to predicting the separation of dispersions and therefore the application to design of equipment, mixer–settler extraction columns etc. Before examining this it is perhaps worth reviewing the results of experiments.

Numerous workers have reported data on the coalescence of single drops. A conclusion common to all of them is what when all conditions are kept constant, temperature, pressure, composition of the phases, drop size (within experimental limits) there is no one value for the coalescence time.

It is now realized that this is due to several factors. The thickness of film separating the drop from the homophase is not uniform and there is no single value when this film will rupture. In addition although all workers have recognized the need to prepare the liquids carefully so as to remove impurities and to design the apparatus such that wetted surfaces are made of materials which can be thoroughly cleaned, it is extremely difficult to remove all traces of materials which might adsorb at the interfaces and thus affect the drainage of the film. Hodgson and Lee (1969) demonstrated this and by significantly modifying the experimental techniques were able to obtain reproducible coalescence times. Stochastic variations in the system due to the stability of the film, perhaps chemical impurities in the system and random vibrations from the surroundings all contribute to the measured variation in coalescence time. Workers have reported coalescence time distributions from repeated measurements to produce data on the mean coalescence time $\bar{\tau}$ or the half-life time $\tau_{1/2}$ defined as the time taken for half of the drops in the sample studied to coalesce, Nielsen *et al.* (1958), Konnecke (1959), Gillespie and Rideal (1956), Jeffreys and Hawksley (1962). An example of such data is shown in Fig. 5.4. In this N is defined as the number of drops which have not coalesced at time t and N_0 is the total number coalesced. Gillespie and Rideal (1956) suggested that the coalescence times could be correlated by the equation

$$\ln(N/N_0) = -K(t - t_0)^{3/2} \tag{5.29}$$

where K, designated as the coalescence constant for the system, was given by

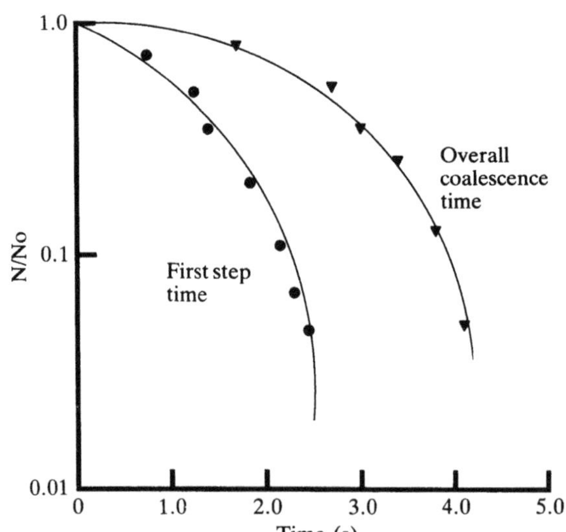

Fig. 5.4. Coalescence time data.

$$K = fC_0A_0\left(\frac{6\sigma}{r\mu}\right)^{1/2} \tag{5.30}$$

They reported that there appeared to be a dwell or delay time, t_0, during which no coalescence took place. Other workers have considered that this is related to the settling time of the drop at the interface where initially some of the momentum of the drop is dissipated by oscillation at the interface. Film rupture was assumed to be initiated by local periodic fluctuations in the film arising from mechanical vibration and 'thermal' effects. To account for these the parameters fC_0A_0 were introduced into eqn (5.30). There is no method however of calculating these a priori so that they can only be deduced from experiment. Nevertheless this early work was important in demonstrating the fact that a distribution of coalescence times is observed, that it may be possible to use mean coalescence time data and that film rupture is probably initiated by random fluctuations. The latter conclusion later led to an interesting study by Lang (1962) and Ivanov *et al.* (1970). Other workers used this approach to analyse experimental data, Elton and Pickett (1957), Sawistowski and James (1963), Lang (1962), Jeffreys and Hawksley (1962). The data were analysed by equations of the form

$$\ln(N/N_0) = -c(t - t_0)^n \tag{5.31}$$

Some workers found that $t_0 = 0$, that is that coalescence may take place instantaneously. There may be a good explanation for this if the phases were not at physical and chemical equilibrium. The value of n was reported to vary from 1.5 to 2.5.

A further important observation was made in several of these studies which in fact has significant implications for the operation of commercial extraction and separation equipment. It was noted that after the film ruptures not all of the drop contents are necessarily transferred into the homophase. When the film ruptures there is set-up immediately a pressure differential between the liquid in the 'drop' and the homophase. This is a result of the capillary forces acting around the curved surface of the 'drop', Fig. 5.5. The capillary pressure, Δp, will vary around this surface because the curvature of the surface, H, varies and from Laplace's equation

$$\Delta p = 2H\sigma \tag{5.32}$$

The resultant computed around the surface is positive so that the liquid is expelled into the homophase and the drop forms a column of liquid which may decay as shown in Fig. 5.5. Detailed high speed photographic studies of this decay process have shown that small surface waves appear on the collapsing column of liquid and in some cases the amplitude of these waves increase so that as the diameter of the liquid column decreases the nodes of the surface curves meet and at this point the column separates to form a

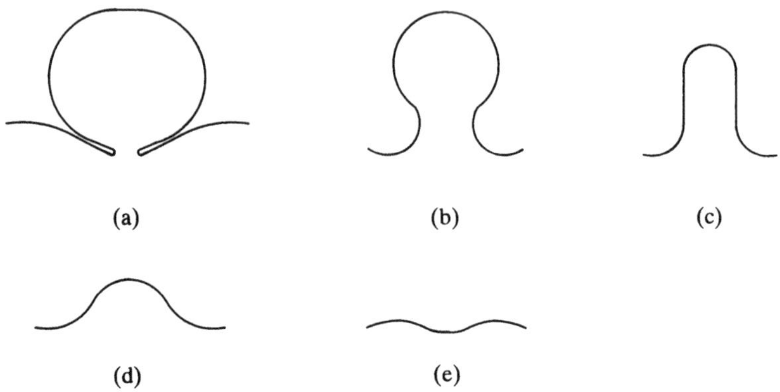

Fig. 5.5. Coalescence of a drop at an interface.

secondary drop, Gillespie and Rideal (1956), Jeffreys and Lawson (1967), and Davies *et al.* (1970), were amongst the first workers to provide quantitative coalescence time data on this process which is referred to as stepwise coalescence although there was evidence of this phenomena in the now classic early photographs of Worthington (1895). The data shown in Fig. 5.4 includes the case of step coalescence. The curve labelled first step time refers to the coalescence time of the primary drop whilst the other curve refers to the overall coalescence time, that is the time for all the material in the primary droplet to transfer to the homophase. This may take place by a sequence of coalescence steps. The primary drop producing a secondary and so on. Sequences involving up to as many as 10 stages have been reported. Although this is an interesting scientific observation it does in fact have significance to the operation of commercial equipment. The stability of very small droplets increases as the diameter is reduced, and if step wise coalescence takes place within dispersions it can produce a large number of small drops which may not coalesce but be swept out of the separator zone of the equipment. This would give rise to an increase in entrainment.

Many workers have tried to correlate the mean coalescence times, measured for various physical systems, with the system properties. The objective of this was to provide a framework from which data could be applied to dispersions. A review of work up to 1970 has been provided by Jeffreys and Davies (1971). The major conclusions from this and later work will be reviewed since not only does it provide a basis from which the application to subsequent problems can be assessed but also it emphasises the difficulties encountered in carrying out experimental work. We should separate the factors controlled in the experiments from the physico-chemical properties of the system. Thus in the first group it is now accepted that the drop size, normally defined by the volume-surface diameter d_{vs}, the distance of fall to the interface, the curvature of interface of the homophase, temperature and temperature gradients within the system, external vibrations and applied electrical fields can all effect coalescence.

The effect of drop size is of particular interest since if we are to apply results to analyse the separation of dispersions we must be able to quantify the coalescence time with diameter. As was shown earlier the dispersion may increase as a result of binary drop coalescence. It is now generally accepted that for primary dispersions the coalescence time of a single drop with its bulk phase increases with diameter. The exact relationship is however not clear. Large droplets tend to flatten, the deforming force is of the order of $O(\Delta\rho g d^3)$ whereas the resisting surface forces are of the order $O(\sigma/d)$. The area of the continuous phase film is then larger. However this effect is difficult to quantify between physical systems since the force pressing on the film is also larger, the difficulty arises in computing the result of these two conflicting factors without a knowledge of the drop dimensions and the shape of the interface. The mean coalescence time, $\bar{\tau}$, varies with d according to the equation

$$\bar{\tau} \propto d^n \tag{5.33}$$

where values of n ranging between 0.6–3.15 have been reported, Burrell and Woods (1973), Jeffreys and Hawskley (1965), Komasava and Otake (1970), Charles and Mason (1960). These values were obtained using carefully purified chemical systems in the absence of surface active agents. The presence of these complicate the picture significantly. Surface active agents will influence the shape of the interface and may produce a surface shear and dilational viscosity which would effect the flow within the film. Therefore it is unlikely that the conditions will be clear. This is confirmed by the results of Cockbain and McRoberts (1953) who, working with oil–water systems containing surfactants, reported $n = 0$ for $0.98\,\text{mm} < d_{vs} < 2.68\,mm$ *and* $n < 0$ for $d_{vs} < 0.98\,\text{mm}$. Neilsen *et al.* (1958) and Lang and Wilke (1971) also found it difficult to predict the effect of drop size on τ when surfactants were present.

The distance of droplet fall from the point of formation to the interface has also been studied in some detail, Jeffreys and Hawksley (1965) and Lawson (1967) both reported that the coalescence time increased with distance of fall and attributed this to disturbances and oscillations set up at the interface which results in an increase in the film thickness at which drainage begins. Results of Edge and Greaves (1967) were interesting since these tend to support this but additionally they found that where stepwise coalescence took place large secondary and ternary drops were formed as the distance of fall increased. Davies and Smith (1971) carried out experiments in a larger apparatus whereby drops could achieve the terminal settling velocity and under these conditions $\bar{\tau}$ was independent of the distance of fall.

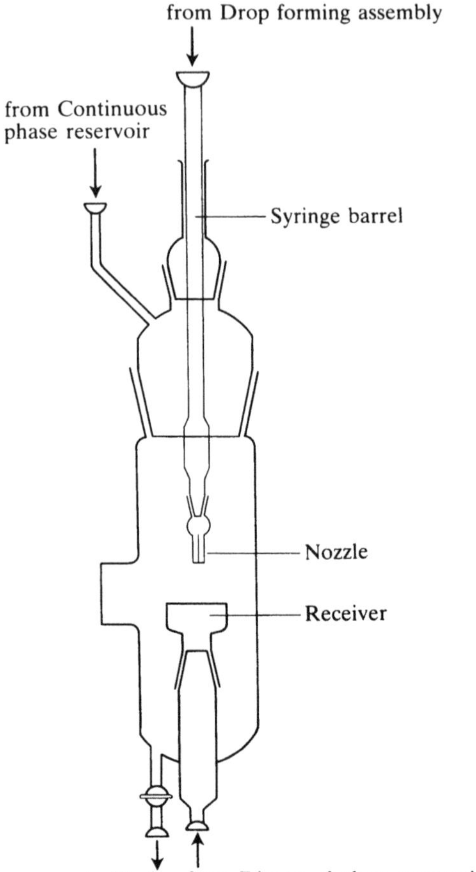

Fig. 5.6. Diagram of an apparatus to measure coalescence time of droplets at an interface.

Burrell and Woods (1973) and Neilsen also reported that $\bar{\tau}$ was independent of fall distance. It is an important factor to assess in examining results from single drop experiments but in commercial equipment the distance of fall (or rise) is either very small when large dispersion bands are present or large in relation to the drop size, and terminal velocities are then likely to be achieved.

The curvature of the interface of the homophase can be controlled by changing the dimensions of the cell, Fig. 5.6, and by adjusting the level of the liquid–liquid interface. This was first examined by Nielsen *et al.* (1958) who showed that the coalescence time could be varied by changing the curvature and that if the curvature of the homophase was concave so that it curved around the drop then the mean coalescence time increased significantly. This would be expected since under these conditions the area of the draining film is larger than when the interface is convex relative to the drop. More interesting studies in relation to the application of data to interpret coalescence in dispersions involves the coalescence between pendant drops. This can be carried out using two nozzles designed co-axially with a controlled gap between them, Fig. 5.7a or with the axes parallel, Fig. 5.7b. By adjusting the liquid flow to each nozzle the approach of the coalescence of the drops can be studied. McAvoy *et al.* (1965) used this technique to study the effect of curvature on coalescence. Similar techniques were employed by Dergaguin and Titijerskaya (1954) and Van den Temple (1958) to examine the suitability of liquid films.

An important factor to which a considerable amount of experimental effort was directed was the study of the effect of vibration. Many workers had experience with the effect of vibration on thin liquid films and it would be obvious then to examine this in relation to coalescence. An important feature in the analysis of data performed by Gillespie and Rideal (1956) was the inclusion of terms to allow for disturbances in the continuous phase film. They suggested that external vibrations transmitted through laboratory floors and walls would tend to stabilize the draining film. Only minor effects were reported by Neilsen *et al.* (1958). Lang and Wilke (1971) introduced vibrations to the system and showed that at certain amplitudes and frequencies these vibrations reduced the stability of the draining film sufficiently to promote coalescence and for those systems where drainage was slow and rate controlling external vibrations could indeed promote separation. Similar results were obtained by Smith (1967) who applied external vibrations to a settling dispersion band and was able to demonstrate rapid coalescence within the band as the vibrations were applied. This seems an important factor which has not been scaled up or examined on commercial plant. Although there would be capital and operating costs associated with such an application, the effects on the small scale are significant enough to warrant further examination.

Electrical fields are used to promote coalescence of aqueous droplets in an oil and underpin the application of electrostatic coalescers. Both a.c. and d.c. steady and pulsed fields are used. The discussion on this will be deferred to the section reviewing the design of equipment. It is however worth discussing the influence of electrical fields on droplet coalescence. This is easily studied in the type of cell shown in Fig. 5.6. The nozzle is insulated and forms an electrode which is then connected to a battery or high tension source. A grid is positioned in the cell and is submerged in the homophase. The continuous phase must be non-conducting. The drop formed at the nozzle is then charged. As a result of this charging its shape and size will change and depend on the surface charge density. When the drop detaches from the nozzle it is subject to an electrical force which can greatly exceed

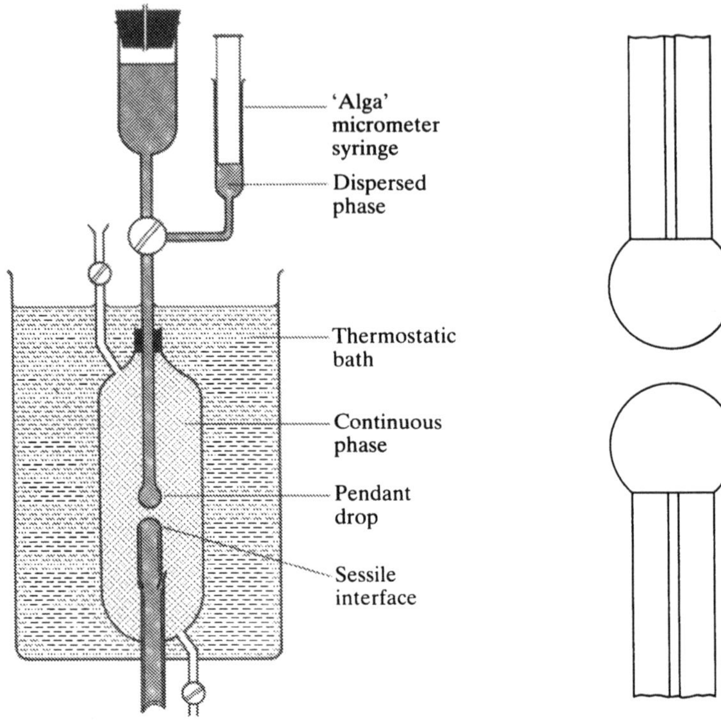

Fig. 5.7. Techniques to simulate drop–drop coalescence; (a) coalescence of coaxial surfaces; (b) coalescence of pendant drops.

the buoyancy force so that as this approaches the interface with the homophase the rate of thinning of the film is greatly accelerated. The early work on single drops was carried out by Charles and Mason (1960), and Allan and Mason (1961) using DC fields. This group showed that in systems in which step wise coalescence was normally observed the effect of the electric field was to eliminate this, no secondary drops were formed. This conclusion was confirmed by Brown and Hanson (1965, 1968) who used AC fields. The additional benefits ensuing from using electrical fields have been expounded by Thornton (1966) who has demonstrated that the rate of extraction from drops is enhanced by the rapid acceleration of drops through the continuous liquid prior to coalescence. The application of external fields, centrifugal, electrical and in some cases magnetic offer significant benefits in reducing equipment size in extraction and phase separation processes. In this case providing the dielectric properties of the liquids are suitable and providing the phase continuity, *W/O*, is established, external electrical fields

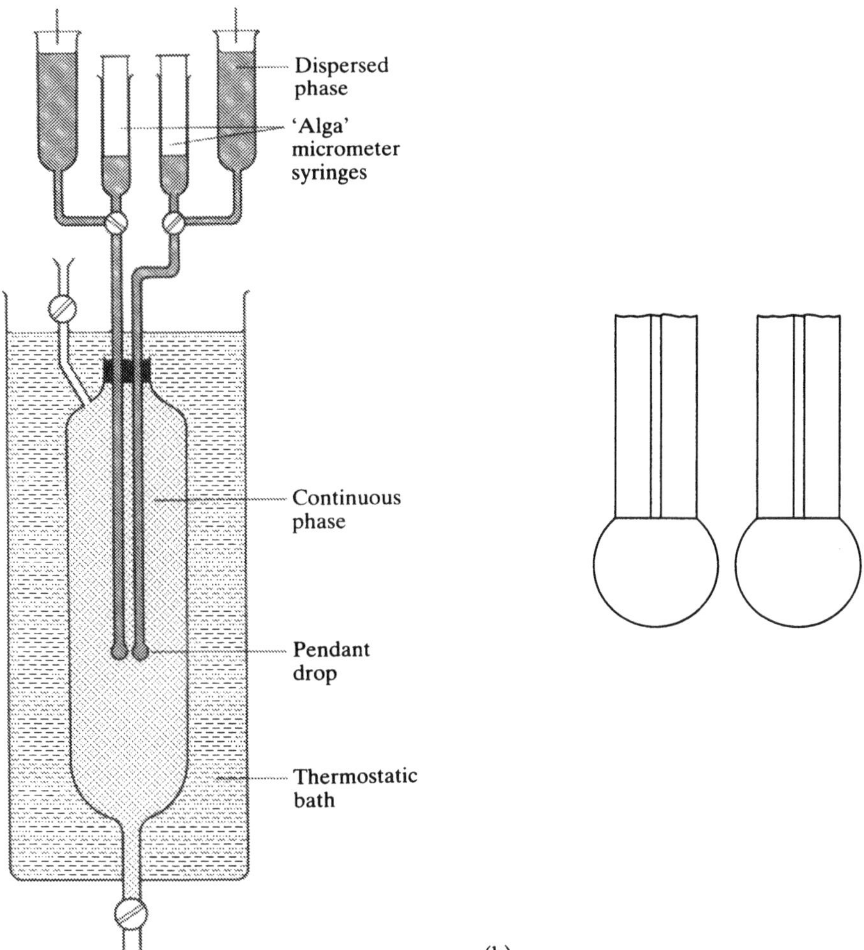

(b)

Fig. 5.7. *cont'd.*

can promote almost instantaneous coalescence coupled with low entrainment. The effect can be seen in Fig. 5.8. In Fig. 5.8a droplets of water are flowing through decane in a 50 mm diameter glass spray column. In Fig. 5.8b the outer walls of the column were gently rubbed with a polypropylene mat. The electrical field set up as a result of charging of the walls was sufficient to cause the dispersion to coalesce.

The physico-chemical properties which effect droplet coalescence are the phase densities (or density difference), phase viscosities, interfacial tension and surface viscosities if adsorbed components are present. Additionally if non-equilibrium conditions exist so that component(s) may be transferring

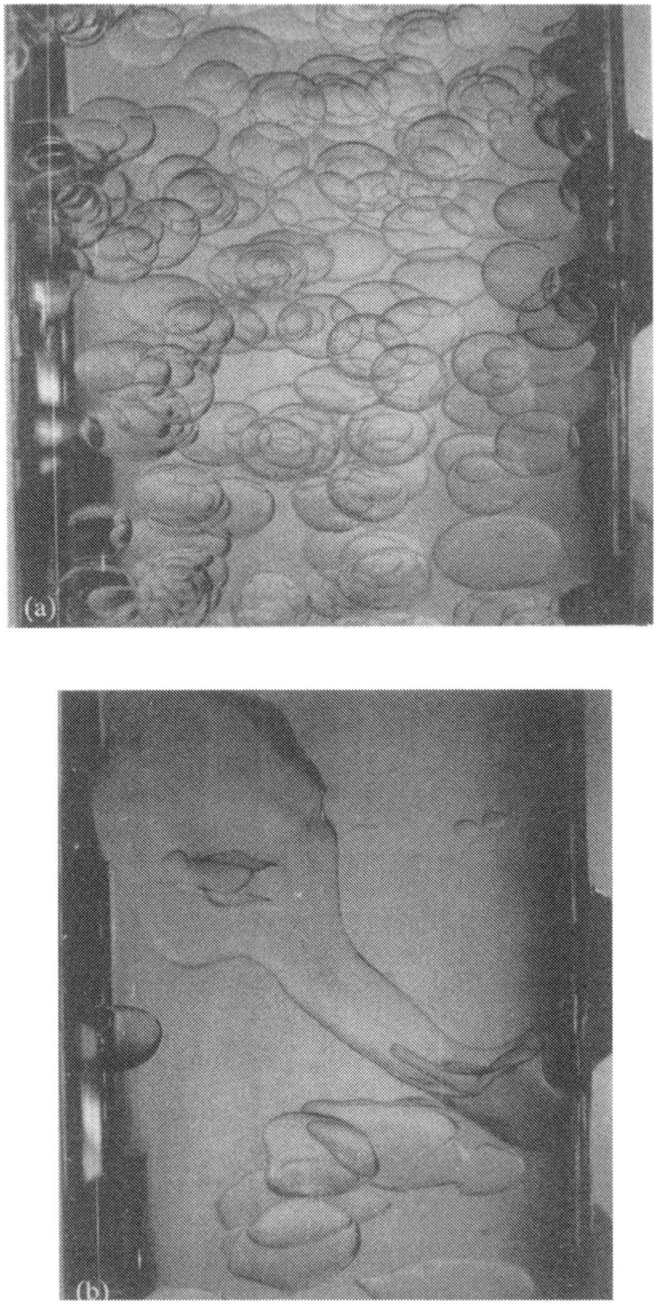

Fig. 5.8. Coalescence of droplets in a spray column: (a) water droplets settling in *n*-decane; (b) coalescence induced by charging the walls of the vessel.

between the drop and homophase this can have profound effects on coalescence.

A problem of examining these in isolation is that although a large number of measurements have been made and reported, in fact the data base is still small and for any factor a limited range may have been studied. It is clear that as the density difference increases or, in terms of a significant dimensionless parameter, as the Bond number increases [$N_B = \Delta\rho g d^2/\sigma$] the deformation of the drop and interface increases. The drop tends to flatten so that the area of the draining film increases whilst the hydrostatic forces, the principal driving force for draining, does not increase proportionally. The expected result from this is that the coalescence time will increase. Jeffreys and Hawksley (1965) showed that half life time $t_{1/2}$ increased with $\Delta\rho$ as did Davies and Jeffreys (1971) and Hanson (1971).

The viscosity of both phases can influence coalescence by their influence on film drainage. Increasing the viscosity of the continuous phases will reduce the film drainage rate thereby increasing coalescence times. This has been confirmed in all experimental work. Additionally the viscosity of the dispersed phase effects coalescence time since, as will be shown later, in the absence of surfactants the liquid in the drop is circulating so that at the drop film interface the fluid must be moving. The surface velocity is determined in part by the viscosity of the dispersed phase μ_d and the ratio μ_c/μ_d. As μ_c is reduced the coalescence time is reduced.

The variations of film drainage with the parameters discussed above $\Delta\rho$, μ_c and μ_d can be examined also by analyses of equations describing the flow of the continuous phase film. What is more difficult to study quantitatively is the effect of transferring solute(s) during coalescence and the effect of trace impurities in the system. The early experimental work suffered from severe problems of contamination. The presence of a third component or phase may accelerate or retard coalescence. One has only to measure separation rates in practical equipment to see that the formation of crud [an acronym for Chalk River Unidentified Deposit to describe solid materials which accumulated in uranium solvent extraction plant operated at Chalk River, Canada and reduced equipment capacity] in a system as a third phase at the liquid–liquid boundary can significantly reduce the rate of coalescence. Under controlled experiments with single drops Charles and Mason (1960) demonstrated that small glass beads as a third phase introduced at the interface promoted spontaneous coalescence of water drops in benzene. This can be explained in terms of the surface wetting of the solid by the dispersed phase. Had the glass beads been treated say with dimethyl dichlorosilane to make them hydrophobic then the effect would have been reversed. This is a rather simple case to examine, in practice the situation is more complicated. When the third component is present in solution in one of the phases and if the Gibbs adsorption, Γ, of this component is finite then significant effects usually ensue. Surface active agents retard coalescence. This has been demonstrated for both catonic, anionic and non-ionic materials and is evidenced in the work of Nielsen *et al.* (1958), Gillespie and Rideal (1956), Allen (1961), Hartland (1968), Hodgson and Lee (1969), Hodgson and Woods (1969). If the phases are not at thermodynamic equilibrium during coalescence then the transfer of any of the components may have a profound effect on the coalescence. Two cases should be considered. One when the transferring component has a positive Gibbs adsorption. The presence of an excess concentration of this at the interface between the liquids will then cause a local reduction in interface tension. The second case is the reverse of this, that is a transferring component with a negative Gibbs adsorption. This again will effect coalescence. The simple case where the third component has a zero Gibbs adsorption can be neglected, this generally has no effect. Consider Case 1. If this material is transferred from the drop to the continuous phase and if, for simplicity we assume that the mass flux is uniform around the drop, then initially the surface concentration will be uniform. However the concentration in the film will rise relative to the surrounding liquid since it can only be dissipated by diffusion from the boundary edge of the film and by convection as the film drains. Then since $\Gamma > 0$ the interface tension in the area of the film will be lower than the surrounding surface of the drop. This will lead to an unstable condition and the interface area of the film will expand and dilate rapidly. This will greatly accelerate the process of drainage and promote coalescence. If the direction of mass transfer were reversed i.e. from the continuous phase into the drop then the opposite result would be found since now the film region would be deficient, relative to the surrounding drop surface, of the absorbing component and the interface tension of the film region could be higher. This would result in a surface flow which would draw material into the film and oppose film drainage. If we consider Case 2 the opposite results would be obtained. These are summarized in Table 5.2. The results are all derived from surface flows promoted by changes in local surface concentrations of adsorbed components. The first scientific observations of spreading and surface flows of oil on water were reported by Marangoni (1865, 1871) and these phenomena are often referred to as Marangoni effects. The experimental confirmation was reported in the now classical work of Linde and Winkler (1964, 1965) and Sawistowski (1971). Some quantitative data on coalescence times may be found in Charles and Mason (1960), MacKay and Mason (1963), Jeffreys and Lawson (1965). A very interesting experiment was reported on the operation of spray towers by Harding, Bliss and Johnson (1946) which can only be properly explained by the effect of mass transfer on droplet coalescence in the flocculating section of the tower.

The effect of temperature gradient on coalescence, referred to earlier, can now be explained since the interfacial tension is a function of temperature.

TABLE 5.2

$\frac{\partial\sigma}{\partial c} < 0$: +ve Gibbs adsorption						$\frac{\partial\sigma}{\partial c} > 1$: −ve Gibbs adsorption					
Direction of mass transfer						Direction of mass transfer					
1 ⟶ 2			2 ⟶ 1			1 ⟶ 2			2 ⟶ 1		
$\frac{v_1}{v_2}$	$\frac{D_1}{D_2}$	Result	$\frac{v_1}{v_2}$	$\frac{D_1}{D_2}$	Result	$\frac{v_1}{v_2}$	$\frac{D_1}{D_2}$	Result	$\frac{v_1}{v_2}$	$\frac{D_1}{D_2}$	Result
< 1	1	Convection cells	> 1	1	Convection cells	< 1	1	Stable or oscillations	> 1	1	Stable or oscillations
1	1	Stable	1	1	Stable	1	1	Stable	1	1	Stable
> 1	1	Stable or oscillations	< 1	1	Stable or oscillations	> 1	1	Convection cells	< 1	1	Convection cells
1	> 1	Convection cells	1	< 1	Convection cells	1	> 1	Stable	1	> 1	Stable
1	< 1	Stable	1	> 1	Stable	1	< 1	Convection cells	> 1	< 1	Convection cells
< 1	> 1	Convection cells	> 1	< 1	Convection cells	< 1	> 1	Stable or oscillations	> 1	< 1	Stable or oscillations
> 1	< 1	Stable or oscillations	< 1	> 1	Stable or oscillations	> 1	< 1	Convection cells	< 1	> 1	Convection cells

Stable conditions infer that mass transfer coefficients can be computed from theory based on transport through boundary layers. There will be little effect on drop coalescence in these systems.

Oscillatory or convection cells infers a break-down of conventional mass transfer theory and correlations. Drop coalescence is likely to be influenced by mass transfer in these systems.

If temperature gradients exist in equipment which might influence the interfacial tension in the coalescence section then 'Marangoni' flows may be set up which would influence coalescence in a way analogous to the results in Table 5.2. Generally the coefficient $\partial\sigma/\partial T$ is less than $\partial\sigma/\partial C$ and temperature effects are of second order with respect to concentration effects. It is however an interesting problem since the solubility is dependent on temperature therefore there may be a coupling which can only be resolved by application of the appropriate equations from irreversible thermodynamics (Standart 1965; Davies and Thornton 1977).

5.4.3 Analysis of film drainage

The factors discussed in Section 5.4.2 (p. 257) all affect film drainage and in principle can be investigated if the approach of a drop to an interface with the homophase is analysed. The problem is quite complicated however since forces acting on the droplet even at large separation distances will cause deformation which in turn will affect the flow. As the interfaces approach, this deformation will be very significant and must be considered in the solution to the equation of motion. In addition as the separation distance is reduced, the effect of any interfacial tension gradient, adsorbed components influencing the surface shear and dilational viscosities must be introduced, as must be the effect of the electrical double layer if one is present. The process can be considered in a number of regions where particular forces are dominant and for two cases rigid and deformable interfaces. Then from these solutions the influence of various properties can be deduced.

5.4.3.1 *Droplet approach leading to drainage*

We will consider first the conditions where the distance of separation is of the order of the drop radius, r, down to 0.1 r. The problem can be solved for the case of a spherical drop approaching a plane or spherical interface. The curvature of the interface can be convex or concave relative to the drop. Conditions are outlined in Fig. 5.9. If we take the curvature of the interface as convex then the solution can also be applied to the approach of two droplets along their lines of centre leading to binary coalescence. If the liquids are Newtonian and are incompressible the equations of motion and continuity are:-

$$\frac{\partial \mathbf{v}}{\partial t} + \mathbf{v}\cdot\nabla\mathbf{v} + \frac{1}{\rho}\nabla p = \mathbf{v}\nabla^2\nu^2 \tag{5.34}$$

$$\nabla\cdot\mathbf{v} = 0 \tag{5.35}$$

The drop velocities in this region are low, in commercial equipment any turbulence must be damped for droplets to settle towards an interface. Then for sufficiently slow motions it is permissible to neglect the quadratic terms, $\mathbf{v}\cdot\nabla\mathbf{v}$, compared with the viscous terms $\nabla^2\mathbf{v}$. The fluid motion is inherently unsteady when a liquid droplet approaches a plane interface or another droplet. However the time dependent term $\partial\mathbf{v}/\partial t$ becomes negligibly small by comparison to the various terms for sufficiently small values of the dimensionless group $(U_r/\mathbf{v})$ (a/h), (Happel and Brenner 1965), h being a characteristic distance in the direction parallel to the flow. The term h can be taken as the minimum distance of separation. The equations then simplify to

$$\nabla p = \mathbf{v}\nabla^2\mathbf{v} \tag{5.36}$$

The equations must be solved for the three regions, the droplet, the continuous phase and the homophase or second droplet. If the flow is assumed to be axisymmetric then the equation can be simplified by introducing the Stokes' stream function ψ_i the subscript refers to the flow region $i, i = 1$, 2 or 3. The stream function is defined for this case as

$$\mathbf{v}_R = \frac{1}{R}\frac{\partial\psi}{\partial z};\ \mathbf{v}_z = -\frac{1}{R}\frac{\partial\psi}{\partial R};\ \mathbf{v}_\phi = 0 \tag{5.37}$$

Then eqn. (5.36) reduces to

$$E^4\psi = 0 \tag{5.38}$$

where,

$$E^2 = \frac{\partial^2}{\partial z^2} + R\frac{\partial}{\partial R}\left(\frac{1}{R}\frac{\partial}{\partial R}\right)$$

The boundary conditions are defined at the surface of the droplet and homophase. Since these are both curved and the distance of separation h depends on ϕ, the problem is tedious unless a transformation is applied. If bipolar coordinates, defined by the conformal transformation

$$Z + iR = iC\cot\frac{1}{2}(\eta + i\xi) \tag{5.39}$$

or equivalently

$$R = \frac{C\sin\eta}{\cosh\xi - \cos\eta} \text{ and } Z = \frac{C\sinh\xi}{\cosh\xi - \cos\eta} \tag{5.40}$$

are used, the problem in the new coordinate frame is equivalent to the approach of two planes. It is only necessary to consider in this problem the case for $R > 0$ which corresponds to a range of values $-\infty < \xi < \infty$ and $0 \le \eta \le \pi$. In eqn (5.39) C is a positive constant.

For the condition shown in Fig. 5.9, the surfaces $\xi = \xi_1 (> 0)$ and $\xi = \xi_2 (< 0)$ are defined by

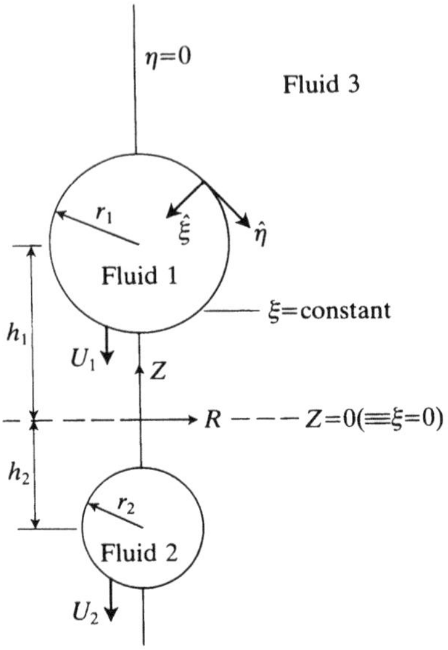

Fig. 5.9. Conditions of a drop approaching a curved interface.

$$\xi_1 = \cosh^{-1}(h_1/r_1) \qquad \xi_2 = \cosh^{-1}(h_2/r_2) \tag{5.41}$$

The solution in terms of the stream fuction is (Rushton and Davies 1970, 1973),

$$\psi_i = (\cosh\xi - \xi)^{-2/3} \sum_{n=0}^{\infty} u_{ni}(\xi)\, C_{n+1}^{-1/2}(s) \tag{5.42}$$

where $s = \cos\eta$ and $C_{n+1}^{-1/2}(s)$ is a Gegenbaur polynomial of order $(n + 1)$ and degree $-1/2$ and $u_{ni}(\xi) = a_{ni}\cosh(n - 1/2)\xi + b_{ni}\sinh(n - 1/2)\xi + c_{ni}\cosh(n + 3/2)\xi + d_{ni}\sinh(n + 3/2)\xi$.

The constants a_{ni}, b_{ni}, c_{ni}, d_{ni} $(i = 1, 2, 3)$ must be determined by the boundary conditions. The boundary conditions, in the absence of surface active agents or in the absence of concentration gradients of solute(s) in the interface, express the continuity of velocity and shear stress at each fluid interface, Batchelor (1967). For normal stresses it was assumed that any discontinuity in the normal stress is balanced by the interface tension forces. This requires that these forces which are of the order of, $O(\sigma_j/r_j)$ should be large compared to the deforming viscous forces of the normal stress due to motion (of order $\mu_3 U_j/\epsilon$). Thus

$$\frac{\mu_3 U_j}{\sigma_j} \cdot \frac{r_j}{\epsilon} \ll 1 \; (j = 1, 2) \tag{5.43}$$

The solution to this problem was presented by Rushton and Davies (1973a). This provided expressions for the drag force acting on the droplets which now are given by

$$F_i = 6\pi \mu_3 r_i U_i \lambda_i \qquad i = 1, 2 \tag{5.44}$$

where for $i = 1$

$$\lambda_1 = a_{11} U_1 + a_{12} U_2 \tag{5.45}$$

a_{11} and a_{12} are functions of the instantaneous position of the drops. The velocity components can be computed from the equations.

$$u_{xi} = \frac{1}{(\cosh\xi - \xi)}\left(\left(u_{\xi i}\sin\eta\sinh\xi + u_{ni}(1 - s\cosh\xi\right)\right) \tag{5.46}$$

$$u_{zi} = \frac{1}{(\cosh\xi - \xi)}\left(u_{\xi i}(1 - \xi\cosh\xi) - u_{\eta i}\sin\eta\sinh\xi\right) \tag{5.47}$$

In the general case for two droplets approaching

$$\frac{U_1}{U_2} = \frac{a_{22}(r_1/r_2)^2\Delta\rho - a_{12}}{a_{11} - a_{12}(r_1/r_2)^2\Delta\rho} \tag{5.48}$$

and since U_1 and U_2 must be related to the drag forces through (5.44)

$$\frac{U_1}{U_2} = f(\Delta\rho, \mu_1/\mu_3, \mu_2/\mu_3, \alpha, \beta) \tag{5.49}$$

The implications for film drainage can be considered by examining the streamlines in a system, take the case of a droplet approaching a second drop (if $r_2 \to \infty$ then this reduces to the drop–plane interface problem). An example is shown in Fig. 5.10, Rushton and Davies (1978). This shows the streamlines generated by motion of a drop towards a stationary drop at velocity U_1. The streamlines are mapped relative to the absolute motion. The approach of the drop sets up a secondary flow inside the stationary drop, this is a direct result of the surface boundary conditions applicable to the fluid–fluid boundaries. Unlike the conditions for an isolated drop settling in a fluid, the centre of circulation in the stationary drop is no longer co-incident with the plane $z = -h_2/c$ but has moved upwards. There is of course fluid circulation in the moving droplet so that as the interfaces approach, liquid in the dispersed and homophase adjacent to the interfaces is in motion. This will increase the drainage rate and the quantitative effect will depend on the magnitude of the dimensionless viscosity ratios, μ_1^* and μ_2^*, defined as

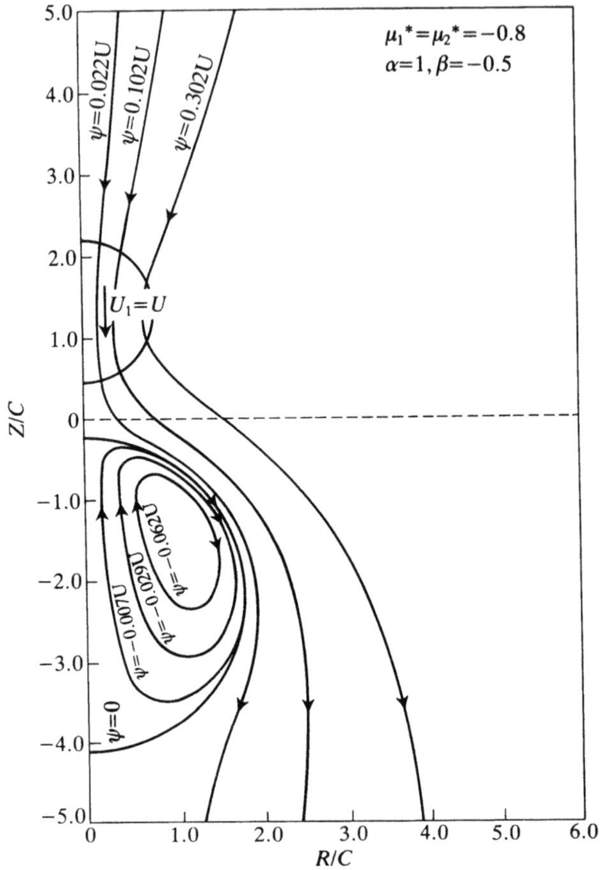

Fig. 5.10. Streamlines generated by the motion of a droplet towards a stationary droplet.

$$\mu_i^* = \frac{\mu_i - \mu_3}{\mu_i + \mu_3} \qquad i = 1, 2 \tag{5.50}$$

$$-1 \leq \mu_i^* \leq 1$$

The film drainage rate is computed from the radial velocity profile. When $\mu_1^* = \mu_2^* = 1$ this is the solid sphere–solid interface case and a parabolic velocity profile is obtained, a result which was reported by Bart (1968) and used in a simplified analysis for film drainage prior to coalescence by Charles and Mason (1960). As μ_i^* decreases the velocity profile flattens until at $\mu_i^* \simeq 0$ plug flow is achieved within the film. If μ_3 is constant then as the viscosity of the drop decreases the drainage rate will increase, a result which has also been confirmed by Reed *et al.* (1974). Although the analysis is restricted to spherical drops the results on the effect of viscosity and the importance of the surface flow set up as a result of the motion are likely to be applicable to larger droplets which exhibit deformation. Experimental confirmation of internal circulation in a droplet during film drainage has been obtained by Rushton and Davies (1973b). In this work they studied the film drainage process at a planar boundary with the homophase, during coalescence of a chlorobenzene drop in water as the continuous phase. A phototropic dye was dissolved in the chlorobenzene and this was triggered to produce the coloured complex by firing a UV laser beam through the drop parallel to the interface while drainage was taking place. A coloured line of dye was formed in the path of the light beam. The subsequent deformation of this line of dye confirmed that the drop internals were in motion, not stagnant as was assumed in all the early models used to examine film drainage. Photographs showing circulation are displayed in Fig. 5.11.

5.4.3.2 *Film drainage*

The early attempts to analyse the drainage in the film assumed axisymmetric flow and applied boundary conditions for immobile interfaces which will inevitably eliminate any influence of the viscosity of the dispersed phase. Various assumptions were made on the geometry of the liquid film.

Early attempts assumed that there was no deformation, the problem then reduces to the sphere–plane problem but with a simple analysis of the flow. This was followed by models in which deformation was considered, but the film was assumed to be uniform in thickness (Gillespie and Rideal 1956; Charles and Mason 1960). Experimental work in this period demonstrated that the film was not uniform and that due account must be taken of the shape of the boundary surfaces if more realistic analyses were to be developed. This has led to a series of papers from Hartland and his co-workers in Zurich, Slattery and others.

The development of this work can be followed from the simpler models. Using the notation in Fig. 5.12, if a differential element in the film is considered, the rate of change of volume of the element is

$$\frac{\mathrm{d}V}{\mathrm{d}t} = 2\pi u \mathrm{d}R \tag{5.51}$$

This must also equal the net outflow of fluid from the element $\mathrm{d}V$. If the thickness of the film at R is ϵ then the drop approach velocity u is

$$u = -\frac{\delta\epsilon}{\mathrm{d}t} \tag{5.52}$$

If the flow in the film is laminar then neglecting gravitational effects

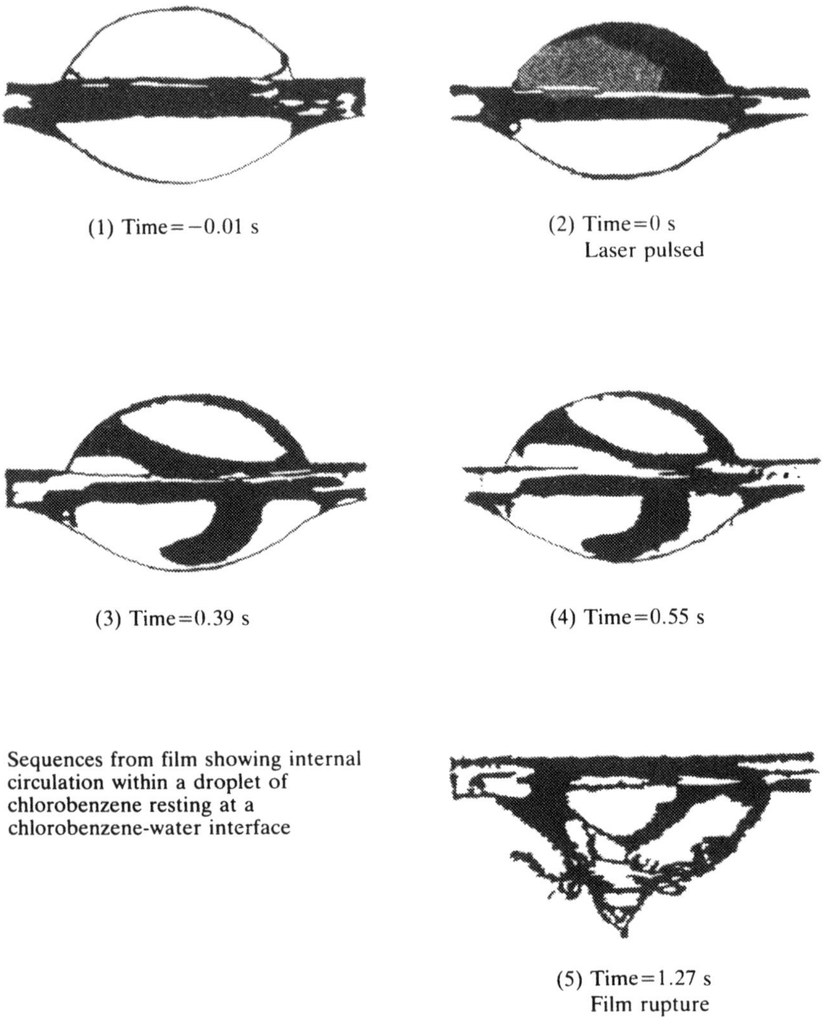

Fig. 5.11. Photographs using a photochromic dye trace to show circulation in a liquid droplet during film drainage.

$$V = \frac{\pi R}{6}\left(-\frac{dp}{dR}\right)\epsilon^3 \tag{5.53}$$

dp/dR is the radial pressure gradient in the film.

Equation (5.53) assumes that the boundary surfaces are immobile. If one or other surfaces were allowed to have a finite surface velocity which in this analysis would have to be assumed constant, cf. Section 5.4.3.1 (p. 270), then this could be accommodated,

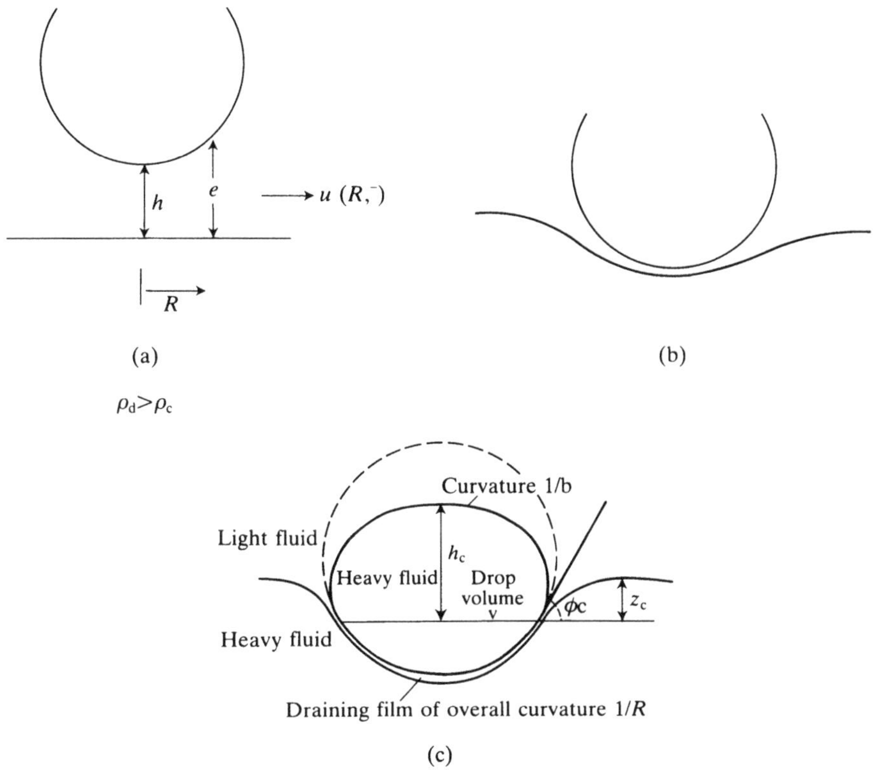

Fig. 5.12. Simplified geometric models to illustrate film draining prior to coalescence: (a) rigid spherical drop, flat interface; (b) rigid drop-deformable interface; (c) deformable drop-deformable interface.

$$V = \frac{2}{3}\pi R\left(-\frac{\mathrm{d}p}{\mathrm{d}R}\right)\frac{\epsilon^3}{\beta^2} \tag{5.54}$$

where β is the number of mobile surfaces. If the droplet velocity is independent of R then from eqn (5.51), $V = \pi r^2 u$ and

$$-\frac{\mathrm{d}p}{\mathrm{d}R} = \frac{3\beta^2\mu R u}{\epsilon^3} \tag{5.55}$$

From which

$$(p_R - p_0) = 3\beta^2\mu u\int_R^{R_f}\frac{R\,\mathrm{d}R}{\epsilon^3} \tag{5.56}$$

p_0 is the pressure in the liquid at the edge of the film. This will be equal to the local hydrostatic pressure so that $p' = (p - p_0)$ is the dynamic pressure in the film. The force acting on the film due to this pressure

$$F = \int_0^{R_f} 2\pi r p'_R \, dR \tag{5.57}$$

from which

$$F = \pi\beta^2 \mu u \int_0^{R_f} \int_R^{R_f} \frac{R\,dR}{\epsilon^3} \cdot R\,dR \tag{5.58}$$

This force is related to the approach velocity. Equation (5.58) can be evaluated for the various geometries assumed for the film.

An advance on this work was made when more realistic descriptions of the shape of the film were introduced. The shape of the interfaces at equilibrium are described by Laplace's equation:

$$\Delta p = \sigma H \tag{5.59}$$

where H is the sum of the principal radii of curvature of the surface

$$H = \left(\frac{1}{R_1} + \frac{1}{R_2}\right) \tag{5.60}$$

and

$$\frac{1}{R_1} = \frac{y'/x}{[1 + (y')^2]^{1/2}} \tag{5.61}$$

and

$$\frac{1}{R_2} = \frac{y''}{[1 + (y')^2]^{3/2}} \tag{5.62}$$

y' and y'' are derivatives dy/dx and d^2y/dx^2 respectively.
The term Δp is the pressure difference across a fluid–fluid interface due to surface forces. Princen (1963) and Princen and Mason (1965) introduced this to define the pressure at any vertical position y, Fig. 5.13, in the droplets.

$$\Delta p = g|\rho_d - \rho_0|y + c \tag{5.63}$$

where c is a constant.

Introducing eqn (5.59) into eqn (5.63) if R_1 is taken as the radius at the point $(x_1 y_1)$, say equal to a, then R_2 becomes $(x/\sin\phi)$ where ϕ is the angle which the normal at $x_1 y_1$ makes with the axis of revolution and

$$\frac{1}{a} + \frac{\sin\phi}{x} = \frac{C + g|\rho d - \rho_c|y}{\sigma} \tag{5.64}$$

If b is the radius of curvature at the lowest point of the drop

$$\lim_{y\to 0} a = b \text{ and } \lim_{y\to 0} \frac{x}{\sin\phi} = b$$

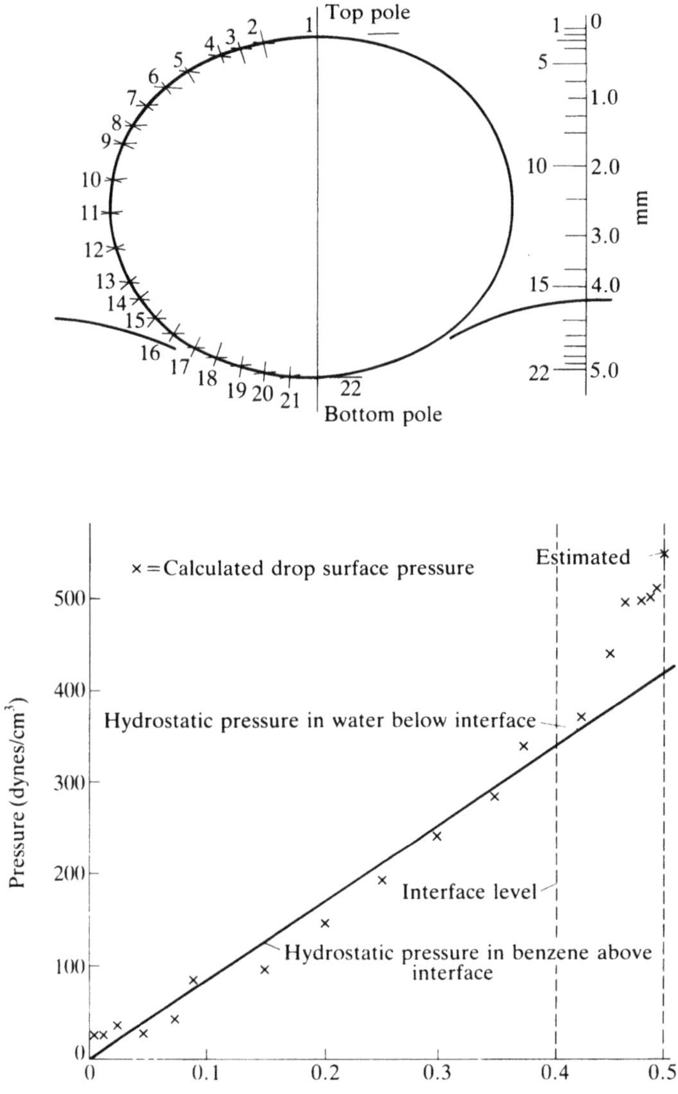

Fig. 5.13. Above – profile of water drop at benzene-water interface. Below – calculated surface and hydrostatic pressure outside drop.

then

$$\frac{C}{\sigma} = \frac{2}{b} \tag{5.65}$$

Then eqn (5.64) becomes

$$\frac{1}{a/b} + \frac{\sin\phi}{x/b} = \frac{C'y}{b} + 2 \tag{5.66}$$

where $C' = \dfrac{g|\rho_d - \rho_c|b^2}{\sigma}$ is the Bond number.

The solutions to this equation for pendant and sessile interfaces were first tabulated by Bashforth and Adams (1883). The values from these tables were used by Prince to compute the film shape. A major study of the topic has subsequently been undertaken by Hartland and his co-workers. The equations to describe drainage in axisymetric films and the pressure distributions therein were set out by Hartland (1971a,b). Methods of obtaining exact solutions were also reviewed. This represents a significant advance on the earlier work involving considerable effort. Extensions of the numerical solutions to the equations describing the shape of menisci and interfaces first considered by Bashford and Adams have been provided in a treatise by Hartland and Hartley (1976). However from the scientific point of view the models still fall short of the real conditions. Practically all of the models published and certainly those discussed here, assume symmetrical drainage. There is ample experimental evidence to show that this is not the case. This evidence comes from photographs of the draining film, Lawson (1967), Hartland (1969a,b), Hartland and Robinson (1970) and of the rupture of the film Davies *et al.* (1970). The film thickness can vary with position, even the shape of the film can change with time, Lin and Slattery (1982a,b). The qualitative effect of these changes which promote unsymmetric drainage is to increase the drainage rate. To describe this quantitatively the process must be described in 3 space. This provides major additional problems, analytical solutions are ruled out making numerical solutions the only possibility. This work will continue to tax the minds of researchers. However until we can produce a quantitative understanding of film rupture the application to predict the overall coalescence is some way off. Attempts have been made to move forward. Jeelani and Hartland (1983) have analysed a problem of unsymmetric flow produced by the approach of two inclined plates. This type of approach taking simple geometries, in much the same way as simple film models did, may well provide a spring board for advance.

5.4.3.3 *Critical film thickness and rupture*

The methods discussed above allow the time taken for a film to reduce in thickness from some initial value ϵ_0 to another value ϵ. Notwithstanding all the difficulties associated with predicting this time the critical thickness at which rupture occurs is not defined in these models. Indeed it is now realised that there is not a single value for this, rather the probability for rupture increases as the film thickness decreases. Attention has been paid to this problem taking as a simple approach the time for the film to attain a thickness defined as the critical thickness ϵ_c where $\epsilon_c \ll \epsilon_0$. The term E_c can be calculated using many of the methods in Section 5.4.3.2. (p. 275) If a uniform thickness model is used and a film area a is assumed, the time to reach ϵ_c is

$$t = \frac{3\mu}{4\pi} \frac{a^2}{F} \frac{1}{\epsilon_c^2} \tag{5.67}$$

For a planar film $a = \pi r_f^2$ where r_f is the radius of the periphery of the film. If the force per unit area acting on this film, $F/\pi r_f^2$, is much greater than the van der Waal surface pressure Π then Vrij (1966) suggested that the critical film thickness is given by

$$\epsilon_c = 0.268(\pi\eta^2 r_f^4/F\xi\sigma)^{1/7} \tag{5.68}$$

where $\eta = 6\pi\omega$, ω being the Hamaker constant and ξ being a constant, $\xi = 6.5$. The other limiting case, that is when $F/\pi r_f^2 \ll \Pi$, is:

$$\epsilon_c = 0.222(\eta r_f^2/\xi\sigma)^{1/4} \tag{5.69}$$

Other studies have been made into this problem, Scheldulko and Manev (1968) proposed that $\epsilon_c \propto r_c^{2/5}$ (Ivanov *et al.* 1974; Ruckenstein and Jain 1974, 1976). This is a start on what is a difficult problem to analyse. It will be some time before any approach to link these results with drainage models can be made to predict accurately coalescence times.

Apart from the challenge in understanding the occurence of film rupture the process of hole expansion in the film and retraction of the film to the surrounding continuous phase fluid is interesting and has some consequences for practical separations.

In the majority of extraction systems where the phase viscosities are not high, $\lesssim$ 20 cP the time taken for the hole in the film to expand after rupture and the film retract is of the order of 0.1 ms. As the hole expands the contents of the drop are propelled rapidly into the homophase forming a trumpet vortex. The retracting film is sometimes unstable and breaks up into small satellite drops of the order of 20–30 μm diameter. An example is shown in Fig. 5.14a. These droplets are entrained in the vortex into the homophase

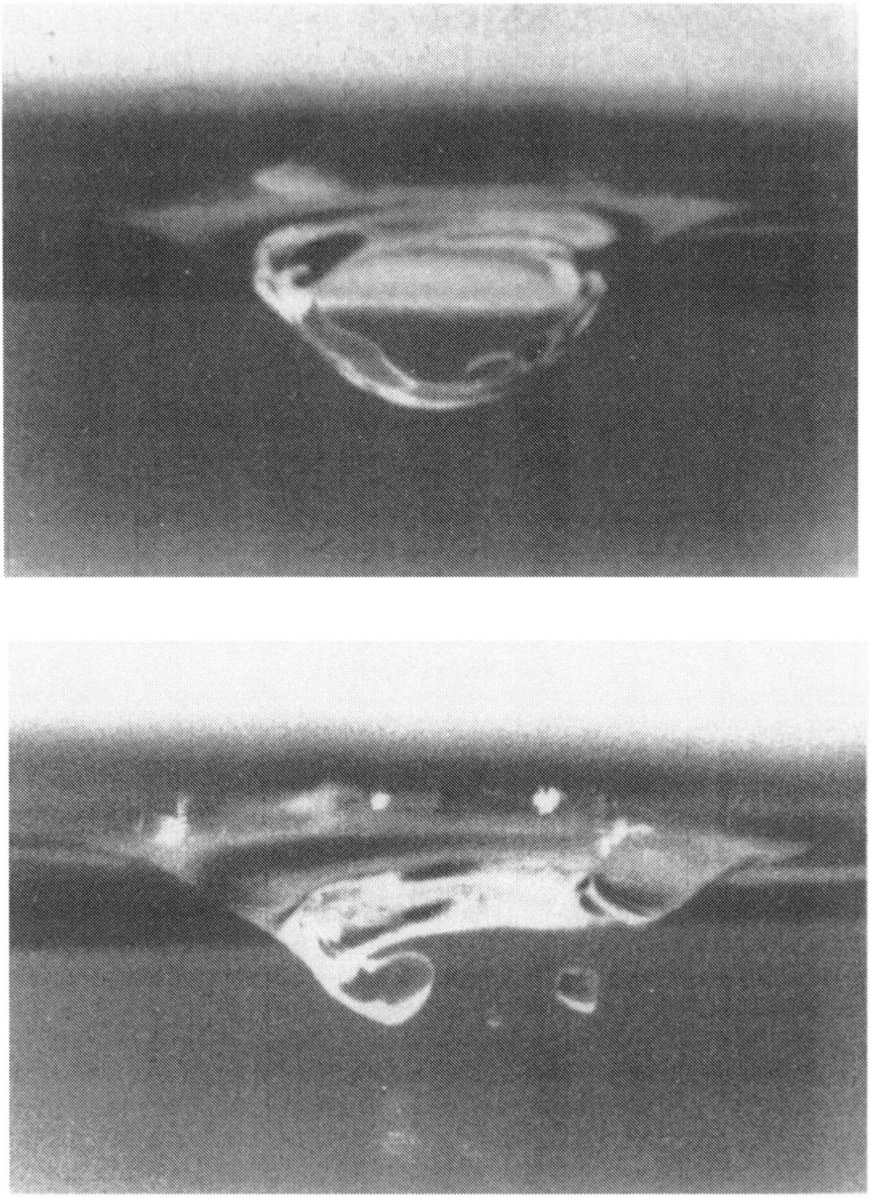

(a)

Fig. 5.14. Photographs showing the formation of small secondary droplets of the continuous phase formed by asymptotic rupture of the film during coalescence: (a) rupture of the film; (b) collection of the droplets at a concave interface.

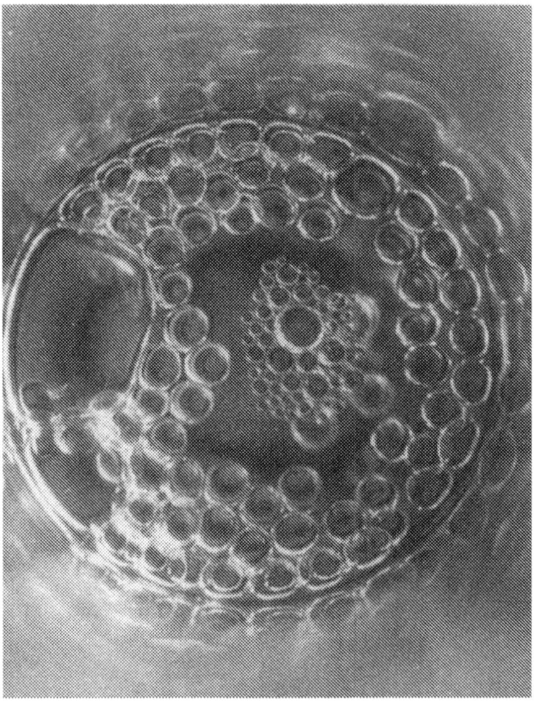

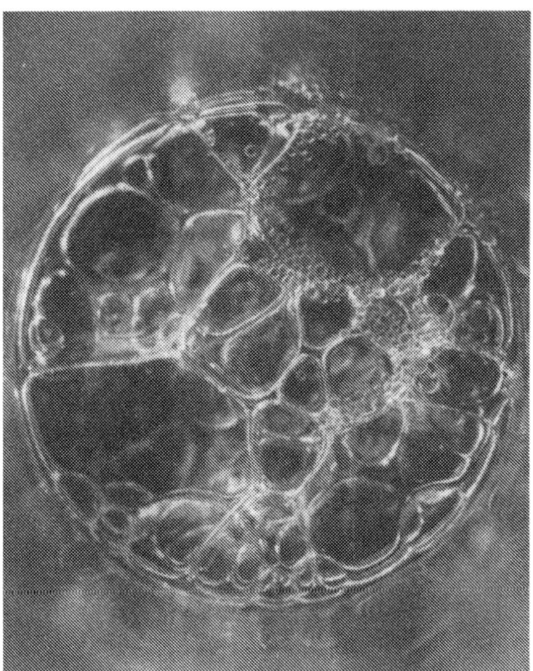

(b)

Fig. 5.14. *cont'd.*

and may be propelled up to 10 cm into the phase. This process has been observed in continuous separators and gives rise to a haze in the homophase and entrainment, Davies *et al.* (1970). It is particularly severe in systems with low interfacial tension and density difference. Under these conditions the film is less stable and settling of the droplets back to the interface is much slower. An interesting demonstration of this can be produced in a small diameter spray column set up to have a convex interface, relative to the arriving droplets, between the continuous and homophase. When this is operated at very low flows so that a small raft of droplets are present at the interface at any time, the curvature of the interface results in these droplets rolling across the interface to the wall of the column. After a short time droplets of continuous phase appear at the centre of the column in the homophase. These have been formed by the process of film rupture. A photograph taken below such an interface is shown in Fig. 5.14b.

5.4.4 Coalescence and dispersions

5.4.4.1 *Drop–drop coalescence*

Apart from the scientific interest, the work on drop coalescence should be directed towards improving our ability to design commercial separation equipment. Some thirty years ago it was thought that if a body of data on single drops was available then it should be possible to predict the separation of dispersions. This is now recognized not to be the case but studies between single drops on one hand and deep dispersion beds on the other may be fruitful in advancing knowledge. In a move towards this, studies have been made on rafts of drops at an interface and shallow dispersion beds comprising two or three droplet layers, Smith (1970). If drops settle towards and rest upon a concave interface then the component of gravitational force acting to press the adjacent surfaces together is often sufficient to cause deformation. Examples are shown in Fig. 5.15. The drainage of the intervening film can be described by one of the simplified models referred to earlier. The drainage rate is dependent on the force acting on the film which is the resultant of the force driving the interfaces towards one another and the capillary forces. The effect of the latter through changes in interfacial tension can be seen in Fig. 5.15. The interfacial tension between the systems shown increases from Fig. 5.15a showing large deformations to Fig. 5.15d where the droplets show little or no deformation and have attained almost the maximum packing efficiency for monodispersed spheres on a plane. These photographs were taken from beneath the interface of the homophase in a cell. A continuous flow of droplets was maintained in the cell, the flow rate was such that the rate of coalescence was of the order of the drop arrival rate so that only a small group of drops remained at the

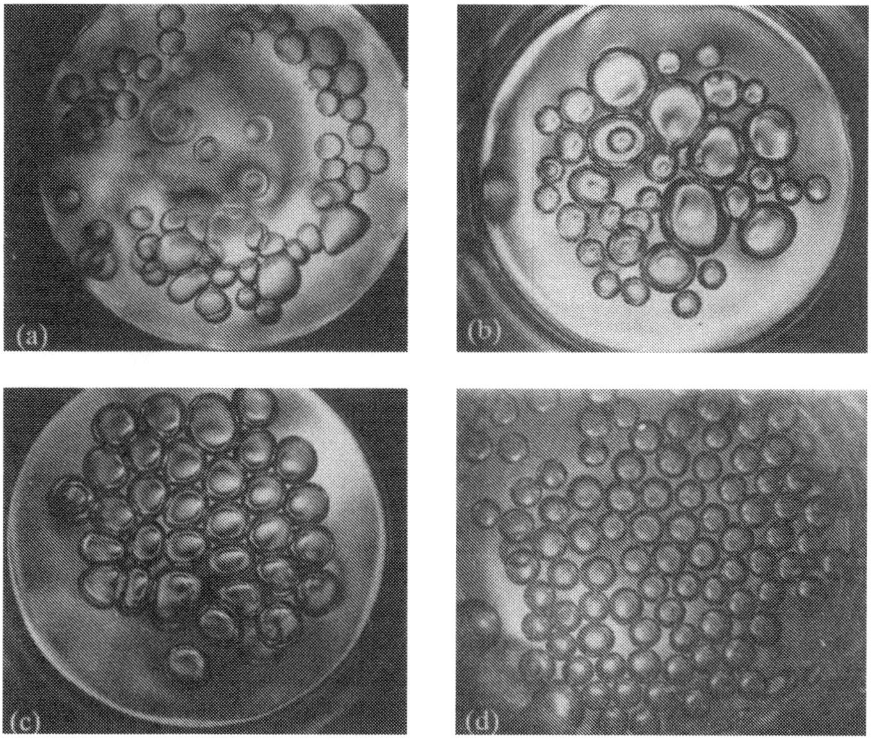

Fig. 5.15. Photograph showing droplets coalescing at a concave interface. The gravitational force causes deformation of adjacent droplets along the lines of contact. (a) water droplets in amyl alcohol; (b) water droplets in cyclohexanane; (c) water droplets in methyl carbonate; (d) water droplets in toluene.

interface. An immediate conclusion from the photographs is that the extent of drop–drop coalescence varies between these systems. Attempts have been made to quantify the times for binary coalescence, $\bar{\tau}_{ij}$, and drop-interface coalescence $\bar{\tau}_i$ at a boundary with the homophase. The binary coalescence time will be dependent on the initial size of the drops i and j. This has not been a very successful exercise, the difficulties of predicting $\bar{\tau}$ for single drops have still not been overcome and the added complication of introducing neighbouring drops has compounded the problem. Nevertheless these studies lead onto detailed investigations into the structure of dispersion bands which are normally present in commercial equipment.

5.4.4.2 *Drop–interface coalescence*

In the experiments in a continuous flow separator when the flow of dispersed phase is low, results similar to Fig. 5.15 are obtained. A raft of drops exists

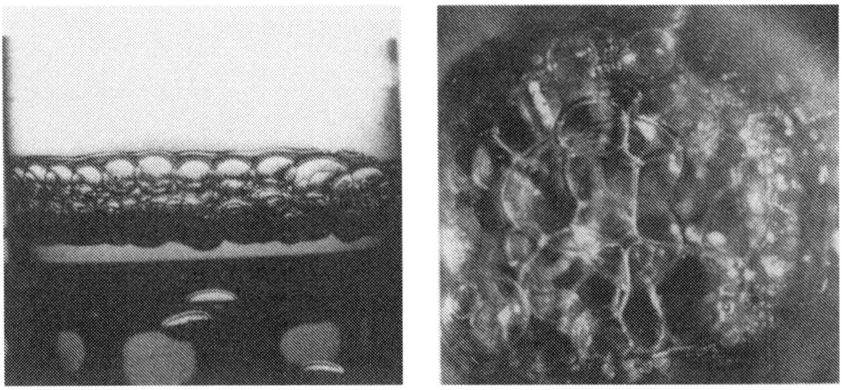

Fig. 5.16. Droplet dispersion band at an interface.

at the interface, the actual number present at any one time will not be constant. There will be a variation since the coalescence time for a given sized drop is not constant and binary coalescence will produce a dispersion in diameters. The results which are generally observed are that the disturbances generated at the interface as drops coalesce tend to descrease—stabilize drops—rather than promote coalescence. As the drop arrival rate increases, a band of drops form at the interface. The gravitational force transmitted through the contacting surfaces in this band causes the droplets and intervening films to deform. Examples are seen in Fig. 5.16. The study of the behaviour of dispersion bands with constant flow of dispersed phase into the band is important, and has proved a fruitful method to develop design procedures for commercial separators. From the engineering standpoint the important relationship is the variation in the depth of the dispersion with the flow rate of dispersed phase. At this flow rate increases the depth of the band increases. It is however dependent not only on the flow rate but also on the drop size of the dispersion entering the band, $f(\phi_0)$:

$$H = H[V, f(\phi_0)] \tag{5.70}$$

As drops entering the band get smaller the depth increases. Some typical results are shown in Fig. 5.17. In these experiments the conditions were carefully controlled so as to produce as near as possible a monodispersion of the droplets at the inlet to the bed. Several features can be seen from these results. The variation of H with V is non-linear and a maximum flow rate is achieved at which there is insufficient height in the vessel of settling to occur, flooding is taking place at this flow rate. The height is seen to vary with inlet drop size as described above. Within this band, binary coalescence takes place which produces a distribution in drop sizes. The flow of the dispersed phase within the band is approximately piston or plug flow (Smith

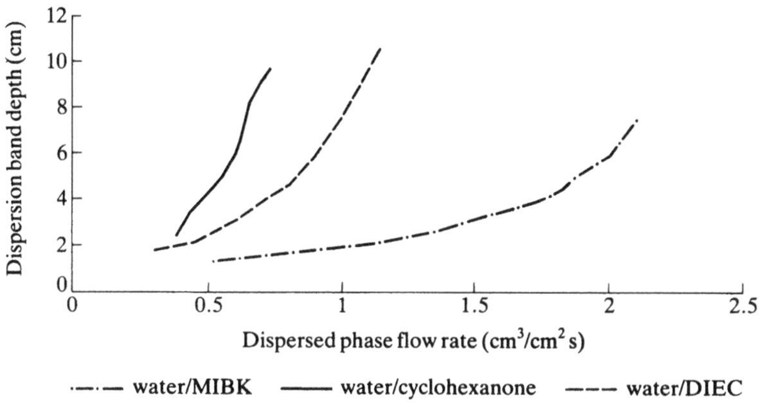

Fig. 5.17. Experimental results for the depth of the dispersion band at different inlet conditions.

and Davies 1970; Allak and Jeffreys, 1974). For the conditions in Fig. 5.17 binary coalescence produces a distribution of drop sizes increasing from the inlet to the disengaging interface. For steady state to be achieved at any flow (less than the flooding condition) then the volume flux of dispersed phase entering the bed must be equal to the volume rate of coalescence per unit area at the disengaging interface. This means that as the volume flux increases the coalescence at the disengaging interface must increase. To understand this, the structure of a dispersion band must be examined. In shallow beds and under certain conditions in small settlers where wall effects are important the hold-up of continuous phase liquid in the dispersion is relatively high ~20 per cent and the deformation of the drops is relatively low. Under these conditions the dispersion resembles the structure of a froth, Smith and Davies (1970). A schematic is shown in Fig. 5.18a. The droplets increase in size within the band. Within a dispersion band the continuous phase must drain towards the inlet. This flow is cumulative and increases towards the inlet. Thus the hold-up of continuous phase in a band is a function of depth and is maximum at the inlet. Very close to the inlet the back flow of the continuous phase can be so great that the droplets can move relative to one another and were described by Allak and Jeffreys to be fluidized. As the depth increases the drops settle together and are then constrained and cannot move freely. Thus back flow helps to maintain the froth structure. In deep bands as drainage continues and the drops move towards the disengaging interface, the hold-up of dispersed phase increases towards 90 per cent where the dispersions take on the structure of a foam. The 'droplets' appear as dodecahedra separated by thin films connected at plateau borders. A photograph of this is shown in Fig. 5.19. The structure of a deep dispersion

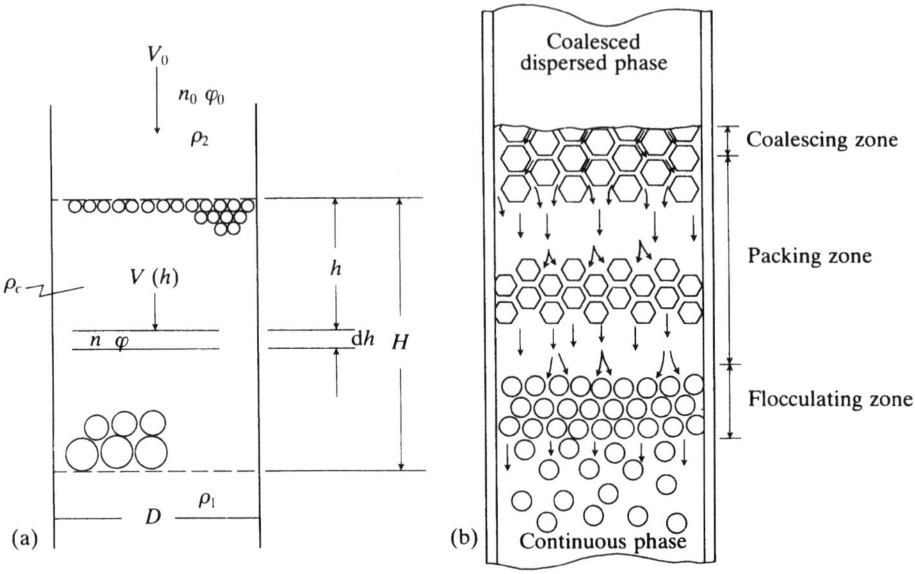

Fig. 5.18. Droplet dispersion bands at an interface; (a) shallow bands with a froth structure; (b) deep bands showing a close packed foam structure at the phase boundary.

band showing the transition from a froth to a foam structure is shown diagrammatically in Fig. 5.18b.

Three regions can be defined, one close to the inlet where the drops can move relative to neighbours, here the hold-up of the continuous phase is high ~20–40 per cent. This is followed by a second region where the drops are constrained but still resemble a froth, the hold-up of continuous phase is lower than in region 1. The final region close to the disengaging interface is a foam in which the dispersed phase has taken the form of dodecahedra and the hold-up of continuous phase is very low ~5–10 per cent.

It has been shown that when foams are formed the depth of the dispersion band is much greater than for a froth. In large equipment it is not generally possible to sustain a froth throughout the dispersion.

Returning to the conditions at steady flow the volume rate of coalescence can increase if either:

1. The coalescence time of the disengaging interface is constant, the volume flux will then increase as a result of binary coalescence within the band giving an increase in drop size. As the depth of the band increases the residence time in the band will increase allowing more binary coalescence to take place.

2. The coalescence time of droplets at the disengaging interface is effected

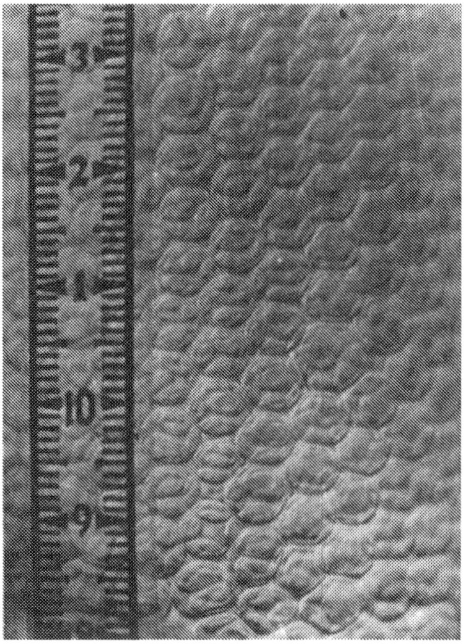

Fig. 5.19. Photograph of a close packed foam structure of a dispersion band.

by the dispersion height. In this region where the packing density of the dispersed phase is high, any relative movement is constrained so that the area of the dividing films cannot extend. The gravitational force transmitted through the dispersion by drop–drop contact will cause the film drainage to increase. Thus coalescence will be influenced by the dispersion height. This has been demonstrated by Hartland and Vohra (1980).

The behaviour of a dispersion band can be modelled by considering binary coalescence and disengagement at the interface, Smith and Davies (1972). Consider the conditions shown in Fig. 5.18a. If the number of drops entering the band per unit time is n_0 and mean diameter ϕ_0 binary coalescence will reduce the number flux at any position h and increase the diameter. The hold-up of dispersed phase η will change with position. If the flow of dispersed phase in the bed is approximately plug flow then $\phi = \phi(h)$. The mean drop diameter at any height will be defined as

$$\phi_h = \left[\frac{\Sigma n_i \phi_i^3}{\Sigma n_i}\right]_h^{1/3} \tag{5.71}$$

At any position h, the volume flux of the dispersed phase, V, is:

$$V = \frac{\pi}{6} n \phi^3 \tag{5.72}$$

where n is the number flux of drops with a mean diameter ϕ. Within any differential element dh

$$\frac{\mathrm{d}V}{\mathrm{d}h} = \frac{\pi}{6}\left[\phi^3 \frac{\mathrm{d}n}{\mathrm{d}h} + 3n\phi^2 \frac{\mathrm{d}\phi}{\mathrm{d}h}\right] \tag{5.73}$$

If wall effects giving rise to coalescence can be ignored (as in large vessels) then d$V = 0$ and

$$\frac{\mathrm{d}\phi}{\mathrm{d}h} = -\frac{\phi}{3n}\frac{\mathrm{d}n}{\mathrm{d}h} \tag{5.74}$$

A number balance for drops over this element can be expressed in terms of binary coalescence so that the total number is reduced by one at each event

$$-\frac{\mathrm{d}n}{\mathrm{d}h} = \xi \frac{A}{2}\frac{1}{\tau(\phi)} \tag{5.75}$$

Here ξ is the number of drops in the element and A is the coordination number. The term $\tau(\phi)$ is the mean coalescence time between drops of diameter ϕ. The term ξ can be expressed in terms of volumetric packing efficiency η within the element

$$\xi = \frac{\eta}{\pi\phi^3/6} \tag{5.76}$$

If the drops are assumed to be in the most stable nested arrangement each drop will on average be in contact with 12 neighbours, thus $A = 6$. Then eqn (5.75) becomes

$$-\frac{\mathrm{d}n}{\mathrm{d}h} = \frac{36}{\pi}\frac{\eta}{\phi^3}\frac{1}{\tau(\phi)} \tag{5.77}$$

At the disengagement boundary, the volume flux V_0 must be balanced by the outflow due to drop-interface coalescence. Then using a similar nomenclature

$$V_0 = \frac{\eta'}{\pi\phi_H^2/4}\cdot\frac{\pi\phi_H^3}{6}\cdot\frac{1}{\tau'(\phi)_H} \tag{5.78}$$

where η' is the area packing efficiency at this boundary and $\tau'(\phi_H)$ is the drop-interface coalescence time for drops of mean diameter ϕ_H. If eqns (5.77) and (5.78) can be solved then an expression for $H = f(V)$ can be obtained. The problem is to obtain realistic data on τ and τ'; from the experimental data discussed earlier this is not a straight-forward task. The simple case described above, that of assuming that the coalescence times τ and τ' are independent of diameter, can be obtained. Then introducing the conservation of volume of dispersed phase in the bed

$$V_0 = \frac{\pi}{6} n_0 \phi_0^3 = \frac{\pi}{6} n \phi^3$$

Combining (5.74 and 5.77) and the above gives

$$\frac{d\phi}{dh} = \frac{2\eta}{V_0}\frac{\phi}{\tau} \tag{5.79}$$

and integrating between the limits $h = 0$, $\phi = \phi_0$; $h = h$, $\phi = \phi$

$$\phi = \phi_0 e^{(2\eta h/V_0 \tau)} \tag{5.80}$$

or in terms of the number of drops

$$n = n_0 e^{-6\eta h/V_0 \tau} \tag{5.80b}$$

If the boundary condition, eqn (5.78) is introduced assuming $\tau' \neq f(\phi_H)$ then

$$H = \frac{1}{2}\frac{\tau}{\eta} V_0 \text{In}\left[\frac{3}{2}\frac{\tau'}{\eta'}\frac{V_0}{\phi_0}\right] \tag{5.81}$$

A similar result was obtained by Hartland and Vohra (1978).
Although this equation quantifies condition (a) above, experimental data on

(i) dispersion bands, Ryon *et al.* (1960), Gondo and Kusanoki (1969), Smith and Davies (1970), Mizrahi and Barnea (1973, 1975); and on

(ii) drop coalescence

have shown that $H = f(V)^x$. This cannot be obtained from the model if τ and τ' are independent of diameter, i.e. $\tau \neq f(\phi)$ and $\tau' \neq f(\phi)$. The experimental work on single drops and binary coalescence suggest that

$$\tau = f(\phi)^m \quad \text{and} \quad \tau' = f(\phi)^n \tag{5.82}$$

This can be introduced into the model but allowance must be made for the conditions the drops experience within the bed. Drainage is taking place continually in zones 2 and 3 of the bed. This can be allowed for by introducing a correction factor related to the residence time in the bed. In addition the gravitational force transmitted to particular drops in the bed which govern drainage will be lower than that experienced for single drops or doublets at an interface. To allow for both factors the drop–interface coalescence time τ' is modified

$$\tau'(\phi_H, H) = \tau'(\phi_H, 0) - \frac{1}{K}\Theta_H \tag{5.83}$$

where Θ_H is the residence time in the bed and K is a factor to allow for reduced drainage

$$K = 1 - \eta$$

For plug flow:

$$\Theta_H = \frac{\eta h}{V_0} \tag{5.84}$$

Then introducing eqns (5.83 and 5.84) into (5.82)

$$\tau'(\phi, H) = \tau_1'[\phi_H]^n - \frac{1}{K}\left[\frac{\eta h}{V_0}\right] \tag{5.85a}$$

and

$$\tau(\phi, h) = \tau_1[\phi(h)]^m - \frac{1}{K}\left[\frac{\eta h}{V_0}\right] \tag{5.85b}$$

the subscript on τ' and τ refers to a reference drop at an interphase. Then in eqns (5.77 and 5.78)

$$H = \frac{1}{\eta/K}\left[t_1' V_0 \phi_H^n - \frac{2}{3}\eta' \phi_H\right] \tag{5.86}$$

and

$$h = \frac{\tau_1 V_0}{\eta}\frac{1}{2(n + 1/2K)}\left[\phi^m - \frac{\phi_0^{m + 1/2K}}{\phi_0^{1/2K}}\right] \tag{5.87}$$

The solutions depend on the values of m and n used. There is still a large variation in values reported in the literature which relates to the difficulties in carrying out the experiments and maintaining constant conditions, particularly levels of impurities which can have catastrophic effects on results. Smith and Davies (1972) assumed that $m = n$ and obtained a solution to the equation

$$\frac{H}{\phi_0} = 0.24\left[\frac{\tau_0 V_0}{\phi}\right]^{1.10} \tag{5.88}$$

τ_0 being the binary coalescence time for drops of diameter ϕ_0. The experimental data in deep dispersion bands show $H \propto V^a$. Ryon *et al.* (1959) studied separation in a deep vertical settler and obtained a relation:

$$H = K_1 V_c^a \tag{5.89}$$

where V_c is the flux of continuous phase liquid. This can be related to the dispersed phase flux V through the total flow flux, V_F, and the phase ratio Φ. Gondo and Kusanoki (1969) suggested a similar equation:

$$H = K_2 \epsilon^b V^a \tag{5.90}$$

where ϵ is the fractional hold-up of dispersed phase. The work was restricted to water in kerosene dispersions for which $a = 3.1$, $b = 4.9$. The dispersions were produced in a mechanical mixer and additionally they found that H was dependent on the drop size in the entering dispersion, which they correlated in terms of the mixer speed, N and showed that $H \propto N^{2.8}$. The dependency of H on ϕ_0 is expected, but surprisingly this was not found by Ryon *et al.* (1959) who used similar experimental techniques to Gondo and Kusanoki. If surfactants were present in the system this might account for the insensitivity of H to N over the range of experiment. Similar results to eqn (5.90) were reported by Gondo and Modhi (1977) who, using an organic phase of alamine 336, n decanol and Shell Sol K and and aqueous phase of sodium sulphate at pH 1.3, obtained values of $a = b = 3.1$ and $K_2 = 1.45$.

The application of theoretical models based on binary and interface coalescence, which require data to be used from single and binary drop studies to equipment design, is unlikely to be successful. This is due to the uncertainty in the basic data and the fact that tests must be carried out on truly representative samples since trace contaminants can have profound effects on such quantitative data; see for example Smith and Davies (1970). More success may be achieved by studying the decay of dispersions in test settlers. This is introduced here since this too can be modelled. If, in the experiments described on continuous flow settling the feed is suddenly swithed off, then the dispersion in the settler will collapse. Again, the same basic mechanisms for interface and binary coalescence are responsible for this. Under such conditions it would be important to be able to predict $H = f(t)$, and from this the total collapse time for a given initial condition $H = H_0$, $t = 0$. In the continuous flow system when steady state is achieved, the position of both boundaries of the respective bands are stationary, but when the flow is stopped these move. If $\rho_d < \rho_c$ then the lower boundary—the disengagement interface—moves up rapidly. The upper boundary moves down as the continuous phase drains out of the band and the drops settle together within the band. A typical curve showing the position of these boundaries relative to the final undisturbed interface between the two liquids is shown in Fig. 5.18. This is for a system where $\rho_d > \rho_c$, the lower curve represents the coalescing interface, the upper curve the settling interface. When the flow is stopped, zone 1 in which the drops are in free relative motion rapidly decays as the continuous phase liquid drains from the bed. Droplets within the band then settle towards one another as further drainage takes place. The hold-up of continuous phase in the band $(1 - \epsilon)$ is reducing as drainage from the band takes place. This is

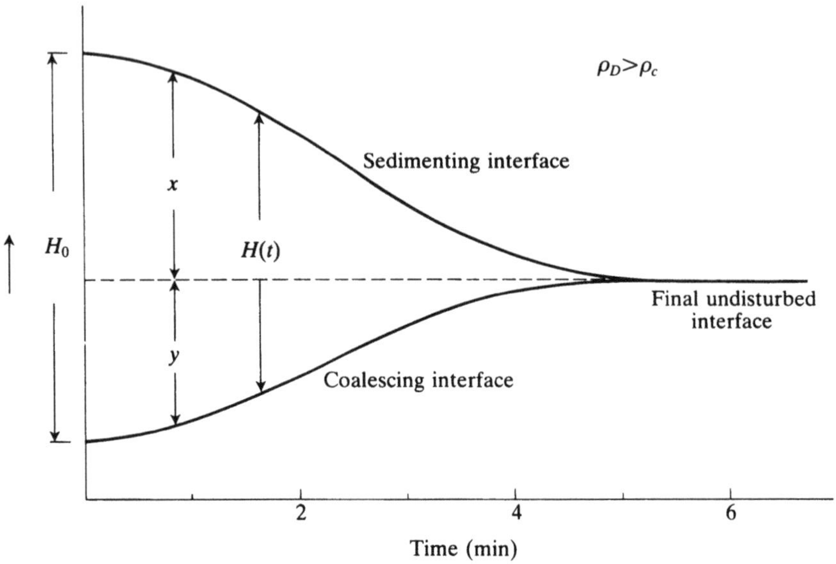

Fig. 5.20. The collapse of a dispersion band with time. Experimental results for a kerosine dispersion in water (Smith 1970). ($\bar{d}_0$ at $t < 0 = 0.35$ cm.)

further affected by binary coalescence, which causesincreased settling and helps to expel continuous phase. At the coalescing interface the packing efficiency increases. In the system the initial collapse is due mainly to drainage and settling within the dispersion phase, and the final stages are controlled by coalescence at the interface. At any time $t > 0$, the volumes of the continuous and dispersed phase discharged from the band are proportional to $x_0 - x(t)$ and $y_0 - y(t)$. The average hold-up of the dispersed phase $\bar{\epsilon}(t)$ is then

$$\bar{\epsilon}(t) = \frac{y}{(x+y)_t} = \frac{y}{h(t)} \tag{5.91}$$

and the flux of dispersed phase from the band is:

$$V = -\frac{\mathrm{d}y}{\mathrm{d}x} = \frac{-d[\bar{\epsilon}(t)h(t)]}{\mathrm{d}t} \tag{5.92a}$$

The collapse of a dispersion band is relatively easy to observe and measure. The importance of such measurements in relation to the studies of continuous systems was recognized by Barnea and Mizrahi (1975), Golob and Modic (1977), and Vieler *et al.* (1977). In the first paper, Barnea showed that a correction could be obtained between the nominal settler capacity V corresponding to a given band height H and the batch settling time, t_B, measured in a standard test. The term t_B is sometimes called the primary break time. Thus:

$$V = Kt_B^n \tag{5.92b}$$

for the system amyl alcohol–HCl–water $n = -0.87$. Further work, based on a wide range of physical systems and phase ratios ϕ of 0.25–0.66 that will cover many practical systems, showed that $n = -0.905$ for O/A dispersions and -0.87 for A/O dispersions. The separation flux of dispersed phase V was measured in units $m^3m^{-2}h^{-1}$, corresponding to a band height of 0.5 m and a standard apparatus measured for batch time measurements using $h_0 = 0.3$ m. Under these conditions Barnea and Mizrahi reported values of K of 1064 and 888 for O/A and A/O dispersions, respectively. The values of these constants and the applications to other systems must be carefully judged, nevertheless this was important work since it provided data that will include the effects of any impurities and so on in the process fluids and is a simple technique. Golob and Modic (1977) confirmed the general relation and attempted to extend the technique by correlating the coalescence rate in terms of the physical properties of the mutually saturated phases. Their experimental work revealed that the properties shown in eqn (5.93) were important

$$V_d = f(\mu_c, \mu_d, \rho_c, \Delta\rho, \sigma, \sigma_r) \tag{5.93}$$

where σ_r is the surface tension of water at 20°C, 72.3 $Nm^{-1} \times 10^{-3}$, which was used as a reference. The correlation proposed was

$$\frac{V\mu_c}{\sigma} = 0.098\left(\frac{\mu_c}{\mu_d}\right)^{1.05}\left(\frac{\sigma_r}{\sigma}\right)^{0.85}\left(\frac{\Delta\rho}{\rho}\right)^{0.41} \tag{5.94}$$

The correlation is interesting since it implies that in their work μ_d is more significant than μ_c, a result which is found in single drop coalescence, providing the Gibbs adsorption of all components in the system is low (see Section 5.43).

A similar model to that discussed above for continuous flow can be developed for a collapsing dispersion band. In analysing binary coalescence and collapse of a dispersion band the initial condition should include the drop size distribution in the band formed during continuous flow, solving eqn (5.74). If the flow is stopped at time $t = 0$ then

$$\phi = \phi(h, t) \qquad 0 \le h \le H, t \ge 0$$

and

$$n = n(h, t)$$

with $\phi = \phi(h, 0)$ and $n = n(h, 0)$ given from the solution of the continuous flow case. If the drop distribution within the band is ignored and a mean value of ϕ is used, $\phi = \phi(t)$. Then a differential balance over a time interval dt gives in eqn (5.73)

$$-\frac{1}{n}\frac{\mathrm{d}n}{\mathrm{d}t} = \frac{3}{\phi}\frac{\mathrm{d}\phi}{\mathrm{d}t} \tag{5.95}$$

The change of dispersion volume *within* the band must be zero. The number change due to binary coalescence will then be

$$-\frac{\mathrm{d}n}{\mathrm{d}t} = \frac{n}{2\tau} \tag{5.96}$$

Dispersed phase will leave the band by coalescence at the disengagement interface and

$$V = -\frac{\mathrm{d}(\bar{\epsilon}h)}{\mathrm{d}t} = \frac{2\eta'\phi_H}{3\tau'} \tag{5.97}$$

The solution of eqn (5.95) and (5.96) for constant binary coalescence time τ produces, using the initial conditions $\phi = \phi_0$, $t = 0$

$$\phi(t) = \phi_0 \mathrm{e}^{t/6\tau} \tag{5.98}$$

This simplification is not generally valid since

$$\tau = \tau(\phi)$$

Substituting eqn (5.82) gives, after integration

$$\left(\frac{\phi}{\phi^*}\right)^m = 1 + \frac{n}{6\tau}(t - t^*) \tag{5.99}$$

where $\phi = \phi^*$ at a reference time $t = t^*$. If eqn (5.82) is substituted into (5.96) then

$$V = \frac{2}{3}\eta' \frac{\phi_H^{1-n}\phi^{*n}}{\tau'^*} \tag{5.100}$$

and if the diameter distribution in the dispersion is ignored so that $\phi = \phi_H$, eqn (5.99) can be substituted into (5.100). Then:

$$V = \frac{2}{3}\frac{\eta'\phi^*}{\tau'^*}\left\{1 + \frac{n}{6\tau}(t - t^*)\right\}^{(1-n)/m} \tag{5.101}$$

t is now the average time taken for a drop in the band to reach the coalescence interface and coalesce. The bracketed term in eqn (5.101) can be rewritten as

$$\left(\frac{n}{6\tau}\right)^{(1-n)/m}[t - t']^{(1-n)/m}$$

where

$$t' = \frac{6\tau}{n} - t^*$$

The model equations cannot be solved unless some relationship between H and t is introduced. At $t = 0$ the average lifetime of a drop in the dispersion band is $\bar{t} = H\bar{\epsilon}/V$. If $\tau \gg t - t^*$ then substituting for V from eqn (5.101) into the mean residence time gives

$$H = \frac{\bar{t}V}{\bar{\epsilon}} = \frac{V^{1+m/(1-n)}}{\epsilon}\left(\frac{3\tau'^*}{2\eta'\phi^*}\right)^{m/(1-n)} \cdot \frac{6\tau}{n} \tag{5.102}$$

or

$$H \alpha V^w K \tag{5.103}$$

where $$w = \left[1 + \frac{m}{1-n}\right]$$

This is similar in form to the equation proposed by Golob and Modic (1977) and to the equations used to correlate experimental results for continuous flow systems. The difficulty in predicting the separation of dispersions from binary and single drop coalescence data has already been emphasized.

A problem with differential models of the type described above for both steady state and collapsing dispersions is that the dispersed phase is still defined as droplets or a froth since, although a packing efficiency in both the band and at the interface is introduced, η is considered not to vary with height. The different structures of the dispersion are not properly accounted for. Barnea and Mizrahi (1978) recognized this in analysing deep layer gravity settlers. An insight into the problem was made by Allak and Jeffreys (1974). They employed a neat experimental method to study the structure of the dispersion. A liquid–liquid system was chosen, the phases of which when equilibrated had the same refractive index, that is they were iso-optic. Many systems have these properties, some are listed by Francis (1963). A spray column was used for the work. When the dispersion was produced and viewed with background illumination from a sodium lamp the interfaces were invisible. Dyes were introduced into first the continuous phase and then, in separate experiments, into the dispersed phase. By photographing the column, the structure of the dispersion band was determined. This work distinguished the three zones described earlier. The rapid deceleration of droplets as they entered the band was observed and measured and the froth–foam transition was determined. The coalescence in deep dispersion bands was shown to be controlled by the drainage within the foam. To describe this, Allak and Jeffreys employed equations to describe the

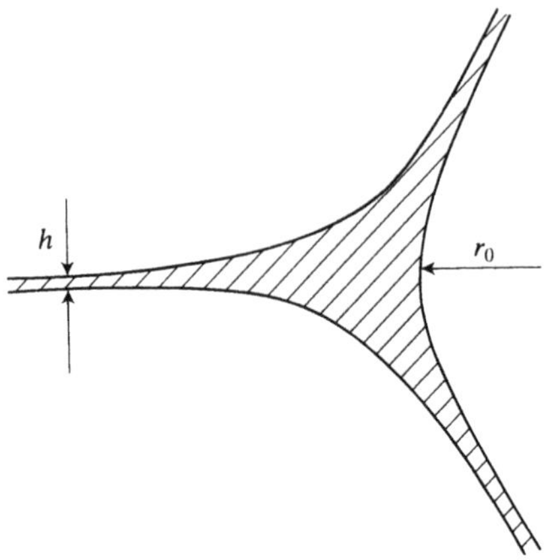

Fig. 5.21. Idealized section of a foam in which the dispersed phase is assumed to be a regular dodecahedron.

drainage within gas–liquid foams proposed by Lemlich and Leonard (1965). A diagram of a section of foam is shown in Fig (5.21). In this the dispersed phase is the less dense phase so that the continuous phase in the foam drains downwards towards the inlet.

The foam was assumed to be formed from deformed droplets arranged as pentagonal dodecahedra. Drainage of the liquid between adjacent surfaces can then be expressed using the equations in Section 5.43, for advancing parallel interfaces. Thus:

$$-\frac{\mathrm{d}h}{\mathrm{d}t} = \frac{2Fh^3}{3\pi\mu_c R^4} \tag{5.104}$$

The liquid drains into the plateau borders and for the regular structure assumed, each drop having 12 faces with 5 edges per face, there are 10 plateau borders per drop. The number of plateau borders per unit volume is

$$N = \frac{60\,\eta}{\pi d^3} \tag{5.105}$$

The flow rate through a plateau border was derived by Leonard and Lemlich (1965) as

$$q = \frac{0.00677}{12.86}\frac{R_0^4}{\mu_c}\frac{\Delta p}{l} \tag{5.106}$$

and for a plateau border inclined at an angle Θ

$$\frac{\Delta p}{\Delta l} = \eta\rho g \sin\Theta$$

so that total flow per unit area of foam draining through the plateau borders

$$Q_d = \int_0^{\pi/2} qN \cos \Theta \, d\Theta \tag{5.107}$$

from which

$$Q_d = \frac{0.00412}{3} \frac{R_0^4 \rho g \eta^2}{\mu d^2} \tag{5.108}$$

At steady state the drainage of continuous phase from the foam through the plateau borders must be equal to the flow into the dispersion band. The rate of arrival of drops is $N' = 6V/\pi d^3$ and the upflow of continuous phase Q_u is:

$$Q_u = \frac{3.32\, Vh}{d} \tag{5.109}$$

or alternatively

$$Q_u = \left(\frac{1-\eta}{\eta}\right) V \tag{5.110}$$

Thus combining eqns (5.109) and (5.110) an expression for h the film thickness can be obtained which can be substituted into eqn (5.104). When this is carried out the force F can be estimated from eqn (5.104)

$$R_0 = \left[\frac{3 \times 3.32\, Vh\mu d}{0.00412\, \rho g \eta^2}\right]^{1/4} \tag{5.111}$$

and introducing Laplace's equation for the pressure in the film

$$\Delta p = \sigma\left[\frac{1}{R_1} + \frac{1}{R_2}\right] \tag{5.112}$$

and $F = a\Delta p$ where $a = 0.29\, d^2$.

$$F = 0.29\, d^2 \sigma \left[\frac{0.00412\, \rho g \eta^2}{9.96\, Vh\mu d}\right]^{1/4} \tag{5.113}$$

from which an expression for the rate of film thinning can be obtained as

$$\frac{dh}{dt} = -\,4.969\, h^{11/4} \left[\frac{\rho g \eta^2 \sigma^4}{V \mu^5 d^9}\right]^{1/4} \tag{5.114}$$

Coalescence in the film was assumed to take place when a critical film thickness h_c was achieved from which the drainage time can be evaluated:

$$t = 0.115 \left[\frac{V\mu^5 d^9}{\rho g \eta^2 h_c^7 \sigma^4}\right]^{1/4} \tag{5.115}$$

These equations formed the basis of a computer model to predict the total bed height as a function of the flowrate. A critical film thickness at rupture was taken as 2 μm and the time taken for drainage from $h_0 = [(1-\eta)/3.32\eta]d$ down to 2.0 μm was calculated from equation (5.115). The thickness of the increment of the coalescence band was then calculated from $H_1 = Vt$. On coalescence the drop size was adjusted since $d_2 = (2d^3)^{1/3}$. The term Q_u was calculated at the new mean drop size and the procedure repeated to calculate H_2. By repeating the procedure until $Q_u \to 0$ and then summing the increment heights the total band height H can be calculated

$$H = \sum_i H_i$$

The procedure in translating the film drainage time in the foam to an effective steady state band height is clearly somewhat idealized and simplified. Nevertheless the model was one of the first to make a real attempt at computing the drainage of continuous phase within the foam structure which controls the separation of most dispersions.

All the attempts to model dispersions have imposed constraints on the structure of the dispersion, either by way of describing the band in terms of a mean drop size at any critical position in the bed, or by assuming a constant drop size at any position in the bed so that the local structure can be defined; in a foam for example, by assuming a pentadodecahedron structure. In practice this is not the case. Even if the dispersion entering the separator were a monodispersion the random coalescence times in the band will mean that a polydispersion giving a somewhat irregular packing structure at any position will be generated. Although it may be justified to consider the mean drop size as a function of position in the bed from the inlet, the structure in the band will change as the standard deviation of the drop size distribution changes and as the forces transmitted to the dispersion increases in the bed. For models to accommodate these properties, a stochastic approach will be necessary and deterministic methods of the type reviewed will at best be first approximations. To date there have been few attempts to approach the problem from this standpoint. There is some understanding of methods for dividing 3 space to form polyhedra when the average coordination number of nodes is equal to that observed in liquid foams (Meijering 1953), so that stochastic models may be developed to simulate the structure of dispersion bands. The drainage process could then be superimposed on this to provide perhaps more accurate descriptions of the separation process. There is still some way to go in this if better quantitative predictions are to be made using structural models. A much simplified version in attempting to introduce stochastic models to simulate steady state dispersion band behaviour was proposed by Doulah and Davies (1974). This was based on queue theory and expressed the separation processes in terms of birth–death equations. The difficulty in this was again in obtaining data to insert into the solution for the respective service (coalescence) times which appear as coefficients in the equations.

From a pragmatic view, the most promising approach now available is based on batch settling and the interrelation of batch to continuous behaviour. This has the merit, if data is obtained on representative solutions taken from the process, of incorporating the effects of any impurities present under realistic separation conditions.

A most interesting approach to the subject was presented by Vieler *et al.* (1977). In this they related the volume of dispersion in the settler with age between t and $t + dt$, $\psi(t)dt$ to the phase disengagement rate $g(x, t)$ so that:

$$\frac{d\psi(t)}{dt} = -g(x, t)\,\psi(t) \tag{5.116}$$

In the case of batch settling all the dispersion is of the same age and if the settler is vertical with a constant cross section area

$$\frac{dh(t)}{dt} = -g(x, t)\,h(t) \tag{5.117}$$

or

$$\frac{d\ln(h(t))}{dt} = -g(x, t) \tag{5.118}$$

From which in principle it may be possible to obtain data for the separation rate from batch data. The importance of this work was the formal relationship derived between the batch case and continuous flow conditions. The separation rate under steady flow conditions was defined in an analogous way as $G(x, \tau)$ and the volume fraction with a residence time τ being

$$\psi(\tau) = \frac{\nu \exp \int_0^\tau G(x, \tau)\,dt}{\int_0^\infty \exp - \int_0^\tau G(x, \tau)\,d\tau} \tag{5.119}$$

where ν is the volume of the dispersion. Then the volume flowrate of the dispersion is

$$\begin{aligned} Q_d &= \int_0^\infty G(x, \tau)\,\psi(\tau)\,d\tau \\ &= \psi(0) - \psi(\infty) \\ &= \psi(0) \end{aligned} \tag{5.120}$$

since $\psi(\infty) = 0$ at steady state. Therefore the mean residence time in the dispersion, (ν/Q) is:

$$\frac{\nu}{Q} = \int_0^{\infty} \exp\left(-\int_0^{\tau} G(x, \tau)\, d\tau\right) d\tau \tag{5.121}$$

and

$$\frac{\nu}{A} = H$$

The analysis shows that any volume element with a small value of G will make a large contribution to the mean residence time. In relation to actual systems any slowly disengaging (coalescing) droplets will tend to build up in a settler and form a large volume fraction in the settler, cf. Davies *et al.* (1970).

Vieler *et al.* considered two cases to examine the relationship between batch and continuous separation. In the first they assumed that the properties were the same, i.e. $X = x$, $\tau = t$. Additionally the properties were not a function of time so that $x \neq x(t)$. Then

$$G(x, \tau) = g(x, t) \tag{5.122}$$

for which

$$\ln \frac{h}{h_0} = -\int_0^{\infty} G(x, \tau)\, d\tau \tag{5.123}$$

and

$$\frac{V}{A} = \int_0^{\infty} \frac{h(t)}{h_0}\, dt \tag{5.124}$$

where $h_0 = H$

The second case which can be solved is when the fractional settling rate is independent of time

$$G = G(x) \tag{5.125}$$

Then

$$\frac{V}{a} = \frac{1}{G(x)} \tag{5.126}$$

and for the batch settler

$$\frac{d \ln h}{dt} = -g(a) \tag{5.127}$$

from which the mean residence time in continuous settlers is obtained from a knowledge of the settling rate in the batch system under the same conditions $x = X$. This is the procedure which has been advocated by many workers in scaling up equipment. It is however a simplification, a more accurate relation can only be obtained by integration of eqn (5.119).

When comparing the operation of batch and continuous settlers there is one important fundamental difference. In a continuous settler operating at steady state the position of the disengaging or coalescence interface is essentially stationary whereas in the case of batch separation this interface is moving. For batch separation, the dispersion band depth decreases with elapsed time, whereas for continuous settlers the dispersion band depth increases rapidly, in some cases exponentially, with the loading rate. The variation of h with time in a batch system is shown in Fig. 5.20, from this the volume flux of continuous phase from the band is $V_c = -\mathrm{d}x/\mathrm{d}t$ and that from the dispersed phase is $V_d = -\mathrm{d}y/\mathrm{d}t$. The total volume flux from the band is then

$$-\frac{\mathrm{d}(x+y)}{\mathrm{d}t} = \frac{\mathrm{d}h}{\mathrm{d}t} \qquad (5.128)$$

The average residence time of an element in the band is given by $\bar{t} = h/(-\mathrm{d}h/\mathrm{d}t)$. This compares to the continuous flow problem where $\bar{\tau} = \mathrm{H}A/Q_{\mathrm{F}}$.

In a batch system, h decreases with time, initially the decay is determined by both sedimentation and coalescence. The curve may show an inflection point where the sedimentation is effectively terminated. The separation is then controlled by coalescence and $h \propto \mathrm{e}^{-\alpha t}$. This was recognized by Stönner and Wöhler (1975) who proposed the equation

$$-\frac{h}{\mathrm{d}h/\mathrm{d}t} = \frac{1}{K_1} + \frac{h}{K_2} \qquad (5.129)$$

to describe the unsteady collapse of a dispersion band. The term $1/K_1$ was used to describe the coalescence at the disengaging interface and was termed 'active' separation, K_1 having units of reciprocal time T^{-1}. The term h/K_2 was used to describe 'passive' separation accounting for sedimentation and drop–drop coalescence within the band. K_2 has units of LT^{-1}. A corresponding equation was suggested for continuous separation

$$\frac{H}{Q/A} = \frac{1}{K_1} + \frac{H}{K_2} \qquad (5.130)$$

Considering the conclusions from Vieler, steady state behaviour can only be successfully interpreted if conditions are such that the relative stages defined by K_1 and K_2 are similar for both processes, and if K_1 and K_2 can be quantified from batch experiments. According to eqns (5.129) and (5.130), $-\mathrm{h}/(\mathrm{d}h/\mathrm{d}t)$ and H/Q_A should be linear with h or H respectively. Data from Sönner and Vieler's work shown in Fig. 5.22 support this.

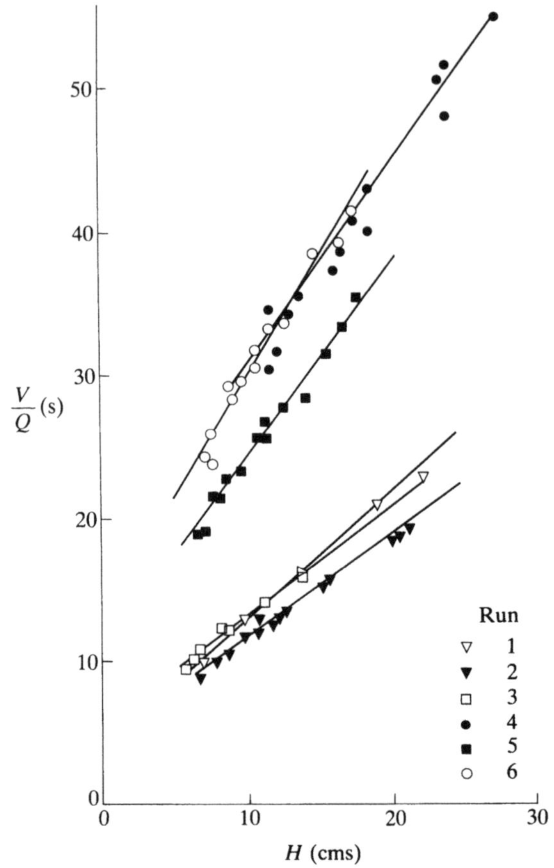

Fig. 5.22. The change in dispersion band height with flow rates, a comparison of experimental results with eqn (5.129) (Vieler 1977).

The approach has subsequently been extended by other researchers, notably Hartland and his co-workers. One of the major thrusts in this work has been to predict the heights of the sedimentation and dense packed layers in the dispersion band. In batch separations $x(t)$ and $y(t)$ are determined by the respective values of separation $-\mathrm{d}x/\mathrm{d}t$ and $-\mathrm{d}y/\mathrm{d}t$. Both of these go through a maximum, with time. Barnea and Mizrahi (1975) used the relation for the settling velocity of a drop to relate the sedimentation velocity in the dispersion band

$$V_s = K_s \phi^\alpha \tag{5.131}$$

and K_s is constant which will be related to the physical properties ρ_c, ρ_d, μ_c, μ_d and $\bar{\eta}_s$. Barnea found that the index $\alpha = 2$. This would correspond to

Stokesian flow then applying Hadamand–Rybczynski's solution for spherical drops $K_s = \Delta\rho g/18\mu_c\lambda$ and $\lambda = (2\mu_c + 3\mu_d)/3(\mu_c + \mu_d)$. Kumar *et al.* (1980) from experiments in a spray column found $\alpha = 0.5$ which would infer that the drag coefficient is independent of the Reynolds number. The prediction of the sedimentation velocity

$$-\frac{dx}{dt} = V_s = K_s\phi^\alpha$$

must take into account the fact that

(a) drop–drop coalescence is simultaneously taking place which will influence V_s through ϕ and;

(b) the hold-up of dispersed phase is of course changing. Hartland has considered this by introducing an equation for the coalescence time for binary coalescence, eqn (5.99) and assuming that the dispersed phase hold-up is constant. A graphical interpretation can be found in the work of Godfrey *et al.* (1977).

The total height $h(t)$ is the sum of the sedimentation and dense packed zones

$$h = h_s + h_p \tag{5.132}$$

Experimentally the problem is in distinguishing this demarkation in the band so that comparison of theory with experiment is difficult. Hartland and co-workers have developed methods to relate measurements to the undisturbed interfaces. Thus if the instantaneous average volume hold-up for the dispersed phase is introduced a volume balance yields:

$$\bar{\epsilon}_s h_s + \bar{\epsilon}_p h_p = \bar{\epsilon} h \tag{5.133}$$

At any limit the volumes of the dispersed and the continuous phase which have exited the bed are proportional to $(y_0 - y)$ and $(x_0 - x)$ respectively. The volumes of the dispersed and continuous phase in the dispersions at any time t are proportional to y and x and then the average hold-up over the band is

$$\bar{\epsilon} = \frac{y}{x + y}$$

so that in eqn (5.133)

$$\bar{\epsilon}_s h_s + \bar{\epsilon}_p h_p = y \tag{5.134}$$

and

$$x + y = h$$

Eliminating h_s from eqn (5.134)

$$h_p = \frac{y - \bar{\epsilon}_s h}{\bar{\epsilon}_p - \bar{\epsilon}_s} \tag{5.135}$$

The locus of the boundary between the two zones in the dispersion band $Z = y - h_p$ is:

$$Z = \frac{\bar{\epsilon}_s x - y(1 - \bar{\epsilon}_p)}{(\bar{\epsilon}_p - \bar{\epsilon}_s)} \tag{5.136}$$

By noting that the maximum value of h_p occurs when

$$\bar{\epsilon}_s = (-\mathrm{d}y/\mathrm{d}t)/(-\mathrm{d}h/\mathrm{d}t) \tag{5.137}$$

Hartland was able to show that

$$h_p = \frac{yh_0 - y_0 h}{(\bar{\epsilon}_p h_0 - y_0)} \tag{5.138}$$

and

$$Z = \frac{y_0 x - yh_0(1 - \epsilon_p)}{\bar{\epsilon}_p h_0 - y_0} \tag{5.139}$$

By combining these equations with eqns (5.129 and 5.130), data from measurements on batch dispersions can be used to predict continuous flow separation. Hartland has taken this further to incorporate expressions for binary coalescence and drop-interface coalescence and the equations for $-\mathrm{d}x/\mathrm{d}t$ and $-\mathrm{d}y/\mathrm{d}t$, and by using correlations of coalescence times has obtained equations relating $\bar{\epsilon}H$ to Q/A, an important step in determining the dispersion band depth in the design of settlers.

5.4.5 Application to the design of industrial settlers

[See also Pratt and Stevens, Ch. 8]

There are two basic types of gravity settlers, vertical flow and cross flow or horizontal settlers. Many modifications to these basic types have been proposed with the object of increasing the separation fluxes but these remain the two general types. Both may be used in single or multi-stage plant.

5.4.5.1 *Vertical settlers*

Vertical separators are straightforward in concept and can be operated with either phase continuity. The design methods are based on either residence time, sedimentation velocities or the rate of coalescence in dispersion bands. In the first method a residence time for the mixture in the vessel is defined. This can vary by 2 orders of magnitude, the lowest value is quoted as being

3 min. There is little or no scientific method in this design. It is based merely on previous experience assuming plug flow and takes no account of the detailed design of the vessel in relation to flow distribution and residence time distribution (Treybal 1963, Perry 1963). The second method has little scientific basis. This assumes that the rate determining step is the sedimentation of the dispersion to form the dispersion band at the interface. This method calculates the critical superficial velocity in the continuous phase at which sedimentation would not take place before the dispersion reached the outlet from the vessel. This critical velocity is a function of the drop size so that a minimum or 'cut off' drop size is defined. The velocity is calculated using either Stokes', Hadamard–Rybczynski's, or Newton's equation to compute the drag coefficient on a single drop. The method uses single drop calculations although of course dispersions are present. This method is used in American Petroleum Institute design standards for oil separators and is only applicable for dilute dispersions at $\Phi \leq 0.1$.

For liquid extraction, the more useful method is based on coalescence since this is almost always the rate determining step. Early design methods used in Petroleum and Petrochemical solvent extraction/separator designs were based on a nominal separation flux, V, or 1 US gallon/ft^2 mm or 24 m^3 m^2 h. This often has no resemblance to practice and results in over-design of equipment. From this flux the area of the disengagement interface is determined for a given total volume flow. The scientific basis for three designs based on coalescence and separation fluxes commenced with the work of Ryon *et al.* with recognition that the dispersion band height was related to the separation flux via equations of the form of (5.89). Data on continuous settling in vertical separators have been published by many workers in addition to those cited in Section 5.4.4, Lowes and Larkin (1967), Naylor and Larkin (1971), Drown and Thompson (1977) and Stönner and Wöhler (1975). Most of the work can be correlated by eqn (5.89) although Stönner derived a modified equation (cf. eqn (5.130))

$$H = \frac{1}{K_1} \frac{V/s}{(1 - 1/K_2)\, V/\epsilon} \tag{5.140}$$

where K_1 and K_2 are constants and V/ϵ has an upper limiting value equal to K_2 corresponding to flooding.

In the design of a commercial separator the maximum dispersion bed depth is defined and then the separation flux calculated from a suitable correlation equation. The diameter of the separator can then be determined for a given total flow. The problem however is to confirm the coefficients in eqn (5.89) (or variant). This has been done in the past by carrying out pilot plant experiments. Barnea and Mizrahi's work relating batch and continuous separator studies attempted to reduce the need for pilot testing and reduce development costs. There is significant potential in this if it is recognized that the dispersed phase hold-up drop diameters may not be the same in these two cases. In the basic retention equation (5.92b), the two coefficients K and n are specific to the systems studied. To avoid experimentation it is necessary to determine these parameters. Golob and Modic (1977) introduced a correlation in terms of the physical properties of the system. They modified the concept of a standardized batch test proposed by Barnea and suggested that the test should be performed under the same conditions as for the continuous flow design, i.e. temperature, phase ratio H and mixing intensity. The batch separation time t_B was defined as

$$t_B = \frac{h_{y0}}{r} \tag{5.141}$$

where h_{y0} is the initial height of the coalescing interface relative to the undisturbed interface (Fig. 5.19) and r is the average coalescence rate obtained by measuring the position of the coalescence front $y(t)$ as a function of time. The separation flux for continuous operation is then computed from

$$V = Ht_{BS}^{-1} \tag{5.142}$$

A more promising method appears to arise from the recent work of Hartland outlined in the previous section, to predict H/Q_A from measurements of $h/(-\mathrm{d}h/\mathrm{d}t)$.

5.4.5.2 *Crossflow settlers*

The majority of large scale solvent extraction plant used in hydrometallurgy applications in this technological area represent plant with the largest throughputs and use crossflow rather than vertical separators. A typical arrangement is shown in Fig. 5.23. The problems in the design are to determine the area of the disengagement boundary, this is normally the plan area of the settler, and the settler depth. As for vertical settlers, different design methods have been developed based on either sedimentation or coalescence. The 'sedimentation' method is based on similar principles to that described in Section 5.4.5.1. A critical droplet trajectory is computed for which the droplet just arrives at the disengagement interface before reaching the liquid outlet. The settling velocity is again computed from Stokes or Hadamard–Rybzynski equations. For crossflow the critical or 'cut-off' drop size is a function of the crossflow velocity as well as the physical properties defining the drag coefficient, viz. $\Delta\rho$, μ_c, μ_d. This approach assumes that settling is the rate determining step which might approximate to the conditions for dilute dispersions but is not acceptable for extraction conditions. For these coalescence is normally the rate limiting step.

Fig. 5.23. Photograph of a crossflow settler. Courtesy of Davy MacKee Ltd.

Models have been developed to describe coalescence in horizontal crossflow settlers. Consider the conditions outlined in Fig. 5.24. Droplets entering the vessels form a close packed dispersion band. To avoid flow disruptions within the settler the dispersion is usually introduced at or near to the position of the bulk interface between two phases. The dispersion flows with the bulk separated phases across the settler. At the disengagement interface drop–interface coalescence takes place and if the system is to operate without flooding the total volume coalescence rate at this interface, must again equal the volume flow rate of dispersed phase into the settler. As the flow rate increases the band at the interface will increase in depth until at some point the coalescence rate is less than the inlet flow. Crossflow settlers usually operate with a dispersion band across the entire bulk interface. At low flows this continuous band does not establish and then a wedge shape dispersion band is formed which is sketched in Fig. 5.24. The analysis of this is important as it provides a sensitive test of any model, the dimensions of the wedge are sensitive to flow and coalescence. At any point in the band, vertical settling and drop–drop coalescence take place so that these two processes produce a variation in drop size, $\phi = \phi(h, l)$. The problem is to determine the total wedge length L as a function of the inlet flow and drop size distribution. From this the settler dimensions can be determined. The first attempt to model this system was published by Jeffreys *et al.* (1967). In this model they assumed that the droplets were packed

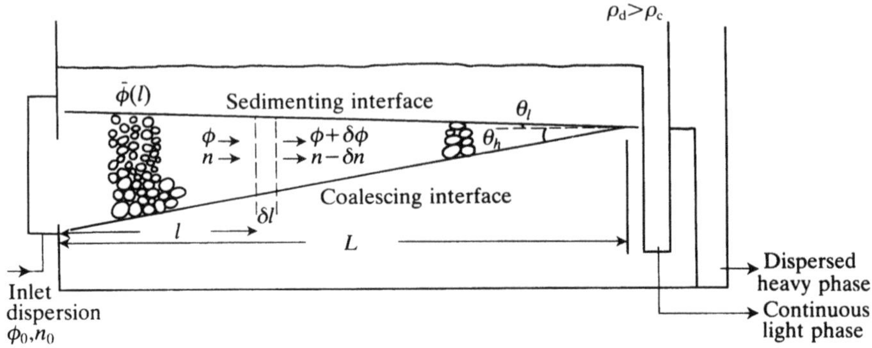

Fig. 5.24. A model of a dispersion 'wedge'.

together in the band as spheres and allowed for drop–drop coalescence in the dispersion by drops in any column coalescing with drops in an adjacent column, thereby producing an increase in mean drop size along the wedge from the inlet. Drop–interface coalescence was allowed for at the coalescing interface. The basic ideas undepinning the model were sound in that a finite difference solution was sought, the time step being related to the coalescence times involved. However the major problem with the model as developed was the rather idealized structure of the dispersion assumed in the wedge. This was necessary to describe the two coalescence stages involved and severely reduced the physical reality.

A different approach was adopted by Jeffreys *et al.* (1970a,b). In this work the processes involved were described by a differential rather than a finite difference model. It was assumed that at any position in the wedge l from the inlet the drop diameter could be defined and that $\phi = \phi(l)$. Then if a differential element is considered, δl the net flux of dispersed phase is equal to the volume rate of coalescence at the interface. If the more dense phase is dispersed, the coalescence interface will be included at an angle of Θ_h to the horizontal. The upper interface may also be inclined, Θ_l, as a result of sedimentation within the wedge following drop-coalescence. In most systems $\Theta_h > \Theta_l$.

The volume balance for dispersed phase within δl can be expressed:

$$\frac{n\pi\phi^3}{6} = \left(n + \frac{\mathrm{d}n}{\mathrm{d}l}\,\delta l\right)\frac{\pi}{6}\left(\phi + \frac{\mathrm{d}\phi}{\mathrm{d}l}\,\delta l\right)^3 + \frac{\pi\phi^3}{6}\left(\frac{1}{\tau^*}\,\frac{4\eta^*}{\pi\phi^2}\,\frac{\delta l}{\cos\Theta_l}\right) \tag{5.143}$$

The nomenclature used follows that in Section 5.4.4.1.

In this equation the second term represents the net flow out of the vertical slice of the element and the third term expresses the volume coalescence from the element at the interface. The equation can be simplified if first order terms are retained

$$\frac{3}{\phi}\frac{d\phi}{dl} + \frac{1}{n}\frac{dn}{dl} + \frac{4\eta^*}{\pi\tau^*}\frac{1}{n\phi^2} = 0 \tag{5.144}$$

A number balance can be formulated for the element and considering binary coalescence

$$n = (n + \frac{dn}{dl}\delta l) + \left(\frac{1}{\tau^*}\frac{4\eta^*}{\pi\phi^2}\frac{\delta l}{\cos\Theta_h}\right) + \frac{1}{2}\frac{h\eta}{\pi\phi^3/6}\frac{\delta l}{\tau} \tag{5.145}$$

which gives

$$-\frac{dn}{dl} = \frac{1}{\phi^2}\left[\frac{3}{\pi}\frac{\eta}{\tau}\frac{h}{\phi} + \frac{4\eta^*}{\pi\tau^*\cos\Theta_h}\right] \tag{5.146}$$

and $$h = (L - l)\,[\tan\Theta_h + \tan\Theta_l]$$

The overall process taking place in the section can be expressed in terms of the total volume coalescence rate at the disengaging interface. Then if the mean drop size entering the settler is ϕ_0

$$Q' = \frac{\pi n_0\phi_0^3}{6} = \int_0^L \frac{\pi\phi^3}{6}\frac{1}{\tau^*}\frac{4\eta^*}{\pi\phi^2}\frac{dl}{\cos\Theta_h} \tag{5.147}$$

where Q' is the volume flow rate of dispersed phase per unit width of the settler. Simplifying this equation gives:

$$n\phi_0^3 = \frac{4}{\pi\cos\Theta_h}\int_0^L \frac{\eta^*}{\tau^*}\phi\, dl \tag{5.148}$$

The boundary conditions for solution of these equations are:

$$l = 0;\ n = n_0;\ \phi = \phi_0.$$

$$\lim_{l\to L} n = 0 \qquad \lim_{l\to L} \phi = \phi_L$$

or

$$\lim_{l\to L} h = \phi_L$$

Eqns (5.144 and 5.146) can be combined to give

$$\frac{d\phi}{dl} = \frac{\eta h}{\pi\tau n\phi^2} \tag{5.149}$$

In this equation $\eta = \eta(l)$ *and* $\tau = \tau(\phi)$. The variation of the packing efficiency along the wedge was related to the crossflow velocity within the wedge, ν.

$$\nu h\eta = \frac{\pi n\phi^3}{6} \tag{5.150}$$

Then

$$\frac{d\phi}{dl} = \frac{\phi}{6\nu\tau} \tag{5.151}$$

The velocity can be estimated from Bernoulli's equation applied to the system. The major problem in solving equations is the relation $\tau = \tau(\phi)$. In the original work Davies *et al.* (1970a) assumed that the functional relationship, $\tau = \tau(\phi)$, was identical in form to $\tau^* = \tau^*(\phi)$ and used a correlation based on eqn (5.33)

$$\tau = f(\phi)^{1.5} \tag{5.152}$$

Then eqn (5.146) can be integrated to give

$$\phi^{1.5} - \phi_0^{1.5} = \frac{1}{4\nu f} \tag{5.153}$$

which when combined with the overall volume balance equation for small wedge angles gives

$$Q' = 2\eta^*\nu\left[\left(\phi_0^{3/2} + \frac{L}{4\nu f}\right)^{2/3} - \phi_0\right] \tag{5.154}$$

The comparison between the predictions of the model with a limited number of physical systems in a small batch settler was reasonably good. Further confirmation was published by Vijayan and Ponter (1974, 1976). An ambitious attempt to obtain quantitative data from a commercial settler was made by Davies *et al.* (1970b). These workers placed a video camera into a pressure separator in a kerosene wash system and although the data on the dispersion within the settling zone was sparse and insufficient for a proper validation of the model, important evidence was obtained of the significance of the design of inlet distributors to efficient phase separation. These workers were able to show that when a simple pipe inlet was used, the subsequent disruption produced in the vessel could result in a reduction in capacity by up to 50 per cent from that achieved if the phases are introduced carefully across the total width of the vessel. This problem has been studied extensively in settlers for hydrometallurgical systems by Warwick *et al.* (1971), Lewis (1977), Barnea and Mizrahi (1975, 1978). The importance of the design of the inlet to settlers was well recognized in these papers. Methods to distribute the mixture across the entire vessel using 'diffusers' and picket fences to suppress turbulence were used on commercial plant, Roberts *et al.* (1977).

The application of models based on coalescence to describe settler

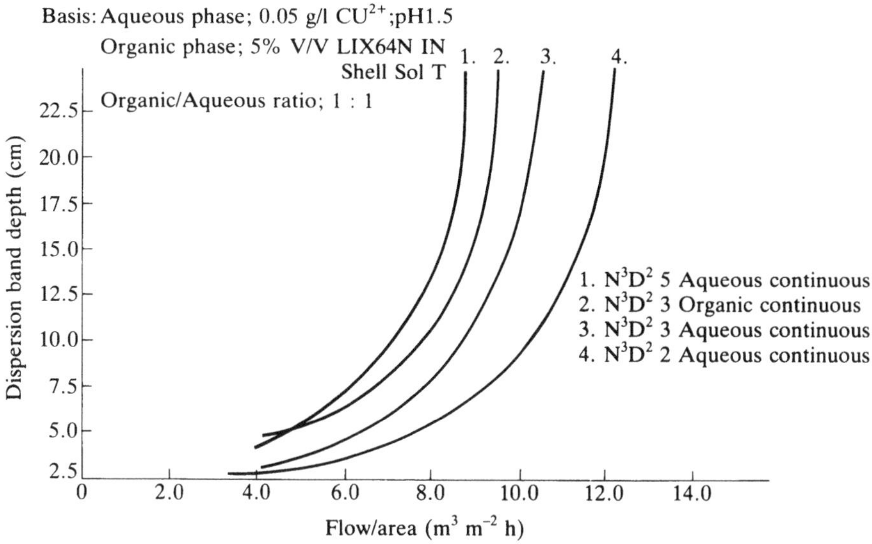

Fig. 5.25. Experimental data of dispersions in crossflow settlers. The effect of variation of phase continuously and mixer N^3D^2 on settler dispersion band depth.

performance was confirmed in a series of papers published by Warwick, Scuffham and co-workers (1971, 1974, 1975). Working on solvent extraction of metal ions Cu^{2+}, Ni^{2+}, U, from aqueous leach solutions, Warwick showed that the dispersion band depth in crossflow separators was strongly dependent on the inlet drop size distribution and phase continuity. For copper extraction systems typical results are shown in Fig. 5.25. The band depth H increases rapidly as flooding is approached. The settler loading, based on the total mixture flow into the separator and taken as 75 per cent of the flooding velocity was typically between 2–5 m^3h. The crossflow velocity was between 1 and 5 cm. As the power input to the mixer increased giving a reduction in mean drop size the settler loading was reduced. Additionally it was found, that in accordance with predictions from coalescence measurements, the loading was lower for aqueous in organic dispersions as shown in Fig. 5.25. This work was significant in that, although it was not the first detailed experimental evaluation of the performance of continuous flow settlers (being preceded by Ryon *et al.* (1959)) it did recognize the effect of phase continuity in settler design and thereby ensured that problems encountered in the first plants designed for copper solvent extraction of unexpected flooding as a consequence of a spontaneous phase inversion Powers *et al.* (1971) were avoided. Additionally Warwick *et al.* showed conclusively that the settler loading was dependent on the inlet dispersion drop size, a fact which would seem obvious from the discussions on coalescence but was confused in the earlier work by Ryon. There was evidence also in the work by Rowden *et al.* (1974) and confirmed by Barnea and Mizrahi (1975) that the smallest droplets in the drop size distribution have an important effect on the phase disengagement. The experimental finding is explained in Viegler's (1977) theoretical work and emphasizes the need for proper selection and design of mixing devices. Following Rowden's results there was considerable interest generated in using static mixing devices as these generally produce more uniform dispersions. Their application to large scale hydrometallurgical solvent extraction plant was thwarted by the relatively long contact times necessary for mass transfer (order 1–3 min) which required long pipes and by the energy costs from the then high pressure differentials across the mixing elements. In the intervening years with improved extraction reagents, contact times have reduced to order 30 s. This might now indicate that the design of mixer and settler could be radically changed.

A major factor in operating solvent extraction plant is that of cross entertainment in the phases leaving the individual mixer settler stages. This was studied in detail by Warwick and co-workers (1971), Barnea and Mizrahi (1975), Jackson *et al.* (1974, 1975), Lewis (1977, 1984), Godfrey *et al.* (1978) and others. In this work it was shown that the entrainment of dispersed phase from settlers increases with the power input to the mixer. In other words as the inlet drop size distribution decreases the entrainment increased. This is due to the increased stability of the dispersion and to the fact that at high power inputs the fraction of small droplets of secondary dispersions increase. These will generally pass through the settler. This entrainment will increase as the density difference and interfacial tension between the phases decreases. The possibility of cross entrainment of both phases was recognized in fundamental studies arising from stepwise coalescence leading to dispersed phase entrainment and in film breakdown leading to continuous phase entrainment. In addition to being influenced by $\Delta\rho$ and σ, the latter is effected by the ratio of the phase visocsities μ_c/μ_d, increasing as the ratio increases. Typical results for entrainment from mixer–settlers are shown in Fig. 5.26.

It can be seen that there is a significant difference between the levels of entrainment of organic in aqueous, *O/A*, and *A/O* in the system studied. The results were obtained from studies of a hydrometallurgical extraction process. An important factor in these processes is that although the phase ratio *A/O* across the entire process may be high ~10/1, internal recycle of the organic phase may be necessary to maintain extraction conditions. Typical phase ratios of 1/1 may be required across mixer–settlers. The operating value usually lies within the ambivalence range of the system when spontaneous phase inversion may take place. An example of a typical phase map is shown in Fig. 5.27. The design must consider the worst case and for

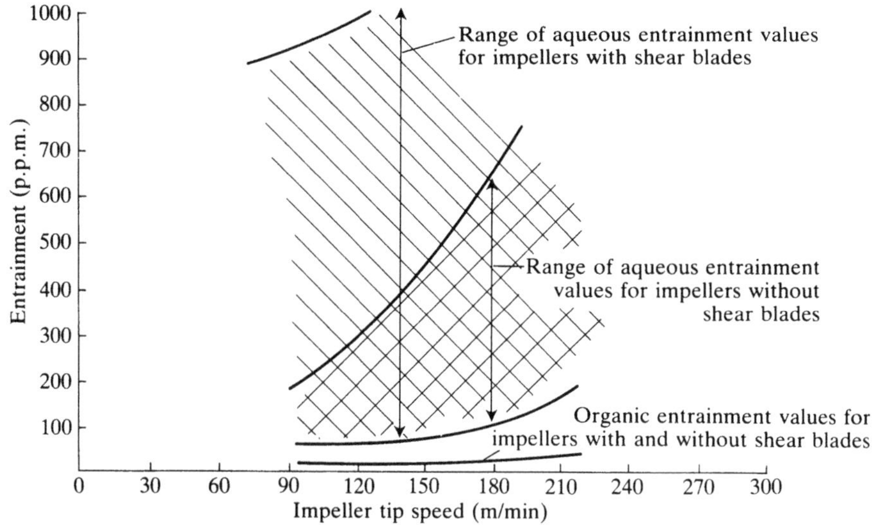

Fig. 5.26. Experimental results for cross entrainment from gravity settlers. Graph of entrainment against tip speed for organic continuous operation.

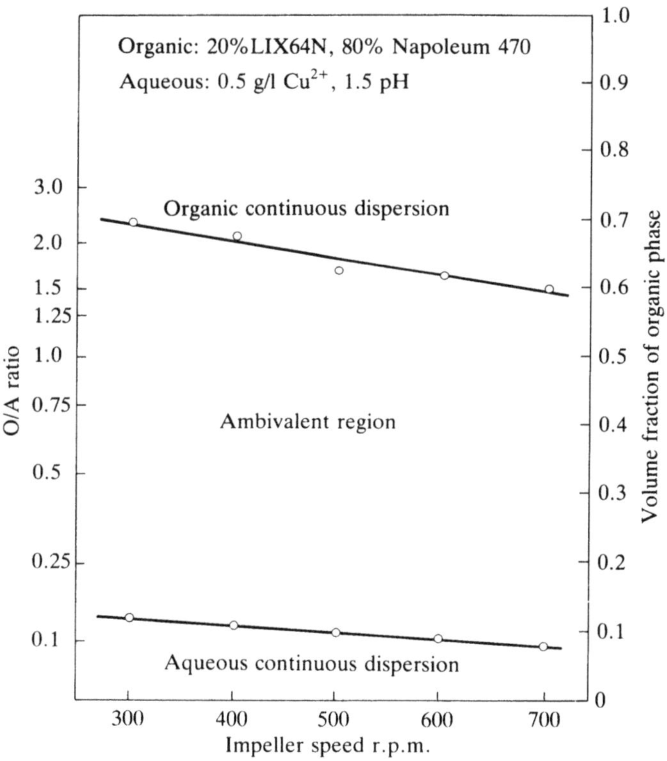

Fig. 5.27. The ambivalence range for a typical solvent extraction system.

the settler this will involve the design for the lower rate of coalescence, A/O or O/A, and the highest entrainment. Most measurements to determine the phase maps and the drop size in the suspension are made on equilibrated phases. These may prove in error since it has been shown that Marangoni effects produced during mass transfer can promote coalescence within a dispersion and give rise to phase inversion, Jackson *et al.* (1975).

5.4.6 Phase entrainment in mixer-settlers

Experimental work on pilot plant equipment and full scale plant experience has shown that cross entrainment can be a most serious problem affecting operating costs through loss of solvent, contamination of product streams and environmental pollution. It warrants a section to review the methods of production and treatment. Normally the entrainment from separators is in the form of secondary dispersions which are produced during the mixing and mass transfer stages of the process and usually to a lesser extent during coalescence. On the premise that prevention is better than cure the design of mixing and transport equipment is particularly important. Much effort has been placed in this in all modern plant. Multistage mixers operated at lower power input, static or pressure mixers have been incorporated into plant design in an attempt to control the dispersion drop size distribution and eliminate small drops less than approximately 60 μm which will give problems during separation. One major item of equipment which will give problems if liquid-liquid mixtures are present is a centrifuge pump. Therefore multistage processes which may require inter-stage pumping of one or both phases should be designed carefully so as to eliminate the second phase before the pump and to specify low shear pumps. A cost reduction for the process might be effected at the design stage by eliminating pumps on one of the streams by elevating the stages, Lott *et al.* (1972). In extraction processes another cause of secondary dispersions is cooling of a saturated solution. This may take place for instance in storage tanks between extraction and stripping processes and because of the mechanism of the nucleation process, leading to the formation of a second liquid phase as droplets, a secondary haze is always formed. Therefore if the process is carried out at temperatures above ambient heat losses should be minimized.

Within settlers several techniques have been employed to reduce entrainment. These involve either the application of surfaces or packing to promote coalescence or the application of external force fields to augment gravity, electrostatic or centrifugal fields. These will be briefly considered.

5.4.6.1 *Surfaces and packings*

The simplest method is to introduce baffles or plates into the settling zone. These are only really effective in promoting coalescence of primary disper-

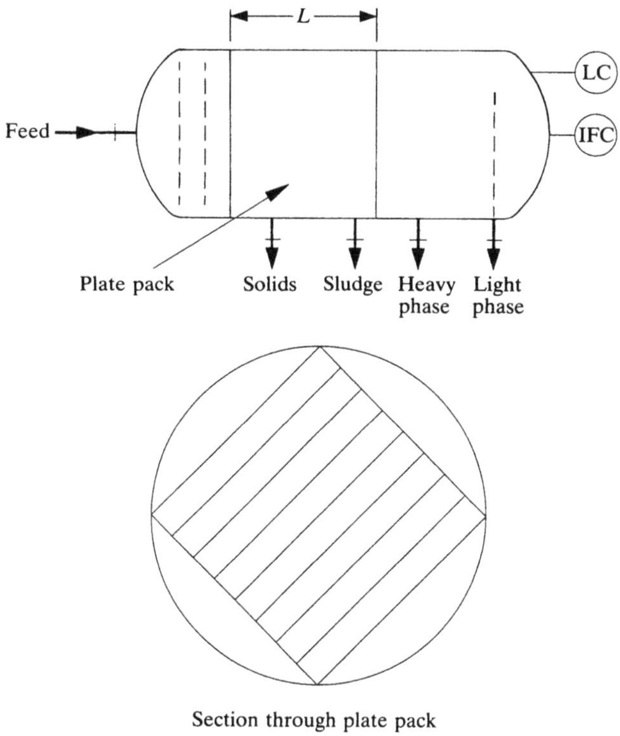

Fig. 5.28. A crossflow inclined plate separator.

sions, but nevertheless enable improved operation in relation to entrainment levels, *and* in increasing the loading rate of the separator. The technique is used extensively in the separation of oil and water phases in the petroleum industry and in treating aqueous effluents. A simple diagram showing a typical installation in a crossflow settler is shown in Fig. 5.28. An improvement in the rate of phase separation and some reduction in entrainment is achieved by (i) acting as a flow 'straightener', (ii) reducing the flocculation distance, (iii) promoting surface coalescence onto the plate. In this type of settler the deep dispersion band formed in conventional gravity separators is eliminated. The plates are designed to be between 2–10 cm apart and as the mixture flows in the channels formed by adjacent plates, the droplets settle towards the plate surface. The plate is constructed of a material which is preferentially wetted by the dispersed phase so that when droplets arrive at the surface, coalescence takes place to form a liquid film of dispersed phase on the surface, Hazlett (1969), Davies and Jeffreys (1969). This film drains from the surface to establish final phase disengagement. In some systems shallow dispersion bands are produced in each channel when the interface coalescence from a raft of drops is lower than the local volume arrival rate at the surface. Generally however flocculation in the channel controls the separation rate. Simplified design equations are used under these conditions. If Stokesian settling is assumed, the critical drop size which can be separated is given by

$$d_c = \left(\frac{18\mu c YV\lambda}{L\Delta\rho g \cos\Theta}\right)^{1/2} \qquad (5.155)$$

where Y and L are the perpendicular distance between adjacent plates and the length of the plates respectively, V_c is the volume flow rate of continuous phase per unit cross sectional area of a channel formed between adjacent plates and Θ is the inclination of the plates. This equation gives the maximum diameter of droplets which will be separated and is usually designed for a minimum of 30 μm. The derivation of eqn (5.155) ignores the effect of the liquid film which is formed on the surface of the plate at the disengaging interface. The effect of this will be to reduce the effective flocculation distance and if coalescence can be maintained the critical drop size will be reduced thereby improving separation. If the drop size distribution in the feed is known, the entrainment leaving the separator can be computed to give an estimate of the overall separation efficiency and the drop size distribution in the effluent, Rowley and Davies (1988). Many different proprietary forms of plates have been developed, corrugated surfaces are used which are useful when solids are present in the system likewise chevron baffles have been developed for three phase, liquid–liquid–vapour and liquid–liquid–solid systems.

Another approach is to use packed beds in which again the surface of the packing is preferentially wetted by the dispersed phase. To avoid the problems of channeling often experienced in random packed beds, regular or structured packings are used. These may be reticulated foam, knitted or woven mesh or formed sheet. The regular structure will produce more uniform flow and the extended surface provides a high surface area per unit volume for coalescence. The mechanism of separation is similar to that above, droplets are intercepted by the surfaces and interface coalescence takes place. Drainage of the coalesced phase by gravity within the packing completes the separation. The technique using these types of packing is only applicable for primary dispersions and, for it to work, the packing surface must be preferentially wetted by the dispersed phase. This presents a potential problem if the system is operated within the ambivalence range when, if the phases invert, preferential wetting by the dispersed phase would not be achieved. To overcome this Davies *et al.* (1972) developed a bi-functional material which could operate in aqueous–oil systems for either phase continuity. This was used in hydrometallurgical solvent extraction systems by Jackson *et al.* (1974). In order to exploit the coalescer and ensure

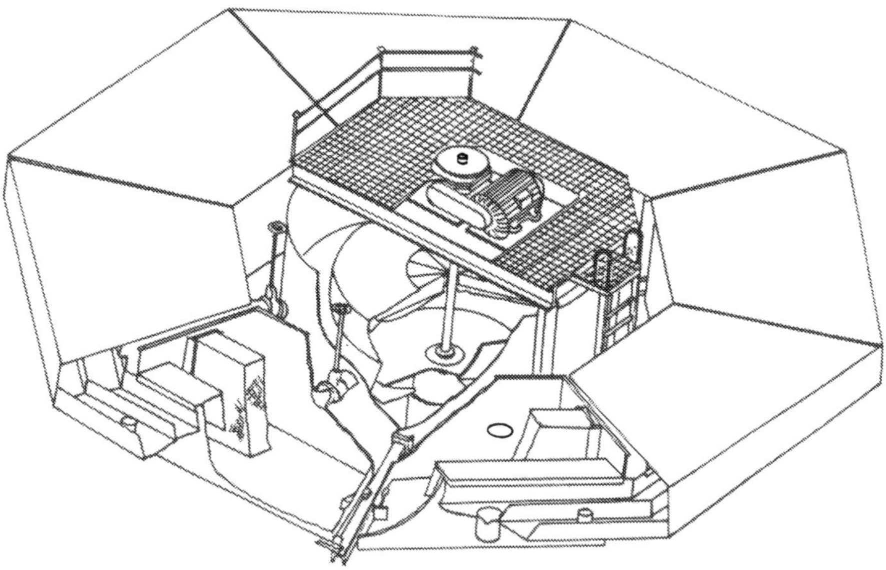

Fig. 5.29. A mixer-settler designed for use with a coalescer packing (after Jackson 1974).

continuous operation a redesign of the mixer-settler was necessary. The design utilized a segmented annular settler to allow for routine maintenance of the packing without stopping the flow. The design was described by Jackson (1974) and a diagram is shown in Fig. 5.29. Typical results showing cross entrainment are given in Fig. 5.30. The results were obtained for similar conditions to those for gravity settling shown in Fig. 5.26 so that comparison between these can be made to assess the improvement in

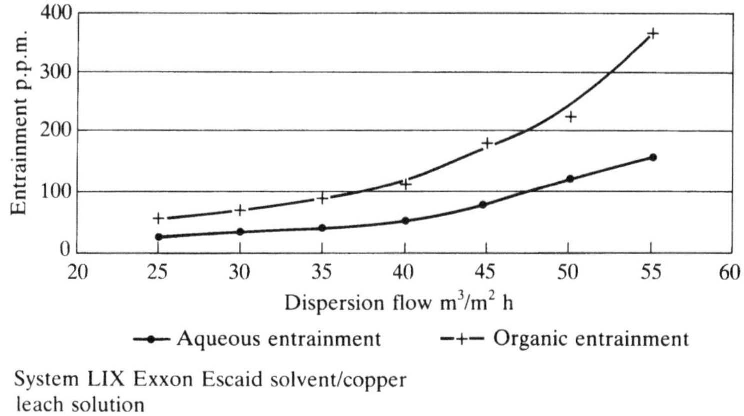

Fig. 5.30. Data on cross entrainment from mixer-settler units fitted with a coalescer packing.

performance. It must be stated however that any form of packing used to promote primary dispersion separation will be subject to blockage by entrained solids. In these packings the mean effective pore diameter is of the order of 800 μm and Jackson demonstrated that blockage could take place when products in the range 60 μm and larger were present. To use these materials in commercial systems a properly designed mixer settler is necessary in which the settler is sectioned and intermittent backflushing can be used to clean the packings. Such systems have been developed by Jackson and co-workers (1975).

When secondary dispersions are present, different types of packings are necessary in order to remove the entrained droplets. By definition these droplets will not settle under gravity. To remove these the dispersion must be pumped through a porous bed of finely divided solids or fibres. The most widely used materials in modern systems are non-woven fibre mat and polymeric membranes. The mechanism of operation is similar for both types of media, a prerequisite to understanding the processes taking place being a knowledge of the structure of the media. Schematic diagrams of a typical non-woven material used in coalescers are shown in Fig. 5.31. Fig. 5.31a shows a thin section taken in the direction perpendicular to the flow whilst 5.31 shows a section normal to this in the direction parallel to flow. First it is evident that the material is structurally anisotropic. It is formed from layers of fibres in which in any layer, the fibres are arranged in a random order. The material shown is a polyester, the filaments in 5.31a can be assumed to be cylindrical rods which are flexible so that where they overlap they bend around to maintain a layered structure. In 5.31a the areas enclosed by intersecting fibres form the position of the pores through which the liquid dispersion must flow. The section of these pores for this particular material are irregular external polygons. In the 3-D structure formed by superimposition of random layers, the pores are made up by inter-connection of the larger pores to produce a network with varying cross-section and shape. Not all non-woven material is the same, two major differences may exist. First the fibres may not be straight rods but might be curved with constant or variable curvature as shown in Fig. 5.32a and b. Otherwise the basic structure remains the same; the pore sections now would have a more complex shape the simplest being circular external polygons. In the second case the layer structure might be disturbed by a process known as needling whereby to increase the tear resistance of the material, needles are forced through which displace filaments between layers so that some run in the direction of flow. These types of material are used widely in filtration, in gas filtration for bag filters, etc. where strength is important, but are little used for coalescers since the needling process produces holes in the material which would result in short-circuiting and bleed through.

The mechanism of operation is that droplets flowing through the matrix

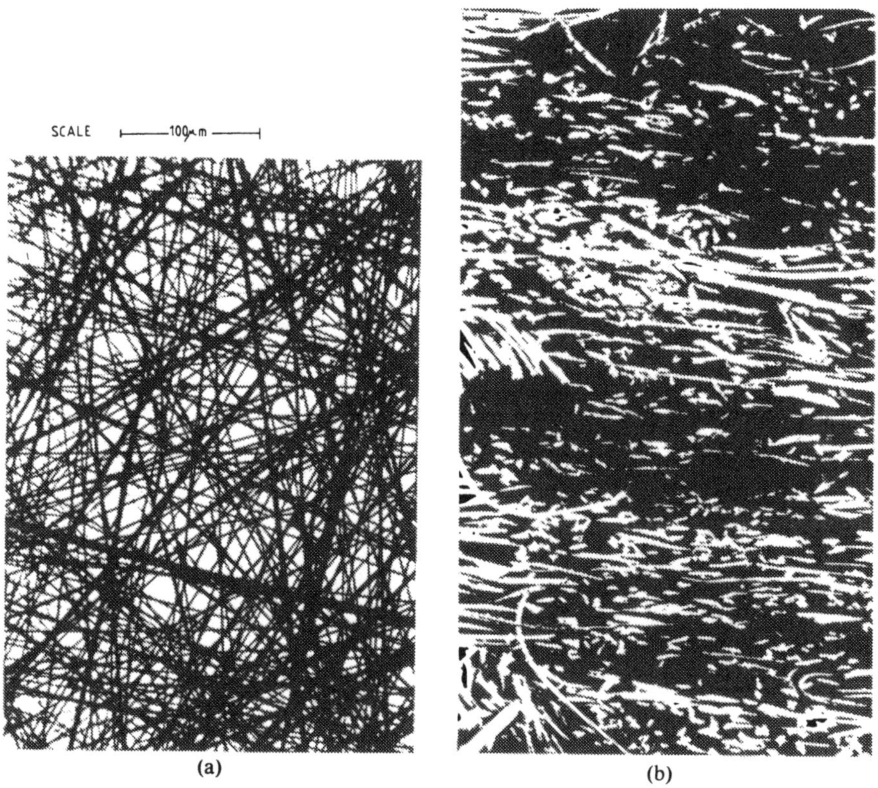

Fig. 5.31. Schematic diagram of a non-woven fibre material used for secondary coalescers; (a) of a sector in the direction perpendicular to flow; (b) of a section in the direction parallel to flows.

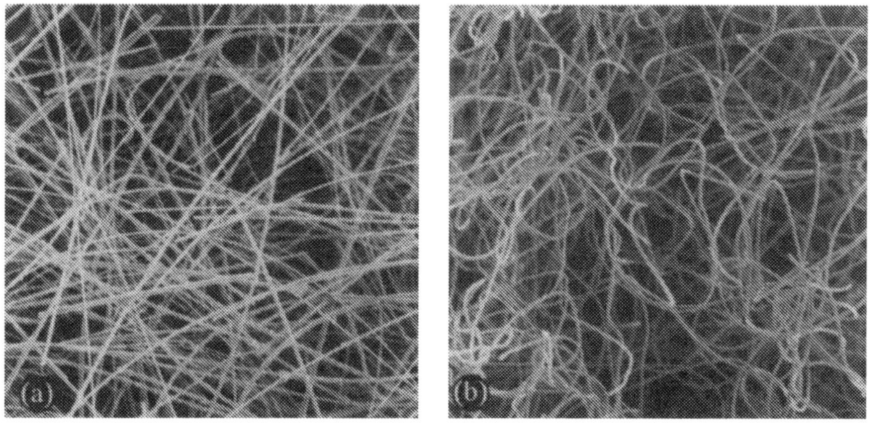

Fig. 5.32. Photomicrograph of non-woven materials used for secondary coalescers; (a) SEM of a polypropylene mat; (b) SEM of a polyester mat.

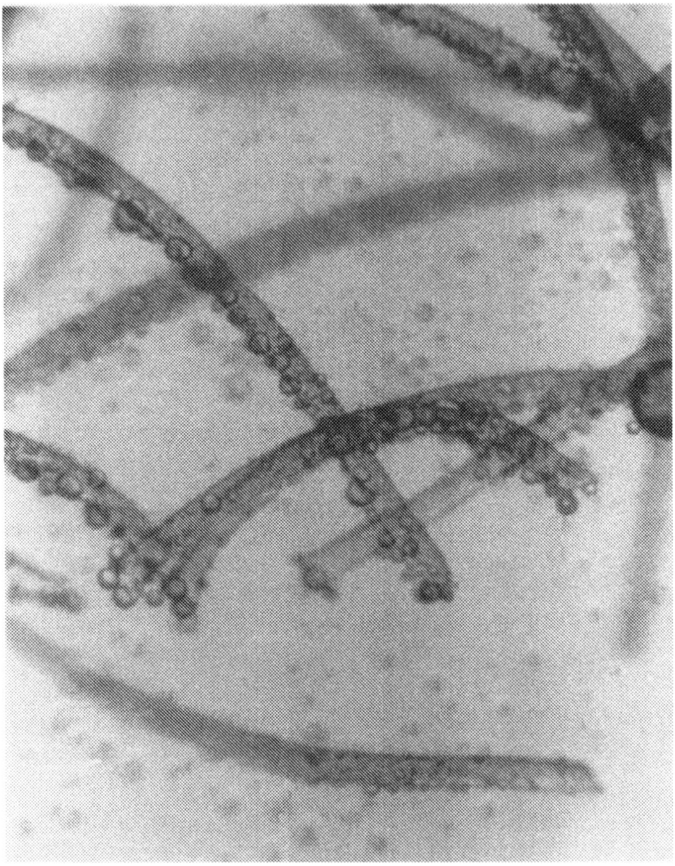

Fig. 5.33. Photomicrograph of drops adhering to a fibre in a secondary coalescer.

are intercepted by and adhere to the fibres. The adhesion force between the droplet and fibre is greater than the viscous drag force transmitted by the flowing fluid. On the surface of the fibre droplets coalesce with neighbouring captured droplets and by direct interception of droplets from the free stream. A photograph of the process is shown in Fig. 5.33. As the droplets grow by coalescence on the fibres the viscous force increases, proportional to the drop area, whilst the force of adhesion remains almost constant. Therefore a stage is reached where the drag forces on a particle overcome the adhesion force and then the drop is re-entrained with the stream. The droplet may be captured again in the next layers of the bed. In this way the mean drop size increases as the dispersion passes through the bed. At the outlet the mean drop size is large enough to allow the droplets to settle in the vessel, i.e. a primary dispersion has been produced. These devices were first developed in the 1950's to remove dispersed water from aviation jet fuel where glass fibre was a popular coalescer material. The analysis of the operation was first centred around the description of flow through the media using the Carman–Kozeny equation, Rose (1963), Sherony and Kintner (1971), Spielman and Goren (1970,1972), Scheidegger (1960). Thus

$$f = \frac{\Delta P}{L} \frac{d_c}{\rho_c u^2} \frac{\epsilon_1^3}{1 - \epsilon_1} = \frac{16 K_1 (1 - \epsilon)}{Re} \tag{5.156}$$

where f is the friction factor, ϵ_1 is the free volume, d_f is the fibre diameter and u the superficial velocity. When this was applied to the two phase flow problem of the coalescer, the pressure drop was consistently higher than that calculated for the single (continuous) phase flow. To accommodate this Rose and Kintner argued that the free volume was reduced for two phase flow due to retention of the dispersed phase in the bed and when eqn (5.156) was applied to these conditions, ϵ should be reduced. Then defining ϵ_2 as the free volume for two phase flow

$$\frac{\Delta P_1}{\Delta P_2} \frac{k_2}{k_1} \frac{\epsilon_1^3}{(1 - \epsilon_1)^2} = \frac{\epsilon_2^3}{(1 - \epsilon_2)^2} \tag{5.157}$$

the suffixes 1 and 2 referring to single and two phase flow respectively. The term ϵ_2 is related to the average saturation of the bed, $\bar{S}$, defined as the fraction of the voids occupied by the dispersed phase

$$\bar{S} = 1 - \frac{\epsilon_2}{\epsilon_1} \tag{5.158}$$

or in terms of the hold-up Φ

$$S = \Phi \left(\frac{1 - \epsilon_1}{\epsilon_1} \right) \tag{5.159}$$

ϕ varies with the Reynolds number and with the ratio of the mean inlet drop diameter to fibre diameter. Sherony *et al.* (1978) defined the overall coalescence efficiency E of a fibre bed in terms of the ratio of the outlet to inlet drop diameters

$$\frac{d_o}{\bar{d}_i} = \exp \left\{ -\frac{3}{4} \frac{\bar{S}}{1 - \bar{S}} \frac{(1 - \epsilon_1)(1 + \bar{d}_i/d_f)}{d_f} EL \right\} \tag{5.160}$$

The conclusions from the work of Kintner's group and Goren is that the coalescence efficiency E is dependent on (i) the bed properties d_f, ϵ_1, L, (ii) the interaction between the dispersed phase, the fibre material and the pore structure which control S, and on (iii) the operating conditions $\bar{d}_i$ and the superficial velocity of the continuous phase. Generally the complex behaviour does not allow the bed saturation S or the outlet drop size $\bar{d}_o$ to be predicted a priori. The computation of the separation efficiency depends on these parameters so that unless these can be predicted the analysis will depend upon the availability of experimental data.

An alternative approach to the problem can be made, this involves describing the microscopic structure of the bed and the flow of dispersion within the pores or around fibres in the bed. The local capture efficiency can be computed and, by integrating this across the entire bed an estimate of the separation efficiency can be made. For fibre beds the problem has been approached by first describing the flow around a single fibre. A model is then assumed to describe the fibre mat and flow equations are solved for the assumed model. This gives expressions for the liquid velocity in the mat. Dispersed droplets are introduced and the probability of capture on individual fibres is computed and estimates of the local separation efficiency can be computed.

The flow around a single cylindrical rod, Fig. 5.34, was first considered by Lamb (1932) and Thorn (1933). Many workers followed this approach and adopted the solution to filtration applications, Davies (1950), Tomotika and Aoi (1950), Imai (1951), Tomotika (1953) and Kaplan (1957). The stream function $\psi(r, \theta)$ derived from this work was

$$\psi(r,\theta) = \frac{a\,U_0 \sin\theta}{2(2 - \ln Re)}\left[\frac{a}{r} - \frac{r}{a} + \frac{2}{a}\ln\frac{r}{a}\right] \tag{5.161}$$

The velocity components u_r and u_0 are related to the stream function $\psi(r, \theta)$ as

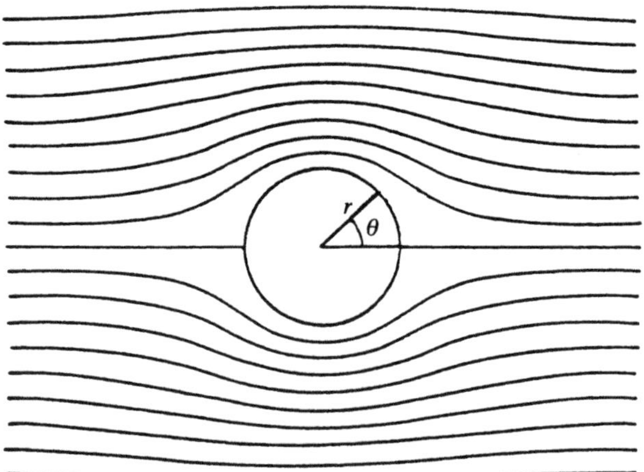

Fig. 5.34. Flow around an isolated cylindrical fibre.

$$u_r = \frac{1}{r}\left(\frac{\partial\psi}{\partial\theta}\right) : u_0 = -\frac{\partial\psi}{\partial r}$$

u_0 is the undisturbed flow velocity. The Reynolds number Re is given by $Re = d_f U_0 \rho_c / \mu_c$. The extension of the analysis to groups of fibres was considered first by Kuwabara (1959) and Happel (1959). They assumed that the fibres in the bed were parallel with the centres arranged on a cell pattern, square and triangular lattices were both used to define the cell patterns. This enables the complex flow around a matrix of fibres to be simplified to the analysis of the flow within a cell. Although the original authors applied different boundary conditions to describe the conditions at the cell surfaces the results were similar. The radius of the cell was defined as b. The free area of a section of filter in the plane of the flow is then $\epsilon_a = [1 - (a/b)^2]$. For high porosities corresponding to $(a/b)^2 \ll 1$ the stream function for this multi-fibre model was given as:

$$\psi(r, \theta) = \frac{aU_0 \sin\theta}{2\left(-C - \frac{1}{2}\ln(1 - \epsilon_A')\right)}\left[\frac{a}{r} - \frac{r}{a} + \frac{2r}{a}\ln\frac{r}{a}\right] \tag{5.162}$$

C is a constant which for Kuwabara's solution was 0.75 and from Happel 0.5. The velocity distribution was derived for a cell model by Spielman and Goren (1968). The transverse flow velocity was described by a similar but more complex equation than (5.162) which includes the resistivity X, where from D'Arcy's equation

$$X = \frac{a^2 \Delta P}{\mu_c U_0 L} \tag{5.163}$$

Once the stream function is defined then the drag coefficient and pressure drop can be computed. If the drag force acting on a unit length of fibre is defined as F, then

$$F = C_D d_f (\rho_c U_0^2 / 2) \tag{5.164}$$

F can be made dimensionless. Thus

$$F^* = \frac{F}{\mu U_0} = \frac{1}{2} C_D Re$$

Using eqn (5.161)

$$F^* = \frac{4\pi}{2 - \ln Re} \tag{5.165}$$

whilst the cell models give:

$$F^* = \frac{4\pi}{-C - \frac{1}{2}\ln\beta + \beta - \beta^2/4} \tag{5.166}$$

$$\beta = 1 - \epsilon'_A$$

In the coalescers, droplets suspended in the continuous phase are first separated by capture onto the surface of the fibres. The motion and trajectory of droplets within the coalescer depends on the forces acting on the individual drops. When the distance between the surface of a drop and fibre, δ, is large, $\delta \gg \delta_p$ then the forces which act on the drop are gravitational force, viscous and inertial forces, and diffusion force (transmitted via Brownian motion in the liquid). The influence of these forces leading to interception of droplets by a fibre are shown in Fig. 5.35. Fibre bed coalescers are used to remove droplets in the range 0.5–30 μm diameter. In practical systems the diffusion force is small and can be neglected. Likewise the gravitational force is small, by definition a secondary dispersion will not separate under gravity, and similarly can be neglected. The dominant mechanisms are inertial and direct interception. The contributions of each mechanism to the overall capture efficiency E_0 on a fibre can be expressed as E_I and E_D respectively where

$$E_0 = E_I + E_D \tag{5.167}$$

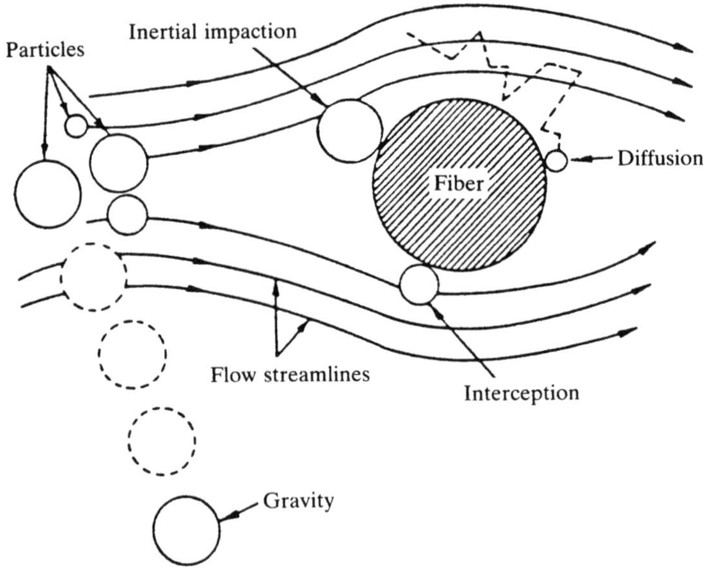

Fig. 5.35. Diagram of interception of droplets by a fibre in a coalescer showing: gravity interception; inertia interception; direct interception; and Brownian diffusion interception.

The individual efficiencies are defined in terms of number flux of droplets which pass through the an area of the fibres projected into the plane perpendicular to the fluid flow, $n_0 U_0 A_M$. The term n_0 is the number concentration of particles. Then if the number of droplets collected per unit time by mechanism i is ϕ_i

$$\phi_i = E_i n_0 U_0 d_f \tag{5.168}$$

Direct interception is characterized by the dimensionless ratio $N_D = d_p/d_f$ and $E_D = E_D(N_D)$. The determination of this function is straight forward when the Stokes number, $St = (dp^2 \rho d v / 9\mu_c d_f)$, tends to infinity. This implies that the inertia of the drops is such as to make them travel directly towards the fibre(s). Then $E_D = N_D$. In the other limiting case, $St \to 0$, the particles have essentially no inertia and travel along the streamlines of the flowing fluid. Then using eqn (5.162)

$$E_D = \frac{1}{2[-C - (1/2)\ln\beta + \beta - \beta^2/4]} [(1 + N_D)^{-1} - (1 + N_D) + 2(1 + N_D)\ln(1 + N_D)] \tag{5.169}$$

This can be approximated to:

$$E_D = \frac{1}{2[-C - 1/2\ln\beta - 1/2\ln\beta + \beta(\beta^2/4)} \tag{5.170}$$

Natanson (1962), Pich (1966).

For inertia interception, E_I is a function of the flow velocity

$$E_I = E_I(Re, N_D, St) \tag{5.171}$$

Davies (1952) derived an expression for E_I:

$$E_I = 0.16[N_D + (0.5 + 0.8\,N_D)\,St - 0.105\,N_D St^2] \tag{5.172}$$

The final problem is to translate this overall capture efficiency of a fibre to the efficiency for the coalescer bed. For the model systems consisting of a series of fibre layers arranged geometrically on a square lattice, Zebal (1965) proposed that

$$E = 1 - (1 - E_0\beta^{\frac{1}{2}})^n \tag{5.173}$$

where n is the number of layers. For a square lattice, n is related to the depth of the fibre bed L, $n = L/2b$.

Although the cell model first introduced by Kuwabara and Happel allows the effect of adjacent fibres on the velocity field to be computed in the stream function, when the conditions are translated to the local bed the geometrical structure produced is unrealistic. Parallel fibres would mean that the pores

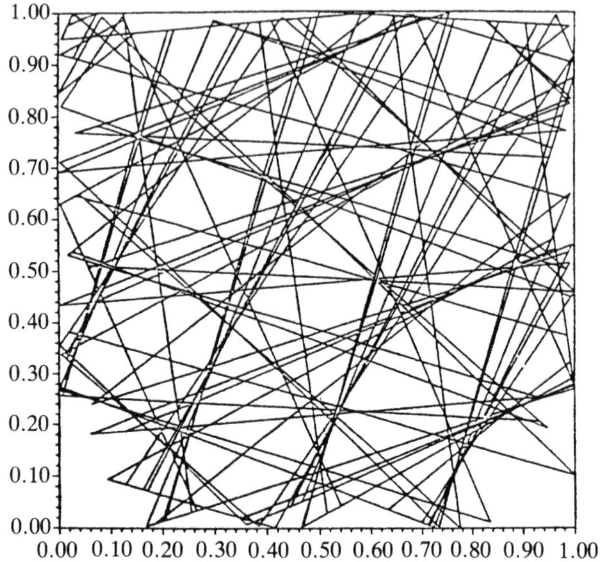

Fig. 5.36. A simulation of a non-woven fibre mat using the 'fan' model of Fuchs and Stechkina (1963) (Fairclough 1986).

produced, would be rectangular, (cf. Fig. 5.31a) and sections as in Fig. 5.31b would be regular on square or triangular particles. Fuchs and Stechkina (1963) recognized this problem and introduced a degree of randomness into the structure, as is observed in real materials, by maintaining the layer structure and rotating each layer through a random angle θ such that $0 \leq \theta \leq \pi$. The fibres in each layer were again assumed to be parallel. This resulted in the so-called fan model, an example of which is shown in Fig. 5.36. The equations for flow u_r, u_0, and capture efficiencies E_D, E_I were re-worked to conform with the assumptions of the structural model. From the standpoint of the pore properties, shapes areas etc. this was an improvement over the cell structures but falls short on comparison with real systems where fibres are not in general parallel even in a single layer.

In recent years, spurred on by the need to improve the quantitative predictions of the performance of coalescers and especially to move towards predicting the concentration of the dispersed phase and drop size distribution in the bulk phase leaving a coalescer, more work has been done on the structure. The ideas of incorporating a cell structure have been abandoned in this work, instead the fibres in any layer are allowed to have a random position in 2 space. As in the example shown in Fig. 5.31, the fibres are assumed to be straight cylinders but are now flexible in the z-plane so that where they overlap they can bend around to maintain the layer structure. The problem of defining a layer then reduces to that of defining a Poisson line process in 2 space where each line will represent a fibre. The individual fibres can have different diameters which can be chosen at random from a defined distribution function. The detailed description of algorithms to produce simulations of fibre media have been given by Davies and co-workers (1985a,b, 1986). Examples of structures produced are shown in Fig. 5.37a and b. The performance of a coalescer using this description of the structure can be calculated by computing the paths of droplets from the inlet into the fibre bed. The position at entry and drop size are chosen, the entry being a random position and the diameter being selected from a defined distribution function. The flow velocity of continuous phase through the coalescer is fixed and the approach of the drop towards fibres in each layer is computed. In the near proximity of the fibres the equations describing the flow, (eqn 5.161), are used. If the droplet is intercepted by the fibre the probability of being retained is calculated. If it is captured then the position on the fibre is recorded and trajectories of subsequent drops calculated. The penetration of drops into the coalescers can be computed as well as drops which escape without capture. Experiments have demonstrated that droplets can be captured by direct interception with drops already adhering to the fibre surfaces, Langden *et al.* (1972). The stochastic trajectory models described by Abdel-Ghani and Davies (1985) take this into account. In addition rearrangement on the fibres due to capillary forces, are also considered. The results predicted by the model for the penetration of droplets into a coalescer before capture give

$$m = m_0 e^{-\lambda z} \tag{5.174}$$

where m is the mass flow of dispersed phase at any point z. The term λ can be viewed as a measure of the overall capture efficiency.

$$\lambda = \lambda(d_f, \sigma_f, \epsilon_A, d_p, \sigma_p, U_0, \mu_c, \rho_c) \tag{5.175}$$

Equation (5.174) is similar to an empirical equation derived from experimental results for the filtration of fine particles in fibre filters, Iwasaki (1937), Witzmann (1940), Radushkevich (1946) and droplets in coalescers, Wasan *et al.* (1978). If the conditions are considered over layer i of the coalescer then the net mass flux of dispersed phase over layer i can be related to the capture efficiency η_i. Thus

$$\eta_i = \frac{m_{i-1} - m_i}{m_{i-1}} \tag{5.176}$$

Extending this to a coalescer of N layers, the overall capture efficiency E is

$$E = \frac{m_0 - m_N}{m_0} = 1 - \prod_{i=1}^{N} (1 - \xi_i) \tag{5.177}$$

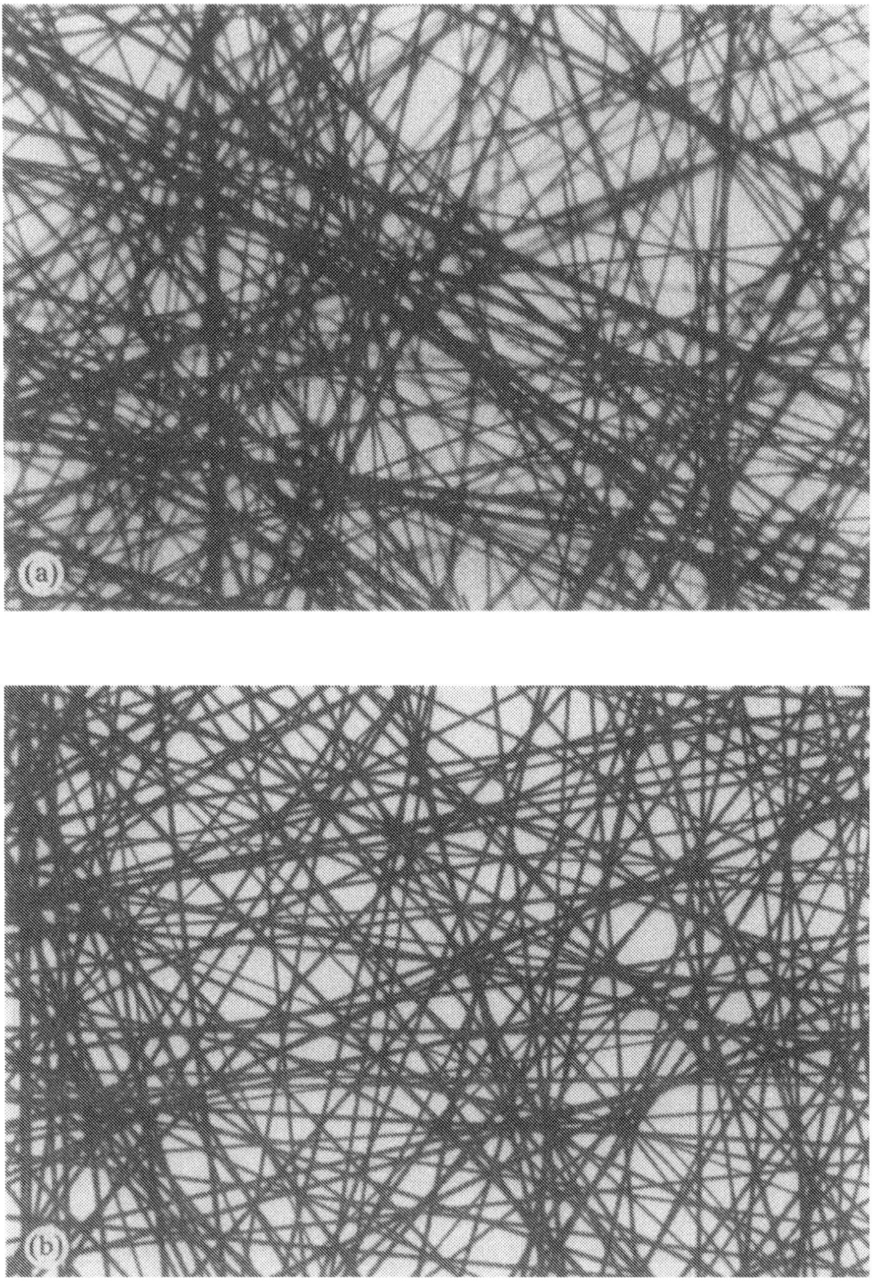

Fig. 5.37. A non-woven carbon coated polyester fibre mat used for coalescers; (a) SEM photomicrograph; (b) simulation using a random network model after Wilkinson and Davies (1986).

If the layer efficiency, ξ_i, is constant throughout the bed then

$$E = 1 - (1 - \bar{\xi})^N \tag{5.178}$$

from which

$$\ln \frac{m_i}{m_0} = N \ln (1 - \bar{\xi}) \tag{5.179}$$

and λ in eqn (5.174) equals $\ln (1 - \bar{\xi})$.

$$\xi = \xi (d_f, \sigma_f, \epsilon_A, d_p, \sigma_p, U_0, \mu_c, \rho_c) \tag{5.180}$$

Using random network models of the type proposed by Abdel-Ghani and Davies (1985) and Piekaar and Clarenburg (1967) in conjunction with stochastic trajectory models to simulate capture and coalescence of droplets, the separation efficiency of fibre beds can be computed and the concentration and droplet size in the outlet predicted. Examples are shown in Figs 5.38–5.40.

In Fig. 5.38 the penetration of droplets into a fibre mat is shown, the slope of the graphs being λ. In Fig. 5.39, the variation in the mean capture efficiency $\bar{\xi}$ is plotted as a function of the free area, ϵ_A, for various dispersions characterized by the mean drop diameter, $\bar{d}_p$. As $\bar{d}_p$ decreases the capture efficiency decreases. It is of course more difficult to separate finer drops! Finally coalescers, unlike filters, operate in co-current flow, that is the dispersed phase after capture and coalescence on the fibre surfaces

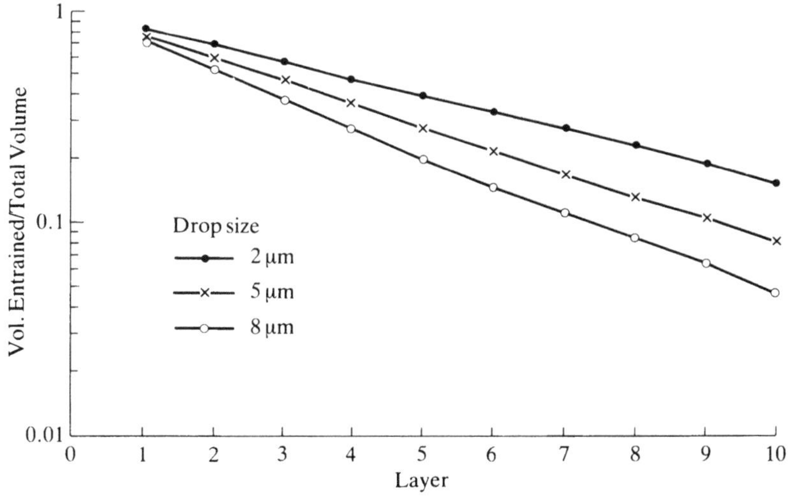

Fig. 5.38. The variation in the concentration of the dispersed phase in flow through a fibre coalescer. Layer free area = 0.7; fibre diameter 7 μm.

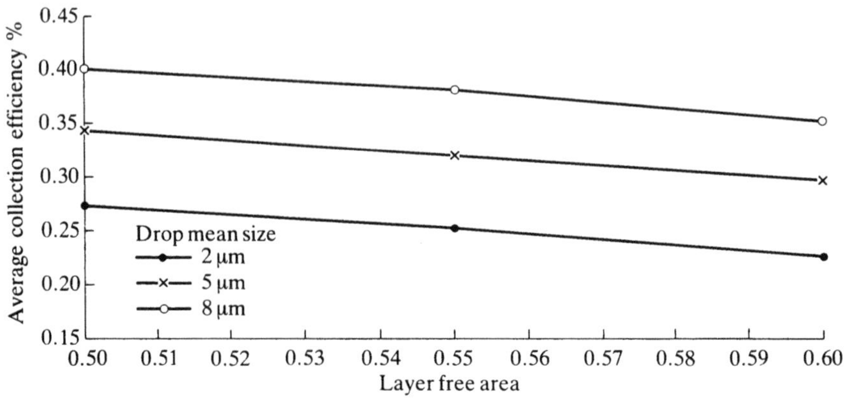

Fig. 5.39. The variation in the mean capture efficiency with the free area of a fibre coalescer.

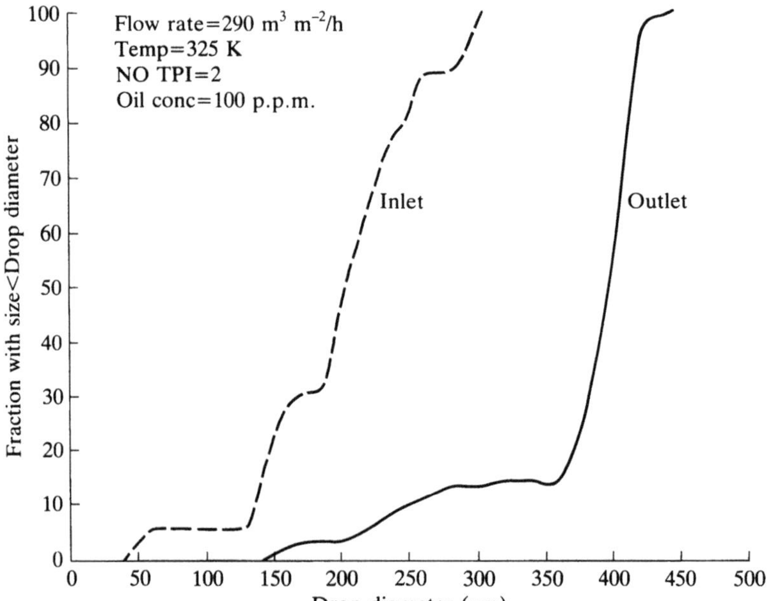

Fig. 5.40. Predictions of the outlet drop size distribution from a coalescer (after Braithwaite 1990).

must then drain through the bed to exit at the outlet as a primary dispersion. The predictions for this from these models allow the outlet drop size distribution to be computed. The results are shown in Fig. 5.40.

The application of the theory to the design of a coalescer is embodied in the coalescer cartridge. The inlet dispersion enters the centre of the cylinder and then flows radially outwards through the media. The fibre media

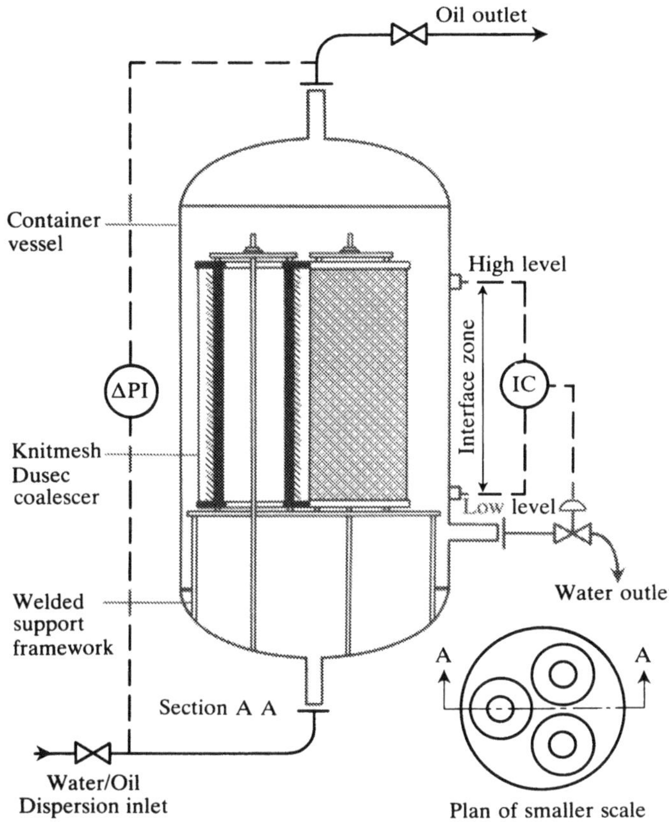

Fig. 5.41. A secondary coalescer unit.

is wound around a central perforated tube making up the inner cylinder. Theory demonstrates that the efficiency of capture and coalescence, ξ, is determined by the bed properties fibre diameter and free volume. Thus:

$$\xi = \xi(d_f/d_p, \epsilon_v) \qquad (5.181)$$

As the dispersion passes through the bed (in the radial direction) the mean drop size will increase due to coalescence. To maintain the separation efficiency and to prevent redispersion the pore volumes within the bed should increase and the fibre diameter should also increase. This is achieved in practice by using successive layers of material in the cartridge for which d_p and ϵ_v increase from layer to layer in the radial direction. Flow being from the centre to the outer cage of the cartridge enables this increase in ϵ_v to be simply achieved. In addition since the inlet layers define the flow capacity, the design based on the inner surface area of the cartridge ensures that flooding will not take place in subsequent layers. The flow scheme is exactly opposite to that adopted for a filter where the flow is radially from the outer surface to the inner tube of a cartridge so that surface filtration takes place on the highest exposed surface to the flow. An overall system where the more dense phase is the dispersed phase is shown in Fig. 5.41.

Nomenclature

A	coordination number
a	interfacial area
a_p	average pore area in a fibre bed
C	concentration
C'	bond number $C' = (g\Delta\rho d^2/\sigma)$
d	drop diameter
D	diffusion coefficient
D	impeller diameter
d_c	'cut off' drop diameter, eqn (5.155)
d_f	fibre diameter
d_i	mean diameter of drops entering fibre bed coalescer
d_{max}	maximum stable drop diameter
d_0	mean diameter of drops leaving a fibre bed coalescer
d_{VS}	Sauter mean drop diameter
d_{VS}	Sauter mean drop diameter at zero hold-up
E	coalescence efficiency of dispersed phase in a porous bed
E_D	capture efficiency due to diffusion interception at a fibre surface
E_i	capture efficiency due to inertia interception at a fibre surface
E_0	overall capture efficiency
F	drag force
F	force causing draining of a liquid film during coalescence or
f	friction factor
Fr	Froude number
F^*	dimensionless drag force
g	gravitational constant
$G(x, t)$	separation rate under steady flow in a dispersion band in a continuous flow settler.
$g(x, t)$	disengagement rate at time t
H	dispersion band depth
H	curvature of a surface
H_i	increment depth of coalescence zone in a foam
h	position in dispersion band measured from the inlet
K	$= 1 - \eta$ eqn(5.83)
K_1, K_2	coefficients eqn(5.129)
L	length of a dispersion wedge
L	depth of packing
L	length of a plate in a plate separator

m	mass flow of dispersed phase
N	mixer speed
n	number of layers in a fibre bed separator, eqn(5.173)
N	number of plateau borders in a foam per unit volume
N	number of drops in a population which have not coalesced (Section 5.4)
N_D	dimensionless ratio dp/df, eqn(5.169)
N_f	number of fibres
N_i	molar flux of component i
N_0	total number of drops in a population
n_0	number flux of droplets at $t = 0$
N_p	power number $N_p = Pg/\rho N^3 D^5$
N_p	number of pores in a fibre bed
P	power input
P_R	pressure at a radial position R
Q_d	volume flow of dispersed phase
R	radius of curvature
r	drop radius
R_1, R_2	principal radii of curvature of a surface
Re	Reynolds number
S	average saturation by the dispersed phase of a porous bed
s	standard deviation
St	Stokes number
T	temperature
t	time
t	average time
t_B	primary break time
t_0	dwell or delay time during coalescence
t^*	reference time
U	drop setting velocity
u	fluid velocity
u	superficial velocity
V	volume flow rate
V	volume flux of dispersed phase
v	velocity
V_c	volume flux of continuous phase
V_s	sedimentation velocity
We_{crit}	critical Weber number
We_d	Weber number based on a drop
We_I	impeller Weber number $We_I = N^2 D^3 \rho_c/\sigma$
x	vertical distance of the sedimentation interface from the final
Y	perpendicular distance between adjacent plates in a plate separator
y	vertical distance of the coalescing mixture from the final
Z	locus of the boundary between two zones in a dispersion band
Z, R, θ	cylindrical coordinates

Subscripts

E	extract phase
R	raffinate
c	continuous phase
d	dispersed or drop phase
aq	aqueous phase

Greek symbols

ϵ	*critical film thickness*
$\epsilon(r)$	distance between the surface of a drop at a given radial position r from an interface
ϵ_v	free volume
ϵ	energy dissipation per unit mass
$\Phi(t)$	fractional hold-up of dispersed phase at time t
Φ	fractional hold-up of dispersed phase
ϵ_A	free area
η'	area packing efficiency of droplets at the coalescing interface of dispersion bed
η	volumetric packing efficiency of droplets in the dispersion bed
Θ	residence time of droplets in a dispersion bed
Θ	turbulence intensity
θ	angle of inclination of plates in a plate separator
θ_1	angle subtended of the sedimenting interface in a dispersion band from the horizontal
λ	Hadamard–Rybzynski factor
μ_i	chemical potential of component i in a mixture
μ	viscosity
ν	kinematic viscosity
$\underline{\nu}$	crossflow velocity in a dispersion wedge
$\bar{\xi}$	average capture efficiency
ξ_i	capture efficiency for layer i of a fibre bed
ξ	coefficients in eqn(5.68)
ξ	number of drops in a unit volume of dispersion bed
π	pressure
$\Delta\rho$	difference in density between continuous and dispersed phases
ρ	density
σ_r	surface tension of water
σ	interfacial tension
τ	drop–drop coalescence time
τ_i	coalescence time of a drop at a phase boundary or interface
$\tau(d)$	coalescence time of a drop of diameter d
τ	shear stress
τ'	drop interface coalescence time
$\bar{\tau}$	mean coalescence time
Φ	hold-up of dispersed phase
ϕ'	hold-up at inversion from organic to aqueous dispersions
ϕ''	hold-up at inversion from aqueous to organic dispersions
ϕ	drop diameter

$\psi(r, \theta)$	stream function, eqn(5.161)
θ_h	angle subtended by the coalescing interface in a dispersion band
$\psi(g(x), t)$	volume of a dispersion band with age between t and $t + dt$
ω	Hamaker constant

References

ABDEL-GHANI, M.S. and DAVIES, G.A. (1985). *Chemical Engineering Science*, **40**, 117.

ALLAK, A.M.A. and JEFFREYS, G.V. (1974). *American Institute of Chemical Engineers Journal*, **20**, 564.

ALLAN, R.S. and MASON, S.G. (1961). *Transactions of the Faraday Society*, **57**, 2027.

BARNEA, E. (1978). *Transactions of the Institution of Chemical Engineers*, **56**, 73.

BARNEA, E. and MIZRAHI, J. (1975). *Transactions of the Institution of Chemical Engineers*, **53**, **61**, 70, 75.

BARNEA, E. and MIZRAHI, J. (1978). *Transactions of the Faraday Society*, **66**, 1262.

BART, E. (1968). *Chemical Engineering Science*, **23**, 193.

BASHFORTH, F. and ADAMS, J.C. (1883). *An attempt to test the theories of capillary action*. CUP, Cambridge.

BATCHELOR, G.K. (1951). *Proceedings of the Cambridge Philosophical Society*, **47**, 359.

BATCHELOR, G.K. (1967). *An introduction to fluid mechanics*. CUP, Cambridge.

BAYATIOTIS, B.A. and THORNTON, J.D. (1966). *Institution of Chemical Engineers Symposium Series* No.26, 43.

BRAITHWAITE, J. (1990). Ph.D. thesis. University of Manchester.

BROWN, A.H. and HANSON, C. (1965). *Transactions of the Faraday Society*, **61**, 1754.

BROWN, A.H. and HANSON, C. (1968). *Chemical Engineering Science*, **23**, 841.

BROWN, D.E. and PITT, K. (1972). *Chemical Engineering Science*, **27**, 577.

BURRELL, K.A. and WOODS, D.R. (1973). *Journal of Colloid and Interface Science*, **42**, 15, 35.

CALDERBANK, P.H. (1958). *Transactions of the Institution of Chemical Engineers*, **36**, 443.

CHEN, H.T. and MIDDLEMAN, S. (1967). *American Institute of Chemical Engineers Journal*, **13**, 989.

CHARLES, G.E. and MASON, S.G. (1960). *Journal of Colloid Science*, **15**, 105, 236.

COCKBAIN, E.G. and MCROBERTS, T.S. (1953). *Journal of Colloid Science*, **8**, 440.

COULALOGLOU, C.A. and TAVLARIDES, L.L. (1976). *American Institute of Chemical Engineers Journal*, **22**, 289.

CUTLER, L.A. (1966). *American Institute of Chemical Engineers Journal*, **12**, 35.

DAVIES, C.N. (1952). *Proceedings of the Institute of Mechanical Engineering*, **B1**(5), 185.

DAVIES, G.A. and JEFFREYS, G.V. (1969). *Filtration and Separation*, **6**, 349.

DAVIES, G.A. and THORNTON, J.D. (1977). *Letters in Heat and Mass Transfer*, **4**, 287.

DAVIES, G.A., JEFFREYS, G.V., and ALI, F.A. (1970a). *Chemical Engineer*, **243**, 378.

DAVIES, G. A., JEFFREYS, G. V., SMITH, D. V., and ALI, F. A. (1970b). *Chemical Engineer*, **243**, 378.
DAVIES, G. A., JEFFREYS, G. V., and SMITH, D. V. (1971). *International Solvent Extraction Conference*, **7**, 385.
DAVIES, G. A., JEFFREYS, G. V., and ATZAL, M. (1972). *British Chemical Engineering*, **17**, 709.
DERJAGUIN, B. V., and TITIJERSKAYA, A. S. (1954). *Discussions of the Faraday Society*, **18**, 27.
DOULAH, S. and DAVIES, G. A. (1974). *Proceedings of the International Solvent Extraction Conference*, **3**, 243.
DROWN, D. C. and THOMPSON, W. J. (1977). *Industrial Engineering Chemistry Process Design and Development*, **16**, 197.
EDGE, R. M. and GREAVES, M. (1967). Institution of Chemical Engineers Symposium on Liquid Extraction, Newcastle.
ELTON, G. A. H. and PICKNETT, R. G. (1957). *Proceedings of the 2nd International Congress on Surface Activity*, **1**, 288. Butterworths, London.
FAIRCLOUGH, A. R. N. (1986). Ph.D. Thesis. University of Manchester.
FAIRCLOUGH, A. R. N. and DAVIES, G. A. (1987). *Institution of Chemical Engineers Symposium Series*, **103**, 331.
FAIRCLOUGH, A. R. N., CHAN, K. L., and DAVIES, G. A. (1987). *Chemical Engineering Research and Design*, **65**, 396.
FRANCIS, A. W. (1963). *Liquid–liquid equilibriums*. Interscience, NY.
FUCHS, N. A. and STECHKINA, I. B. (1963). *Annals of Occupational Hygiene*, **6**, 27.
GILLESPIE, T. and RIDEAL, E. K. (1956). *Transactions of the Faraday Society*, **52**, 173.
GODFREY, J. C. (1978). *Transactions of the Institution of Chemical Engineers*, **55**, 207.
GODFREY, J..C. and GRILIE, V. (1977). European Conference on Mixing, BHRA, C1, Cambridge.
GODFREY, J. C., CHAN-KAKOTI, D. K., SLATER, H. J., and THARMALINGAM, S. (1977). *Proceedings of the International Solvent Extraction Conference*, **1**, 406.
GODFREY, J. C., OBI, F. I. N., and REEVE, R. N. (1989). *Chemical Engineering Progress*, **61**.
GOLOB, J. and MODIC, R. (1977). European Conference on Mixing, BHRA, Cambridge.
GONDO, S. and KUSANOKI, K. (1969). *Hydrocarbon Proceedings* **48**, (9), 209.
GONDO, S. and MODHI, (1977). *Advances in separation techniques*, American Institute of Chemical Engineers/Institution of Chemical Engineers Symposium, **1**, 16.
HANSON, C. (1971). *Recent advances in liquid/liquid extraction*. Pergamon Press, Oxford.
HAPPEL, J. (1959). *American Institute of Chemical Engineers Journal*, **5**, 174.
HAPPEL, J. and BRENNER, H. (1965). *Low Reynolds number hydrodynamics*. Prentice Hall, Englewood Cliffs, NJ.
HARDING BLISS, and JOHNSON, H. (1946). *Transactions of the American Institute of Chemical Engineers*, **42**, 331.
HARLETT, R. N. (1969). *Industrial and Engineering Chemistry: Fundamentals*, **8**, 625.
HARTLAND, S. (1968). *Transactions of the Institution of Chemical Engineers*, **46**, 275.

HARTLAND, S. (1969a). *Chemical Engineering Progress Symposium Series*, **65**, 82.
HARTLAND, S. (1969b). *Chemical Engineering Science*, **24**, 987.
HARTLAND, S. (1971a). *Journal of Colloid and Interface Science*, **35**, 2.
HARTLAND, S. (1971b). *Chemical Engineering Science*, **26**, 257.
HARTLAND, S. (1974). *Transactions of the Faraday Society*, **57**, 2027.
HARTLAND, S. and HARTLEY, R.W. (1976). *Axisymmetric fluid interfaces*. Elsevier, Amsterdam.
HARTLAND, S. and JEELANI, S.A.K. (1987). *Chemical Engineering Science* **42**, 1927.
HARTLAND, S. and JEELANI, S.A.K. (1987). *Chemical Engineering Science* **43**, 2421.
HARTLAND, S. and ROBINSON, J.D. (1970). *Chemical Engineering Science*, **25**, 277.
HARTLAND, S. and VOHRA, D.K. (1978). *American Institute of Chemical Engineers Journal*, **24**, 5.
HARTLAND, S. and VOHRA, D.K. (1980). *Journal of Colloid and Interface Science*, **II**, 295.
HODGSON, T.D. and LEE, J.C. (1969). *Journal of Colloid and Interface Science*, **30**, 94.
HODGSON, T.D. and WOODS, D.R. (1969). *Journal of Colloid and Interface Science*, **30**, 429.
HAZLETT, R.N. (1969). *Industrial Engineering Chemistry: Fundamentals*, **8**, 625.
IMAI, I. (1957). *Proceedings of the Royal Society*, **A208**, 487.
IVANOV, I.B., RADOEV, B., MANEV, E., and SCHELUDKO, A. (1970). *Transactions of the Faraday Society*, **66**, 1262.
IWASAKI, (1937). *J.A.W.W.A.*, **29**, 1591.
JACKSON, I.D. (1974). British Patent. Appl. 19266.
JACKSON, I.D., SCUFFHAM, J.B., WARWICK, G.C.I., and DAVIES, G.A. (1974). *Proceedings of the International Solvent Extraction Conference*, **1**, 567.
JEELANI, S.A.K. and HARTLAND, S. (1984). *Canadian Journal of Chemical Engineering*, **64**, 13.
JEELANI, S.A.K. and HARTLAND, S. (1985). *American Institute of Chemical Engineers Journal* **31**, 711.
JEFFREYS, G.V. and DAVIES, G.A. (1971). *Recent advances in liquid-liquid extraction* (ed. C. HANSON), Ch. 14, 495. Pergamon, Oxford.
JEFFREYS, G.V. and HAWKSLEY, J.W. (1962). *Journal of Applied Chemistry*, **12**, 329.
JEFFREYS, G.V. and HAWKSLEY J.L., (1965). *American Institute of Chemical Engineers Journal*, **11**, 413.
JEFFREYS, G.V., DAVIES, G.A., and PITT, K. (1970a). *American Institute of Chemical Engineers Journal*, **16**, 832.
JEFFREYS, G.V., DAVIES, G.A., and SMITH, D.V. (1970b). *American Institute of Chemical Engineers Journal*, **16**, 827.
JEFFREYS, G.V. SMITH, D.V., and PITT, K. (1967). *Institution of Chemical Engineers Symposium Series*, No. 26, 93.
KAPLAN, S. (1957). *Journal of Mathematics and Mechanics*, **6**, 595.
KOLMOGOROFF A.N. (1949). *Doklady Akademii Nauk SSSR*, **66**, 825.
KOMOSAVA, I. and OTAKE, T. (1970). *Journal of Chemical Engineering of Japan*, **3**, 243.

KONNECKI, H.G. (1959). *Zeitschrift fuer Physikalische Chemie (Leipzig)*, **211**, 208.
KUMAR, A., VOHRA, D.K., and HARTLAND, S. (1980). *Canadian Journal of Chemical Engineering*, **58**, 154.
KUWABARA, S. (1959). *Journal of the Physical Society of Japan*, **19**, 527.
LAMB, H. (1932). *Hydrodynamics* (6th edn). Cambridge University Press.
LANGDEN, W.M., NAIK, P.P., and WASAN, D.T. (1972). *Environmental Science and Technology*, **6**, 10.
LANG, S.B. (1962). Ph.D. Thesis. University of California.
LANG, S.B. and WILKE, C.R. (1971). *Industrial and Engineering Chemistry: Fundamentals*, **10**, 341.
LAWSON, G.B. (1967). Ph.D. Thesis. University of Manchester.
LEE, J.C. and TASAKORN, P. (1979). *Proceedings of the 3rd European Conference on Mixing*, 157, BHRA, York.
LEMLICH, R. and LEONARD, R.A. (1965). *Chemical Engineering Science*, **20**, 790.
LEONARD, R.A. and LEMLICH, R. (1965). *American Institute of Chemical Engineers Journal*, **11**, 18.
LEWIS, I.E. (1977). *Proceedings of the International Solvent Extraction Conference*, **1**, 325.
LINDE, H. and WINKLER, K. (1964). *Zeitschrift fuer Physikalische Chemie*, **225**, 223.
LINDE, H. and WINKLER, K. (1965). *Zeitschrift fuer Physikalische Chemie*, **230**, 207.
LIN, C.Y. and SLATTERY, J.G. (1982a). *American Institute of Chemical Engineers Journal*, **28**, 147.
LIN, C.Y. and SLATTERY, J.G. (1982b). *American Institute of Chemical Engineers Journal*, **28**, 786.
LOTT, J.B., WARWICK, G.C.I., and SCUFFHAM, J.B. (1972). *Transactions of the Society of Mining Engineers*, **252**, 27.
LOWES, L. and LARKIN, H.J. (1967). *Institution of Chemical Engineers Symposium Score*, **26**, 111.
MACKAY, G.D.H. and MASON, S.G (1963). *Canadian Journal of Chemical Engineering*, **41**, 203.
MARANGONI, C. (1865). *Sull expansione delle gocie di un liquido galleggiante sulla superficie di altro liquido*. Fusi, Pavia.
MARANGONI, C. (1871). *Annalen der Physik*, **143**, 337.
MCAVOY, R.M., WIEGARD, W.A., TOMKIN, E.E., and KINTNER, R.G. (1965). *Advances in Separation Techniques, American Institute of Chemical Engineers/Institution of Chemical Engineers Symposium*, **1**, 16.
MEIJERING, J.L. (1953). Phillips Research Rept. 8, 270.
MIZRAHI, J. and BARNEA, E. (1973). *Process Engineering*, **1**, 60.
MLYNECK and RESNICK (1972). *American Institute of Chemical Engineers Journal*, **18**, 122.
NATANSON, G.L. (1962). *Kolloidnyi Zhurnal*, **24**, 52.
NAYLOR, A. and LARKIN, M.J. (1971). *Proceedings of the International Solvent Extraction Conference*, **2**, 1356.
NIELSEN, L.E., WALL, R., and ADAMS, G. (1958). *Journal of Colloid Science*, **13**, 441.
PERRY, J.H. (ed) (1963). *Chemical engineers handbook* (4th edn), McGraw-Hill, New York.

PICH, J. (1966). *Staub*, **25**, 2.
PIEKAAR, H. W. and CLARENBURG, L. A. (1967). *Chemical Engineering Science*, **22**, 1399.
PRINCEN, H. M. (1963). *Journal of Colloid Science*, 18, 178.
PRINCEN, H. M. and MASON, S. G. (1965). *Journal of Colloid Science*, **20**, 156.
RADUSHKEVICH, L. V. (1946). *Russian Journal of Physical Chemistry*, **40**, 4, 520.
REED, X. B., RIOLO, E. and HARTLAND, S. (1974). *Journal of Multiphase Flow*, **1**, 437.
REYNOLDS, O. (1881). *Chemical Newsletter*, **44**, 221.
ROBERTS, J. (1977). *Proceedings of the International Solvent Extraction Conference*, **1**, 428.
ROBERTS, J., LYNE-SMITH, K., COLLIER, D. E. and MCGEE, J. (1977). *Proceedings of the International Solvent Extraction Conference*, **1**, 418.
ROSE, P. M. (1963). M. Sc. Thesis. Illinois Institute of Technology, Chicago.
ROWDEN, G. A., SCUFFHAM, J. B., and WARWICK, G. C. I. (1974). *Proceedings of the International Solvent Extraction Conference*, **1**, 81.
ROWLEY, M. E. and DAVIES, G. A. (1988). *Chemical Engineering Research and Design*, **66**, 313.
RUCKENSTEIN, E. and JAIN, (1974). *Faraday Transactions* II, **70**, 132.
RUCKENSTEIN, E. and JAIN, (1976). *Journal of Colloid and Interface Science*, **54** 108.
RUSHTON, E. and DAVIES, G. A. (1970). *International Fluid Mechanics Symposium*, 3, McMaster University.
RUSHTON, E. and DAVIES, G. A. (1973a). *Applied Science Research*, **28**, 37.
RUSHTON, E. and DAVIES, G. A. (1973b). *Interfacial phenomena in two phase flow*, American Institute of Chemical Engineers Symposium.
RUSHTON, E. and DAVIES, G. A. (1974). *International Solvent Extraction Conference*, **1**, 289.
RUSHTON, E. and DAVIES, G. A. (1978). *International Journal of Multiphase Flow*, **4**, 357.
RYON, A. D., DALEY, F. L., and LOWRIE, R. A. (1959). *Chemical Engineering Progress*, **55**, 70.
RYON, A. D., DALEY, F. L., and LOWRIE, R. A. (1959). *Chemical Engineering Progress*, **55**, 70.
RYON, A. D., DALEY, F. L., and LOWRIE, R. A. (1960). Rept. No. O R N L 2951, Oak Ridge National Laboratory, Tennessee.
SAWISTOWSKI, H. (1971). *Recent advances in liquid–liquid extraction* (ed. E. C. HANSON), Ch. 11. Pergamon, Oxford.
SAWISTOWSKI, H and JAMES, B. R. (1963). *Chemie-Ingenieur-Technik*, **35**, 175.
SCHEIDEGGER, A. E. (1960). *Physics of flow through porous media*. Univ. Toronto Press, Toronto.
SCHELDULKO, A. and MANOV, E. (1968). *Transactions of the Faraday Society*, **64**, 1123.
SHERONY, D. F. and KINTNER, R. C. (1971). *Canadian Journal Chemical Engineering*, **49**, 314, 321.
SHERONY, D. F., KINTNER, R. C. and WASAN, D. T. (1978). Coalescence of Secondary Emulsions in Fibrous Beds in Surface and Colloid Science.
SHINNAR, R. and CHURCH, J. M. (1960). *Industrial and Engineering Chemistry*, **52**, 253.

SMITH, D.V. (1967). M. Sc. Thesis. University of Manchester.
SMITH, D.V. (1970). Ph.D. Thesis. University of Manchester.
SMITH, D.V. and DAVIES, G.A. (1970). *Canadian Journal of Chemical Engineering*, **48**, 628.
SMITH, D.V. and DAVIES, G.A. (1972). *American Institute of Chemical Engineers Symposium Series*, **68**, 124.
SMOLUCHOWSKI, J. (1917). *Zeitschrift fuer Physikalische Chemie*, **92**, 129.
SPIELMAN, L. and GOREN, S.L. (1968). *Environmental Science and Technology*, **2**, 179.
SPIELMAN, L.A. and GOREN, S..L. (1970). *Industrial and Engineering Chemistry*, **62**(10), 10.
SPIELMAN, L.A. and GOREN, S.L. (1972). *Industrial Engineering Chemistry: Fundamentals*, **11**, 66.
SPROW, F.B. (1969). *Chemical Engineering Science*, **22**, 435.
STANDARD, G.L. (1964). *Chemical Engineering Science*, **19**, 227.
STEWART, G. and THORNTON, J.D. (1966). *Institution of Chemical Engineers Symposium Series*, No. 26, 29, 37.
STÖNNER, H.M. and WÖHLER, F. (1975). *Institution of Chemical Engineers Symposium Series*, 42, 14, 1.
THOM, A. (1933). *Proceedings of the Royal Society*, **A141**, 651.
TOMOTIKA, S. and AOI, T. (1950). *Quarterley Journal of Mechanics and Applied Mathematics*, **3**, 140.
TREYBAL, R.E. (1963). *Liquid extraction* (2nd edn). McGraw-Hill, New York.
VAN DER TEMPLE, M. (1958). *Journal of Colloid Science*, **13**, 125.
VAN HEUVEN, J.W. and BECK, M.S. (1971). *Proceedings of the International Solvent Extraction Conference*, **1**, 70.
VERMEULEN, T., WILLIAMS, G.M., and LANGLOIS, G. (1955). *Chemical Engineering Progress*, **57**, 85F.
VIELER, A.M.S., GLASSER, D., and BRYSON, A.W. (1977). *Proceedings of the International Solvent Extraction Conference*, **1**, 377.
VIJAYAN, S. and PONTER, A.B. (1974). *Proceedings of the International Solvent Extraction Conference*, **1**, 591.
VIJAYAN, S. and PONTER, A.D. (1976). *Tenside Detergents*, **13**, 193.
VRIJ, A. (1966). *Discussions of the Faraday Society*, **42**, 23.
WARWICK, G.C.I. (1974). *Proceedings of the International Solvent Extraction Conference*, **1**, 81.
WARWICK, G.C.I., SCUFFHAM, J.B., and LOTT, J.B. (1971). *Proceedings of the International Solvent Extraction Conference*, **2**, 1373.
WASAN, D.T., SHERONY, D.F., and KINTNER, R.C. (1978). *Surface and Colloid Science*, **10**, 99.
WILKINSON, E.T. and DAVIES, G.A. (1985). *Canadian Journal of Chemical Engineering*, **63**, 891.
WILKINSON, E.T., FAIRCLOUGH, A.R.N. and DAVIES, G.A. (1986). *Proceedings of the 4th World Filtration Congress*, **1**, 7.
WORTHINGTON, A.M. and COLE, R.S. (1895). *Proceedings of the Royal Society*, **33**, 137.
ZEBAL, (1965). *Journal of Colloid Science*, **20**, 522.

INTRODUCTION TO CHAPTERS 6, 7, AND 8

H.R.C. PRATT and G.W. STEVENS

Design procedures for countercurrent extractors

Types of operation and equipment

Liquid–liquid extraction, as normally practised, is a member of the class of countercurrent diffusional separation processes, in which it ranks as second in importance only to distillation. Its applications cover a wide range of industries, e.g. chemical, petroleum, metallurgical, pharmaceutical and food processing, and its scale of operation can range from that of the laboratory to throughputs of many thousands of tonnes per week.

In its simplest from liquid extraction involves the separation of one of the components of a binary mixture (the solute) using a solvent which is only partially miscible with the other component. Thus a minimum of three components is required, although it is apparent that more components may be present in either the solvent, the mixture being separated, or both. In an alternative form a mixture of two solutes is separated by fractional distribution between two partially miscible solvents. This method of operation, known as dual solvent extraction, requires a minimum of four components, although again more may be present in either the solvents, the feed mixture, or both.

Various types of extraction process are available, the simplest being single-contact batch extraction, analogous to the laboratory separating funnel, in which the feed mixture is agitated with solvent, allowed to settle and the phases separated. Continuous operation is more usual, however, and includes both single and multistage processes, the latter generally in countercurrent form. These processes will be described in the next chapter. More complex batch processes have also been proposed, but have achieved little importance and will therefore receive only brief treatment.

A variety of extractor types is available, comprising discrete-stage mixer–settlers in which the phases are repeatedly mixed and separated; differential columns in which the concentration gradients are continuous; multi-compartment cell-type columns with or without agitation; and centrifugal contactors, which may be of either mixer–settler or differential type. It should be noted that the column types have only a single interface, in contrast to mixer–settlers which have one per stage.

Basis of Modelling Procedures

The process design of liquid extraction equipment requires the determination of two independent parameters, viz.(i) the cross-section required to accommodate the desired flows without 'flooding', and (ii) the number of mixer-settler stages, or height of column required to achieve the desired performance (i.e. degree of extraction of solute(s) from the feed). True mixer-settlers achieve a high efficiency per stage—generally over 90 percent—and their performance is usually evaluated in terms of the number of theoretical (i.e. 'ideal') stages together with the stage efficiency; however, an alternative approach for calculating the number of non-ideal stages will also be described. Countercurrent columns of both differential and compartment types mostly suffer from axial dispersion of one or other, or both phases, which reduces efficiency. Axial dispersion models covering both types are available, and in view of their complexity will be described in Chapter 7 (p. 416), following the traditional 'plug flow' models in Chapter 6 (p. 345).

The applications of the theoretical models in practice are threefold, viz.(i) design, for predicting number of stages (or compartments), or height to achieve a desired performance, (ii) performance, to evaluate a suitable scale-up parameter from pilot scale data, and (iii) operational, to predict the exit concentrations for a given column. Although it would be desirable to predict column height from first principles, in fact the mass transfer process involving droplet 'swarms' has hitherto defied simple interpretation. It is therefore generally necessary in practice to conduct tests with the given system on a pilot scale column. The performance can then be interpreted as in (ii) above, and the resulting value of the parameter used in (i), to determine the design, or (iii), to determine the operational performance for a given column.

Following the accounts of the plug flow and backmixing models in the next two chapters, a description will be given in Chapter 8 (p. 492) of the more common extractor types together with methods for their selection, design, and scale-up. This will necessarily include also the consideration of the fluid dynamical factors, mentioned earlier, which determine extractor cross-section and flood-point.

6

PLUG FLOW ANALYSIS

G.W. STEVENS and H.R.C. PRATT

6.1 Introduction

6.1.1 General

The number of stages, or height of column required for a given separation, is obtained from a combination of the equilibrium relationship, material balance, and mass transfer rates. Contactor performance is normally expressed either in terms of the number of ideal (i.e. theoretical) stages required together with the stage efficiency, or as the number of transfer units. The theoretical stage approach is strictly applicable only to staged contactors such as mixer–settlers, and the transfer unit approach to differential contactors such as packed columns; however, many contactor types used in practice are of an intermediate class and can be treated by either approach. Both the theoretical stage and the transfer unit methods will be described in this chapter. However, as the former cannot be realized in practice, an alternative approach based on non-ideal stages will also be described.

The analyses presented in this chapter are based on the assumption that both phases flow through the contactor in plug flow; that is, every fluid element moves with the same velocity and all elements have the same residence time in the column. This restriction will be relaxed in the next chapter, when axial dispersion is considered.

6.1.2 Concentration units

6.1.2.1 *Frames of reference*

The most suitable frame of reference for mass transfer calculations is that of no net volume flux; however, in practice it is inconvenient to use concentrations based on volume and a mass or molar basis is used. If the mass density is constant with composition it can be shown that the mass average velocity is equal to the volume average velocity, and this can be used as a reference basis. A similar argument applies to the molar average velocity; however, the mass density rather than the molar density is normally found to vary less with concentration in the case of the liquid phase, and so more closely approximates the ideal case of no net volume flux in liquid extraction calculations.

The most suitable concentration units are therefore kg/m^3, although $kmol/m^3$ is sometimes used. Weight or mole fraction units have also been used in the past, and the former are still necessarily used in the presentation of some phase equilibrium diagrams (see Section 6.1.3, p. 348), as these require the concentrations of all the components to sum to unity. They are not to be recommended, however, for computations involving mass transfer rates.

6.1.2.2 *Dimensionless concentrations*

In the case of the more complex treatments required when axial dispersion, especially backmixing, occurs it has been found convenient to express concentrations in dimensionless form by a change of coordinates. This enables generalized solutions of extractor performance to be obtained and applied to a variety of operational modes. Since the generalized solutions reduce to those for the plug flow cases in the limit, it is convenient to express the present solutions in dimensionless form also.

A typical operating (i.e. equilibrium distribution) diagram for counter-current extraction is shown in Fig. 6.1(a), where AB represents a linear approximation to the equilibrium line, defined by eqn (6.1), and CD the operating line

$$c_y^* = mc_x + q \qquad (6.1)$$

The coordinate change is conducted as follows: (i) the exit feed ($X-$) phase composition $c^*_{x,N+1}$, in equilibrium with the inlet solvent ($Y-$) phase, of composition $c_{y,N+1}$, is located at A, to which the coordinate origin is moved; (ii) the coordinate scales are expanded or contracted so that the inlet X-phase composition $c_{x,0}$ and the Y-phase composition in equilibrium with this have values of 1.0. In this way the coordinates of a point such as P are defined by $X = AH/AJ$ and $Y = AE/AG$. The modified operating diagram thus obtained is shown in Fig. 6.1(b), where the equilibrium line, of unit slope, is represented by $Y^* = X$.

The dimensionless concentrations, defined in terms of the measured concentrations c_x and c_y in units of mass per unit volume are as follows

$$X = \frac{c_x - c^*_{x,N+1}}{c_{x,0} - c^*_{x,N+1}} = \frac{c_x - (c_{y,N+1} - q)/m}{c_{x,0} - (c_{y,N+1} - q)/m} \qquad (6.2)$$

$$Y = \frac{c_y - c_{y,N+1}}{c^*_{y,0} - c_{y,N+1}} = \frac{c_y - c_{y,N+1}}{mc_{x,0} - (c_{y,N+1} - q)} \qquad (6.3)$$

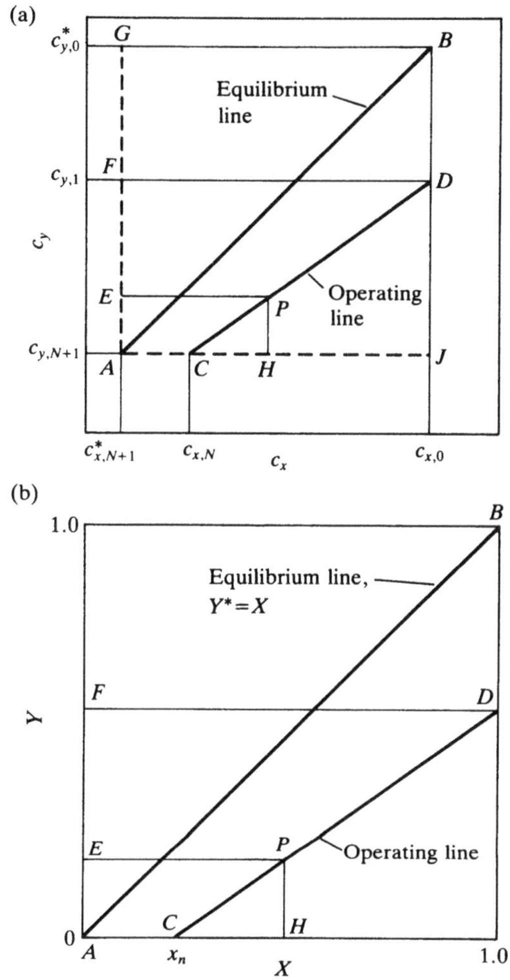

Fig. 6.1. Operating diagrams for countercurrent extraction: (a) on c_x-c_y coordinates; (b) using dimensionless concentrations.

The corresponding expressions for differential extractors are obtained by replacing the subscript $N + 1$ by I.

6.1.2.3 *High solute concentrations*

Solute transfer within an extractor results in changes in the flows and thus to non-linearity of the mass balance (i.e. operating line) relationship. A method of allowing for this, applicable to relatively low solute concentrations, is described in Section 6.3.3 (see p. 00). However, in some circumstances, especially in dual-solvent extraction (Section 6.7, see p. 00), the change in solute concentration is much greater, and it is then necessary to express the solvent flows on a solute-free basis, and the concentrations as kg of solute per m^3 of pure (i.e. solute-free) solvent. By this means the operating line is rendered strictly linear, provided that the solvent miscibility remains unchanged; on the other hand, the linearity of the equilibrium relation will generally be decreased.

These concentration units can also be expressed in dimensionless form, using eqns (6.2) and (6.3) with the appropriate values of m and q. This is particularly convenient when using the generalized design equations for the calculations of dual-solvent extractors, as described in Section 6.7.5 (see p. 00) and in Section 7.4.5 (see p. 00).

6.1.3 Graphical representation of equilibrium data
[See also Newsham, Ch. 1]

The extent of extraction of a single solute from a binary mixture by an appropriate solvent can be determined from a ternary equilibrium diagram. Ternary equilibrium data are commonly represented either on equilateral or right-angled triangular diagrams, on the rectilinear solvent-free coordinate system of Janecke (1906), or on rectangular equilibrium distribution or selectivity diagrams. A ternary system with only one partially miscible pair of components is known as a Type I system (Treybal 1963), an example of which is shown in Fig. 6.2(a). The term Type II(b) is used to describe a ternary system with two partially miscible pairs and a single two-phase region in a band across the equilibrium diagram (Fig. 6.2(b)). Type III ternary systems, with two partially miscible pairs forming two separate two-phase regions, are comparatively rare.

Figure 6.3 shows the relationship between the two triangular coordinate systems and the equilibrium distribution diagram, with concentratiohs expressed in mass fraction units. The curve DRHPECG in Figs 6.3(a) and

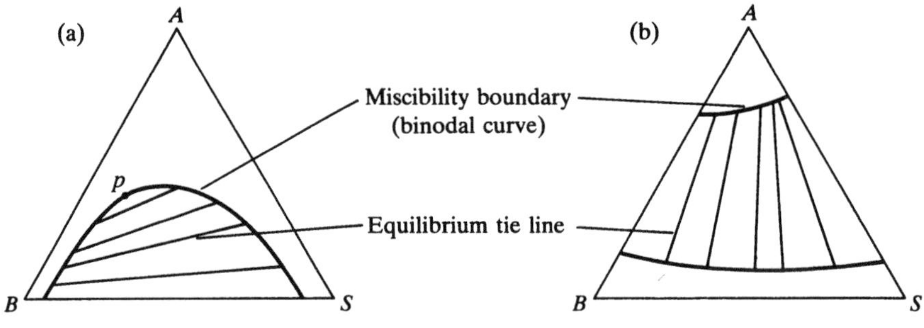

Fig. 6.2. Ternary equilibria: (a) Type I; (b) Type II.

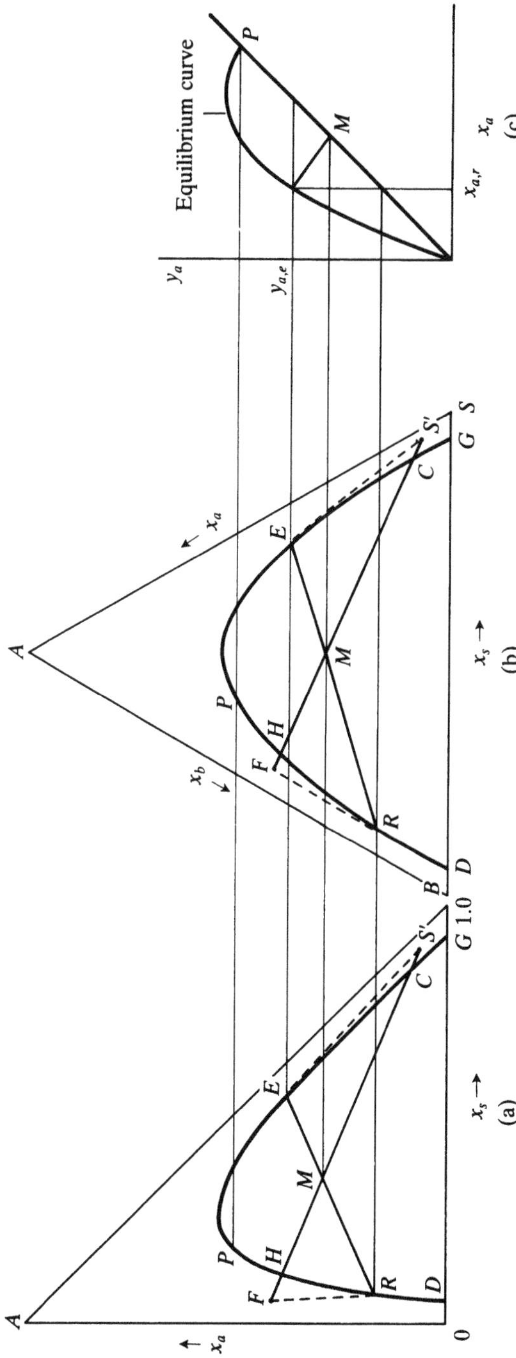

Fig. 6.3. Single-stage extraction on: (a) rectangular (right-angled triangular) coordinates; (b) equilateral traingular coordinates; (c) projection on to equilibrium distribution diagram.

(b) represents the miscibility boundary and point P the 'plait point', where the two conjugate phases have the same composition. The line RME represents a typical tie line, the ends of which represent the composition of the conjugate phases for a mixture of composition M within the two-phase region. In the equilibrium distribution diagram (Fig.6.3(c)), x_a, the concentration of solute A in the feed phase, is plotted against that in the solvent phase, y_a. For the system in Fig. 6.3 the solvent S has a greater affinity for the solute A so that the equilibrium line lies above the diagonal, i.e. $y_{a,E}/y_a$ is greater than unity; if the feed solvent, B, had a greater affinity for the solute, the equilibrium would lie below the diagonal. The plait point always lies on the diagonal as the composition of the two phases are identical.

The Janecke diagram is constructed by plotting the solvent ratio $S/(A+B)$, that is, the mass ratio of pure solvent to total solute, against the extract and raffinate compositions on a solvent-free mass fraction basis, as shown in Fig. 6.4(a) for a Type I system. These compositions are defined as follows

$$\bar{x}_a = x_a/(x_a + x_b) \tag{6.4}$$

$$\bar{y}_a = y_a/(y_a + y_b) \tag{6.5}$$

The equilibrium selectivity diagram consists of a rectilinear plot of $\bar{x}_a$ vs $\bar{y}_a$. This is shown in Fig. 6.4(b) as a projection from the Janecke diagram.

6.2 Single and multiple contact extraction

6.2.1 Single-stage extraction

The extract and raffinate compositions leaving a single stage can be obtained by combining the equilibrium relationship with mass balances taken around the stage. The latter are as follows, using mass fraction units

Total mass: $$F + S = R + E = M \tag{6.6}$$

Solute A: $$Fx_{a,F} + Sy_{a,S} = Rx_{a,R} + Ey_{a,E} = Mx_{a,M} \tag{6.7}$$

Solvent S: $$Fx_{s,F} + Sy_{s,S} = Rx_{s,R} + Ey_{s,E} = Mx_{s,M} \tag{6.8}$$

In these equations M, $x_{a,M}$, and $x_{s,M}$ refer to the mixture M. These equations can be rewritten as follows

$$F + S - M = 0 \tag{6.9}$$

$$Fx_{a,F} + Sy_{a,S} - Mx_{a,M} = 0 \tag{6.10}$$

$$Fx_{s,F} + Sy_{s,S} - Mx_{s,M} = 0 \tag{6.11}$$

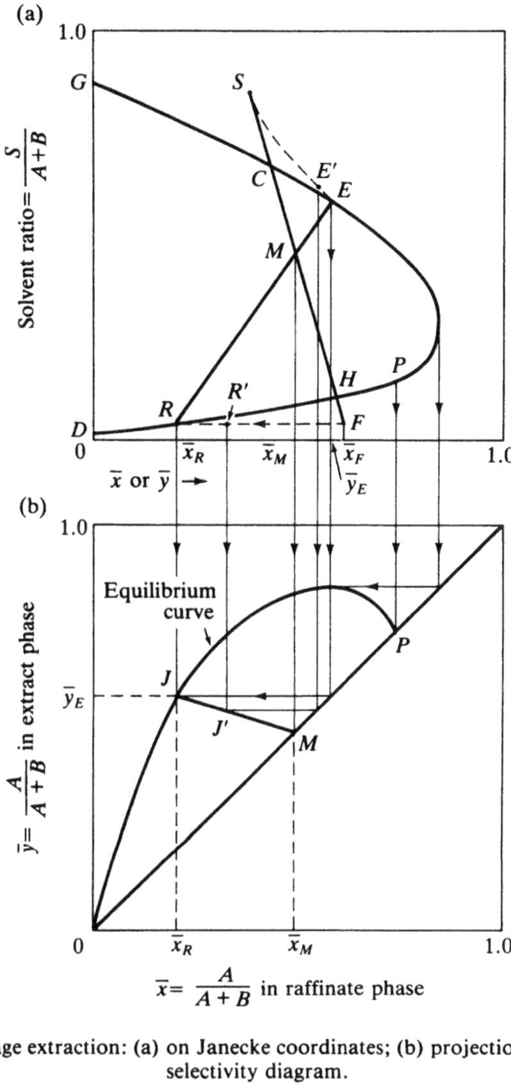

Fig. 6.4. Single-stage extraction: (a) on Janecke coordinates; (b) projection on to equilibrium selectivity diagram.

Equations (6.9–6.11) comprise a set of linear homogeneous equations in F, S, and M which have a nontrivial solution only if the determinant of the coefficients is zero; thus

$$\begin{vmatrix} 1 & 1 & 1 \\ x_{a,F} & y_{a,S} & x_{a,M} \\ x_{s,F} & y_{s,S} & y_{s,M} \end{vmatrix} = 0 \tag{6.12}$$

Equation (6.12) expresses the condition that the three points $(x_{a,F}, x_{s,F})$, $(y_{a,S}, y_{s,S})$, and $(x_{a,M}, x_{s,M})$ are collinear,* that is, M lies on the line FS′ in Fig. 6.3a. The location of M is found by elimination of M from eqns (6.9 and 6.10), giving

$$\frac{F}{S} = \frac{x_{a,M} - y_{a,S}}{x_{a,F} - x_{a,M}} = \frac{MS'}{MF} \tag{6.13}$$

This is an example of the *lever arm rule*, and is applicable to all conserved property diagrams, including the Janecke solvent-free diagram. Thus, F and S′, representing the feed and the solvent compositions, are located on Figs 6.3(a) or (b) and joined. The compositions of the extract and raffinate products are then given by the intersections with the phase envelope of the equilibrium tie line RE through mixture point M, obtained from eqn (6.13). The amount of extract is obtained by eliminating R from eqns (6.6 and 6.7), and that of raffinate from eqn (6.6). Finally, elimination of M from these two equations gives R/E = EM/MR. The equilibrium distribution diagram can be obtained by projection from either of the triangular diagrams as shown in Fig. 6.3. The analogous construction on the Janecke and equilibrium selectivity diagrams is shown in Fig. 6.4.

6.2.2 Multistage crossflow extraction

6.2.2.1 *Graphical methods*

Multistage crossflow extraction can be considered as an extension of single-contact extraction in which the raffinate from each stage is re-treated with fresh solvent. The concentrations of each of the streams can be determined in a similar manner to that for single-contact extraction on the triangular or Janecke diagram. An example of the latter is shown in Fig. 6.5, in which the successive raffinates are indicated by R_1, R_2, etc. Since the raffinate from any stage provides the feed to the next, eqns (6.6–6.8) are written in terms of stage n, replacing R and E by $\bar{R}_n$ and $\bar{E}_n$ respectively, and similarly F and S are replaced by $\bar{R}_{n-1}$ and $\bar{S}_n$, where the overbar refers to solvent-free values. In Fig. 6.5, the case where impure solvent of composition $\bar{S}$ is used is shown on the right, and that of pure solvent on the left; in the latter case

*The determinant

$$\begin{vmatrix} 1 & 1 & 1 \\ x_1 & x_2 & x_3 \\ y_1 & y_2 & y_3 \end{vmatrix} = 0$$

is equal to twice the area of a triangle with vertices (x_1, y_1), (x_2, y_2), (x_3, y_3). If this is equal to zero the points are collinear.

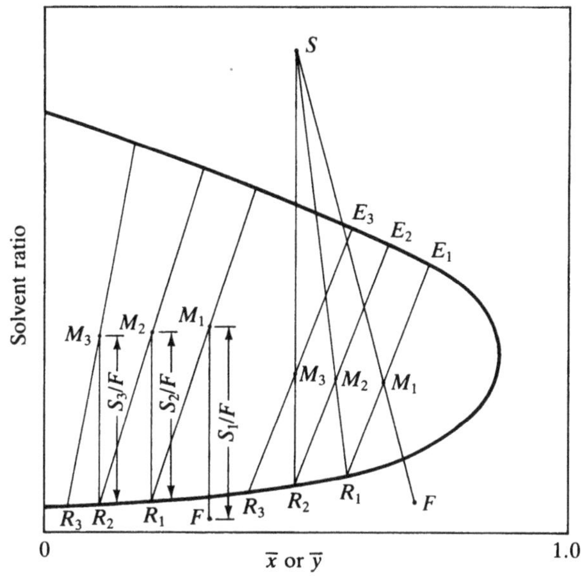

Fig. 6.5. Multistage crossflow extraction on Janecke coordinates. The left-hand construction applies to pure solvent, the right-hand one to impure solvent.

point S is at infinity so that the tie lines through M_n are vertical. The Janecke diagram can be projected on to the selectivity diagram if desired, in the same way as for single stage extraction.

The various extracts are usually combined, giving a mixture of composition

$$\bar{y}_{av} = \sum_n \bar{E}_n \bar{y}_n / \sum_n \bar{E}_n \tag{6.14}$$

Recovery of solvent, if required, is normally carried out on the combined extracts and on the final raffinate. In general, unless the solvent is solute-free, one of the tie lines on extension will pass through the point S on Fig. 6.5; this corresponds to the maximum possible degree of extraction and requires an infinite number of stages.

A simpler method, applicable when the solvents are effectively immiscible, has been described by Hunter and Nash (1932) and Evans (1934); see also Pratt (1983*a*, Section 2.4.1(b)).

6.2.2.2 *Analytical method*

An analytical solution is available for the case of effectively immiscible solvents if the solute-free solvent ratio, $\bar{R}/\bar{E}$, on each stage is constant. By expressing the concentrations in dimensionless units (Section 6.1.2.2, see p. 00), the initial feed concentration $X_F = 1.0$ and the inlet solvent concentration $Y = 0$.

The material balance for any stage is given by

$$Y_n = E(X_{n-1} - X_n) \tag{6.15}$$

where $E = \bar{R}/m\bar{E}$, the extraction factor. The equilibrium relationship in the same units is $Y_n^* = X_n$ so that eqn (6.15) becomes, for an equilibrium stage,

$$X_N(1 + 1/E) - X_{n-1} = 0 \tag{6.16}$$

The solution to eqn (6.16), with boundary condition $X_n = X_N$ when $n = N$, is then

$$X_N = (1 + 1/E)^{-N} \tag{6.17}$$

In deriving eqn (6.17) it is assumed that the same amount of solvent is used in each stage. If this is not the case, the right hand side of this equation must be replaced by the reciprocal of the products of $(1 + 1/E)$ for each stage.

Equation (6.17) also shows that, for a given number of stages, the degree of extraction from the feed always increases (i.e. X_N decreases) as the solvent quantity is increased. However, the proportion removed by a given amount of solvent relative to that removed by the same amount using infinite stages always passes through a minimum as the solvent quantity is increased; Evans (1934) and Tiller (1949) have shown that this occurs at a value of E between 0.56 and 0.63, depending upon the number of stages. The relative efficiency, defined in this way, does not increase very rapidly when the number of stages is greater than four.

6.3 Simple countercurrent extraction. (I) Ideal stages

6.3.1 Graphical methods

In countercurrent extraction the feed and solvent enter at opposite ends of the extractor so that the raffinate and extract flow countercurrently, as shown in Fig. 6.6. By this means the outgoing extract is brought into contact with entering feed containing the highest concentration of solute, thus achieving a low solvent consumption and a high extract concentration.

6.3.1.1 *General solution*

Either of the two triangular coordinate systems or the Janecke method described in Section 6.2.1 (see p. 00) for single-contact extraction can be extended to cover continuous countercurrent extraction in stagewise contactors. Treybal (1963, 1981) has described the triangular coordinate methods; in most respects, however, Janecke coordinates are more convenient and their use is described below.

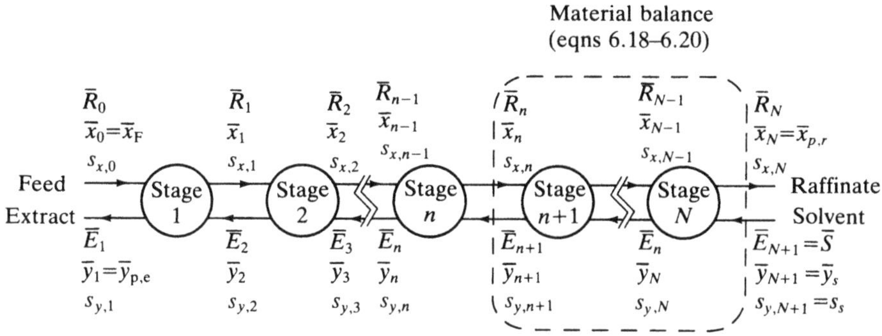

Fig. 6.6. Countercurrent extraction: stage material balance in solute-free concentration units.

The required relationships are obtained by taking material balances between any two stages n and $n + 1$ and around the solvent inlet end in terms of solvent-free flows and compositions as shown in Fig. 6.6; thus

Non-solvent $A + B$: $\bar{R}_n - \bar{E}_{n+1} = \bar{R}_N - \bar{S} = \bar{\Delta}_R$ (6.18)

Solute A: $\bar{R}_n\bar{x}_n - \bar{E}_{n+1}\bar{y}_{n+1} = \bar{R}_N\bar{x}_N - \bar{S}\bar{y}_s = \bar{\Delta}_R\bar{x}_{\Delta,R}$ (6.19)

Solvent S: $\bar{R}_n s_{x,n} - \bar{E}_{n+1}s_{y,n+1} = \bar{R}_N s_{x,N} - \bar{S}s_s = \bar{\Delta}_R s_{\Delta,R}$ (6.20)

The terms containing $\bar{\Delta}_R$ refer to the net flows of non-solvent, solute and solvent, which are constant throughout the extractor. On subtracting the terms containing $\bar{\Delta}_R$ from those on the left-hand side, a set of linear homogeneous equations in $\bar{R}_n$, $\bar{E}_{n+1}$, and $\bar{\Delta}_R$ is obtained; these have a non-trivial solution given by

$$\begin{vmatrix} 1 & 1 & 1 \\ \bar{x}_n & \bar{y}_{n+1} & \bar{x}_{\Delta,R} \\ s_{x,n} & s_{y,n+1} & s_{\Delta,R} \end{vmatrix} = 0 \qquad (6.12)$$

This expresses the condition that the points $(\bar{x}_n, s_{x,n})$, $(\bar{y}_{n+1}, s_{y,n+1})$ and $(\bar{x}_{\Delta,R}, \bar{s}_{\Delta,R})$ are collinear; consequently a line drawn through the net flow point Δ_R on the Janecke diagram [Fig. 6.7(a)] and the miscibility boundary will intersect the latter at points that satisfy the material balance.

By combining the terms in $\bar{R}_N$, $\bar{S}$ and $\bar{\Delta}_R$ in eqns (6.18–6.20) in the same way and writing in determinant form it can be shown that the points $(\bar{x}_N, s_{x,N})$, $(\bar{y}_s, s_s)$ and $(\bar{x}_{\Delta,R}, s_{\Delta,R})$ are also collinear. Consequently Δ_R lies on an extension of SR_N on Fig. 6.7(a), where R_N has the coordinates $(\bar{x}_{x,N}, s_{x,N})$; the coordinates of Δ_R are obtained as follows from eqns (6.18–6.20)

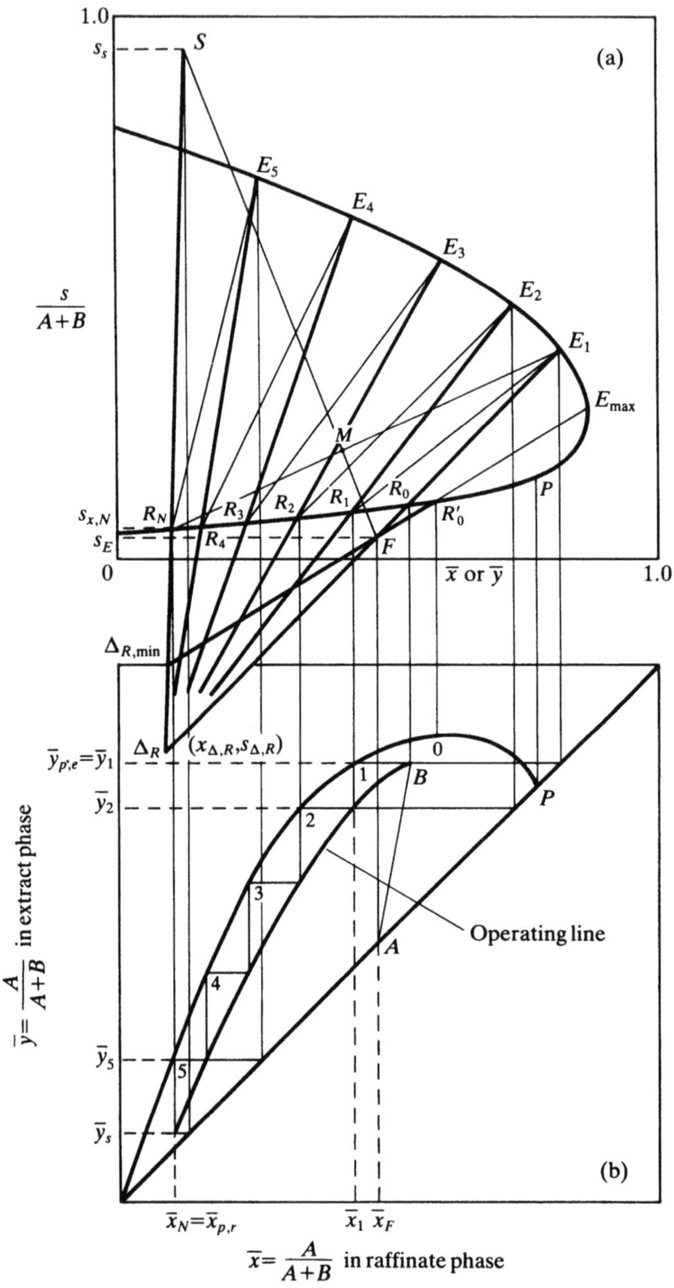

Fig. 6.7. Countercurrent extraction: (a) on Janecke coordinates; (b) projection on to selectivity diagram.

$$\bar{x}_{\Delta,R} = \frac{\bar{R}_N\bar{x}_N - \bar{S}y_s}{\bar{R}_N - \bar{S}}\ ;\ s_{\Delta,R} = \frac{\bar{R}_N s_{x,N} - \bar{S}s_s}{\bar{R}_N - \bar{S}} \tag{6.22}$$

If the entering solvent is solute-free, $\bar{S} = 0$ and $\bar{x}_{\Delta,R} = \bar{x}_N$; also, s_s is infinite and $\bar{S}s_s = S$, the quantity of pure solvent fed. Consequently $\bar{s}_{\Delta,R} = (s_N - S/R_N)$, so that Δ_R is located vertically below R_N at a distance $S/\bar{R}_N$, with S at infinity. This distance corresponds to the solvent ratio, that is, the ratio of solvent fed to raffinate product.

The number of theoretical stages can now be determined, using the construction shown in Fig. 6.7(a). The net flow point Δ_R is located using eqns (6.22), or from the solvent ratio for pure solvent, and is joined to feed point F with coordinates $(\bar{x}_0,\ s_{x,0})$; this line, on extension, meets the miscibility boundary at E_1, giving the composition and solvent content of the extract product. Equilibrium tie-line E_1R_1 is then drawn through E_1 and $\Delta_R R_1$ is extended to meet the miscibility boundary at E_2, giving the composition of the second stage extract. This procedure, drawing alternate equilibrium tie-lines and material balance lines is repeated until the specified exit raffinate composition is reached, requiring five stages for the example shown in Fig. 6.7(a).

The quantities of solvent-free extract and raffinate products are obtained from the overall material balance, viz.

$$\bar{R}_N + \bar{E}_1 = \bar{R}_0 + \bar{E}_{N+1} \tag{6.23}$$

The overall mixing point M is located on Fig. 6.7(a) at the intersection of FS and R_NE_1 so that, by the lever arm rule, $\bar{E}_1/\bar{R}_N = R_NM/ME_1$. Substitution of this value in eqn (6.23) then gives $\bar{R}_N$ and $\bar{E}_1$.

As the solvent rate is decreased, the net flow point Δ_R approaches R_N and the number of stages required increases until finally a point $\Delta_{R,\min}$ is reached at which the number becomes infinite. This limit, known as the minimum solvent ratio, occurs when an equilibrium tie line coincides with a balance line through Δ_R. Generally, unless there is an inflection in the tie line slopes, this occurs when the tie line for the first stage, on extension, passes through F and Δ_R, as shown by line $E_{\max}\ F\Delta_{R,min}$ in Fig. 6.7(a).

On increasing the solvent rate beyond the minimum, the number of stages decreases; however, at the same time the extractor cross-section and the cost of solvent recovery from the extract product both increase. In practice, a solvent ratio of about 20–40 per cent above the minimum would be used, but the exact value must be determined on economic grounds at the design stage. Account must also be taken during design of the stage efficiency; for this purpose the Murphree efficiency, described in Section 6.4.3, (see p. 00), is normally most appropriate.

If preferred, the number of stages can be obtained by the McCabe–Thiele method using the selectivity diagram, obtained by projection from the

Janecke diagram as shown in Fig. 6.7(b). For this purpose the equilibrium curve is obtained by projection of the tie-lines, and the operating lines by drawing lines at random through Δ_R and projecting the points of intersection with the miscibility boundary. Line *AB* represents the composition change of the feed on saturation with solvent in the first stage, and is analogous to the q-line in distillation.

With some systems the solvents are effectively immiscible over the range of solute concentrations to be used. For such cases a simplified method using mass ratio coordinates, described by Hunter and Nash (1932) and Evans (1934) can be used [see also Pratt 1983a, Section 2.5.1(c)].

6.3.1.2 *Dual feeds*

It is sometimes necessary to operate an extractor with a second feed of lower strength introduced at an intermediate stage. In this case the extractor must be considered as having two regions, between the first and second feeds and between the latter and the raffinate outlet respectively. The two regions have separate net flow points which are collinear with the second feed concentration, and the stagewise construction is carried out as described above for each region. Details of the construction have been given by Treybal (1963).

6.3.2 Analytical solutions

Analytical solutions for the number of stages are in principle more convenient and more accurate than graphical solutions, provided that changes in solvent miscibility can be disregarded. The early solutions presented by Colburn (1939) were restricted to cases where the distribution coefficient is constant. Colburn (1941) and Scheibel and Othmer (1942) later described methods which permitted some curvature of the equilibrium line.

The solutions presented below are somewhat less restrictive in that a curved equilibrium line can be approximated by a straight line defined by eqn (6.1), which can possess an intercept q and so need not pass through the origin. In a further development to be described, a strongly curved equilibrium relation is approximated by subdivision of the extractor into two or more sections, within each of which the equilibria are linearized. These solutions are presented in terms of the dimensionless concentration units defined by eqns (6.2 and 6.3). This has advantages in that the solutions are simpler in form, and also that they can be related to the more complex solutions for axial dispersion presented in Chapter 7.

A solute mass balance taken around the solvent inlet end of the extractor [Fig. 6.8(a)] is as follows

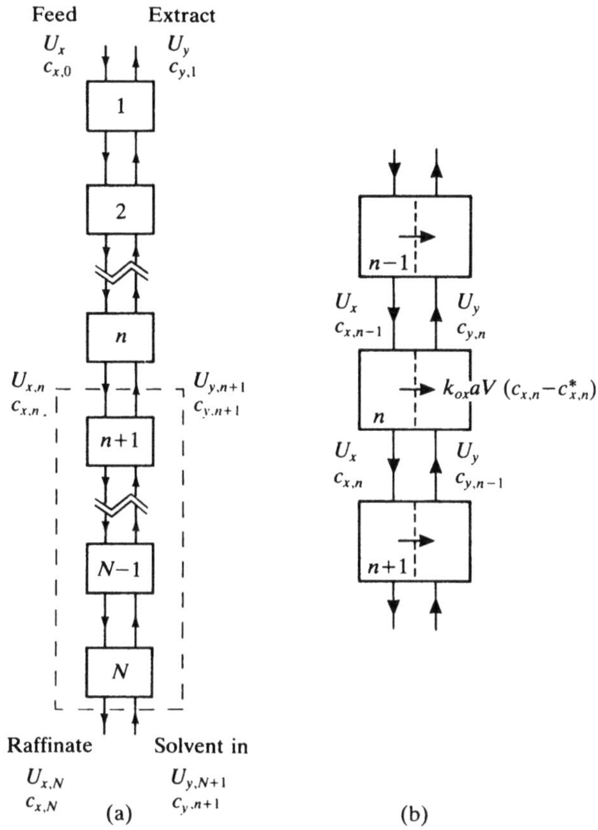

Fig. 6.8. Material balance for countercurrent extraction: (a) around solvent inlet end; (b) across stage.

$$U_{y,n+1}c_{y,n+1} + U_{x,N}c_{x,N} = U_{y,N+1}c_{y,N+1} + U_{x,n}c_{x,n} \tag{6.24}$$

Solving for $c_{y,n+1}$ neglecting changes in U_x and U_y, gives the equation to the operating line, as follows

$$c_{y,n+1} = \frac{U_x}{U_y}c_{x,n} + c_{y,N+1} - \frac{U_x}{U_y}c_{x,N} \tag{6.25}$$

For mixer–settlers the superficial velocities U_x and U_y must be replaced by the flowrates F_x and F_y.

Substitution for c_x and c_y in terms of the dimensionless concentrations X and Y, using eqns (6.2 and 6.3), gives

$$Y_{n+1} = E\ (X_n - X_N) \tag{6.26}$$

where $E = U_x/mU_y$ is the *extraction factor*. The equilibrium relation in dimensionless units is $Y_n^* = X_n$; dropping the superscript on Y, as the phases within each ideal stage are at equilibrium, and substituting in eqn (6.26) gives

$$X_{n+1} - EX_n = -EX_N \qquad (6.27)$$

This is a linear, first-order difference equation, the solution to which is

$$X_n = CE^n - EX_N/(1 - E) \qquad (6.28)$$

where the terms on the right hand side are the general and particular solutions respectively. Evaluating the constant C using the boundary condition $X_0 = 1.0$ at $n = 0$ and substituting for X_N gives the final solution as follows

$$X_n = Y_n = \frac{E^{N+1} - E^n}{E^{N+1} - 1} \qquad (E \neq 1.0) \qquad (6.29)$$

$$X_n = Y_n = (1 + N - n)/(1 + N) \qquad (E = 1.0) \qquad (6.30)$$

The number of theoretical stages is obtained by putting $n = N$, $X_n = X_N$ in eqns (6.29 and 6.30) and solving for N, giving

$$N = \frac{\log\left[(1 - E)/X_N + E)\right]}{\log(1/E)} \qquad (E \neq 1.0) \qquad (6.31)$$

$$N = 1/X_N - 1 \qquad (E = 1.0) \qquad (6.32)$$

6.3.3 Allowance for operating line curvature

The transfer of solute results in changes in flow rate of the phase within an extractor, so that the extraction factor E also varies. The mean of the values of E at the two ends of the extractor must therefore be used in the analytical expressions, so that the overall material balance in dimensionless form becomes

$$Y_1 = E_{av}(1 - X_N) \qquad (6.33)$$

Using dimensionless coordinates, this corresponds to a straight operating line of slope E_{av}, whereas the true operating line is generally curved; thus if the flow ratio $U_x/U_y < 1.0$ it is convex downwards, as shown in Fig. 6.9. A line of slope E_{av} through X_N will therefore give an adjusted exit extractant phase composition $Y_{1,adj} < Y_1$, while one through Y_1 will give an exit feed phase composition $X_{N,adj} < X_N$. It is apparent from eqn (6.31) that the latter, requiring the use of $X_{N,adj}$ in eqn (6.33), will give a higher value for N and hence a 'safer' design. On the other hand, the opposite will apply if

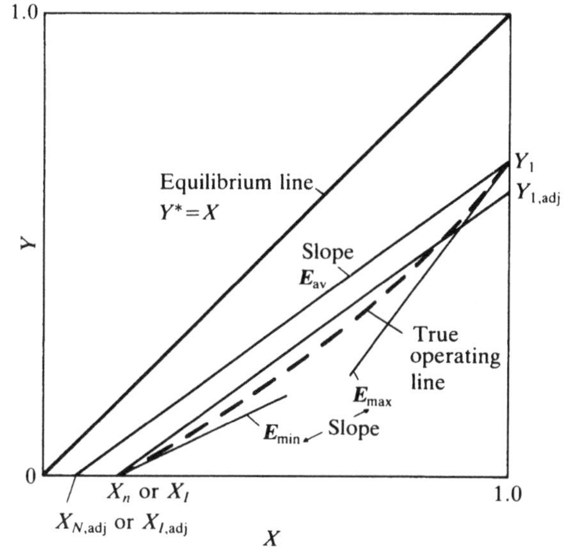

Fig. 6.9. True and linearized operating lines on X–Y coordinates.

the operating line is convex upwards and the use of X_N will give the safer design.

The correct value of $c_{y,0}$, assuming the solute-free flow rates of the phases to be constant and the mixture law to apply, can be obtained from a solvent balance on the feed phase around the extractor, as follows

$$U_{x,N} = U_{x,0}\left[1 - \frac{c_{x,0}}{\rho_s}\right] \Big/ \left[1 - \frac{c_{x,N}}{\rho_s}\right] \tag{6.34}$$

where ρ_s is the solute density. The overall rate of solute transfer per unit cross-section of the extractor is given by

$$\bar{N}_a = U_{x,0}\,c_{x,0} - U_{x,N}\,c_{x,N} \tag{6.35}$$

and the extractant phase composition at the exit is therefore

$$c_{y,0} = (U_{y,N}\,c_{y,N} + \bar{N}_a)/(U_{y,N} + \bar{N}_a/\rho_s) \tag{6.36}$$

Substitution of the values of $U_{x,N}$ and $\bar{N}_a$ from eqns (6.34 and 6.35) into eqn (6.36) gives $c_{y,0,\text{adj}}$ from which Y_0 is obtained. Substituting this in eqn (6.33) then gives $X_{N,\text{adj}}$, which should be used in eqns (6.31 and 6.32) in place of X_N if lower.

6.4 Stage efficiency
[See also Skelland, Ch. 2]

6.4.1 General

The concept of a theoretical stage can sometimes be applied to batch extraction provided the phases are contacted closely for an adequate period of time. However, equilibrium is seldom reached, and in continuous flow extractors it is therefore usual to define a stage efficiency, analogous to the plate efficiency in distillation, to relate the performance of a real stage to that of an ideal theoretical stage.

As Standart (1956) has pointed out, two separate aspects of efficiency must be considered, viz. (i), its use to predict the performance of a real extractor from that of an ideal one, and (ii), the interpretation of the efficiency in terms of the actual mechanisms of mass transfer which occur. A number of definitions of stage efficiency have been proposed, as described below.

The following discussion is primarily concerned with the case where a single solute only is present. When two or more solutes are present the situation is complicated by coupling between the motions of the diffusing species (Krishna and Standart 1976).

6.4.2 Overall efficiency

The overall efficiency, E_o, of an extractor is defined as the ratio of the number of ideal to real stages required to accomplish the same duty, i.e. the same concentration change with the given flows; thus

$$E_o = \frac{N(\text{ideal})}{N(\text{real})} \tag{6.37}$$

Although convenient for design purposes, this efficiency is only meaningful for a linear equilibrium relationship. Its relation to other efficiencies under such circumstances is given in Table 6.1.

6.4.3 Murphree efficiencies

Murphree (1925) defined efficiency as the ratio of the actual concentration change within a stage to the change which would have occurred if equilibrium had been reached. This can be expressed in terms of either the X- or Y-phase, as illustrated in Fig. 6.10 in terms of volumetric concentration units; thus

$$E_{Mx} = \frac{c_{x,n-1} - c_{x,n}}{c_{x,n-1} - c^*_{x,n}} \tag{6.38}$$

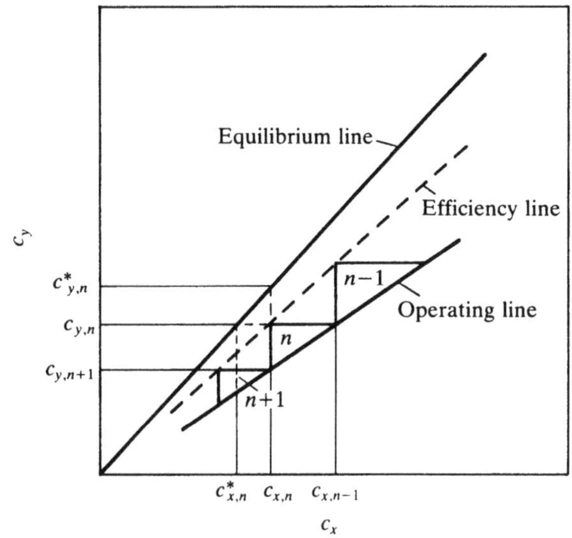

Fig. 6.10. Murphree efficiency.

$$E_{My} = \frac{c_{y,n} - c_{y,n+1}}{c^*_{y,n} - c_{y,n+1}} \tag{6.39}$$

These efficiencies are defined in terms of overall stage values, so that the concentrations $c_{x,n-1}$ and $c_{y,n+1}$ refer to the streams entering the stage, and $c_{x,n}$, $c_{y,n}$ to those leaving. A pseudo-equilibrium (i.e. efficiency) line can therefore be drawn using eqns (6.38 or 6.39) to obtain the coordinates $c_{x,n}$, $c_{y,n}$, and the stages can be stepped off against this as shown.

In most mixer–settlers the phases are either fully mixed, or pass through the mixer in cocurrent flow; the end result is the same in both cases and eqns (6.38) and (6.39) both apply. With some column extractors, for example, perforated plate columns, an additional enrichment can occur due to crossflow of continuous phase with respect to the dispersed phase. In such cases it is usual to define local (or point) efficiencies as follows

$$E'_{Mx} = \frac{c_{x,n-1} - c_x}{c_{x,n-1} - c^*_x} \tag{6.40a}$$

$$E'_{My} = \frac{c_y - c_{y,n+1}}{c^*_y - c_{y,n+1}} \tag{6.40b}$$

where c_x and c_y refer to concentrations within the stage, and c^*_x and c^*_y to the corresponding equilibrium values. Lewis (1936) obtained theoretical relations between point and overall efficiencies for distillation plates which are applicable also to extractors of this type. For the important case of a fully mixed dispersed extractant phase and unmixed continuous phase in crossflow (Lewis, Case I), the result is as follows

$$E_{My} = E \exp\,[E^{1}_{My} - 1] \tag{6.41}$$

6.4.4 Hausen efficiency

Hausen (1953) proposed an alternative definition of plate efficiency in distillation based upon the individual plate operating line. In terms of liquid extraction, this is obtained by taking a balance on solute around the stage. Neglecting changes in flow, this gives

$$N_a = U_x(c_{x,n-1} - c_{x,n}) = U_y(c_{y,n} - c_{y,n+1})$$

or

$$\frac{c_{y,n} - c_{y,n+1}}{c_{x,n} - c_{x,n-1}} = -\frac{U_x}{U_y} \tag{6.42}$$

This is the equation to the line *AB* of slope $-(U_x/U_y)$ in Fig. 6.11. If the exit streams had come to equilibrium, they would have had solute concentrations c_x^* and c_y^*, corresponding to point *C*, with an amount N_a^* of solute transferred. The Hausen stage efficiency E_H is then defined as

$$E_H = \frac{N_a}{N_a^*} = \frac{c_{x,n-1} - c_{x,n}}{c_{x,n-1} - c^*_{x,n}} = \frac{c_{y,n} - c_{y,n+1}}{c^*_{y,n} - c_{y,n+1}} = \frac{AB}{AC} \tag{6.43}$$

The Hausen efficiency therefore has the advantage that it is independent of the phase in which it is expressed. On the same diagram it is seen that E_{Mx} is represented by *PK/PS* and E_{My} by PB/PL.

The Hausen efficiency suffers from a disadvantage when applied to a countercurrent cascade. This is shown in Fig. 6.11, where the streams $c_{x,n-1}$ and $c_{y,n+1}$ enter the real stage *PBQ* at *A* and leave at *B*. To retain the same inlet concentrations the corresponding equilibrium step must be displaced to *GCF* and the extractor operating line to *FG*, thus necessitating a change in the raffinate (or extract) product composition. Alternatively, if the true operating line is retained so that the equilibrium stage is represented by *ECD*, the stage efficiency becomes *AB/MC*; in this case the driving force at the stage inlet on an *X*-phase basis corresponds to *MR* instead of *AS* and is incorrect. Consequently, although the Hausen efficiency is satisfactory for a single stage, it is unrealistic when applied to countercurrent multistage processes.

Relationships between the Hausen and other efficiencies have been given by Treybal (1963) and Pratt (1983a), and are summarized in Table 6.1.

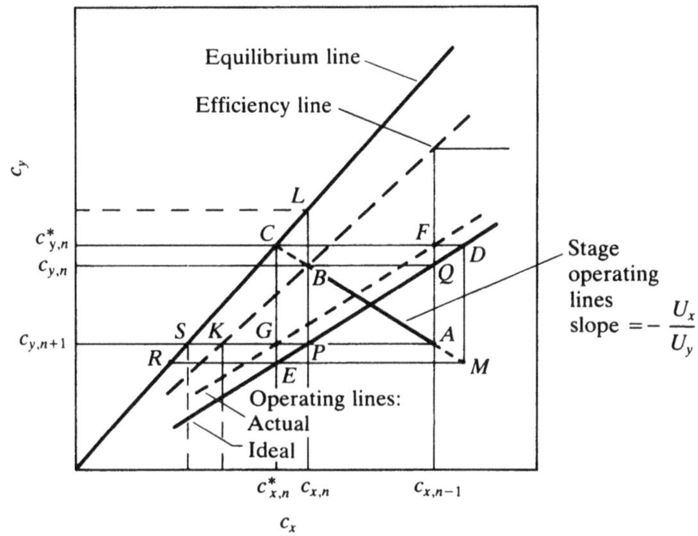

Fig. 6.11. Hausen efficiency for a countercurrent extractor.

6.5 Simple countercurrent extraction (II)

6.5.1 Non-ideal stages

The analysis presented in Section 6.3 (see p. 00) was based on the assumption that the phases leaving a given stage are in equilibrium, that is, the stages are ideal. A more realistic approach takes into account the rate of mass transfer within the stages; as this is necessarily finite, the phases do not come to equilibrium and the stages are non-ideal. This case is of particular relevance to the design of mixer-settlers.

The concentration profile in a countercurrent extractor may be determined from a material balance around the nth stage, as shown in Fig. 6.8(b); thus for the X-phase (cf. eqn (7.39) with $\alpha_x = 0$, see p. 00)

$$U_x(c_{x,n-1} - c_{x,n}) = k_{ox} a V_c (c_{x,n} - c^*_{x,n}) \tag{6.44}$$

This becomes, in dimensionless form

$$X_{n-1} - X_n = N^1_{ox}(X_n - Y_n) \tag{6.45}$$

where $N^1_{ox} = k_{ox} a V_c / F_x$ is the number of totally backmixed overall transfer units per stage, based on the X-phase. Similarly for the Y-phase,

$$Y_{n+1} - Y_n = -N^1_{ox} E (X_n - Y_n) \tag{6.46}$$

TABLE 6.1 *Relationship between efficiencies*

	Murphree		
Efficiency	*X*-phase	*Y*-phase	Hausen
Individual stages			
E_{Mx}, E_{My}	$E_{Mx} = \frac{E_{My}}{E + E_{My}(1 - E)}$	$E_{My} = \frac{E\,E_{Mx}}{1 + E_{Mx}(E - 1)}$	
	$E_{Mx} = \frac{E_H}{E(1 - E_H) + 1}$	$E_{My} = \frac{E\,E_H}{(1 - E_H) + E}$	
E_H	$E_H = \frac{E_{Mx}(1 + E)}{E\,E_{Mx} + 1}$	$E_H = \frac{E_{My}(1 + E)}{E_{My} + E}$	
Cross-flow extractors			
E_0	$E_0 = \frac{\log[E(E_{Mx} + 1)/(E_{Mx}(E - 1) + 1)]}{\log(E + 1/E)}$	$E_0 = \frac{\log[1 + E_{My}/E]}{\log(1 + 1/E)}$	$E_0 = \frac{\log[1 + (1 - E_H)/E]}{\log[1 + 1/E]}$
Countercurrent extractors			
E_0	$E_0 = \frac{\log[1 + E_{Mx}(E - 1)]}{\log E}$	$E_0 = \frac{\log[1 - E_{My}(1 - 1/E)]}{\log 1/E}$	$E_0 = \frac{\log[(E + 1 - E_H E)/(E + 1 - E_H)]}{\log 1/E}$

Elimination of Y between eqns (6.45 and 6.46) gives a second-order finite difference equation in X which is subject to the boundary conditions $X = 1$ when $n = 0$ and $X = X_N$ when $n = N$. This has the solution

$$X_N = \frac{E\mu_4^N - \mu_4^n}{E\mu_4^N - 1} \qquad (E \neq 1.0) \qquad (6.47)$$

where

$$\mu_4 = (1 + N_{ox}^1 E)/(1 + N_{ox}^1) \qquad (6.48)$$

The total number of stages is obtained as follows on setting $n = N$ and solving for N

$$N = \frac{\log X_N/[1 - E(1 - X_N)]}{\log \mu_4} \qquad (6.49)$$

The full set of solutions for this case are included in Table 6.2.

In practice, the extent of approach to equilibrium in a stage is expressed in terms of a Murphree efficiency, e.g. E_{Mx}. This is related as follows to N_{ox}^1 provided that both phases are fully mixed (Section 6.5.2.5, p. 375, and Case 4 of Table 6.3).

$$E_{Mx} = \frac{N_{ox}^1}{1 + N_{ox}^1} \qquad (6.50)$$

It is apparent that $E_{Mx} \to 1.0$ as $N_{ox}^1 \to \infty$, and consequently the above solutions approach those given in Section 6.3.2 (p. 358) for ideal stages as N_{ox}^1 becomes very large.

6.5.2 Differential extractors

[See also Skelland, Ch. 2]

Countercurrent differential extractors are characterized by a continuous change in the concentration profile from one end to the other, rather than step changes; typically the two phases are in continuous contact throughout the extractor. A simplified representation of a column extractor of this type is shown in Fig. 6.12; for simplicity, the two phases are shown as occupying separate vertical channels with an interface between.

6.5.2.1 *Transfer unit concept*

Considering first the X-phase, the flux, dN_a, of solute through the interface in the differential height δz is related as follows to the change in solute concentration across the section

TABLE 6.2 *Summary of design equations in dimensionless units*

Equations for	Staged contactors	Differential contactors
Equilibrium	$Y_n^* = X_n$	$Y^* = X$
Operating line	$Y_{n+1} = E(X_n - X_N)$	$Y = E(X - X_1)$
Overall material balance	$Y_1 = E(1 - X_N)$	$Y_o = E(1 - X_1)$
Concentration profiles	$X_n = Y_n = \dfrac{E^{N+1} - E^n}{E^{N+1} - 1}$	$X = \dfrac{\exp[\lambda L(1-z)] - E}{\exp(\lambda L) - E}$
$E \neq 1.0$		$Y = \dfrac{E\exp[\lambda L(1-z)] - E}{\exp(\lambda L) - E}$
		where $\lambda = \dfrac{1-E}{H_{Tox}} = \dfrac{1-E}{EH_{Toy}}$
$E = 1.0$	$X_n = Y_n = \dfrac{1 + N - n}{1 + N}$	$X = \dfrac{L(1-z) + H_{ox}}{L + H_{ox}}$
		$Y = \dfrac{L(1-z)}{L + H_{oz}}$

Number of stages $E \neq 1.0$	$N_S = \dfrac{\log[E + (1 - E)/X_N]}{\log(1/E)}$	$N_{Tox} = \dfrac{1}{1 - E} \ln[(E + (1 - E)/X_I]$ $N_{Toy} = \dfrac{E}{1 - E} \ln[E + (1 - E)/X_I]$
$E = 1.0$	$N_S = \dfrac{1}{X_N} - 1$	$N_{Tox} = N_{Toy} = \dfrac{1}{X_I} - 1$
Extractor efficiency	$E_{sx} = (1 - X_N) = Y_1/E$	$E_{oX} = E_{oY} = (1 - X_I) = Y_o/E$
$(E \neq 1.0)$	$= \dfrac{E^N - 1}{E^{N+1} - E}$	$= \dfrac{\exp(\nu) - 1}{\exp(\nu) - E}$
$(E = 1.0)$	$\dfrac{N}{N + 1}$	$= \dfrac{N_T}{1 + N_T}$ where $\nu = N_{Tox}(1 - E)$ $= N_{Toy}\left(\dfrac{1}{E} - 1\right)$ and $N_T = N_{Tox}$ or N_{Toy}

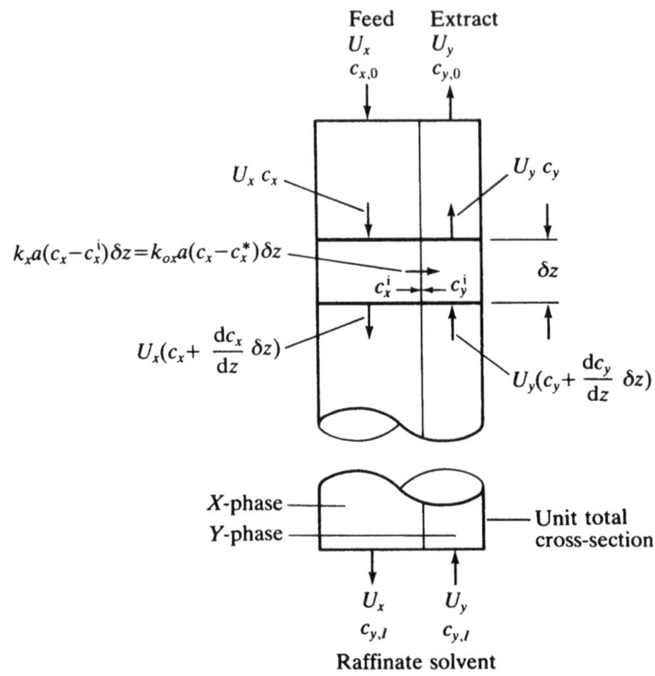

Fig. 6.12. Material balance for differential extractor over differential height δz. The phases are shown for simplicity as occupying separate channels.

$$dN_a = -U_x dc_x = k_x a\,(c_x - c_x^i)\,dz \tag{6.51}$$

where c_x is the composition of the bulk phase and c_x^i is that at the interface. Rearranging and integrating between the limits $c_x = c_{x,0}$ at $z = 0$ and $c_x = c_{x,I}$ at $z = H$, assuming the flow rate U_x to be constant, gives

$$H = \frac{U_x}{k_x a} \int_{c_{x,I}}^{c_{x,0}} \frac{dc_x}{(c_x - c_x^i)} \tag{6.52}$$

This may conveniently be expressed in the form

$$H = H_x N_x \tag{6.53}$$

where N_x, termed the *number of X-phase transfer units*, is given by the integral in eqn (6.52), viz.

$$N_x = \int_{c_{x,I}}^{c_{x,0}} \frac{dc_x}{(c_x - c_x^i)} \tag{6.54}$$

and H_x, the *height of an X-phase transfer unit* is given by

$$H_x = U_x / k_x a \tag{6.55}$$

Eqns (6.53–6.55) indicate that H_x can be interpreted as the height of column over which the change in concentration, Δc_x, is numerically equal to the mean driving force $(c_x - c_x^i)$. Similarly, N_x can be considered as a dimensionless measure of the difficulty of separation.

The Y-phase can be treated in the same way, giving

$$H = \frac{U_y}{k_y a} \int_{c_{y,I}}^{c_{y,0}} \frac{dc_y}{(c_y^i - c_y)} \tag{6.56}$$

$$= H_y N_y \tag{6.57}$$

where

$$H_y = U_y / k_y a \tag{6.58}$$

and

$$N_y = \int_{c_{y,I}}^{c_{y,0}} \frac{dc_y}{(c_y^i - c_y)} \tag{6.59}$$

6.5.2.2 *Overall transfer units*

In practice, it is difficult to obtain reliable values of the individual mass transfer coefficients, and in most cases the interfacial compositions are not known. Mass transfer data are therefore generally based upon the overall coefficients k_{ox} and k_{oy}; thus, using the former, eqn (6.51) becomes

$$dN_a = -U_x dc_x = k_{ox} a (c_x - c_x^*) dz \tag{6.60}$$

where c_x^* is the X-phase composition in equilibrium with c_y. Integration then gives, as before,

$$N_{ox} = \int_{c_{x,I}}^{c_{x,0}} \frac{dc_x}{(c_x - c_x^*)} \tag{6.61}$$

and

$$H_{ox} = U_x / k_{ox} a \tag{6.62}$$

Similarly, in terms of the Y-phase

$$N_{oy} = \int_{c_{y,I}}^{c_{y,0}} \frac{dc_y}{(c_y^* - c_y)} \tag{6.63}$$

$$H_{oy} = U_y / k_{oy} a \tag{6.64}$$

The overall height is then given by

$$H = H_{ok} N_{ok} \tag{6.65}$$

where subscript k refers to the X or Y phase. These integrals are simple to evaluate since c_x^* and c_y^* are the equilibrium values corresponding to c_y and c_x respectively and can be obtained directly from the equilibrium relationship.

The individual and overall values of the height of a transfer unit (HTU) are related by expressions for the sum of resistances. Thus, for the X-phase

$$\frac{1}{k_{ox}} = \frac{1}{k_x} + \frac{1}{mk_y} \tag{6.66}$$

and multiplying through by U_x/a gives

$$H_{ox} = H_x + EH_y \tag{6.67}$$

Similarly, for the Y-phase

$$H_{oy} = H_y + H_x/E \tag{6.68}$$

An important restriction is involved in the use of overall transfer units if the equilibrium relationship is non-linear, since then the slope m, and hence the extraction factor E, is a function of concentration. This is illustrated in Fig. 6.13(a) in which points B and I represent the bulk and interfacial concentrations respectively. The value of m in eqn (6.66), used to calculate E in eqn (6.67), is given by the slope of chord IC, whereas that used for E in eqn (6.68) is given by the slope of ID. These are not equal unless the equilibrium relation is linear.

6.5.2.3 *Height of extractor*

6.5.2.3.1 Graphical method

To use the foregoing transfer unit expressions, it is first necessary to plot the equilibrium and operating lines on c_y–c_x coordinates, as shown in Fig. 6.14. The former is obtained from the appropriate equilibrium data, and may or may not be straight. The operating line equation is obtained from a solute mass balance around the solvent inlet end and has the same form as eqn (6.25) with the subscripts n and $n+1$ dropped, and with N and $N+1$ both replaced by I. This can be put into the form

$$c_y = pc_x + b \tag{6.69}$$

where $p = U_x/U_y$ and $b = c_{y,I} - pc_{x,I}$. In Fig. 6.14, the point A on the operating line, with coordinates c_x, c_y correspondes to interfacial composi-

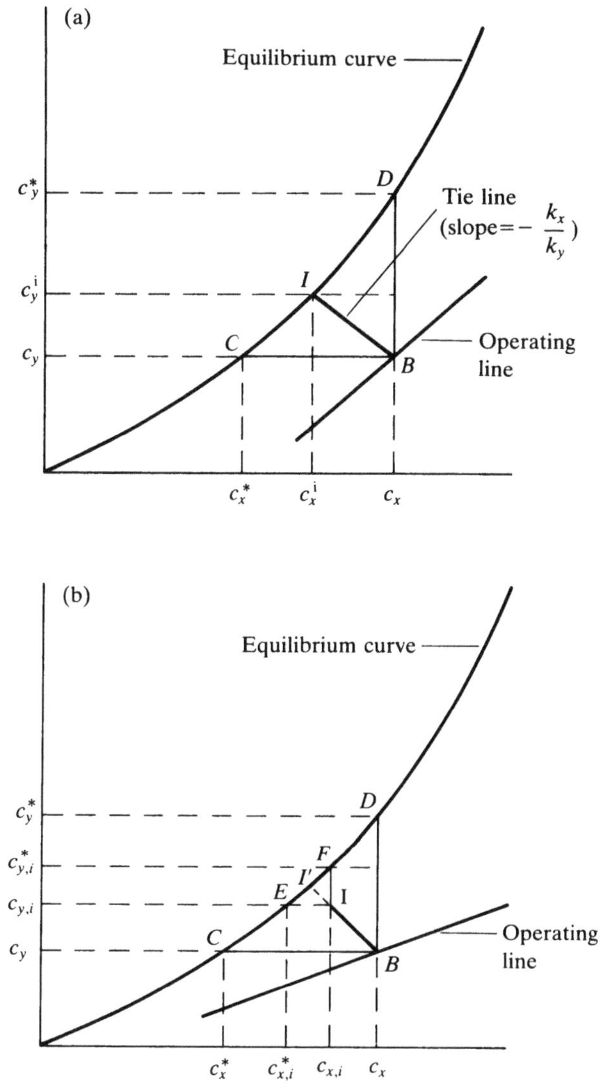

Fig. 6.13. Relationship between driving forces for mass transfer: (a) physical extraction; (b) with heterogeneous reaction.

tions c_x^i and c_y^i at I, where the slope of AI is given by $-k_x/k_y$, c_y is the ordinate of point B, and c_x the abscissa of point C.

To determine the number of individual-phase transfer units, tie lines such as AI, of slope $-k_y/k_x$, are drawn at a number of points along the operating line to intersect the equilibrium line at compositions c_x^i and c_y^i.

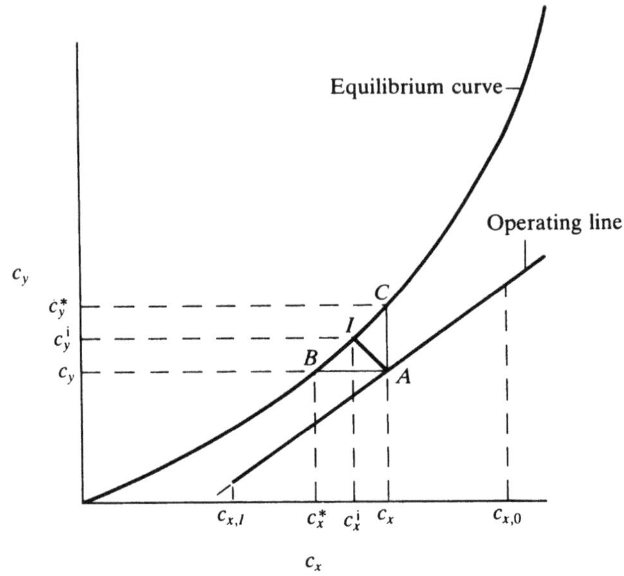

Fig. 6.14. Graphical determination of transfer units; A is a typical point on the operating line.

The area under a plot of $1/(c_x - c_x^i)$ vs c_x, or of $1/(c_y^i - c_y)$ vs c_y, over the range of c_x or c_y values between those at the extractor terminals, then gives N_{ox} or N_{oy}, and substitution in eqns (6.53 or 6.57) gives the required height.

The procedure is similar in the case of overall transfer units, except that the equilibrium compositions, c_x^* or c_y^*, replace the interfacial compositions c_x^i or c_y^i in the integration. The extractor height is then obtained by substitution in eqn (6.65).

6.5.2.3.2 Analytical solution

The analytical solutions can conveniently be expressed in terms of dimensionless concentration units, as for the stagewise cases. For this purpose the operating line becomes [cf. eqn (6.26) for the stagewise case]

$$Y = E(X - X_l) \tag{6.70}$$

Eqn (6.61) therefore becomes, in dimensionless units

$$N = \int_{X_l}^{1.0} \frac{\mathrm{d}X}{(X - X^*)} \tag{6.71}$$

and substitution of $X^* = Y$ from the equilibrium relation, followed by elimination of Y using eqn (6.70) gives

$$N_{ox} = \int_{X_l}^{1.0} \frac{\mathrm{d}X}{X(1-E)+EX_l} \quad (6.72)$$

Integration then gives

$$N_{ox} = \frac{1}{(1-E)} \ln\left[(1-E)/X_l + E\right] \qquad (E \neq 1.0) \quad (6.73)$$

The solutions for all cases are summarized in Table 6.2.

When using these expressions it is generally desirable to adjust the specified value of X_l, as described in Section 6.3.3 (p. 360), to allow for curvature of the operating line.

6.5.2.4 *Relationships between stages and transfer units*

Combination of the expressions for N_{ideal}, N_{ox} and N_{oy} in Table 6.2 gives

$$\frac{N_{ox}}{N_S} = \frac{\ln(1/E)}{1-E} \qquad (E \neq 1.0) \quad (6.74)$$

$$\frac{N_{oy}}{N_S} = \frac{E\ln(1/E)}{1-E} \qquad (E \neq 1.0) \quad (6.75)$$

$$N_{oy}/N_{ox} = E \quad (6.76)$$

6.5.2.5 *Transfer units and mixing on a stage*

The performance of a single stage can also be expressed in transfer units. The operating diagram for such a stage is shown in Fig. 6.15, where *AB* represents the extractor operating line and *CD*, *C'D'* the equilibrium and pseudo-equilibrium lines respectively. The feed enters stage n at point *F*, comprising X-phase of composition $c_{x,1}$ from stage $(n-1)$ and Y-phase of composition $c_{y,1}$ from stage $(n+1)$, and the compositions within the stage lie on the stage operating line *FI* of slope $-(U_x/U_y)$, as given by eqn (6.42).

The rate of mass transfer for such a cocurrent stage is given by eqn (6.60), where $\mathrm{d}z$ now represents a height increment within the stage. As both operating and equilibrium lines are linear, integration over the stage gives

$$N^1_{ox,P} = \int_{c_{x,2}}^{c_{x,1}} \frac{\mathrm{d}c_x}{c_x - c_x^*} = \frac{c_{x,1} - c_{x,2}}{(c_x - c_x^*)_{\text{lm}}} = \frac{k_{ox} a h_c}{U_x} \quad (6.77)$$

where $(c_x - c_x^*)_{\text{lm}}$ is the logarithmic mean of $(c_{x,1} - c^*_{x,1})$ and $(c_{x,2} - c^*_{x,2})$, $N^1_{ox,P}$ is the number of overall X-phase transfer units for plug flow, and h_c is the stage (i.e. compartment) height.

In practice, considerable mixing of the phases is likely to occur within the stage. If the X-phase alone is completely mixed, its composition remains constant at $c_{x,2}$, and integration gives

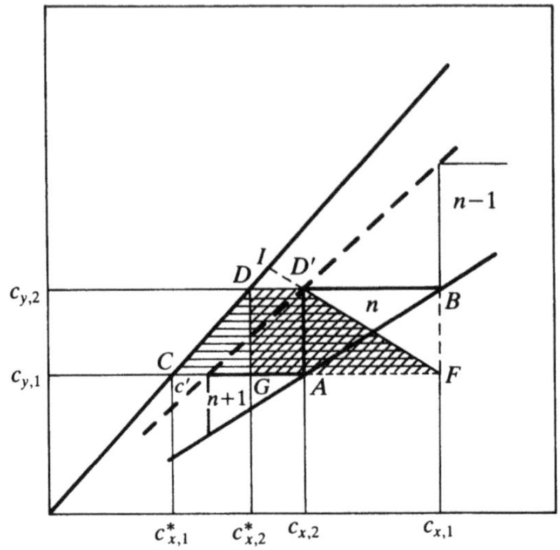

Fig. 6.15. Range of driving force for: (a) plug flow stage (area *D'DCF*); (b) complete mixing of *X*-phase (area *ACDD'*); and (c) complete mixing of *Y*-phase (area *FGDD'*).

$$N^1_{ox,X} = \frac{c_{x,1} - c_{x,2}}{(c_{x,2} - c^*_x)_{\mathrm{lm}}} = \frac{k_{ox} a h_c}{U_x} \tag{6.78}$$

where $(c_{x,2} - c^*_x)_{\mathrm{lm}}$ is the logarithmic mean of $(c_{x,2} - c^*_{x,1})$ and $(c_{x,2} - c^*_{x,2})$. This can be put into an alternative form by expressing c^*_x in eqn (6.60) in terms of c_y by means of eqn (6.1) and putting $\mathrm{d}c_y = (U_y/U_x)\,\mathrm{d}c_x$; integration then gives

$$N^1_{ox,X} = -\frac{1}{E} \ln \left[\frac{c^*_{y,2} - c_{y,2}}{c^*_{y,2} - c_{y,1}}\right] \tag{6.79}$$

$$= -\frac{1}{E} \ln (1 - E_{My}) \tag{6.79a}$$

Similarly, if the *Y*-phase is fully backmixed its composition remains constant at $c_{y,2}$, giving an average driving force of $(c_x - c^*_{x,2})_{\mathrm{lm}}$, the logarithmic mean of $(c_{x,1} - c^*_{x,2})$ and $(c_{x,2} - c^*_{x,2})$.

Finally, if both phases are completely backmixed the driving force is constant at $(c_{x,2} - c^*_{x,2})$, giving

$$N^1_{ox,XY} = \frac{c_{x,1} - c_{x,2}}{c_{x,2} - c_{x,2}} = \frac{k_{ox} a h_c}{U_x} \tag{6.80}$$

$$= \frac{E_{Mx}}{1 - E_{Mx}} \tag{6.81}$$

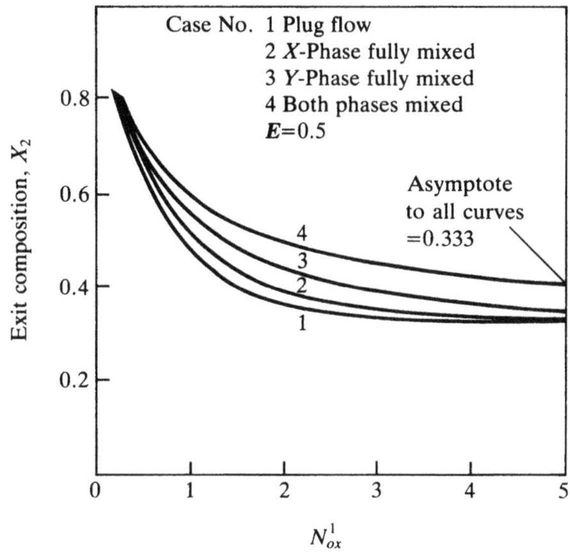

Fig. 6.16. Exit composition of X-phase from single stage as a function of N^1_{ox} and degree of mixing (inlet composition $X_0 = 1.0$).

These expressions are summarized in Table 6.3 in terms of the dimensionless concentrations, together with the corresponding ones for the Y-phase; use of the latter is to be preferred when the Y-phase resistance is controlling. The effect of mixing is shown in Fig. 6.16 for the X-phase, where the dimensionless exit composition is plotted against N^1_{ox}, with $E = 0.50$, for each of the four possible modes; as expected, this shows that plug flow gives the best performance and totally mixed flow the worst.

It is to be noted that right-hand terms in eqns (6.77, 6.78, and 6.80), i.e. $k_{ox}ah_c/U_x$, are identical for all cases. The number of X-phase transfer units given by a stage is therefore constant, but the degree of enrichment achieved depends upon the extent of mixing. The parameters N^1_{ox} and N^1_{oy} provide useful measures of stage effectiveness, but as they vary from zero to infinity it is convenient to express them in terms of an arbitrary transfer unit efficiency E_T which varies from zero to unity; thus for phase k, from eqn (6.81)

$$E_T = \frac{N^1_{ok}}{N^1_{ok} + 1} \tag{6.82}$$

This is identical with the Murphree efficiency if both phases are fully backmixed, but in other cases the performance will be greater, as shown in Fig. 6.16.

TABLE 6.3 *Relationship between NTU per stage, exit compositions, and efficiencies*

Case number and description	X-phase	Y-phase
Case 1: Plug flow	$N^1_{oxP} = \dfrac{-\ln[X_2(E+1)-E]}{E+1}$	$N^1_{oyP} = \dfrac{-E\ln[1-Y_2(1+E)/E]}{1+E}$
	$X_2 = \dfrac{E+\exp[-N^1_{oxP}(E+1)]}{E+1}$	$Y_2 = \dfrac{E-E\exp[-N_{oyP}((E+1)/E)]}{1+E}$
Case 2: X phase completely backmixed	$N^1_{oxX} = \dfrac{1}{E}\ln\left[\dfrac{X_2}{X_2(E+1)-E}\right]$	$N^1_{oyX} = \ln\left[\dfrac{E-Y_2}{E-Y_2(1+E)}\right]$
	$= -\dfrac{1}{E}\ln(1-E_{My})$	$= -\ln(1-E_{My})$
	$X_2 = \dfrac{E\exp(EN^1_{oxX})}{(1+E)\exp(EN^1_{oxX})^{-1}}$	$Y_2 = \dfrac{E[\exp(N^1_{oyX})-1]}{(1+E)\exp(N^1_{oyX})-1}$
Case 3: Y phase completely backmixed	$N^1_{ox,Y} = \ln\left[\dfrac{1-E(1-X_2)}{X_2(E+1)-E}\right]$	$N^1_{oyY} = E\ln\left[\dfrac{E(1-Y_2)}{E-Y_2(1+E)}\right]$
	$= -\ln(1-E_{Mx})$	$= -E\ln(1-E_{Mx})$
	$X_2 = \dfrac{1+E[\exp(N^1_{oxY})-1]}{(1+E)\exp(N^1_{oxY})-E}$	$Y_2 = \dfrac{E[\exp(N^1_{oyY}/E)-1]}{(1+E)\exp(N^1_{oyY}/E)-E}$
Case 4: Both phases completely backmixed	$N^1_{ox,B} = \dfrac{1-X_2}{X_2(E+1)-E}$	$N^1_{oyB} = \dfrac{EY_2}{E-Y_2(1+E)}$
	$= \dfrac{E_{Mx}}{1-E_{Mx}}$	$= \dfrac{E_{My}}{1-E_{My}}$
	$X_2 = \dfrac{E\,N^1_{oxB}+1}{N^1_{oxB}(E+1)+1}$	$Y_2 = \dfrac{E\,N^1_{oyB}}{N^1_{oyB}(1+E)+E}$

6.5.3 Non-linear equilibria

The analytical methods described above are not directly applicable if the equilibrium line has appreciable curvature. However, in such cases it is generally possible to subdivide the extractor into two or more sections within each of which the equilibrium relationships can be linearized, and to calculate the height or number of stages separately for each section.

A method of doing this is described in Section 6.2.5.2 (p. 367) using a generalized expression for backmixed extractors in terms of both transfer units and non-ideal stages. This reduces to the plug flow cases when the backmixing parameters are set to zero.

6.6 Countercurrent extraction with reflux (single-solvent fractional extraction)

6.6.1 Process description

The richest possible extract composition obtainable in simple countercurrent extraction corresponds to equilibrium with the entering feed. However, higher concentrations are obtainable by provision of an enriching section at the extract end of the extractor to which reflux is supplied by means of a suitable solvent removal unit (e.g. by evaporation or fractional distillation), as shown in Fig. 6.17. This process can be carried out either in columns, as shown, or in mixer–settler cascades.

With Type I systems the maximum extract composition obtainable corresponds to that at the plait point, and in practice is considerably less due to the need to provide an adequate density difference of the phases. With banded Type II systems, however, a 'complete' separation is possible, as with distillation; unfortunately, such systems are rare, so that applications of the process are limited. An important example of its use is in aromatics separation from petroleum reformates using solvents such as tetraethylene glycol and Sulfolane.

6.6.2 Graphical solution

This process can be treated on a Janecke diagram, as shown in Fig. 6.18, by an extension of the method described in Section 6.3.1, p. 354 (Maloney and Schubert 1940). Considering a banded Type II system, material balances for the enriching section, taken around the solvent recovery unit adjacent to stage 1 and between stages m and $m + 1$, are as follows

Nonsolvent $A + B$: $\bar{E}_{m+1} - \bar{R}_m = \bar{P}_E + \bar{S}_E = \bar{\Delta}_E$ (6.83)

Solute A: $\bar{E}_{m+1}\bar{y}_{m+1} - \bar{R}_m\bar{x}_m = \bar{P}_E\bar{y}_{p,E} + \bar{S}_E\bar{y}_{E,s} = \bar{\Delta}_E y_{\Delta,E}$ (6.84)

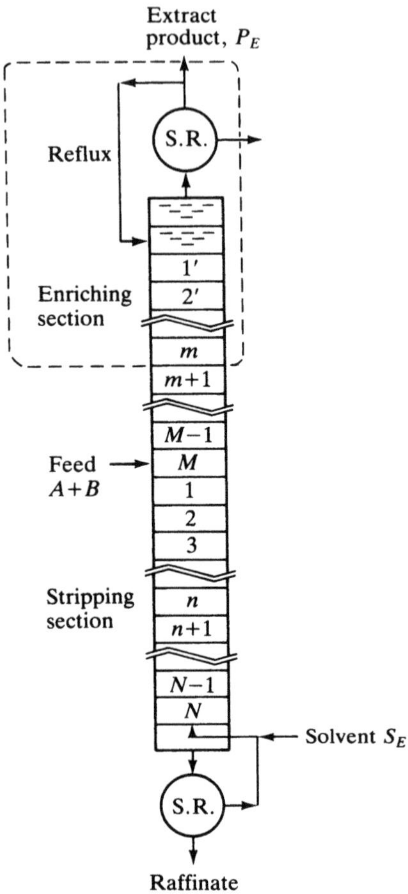

Fig. 6.17. Single-solvent fractional extraction in column extractor, showing material balance around extract phase solvent removal unit (S.R.).

Solvent S: $$\bar{E}_{m+1}s_{y,m+1} - \bar{R}_m s_{x,m} = \bar{P}_E s_{p,E} + \bar{S}_E s_{E,s} = \Delta_E S_{\Delta,E} \quad (6.85)$$

The values of $\bar{S}_E$, $\bar{y}_{E,s}$ and $s_{E,s}$ are the same as $\bar{S}_R$, $\bar{y}_{R,s}$ and $s_{R,s}$ when the solvent is recycled. Elimination of E, R and Δ_E by writing in determinantal form [cf. eqn (6.12)] or use of the lever arm rule shows that the points $(\bar{x}_m, s_{x,m})$, $(\bar{y}_{m+1}, s_{y,m+1})$ and $(\bar{y}_{\Delta,E}, s_{\Delta,E})$ are collinear with $\bar{y}_{\Delta,E}$ and $S_{\Delta,E}$ given by

$$\bar{y}_{\Delta,E} = \frac{\bar{P}_E \bar{y}_{p,E} + \bar{S}_E \bar{y}_{E,s}}{\bar{P}_E + \bar{S}_E} \quad (6.86)$$

$$s_{\Delta,E} = \frac{\bar{P}_E s_{p,E} + \bar{S}_E s_{E,s}}{\bar{P}_E + \bar{S}_E} \tag{6.87}$$

When the solute content of the solvent leaving the solvent separator is zero, these expressions become $\bar{y}_{\Delta,E} = \bar{y}_{p,E}$ and $\bar{s}_{\Delta,E} = (\bar{s}_{p,E} + \bar{S}/\bar{P}_E)$.

The equations for the stripping section are the same as for simple countercurrent extraction (Section 6.3.1, p. 354). Similar balances around the whole extractor, i.e. on $\bar{F}$, $\bar{R}_N$, and $\bar{P}_E$, show that the points $\Delta_R F \Delta_E$ are collinear. The construction can now be completed on the Janecke diagram (Fig. 6.18) as follows. The extract product point P_E, with coordinates ($\bar{x}_{p,E}$, $s_{p,E}$), is joined to S_E and equilibrium tie lines to the right of F are extended to intersect the line $S_E P_E$; the intersection most remote from S_E locates the point $\Delta_{E,\min}$ which defines the minimum extract reflux ratio, given by $\Delta_{E,\min} E_1/E_1 P_E$. A somewhat higher reflux ratio, corresponding for example to Δ_E, is chosen and extended through F to intersect the extension of the line $S_R R_N$ at Δ_R, thus satisfying the overall material balance. The usual stage construction is then carried out by drawing equilibrium tie-line $E_1' R_1'$ through E_1', the intersection of $P_E \Delta_E$ with the Y-phase boundary, followed by stage operating line $R_1' \Delta_E$ to locate E_2'. This process is continued until the feed point is reached, after which the operating lines are drawn through the lower net flow point Δ_R and the procedure continued until the required raffinate composition is reached.

The equilibrium tie lines can be projected if desired on to the selectivity diagram and the stages stepped off by the McCabe–Thiele method, as shown in Fig. 6.18. This illustrates the similarity of this process to distillation. As the reflux ratio is increased, point Δ_E moves further from P_E until finally, if the solvent is pure, it approaches infinity and the stage operating lines become vertical; at the same time the operating lines on the selectivity diagram coincide with the diagonal. This is the total reflux condition, which requires infinite heat for solvent removal, as the production rate is zero. On the other hand, the heat requirement is least at the minimum reflux ratio, but the number of stages and hence the capital charges are infinite. Clearly therefore there is an optimum reflux ratio which corresponds to a minimum in the relation between total cost (i.e. operating plus capital charges) and reflux ratio.

6.7 Dual-solvent fractional extraction

6.7.1 Process description

In dual-solvent extraction two separate, preferably immiscible solvents are used to separate at least two solutes from a feed which is usually, although not necessarily dissolved in one of the solvents (Scheibel 1948). The process

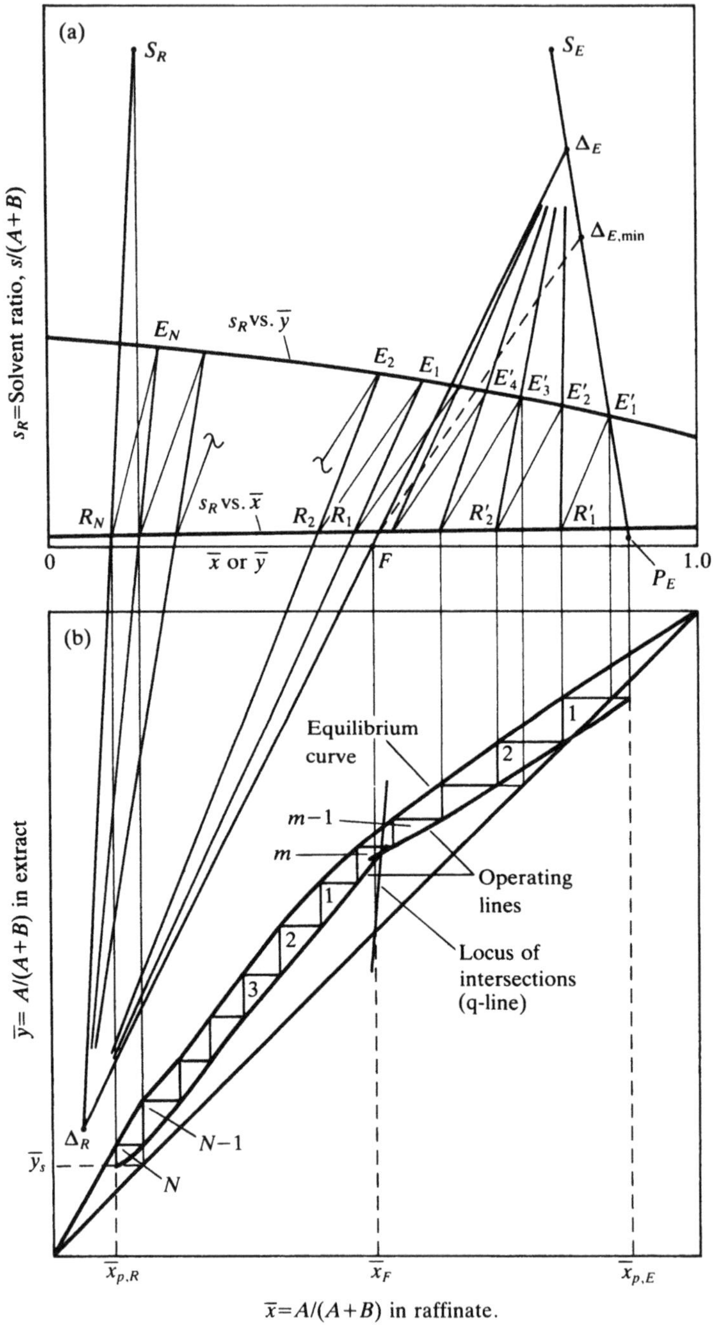

Fig. 6.18. Single-solvent fractional extraction on (a) Janecke coordinates; (b) selectivity diagram.

can be carried out either in mixer–settlers or in columns; a schematic of the former is shown in Fig. 6.19, and of the latter in Fig. 6.26(a), where it is seen that the two solvents enter at opposite ends and the feed at some intermediate point. The two sections of the extractor are conventionally termed the scrubbing and extraction sections, the former because, in inorganic separations, of the need to treat the organic extract with aqueous 'scrub' solution to remove impurities. A minimum of four components is involved in the process; of the solutes, that with the higher distribution coefficient appears predominantly in the extract product leaving the scrub section, and the other predominantly in the raffinate.

In normal practice the solute concentrations in this process are appreciably higher at the feed point than at the two ends. As a consequence, the flow rates of the two phases change throughout the extractor, even though the mutual solubilities of the solvents themselves may be unaltered, leading to curvature of the operating lines. To allow for this, therefore, the flow rates and concentrations are normally expressed on a solute-free basis; thus either solute-free mass flow rates with mass ratio concentrations, or solvent-free volumetric flow rates, with concentrations expressed as mass per unit volume of solute-free solvent, can be used. In the following treatment the latter will be used.

Although such units results in straight operating lines, the equilibrium relationship is usually non-linear. Thus for two solutes J, where $J = A$ or B, the linear distribution coefficient relationshi]p $c_{y,J} = K_i c_{x,J}$ becomes, assuming the simple mixture law to apply

$$\bar{c}_{y,J} = \bar{K}_i \bar{c}_{x,J} \tag{6.88}$$

where

$$\bar{K}_J = K_J \frac{(1 + c_{y,a}/\rho_a + c_{y,b}/\rho_b)}{(1 + \bar{c}_{x,a}/\rho_a + \bar{c}_{x,b}/\rho_b)} \tag{6.89}$$

and $\bar{c}_{x,J}$, $\bar{c}_{y,J}$ refer to concentration units in terms of solute-free solvent. Despite this, there is considerable advantage in practice in expressing the concentrations in this way.

6.7.2 Graphical solution

6.7.2.1 *Number of stages*

Phase equilibria for quaternary systems can be represented in three-dimensional form on equilateral or right-angled tetrahedra or on a triangular prism. However, such models are difficult to construct for general use, and although Brancker *et al.* (1940, 1941) have described methods of two-dimensional projection, these are very tedious to use. It is generally assumed,

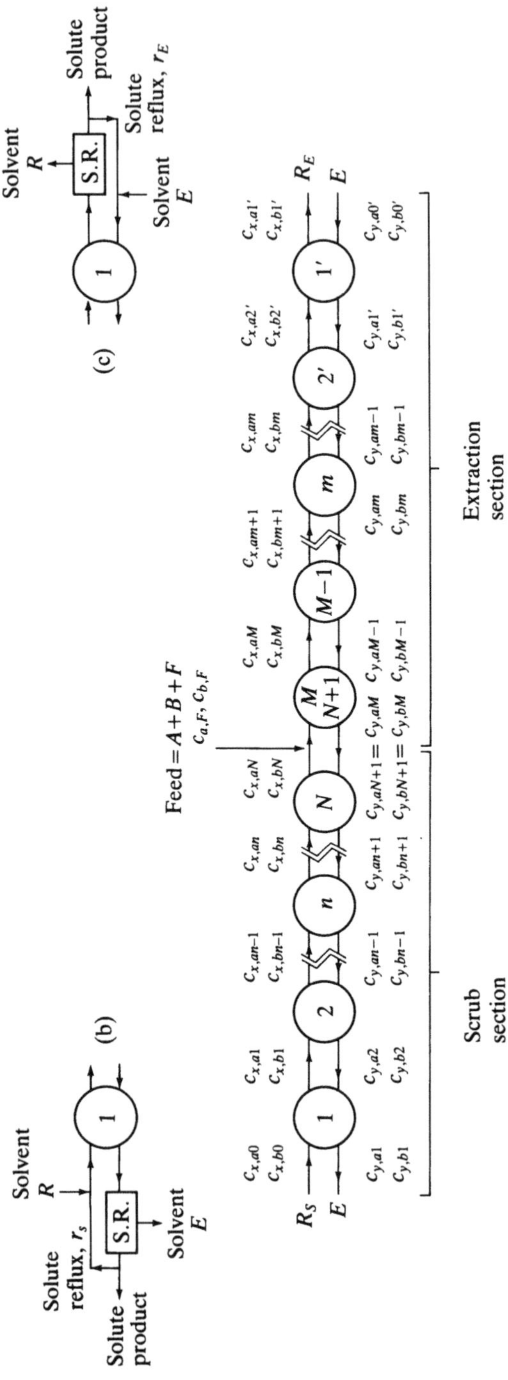

Fig. 6.19. Dual-solvent fractional extraction in staged extractor: (b) and (c) show extract and raffinate phase reflux.

therefore, that the solvents themselves are immiscible; this is normally acceptable for inorganic systems, but may be less so for organic solutes, particularly in the region of the feed point where total solute concentrations can be high.

The graphical design of a dual-solvent fractional extractor will be described for the case shown in Fig. 6.19, where two immiscible solvents E and R are used to separate solutes A and B which enter as feed dissolved in the raffinate solvent R; solute A will be assumed to have the higher distribution coefficient. A material balance for either solute in the scrub section, taken around stage n and the R phase inlet gives, in solute-free units

$$E\bar{c}_{yS,n+1} + R_S\bar{c}_{x,0} = E\bar{c}_{yS,1} + R_S\bar{c}_{xS,n} \tag{6.90}$$

Solving for $\bar{c}_{y,S,n+1}$,

$$\bar{c}_{yS,n+1} = \frac{R_S}{E}\bar{c}_{xS,n} + \left[\bar{c}_{yS,1} - \frac{R_S}{E}\bar{c}_{xS,0}\right] \tag{6.91}$$

Similar balances for the extraction section around stage m and the $E-$ phase inlet give

$$\bar{c}_{yE,m} = \frac{R_E}{E}\bar{c}_{xE,m+1} + \left[\bar{c}_{yE,0} - \frac{R_E}{E}\bar{c}_{xE,1'}\right] \tag{6.92}$$

Equations (6.91 and 6.92) represent the operating lines for either solute in the scrub and extraction sections respectively. The raffinate solvent flow rates in the two sections of the column are related by

$$R_E = R_S + F \tag{6.93}$$

The operating lines and corresponding equilibrium curves are plotted separately for each solute in Figs 6.20(a) and (b), and theoretical stages are stepped off in the usual manner from the two ends, i.e. from stages 1 and 1′. The feed stage compositions must then be matched, assuming that the same number of stages is required for each component. For this purpose plots are made on the same graph of the concentrations of each solute against stage number for each section, as shown in Fig. 6.21; the feed stage, corresponding to the rectangle, is then found by trial. If the concentrations are plotted in terms of $\bar{c}_x$, that is, for the raffinate phase, the match occurs on the Mth stage of the extraction and the $(N + 1)$th of the scrubbing section, so that the total number of stages is $(M + N)$. If a match cannot be obtained the flow rates must be changed, noting that there are strict upper and lower limits to the flow ratio R_S/E, beyond which the desired product specifications cannot be met. This subject will be discussed further in Section 6.7.3 (p. 389).

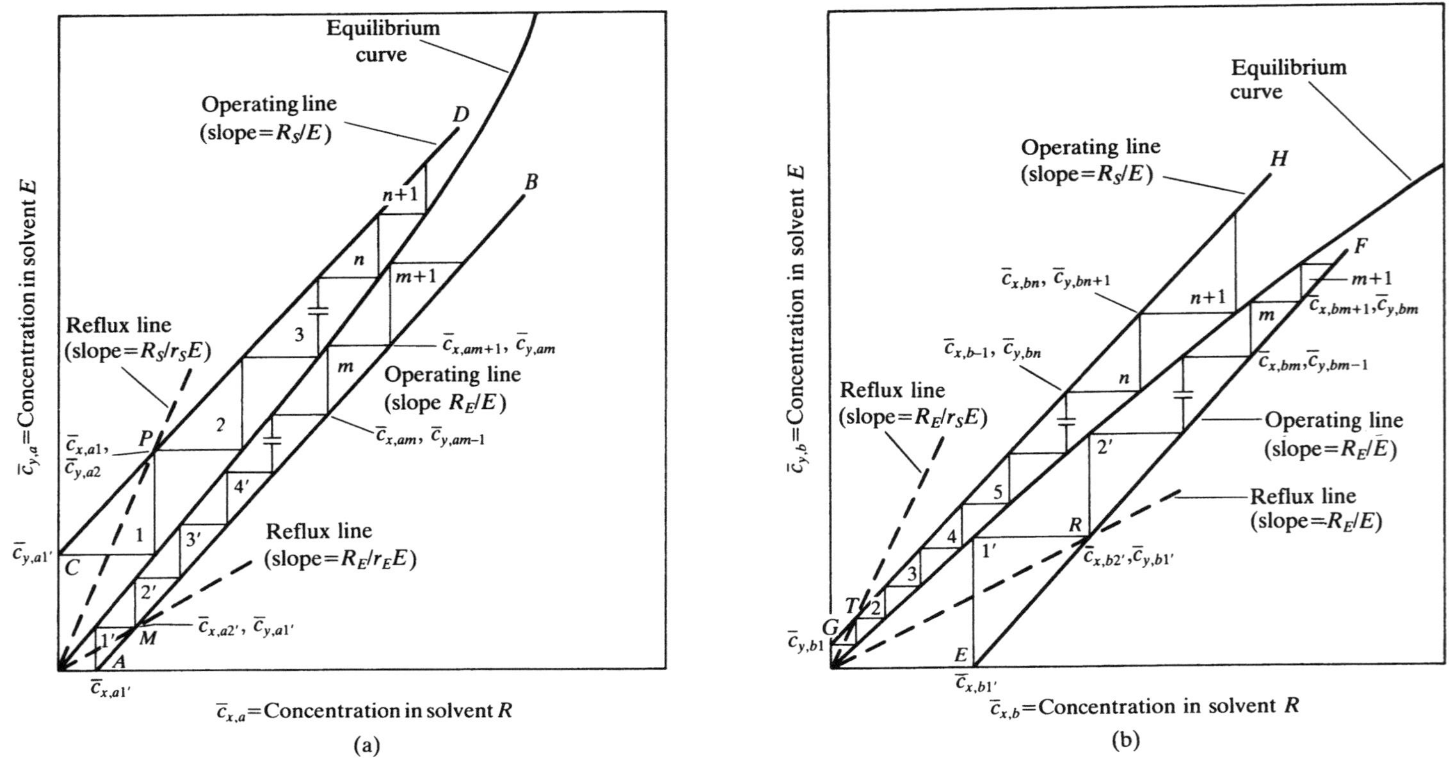

Fig. 6.20. Dual-solvent fractional extraction: operating diagrams for (a) solute *A*; (b) solute *B*.

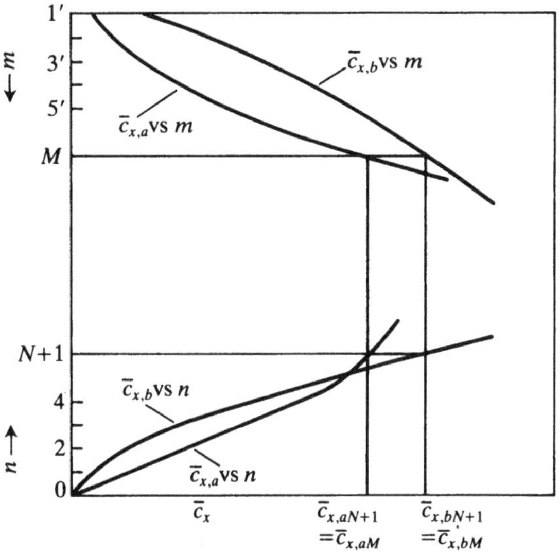

Fig. 6.21. Dual-solvent extraction: matching of feed-point compositions.

6.7.2.2 *Use of reflux*

External reflux can be used as shown at (b) and (c) in Fig. 6.19 to increase the solute concentrations at the ends of the extractor. Allowance for this can be made by taking balances for each solute around the solvent removal units; thus if a fraction r_s of the product is returned as reflux to the solvent end of the scrubbing section inlet, then

$$R_S \bar{c}_{x,0} = E r_S \bar{c}_{y,1}$$

therefore

$$\bar{c}_{y,1} = R_S \bar{c}_{x,0} / r_S E \tag{6.94}$$

Equation (6.94) represents straight lines of slope $R_S/r_S E$ through the origin, shown by the broken lines in Figs 6.20(a) and (b). Similarly for the extraction end of the column where the fraction r_E is refluxed,

$$\bar{c}_{y,0} = r_E R_E \bar{c}_{x1} / E \tag{6.95}$$

This represents lines of slope $r_E R_E / E$ on the same diagrams. The operating lines terminate on the reflux lines and the stages are stepped off from the intersection points, as shown.

Using reflux, six different cases can arise according to the relative slopes of the reflux and equilibrium lines, as follows (Asselin and Comings 1950):

1. Scrubbing section

(a) If a reflux line lies above the corresponding equilibrium line, the exit solute concentration in solvent E will be less than that in the feed stage.

(b) If a reflux line lies below the equilibrium line, the exit solute concentration in solvent E will be greater than in the feed stage.

(c) If the reflux and equilibrium lines have the same slope the solute concentration will remain constant throughout the section.

2. Extraction section

(a) If a reflux line lies above the equilibrium line, the exit solute concentration in solvent R will be greater than in the feed stage.

(b) If a reflux line lies below the equilibrium curve, the exit solute concentration in solvent R will be less than in the feed stage.

(c) If the reflux and equilibrium lines have the same slope, the solute concentration will remain constant throughout the section.

Thus it is possible, by suitable selection of the reflux ratios, to cause a solute to increase or decrease progressively in concentration from one end of the extractor to the other, to pass through a maximum or minimum at the feed stage, or to remain constant in one or both sections. An example is shown in Fig. 6.22 in which the reflux lines lie between the two equilibrium lines so that the concentration of component A increases continuously from the extract solvent inlet to the scrub inlet, while the concentration of B continuously decreases.

At total reflux the operating lines pass through the origin and are coincident with the reflux lines. The output of the extractor is then zero, and the number of theoretical stages is a minimum; the latter is given as follows by the Fenske–Underwood equation if the distribution coefficients are constant:

$$(M+N)_{\min} = \frac{\log(P_A/P_B)}{\log(K_a/K_b)} \tag{6.96}$$

where P_A, P_B are the *rejection ratios* for solutes A and B, that is, the ratios of the amounts in the extract product to those in the raffinate.

In most cases, the use of reflux reduces the number of stages required for a given separation; on the other hand the solvent requirement may be increased, requiring a greater cross-section, and the cost of solvent recovery must be taken into account. The problem of optimization is therefore complex and can best be investigated using an appropriate computer program.

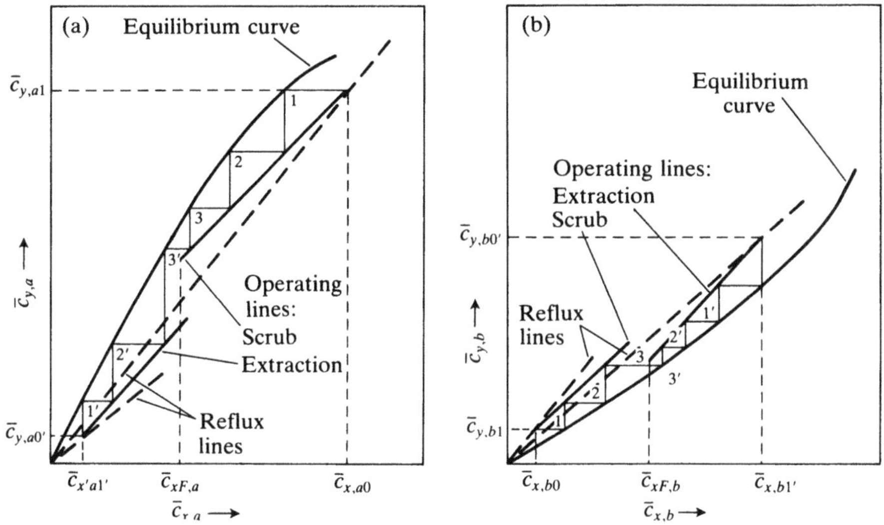

Fig. 6.22. Dual-solvent extraction with reflux lines between equilibrium lines: (a) solute A; (b) solute B.

6.7.3 Material balances and limiting flows

It is convenient to specify the product compositions in terms of the ratios E_{ab} of solute A to B in the extract product, and R_{ba}, of B to A in the raffinate. On this basis the overall material balances become

$$F\bar{c}_{aF} = A_E + B_R/R_{ba} \tag{6.97}$$

$$F\bar{c}_{bF} = A_E/E_{ab} + B_R \tag{6.98}$$

where A_E and B_R are the mass flow rates of A and B in extract and raffinate products respectively.

Elimination of B_R gives

$$A_E = \frac{F(\bar{c}_{aF} - \bar{c}_{bF}/R_{ba})}{1 - (E_{ab}R_{ba})^{-1}} \tag{6.99}$$

and similarly,

$$B_R = \frac{F(\bar{c}_{bF} - \bar{c}_{aF}/E_{ab})}{1 - (E_{ab}R_{ba})^{-1}} \tag{6.100}$$

Equations (6.99 and 6.100) show that limits exist to the possible range of values of E_{ab} and R_{ba}, as follows

$$E_{ab} > \bar{c}_{aF}/\bar{c}_{bF}\ ;\ R_{ba} > \bar{c}_{bF}/\bar{c}_{aF} \tag{6.101}$$

together with $E_{ab}R_{ba} \gg 1.0$.

The presence of raffinate solvent in the feed results in the operating lines for the extraction and scrub sections having different slopes. They therefore intersect at coordinates ($\bar{c}_{x,i}$, $\bar{c}_{y,i}$), which may be obtained by solving eqns (6.91 and 6.92) for these concentrations and combining with the overall material balance to eliminate the bracketed terms in these equations. This gives $\bar{c}_{x,i} = \bar{c}_F$ for both solutes together with a value of $\bar{c}_{y,i}$, obtained from eqn (6.92) assuming the entering solvents to be solute-free, as follows

$$\bar{c}_{y,i} = \frac{R_E}{E}(\bar{c}_F - \bar{c}_{x,i}) \tag{6.102}$$

A solution to the problem of defining the permissible range of flow rates has been obtained by Prochazka and Jiricny (1976), based on the following parameters:

$$T = \frac{\bar{c}_{xb,F}}{\bar{c}_{xa,F}} ; \; S = \frac{\bar{c}_{yb,1}}{\bar{c}_{ya,1}} ; \; Z = \frac{R_E \bar{c}_{xa,R_p}}{F\bar{c}_{xa,F}}$$

together with

$$P = R_S/F \; ; \; 1 + P = R_E/F \; ; \; R = E/F$$

Parameter Z corresponds to the fractional loss of solute A in the raffinate product, and $S = 1/E_{ab}$. Using these definitions eqn (6.102) becomes

$$\bar{c}_{ya,i} = \bar{c}_{xa,F}(1 + P - Z)/R \tag{6.103}$$

$$\bar{c}_{yb,i} = \bar{c}_{xa,F}[TP + S(1 - Z)]/R \tag{6.104}$$

A case of particular interest arises when the two operating lines for a solute intersect on the equilibrium line for that solute. This is obtained by substituting for the $\bar{c}_{y,i}$ in eqns (6.103 and 6.104) in terms of the K values, thus

$$\bar{c}_{ya,i} = K_a\bar{c}_{xa,F} = \bar{c}_{xa,F}(1 + P_K - Z)/R_K \tag{6.105}$$

$$\bar{c}_{yb,i} = K_b T\bar{c}_{xa,F} = \bar{c}_{xa,F}[TP_K + S(1 - Z)]/R_K \tag{6.106}$$

The values of P and R in these equations, i.e. P_K and R_K, are termed the 'critical values'; solving for these gives

$$P_K = (1 - Z)(\beta - S/T)/(1 - \beta) \tag{6.107}$$

$$R_K = (1 - Z)(1 - S/T)/[K_a(1 - \beta)] \tag{6.108}$$

where

$$\beta = K_b/K_a < 1.0.$$

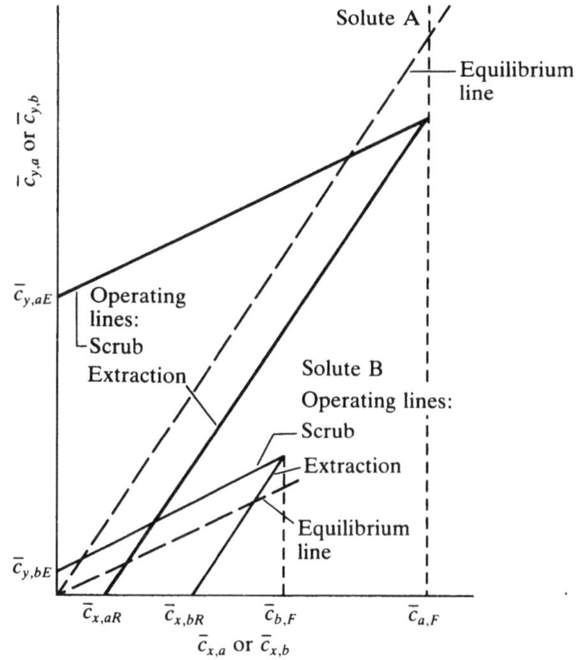

Fig. 6.23. Intersections of operating lines in dual-solvent extraction.

As P_K must be positive, eqn (6.107) requires that $\beta > S/T$, i.e. $E_{ab} > K_a\bar{c}_{xa,F}/(K_b\bar{c}_{xb,F})$, which is greater than the value given by eqn (6.101).

The value of P_K corresponds to the lowest possible flow of raffinate solvent in the scrub section; with the corresponding extract solvent flow, defined by R_K, the cross-over from scrub to extraction section operating line occurs on the equilibrium line itself for each solute at its feed concentration, and requires infinite stages in each section of column. Without reflux, the feed stage solute concentrations cannot exceed those of the feed itself, so that the cross-over must occur at somewhat lower concentrations, that is, both pairs of operating line intersections must occur between the equilibrium lines as shown in Fig. 6.23. (An exception to this could occur at high reflux ratios, such that the solute concentration of the entering solvent exceeds that of the feed). This requires the value of P to be increased; thus, for a given value of $P > P_K$, eqns (6.105 and 6.106) give two limiting values for R, denoted by R_A and R_B, as follows

$$R_A = (1 + P - Z)/K_A \tag{6.109}$$

$$R_B = [P + S(1 - Z)/T]/K_B \tag{6.110}$$

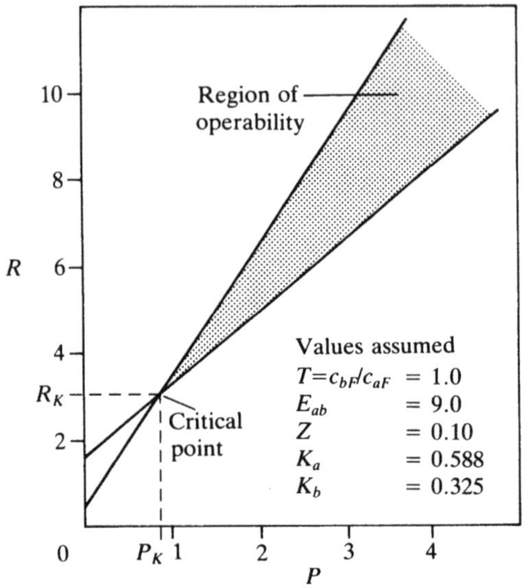

Fig. 6.24. Limiting lines in dual-solvent extraction.

Plots of eqns (6.109 and 6.110) in the form R vs P give two straight lines, termed 'limiting lines', which intersect at the critical point (P_K, R_K) as shown in Fig. 6.24. These lines correspond to the intersection of one or other pair of operating lines on the corresponding equilibrium line, so that a range of extract solvent flows between R_A and R_B can be used with a given raffinate flow $P > P_K$. Under these conditions the number of stages required is infinite when $R = R_A$ or R_B, and between these there is a well-defined minimum for the total stages, as shown in Fig. 6.25. A sufficient condition for the existence of a solution, as first obtained by Prochazka and Jiricny (1976) is therefore that

$$P > P_K; R_A > R > R_B \tag{6.111}$$

As the value of P increases, the solvent flows increase and the solute concentrations in the two products decrease, requiring greater energy for solvent removal. The optimization problem therefore involves the determination of the value of P, with associated value of R to give minimum total stages, for which the total of energy costs and capital charges is a minimum.

6.7.4 Analytical solutions

When the distribution coefficients are constant, the stage calculations can be carried out analytically using suitable forms of the expressions given in

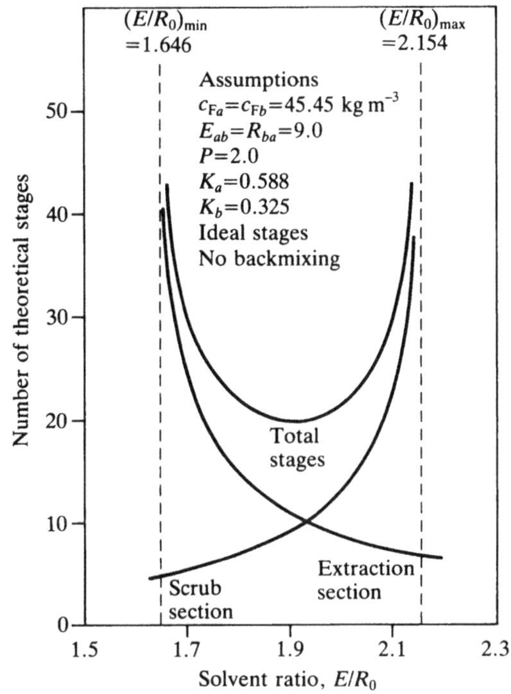

Fig. 6.25. Dual-solvent extraction: effect of solvent ratio on number of stages.

Table 6.2. These are given in detail by Treybal (1963) who covers both the general case, and special cases with no reflux and with no feed solvent present. However, matching of the feed stage compositions involves difficulty and must be done graphically or numerically. A numerical method of accomplishing this will be described in the next section.

For the case of no reflux and no solvent present in the feed, Klinkenberg (1951) and Klinkenberg *et al.* (1951) described six cases which permit simple calculation of the plot of total stages against solvent ratio shown in Fig. 6.25. This involves the use of the following expression

$$P_J = \frac{(E_J^{-m} - 1)}{E_J^{N+1}(E_J^{-(N+1)} - 1)} \tag{6.112}$$

where P_J is the rejection ratio for solute J (i.e. A or B). Of the six cases studied, their case 4 for $M = N$, corresponding to $1/P_A = E_a$ and $P_B = E_b^N$, is of particular interest as it gives a value as follows for total stages which is close to the minimum in Fig. 6.25

$$M + N = \frac{2\log(P_A/P_B)}{\log\beta} - 1 \tag{6.113}$$

This is twice the minimum number of stages at total reflux, as given by eqn (6.96), less one and is the same as the optimum number of stages for an ideal tapered cascade of separating elements (Pratt 1967).

6.7.5 Numerical methods

Prochazka and Jiricny (1976) described a numerical method for the computation of stage requirements in dual-solvent extraction, based on the arrangement shown in Fig. 6.26(b) in which the scrub and extraction sections are treated as separate units. This was based on the use of an assumed value for the concentration of solute B in the extract phase between the column sections and iterating on this, computing the stages for each solute in each section until convergence is obtained. Stage-to-stage computations were used for this purpose, but as will be shown, these can be replaced by analytical expressions.

The steps involved in the computer program are as follows (Dongaonkar *et al.* 1987):

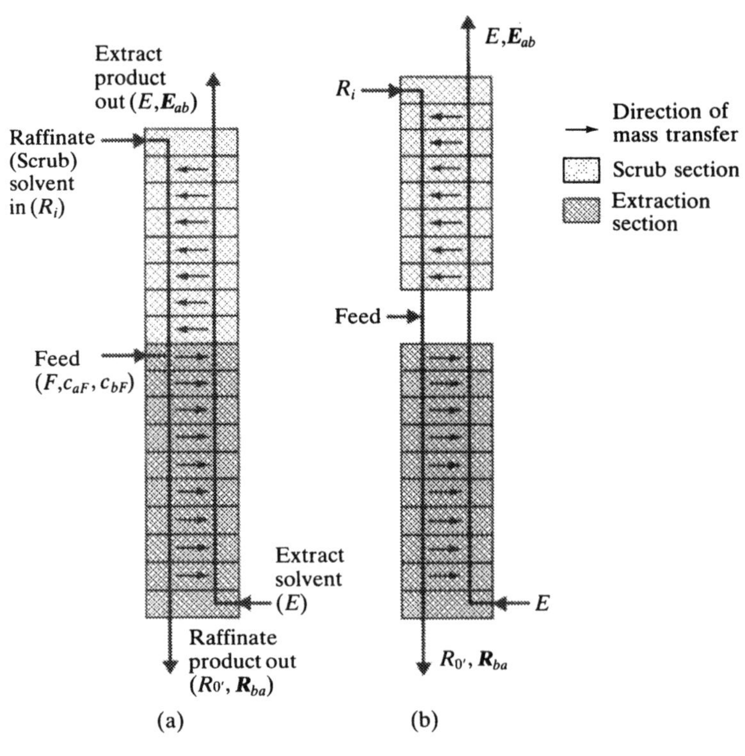

Fig. 6.26. Dual-solvent extraction in column: (a) normal arrangement; (b) as assumed for numerical solution.

1. Input the feed rate and solute concentrations and convert these if necessary to a solute-free basis using the solute densities.

2. Input the desired product ratios $\boldsymbol{E}_{ab}$ and $\boldsymbol{R}_{ba}$ and the distribution coefficients K_a and K_b.

3. Compute P_K using eqn (6.107); select a value for P_K, compute R_A and R_B using eqns (6.109 and 6.110) and select a value for R between these. Hence calculate the flows R_S, R_E, and E followed by the extract and raffinate product compositions.

4. Assume a value for $\bar{c}_{yb,M} > \bar{c}_{yb,S}$ (out) and compute the number of stages required in the extraction section for solute B, i.e. $N_{E,b}$.

5. Compute the number of stages required in the scrub section for solute B, i.e. $N_{S,b}$.

6. Compute $\bar{c}_{ya,N}$ for solute A in the extract phase leaving the scrub section assuming that $N_{S,a} = N_{S,b}$.

7. Assuming that $\bar{c}_{ya,M} = \bar{c}_{ya,N}$, compute the number of stages in the extraction section for solute A, i.e. $N_{E,a}$.

8. Compare $N_{E,a}$ with $N_{E,b}$; if these differ by more than, say, 0.01 units then let $N_{E,b} = N_{E,a}$, compute $\bar{c}_{yb,M}$ and return to step 5.

In Section 7.4.4 (p. 468), a modification of this method is described using generalized design equations which give the solutions in terms of either height (using transfer units) or non-ideal stages, with either plug flow or backmixing of the phases, as desired (Pratt 1983b). An alternative method employing matrix algebra is also described.

6.8 Effect of Chemical Reaction
[See also Skelland, Ch. 2]

6.8.1 Mass transfer with homogeneous reaction

Processes in which the extracted solute undergoes homogeneous chemical reaction in the extract phase range from simple washing operations, e.g. removal of acid from an organic liquid by alkali treatment, to the nitration or sulphonation or aromatics. However, processes such as the latter are properly classified as forms of chemical reaction engineering rather than as separation processes and they are therefore outside the scope of this book; a review of such processes has been given by Hanson (1971).

Homogeneous reaction, unlike heterogeneous reaction (Section 6.8.2) has the effect of increasing the mass transfer rate due to (i), an increase in driving force, $(c_i - c_y)$, resulting from a reduction in the value of c_y in the bulk

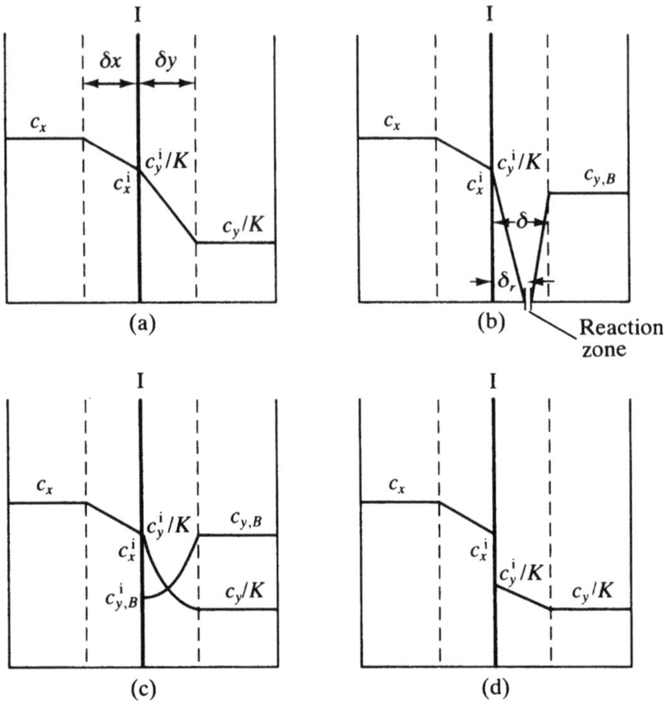

Fig. 6.27. Interfacial concentration profiles based on film theory: (a) physical extraction; (b) and (c) fast and slow homogeneous reaction respectively; (d) heterogeneous reaction.

phase, often to zero, and (ii), an increase in mass transfer coefficient, k_y, on the reaction side of the interface. This is illustrated in Fig. 6.27 in terms of simple two-film theory, where case (a) relates to simple physical extraction, and (b) and (c) to fast and slow homogeneous reactions respectively of the following type

$$A + bB \rightarrow rR$$

In case (b) the reactant B, which is insoluble in the feed phase, has a concentration c_B in the bulk extractant phase and zero in the reaction zone within the film. Solute A diffuses from the interface to the latter where its concentration becomes zero, so that its diffusion path is reduced from δ to δ_r and its mass transfer coefficient is increased accordingly. In the limit, if the concentration of B is sufficiently high in the bulk phase its concentration at the interface will be greater than zero; in this case the reaction zone will coincide with the interface itself and the rate of mass transfer will be controlled entirely by that of the feed phase.

In Fig. 6.27(c) the reaction rate is much slower so that solute A does not react fully within the film and is present in the bulk extractant phase, where the reaction continues. Similarly, reactant B is not fully consumed in the film and has a concentration c_{yB}^{i} at the interface. It is apparent that reaction rates can vary over a very wide range, from the effectively instantaneous as with ionic reactions in an aqueous extractant phase, to extremely slow as may occur with reactions in an organic extraction solvent.

Theoretical treatments in terms of both film and penetration models lead to expressions for an enhancement (or reaction) factor, defined as the ratio of the mass transfer coefficient with reaction to that for physical extraction only. The solutions for the film theory are simpler than for the penetration models since they involve ordinary instead of partial differential equations; however, both give results which agree to within a few percent, so that the former is to be preferred in practice.

Published work on this subject has been mainly concerned with the parallel case of gas absorption with reaction. Comprehensive accounts have been given by Danckwerts (1970), Astarita (1967) and Sherwood *et al.* (1975), to which reference should be made for further details.

6.8.2 Mass transfer with heterogeneous reaction

Solutes may exist in different chemical forms in different solvents; for example, simple carboxylic acids have been shown to exist as single molecules in an aqueous phase but as dimers in non-polar solvents such as benzene. Similarly, amines such as diethylamine in water have water of hydration which is shed on transfer into organic solvents. The transfer of solute from one phase to another in such cases involves a reaction at the interface.

Inorganic solvent extraction systems provide frequent examples of heterogeneous reaction between metal solutes and solvent. This is particularly important in hydrometallurgical extractions where the metal cations and the anions can only exist in an aqueous phase. It is therefore necessary for them to react with an appropriate organic compound to form a neutral complex before they can enter a solvent phase. Since such reagents normally have only a very slight solubility in water, especially when diluted with a solvent such as kerosene, the reaction generally takes place at the interface itself. Such reactions, which are often slow, are exemplified in the extraction of uranium with tributyl phosphate, and of copper with oxime reagents, diluted with kerosene in both cases.

The effect of heterogeneous reaction can conveniently be described in terms of the two-film theory, as shown in Figs 6.27(a) and (d) where the cases of physical extraction and heterogeneous reaction are compared. In the latter case the points c_x^i and c_y^i/K, representing the concentrations on the two sides of the interface, are no longer coincident. This can be expressed quantitatively for the case of an mth order forward and nth order reverse reaction, with rate constants k_m^+ and k_n^-, as follows, noting that the net rate of reaction is equal to the flux of solute A through the interface

$$N_a = k_m^+ c_x^{im} - k_n^- c_y^{in} \tag{6.114}$$

$$= k_m^+ \left[c_x^{im} - \frac{k_n^-}{k_m^+} c_y^{in} \right] \tag{6.115}$$

At equilibrium, the net reaction rate is zero so that the interfacial compositions are related by

$$K_a' = k_n^- / k_m^+ = c_x^{im*} / c_y^{in} \tag{6.116}$$

where K_a' is the ratio of the forward and reverse reaction rates and c_x^{i*} is the concentration on the X-phase side of the interface in equilibrium with c_y^i on the Y-side. This expression can be combined with eqn (6.51) for the X-phase and the corresponding one for the Y-phase to give [cf. eqn (6.66)]

$$\frac{1}{k_{ox}} = \frac{1}{k_x} + \frac{1}{m_y k_y} + \frac{1}{k_m^+} \left[\frac{c_x^i - c_x^{i*}}{c_x^{im} - c_x^{im*}} \right] \tag{6.117}$$

$$\frac{1}{k_{oy}} = \frac{1}{k_y} + \frac{m_x}{k_x} + \frac{1}{k_n^-} \left[\frac{c_y^{i*} - c_y^i}{c_y^{in*} - c_y^{in}} \right] \tag{6.118}$$

The operating diagram for heterogeneous reaction is compared with that for physical extraction in Fig. 6.13(b), where it is seen that in the chemical reaction case the tie line BI no longer terminates on the equilibrium curve as it does with physical extraction; consequently m_x and m_y correspond to the slopes of chords FD and CE respectively, as compared with those of DI' and CI' for physical extraction.

It is to be noted that the right-hand terms in eqns (6.117 and 6.118), which represent the chemical reaction resistance, are concentration-independent only if m or $n = 1$, that is, for first-order reactions. Although such a reaction order is rare, in fact many reactions are pseudo first-order when one of the reactants is in considerable excess. It is not possible to predict rate constants theoretically for the liquid phase, and values must therefore be obtained experimentally, preferably over a range of temperatures in order to evaluate the activation energy using the Arrehenius equation.

6.9 Multicomponent separations

6.9.1 Types of operation

Multicomponent separations can comprise those in which (i), the added solvent (or solvents) are mixtures, (ii), the feed solvent in single-solvent operation is a mixture, or (iii), more than one solute is present (more than two in dual-solvent processes). However, the use of mixed solvents is relatively trivial in terms of design, and for the present purposes multicomponent separations will be defined as those in which the feed contains two or more solutes in single-solvent processes, and at least three in the dual-solvent case. Thus, the systems involved contain a minimum of four components in the former case and five in the latter.

Although quaternary equilibrium data can be represented using solid models or two-dimensional projections from these (Treybal 1963), this is generally inconvenient and cannot be applied to systems with five or more components. Equilibrium data for multicomponent systems are therefore generally represented by analytical expressions such as those obtained from the UNIQUAC or NRTL equations, etc. or by empirical correlations of numerical data. For systems where individual molecular species cannot be identified, some workers have suggested replacing one of the components by a relevant physical property. For example, in the separation of saturated and unsaturated triglycerides from vegetable oils the iodine number is used as a measure of unsaturation, and this can be treated as a pseudo-component on the Janecke diagram with iodine number on the horizontal axis (Ruthruff and Wilcock 1941). Similarly the viscosity–gravity constant has been used as a measure of aromaticity in lubricating oil refining (Hunter and Nash 1932). The data obtained in this way are only applicable to the particular feed mixture used and cannot be applied to crossflow or countercurrent extraction processes (Treybal 1963; Alders 1959). In such cases a complete pilot scale simulation is therefore required.

6.9.2 Crossflow extraction

The graphical and analytical methods described in Section 6.2.2 (p. 352) can be applied to each solute separately in multistage crossflow extraction if the distribution coefficients are constant and independent of the concentrations of the other solutes. In such cases eqn (6.17) takes the following form for the fraction of a feed solute, say A, remaining in the raffinate after n stages

$$q_{a,n} = 1/(1 + 1/E_a)^{-n} \tag{6.119}$$

and $p_{a,n}$, the fraction in the nth stage extract, is $q_{a,n}E_a$.

Using this approach Treybal (1963) has shown that, for solute-free solvent, the optimum number of stages and optimum solvent ratio for crossflow extraction are as follows:

$$N_{a,\mathrm{opt}} = \frac{\log[\log(1 + 1/E_a)/\log(1 + 1/E_b)]}{\log[(1 + 1/E_a)/(1 + 1/E_b)]} \tag{6.120}$$

$$E_{a,\text{opt}} = \frac{\bar{R}}{K_a\bar{E}} = \frac{\beta - \beta^{1/(N+1)}}{\beta(\beta^{1/(N+1)} - 1)} \tag{6.121}$$

These optima correspond to the case where the difference in the fraction of the two components remaining in the raffinate, i.e. $(q_{a,n} - q_{b,n})$ is a maximum. These equations allow the determination of the optimum number of stages, given the solvent ratio, or the optimum solvent ratio given the number of stages. Solving eqns (6.120 and 6.121) simultaneously results in an infinite number of stages and a solvent ratio of zero, indicating that differential extraction gives the best separation.

When the distribution coefficients are interdependent an iterative graphical procedure can be used to determine the distribution of solutes. This is cumbersome, however, particularly if the number of distributed solutes exceeds two, and it is more appropriate to use a numerical method.

6.9.3 Countercurrent extraction

This process in its simplest form involves the simultaneous extraction of two or more solutes from a feed solution using an immiscible solvent. If the distribution coefficients are constant and independent of the concentration of the other solutes, the analyses in terms of ideal stages, in Section 6.3 (see p. 354) for ternary systems, can be applied to each solute independently. The graphical method is best carried out on the distribution diagram as shown in Fig. 6.28, or on the solute-free selectivity diagram if the total solute concentration is high. It is convenient in practice to designate one of the solutes, of intermediate distribution coefficient, as a key component and to select a solvent ratio U_x/U_y; this would generally correspond to a value of the extraction factor E_{key} $(= U_x/U_y m_{\text{key}})$ for this component within the range 1.3–2.0, although this may require revision later.

An operating line of slope R/E (or R_i/E_i on the selectivity diagram) is drawn through the specified key component concentration in the extract or raffinate and the stages are stepped off in the usual way. Parallel operating lines are then drawn by trial for the other solutes and their position adjusted to give the same number of stages. In Fig. 6.28, component B is selected as key and A, with a higher distribution coefficient, is stripped virtually completely from the raffinate so that its operating line passes close to the origin. Component C, on the other hand, has a low distribution coefficient so that its operating line 'pinches in' to its equilibrium line at the feed concentration. If the distribution coefficients of the solutes are dependent on the concentration of the other solutes present, an iterative procedure must be used and the solution becomes increasingly complex as the number of solutes increases beyond two.

It is apparent that the analytical method described in Section 6.3.2 (p. 358)

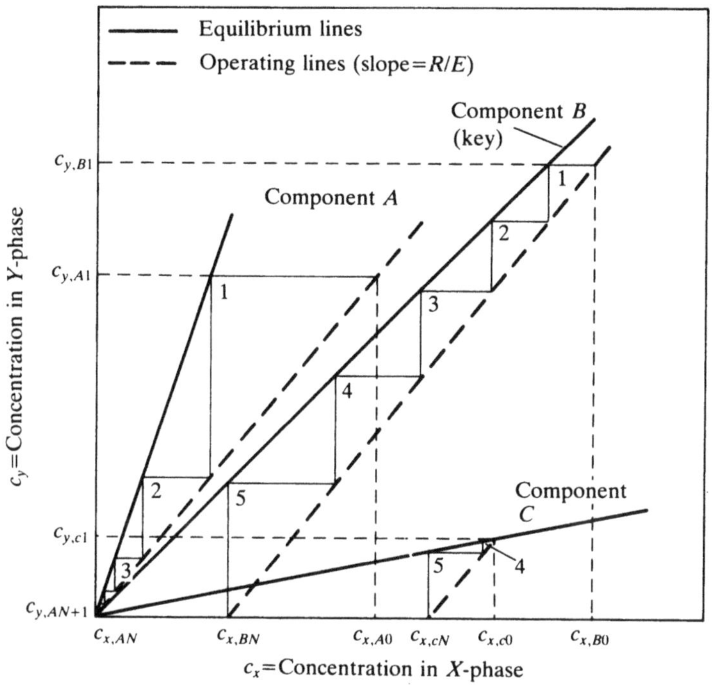

Fig. 6.28. Graphical design of mutlicomponent countercurrent extraction (three solutes).

can also be used, on a trial-and-error basis, to achieve the same result.

Various numerical algorithms have been described for the computation of the column performance, given the number of stages. Basically the procedures can be characterized either as a stage-to-stage calculation, or a matrix formulation. Both methods require the initial specification of the following parameters:

1. The flow rate and composition of feed and extraction solvents.
2. The number of theoretical stages and the location of the feed point.
3. The solvent ratio.
4. The extract reflux ratio (if any).
5. The equilibrium data for each solute, and the temperature profile if appropriate.

The programs then compute the raffinate and extract flow rates and compositions at each stage.

Details of the stage-to-stage method are given by Smith (1963), and matrix formulations have been described by Tierney *et al.* (1971), Tierney (1977) and Roche (1969, 1971, 1973). The matrix method described in Section 7.4.3 (p. 460) is also applicable with the axial dispersion parameters set to zero.

The above methods are all based on the concept of an ideal stage, which is not realizable in practice. In practical terms this can be allowed for by the introduction of the Murphree efficiency (Section 6.4.3, p. 362); this however is likely to vary for each solute, so that the value for the key component should normally be used. A similar situation arises with the methods based on non-ideal stages and transfer units described in Section 6.5, (p. 365) and with these care must be taken in selecting appropriate values of the mass transfer parameters.

6.9.4 Dual-solvent systems

Dual-solvent extraction is capable in principle of giving a complete separation of each of the components of a multicomponent mixture. As in distillation, the separation of a C-component mixture requires $C - 1$ extractors; however, as the solvent effectiveness often depends on complex formation, especially in metal separations, the same solvent pair is not necessarily used in each of the extractors.

To specify the performance of a multicomponent dual-solvent extraction column it is necessary to select two key components between which a separation is desired. Thus the problem reduces to a pseudo dual-solvent extraction system, for which a solution is found using the methods described in Section 6.7 (p. 381). The choice of keys is based on similar considerations to those for multicomponent distillation, that is, both must be distributed in such a way that a major amount of one appears in the extract and a major amount of the other in the raffinate, with small but quantifiable amounts of each in the other phase outlets. It is usually desirable to select as keys, components with adjacent solubilities so that all solutes of lower partition coefficient than the less soluble key can be assigned to the raffinate product and all those of higher partition coefficient to the extract.

The distribution coefficients of the various solutes are often concentration-dependent and influenced by the presence of the other solutes; further, their mass transfer fluxes are interactive to varying degress (Krishna and Standart 1976). The design problems for such systems can therefore be very complex, and are best solved by the matrix method described in Section 7.4.3 (p. 460), provided that the relevant data are available.

6.10 Batch processes

Batch extraction processes take a variety of forms, of which treatment of a solution with an immiscible solvent in a separating funnel is the most basic. Complex operating schemes of various types can be used to simulate countercurrent processes in the laboratory, and to obtain powerful separations of complex mixtures. A number of types of differential extraction process, akin to batch distillation, have also been described but these do not appear to have been used other than on a laboratory scale.

6.10.1 Single and multiple contact batch extraction

This process, the basic laboratory method of extraction, consists of adding a solvent to a feed mixture in a separating funnel, shaking and allowing the phases to separate. After removal of the extract phase the treatment can be repeated with the addition of fresh solvent as many times as desired. On the larger scale the process is carried out in vessels provided with agitators, and with sight glasses to enable the phases to be run off separately after disengagement.

These processes can be treated on the triangular or Janecke diagrams in a similar manner to the corresponding continuous processes described in Sections 6.2.1 (p. 350) and 6.2.2 (p. 352), by substituting mass for flow rate. The phases are likely to reach very close to equilibrium if agitation is continued for a sufficient time. The analytical method of Section 6.2.2.2 (p. 353) can also be used if the phase miscibility is small.

6.10.2 Simulation of countercurrent extraction

It is sometimes necessary to simulate continuous countercurrent processes in the laboratory using separating funnels. Fig. 6.29(a) shows an operating scheme for such a purpose, simulating a 5-stage simple countercurrent extraction in which the circles represent shake-outs. Starting at stage 1, feed and solvent are contacted and extract E_1 is withdrawn; the raffinate then passes in turn through stages 2–5 in each of which it is re-extracted with fresh solvent, finally giving raffinate product $R_{p,1}$. The extracts from stages 2–5 are transferred to stages 1′–4′ of a second set of separating funnels and fresh solvent is added to stage 5′. The process then continues as shown, using the original set of separating funnels as the third set; thus, extracts move to the left and raffinates to the right. It is important to ensure that the quantities of feed and solvent used in each stage are identical throughout.

It is apparent, in the first set of shake-outs, that the first feed is subjected to five stages of cross flow extraction. However, as the extraction proceeds the results approach asymptotically to the continuous process, as is found

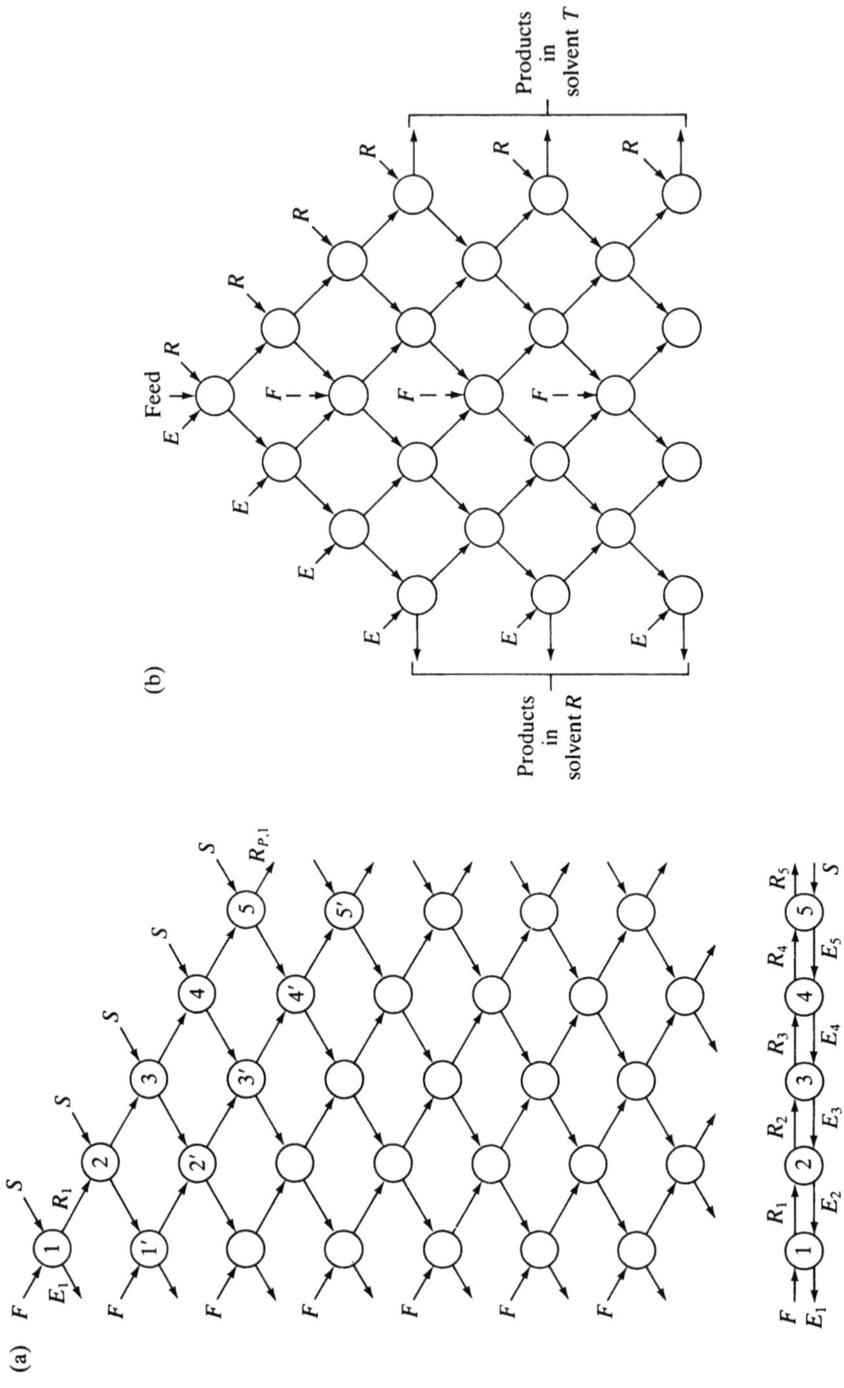

Fig. 6.29. Operation schemes for simulation of (a) countercurrent extraction; (b) dual-solvent extraction.

on analysing the raffinates on the right of the diagram. When the concentrations of these become effectively constant it can be assumed that conditions approximate to those of the continuous process, and the extracts and raffinates along the bottom of the diagram are then analysed. These can be used to plot the operating diagrams, giving points on the equilibrium and operating (i.e. mass balance) lines, using either distribution or selectivity coordinates as desired; if more than one solute is present, one diagram must be constructed for each of these.

A similar scheme has been devised for single-solvent fractional extraction (Scheibel 1954), although this is more difficult to implement in practice due to the need to separate the solvent from the extract at each stage to provide the reflux. This problem does not arise with the usual form of dual-solvent fractional extraction, the operating scheme for which is shown in Fig. 6.29(b).

Most of these procedures were first described by Jantzen (1932) and calculation procedures have been given by Stene (1944), and Compere and Ryland (1951, 1954). These are similar in principle to those described for countercurrent distribution, a similar type of process, in the next section.

6.10.3 Countercurrent distribution

6.10.3.1 *Process description*

This process is used in the laboratory for the resolution of complex natural products and synthetic mixtures into their components. It is, however, primarily a small scale process suitable for handling quantities of the order of grams and is not amenable to use for large scale operation. The operating scheme, shown in Fig. 6.30, bears a resemblence to that for dual-solvent extraction described above, with the important difference that the feed is introduced only in the first cycle. The solutes in this cycle are distributed between solvents S and T, after which solvent S, containing a fraction p of a given solute, is withdrawn and treated with a fresh quantity of solvent T in the second cycle; the original solvent T, containing a fraction q of the solute, is further treated with fresh solvent S. This procedure is continued, giving the solute distribution shown in the bottom line of Fig. 6.30 after six such cycles.

This process can be carried out in separating funnels; however, Craig has developed a number of ingenious methods which greatly simplify the procedure (Craig and Craig 1956). The most important of these employs a bank of glass equilibration tubes of the type shown in the upper part of Fig. 6.31. These are arranged in a rack in which they are rocked from position (1) to position (2) to mix the two phases and bring the solutes to equilibrium. After a pre-set time they are brought to rest in position (1) to allow phase separation to occur. The tubes are then inverted to position (3)

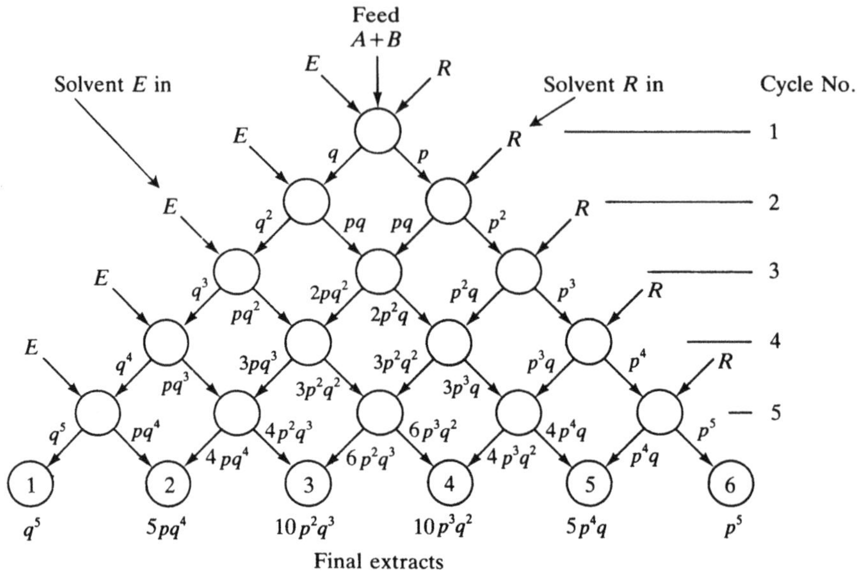

Fig. 6.30. Operating scheme for countercurrent distribution.

so that the light phase flows via side-arms A into the shorter tube B, leaving the heavy phase behind at C. On tilting again to position (1) the light phase flows into the adjacent equilibration tube through connecting tube E. As the extraction continues, therefore, the light phases progress simultaneously from one tube to the next while the heavy phases remain behind.

An extractor of this type consists of many such tubes, often several hundreds, arranged as shown at (4) in Fig. 6.31. The cycle is controlled by a timing mechanism so that it can operate automatically and unattended, often over several days during which a mixture can be subjected to many thousands of extractions, depending upon the number of tubes used. Craig tubes are available in a range of sizes, holding liquid volumes from around 10 cm^3 up to about 1 l.

6.10.3.2 *Calculation of distribution*

The ratio of the quantities of solute passing to solvents S and T respectively within a stage can be obtained from a mass balance; thus, for component A, the rejection ratio P_a is given by

$$P_a = \frac{p_a}{1 - p_a} = \frac{c_{a,S}S}{c_{a,T}T} = \frac{K_a S}{T} = \frac{1}{E_a} \tag{6.122}$$

Solving for p_a and $q_a(= 1 - p_a)$,

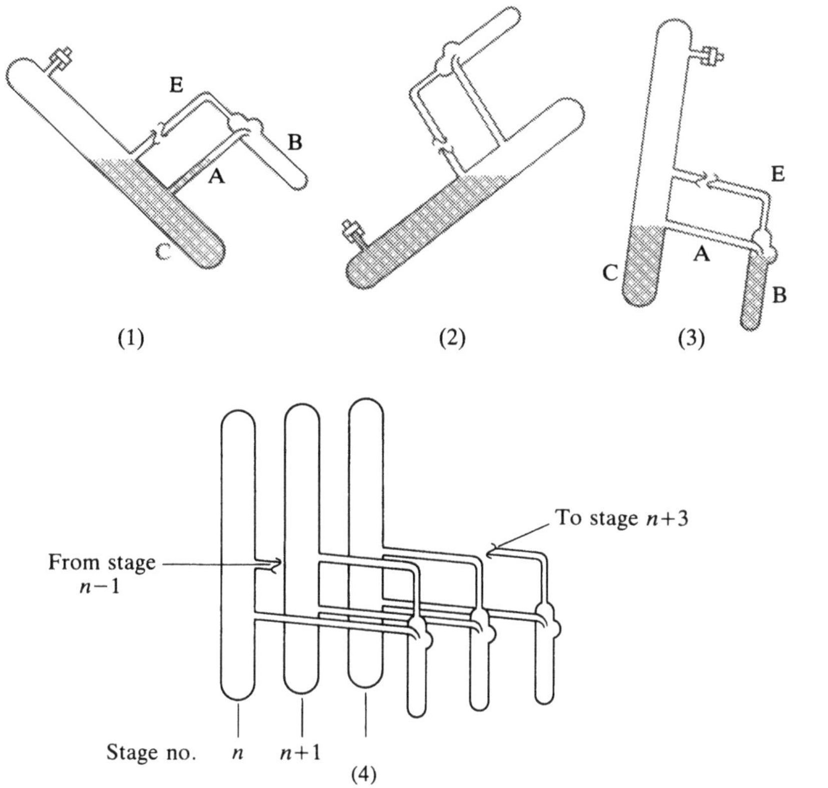

Fig. 6.31. Craig extractor tubes.

$$p_a = \frac{1}{1 + E_a}\,;\; q_a = \frac{E_a}{1 + E_a} \tag{6.123}$$

In Fig. 6.30 the compositions in each cycle are shown in terms of p and q, which are assumed to be constant; these are seen to be the successive terms in the binomial expansion of $(p + q)^t$, where t is the cycle number. Consequently for a component with constant partition coefficient the proportion present in the two combined phases is as follows

$$(p + q)^t_m = \frac{(t - 1)!}{(m - 1)!\,(t - m)!}\left(\frac{E}{1 + E}\right)^{t-1} E^{1-m} \tag{6.124}$$

where m is the tube number and t the cycle number; the numbering starts from $m = 1$ for the tube into which solvent S is introduced, and $t = 1$ for the cycle into which feed is introduced.

The best resolution of two solutes is obtained when the solvent ratio is chosen to give a symmetric separation, i.e. when $E_a = E_b^{-1}$, so that

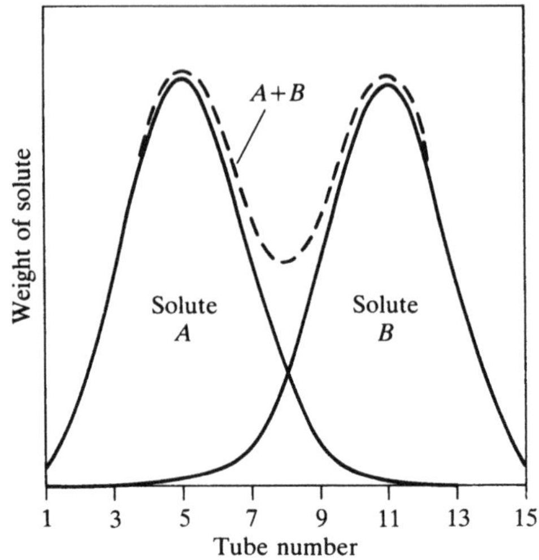

Fig. 6.32. Product distribution from Craig extractor for symmetric separation.

$$\frac{S}{T} = \left(\frac{1}{K_a K_b}\right)^{1/2} \tag{6.125}$$

The product distribution for a typical symmetric separation is shown in Fig. 6.32. The peak value of $(p + q)^t_m$ can be shown to occur at tube $m_{\max}$, given approximately by

$$m_{\max} \approx pt + 1 \tag{6.126}$$

The presence of the two peaks in Fig. 6.32 is a good indication that at least two components were present in the feed. However, if the number of tubes is too small in relation to the relative extraction factors the separate peaks in the plot of total solute against tube number may not be detectable, although there would be a flattening of the curve as compared with that for a single substance.

For a more detailed treatment of this subject reference should be made to the reviews of Craig and Craig (1956) and Hecker (1955).

6.10.4 Differential extraction

6.10.4.1 *Types of process*

In its simplest form this process is analogous to Rayleigh distillation in that is involves the repeated extraction of a charge with infinitesimal quantities of solvent, with removal of the extract as it is formed. It can be approached

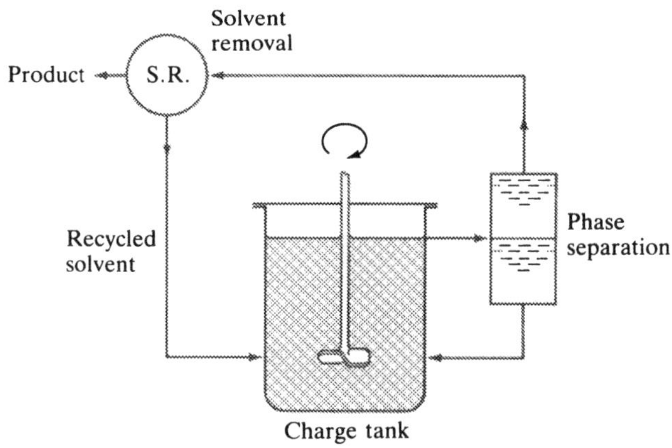

Fig. 6.33. Differential (Rayleigh) extraction.

in practice using the apparatus shown in Fig. 6.33, in which solvent is added continuously to the charge in the mixing tank; the resulting two-phase mixture passes to a phase separator, the extract phase from which enters the solvent removal unit, and the raffinate returns to the mixing vessel.

This process, like Rayleigh distillation, cannot give a high degree of separation unless the distribution coefficient is very large. However, in an extension of the process a multi-stage extraction column can be interposed between the mixing vessel and the solvent removal unit, and a portion of the separated extract product refluxed to the top of the column as shown in Fig. 6.34. By this means, fractions of decreasing solubility are obtained, and the process bears an obvious resemblance to normal batch distillation in which successive fractions of decreasing volatility are withdrawn; the mixing vessel therefore takes the place of the still pot/reboiler. A number of possible alternative forms of this process, involving both normal and inverted operation with single and dual-solvent systems has been described (Pratt 1954).

This process has been used in the laboratory to fractionate lubricating oil with acetone (Cannon and Fenske 1936) and a petroleum fraction with phenol (Kondrat'ev 1964). No large-scale applications have been reported.

6.10.4.2 *Rayleigh extraction*

The general case, including that where solvent miscibility cannot be neglected, can be treated using the Janecke and selectivity diagrams. On this basis a differential mass balance on solute, assuming the entering solvent to be solute-free, gives

$$\bar{y}\,\mathrm{d}\bar{R} = d(\overline{Rx}) = \bar{R}\,\mathrm{d}x + \bar{x}\,\mathrm{d}R \tag{6.127}$$

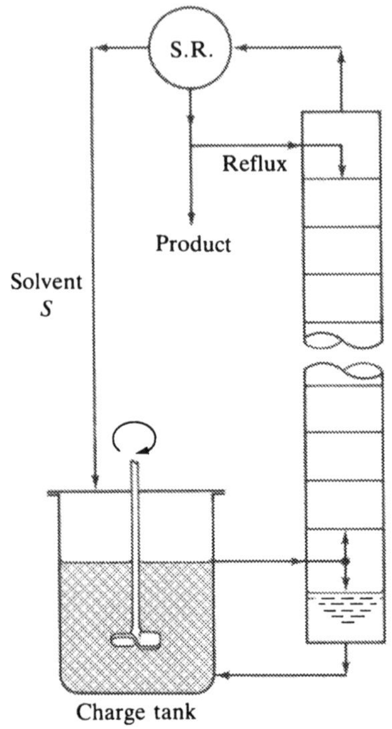

Fig. 6.34. Single-solvent batch fractional extraction.

This may be integrated as follows between the initial charge composition and amount, designated by subscript 0, and the values at any later stage of the extraction (unsubscripted), as follows

$$\int_{\bar{R}}^{\bar{R}_0} \frac{d\bar{R}}{\bar{R}} = \ln \frac{\bar{R}_0}{\bar{R}} = \int_{\bar{x}}^{\bar{x}_0} \frac{d\bar{x}}{(\bar{y} - \bar{x})} \tag{6.128}$$

The integration can be carried out graphically by measuring the area under a plot of $1/(\bar{y} - \bar{x})$ against $\bar{x}$, using values of $\bar{y}$ and $\bar{x}$ obtained from the selectivity diagram after projection from the Janecke diagram. If the stages are non-ideal, as is usually the case, the values of $\bar{y}$ must be adjusted by incorporation of the Murphree or Hausen efficiency.

The course of the extraction can be depicted by means of batch extraction curves, which comprise plots of the solvent-free extract and raffinate compositions against $(1 - \bar{R}/R_0)$, the fraction of the original solvent-free charge extracted. To obtain these it is necessary to perform the integration between the initial charge compositions $\bar{x}_0$ and a range of values of the residue composition $\bar{x}$.

For some banded Type II systems the selectivity ratio β, defined by

$$\beta = \frac{K_a}{K_b} = \frac{\bar{y}\,(1 - \bar{x})}{\bar{x}\,(1 - \bar{y})} \tag{6.129}$$

is effectively constant using either mass or mole fraction units. For such cases the following analytical expression can be shown to apply (Pratt 1967)

$$\ln\frac{\bar{R}_0}{\bar{R}} = \frac{1}{(\beta - 1)}\left[\ln\frac{\bar{x}_0}{\bar{x}} + \beta\ln\frac{1 - \bar{x}}{1 - \bar{x}_0}\right] \tag{6.130}$$

6.10.4.3. *Batch fractional extraction*

In the single-solvent process shown in Fig. 6.34, the single-stage mixer in Fig. 6.33 is replaced by a multistage extractor, so that $(\bar{y} - \bar{x})$ in eqn (6.128) must be replaced by $(\bar{x}_p - \bar{x}_r)$, where subscript p refers to the product composition after solvent removal and r to that of the residue, i.e. raffinate. Values of $\bar{x}_p$ and $\bar{x}_r$ can be obtained from the selectivity diagram by projecting operating lines from the Janecke diagram starting from a number of values of $\bar{x}_p$ and stepping off the stages by the McCabe–Thiele method to obtain the corresponding value of $\bar{x}_r$. A plot can thus be drawn of $\bar{x}_p$ against $\bar{x}_r$, and values read off for the evaluation of the integral in eqn (6.128). The results can be used to plot the batch extraction curves as described in the previous section. This procedure has been described in detail for batch distillation by Pratt (1967).

It is assumed implicitly in deriving eqn (6.128) that the hold-up of the extraction column is negligible in comparison with that of the charge vessel. The effect of such hold-up has been shown by Rose *et al.* (1952) and Rose and Johnson (1953) to be both complex and significant for batch distillation, and the same is most likely to be the case for batch extraction, necessitating detailed computation for obtaining the true batch curves.

Nomenclature

A	component A (solute), mass or mass flow rate, kg or kg/s
a	superficial area of contact of the phases, $m^2/m^3 = m^{-1}$
a_i	concentration of solute A at the interface, kg/m^3 or $kmol/m^3$
B	component B mass or mass flow rate, kg or kg/s
$\bar{c}_k$	solute concentration in phase k on solute-free basis, kg/m or kmol/m
D	diffusivity, m^2/s
E	mass flow of extract phase (except Section 6.7, p. 381), kg/s or kmol/s
E	flow rate of extract phase on a solute-free basis (Section 6.7, p. 381), m^3/s
$\bar{E}$	mass flow of extract phase on a solvent-free basis, kg/s or kmol/s
E	U_x/mU_y, extraction factor
E_H	Hausen efficiency

E_{Mk} Murphree efficiency based on phase k
E_o overall efficiency of extractor
E_T transfer unit efficiency
$\boldsymbol{E}_{ab}$ ratio of solute A to solute B in the extract product
F mass flow rate of feed (except Section 6.7), kg/s
F flow rate of feed on a solute-free basis (Section 6.7, p. 381), m^3/s
H height of column, m
h_c height of compartment, m
H_k height of a transfer unit based on phase k, m
H_{ok} height of an overall transfer unit based on phase k, m
K_J distribution coefficient for solute J
k_{ok} overall mass transfer coefficient based on phase k, m/s
k_k individual phase mass transfer coefficient based on phase k, m/s
k_m^+ mth order heterogeneous rate constant for forward reaction, $m^{3m-2}/s\ kmol^{m-1}$
k_n^- nth order heterogeneous rate constant for reverse reaction, $m^{3n-2}/s\ kmol^{n-1}$
M quantity or mass flow of mixed phase, kg or kg/s
$\bar{M}$ quantity or mass flow of mixed phase on a solute-free basis, kg or kg/s
m slope of equilibrium line on c_x, c_y, or other coordinates, as defined [e.g. eqn (6.1)]
N_a flux of solute A across interface, $kg/m^2\ s$ or $kmol/m^2\ s$
$\bar{N}_a$ rate of solute transfer per unit cross-section of extractor (Section 6.3.3, p. 360), $kg/m^2\ s$ or $kmol/m^2\ s$
N total number of stages
N_k number of transfer units based on phase k
N_{ideal} total number of ideal stages
N_{ok} number of overall transfer units based on phase k
N_{ok}^1 number of overall transfer units per stage or compartment based on phase k
n number of stages within the extractor
P_J $p/(1-p)$, rejection ratio for solute J
$\bar{P}$ product rate on solvent-free basis, kg/s
p fraction of solute in extract phase at equilibrium (Section 6.10.3.2, p. 406)
q fraction of solute in raffinate phase at equilibrium (Section 6.10.3.2, p. 406)
q intercept of equilibrium line on (c_x, c_y) or other coordinates, as defined [e.g. eqn (6.1)]
R mass rate of flow of raffinate (except Section 6.7, p. 381), kg/s
R flow rate of raffinate on a solute-free basis (Section 6.7, p. 381), m^3/s
$\bar{R}$ mass rate of flow of raffinate on a solvent-free basis, kg/s
$\boldsymbol{R}_{ba}$ ratio of solute B to solute A in the raffinate product
r fraction of solute returned as reflux
S mass rate of flow of solvent, kg/s
$\bar{S}$ mass rate of flow of non-solvent (i.e. of $A + B$) in solvent phase on a solvent-free basis, kg/s

s solvent ratio, $S/(A + B)$, i.e. ratio of solvent to total non-solvent on a mass or molar basis
V_c volume of compartment (i.e. stage), m^3
X $[c_x - (c_{y,\mathrm{I}} - q)/m]/[c_{x,\mathrm{o}} - (c_{y,\mathrm{I}} - q)/m]$, dimensionless concentration of feed phase (for staged extractors, $c_{y\mathrm{I}}$ is replaced by $c_{y,N+1}$)
U_k superficial velocity of phase k, m/s
x mass or mole fraction of solute in X-phase
$\bar{x}$ mass or mole fraction of solute in X-phase on a solvent-free basis
Y $(c_y - c_{y,\mathrm{I}})/[mc_{x,\mathrm{o}} - (c_{y,\mathrm{I}} - q)]$, dimensionless concentration of solute in Y phase (for staged extractors, $c_{y\mathrm{I}}$ is replaced by $c_{y,N+1}$)
y mass or mole fraction of solute in Y phase
$\bar{y}$ mass or mole fraction of solute in Y phase on a solvent-free basis
z height or length within extractor, m

Greek Letters
β K_a/K_b (K_b/K_a in Section 6.7)
ρ density, kg/m^3

Subscripts
A,a component A (solute)
adj adjusted
av average
B,b component B [solute in Section 6.7 (p. 381), reactant in Section 6.8.1 (p. 395)]
D inlet
E extract phase
E extraction section (Section 6.7)
f final
F feed phase
J solute A, B, etc
i intersection of operating lines
I solvent inlet end of extractor
k x or y phase
lm logarithmic mean
M feed stage, counted from reflux inlet end (single-solvent extraction), or from scrub section inlet (dual-solvent extraction)
M relates to the mixture (Section 6.2.1, p. 350)
m stage within enriching or scrub section
n stage within stripping or extraction section
N solvent inlet stage (single-solvent extraction); feed stage, counted from extract solvent inlet (dual-solvent extraction)
o feed phase inlet end of extractor
P plug flow
R raffinate phase
R_p raffinate product
S solvent phase
s scrub section (Section 6.7)

s component S (solvent)
T total
x relates to the X-phase
y relates to the Y-phase
Δ relates to the net flow of solute

Superscripts
$*$ at equilibrium
l local value
i at interface

References

ALDERS, L. (1959). *Liquid–liquid extraction* (2nd edn). Elsevier, Amsterdam.

ASSELIN, G.F. and COMINGS, E.W. (1950). *Industrial and Engineering Chemistry*, **42**, 1198.

ASTARITA, G. (1967). *Mass transfer with chemical reaction*. Elsevier, Amsterdam.

BRANCKER, A.V., HUNTER, T.G., and NASH, A.W. (1940). *Journal of Physical Chemistry*, **44**, 683.

BRANCKER, A.V., HUNTER, T.G., and NASH, A.W. (1941). *Industrial and Engineering Chemistry*, **33**, 880.

CANNON and FENSKE, (1936). *Industrial and Engineering Chemistry*, **28**, 1035.

COLBURN, A.P. (1939). *Transactions of the American Institute of Chemical Engineers*, **35**, 91.

COLBURN, A.P. (1941). *Industrial and Engineering Chemistry*, **33**, 459.

COMPERE, E.L. and RYLAND, A.L. (1951, 1954). *Industrial and Engineering Chemistry*, **43**, 239; **46**, 24.

CRAIG, L.C. and CRAIG, D. (1956). *Techniques of organic chemistry*, (2nd edn), (ed. A. WEISSBURGER), Vol. 3, Part 1, p. 149. Interscience, New York.

DANCKWERTS, P.V. (1970). *Gas-liquid reactions*. McGraw-Hill, New York.

DONGAONKAR, K.R., STEVENS, G.W., and PRATT, H.R.C. (1987). *Institution of Chemical Engineers Symposium Series*, No. 103, p. 279.

EVANS, T.W. (1934). *Industrial and Engineering Chemistry*, **26**, 860.

HANSON, C. (1971). *Recent advances in liquid–liquid extraction*, Ch. 12. Pergamon Press, Oxford.

HAUSEN, H. (1953). *Chemische Ingenieur Technology*, **25**, 595.

HECKER, E. (1955). *Verteilungsverfahren im Laboratorium*. Verlag Chemie, Weinheim/Bergstr.

HUNTER, and NASH, (1932). *Journal of the Society of Chemical Industry*, **51**, 285.

JANECKE, Z. (1906). *Anorganische Chemie*, **51**, 132.

JANTZEN (1932). *Das fractioniente Distillieren and das fractioniente Verteilen*, Dechema Monographien V, No. 48. Verlag Chemie, Berlin.

KLINKENBURG, A. (1951). *Chemical Engineering Science*, **1**, 86.

KLINKENBURG, A. LAUWERIER, A.H., and REMAN, G.H. (1951). *Chemical Engineering Science*, **1**, *93*.

KONDRAT'EV, A.A. (1964). *International Chemical Engineering*, **4**, 434.

KRISHNA, R. and STANDART, G. (1976). *American Institute of Chemical Engineers Journal*, **22**, 838.

LEWIS, W. K. (1936). *Industrial and Engineering Chemistry*, **20**, 399.

MALONEY, J. O. and SCHUBERT, A. E. (1940). *Transactions of the American Institute of Chemical Engineers*, **36**, 741.

MURPHREE, E. V. (1925). *Industrial and Engineering Chemistry*, **17**, 747.

PRATT, H. R. C. (1954). *Chemical Engineering Science*, **3**, 189.

PRATT, H. R. C. (1967). *Countercurrent separation processes*. Elsevier, Amsterdam.

PRATT, H. R. C. (1983a). *The handbook of solvent extraction* (eds T. C. LO, M. H. I. BAIRD, and C. HANSON), Ch. 5. Wiley, New York.

PRATT, H. R. C. (1983b). *Solvent Extraction and Ion Exchange*, **1**, 669.

PROCHAZKA, J. and JIRICNY, V. (1976). *Chemical Engineering Science*, **31**, 176.

ROCHE, E. C. (1969). *British Chemical Engineering*, **14**, 1393.

ROCHE, E. C. (1971). *British Chemical Engineering*, **16**, 821.

ROCHE, E. C. (1973). *Computer calculations for chemical engineering education, Vol. VI, Stagewise Operations*, (ed. J. CHRISTENSEN), p. 46. National Academy of Engineering, U.S.A.

ROSE, A. and JOHNSON, R. C. (1953). *Chemical Engineering Progress*, **49**, 15.

ROSE, A., JOHNSON, R. C., and WILLIAMS, T. J. (1952). *Chemical Engineering Progress*, **48**, 16.

RUTHRUFF and WILCOX (1941). *Transactions of the American Institute of Chemical Engineers*, **37**, 649.

SCHEIBEL, E. G. (1948). *Chemical Engineering Progress*, **44**, 681, 771.

SCHEIBEL, E. G. (1954). *Industrial and Engineering Chemistry*, **46**, 16.

SCHEIBEL, E. G. and OTHMER, D. F. (1942), *Transactions of the American Institute of Chemical Engineers*, **38**, 339.

SHERWOOD, T. K., PIGFORD, R. L., and WILKE, C. R. (1975). *Mass transfer*. McGraw-Hill, New York.

SMITH, B. D. (1963). *Design of equilibrium stage processes*, Ch. 12. McGraw-Hill, New York.

STANDART, G. (1956). *Chemical Engineering Science*, **20**, 611.

STENE, S. (1944). *Arkiv Kemi, Mineralogy, Geology*, **184**, No. 18.

TIERNEY, J. W. (1977). *Proceedings of the International Solvent Extraction Conference*, Toronto, (ISEC '77), Vol. 1, 289. Canadian Institute of Mining and Metallurgy.

TIERNEY, J. W., YANOSIK, J. L., Bruno, J. A., and BRAINARD, A. J. (1971), *Proceedings of the International Solvent Extraction Conference*, The Hague, (ISEC '71), Vol. 2, 1051. Society of Chemical Industry, London.

TILLER, F. M. (1949). *Chemical engineering progress*, **45**, 391.

TREYBAL, R. E. (1963). *Liquid extraction* (2nd edn). McGraw-Hill, New York.

TREYBAL, R. E. (1981). *Mass transfer operations* (3rd edn). McGraw-Hill, New York.

7

AXIAL DISPERSION

H.R.C. PRATT and G.W. STEVENS

7.1 Introduction

The treatment in the previous chapter relates specifically to pure plug flow, which in practice occurs only in extractors such as discrete-stage mixer–settlers and perforated plate columns where phase separation is virtually complete between stages. Other contactor types suffer adversely in performance from non-uniform flows in the continuous, and sometimes the dispersed phase; these are referred to collectively as axial dispersion and require the more advanced treatment described in this chapter.

7.1.1 Nature of axial dispersion

The overall flow pattern in extraction equipment is complex, but the following factors can be identified as contributing to a reduced performance:

1. Transport and shedding of continuous phase in wakes attached to the rear of droplets in motion.
2. Circulatory flow of continuous phase in the direction of droplet motion, due to the dissipation of the dispersed phase potential energy.
3. Circulation of continuous and entrained dispersed phase in mechanically agitated extractors.
4. Non-uniform velocity profiles in both forward and backward directions relative to an observer moving with the average velocity of flow; these are caused by frictional drag of packing or other column internals, and lead to a distribution of residence times.
5. Dispersion of droplet velocities due to the range of sizes present ('forward mixing').
6. Channelling and consequent maldistribution of the phases.

The first two factors result in pure backmixing of continuous phase (Letan and Kehat 1968; Anderson and Pratt 1978). Thus if the feed phase is continuous, the countercurrent droplet motion results in transport of a proportion of this phase, from which solute has been transferred, into a region nearer the feed entry where the concentration is higher. As a consequence the concentration gradient is degraded throughout the extractor, and at the feed entry itself the mixing of depleted feed phase with fresh incoming feed results in a sudden downward 'jump' in concentration, as will be shown later. Similarly if the extractant phase is continuous, part of the phase which has been enriched in solute is transported against its normal direction of flow into a region of lower concentration; hence at the extractant entry there will be an upward concentration 'jump'. Of these two factors, the second is of greater importance as only 5–10 per cent of the attached wakes, with a volume of one to four times that of the droplets, is shed from droplets for which $Re_d > 150$–200 (Yeheskel and Kehat 1971; Anderson and Pratt 1978). The third factor also contributes to backmixing in the continuous phase, to an extent determined by the agitation rate.

The fourth factor is more obscure, but it has been shown from a comparison of the solutions for the 'mixing cell' and diffusion models that the passage of tracer in a bed of packed spheres can be described in terms of a uniform forward velocity with eddy diffusion in an axial direction towards the exit (Aris and Amundsen 1957). The contribution due to this factor is probably relatively small for two-phase flow, however.

True backmixing is less common in the dispersed phase, as the droplets normally move only in the forward direction relative to the continuous phase. The main cause of backmixing, if any, in this phase is therefore the third factor, that is, circulation of droplets due to mechanical agitation. The fifth factor, an effect due to droplet polydispersivity, is quite different in nature from backmixing although it also has a deleterious effect on performance; this will be discussed in Section 7.5 (p. 470).

The sixth factor is not strictly relevant to the present treatment, although its effect on the driving force for mass transfer can be important. This will be discussed briefly in the next chapter in relation to the performance of pulsed perforated plate columns.

7.1.2 Axial dispersion models

Two distinct types of models are in use, namely the diffusion model, which assumes turbulent axial diffusion of solute superimposed on plug flow of the phase under consideration (Sleicher 1959; Miyauchi and Vermeulen 1963), and the backflow model, which assumes well-mixed non-ideal stages between which backflow occurs (Sleicher 1960). The two models represent idealized limiting cases, but in practice the diffusion model is approached by differential extractors such as packed and baffle-plate columns. Similarly the backflow model, originally developed for mixer–settlers to account for the effect of inter-stage entrainment, is to be preferred for multicompartment columns such as rotary agitated, pulsed plate and reciprocating plate types. In the past, many workers have applied the diffusion model to such compartment-type extractors, and vice versa. However, this practice is not to be recommended, even though in the limit there is a relatively simple relationship between the two models, as will be shown.

7.2 Development of models

7.2.1 Diffusion model

7.2.1.1 *Assumptions made*

The assumptions made for the purpose of this model are as follows (Sleicher 1959):

1. The backmixing of each phase can be characterized by a constant turbulent axial diffusion coefficient, E_j.
2. The boundary conditions of Wehner and Wilhelm (1956) are applicable.
3. The mean velocity and concentration of each phase is constant within any given cross-section.
4. The volume mass transfer coefficient is constant or can be averaged over the column.
5. The solute concentration gradients in each phase are continuous (except at the phase inlets, as shown later).
6. The solute-free feed and extractant (i.e. X and Y) phases are effectively immiscible, or have a constant miscibility irrespective of solute concentration.
7. The volumetric flow rates of feed and extractant phases are constant throughout the extractor.
8. The equilibrium relation is linear or can be approximated by a straight line.

7.2.1.2 *Derivation of equations*

Mass balances on solute in the two phases taken over a differential height of column, dz, are shown in Fig. 7.1 for one-dimensional flow (see Fig. 12, p. 370, of Chapter 6 for the corresponding plug flow case). These lead to the following equations:

$$E_x \frac{d^2 c_x}{dz^2} - U_x \frac{dc_x}{dz} - k_{ox} a(c_x - c_x^*) = 0 \tag{7.1}$$

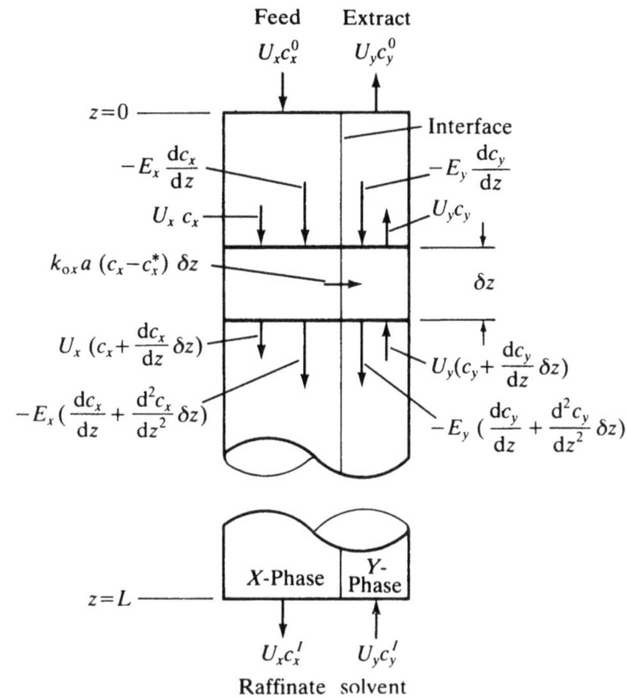

Fig. 7.1. Diffusion model: material balances over differential section.

$$E_y \frac{d^2c_y}{dz^2} + U_y \frac{dc_y}{dz} + k_{ox} a(c_x - c_x^*) = 0 \tag{7.2}$$

With no backmixing, $E_x = E_y = 0$ and these expressions reduce to those for plug flow [cf. eqn (6.60), p. 371 for the X-phase]. Assuming a linear equilibrium relation, viz.

$$c_y^* = mc_x + q, \tag{7.3}$$

eqns (7.1) and (7.2) can be expressed in dimensionless form as follows:

$$\frac{d^2X}{dZ^2} - Pe_x B \frac{dX}{dZ} - N_{ox} Pe_x B(X - Y) = 0 \tag{7.4}$$

$$\frac{d^2Y}{dZ^2} + Pe_y B \frac{dY}{dZ} + EN_{ox} Pe_y B(X - Y) = 0 \tag{7.5}$$

where $Z = z/H$, the dimensionless height; X and Y are the dimensionless concentrations (Section 6.1.2.2, p. 346); $Pe_k = U_k d_c/E_k$, the turbulent Peclet number; $N_{ox} = k_{ox}aH/U_x$, the number of overall X-phase transfer units; $B = H/d_c$; and d_c is a characteristic dimension, e.g. the packing diameter or compartment height. Solving eqn (7.4) for Y, differentiating twice and substituting in eqn (7.5) then gives:

$$\frac{d^4X}{dZ^4} - \alpha\frac{d^3X}{dZ^3} - \beta\frac{d^2X}{dZ^2} - \gamma\frac{dX}{dZ} = 0 \tag{7.6}$$

where

$$\alpha = B(Pe_x - Pe_y) \tag{7.7}$$

$$\beta = N_{ox}B(Pe_x + EPe_y) + Pe_x Pe_y B^2 \tag{7.8}$$

$$\gamma = N_{ox} Pe_x Pe_y B^2(1 - E) \tag{7.9}$$

The four boundary conditions applicable to the solution of eqn (7.6) relate to the inlet and exit concentrations of each phase, and are obtained as follows (Wehner and Wilhelm 1956; Sleicher 1959)

i. X-Phase, $Z = 0$. At the X-phase inlet, solute is transferred away from the inlet of the contacting section by backmixing, i.e. at $Z = 0$ where the concentration is $c_{x,0}$, at a rate given by

$$-E_x\left[\frac{dc_x}{dz}\right]_0 + U_x c_{x,0}$$

The solute enters with the feed at a rate of $U_x c_x^0$, where c_x^0 is the concentration external to the column. Hence equating the flows

$$-E_x\left[\frac{dc_x}{dz}\right]_0 = U_x(c_x^0 - c_{x,0}) \tag{7.10}$$

In dimensionless units this becomes, noting that $X^0 = 1.0$,

$$-\left[\frac{dX}{dZ}\right]_0 = Pe_x B(1 - X_0) \tag{7.11}$$

ii. X-Phase, $Z = 1.0$. At the X-phase exit where the dimensionless internal and external concentration are X_l and X^l respectively, similar reasoning gives

$$-\left[\frac{dX}{dZ}\right]_l = Pe_x B(X^l - X_l)$$

However, $Pe_x B$ is always positive or zero so that the gradient given by the term on the left hand side is of opposite sign to that on the right. The only possible condition is therefore

$$X^l = X_l; \left[\frac{dX}{dZ}\right]_l = 0 \tag{7.12}$$

iii. Y-Phase, $Z = 1.0$. A similar procedure to that used in (i) gives

$$-\left[\frac{\mathrm{d}Y}{\mathrm{d}Z}\right]_I = Pe_y BY_I \tag{7.13}$$

where Y_I relates to the internal concentration at the Y-phase inlet. (The external concentration $Y^I = 0$).

iv. Y-Phase, $Z = 0$. Similar reasoning to that used in (ii) gives

$$Y^0 = Y_0; \left[\frac{\mathrm{d}Y}{\mathrm{d}Z}\right]_0 = 0 \tag{7.14}$$

7.2.1.3 *Solutions for concentration profiles*

The solution of eqn (7.6) for the X-profile in terms of dimensionless height, Z, is as follows

$$X = A_1 + A_2 \mathrm{e}^{\lambda_2 Z} + A_3 \mathrm{e}^{\lambda_3 Z} + A_4 \mathrm{e}^{\lambda_4 Z} \tag{7.15}$$

Combination of this result with eqn (7.4) then gives the Y-profile

$$Y = A_1 + A_2 a_2 \mathrm{e}^{\lambda_2 Z} + A_3 a_3 \mathrm{e}^{\lambda_3 Z} + A_4 a_4 \mathrm{e}^{\lambda_4 Z} \tag{7.16}$$

The λ_i in these expressions ($i = 2$, 3 and 4) are the roots of the characteristic equation

$$\lambda^3 - \alpha\lambda^2 - \beta\lambda - \gamma = 0 \tag{7.17}$$

and the a_i are given by

$$a_i = 1 + \frac{\lambda_i}{N_{ox}} - \frac{\lambda_i^2}{N_{ox} Pe_x B} \tag{7.18}$$

$$= \frac{E(1 - \lambda_i / Pe_x B)}{(1 + \lambda_i / Pe_y B)} \tag{7.19}$$

where the second, more convenient form is obtained by elimination of N_{ox} between eqns (7.17) and (7.18).

The roots of eqn (7.17) are all real and are therefore given by:

$$\lambda_i = \frac{\alpha}{3} + 2p^{1/2} \cos\left(\frac{u}{3} + k\right) \tag{7.20}$$

with $k = 0°$, 120°, and 240° for λ_2, λ_3 and λ_4 respectively, and

$$p = \left(\frac{\alpha}{3}\right)^2 + \frac{\beta}{3}$$

$$q = \left(\frac{\alpha}{3}\right)^3 + \frac{\alpha\beta}{6} + \frac{\gamma}{2}$$

$$u = \cos^{-1}\left(\frac{q}{p^{3/2}}\right)$$

where the angle u lies between 0° and 180°. This solution applies only if

$$q^2 - p^3 < 0$$

which is always the case in the present application.

The values of the A_i are obtained by substitution of eqns (7.15 and 7.16) into the boundary conditions, eqns (7.11–7.14), thus

$$A_i + \sum_{i=1}^{4} A_i\left(\frac{1-\lambda_i}{P_x B}\right) = 1 \tag{7.21}$$

$$\sum_{i=2}^{4} A_i \lambda_i e^{\lambda_i} = 0 \tag{7.22}$$

$$\sum_{i=2}^{4} A_i a_i \lambda_i = 0 \tag{7.23}$$

$$A_i + \sum_{i=2}^{4} A_i a_i e^{\lambda_i}\left(1 + \frac{\lambda_i}{P_y B}\right) = 0 \tag{7.24}$$

The solution to these equations can be expressed conveniently in determinantal form, as follows (Miyauchi and Vermeulen 1963a):

$$A_i = \frac{D_{A_i}}{D_{A_1} - D_A} \tag{7.25}$$

where

$$D_A = \begin{vmatrix} 1 - \dfrac{\lambda_2}{Pe_x B} & 1 - \dfrac{\lambda_3}{Pe_x B} & 1 - \dfrac{\lambda_4}{Pe_x B} \\ a_2\lambda_2 & a_3\lambda_3 & a_4\lambda_4 \\ \lambda_2 e^{\lambda_2} & \lambda_3 e^{\lambda_3} & \lambda_4 e^{\lambda_4} \end{vmatrix}$$

$$D_{A_1} = \begin{vmatrix} a_2\lambda_2 & a_3\lambda_3 & a_4\lambda_4 \\ \lambda_2 e^{\lambda_2} & \lambda_3 e^{\lambda_3} & \lambda_4 e^{\lambda_4} \\ a_2 e^{\lambda_2} f(\lambda_2) & a_3 e^{\lambda_3}(\lambda_3) & a_4 e^{\lambda_4} f(\lambda_4) \end{vmatrix}$$

with $f(\lambda_i) = \left(1 + \dfrac{\lambda_i}{Pe_y B}\right)$

$$D_{A_2} = \lambda_3\lambda_4(a_4 e^{\lambda_3} - a_3 e^{\lambda_4})$$

$$D_{A_3} = \lambda_2\lambda_4(a_2 e^{\lambda_4} - a_4 e^{\lambda_2})$$

$$D_{A_4} = \lambda_2\lambda_3(a_3 e^{\lambda_2} - a_2 e^{\lambda_3})$$

Typical concentration profiles calculated using these equations are shown on X–Y coordinates in Fig. 7.2 together with the curves for the corresponding plug flow case. The effects of backmixing are apparent from the concentration 'jumps' AB and CD in Fig. 7.2 at the X- and Y-phase inlets (Geankoplis *et al.* 1950, 1951; Gier and Hougen 1953) and the zero slopes at the exits, which result from the specified boundary conditions. The operating diagram for the same case is shown in Fig. 7.3, in which the inlet jumps are again clearly indicated.

Figure 7.3 also shows that the true operating line is displaced to a considerable extent from the plug flow operating line; the latter is now termed the balance line, as it represents the overall material balance across the extractor. The driving force for mass transfer on an X-phase basis is therefore reduced by backmixing, typically from the plug flow value of AC to AB, so that the number of transfer units required for a given duty is increased considerably. If backmixing is absent in one or other phase, the profile for that phase does not show the inlet concentration jump or the zero gradient at the exit. Such cases will be discussed later.

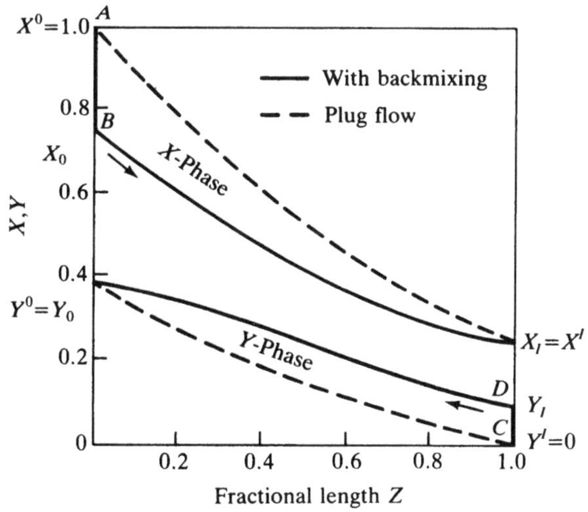

Fig. 7.2. Effect of backmixing on concentration profiles.

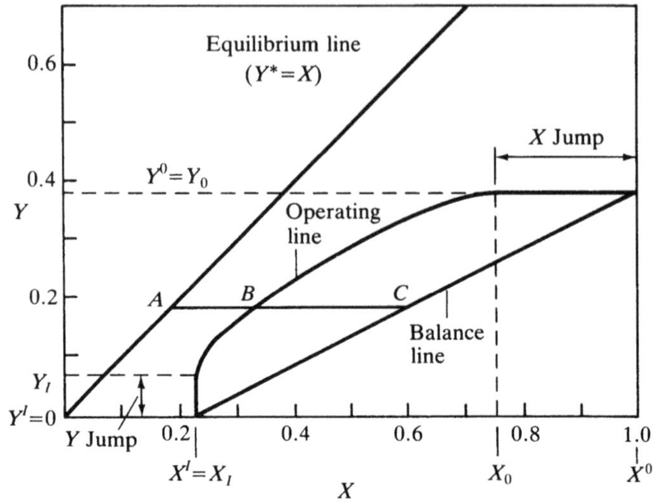

Fig. 7.3. Operating diagram of X–Y coordinates. The parameter values are the same as for Fig. 7.2.

7.2.1.4 *Effect of entrance sections*

Wilburn (1964) extended the diffusion model to allow for the semi-stagnant zones at the ends of a differential column extractor, beyond the feed and extractant inlets (i.e. for $Z < 0$ and > 1.0). This led to a set of 6 differential equations and 12 boundary conditions. The resulting solutions were similar in form to eqns (7.15 and 7.16) and contained 12 constants, the values of which were obtained by numerical inversion of a 12×12 matrix. Typical computed results indicated that the inlet concentration jumps were increased by the presence of the stagnant zones. Some doubt as to the realism of this model remains, however, due to the assumption that the backmixing Peclet numbers for the stagnant zones were the same as for the extractor itself, although with no net convective flow of the phases.

In an extension of this method (Wilburn and Nicholson 1965), allowance was also made for variations in dispersed-phase hold-up and for a non-linear equilibrium relationship by subdivision of the column into n sections, within each of which the hold-up and partition coefficient could be assumed constant. This led to a $4n \times 4n$ matrix which was inverted numerically for solution. An alternative matrix solution will be described in Section 7.4.4 (p. 468).

7.2.1.5 *Transfer units and backmixing*

Integration of eqn (7.1) between the internal concentration at the X-phase inlet, $c_{x,0}$, and the exit concentration, $c'_{x,l}$, gives

$$N_{ox} = \frac{k_{ox}aH}{U_x} = \int_{c'_{x,l}}^{c_{x,0}} \frac{\mathrm{d}\,[c_x - (E_x/U_x)(\mathrm{d}c_x/\mathrm{d}z)]}{(c_x - c_x^*)_B} \tag{7.26}$$

where $(c_x - c_x^*)_B$ represents the 'true' driving force corresponding to, e.g. AB in Fig. 7.3. This may be compared with the expression for the plug flow case [cf. eqns (6.61, 6.62 and 6.65), pp. 371–2], as follows

$$N_{ox} = \frac{k_{ox}aH}{U_x} = \int_{c_{x,l}}^{c_{x,0}} \frac{dc_x}{(c_x - c_x^*)_P} \tag{7.27}$$

It is apparent that the value of the right hand side of eqn (7.26) differs from that in eqn (7.27) in two respects, viz. (i), the amount of solute transferred is reduced by an amount $(E_x/U_x)(\mathrm{d}c_x/\mathrm{d}z)$ due to remixing, and (ii), the driving force is reduced, e.g. from AC to AB (in X, Y coordinates) in Fig. 7.3. As $\mathrm{d}c_x/\mathrm{d}z$ is negative, both effects increase the value of the right hand term; however, the value of N_{ox}, i.e. $k_{0x}aH/U_c$, is the same in both cases, so that the exit concentration $c'_{x,l} > c_{x,l}$ and the performance is therefore reduced by backmixing, as expected; alternatively, the height H must be increased in the backmixing case to make $c'_{x,l} = c_{x,l}$ so that $N_{oxB} > N_{oxP}$ for the same performance. (Subscript B refers to the backflow case, and P to plug flow).

Geankoplis *et al.* (1950, 1951) and Gier and Hougen (1953) have employed a third definition of N_{ox}, viz. N_{oxM}, termed the measured or interior apparent value, using eqn (7.27) with the driving force replaced by $(c_x - c_x^*)_B$, obtained from measured concentration profiles, to evaluate extractor performance. This gives values intermediate between the true backmix and the plug flow values, but there is no theoretical justification for its use.

To simplify extractor design, some workers have attempted to express the effect of backmixing in terms of a parameter N_{oxD}, the number of dispersion transfer units, related as follows to the number of true backmixed, N_{oxB}, and of plug flow transfer units, N_{oxP}, required for a specified duty:

$$N_{oxP}^{-1} = N_{oxB}^{-1} + N_{oxD}^{-1} \tag{7.28}$$

An empirical method of correlation for N_{oxD} in terms of Pe_xB and Pe_yB was proposed by Miyauchi and Vermeulen (1963a). However, an exact relationship for N_{oxD} can be obtained by combining eqn (7.4) with a material balance around one end of the column and with eqn (7.28), as follows (Pratt 1975)

$$\frac{1}{N_{oxD}} = \frac{E}{Pe_x B} - \frac{I_x'}{N_{oxB} N_{oxP}^1 Pe_x B} + \frac{I_y}{N_{oxP}^1 Pe_x B} - \frac{1}{N_{oxP}^1} + \frac{1}{N_{oxP}} \quad (7.29)$$

where N_{oxP}^l is the value of N_{oxP} between X^l (= 1.0) and X_l, and

$$I_x' = \int_{(dX/dZ)_1}^{(dX/dZ)_0} \frac{d(dX/dZ)}{X(1 - E) + EX^l},$$

$$I_y = \int_{Y}^{Y^0} \frac{dY}{X(1 - E) + EX^l}$$

The accuracy of eqn (7.29) was confirmed from computed profiles for a number of cases. It is apparent therefore that the relationship between N_{oxB} and N_{oxP} is too complex to be expressed in terms of N_{oxD} by means of a simple expression such as eqn (7.28).

7.2.1.6 *Simplified solutions*

The solution given by eqns (7.15–7.25), although tedious to use, is satisfactory for direct calculation of the exit concentrations from an extractor of known height, by putting $Z = 1.0$. However an iterative solution is required for design purposes, when the height is to be determined, and this is further complicated by the fact that the roots λ_i are themselves functions of height. This latter problem can be circumvented by substituting $\lambda_i' H$ for λ_i, H/d_c for B and H/H_{ox} for N_{ox} in eqns (7.6–7.9) and (7.15–7.19), thus eliminating H from the characteristic equation, which becomes

$$\lambda'^3 - \alpha'\lambda'^2 - \beta'\lambda' - \gamma' = 0 \quad (7.30)$$

with

$$\alpha' = \frac{Pe_x - Pe_y}{d_c} \quad (7.31)$$

$$\beta' = \frac{Pe_y + EPe_y}{H_{ox} d_c} + \frac{Pe_x Pe_y}{d_c^2} \quad (7.32)$$

$$\gamma' = \frac{Pe_x Pe_y (1 - E)}{H_{ox} d_c^2} \quad (7.33)$$

The values of the λ_i' can be obtained as before using eqn (7.20).

Convenient approximate simplified solutions for the concentration profiles and height were obtained by Pratt (1975) from an examination of the properties of the roots, λ_i, of eqn (7.30) over a wide range of parameter values; these are as follows:

(1) λ_2 is a large and positive, increasing with Pe_x;

(2) λ_3 is negative and $|\lambda_3|$ is large, increasing with Pe_y;

(3) $\lambda_4 = 0$ when $E = 1.0$, and is small and negative when $E < 1.0$, and small and positive when $E > 1.0$.

Consequently terms containing $\exp(-\lambda_2)$ and $\exp(\lambda_3)$ can generally be disregarded in view of (i) and (ii).

Simplified profile equations can therefore be derived from eqns (7.15 and 7.16) by writing them out with the A_i in full, dividing both numerator and denominator by $\exp(\lambda_2)$ and disregarding all terms containing $\exp(-\lambda_2)$ and $\exp(\lambda_3)$. On putting $\lambda_i = \lambda_i' H$, the expression for the X-profile becomes

$$X \simeq \frac{E\{C_1 + C_2 e^{-\lambda_2' H(1-Z)} + \lambda_2'(a_4\lambda_4' e^{\lambda_3' HZ} - a_3\lambda_3' e^{\lambda_4' HZ})\}}{EC_1 + a_3 a_4 \lambda_2'(\lambda_4' - \lambda_3')} \tag{7.34}$$

where

$$C_1 = Ea_3\lambda_3'(\lambda_2' - \lambda_4')\, e^{\lambda_4' H}$$

$$C_2 = a_3\lambda_3'\lambda_4' e^{\lambda_4' H}$$

The Y-profile is obtained from this by multiplying the terms containing $\exp[-\lambda_2' H(1-Z)]$, $\exp(\lambda_3' HZ)$ and $\exp(\lambda_4' HZ)$ by a_2, a_3, and a_4 respectively.

A simplified expression for height can now be obtained from the equation for the Y-profile by noting that the term containing $\exp[-\lambda_2' H(1-Z)]$ can be disregarded (except near the X-phase outlet where $Z \to 1.0$). This gives

$$(1 - Y) \simeq \frac{a_4\lambda_2'(\lambda_4' - \lambda_3') - Ea_4\lambda_2'(\lambda_4' e^{\lambda_3' HZ} - \lambda_3' e^{\lambda_4' HZ})}{E^2\lambda_3'(\lambda_2' - \lambda_4')e^{\lambda_4' H} + a_4\lambda_2'(\lambda_4' - \lambda_3')}$$

On substitution of $Y = Y^0$ at $Z = 0$, this can be solved directly for the column height, H, as follows

$$H \simeq \frac{1}{\lambda_4'} \ln \frac{a_4\lambda_2'(\lambda_4' - \lambda_3')\,(E - Y^0)}{E^2\lambda_3'(\lambda_4' - \lambda_2')\,(1 - Y^0)} \tag{7.36}$$

The corresponding solution in terms of X^l can be obtained by substituting for Y^0 in eqn (7.36) by means of the overall material balance, [cf. eqns (6.33 and 6.70), pp. 360 and 374], viz.

$$Y^0 = E(1 - X^l) \tag{7.37}$$

Equation (7.19) takes the following form when expressed in terms of the modified roots λ_i, although the values of the a_i are unchanged.

$$a_i = \frac{E(1 - \lambda_i' d_c/Pe_x)}{1 + \lambda_i' d_c/Pe_y} \tag{7.38}$$

These results will be further discussed later, in Section 7.2.5 (p. 437), where generalized forms of the solutions will be developed.

7.2.2 Backflow model

7.2.2.1 *Assumptions made*

This model assumes a series of stages interconnected as shown in Fig. 7.4; each stage is mixed by means of an impeller or other device, and may or may not include a settling section in which partial or complete coalescence occurs. The assumptions inherent in this model are as follows (Sleicher 1960);

1. Each stage is well-mixed, and backmixing occurs by entrainment of the phases between stages, after coalescence if appropriate.
2. The backmixing is characterized by backmixing ratios α_k of backmixed to net forward flow.
3. All mass transfer occurs in the mixer in cases where separate settlers are provided.
4. The value of $k_{ox}aV$, the product of volume mass transfer coefficient and stage volume, is constant for each stage.

5–7. Same as 6–8 for the diffusion model.

7.2.2.2 *Derivation of equations*

The interstage flows for the two phases around any stage n are shown in Fig. 7.4; for columns, the total volumetric flows, F_j, can be replaced by the superficial velocities, U_j. The resulting mass balance for the X-phase is

$$(1 + \alpha_x)c_{x,n-1} - (1 + 2\alpha_x)c_{x,n} + \alpha_x c_{x,n+1} = \frac{k_{ox}ash_c}{F_x}(c_{x,n} - c^*_{x,n}) \tag{7.39}$$

and for the Y-phase,

$$\alpha_y c_{y,n-1} - (1 + 2\alpha_y)c_{y,n} + (1 + \alpha_y)c_{y,n+1} = -\frac{k_{ox}ash_c}{F_y}(c_{x,n} - c^*_{x,n}) \tag{7.40}$$

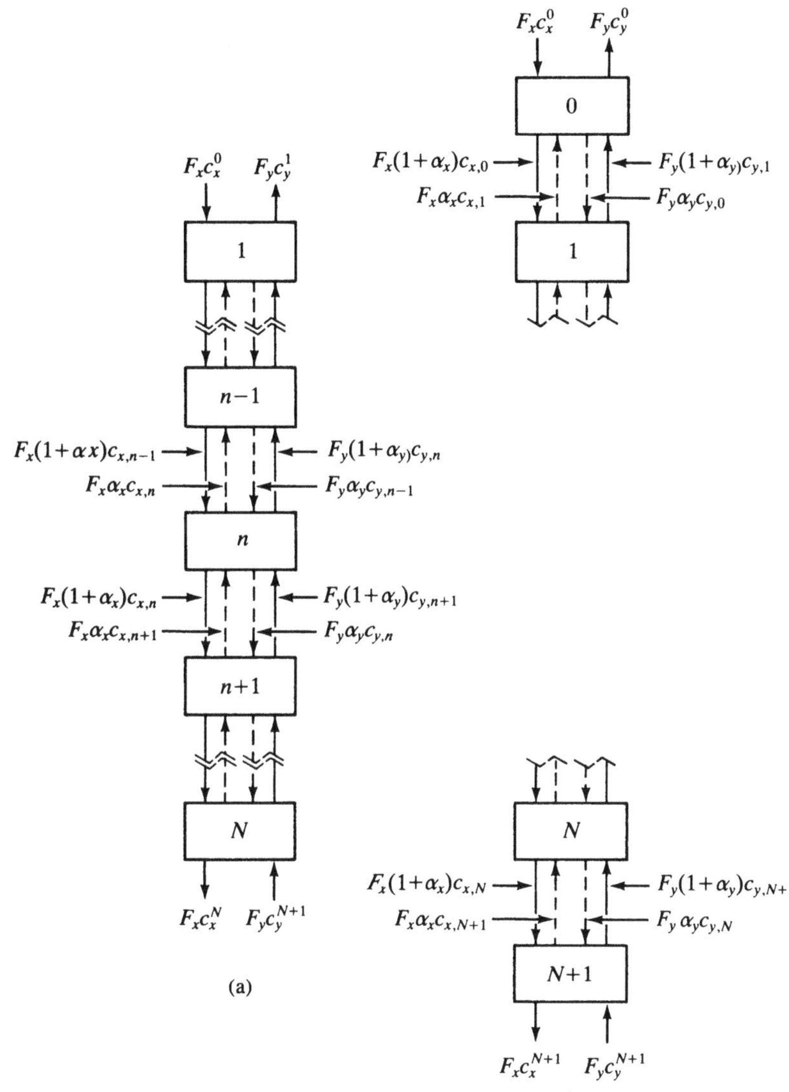

Fig. 7.4. Backflow model: material balances over stage. (a) Normal arrangement; (b) with fictitious and stages.

Assuming, as before, a linear equilibrium relationship given by eqn (7.3), eqns (7.39 and 7.40) can be put into dimensionless form as follows,

$$(1+\alpha_x)X_{n-1} - (1+2\alpha_x)X_n + \alpha_x X_{n+1} = N_{ox}^1(X_n - Y_n) \quad (7.41)$$

$$\alpha_y Y_{n-1} - (1+2\alpha_y)Y_n + (1+\alpha_y)Y_{n+1} = -N_{ox}^1 E(X_n - Y_n) \quad (7.42)$$

where $N_{ox}^1 = k_{ox}aV/F_x = k_{ox}ah_c/U_x$, the number of perfectly mixed transfer units per stage or compartment.

By expressing eqns (7.41 and 7.42) in E operator form (that is, $EX_n = X_{n+1}$), Y can be eliminated to give

$$[(E-1)^4 - \alpha(E-1)^3 - \beta(E-1)^2 - \gamma(E-1)]X_n = 0 \quad (7.43)$$

where

$$\alpha = \frac{1 + N_{ox}^1}{\alpha_x} - \frac{1 - EN_{ox}^1}{(1 + \alpha_y)} \quad (7.44)$$

$$\beta = N_{ox}^1\left[\frac{1}{\alpha_x} + \frac{E}{1 + \alpha_y}\right] + \frac{1 + N_{ox}^1(1 - E)}{\alpha_x(1 + \alpha_y)} \quad (7.45)$$

$$\gamma = \frac{N_{ox}^1(1 - E)}{\alpha_x(1 + \alpha_y)} \quad (7.46)$$

To obtain the boundary conditions, following Sleicher's proposal (1960) it is assumed that fictitious after-settlers designated as stages 0 and $N+1$, in which no mass transfer takes place, are provided at the two ends of the extractor as shown in Fig. 7.4(b). The four boundary conditions required to solve eqn (7.43) are then obtained as follows:

(i) X-phase, $n = 0$. A balance on X-phase around stage 0 gives

$$c_x^0 = c_{x,0} + \alpha_x(c_{x,0} - c_{x,1})$$

In dimensionless concentration units this becomes

$$X_0 + \alpha_x(X_0 - X_1) = 1 \quad (7.47)$$

(ii) X-phase, $n = N + 1$. A balance on X-phase around stage $N + 1$ gives

$$c_x^{N+1} + \alpha_x c_{x,N+1} = (1 + \alpha_x)c_{x,N}$$

Since there is no mass transfer, $c_x^{N+1} = c_{x,N+1}$ so that $c_x^{N+1} = c_{x,N}$, in dimensionless concentration units,

$$X^{N+1} = X_{N+1} = X_N \quad (7.48)$$

(iii) Y-phase, $n = N+1$. A balance on Y-phase around stage $N + 1$ gives

$$Y_{N+1} - \alpha_y(Y_N - Y_{N+1}) = 0 \quad (7.49)$$

(iv) Y-phase, $n = 0$. A procedure similar to that in (ii) gives

$$Y^0 = Y_0 = Y_1 \quad (7.50)$$

7.2.2.3 *Solution of equations*

The solutions for the X- and Y-profiles are obtained from eqns (7.43) and (7.41), as follows

$$X_n = A_1 + A_2\mu_2^n + A_3\mu_3^n + A_4\mu_4^n \tag{7.51}$$

$$Y_n = A_1 + A_2a_2\mu_2^n + A_3a_3\mu_3^n + A_4a_4\mu_4^n \tag{7.52}$$

where the $\mu_i (i = 2, 3 \text{ and } 4)$ are given by the roots, $(\mu_i - 1)$, of the characteristic equation

$$(\mu - 1)^3 - \alpha(\mu - 1)^2 - \beta(\mu - 1) - \gamma = 0 \tag{7.53}$$

and

$$a_i = \frac{1}{\mu_i}\left[1 + \frac{1 + N_{ox}^1}{N_{ox}^1}(\mu_i - 1) - \frac{\alpha_x}{N_{ox}^1}(\mu_i - 1)^2\right] \tag{7.54}$$

$$= \frac{E[1 - \alpha_x(\mu_i - 1)]}{\mu_i + \alpha_y(\mu_i - 1)} \tag{7.55}$$

Equation (7.55), the more convenient form for a_i, is obtained by elimination of N_{ox}^1 between eqns (7.53 and 7.54). The roots of eqn (7.53) are all real and are given by eqn (7.20), as for λ_i.

Finally, substitution of eqns (7.47–7.50) into eqns (7.51 and 7.52) gives the following four equations which determine the A_i:

$$A_1 + \sum_{i=2}^{4} A_i[1 - \alpha_x(\mu_i - 1)] = 1 \tag{7.56}$$

$$\sum_{i=2}^{4} A_i a_i(\mu_i - 1) = 0 \tag{7.57}$$

$$A_i + \sum_{i=2}^{4} A_i a_i \mu_i^N [\mu_i + \alpha_y(\mu_i - 1)] = 0 \tag{7.58}$$

$$\sum_{i=2}^{4} A_i \mu_i^N (\mu_i - 1) = 0 \tag{7.59}$$

The solution for the A_i is given in determinantal form as follows (cf. eqn (7.25) for the diffusion model)

$$A_i = \frac{D_{A_i}}{D_{A_1} - D_A} \tag{7.60}$$

where

$$D_A = \begin{vmatrix} [1-\alpha_x(\mu_2-1)] & [1-\alpha_x(\mu_3-1)] & [1-\alpha_x(\mu_4-1)] \\ a_2(\mu_2-1) & a_3(\mu_3-1) & a_4(\mu_4-1) \\ \mu_2^N(\mu_2-1) & \mu_3^N(\mu_3-1) & \mu_4^N(\mu_4-1) \end{vmatrix}$$

$$D_{A_1} = \begin{vmatrix} a_2(\mu_2-1) & a_3(\mu_3-1) & a_4(\mu_4-1) \\ \mu_2^N(\mu_2-1) & \mu_3^N(\mu_3-1) & \mu_4^N(\mu_4-1) \\ a_2\mu_2^N f(\mu_2) & a_3\mu_3^N f(\mu_3) & a_4\mu_4^N f(\mu_4) \end{vmatrix}$$

with $f(\mu_i) = [\mu_i + \alpha_y(\mu_i - 1)]$

$$D_{A_2} = (\mu_3 - 1)(\mu_4 - 1)(a_4\mu_3^N - a_3\mu_4^N)$$

$$D_{A_3} = (\mu_2 - 1)(\mu_4 - 1)(a_2\mu_4^N - a_4\mu_2^N)$$

$$D_{A_4} = (\mu_2 - 1)(\mu_3 - 1)(a_3\mu_2^N - a_2\mu_3^N)$$

Concentration profiles calculated from these equations are similar in form to those shown in Fig. 7.2 for the diffusion model. The operating diagram is also similar, although in this case the concentrations within the stages follow a stepwise pattern between the true operating and equilibrium lines; in the limit, for ideal stages ($E_{Mx} = 1.0$, corresponding to $N_{ox}^1 \to \infty$) the steps actually touch the equilibrium curve in the usual manner.

7.2.2.4 *Simplified solutions*

As with the diffusion model, the solutions given by eqns (7.51–7.60) can be used for the direct calculation of concentration profiles, but iterative solution is needed for the calculation of the number of stages. However, again simplified approximate solutions can be obtained by making use of the properties of the roots, μ_i, of the characteristic equation. These are as follows (Pratt 1976b):

(1) μ_2 is large and positive, increasing as $\alpha_x \to 0$;

(2) $0 < \mu_3 < 1$, and $\to 0$ as $\alpha_y \to 0$;

(3) $(\mu_4 - 1) = 0$ when $E = 1.0$, and is small and negative when $E < 1.0$, and small and positive when $E > 1.0$.

Consequently terms containing μ_2^{-N} and μ_3^N can be disregarded if N is not too small.

The procedure used is similar to that for the diffusion model. Thus, for the X-profile eqn (7.51) is written out with the A_i in full, and both numerator and denominator are divided by μ_2^N. On disregarding terms containing μ_2^{-N} and μ_3^N, the simplified expression for the X-profile becomes

$$X_n \simeq \frac{E\{C_1 + C_2\mu_2^{(n-N)} + (\mu_2 - 1)[a_4(\mu_4 - 1)\mu_3^n - a_3(\mu_3 - 1)\mu_4^n]\}}{EC_1 + a_3a_4(\mu_2 - 1)(\mu_4 - \mu_3)} \tag{7.61}$$

where

$$C_1 = Ea_3(\mu_3 - 1)(\mu_2 - \mu_4)\mu_4^N$$

$$C_2 = a_3(\mu_3 - 1)(\mu_4 - 1)\mu_4^N$$

The Y-profile is obtained by multiplying the terms containing $\mu_2^{(n-N)}$, μ_3^n and μ_4^n by the a_i term, as given by eqn (7.55).

The expression for number of stages is obtained from the simplified Y-profile expression by substituting $Y = Y^0$ with $n = 0$, and noting that the term containing $\mu_2^{(n-N)}$ can be disregarded. The final result is then

$$N \simeq \log\left(\frac{a_4(\mu_2 - 1)(\mu_4 - \mu_3)(E - Y^0)}{E^2(\mu_3 - 1)(\mu_4 - \mu_2)(1 - Y^0)}\right)\bigg/\log\mu_4 \tag{7.62}$$

Generalized forms of these results will be developed later, in Section 7.2.5 (p. 437).

7.2.3 Comparison of models

7.2.3.1 *Parameters involved*

Both models involve five parameters, as listed in Table 7.1, and this leads to a close similarity of the solutions, which will be exploited in Section 7.2.5 (p. 437). It should be mentioned, in the case of the diffusion model, that the appearance of the parameters Pe_k and B in the equations as a product, Pe_kB, led some early workers to assume that the number of

TABLE 7.1 *Comparison of model parameters*

Parameter		Diffusion model		Backflow model	
No.	Description	Name	Symbol	Name	Symbol
1	Backmixing, X-phase	Peclet no.	Pe_x	Backmix ratio	α_x
2	Backmixing, Y-phase	Peclet no.	Pe_y	Backmix ratio	α_y
3	Extractor efficiency	HTU	H_{ox}	NTU per stage	N_{ox}^1
4	Extractor size	Length ratio	B	No. of stages	N
5	Extraction factor	—	E	—	E

parameters is reduced to four. However, this is unjustified as the local Peclet number, Pe_k, would be expected to be approximately constant throughout a column, whereas $Pe_kB = U_kH/E_k$, termed the column Peclet number, varies in proportion to the height even when the local value is constant. Thus it is clear that Pe_k and B must be treated as separate parameters.

With some extractor types backmixing in one or other phases is negligible or absent, in which case the number of parameters is reduced to four; if absent in both phases, the solutions reduce to those for plug flow. These and other special cases will be discussed in Section 7.2.4 (p. 435).

7.2.3.2 *Relationship between models*

A number of working relationships between the two models have been obtained by Miyauchi and Vermeulen (1963b), of which the most direct is obtained by approximating the differential equations for the diffusion model by the finite difference equivalents. Thus, division of eqn (7.4) by P_xB and replacement of the differential terms by lowest order central differences gives

$$\frac{1}{P_xB}\left[\frac{X_{n+1} - 2X_n + X_{n-1}}{\Delta Z^2}\right] - \frac{(X_{n+1} - X_{n-1})}{2\Delta Z} = N_{ox}(X_n - Y_n) \tag{7.63}$$

For a total of N segments, the size of each is $\Delta Z = 1/N$. Substitution of this value in eqn (7.63), putting $N^1_{ox} = N_{ox}/N$, gives

$$\left[\frac{N}{Pe_xB} + \frac{1}{2}\right](X_{n-1} - X_n) - \left[\frac{N}{Pe_xB} - \frac{1}{2}\right](X_n - X_{n+1}) = N^1_{ox}(X_n - Y_n) \tag{7.64}$$

Equation (7.41) for the backflow model can be arranged as follows:

$$(1 + \alpha_x)(X_{n-1} - X_n) - \alpha_x(X_n - X_{n-1}) = N^1_{ox}(X_n - Y_n) \tag{7.65}$$

A comparison of eqns (7.64) and (7.65) and of the corresponding expressions for the Y-phase shows that these are essentially the same when relationships of the following type are satisfied

$$\frac{1}{Pe_kB} = \frac{1}{N}\left(\alpha_k + \frac{1}{2}\right) \tag{7.66}$$

Hence

$$Pe_kB \simeq \frac{N}{\alpha_k} \quad (N \rightarrow \infty) \tag{7.67}$$

The boundary conditions for the backflow model [eqns (7.47–7.50)] can be obtained from those for the diffusion model [eqns (7.11–7.14)] by substitution of eqn (7.67) in the latter and letting $\Delta Z = 1/N \rightarrow 0$.

The other relations differ in relatively minor ways from eqn (7.66), e.g. with N replaced by $(N - 1)$. These results are used by many workers to justify the use of the diffusion model for compartment-type extractors, and vice versa.

Since $Pe_k B = U_k H/E_k$ and $H/N = h_c$, eqn (7.66) reduces to $Pe_k = 2.0$ if $\alpha_k = 0$, that is, there is no interstage backflow. This is the same as the limiting value predicted theoretically for single phase flow through a bed of spheres (Aris and Amundsen 1957).

7.2.4 Special solutions

Modified solutions for a number of special cases have been given by Miyauchi and Vermeulen (1963a) for the diffusion model, and by Hartland and Mecklenburgh (1966) for both models; see also Pratt and Baird (1983). These are summarized below.

7.2.4.1 *E = 1.0*

When the extraction factor, E, has a value of unity the equilibrium and balance lines are parallel. In this case $\gamma = 0$ in eqns (7.6 and 7.43) so that the characteristic equation reduces to a quadratic, and the solutions for the profiles take different forms. These are given in full by Pratt and Baird (1983) as Case 2 in their Table 1.

In practice, this case rarely arises, and in any case, solutions obtained by the simplified methods described in Section 7.2.5 (p. 437) for $E = 0.99$ and $E = 1.01$, on averaging, were found to give results identical with those obtained using the corresponding expressions for $E = 1.00$ (Pratt 1985).

7.2.4.2 *Backmixing in one or neither phase*

When backmixing occurs in the Y-phase only, the terms containing A_2 in eqns (7.15–7.16) and (7.51–7.52) vanish, as do these containing A_3 when there is backmixing in the X-phase alone; similarly, both terms vanish with no backmixing in either phase. As a consequence, the number of boundary conditions is reduced and the solution for backmixing in one phase only cannot be approximated by the use of a very small value, e.g. 10^{-9}, for the corresponding Pe_k^{-1} or α_k, as the boundary condition at the exit of that phase would still force the profile through a zero slope at the exit. Exact solutions for these cases have been given by Pratt and Baird (1983) in their Table 3 for both models, and simplified solutions of sufficient accuracy for all practical purposes are described in Section 7.2.5 (p. 437).

TABLE 7.2 *Simplified design equations*

Note: See Table 7.3 for definitions of parameters A,B,C, etc.

(I) Height or No. of stages

(a) $\mathbf{E} \neq 1.00$

$$H \text{ or } N = \frac{1}{A} \ln \frac{a_4\left(\dfrac{\kappa_4}{\kappa_3} - 1\right)(\mathbf{E} - Y^0)}{\mathbf{E}^2\left(\dfrac{\kappa_4}{\kappa_2} - 1\right)(1 - Y^0)}$$

(b) $\mathbf{E} = 1.00$

$$H \text{ or } N = \frac{Y^0}{1 - Y^0}(B + C) - B + \left[\frac{1}{\kappa_2} - \frac{1}{\kappa_3}\right]$$

(II) Exit X-phase composition

$$X^I \text{ or } X^N \simeq \frac{\mathbf{E}\left\{F\left[(1 - \mathbf{E})\left(\dfrac{\kappa_4}{\kappa_2} - 1\right)\right]\right\}}{\mathbf{E}^2F\left(1 - \dfrac{\kappa_4}{\kappa_2}\right) - a_4\left(1 - \dfrac{\kappa_4}{\kappa_3}\right)}$$

(III) Fractional height or partial No. of stages ($\mathbf{E} \neq 1.00$)

$$Z(Y_k) \text{ or } n(Y_k) = \frac{1}{D} \ln \frac{\mathbf{E}^2F\left(1 - \dfrac{\kappa_4}{\kappa_2}\right)(1 - Y_k) - a_4Y_k\left(\dfrac{\kappa_4}{\kappa_3} - 1\right)}{\mathbf{E}a_4}$$

(IV) 'Jump' concentrations ($\mathbf{E} \neq 1.00$)

$$X_0 = \frac{\mathbf{E}\left\{F\left[\mathbf{E}\left(1 - \dfrac{\kappa_4}{\kappa_2}\right) + \dfrac{\kappa_4}{\kappa_2}G\right] + \left(\dfrac{a_4\kappa_4}{a_3\kappa_3} - 1\right)\right\}}{\mathbf{E}^2F\left(1 - \dfrac{\kappa_4}{\kappa_2}\right) + a_4\left(\dfrac{\kappa_4}{\kappa_3} - 1\right)}$$

$$Y_I = \frac{\mathbf{E}\left\{F\left[\mathbf{E}\left(1 - \dfrac{\kappa_4}{\kappa_2}\right) + a_2\dfrac{\kappa_4}{\kappa_2}M - a_4N\right] + a_4J\dfrac{\kappa_4}{\kappa_3}\right\}}{\mathbf{E}^2F\left(1 - \dfrac{\kappa_4}{\kappa_2}\right) + a_4\left(\dfrac{\kappa_4}{\kappa_3} - 1\right)}$$

7.2.4.3 *Other cases*

Further solutions are given by Pratt and Baird (1983) in their Table 4 for cases in which perfect mixing of one or other, or both phases occurs and also for the case of an infinitely fast first order reaction taking place in the extractant phase.

7.2.5 Generalized solutions

7.2.5.1 *Linear equilibrium relationship*

A comparison of the simplified expressions for height and number of stages, eqns (7.36 and 7.62), shows that they are similar in form but with the roots λ_i' replaced in the latter by $(\mu_i - 1)$, and the $1/\lambda_4'$ outside the main logarithmic term replaced by $1/\ln \mu_4$. By writing the corresponding characteristic equations as follows:

$$\kappa^3 - \alpha\kappa^2 - \beta\kappa - \gamma = 0 \tag{7.68}$$

where κ represents either λ' or $(\mu - 1)$, the two solutions can be rearranged into the generalized form given as eqn I(a) in Table 7.2 (Pratt 1985). Also given as eqn I(b), for completeness, is the corresponding solution for

TABLE 7.3 *Definitions of parameters in Table 7.2*

Parameter	Diffusion model	Backflow model
A	κ_4	$\ln(\kappa_4 + 1)$
a_i	$\dfrac{\mathbf{E}(1 - \kappa_i d_c/Pe_x)}{(1 + \kappa_i d_c/Pe_y)}$	$\dfrac{\mathbf{E}(1 - \alpha_x \kappa_i)}{1 + (\alpha_y + 1)\kappa_i}$
B	H_{ox}	$1/N_{ox}^1$
C	$d_c\left(\dfrac{1}{Pe_x} + \dfrac{1}{Pe_y}\right)$	$(\alpha_x + \alpha_y + 1)$
D	$\kappa_4 H$	$\ln(\kappa_4 + 1)$
F	$\exp(\kappa_4 H)$	$(\kappa_4 + 1)^N$
G	$\exp(-\kappa_2 H)$	$(\kappa_2 + 1)^{-N}$
J	$\exp(\kappa_3 H)$	$(\kappa_3 + 1)^{N+1}$
M	1	$(\kappa_2 + 1)$
N	1	$(\kappa_4 + 1)$

$E = 1.0$, although as stated earlier this is unlikely to be required in practice. A similar generalized solution for the exit X-phase composition, derived from eqns (7.34 and 7.61) with $Z = 1.0$ or $n = N$ respectively, is also given (eqn II). The definitions of the various parameters used in these equations are listed in Table 7.3.

To apply these solutions to the case where backmixing occurs in only one, or in neither phase it is necessary to examine the properties of the roots, κ_2, κ_3, and κ_4 of eqn (7.68). These are as follows:

1. κ_2, which relates to conditions at the feed inlet end, is relatively large and positive and $\rightarrow \infty$ as $1/P_x$ or $\alpha_x \rightarrow 0$.
2. κ_3, which relates to the extractant inlet end, is negative and numerically large, and $\rightarrow -\infty$ as $1/Pe_y \rightarrow 0$ in the case of the diffusion model; for the backflow model $-1 < \kappa_3 < 0$, and $\kappa_3 \rightarrow -1$ as $\alpha_y \rightarrow 0$.
3. κ_4, which relates to the central part of the extractor, is zero when the extraction factor $E = 1.0$; is small and negative when $E < 1.0$, and small and positive when $E > 1.0$.

When backmixing occurs in one phase only the characteristic equation reduces to a quadratic, the solution to which gives κ_4 together with either κ_2 or κ_3. The remaining root is then the limiting value given under (1) or (2), and substitution of these into eqns (I) or (II) of Table 7.2 gives the desired solution. The various forms of the characteristic equation and roots are summarized in Table 7.4 for the different cases.

The equations given in Table 7.2 reduce to the exact solutions for plug flow, that is, no backmixing in either phase, using the values of the roots given under Case 4 in Table 7.4. Thus, substituting these values into eqn I(a) of Table 7.2 and incorporating the overall material balance [eqn (7.37)] gives for the diffusion model

$$H = \frac{H_{ox}}{(1 - E)} \ln [E + (1 - E)/X_l) \tag{7.69}$$

Similarly for the backflow model

$$N = \frac{\ln [E + (1 - E)/X_l]}{\ln [(1 + N_{ox}^1)/(1 + N_{ox}^1 E)]} \tag{7.70}$$

where the value of N_{ox}^1 for complete backmixing of both phases is given by eqn (6.81), p. 376, in the following form

$$N_{ox}^1 = \frac{E_{Mx}}{1 - E_{Mx}} \tag{7.71}$$

When $E_{Mx} \rightarrow 1.0$, $N^1_{ox} \rightarrow \infty$ and on substitution into eqn (7.70) the denominator $\rightarrow \ln(1/E)$, giving the usual expression for theoretical stages.

The accuracy of the simplified equations was tested by comparing the results obtained over a wide range of parameter values with those computed using the exact solutions, that is, eqns (7.15–7.25) and (7.51–7.60). This indicated for the diffusion model that the error would be well within 5 per cent if the parameter $H(\lambda_2' - \lambda_3')$ exceeds 7–8 with $N_{ox} \geqq 2$, and for most practical purposes the error is negligible (Pratt 1975). A similar comparison for the backflow model showed that the results are of acceptable accuracy provided that N is not less than 4 stages, and that the error becomes quite negligible for larger numbers of stages (Pratt 1976b).

7.2.5.2 *Non-linear equilibria*

The foregoing methods are not directly applicable when the equilibrium line is curved, and the graphical or numerical methods described later must be used to obtain an exact solution. However, for most purposes a solution of adequate accuracy based on the above simplified methods can be obtained by subdivision of the column into two (or more) sections, within each of which the equilibrium curve can be approximated by straight line segments. Two such methods were described by Pratt (1976a), of which the following is the simpler although not always applicable, as will be seen.

The method is illustrated by the operating diagram shown in Fig. 7.5 in

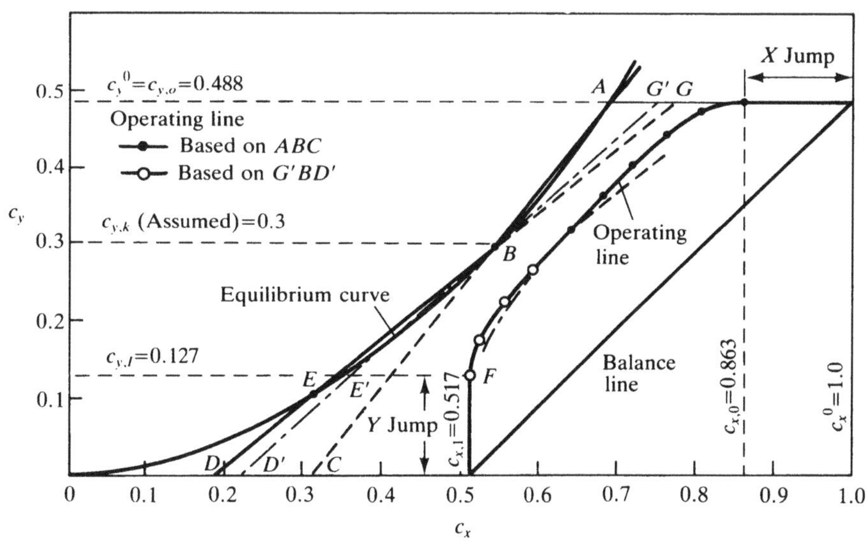

Fig. 7.5. Operating diagram on c_y–c_x coordinates for non-linear equilibrium relationship (Example 4).

TABLE 7.4 *Characteristic equations and roots*

No.	Case	Diffusion model	Backflow model
1	$1/Pe_x$ and $1/Pe_y \neq 0$	*Characteristic equation* $\kappa^3 - \alpha\kappa^2 - \beta\kappa - \gamma = 0$	
	or α_x and $\alpha_y \neq 0$	*Coefficients*:	
	(Backmixing in both phases)	$\alpha = (Pe_x - Pe_y)/d_c$	$\alpha = \dfrac{1 + N^1_{ox}}{\alpha_x} - \dfrac{1 - N^1_{ox}E}{1 + \alpha_y}$
		$\beta = \dfrac{Pe_x + EPe_y}{H_{ox}d_c^2} + \dfrac{Pe_xPe_y}{d_c^2}$	$\beta = \dfrac{N^1_{ox}}{\alpha_x} + \dfrac{N^1_{ox}E}{1 + \alpha_y} + \dfrac{1 + N^1_{ox}(1 - E)}{\alpha_x(1 + \alpha_y)}$
		$\gamma = \dfrac{Pe_xPe_y(1 - E)}{H_{ox}d_c^2}$	$\gamma = \dfrac{N^1_{ox}(1 - E)}{\alpha_x(1 + \alpha_y)}$
		Roots: $\kappa_2 = \alpha/3 + 2p^{1/2}\cos(u/3)$; $\kappa_3 = \alpha/3 + 2p^{1/2}\cos(u/3 + 120)$; $\kappa_4 = \alpha/3 + 2p^{1/2}\cos(u/3 + 240)$	
		where: $p = (\alpha/3)^2 + \beta/3$; $q = (\alpha/3)^3 + \alpha\beta/6 + \gamma/2$; $u = \cos^{-1}(q/p^{3/2})$	
2	$1/Pe_x$ or $\alpha_x \neq 0$	*Characteristic equation* $\kappa^2 - \beta_x\kappa - \gamma_x = 0$	

	$1/Pe_y$ or $\alpha_y = 0$ (Backmixing in X-phase only)	*Coefficients*:	
		$\beta_x = E/H_{ox} + Pe_x/d_c$	$\beta_x = [1 + N^1_{ox}(1 + \alpha_x E)]/\alpha$
		$\gamma_x = Pe_x(1 - E)/H_{ox}d_c$	$\gamma_x = N^1_{ox}(1 - E)/\alpha_x$
		Roots:	
		$\kappa_2 = \beta_x/2 + [(\beta_x/2)^2 + \gamma_x]^{1/2}$	
		$\kappa_3 \to -\infty$	$\kappa_3 = -1$
		$\kappa_4 = \beta_x/2 - [(\beta_x/2)^2 + \gamma_x]^{1/2}$	
3	$1/Pe_x$ or $\alpha_x = 0$	*Characteristic equation*	
		$\kappa^2 + \beta_y\kappa + \gamma_y = 0$	
	$1/Pe_y$ or $\alpha_y \neq 0$	*Coefficients*:	
	(Backmixing in *Y*-phase only)	$\beta_y = 1/H_{ox} + Pe_y/d_c$	$\beta_y = \dfrac{1 + N^1_{ox}(2 - E + \alpha_y)}{(1 + \alpha_y)(1 + N^1_{ox})}$
		$\gamma_y = Pe_y(1 - E)/d_c H_{ox}$	$\gamma_y = \dfrac{N^1_{ox}(1 - E)}{(1 + \alpha_y)(1 + N^1_{ox})}$
		Roots:	
		$\kappa_2 \to \infty$	$\kappa_2 \to \infty$
		$\kappa_3 = -\beta_y/2 - [(\beta_y/2)^2 - \gamma_y]^{1/2}$	
		$\kappa_4 = -\beta_y/2 + [(\beta_y/2)^2 - \gamma_y]^{1/2}$	
4	$1/Pe_x$ and $1/Pe_y = 0$ or α_x and $\alpha_y = 0$	*Roots*:	
		$\kappa_2 \to \infty$	$\kappa_2 \to \infty$
		$\kappa_3 \to \infty$	$\kappa_3 = -1$
	(No backmixing in either phase)	$\kappa_4 = (E - 1)/H_{ox}$	$\kappa_4 = \dfrac{N^1_{ox}(E - 1)}{1 + N^1_{ox}}$

which the equilibrium relationship, represented by $c_y = c_x^2$, corresponds to the curve $ABEO$. The steps involved are as follows:

1. The jump concentration $c_{y,l}$ or $c_{y,N+1}$ is assumed and the equilibrium curve is approximated by the two straight lines AB and BE, covering the range of Y-phase concentrations from exit to jump value. These lines are then extended so that each covers the full concentration range, as shown by ABC and DEBG respectively. (The method is inapplicable if either intersects the operating line.)
2. The dimensionless concentrations Y^0 and Y_k, the latter corresponding to concentration c_{yk} at intersection point B, are calculated; these will differ for the two sections as the two equilibrium lines ABC and DEBG have different slopes.
3. The overall height, H_f^o, or number of stages, N_f, according to model, is calculated for equilibrium line ABC using eqn I(a) of Table 7.2. For the diffusion model, the fraction $Z(Y_k) = Z_f$ of this height required for the feed end of the column down to concentration Y_k is then obtained from eqn III of Table 7.2, so that the actual length of this section is $H_f^oZ_f$. Similarly for the backflow model the same equation gives the number of stages, $n_f = n(Y_k)$ for the feed end down to Y_k.
4. The values of H_e^o and Z_e, or of N_e and n_e for the extractant inlet section are obtained in the same way using equilibrium line DBG. The height or the number of stages for this section is then $H_e^o(1 - Z_e)$ or $N_e - n_e$ respectively.
5. The 'jump' concentrations $c_{x,0}$ and/or $c_{y,l}$ (or $c_{y,N+1}$) at the inlets of the phases which are backmixed are then calculated using eqns IV of Table 7.2, and used if necessary to relocate the straight line approximations ABC and DEBG to the equilibrium curve; if such relocation is required the previous calculations are repeated.
6. The total height or number of stages is then given by $[H_f^oZ_f + H_e^o(1 - Z_e)]$ or $[nf + (N_e - n_e)]$ respectively.

Equation III of Table 7.2 is obtained from the Y-profile equations corresponding to eqns (7.34 and 7.61) by omitting the terms $\exp[-\lambda_2 H(1 - Z)]$ and $\exp(\lambda_3 HZ)$, or $\mu_2(n - N)$ and μ_3^N, and solving for Z or n. Similarly, eqns IV are obtained from the X or Y profile equations by substitution of Z or $n = 0$ and $Z = 1.0$ or $n = N$ respectively.

It is apparent that this procedure could be extended if necessary by subdivision of the equilibrium line into three sections. In this case, using the diffusion model it would be necessary to calculate the height, H_m^o and the values of Z_{m1} and Z_{m2} for the middle section, and to add a length

$H^o_m(Z_{m2} - Z_{m1})$ to that obtained in step 6. Similarly, for the backflow model the number of stages for the middle section, to be added to that obtained in step 6, is $(n_{m,2} - n_{m,1})$.

The need to approximate the equilibrium curve by three straight sections would be likely to arise in cases such as that shown in Fig. 7.5 if there were no backmixing in the Y-phase, as it would then be necessary to take into account the part of the curve between E and the origin. This would involve difficulty, however, if the straight line representing this part of the curve intersected the abscissa for $c^o_x = 1.0$ at a value of c_y below c^o_y, as it would then not be possible to apply eqn I(a) of Table 7.2 over the full concentration range. In such cases a solution could be obtained for the diffusion model by using Pratt's first method (1976a; see also Pratt and Stevens 1991), or alternatively this method could be used for the extractant inlet section.

7.2.5.3 *Programming of solutions*

The foregoing simplified solutions are particularly convenient for solution by programmable calculator or microcomputer, and the outline of a set of three programs written for the former is described by Pratt (1983). Of these programs, the first is used as an option to calculate E_{av} and an adjusted value of the exit feed phase composition $c^I_{x,\mathrm{adj}}$ or $c^{N+1}_{x,\mathrm{adj}}$, from the exact overall material balance allowing for solute transfer, as given in Section 6.3.3 (p. 360). These values are then used in the second, main program to calculate height or number of stages. Finally, if the curved equilibrium line case is required, transfer is effected automatically to the third program, which calculates $Z(Y_k)$ or $n(Y_k)$ together with the 'jump' concentrations. A manual iterative method is also described for estimating H_{ox} or N^1_{ox} from pilot column data, for scale-up purposes.

Using a computer, the above three programs can conveniently be combined and extended to provide a number of options. Typically such a program would provide a choice of three modes, as follows:

1. Design mode, using eqn I(a) of Table 7.2 for calculating height or number of stages, given the desired inlet and exit compositions together with the backmixing and mass transfer parameters. A further option allows for the subdivision of the column into two or three sections to take account of a non-linear equilibrium relation, using also eqns III and IV.

2. Performance mode, for determining H_{ox} or N^1_{ox} from pilot plant performance data by iterative calculation using eqn I(a) of Table 7.2, knowing the backmixing parameters, for scale-up purposes.

3. Operational mode, using eqn II of Table 7.2, for calculating the exit compositions for an existing column, knowing the inlet parameters.

A further option applicable to all three modes would permit the choice either of specifying the extraction factor, E, or of calculating E_{av} from the exact overall mass balance, as before. In the latter case, the value of Y^{I}_{adj}, corresponding to $c^{I}_{x,adj}$, would also be computed and used if higher than the value of Y^{I} given by eqn (7.37), to obtain a 'safer' design, i.e. a greater height or number of stages.

In developing an iterative procedure for mode 2, Newton's approximation method was found to give satisfactory results. Convergence was usually obtained within about 10 iterations using initial values of the gradient $\Delta H_{ox}/\Delta H$ calculated for $H_{ox} = 1.0$ and 1.1 m or ft, or of $\Delta N^{1}_{ox}/\Delta N$ for $N^{1}_{ox} = 0.10$ and 0.11.

7.2.5.4 *Worked examples*

Note: Length units of feet are used in Examples 1, 2, and 4, for comparison with the exact solutions obtained by Mecklenburgh and Hartland (1967).

Example 1
Design mode, diffusion model, linear equilibrium, $E \neq 1.0$.
Specification:

$c_x^0 = 120.00\ \text{kg/m}^3$	$m = 1.60$	$1/P_x = 40$
$c_x^I = 34.43\ \text{kg/m}^3$	$q = -0.80$	$1/P_y = 10$
$c_y^I = 8.00\ \text{kg/m}^3$	$U_x/U_y = 0.800$	$H_{ox} = 2.00$ ft
		$d_c = 0.10$ ft

The solution for this example will be given in detail.

Mass balance. The value of c_x in equilibrium with the inlet Y-phase, i.e. c_x^{*I}, is obtained as follows from the equilibrium relationship, eqn (6.1)

$$c_x^{*I} = (8.00 + 0.80)/1.6 = 5.50$$

From eqn (6.2) (see p. 00).

$$X^I = \frac{34.43 - 5.50}{120.0 - 5.50} = 0.2527$$

Since $E = 0.80/1.6 = 0.500$, eqn (7.37) gives

$$Y^0 = E(1 - X^I) = 0.3737$$

Characteristic equation. The values of the coefficients are given by eqns (7.31–7.33) (see also Table 7.4, Case No. 1), as follows:

$$\alpha' = \frac{0.025 - 0.010}{0.100} = -0.750$$

$$\beta' = \frac{0.025 + 0.500 \times 0.100}{2.00 \times 0.10} + \frac{0.025 \times 0.10}{0.10^2} = 0.625$$

$$\gamma' = \frac{0.025 \times 0.100(1.0 - 0.500)}{2.00 \times 0.10^2} = 0.0625$$

Substitution into the characteristic equation for Case No. 1 in Table 7.4 gives the roots as follows:

$$\kappa_2 = +0.5614525; \kappa_3 = -1.2202247; \kappa_4 = -0.0912278$$

From eqn (7.38), $a_4 = 0.750964$.

Height required. Substitution into eqn I(a) of Table 7.2 gives $H = 7.995$ ft. This is in close agreement with the exact computed value of 8.00 ft (Mecklenburgh and Hartland 1967c).

Example 2

Design mode, diffusion model, linear equilibrium, $E = 1.00$ (specified)

Specification:

$c_x^0 = 1.00\,\text{kmol/m}^3$	$m = 1.00$	$1/P_x = 1/P_y = 25$
$c_x^I = 0.360\,\text{kmol/m}^3$	$q = 0.00$	$H_{ox} = 2.00\,\text{ft}$
$c_y^I = 0.00\,\text{kmol/m}^3$	$E = 1.00$	$d_c = 0.10\,\text{ft}$

Solving by programmable calculator for $E = 0.999$ and 1.001 gave $H = 10.102$ and 10.132 ft respectively, with a mean of 10.12. This agreed exactly with the value given by eqn I(b) of Table 7.2, and compares with an exact computed value of 10.00 ft. (Mecklenburgh and Hartland 1967c)

Example 3

Design mode, backflow model, linear equilibrium, $E \neq 1.0$

Specification:

$c_x^0 = 120.00\,\text{kg/m}^3$	$m = 1.60$	$\alpha_x = 3.00$
$c_x^N = 53.24\,\text{kg/m}^3$	$q = -0.80$	$\alpha_y = 1.50$
$c_y^I = 8.00\,\text{kg/m}^3$	$U_x/U_y = 0.800$	$N_{ox}^1 = 0.160$

Mass balance. Calculating as for Example No. 1 gives $X_m = 0.4169$, $E = 0.650$ and $Y^1 = 0.3790$.

Characteristic equation. The coefficients, as given by the expressions in Table 3 for this case, are as follows

$$\alpha = 0.02827; \beta = 0.23573; \gamma = 0.0074667$$

Solving for the roots and a_4,

$$\kappa_2 = 0.514576 \qquad \kappa_3 = -0.454375$$

$$\kappa_4 = -0.031936 \qquad a_4 = 0.77407$$

No. of stages. Substitution in eqn I(a) of Table 7.2 gives 10.99 stages, or on rounding off upwards, 11 stages. This agrees with the exact computed value (Pratt 1976b).

Example 4

Design mode, diffusion model, non-linear equilibrium

Specification:

$$c_x^0 = 1.00\ \text{kmol/m}^3 \qquad U_x/U_y = 1.00 \qquad H_{ox} = 2.00\ \text{ft}$$

$$c_x^l = 0.512\ \text{kmol/m}^3 \qquad 1/P_x = 1/P_y = 25 \qquad d_c = 0.10\ \text{ft}$$

$$c_y^l = 0.00\ \text{kmol/m}^3 \qquad \text{Equilibrium relation:} \qquad c_y = c_x^2$$

Equilibria. The equilibrium curve and balance line are shown in the operating diagram, Fig. 7.5. From the overall material balance, the exit extractant composition is

$$c_y^0 = \frac{U_x}{U_y}(c_x^0 - c_x^l) = 1.00(1.00 - 0.512) = 0.488$$

As $1/P_y$ is large, the Y jump will be considerable so that the lower part of the equilibrium curve can be disregarded. For a first trial, therefore, the column will be divided at $c_{yk} = 0.300$, with equilibria represented by AB and BE.

Feed entry section. Extending AB to C, the linearized equilibria are represented by $c_y^* = m_f c_x + q_f$, with slope $m_f = 0.488/(0.6986 - 0.3120) = 1.2623$ and intercept q_f' on the c_x axis of 0.3120, corresponding to $q_f = -m_f q_f' = 0.3938$.

Transformation to dimensionless coordinates over the *full* concentration range AC then gives

$$Y_f^0 = (0.488 - 0)/(1.00 \times 1.2623 - 0.3938) = 0.5619$$

$$Y_{kf} = (0.300 - 0)/(1.00 \times 1.2623 - 0.3938) = 0.3454$$

Also, $E = U_x/m_f U_y = 0.7922$.

The roots of the characteristic equation, together with a_4, are then

$$\kappa_2 = 0.735555 \qquad \kappa_3 = -0.703425$$

$$\kappa_4 = -0.032129 \qquad a_4 = 0.930580$$

Substitution in eqn I(a) of Table 7.2 then gives $H_f^0 = 10.540$ ft; also $\exp(\kappa_4 H_f^0) = 0.7123$ so that, from eqn III, $Z_f = 0.5480$ and hence $H_f = H_f^0 Z_f = 5.776$ ft.

Extractant entry section. Line BD on Fig. 7.5 has a slope $m_e = E_e = 0.8264$ and intercept $q_e = -0.1545$, with $E = 1.210$. Similar calculations to those for the feed section give

$$Y_e^0 = 0.7263 \qquad Y_{ke} = 0.4465$$

$$\kappa_2 = 0.761538 \qquad \kappa_3 = 0.027943$$

$$\kappa_4 = -0.789481 \qquad a_4 = 1.051982$$

The height is then obtained as follows

$$H_e^0 = 11.131 \text{ ft and } Z_s = 0.6370$$

therefore

$$H_e = H_e^0(1 - Z_e) = 4.041 \text{ ft}$$

Jump concentrations. Equations IV of Table 7.2 give, on conversion to dimensional concentrations

$$c_{xO,f} = 0.8625; \; c_{yI,e} = 0.1325$$

Total height. $H = 5.776 + 4.041 = 9.82$ ft. This result can be refined by adjusting the equilibrium line for the extractant inlet section to $D'E'G'$, where E' better corresponds to the Y-jump concentration at F, giving $H_e = 4.180$ with $c_{yI,e} = 0.127$. The total length is then 9.96 ft, which agrees well with the exact computed value of 10.0 ft (Mecklenburgh and Hartland 1967c).

Example 5

Performance and design modes, backflow model, linear equilibrium

This example illustrates the method used to design extractors by scale-up from pilot plant data. The column is required for the extraction of acetone from a dispersed phase of toluene with water in a pulsed perforated plate column. The relevant pilot column data are as follows:

$c_x^0 = 50.00$ kg/m^3	m = 1.5699	$F_x = 0.0765$ m^3/h
$c_x^N = 3.917$ kg/m^3	$q = 0.4587$	$F_y = 0.0907$ m^3/h
$c_y^{N+1} = 0.0$ kg/m^3	$d = 0.074$ m	$\alpha_x = 0.00$
No. of stages = 18	$\rho_s = 0.783$ kg/m^3	$\alpha_y = 2.39$

The full scale column, to be of 0.305 m diameter with proportionately higher flow rates of 1.30 and 1.54 m^3/h of X and Y phases respectively, is required to give an exit toluene phase composition, c_x^N, of 1.0 kg/m^3. The value of the continuous phase backmixing ratio, α_x, will be increased to 7.46.

The solution described below was obtained using the calculator program, and was confirmed by computer.

Pilot column. Using the mass balance program it was found that $E_{av} = 0.5090$ and $c_{x,adj}^N = 3.402$; the latter was used as it is lower than the measured value of 3.917. The following results were then obtained by calculator, using the design program iteratively to calculate the number of stages, N, with an initially assumed value of N_{ox}^1 of 0.40.

N_{ox}^1 (assumed)	0.40	0.45	0.430	0.432
N (calculated)	18.72	17.64	18.04	18.00

The value of N_{ox}^1 obtained by computer was 0.43243.

Full scale column. The mass balance program for this case gave $E_{av} = 0.5074$ and $C_{x,adj}^N = 0.6144$ kg/m^3. The value of N_{ox}^1 obtained for the pilot column is assumed to apply to the full scale column, and using these values in the design program gave $N = 57.2$ plates, i.e. 58 on rounding off upwards.

Note: The above values of α_y are not necessarily applicable to pulsed plate columns of the stated diameters.

7.3 Graphical solution methods

An approximate graphical solution method for the calculation of the height of differential columns, involving stepwise integration of the diffusion model equations, has been described by Rod (1964). This was later extended to stagewise contactors, using the backflow model, for the calculation of the number of stages or compartments (Rod 1971).

These methods have the advantage that they are directly applicable when the equilibrium relationship is non-linear. As is often the case with such methods, however, they are tedious to use, especially if backmixing occurs in both phases, as will be seen, and of somewhat limited accuracy. On the other hand they provide a useful insight into the numerical methods described later, and will therefore be given in some detail below.

7.3.1 Differential columns

Using Rod's method (1964), the true operating line is constructed by stepwise integration of eqn (7.1) in conjunction with a material balance on solute. The latter, taken through the upper boundary of the differential section shown in Fig. 7.1 and around the feed phase inlet of the column gives

$$J = U_x c_x^0 - U_y c_{y,0}$$

$$= U_x\left[c_x - \frac{E_x}{U_x}\frac{dc_x}{dz}\right] - U_y\left[c_y + \frac{E_y}{U_y}\frac{dc_y}{dz}\right] \quad (7.72)$$

where J is the net solute flow per unit cross-section. Introducing new variables defined as

$$C_x = c_x - \frac{E_x}{U_x}\frac{dc_x}{dz} \quad (7.73)$$

$$C_y = c_y + \frac{E_y}{U_y}\frac{dc_y}{dz} \quad (7.74)$$

eqns (7.1 and 7.72) become

$$\frac{dC_x}{dz} = \frac{k_{ox}a}{U_x}(c_x - c_x^*) \quad (7.75)$$

$$J = U_x C_x - U_y C_y \quad (7.76)$$

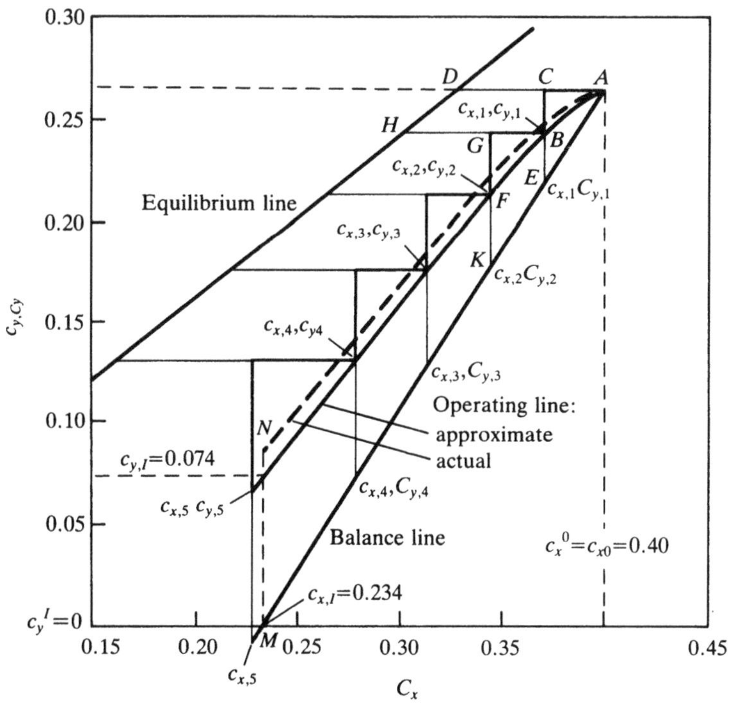

Fig. 7.6. Diffusion model: graphical solution for case of backmixing in Y phase only (Rod 1964) (Example 6).

A plot of eqn (7.76) on C_x–C_y coordinates gives the balance line (i.e. the plug flow operating line) shown in Fig. 7.6. The boundary conditions, eqns (7.11–7.14), become in the new coordinates

$$z = 0\text{: } C_{x,0} = c_x^0 \text{ and } C_{y,0} = c_{y,0} \tag{7.77}$$

$$z = H\text{: } C_{x,l} = c_{x,l} \text{ and } C_{y,l} = c_y^l \tag{7.78}$$

Equations (7.73–7.76) can be approximated by replacing the differentials dc_x, dc_y and dz by the finite differences $\Delta c_{x,j} = c_{x,j+1} - c_{x,j}$, $\Delta c_{y,j} = c_{y,j+1} - c_{y,j}$, and Δz respectively, giving

$$\Delta c_{x,j} = \frac{U_x \Delta z}{E_x}(C_{x,j} - c_{x,j}) \tag{7.79}$$

$$\Delta c_{y,j} = -(c_{y,j} - C_{y,j+1})\Big/\left[1 + (E_y/U_y \Delta z)\right] \tag{7.80}$$

$$\Delta C_{xj} = -\frac{k_{ox} a \Delta z}{U_x}(c_{x,j} - c^*_{xj}) \tag{7.81}$$

$$C_{y,j+1} = \frac{U_x}{U_y} C_{x,j+1} - \frac{J}{U_y} \tag{7.82}$$

In general, the inlet jump concentrations $c_{x,0}$ and $c_{y,l}$ are unknown so that trial and error solution is required. However, if backmixing occurs in only one of the phases the graphical solution is greatly simplified, as the computation can be started at the inlet of the phase which is not backmixed.

7.3.1.1 *Backmixing in one phase only*

If backmixing occurs in the Y-phase only, the variables C_x and c_x are equal so that eqns (7.81 and 7.82) become

$$\Delta c_{x,j} = -\frac{k_{ox} a \Delta z}{U_x}(c_{xj} - c^*_{xj}) \tag{7.83}$$

$$C_{y,j+1} = \frac{U_x}{U_y} c_{x,j+1} - \frac{J}{U_y} \tag{7.84}$$

To start the computation, the compositions $c_{x,0}$ $(= c_x^0)$ and $c_{y,0}$ $(= C_{y,0})$ are located on the balance line at A on a plot of c_y and C_y vs c_x, as shown in Fig. 7.6. A suitable value is selected for the incremental height Δz so that the value of $k_{ox} a \Delta z / U_x$ lies between 0.1 and 0.5. The procedure is then as follows:

1. Draw a horizontal line through A to meet the equilibrium curve at D, and locate C so that $AC/AD = k_{ox} a \Delta z / U_x$, in accordance with eqn (7.83).

2. Draw a vertical line through C to intersect the balance line at E, that is, at $C_{y,I}$ (eqn (7.84)).
3. Locate B on the operating line such that $CB/CE = 1/(1 + E_y/U_y \Delta z)$, giving $c_{y,I}$ (eqn (7.80)).
4. Draw a horizontal line through B to meet the equilibrium curve at H, and repeat steps 1–3 until the desired exit composition $c_{x,I}$ is reached.
5. Multiply the number of height elements obtained by Δz to obtain the overall column height.

The method is illustrated by the following example given by Rod (1964), who assumed a linear equilibrium in order to be able to compare the result with the exact solution.

Example 6
Calculate the contactor height required to meet the following specification

$$c_x^0 = 0.40\,\text{kmol m}^{-3} \qquad U_x = 16\,\text{m h}^{-1} \qquad E_x = 0$$

$$c_{x,I} = 0.234\,\text{kmol m}^{-3} \qquad U_y = 10\,\text{m h}^{-1} \qquad E_y = 25\,\text{m}^2\text{h}^{-1}$$

$$c_y^0 = 0.0\,\text{kmol m}^{-3} \qquad k_{ox}a = 3.2\,\text{h}^{-1}$$

The equilibrium relation is given by $c_y^* = 0.80\,c_x$

Solution. From overall material balance

$$16(0.40 - 0.234) = 10(c_{y,0} - 0.)$$

$$\therefore\ c_{y,0} = 0.266$$

The balance line AM, with coordinates (0.400, 0.266) and (0.234, 0.0) at the two ends, is shown in Fig. 7.6. Assuming $\Delta z = 2.0$ gives $k_{ox}a\Delta z/U_x = 0.40$, so that, from eqn (7.83), $\Delta c_{x,j}/(c_{x,j} - c_{x,j}^*) = -0.40$; also from eqn (7.80), $\Delta c_{y,j}/(c_{y,j} - C_{y,j+1}) = -0.444$.

Using these values, the procedure described above is carried out as shown in Fig. 7.6. This requires 4.9 steps to reach the specified raffinate exit composition at N, with the Y-phase inlet jump given by MN. The height required is therefore $4.9 \times 2.0 = 9.8$ m. This result is in reasonable agreement with the exact value of 10.0 m obtained by Rod (1964). The discrepancy would be even less if a smaller value of Δz were used, bringing the operating line closer to the equilibrium curve as shown by the broken curve in Fig. 7.6, which represents the true operating line.

When backmixing occurs only in the X-phase the construction is started from the Y-phase inlet using the following equations

$$\Delta c_{x,j'} = (C_{x,j'+1} - c_{x,j'})\Big/\left(1 + \frac{E_x}{U_x \Delta z}\right) \tag{7.85}$$

$$\Delta c_{y,j'} = \frac{k_{oy} a \Delta z}{U_y}\,(c^*_{y,j'} - c_{y,j'}) \tag{7.86}$$

$$C_{x,j'+1} = \frac{U_y}{U_x} c_{y,j'+1} + \frac{J}{U_x} \tag{7.87}$$

In these expressions, j' refers to the number of the step counted from the Y-phase entry, that is, from the point $(c_{x,l}, c_{y,l})$, and K_{oy} is the overall mass transfer coefficient based on the Y-phase. In other respects the construction is similar to that for the preceding case.

7.3.1.2 *Backmixing in both phases*

The solution is more difficult in this case since concentration jumps occur at both phase entries, making trial and error calculation necessary. For

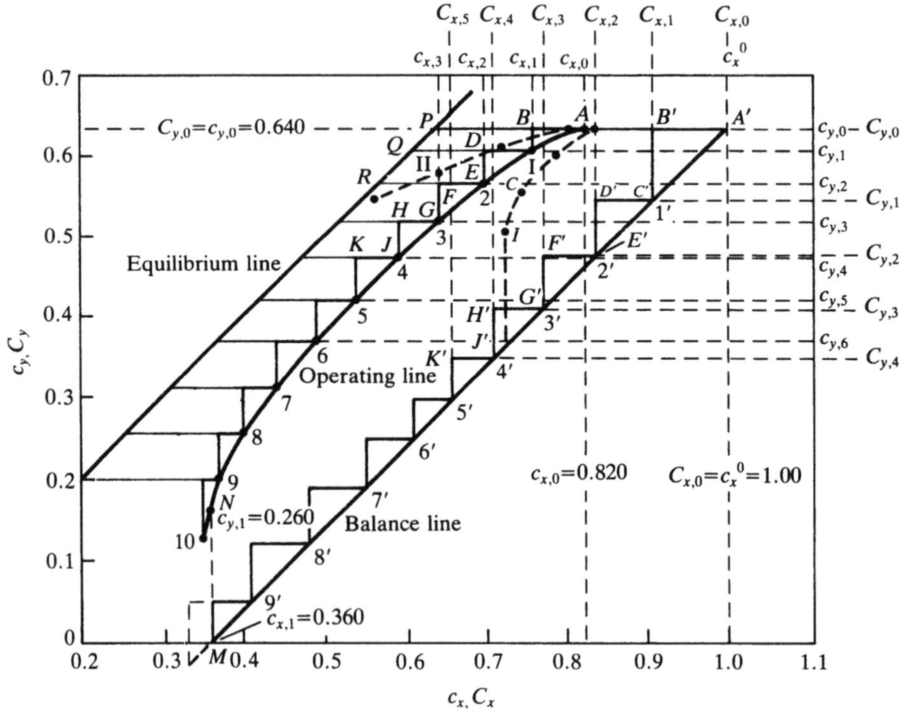

Fig. 7.7. Diffusion model: graphical solution for case of backmixing in both phases (Rod 1964). Curves I and II show effect of incorrect jump concentration $c_{x,o}$ (Example 7).

example, if calculations are started from the X-phase inlet a value must be assumed for $c_{x,0}$ and refined if the specified exit composition cannot be reached. The method, using eqns (7.79–7.82), is carried out by the stepwise procedure detailed below and illustrated in Fig. 7.7. In this diagram point A' on the balance line corresponds to the external concentrations $(c_x^0, c_{y,0})$, which are identical with $C_{x,0}$, $C_{y,0}$.

1. Draw a horizontal line $A'P$ through A' to meet the equilibrium curve at P and assume a value for $c_{x,0}$, e.g. at A.
2. Locate $B'(=C_{x,1})$ on $A'P$ from the relation $A'B'/AP = k_{ox}a\Delta z/U_x$ (eqn (7.81)) and draw a vertical line through B' to meet the balance line at $C'(=C_{y,1}$, from eqn (7.82)).
3. Locate $B(=c_{x,1})$ on $A'P$ so that $AB/A'A = U_x\Delta z/E_x$ (eqn (7.79)).
4. Locate $C(=c_{y,1})$ vertically below B so that $BC/B'C' = 1/(1 + E_y/U_y \Delta z)$ (eqn. (7.80)).
5. Draw a horizontal line through C to meet the equilibrium curve at Q and repeat steps 2–4; if the desired X-phase exit composition is reached after N steps, the required height is $N\Delta z$.
6. If the desired exit composition cannot be reached, repeat from step 1 with a new assumed value for $c_{x,0}$.

Example 7. Calculate the extractor height required to meet the following specification

$c_x^0 = 1.00\ \text{kmol m}^{-3}$	$U_x = 10\ \text{m h}^{-1}$	$E_x = 25\ \text{m}^2\,\text{h}^{-1}$
$c_{x,l} = 0.36\ \text{kmol m}^{-3}$	$U_y = 10\ \text{m h}^{-1}$	$E_y = 25\ \text{m}^2\,\text{h}^{-1}$
$c_y^l = 0.0\ \text{kmol m}^{-3}$	$k_{ox}a = 5.0\ \text{h}^{-1}$	

The equilibrium relationship is given by $c_y^* = c_x$

Solution. From the overall material balance,

$$10.0(1.00 - 0.36) = 10.0(c_{y,0} - 0.0)$$

therefore

$$c_{y,0} = 0.640$$

The balance line A′M is located on Fig. 7.7 as shown. A value of Δz of 1.00 is assumed, giving $k_{ox}a\Delta z/U_x = 0.50$, $U_x\Delta z/E_x = 0.40$ and $1/(1 + E_y/U_y\Delta z) = 0.286$. It was found by trial that a solution was obtainable with $c_{x,0} = 0.820$, requiring 9.6 steps as shown in Fig. 7.7.

If an incorrect value were assumed for $c_{x,0}$ in the first step, the operating line would go to $+\infty$ or $-\infty$ without reaching the desired exit X-phase composition. This is illustrated by the broken lines I and II in Fig. 7.7, which correspond to high and low values respectively of $c_{x,0}$. It is apparent that the computation is very sensitive to small errors in $c_{x,0}$, and this is found to become increasingly so as the value of E_x, that is, the degree of X-phase backmixing, decreases. A similar situation applies to $c_{y,l}$ and E_y when starting from the Y-phase entry. (See Section 7.4.1 for a further discussion).

7.3.2 Stagewise contactors

The auxiliary variables for this case were defined by Rod (1971) as follows

$$C_{x,n} = (1 + \alpha_x)c_{x,n} - \alpha_x c_{x,n+1} \tag{7.88}$$

$$C_{y,n} = (1 + \alpha_y)c_{y,n} - \alpha_y c_{y,n-1} \tag{7.89}$$

These transform eqn (7.39) and the overall material balance as follows

$$C_{x,n-1} - C_{x,n} = N^1_{ox}(c_{x,n} - c^*_{x,n}) \tag{7.90}$$

$$J' = F_x C_{x,n} - F_y C_{y,n+1} \tag{7.91}$$

The boundary conditions, eqns (7.47–7.50), similarly transform to

$$n = 0: \qquad C_{x,0} = c^0_x, \qquad C_{y,1} = c_{y,1} \tag{7.92}$$

$$n = N + 1: \qquad C_{x,N} = c_{x,N} \qquad C_{y,N+1} = c^{N+1}_y \tag{7.93}$$

7.3.2.1 *Backmixing in one phase only*

As before, if backmixing occurs only in, say, the Y-phase, the construction is started from the X-phase inlet, putting $C_{x,n} = c_{x,n}$ throughout. Also N^1_{ox} in eqn (7.90) related by eqn (7.71) to the fully backmixed Murphree efficiency, defined by

$$E_{Mx} = \frac{N^1_{ox}}{1 + N^1_{ox}} = \frac{c_{x,n} - c_{x,n+1}}{c_{x,n} - c^*_{x,n+1}} \tag{7.94}$$

and eqn (7.89) can be expressed as follows:

$$c_{y,n+1} - C_{y,n+1} = \frac{\alpha_y}{1 + \alpha_y}(c_{y,n} - C_{y,n+1}) \tag{7.95}$$

The construction, shown in Fig. 7.8; closely resembles that of Fig. 7.6 for the diffusion model. After locating $(c^0_x, c_{y,1})$ at A on the balance line, the procedure is as follows:

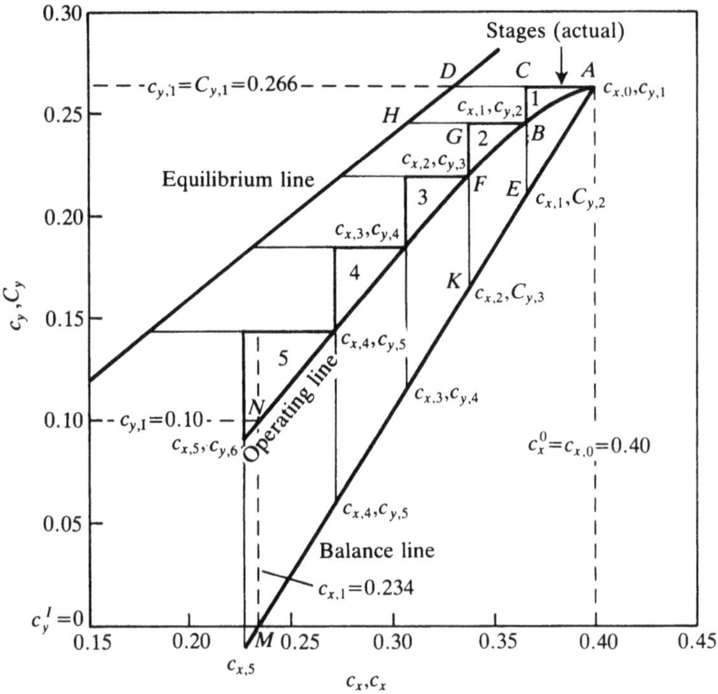

Fig. 7.8. Backflow model: graphical solution for case of backmixing in Y phase only (Example 8).

1. Draw a horizontal line through A to meet the equilibrium curve at D, and locate C so that $AC/AD = E_{Mx}$ [eqn (7.94)].
2. Draw a vertical line through C to meet the balance line at E, that is, at $C_{y,2}$, in accordance with eqn (7.91).
3. Locate B on the operating line so that $BE/CE = \alpha_y/(1 + \alpha_y)$ [eqn (7.95)], thus obtaining $c_{y,2}$.
4. Draw a horizontal line through B to meet the equilibrium curve at H and repeat steps 1–3 until the desired exit composition $c_{x,N}$ is reached.

Example 8. Calculate the number of stages required to meet the same specification as Example 6, given that $N_{ox}^1 = 1.0$, $\alpha_x = 0.0$, and $\alpha_y = 2.0$.

Solution. The construction for this case is shown in Fig. 7.8, and requires 4.8 stages. This compares with a calculated value of 4.79 stages, and is therefore in almost exact agreement.

If backmixing occurs only in the X-phase, a similar procedure is used starting from the Y-phase inlet. The mass transfer rate for this case is expressed on a Y-phase basis, that is, as $N^1_{oy} = k_{oy}ah_c/U_y$, together with the corresponding Murphree efficiency E_{My}.The relevant equations, replacing eqns (7.94, 7.95 and 7.91) are therefore

$$E_{My} = \frac{N_{oy}}{1 + N_{oy}} = \frac{c_{y,n'+1} - c_{y,n'}}{c^*_{y,n'+1} - c_{y,n'}} \tag{7.96}$$

$$C_{x,n'} - c_{x,n'} = \frac{\alpha_x}{1 + \alpha_x}(C_{x,n'+1} - c_{x,n'}) \tag{7.97}$$

$$C_{x,n'+1} = \frac{F_y}{F_x}C_{y,n'} + \frac{J'}{F_x} \tag{7.98}$$

in which n' represents the stage number counting from the Y-phase entry.

7.3.2.2 *Backmixing in both phases*

Again the procedure is more complicated in this case, requiring trial-and-error location of one of the inlet jump concentrations, as with the diffusional model. Assuming the computation to start from the X-phase inlet, eqns (7.88, 7.90, and 7.91) together with (7.95) are used; eqn (7.90) is retained as such, since it cannot be expressed in terms of E_{Mx}.

The detailed construction is shown in Fig. 7.9; the procedure is as follows, starting from point A' with coordinates $(c^0_x, c_{y,1})$ on the balance line:

1. Draw a horizontal line $A'R$ through A' to meet the equilibrium curve at R and assume a value for $c_{x,0}$, e.g. at A.
2. Locate $B(= c_{x,1})$ on $A'R$ such that $AB = A'A/\alpha_x$ [from eqn (7.88)].
3. Locate $B'(= C_{x,1})$ from $A'B' = N^1_{ox}.\ BR$ [eqn (7.90)] and draw a vertical line through B' to meet the balance line at B'', i.e. at $C_{y,2}$ [eqn (7.91)].
4. Locate $C'(=c_{y,2})$ from $C'B''/B'B'' = \alpha_y/(1 + \alpha y)$ [eqn (7.95)], draw a horizontal line through C' to meet the equilibrium curve at S and a vertical line through B to meet $C'S$ at C.
5. Locate $D(=c_{x,2})$ on $C'S$ from $CD = C'C/\alpha_x$ [eqn (7.88)].
6. Repeat from step 3; if the specified X-phase raffinate composition $c_{x,N}$ is not reached the computation is repeated with a new assumed value of $c_{x,0}$.

Example 9. Calculate the number of stages required to meet the same specification as Example 7, given that $N^1_{ox} = 1.0$, and $\alpha_x = \alpha_y = 2.0$.

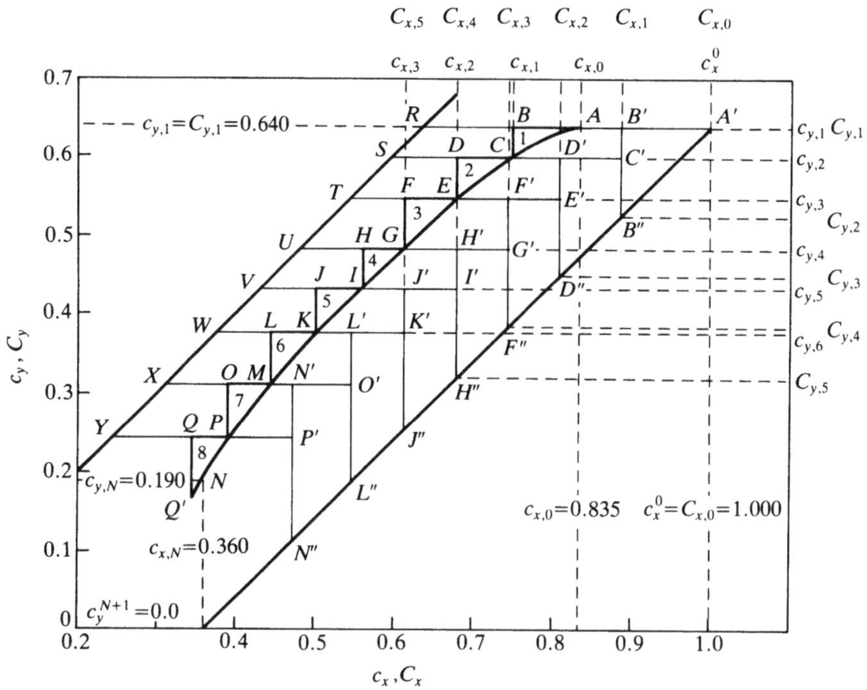

Fig. 7.9. Backflow model: graphical solution for case of backmixing in both phases (Example 9).

Solution. The construction is shown in Fig. 7.9. This shows that slightly less than 8.0 stages are required, as compared with a calculated value of 7.9. Again the computation was extremely sensitive to the assumed value of $c_{x,0}$, which was found to be 0.835, in very close agreement with the calculated value of 0.834; on the other hand, the Y-jump concentration, $c_{y,N}$, was somewhat in error, being 0.190 as compared with the calculated value of 0.166.

7.4 Numerical solution methods

The analytical methods described above are adequate for most cases of simple countercurrent extraction of single solutes, but more complex cases require numerical treatment. These include those in which multiple feeds are to be used; the equilibrium relationship is too complex to be represented by simple straight line segments; or the mass transfer rate varies throughout the extractor due to changes in physical properties. Still more complex cases result when two or more solutes are present, as in dual-solvent and dissociation extraction processes.

A number of alternative numerical methods are available, some of which are applicable to the more complex cases. These are described below under appropriate headings.

7.4.1 Stage-to-stage calculation

Methods of this type are particularly suitable for design purposes, that is, the computation of extractor height or number of stages required to achieve a specified duty. They are, in fact, directly analogous to the graphical methods described previously, and suffer from the same disadvantages, as will be seen.

For differential contactors, the procedure can be carried out in terms of the diffusion model by direct numerical integration of eqns (7.73–7.76) together with the boundary conditions, eqns (7.77–7.78), using, for example, a fourth-order Runge–Kutta method. However, it is more expedient to employ a direct stepwise procedure based on the graphical method of Rod (1964), using eqns (7.79–7.82). Similarly, the backflow model is applicable to staged and plate-type contactors, in which case eqns (7.88–7.93) apply. The backflow model can also be used for differential contactors, in view of the following relationship (cf. eqn 7.66)

$$\alpha_k = \frac{N}{Pe_k B} - \frac{1}{2} \tag{7.99}$$

together with

$$N^1_{ox} = \frac{N_{ox}}{N} = \frac{H}{H_{ox} N} \tag{7.100}$$

There is in fact some advantage in using the backflow model for differential columns, as it was shown by Mecklenburgh and Hartland (1969) that the error in this case is of the order of $1/N^2$, as compared with $1/N$ for Rod's original method where N is the number of steps in each case.

Computer programs for either model can easily be written by analogy with the graphical procedures described in Section 7.3. (p. 448). These are quite straightforward when backmixing occurs in only one of the phases, since the solution is obtained directly by starting the computation at the inlet of the phase which is not backmixed. When there is backmixing in both phases, however, it is necessary to employ a 'boundary iteration' method, in which a value is assumed for one of the inlet jump concentrations, say $c_{x,0}$, and the stepwise concentrations are calculated towards the other terminal. Unless the assumed value of $c_{x,0}$ were correct, the calculated c_y compositions will go to $\pm\infty$, and the computation must be repeated with a new value of $c_{x,0}$.

Unfortunately, the boundary iteration method is extremely sensitive to the assumed value of $c_{x,0}$ (or $c_{y,N}$). A starting value can be obtained by use of the appropriate simplified solution for the 'jump' concentration given by eqns IV of Table 7.2, provided that the equilibrium relation is suitably linearized. Alternatively, solutions obtained for one and two steps, which converge rapidly, can be used as a basis for increasing the number of steps successively until the desired solution is obtained (Mecklenburgh and Hartland 1969). This problem becomes worse as the degree of backmixing decreases, as with the graphical methods. Thus it was found for particular cases, using a 39-bit computer, that convergence was not obtainable with a 1-bit error in $c_{x,0}$, that is, with an accuracy of 1 in 10^{12} (Mecklenburgh and Hartland 1967a). A satisfactory solution should be obtainable with low backmixing in one phase, provided that the computation were started at the inlet of the other phase. With low backmixing in both phases, however, convergence is likely only with a small number of steps, so that the results would most likely be inaccurate.

7.4.2 Unsteady-state method

An alternative method of solution has been devised in which the backflow model equations are put into unsteady state form and integrated until the profiles become steady (Mecklenburgh and Hartland 1969). For this purpose eqn (7.90) and the corresponding expression for the Y-phase are written as follows:

$$\frac{\mathrm{d}c_{x,n}}{\mathrm{d}\theta} = C_{x,n-1} - C_{x,n} - N^{1}_{ox}\,(c_{x,n} - c^{*}_{x,n}) \qquad (7.101)$$

$$\frac{\mathrm{d}c_{y,n}}{\mathrm{d}\theta} = C_{y,n+1} - C_{y,n} + F_R N^{1}_{ox}\,(c_{x,n} - c^{*}_{x,n}) \qquad (7.102)$$

where θ is a variable proportional to time. For differential contactors the values of the α_k in eqns (7.88 and 7.89), and of N^{1}_{ox}, are obtained from the Pe_k and H_{ox} respectively using eqns (7.99 and 7.100).

The procedure consists in first calculating the exit compositions for $N = 1$. These are then used as a starting point for the simultaneous integration of eqns (7.101 and 7.102) for a progressively increasing number of stages, starting from $N = 2$, until the desired degree of accuracy of the exit compositions is reached. The resulting solution gives the operational performance of the extractor; for design purposes the length or number of stages must be assumed, and iterated upon until the desired exit compositions are obtained. Full details of the method are given by Mecklenburgh and Hartland (1969) in their Appendix II.

7.4.3 Matrix solution

As an alternative to the foregoing procedures, the set of mass balance equations for each phase, written around each individual stage, can be solved simultaneously in matrix form. Such a method is strictly applicable only to stagewise contactors, using the backflow model; however, it can also be applied to differential contactors, using the diffusion model, by subdivision into a number of arbitary stages with parameter values given by eqns (7.99 and 7.100).

The formulation described below follows that given by Ricker *et al.* (1981) with minor changes of symbols. This was an extension of earlier methods described by McSwain and Durbin (1966), and by Napthali and Sandholm (1971).

7.4.3.1 *Derivation of equations*

The schematic arrangement of the extractor is shown in Fig. 7.10. This comprises a total of N stages, numbered from 1 to N and arranged as follows:

(i) Stages 1 to $m - 1$: Y-phase settling zone

(ii) Stage m: X-phase (feed) inlet

(iii) Stages $m + 1$ to $n - 1$: Countercurrent mass transfer region

(iv) Stage n: Y-phase (extractant) inlet

(v) Stages $n + 1$ to N: X-phase settling zone

Mass transfer is assumed to occur in all stages except m and n, the feed and extractant inlets. The settling zones are relevant to column-type contactors of either differential or stagewise type rather than mixer-settlers (cf. Wilburn 1964), and may be disregarded if desired.

Referring to Fig. 7.10, an X-phase mass balance for solute i in stage j, assuming the backflow f_x to vary with stage number, is as follows:

$$(F_x + f_x)\, c_{xi,j-1} - (F_x + f_{x,j} + f_{x,j+1})\, c_{xi,j} + f_{x,j+1}\, c_{xi,j+1} - r_{i,j} = 0 \tag{7.103}$$

Similarly for the Y-phase,

$$f_{y,j-1}\, c_{yi,j-1} - (F_y + f_{y,j-1} + f_{y,j})\, c_{yi,j} + (F_y + f_{y,j})\, c_{yi,j+1} + r_{i,j} = 0 \tag{7.104}$$

The steady-state interphase transfer rate of component i is defined as follows in terms of the individual phase mass transfer coefficients and the interfacial compositions c_x^i and c_y^i,

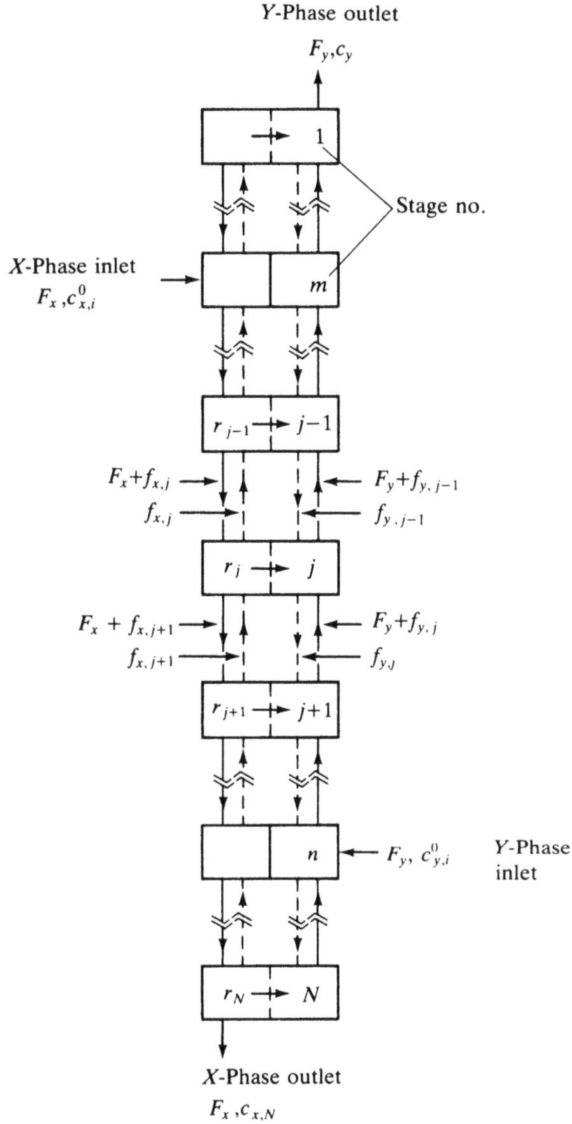

Fig. 7.10. Arrangement of extractor stages for matrix solution.

$$r_{i,j} = k_{xi,j}\, a_j V_j\, (c_{xi,j} - c^i_{xi,j}) = k_{yi,j}\, a_j V_j\, (c^i_{yi,j} - c_{yi,j}) \tag{7.105}$$

Finally, the interfacial compositions are assumed to be at equilibrium:

$$c^i_{yi,j} = K_{i,j}\, c^i_{xi,j} \tag{7.106}$$

For stages m and n, eqns (7.103 and 7.104) must include the feed and extractant inlet terms, viz. $F_x c^0_{xi}$ and $F_y c^0_{yi}$ respectively; also, for these stages, $r_{i,j} = 0$ so that eqns (7.105 and 7.106) are replaced by

$$c^i_{xi,j} = c_{xi,j} \tag{7.105a}$$

and

$$c^i_{yi,j} = c_{yi,j} \tag{7.106a}$$

Equations (7.103 and 7.104) must also be modified for the end stages, 1 and N, to account for the requirement that there be no entrainment of X-phase in Y-product and vice versa.

Equations (7.103 and 7.104) are put into dimensionless form by dividing through by $F_x c^0_{xi}$ and $F_y c^0_{xi}$ respectively, setting $f_x/F_x = \alpha_x$, $f_y/F_y = \alpha_y$, $c_x/c^0_{xi} = \bar{X}$, $c_y/c^0_{xi} = \bar{Y}$, and $F_R = F_x/F_y$. (The dimensionless concentrations $\bar{X}$ and $\bar{Y}$ differ from the X and Y used previously since there is no requirement to linearize the equilibria). Equations (7.103–7.106) then transform as follows:

(i) All stages except 1, m, n, and N

$$F_{1i,j} = (p_{x,j} + \alpha_{x,j})\,\bar{X}_{i,j-1} - (p_{x,j} + \alpha_{x,j} + \alpha_{x,j+1})\,\bar{X}_{i,j} + \alpha_{x,j+1}\bar{X}_{i,j+1} - R_{xi,j} = 0 \tag{7.107}$$

$$F_{2i,j} = \alpha_{y,j-1}\,\bar{Y}_{i,j-1} - (p_{y,j} + \alpha_{y,j} + \alpha_{y,j-1})\,\bar{Y}_{i,j} + (p_{y,j} + \alpha_{y,j})\,\bar{Y}_{i,j+1} + R_{yi,j} = 0 \tag{7.108}$$

$$F_{3i,j} = R_{xi,j} - R_{yi,j}/F_{\mathrm{R}} = 0 \tag{7.109}$$

$$F_{4i,j} = \bar{Y}^i_{i,j} - K_{i,j}\bar{X}^i_{i,j} = 0 \tag{7.110}$$

In these equations, $p_{x,j} = 0$ for stages 1 to $m - 1$ and $= 1$ otherwise; similarly, $p_{y,j} = 0$ for stages $n + 1$ to N and otherwise $= 1$. The dimensionless transfer rates $R_{ki,j}$ are given by

$$R_{xi,j} = N^1_{xi,j}\,(\bar{X}_{i,j} - \bar{X}^i_{i,j}) \tag{7.111}$$

$$R_{yi,j} = N^1_{yi,j}\,(\bar{Y}^i_{i,j} - \bar{Y}_{i,j}) \tag{7.112}$$

in which the number of k-phase transfer units per stage for solute i and stage j is given by

$$N^1_{ki,j} = N_{ki,j}\frac{V_j a_j}{V_t \bar{a}} = \frac{k_{ki,j} a_j V_j}{F_k} \tag{7.113}$$

In eqn (7.113) V_t represents the total extractor volume, $\bar{a}$ its average interfacial area, and $N_{ki,j}$ the total number of k-phase transfer units in the extractor based on the local value of the mass transfer coefficient, $k_{ki,j}$ (i.e. for stage j).

(ii) Feed and extractant inlet stages, m and n
For stage m, eqns (7.107 and 7.108) become

$$F_{1i,m} = \alpha_{x,m}\bar{X}_{i,m-1} - (1 + \alpha_{x,m} + \alpha_{x,m+1})\,\bar{X}_{i,m} + \alpha_{x,m+1}\bar{X}_{i,m+1} - R_{xi,m} + 1 = 0 \tag{7.114}$$

$$F_{2i,m} = \alpha_{y,m-1}\,\bar{Y}_{i,m-1} - (1 + \alpha_{y,m} + \alpha_{y,m-1})\,\bar{Y}_{i,m} + (1 + \alpha_{y,m})\,\bar{Y}_{i,m+1} - R_{yi,m} = 0 \tag{7.115}$$

Similarly for stage n

$$F_{1i,n} = (1 + \alpha_{x,n})\,\bar{X}_{i,n-1} - (1 + \alpha_{x,n} + \alpha_{x,n+1})\,\bar{X}_{i,n} + \alpha_{x,n+1}\bar{X}_{i,n+1} - R_{xi,n} = 0 \tag{7.116}$$

$$F_{2i,n} = \alpha_{y,n-1}\,\bar{Y}_{i,n-1} - (1 + \alpha_{y,n} + \alpha_{y,n-1})\,\bar{Y}_{i,n} + \alpha_{y,n}\,\bar{Y}_{i,n+1} + R_{yi,n} + \bar{Y}_i^0 = 0 \tag{7.117}$$

where $\bar{Y}_i^0 = c_{yi}^0/c_{xi}^0$. As there is no mass transfer in these stages, eqns (7.109 and 7.110) become

$$F_{3i,j} = \bar{X}_{i,j}^i - \bar{X}_{i,j} = 0 \tag{7.118}$$

$$F_{4i,j} = \bar{Y}_{i,j}^i - \bar{Y}_{i,j} = 0 \tag{7.119}$$

(iii) Stages 1 and N
At the phase exists it is assumed that there is no entrainment (i.e. backmixing) of the other phase. For stage 1, eqns (7.107 and 7.108) therefore become

$$F_{1i,j} = \alpha_{x,2}\,(\bar{X}_{i,2} - \bar{X}_{i,1}) - R_{xi,1} = 0 \tag{7.120}$$

$$F_{2i,j} = (1 + \alpha_{y,1})\,(\bar{Y}_{i,2} - \bar{Y}_{i,1}) + R_{yi,1} = 0 \tag{7.121}$$

and for stage N

$$F_{1i,N} = (1 + \alpha_{x,N})\,(\bar{X}_{i,N-1} - \bar{X}_{i,N}) - R_{xi,N} = 0 \tag{7.122}$$

$$F_{2i,N} = \alpha_{y,N-1}\,(\bar{Y}_{i,N-1} - \bar{Y}_{i,N}) + R_{yi,N} = 0 \tag{7.123}$$

7.4.3.2 *Matrix formulation*

The set of equations represented by eqns (7.107 and 7.123) can be expressed in matrix form as follows:

$$\begin{bmatrix} [\mathbf{B}_1][\mathbf{C}_1] & 0 & 0 & 0 & 0 & 0 \\ [\mathbf{A}_2][\mathbf{B}_2][\mathbf{C}_2] & 0 & 0 & 0 & 0 \\ 0 \; [\mathbf{A}_3][\mathbf{B}_3][\mathbf{C}_3] & & & & & \\ - \; - \; - & - & - & - & - & - \\ 0 \; 0 \; 0 \; 0 & [\mathbf{A}_{N-1}][\mathbf{B}_{N-1}][\mathbf{C}_{N-1}] & & & & \\ 0 \; 0 \; 0 \; 0 & 0 & [\mathbf{A}_N] & [\mathbf{B}_N] & & \end{bmatrix} \begin{bmatrix} (\bar{\mathbf{X}}_1) \\ (\bar{\mathbf{X}}_2) \\ (\bar{\mathbf{X}}_3) \\ - \\ (\bar{\mathbf{X}}_{N-1}) \\ (\bar{\mathbf{X}}_N) \end{bmatrix} + \begin{bmatrix} (\mathbf{C}_1) \\ (\mathbf{C}_2) \\ (\mathbf{C}_3) \\ - \\ (\mathbf{C}_{N-1}) \\ (\mathbf{C}_N) \end{bmatrix} = (\mathbf{0}) \tag{7.124}$$

or more compactly, as

$$[\mathbf{M}]\,(\mathbf{X}) + (\mathbf{C}) = (\mathbf{0}) \tag{7.125}$$

In eqn (7.125), $[\mathbf{M}]$ is a block tridiagonal matrix composed of $4M \times 4M$ sub-matrices $[A_j]$, $[B_j]$, and $[C_j]$, where M is the number of solutes; these contain the coefficients of the equations for stage j with respect to the variables for stages $j-1$, j, and $j+1$ respectively.

The vector $(\mathbf{X})$ is composed of corresponding sub-vectors $(\mathbf{X}_j)$, as follows:

$$(\mathbf{X}) = [\,(\bar{\mathbf{X}}_1)\,..(\bar{\mathbf{X}}_j)\,..\,(\bar{\mathbf{X}}_N)\,]^T \tag{7.126a}$$

where

$$(\mathbf{X}_j) = [\,(\bar{X}_{1,j}\,..\,\bar{X}_{M,j})\,(\bar{Y}_{1,j}\,..\,\bar{Y}_{M,j})\,(\bar{X}^{-i}_{1,j}\,..\,\bar{X}'^{i}_{M,j})\,(\bar{Y}^{-i}_{1,j}\,..\,\bar{Y}^{-i}_{M,j})\,]^{\mathrm{T}} \tag{7.126b}$$

The vector $(\mathbf{C})$ contains the constant terms in eqns (7.107–7.119). In the present case with two feed stages, the two sub-vectors for stages m and n (for two solutes) are $[1\,1\,0\,0\,0\,0\,0\,0]^{\mathrm{T}}$ and $[0\,0\,\bar{Y}^0_1\,\bar{Y}^0_2\,0\,0\,0\,0]^{\mathrm{T}}$ respectively; all other elements are zero.

It is assumed in listing the elements of matrix $[\mathbf{M}]$ that the backmixing ratios $\alpha_{x,j}$ and $\alpha_{y,j}$ are independent of composition. The sub-matrices $[A_j]$ and $[C_j]$ are then diagonal, with the following non-zero diagonal elements

$$\begin{array}{lll} a_{i,i} = a_{x,j} + p_{x,j} & : i = 1, M & : j \neq m \neq 1 \\ a_{i,i} = \alpha_{x,j} & : i = 1, M & : j = m \\ a_{i,i} = \alpha_{y,j-1} & : i = M+1, 2M & : j \neq 1 \\ c_{i,i} = \alpha_{x,j+1} & : i = 1, M & : j \neq N \\ c_{i,i} = \alpha_{y,j} + p_{y,j} & : i = M+1, 2M & : j \neq n \neq N \\ c_{i,i} = \alpha_{y,j} & : i = M+1, 2M & : j = n \end{array}$$

The $[B_j]$ sub-matrix is more complex and must be treated as a general $4M \times 4M$ matrix with the following non-zero elements

$$b_{i,i} = -(\alpha_{x,j+1} + N^1_{xi,j}) \quad : i = 1, M \quad : j = 1$$
$$b_{i,i} = -(p_x + \alpha_{x,j} + N^1_{xi,j}) \quad : i = 1, M \quad : j = N$$
$$b_{i,i} = -(p_x + \alpha_{x,j} + \alpha_{x,j+1} + N^1_{xi,j}) \quad : i = 1, M \quad : j \neq 1, N$$
$$b_{i,i} = -(p_y + \alpha_{y,j} + N^1_{yi,j}) \quad : i = M+1, 2M \quad : j = 1$$
$$b_{i,i} = -(\alpha_{y,j-1} + N^1_{yi,j}) \quad : i = M+1, 2M \quad : j = N$$
$$b_{i,i} = -(p_y + \alpha_{y,j} + \alpha_{y,j+1} + N^1_{yi,j}) \quad : i = M+1, 2M \quad : j \neq 1, N$$
$$b_{i,i+2M} = N^1_{xi,j} \quad : i = 1, M$$
$$b_{i+M,i+3M} = N^1_{yi,j} \quad : i = 1, M$$

The following apply to non-feed stages, i.e. $j \neq m, n$

$$\left.\begin{aligned} b_{i+2M,i} &= N^1_{xi,j} \\ b_{i+2M,i+M} &= N^1_{yi,j} \\ b_{i+2M,i+2M} &= -N^1_{xi,j} \\ b_{i+2M,i+3M} &= -N^1_{yi,j} \\ b^*_{i+3M,i+2M} &= K_{i,j} \\ b^*_{i+3M,i+3M} &= -1 \end{aligned}\right\} \quad i = 1, M$$

*These elements, together with certain zero elements, differ in matrix [**M**′] [eqn (7.129)].

The following apply to feed stages only, i.e. $j = m$ or n.

$$\left.\begin{aligned} b_{i+2M,i} &= -1 \\ b_{i+2M,i+2M} &= 1 \\ b_{i+3M,i+M} &= -1 \\ b_{i+3M,i+3M} &= 1 \end{aligned}\right\} \quad i = 1, M$$

All remaining elements in the B matrix are zero.

7.4.3.3 *Method of solution*

The set of equations represented by eqn (7.125) can be solved directly for the concentration vector (**X**) if the distribution coefficients, K_k, are constant. Thus, premultiplication of both sides by $[\mathbf{M}]^{-1}$ gives

$$(\mathbf{X}) = -[\mathbf{M}]^{-1}(\mathbf{C}) \qquad (7.127)$$

If the K_k are not constant, the solution can be obtained iteratively, using the values corresponding to each solution for (**X**) in the next trial.

Frequently, however, the distribution coefficients are complex functions of the concentrations of each component. In such cases it is preferable to solve eqn (7.125) by an iterative Newton–Raphson algorithm in which the equations are linearized about the current values of the dependent variables for each iteration (Ricker *et al.* 1981). By this means the linearization becomes more accurate as the solution is approached, so that the rate of convergence accelerates to the final value (Naphtali and Sandholm 1971).

In linearizing the equations, particular attention should be given to $F_{41,j}$, since this includes the equilibrium distribution coefficients $K_{i,j}$, which are functions of composition. Thus, partial differentiation of eqn (7.110) gives, for solute i (omitting subscript j)

$$\delta F_{4i} = \Delta \bar{Y}_i^i - K_i \Delta \bar{X}_i^i - \bar{X}_i^i \left(\frac{\delta K_i}{\delta \bar{X}_1^i} \Delta \bar{X}_1^i + \ldots \frac{\delta K_i}{\delta \bar{X}_M^i} \Delta \bar{X}_M^i + \frac{\delta K_i}{\delta \bar{Y}_1^i} \Delta \bar{Y}_1^i + \ldots \frac{\delta K_i}{\delta \bar{Y}_M^i} \Delta \bar{Y}_M^i \right) \tag{7.128}$$

Incorporation of these relations into eqn (7.125) gives

$$[\mathbf{M}'] \, (\Delta \mathbf{X}) = -(\mathbf{F}) \tag{7.129}$$

in which matrix $[\mathbf{M}']$ is the first derivative of $[\mathbf{M}]$, $(\Delta\bar{\mathbf{X}})$ is the vector of predicted changes in the dependent variables, obtained by replacing the elements $\bar{X}_1 \ldots \bar{X}_M$, $\bar{Y}_1 \ldots \bar{Y}_M$, etc. of $(\mathbf{X})$ by $\Delta\bar{X}_1 \ldots \Delta\bar{X}_M$, $\Delta\bar{Y}_1 \ldots \Delta\bar{Y}_M$, etc. and $(\mathbf{F})$ is the discrepancy vector, which replaces the zero vector, $(\mathbf{0})$, on the right of eqn (7.125) when this is not exactly balanced.

Matrix $[\mathbf{M}']$ remains block tridiagonal with sub-matrices $[A_j]$ and $[C_j]$ identical wtih those given previously for $[\mathbf{M}]$. Certain elements of of corresponding sub-matrix $[\mathbf{B}']$ differ from those of $[\mathbf{B}]$, however, as follows:

$$b_{i+3M,i'+2M} = X_{i,j}^i \, \delta K_{i,j} / \delta X_{i',j}^i \qquad : i = 1, M;\ i' = 1, M \text{ with } i' \neq i$$

$$b_{i+3M,i+2M} = X_{i,j}^i \, \delta K_{i,j} / \delta X_{i,j}^i + K_{i,j} \qquad : i = 1, M$$

$$b_{i+3M,i'+3M} = X_{i,j}^i \, \delta K_{i,j} / \delta Y_{i',j}^i \qquad : i = 1, M;\ i' = 1, M \text{ with } i' \neq i$$

$$b_{i+3M,i+3M} = X_{i,j}^i \, \delta K_{i,j} / \delta Y_{i,j}^i - 1 \qquad : i = 1, M$$

These elements, $2M^2$ in number, appear in the bottom right-hand corner of $[\mathbf{B}']$, occupying the last M rows and $2M$ columns,

In this form the Newton–Raphson method gives, for the $(m + 1)$th iteration.

$$(\Delta\bar{\mathbf{X}})_m = -(\mathbf{M}')_m^{-1}(\mathbf{F})_m \tag{7.130a}$$

and

$$(\bar{\mathbf{X}})_{m+1} = (\bar{\mathbf{X}})_m + (\Delta\bar{\mathbf{X}})_m \tag{7.130b}$$

The solution is obtained by the following stepwise procedure:

1. Make initial estimates of the concentrations $\bar{X}_j$, $\bar{Y}_j$, $\bar{X}^i_j$, and $\bar{Y}^i_j$ in each stage and arrange these as the combined vector $(\mathbf{X})^0$. Ricker *et al.* (1981) found that satisfactory convergence was obtained by assuming that $\bar{X}_{i,j} = \bar{X}^i_{i,j} = \bar{X}_{i,f}$ and $\bar{Y}_{i,j} = \bar{Y}^i_{i,j} = \bar{Y}_{i,f}$, i.e. as if the extractor were operating with no interphase fluxes.
2. Form matrix $[M]$ using the current values of the equilibrium constants, $K_{i,j}$, and the mass transfer parameters, $N^1_{ki,j}$, in submatrices $[\mathbf{B}]$ and calculate $(\mathbf{F})_0$, as given by the left hand side of eqn (7.125).
3. Form modified sub-matrices $[\mathbf{B}_j']_0$, and hence $[\mathbf{M}']_0$; invert the latter and substitute in eqns (7.130a and 7.130b) to obtain $(\bar{\mathbf{X}})_1$.
4. Repeat from step 2 using the new value of $(\bar{\mathbf{X}})$ until

$$\sum_{k'=1}^{4} \sum_{i=1}^{M} \sum_{j=1}^{N} (F^2_{k'i,j}) < \epsilon$$

   where ϵ is the specified error tolerance of, say 10^{-8}.

(Alternative definitions are given by Spencer *et al.* (1981).)

Spencer *et al.* (1981) described a similar method in which the mass transfer rate was expressed in terms of an overall mass transfer coefficient, based for example on the X-phase as follows

$$r_{i,j} = k_{ox,j}\, a_j\, V_j\, (K_j\, c_{y,j} - c_{x,j})$$

The resulting matrix $[\mathbf{M}]$ was therefore block tridiagonal, made up of sub-matrices of size $3M \times 3M$. A similar method was also described by McSwain and Durbin (1966) who, however, expressed matrix $[M]$ in quindiagonal instead of block tridiagonal form.

7.4.3.4 *Applications of method*

Ricker *et al.* (1981) gave several examples of the use of the method, and found for the case of linear equilibrium that only a single iteration was required, as expected. For the non-linear cases, convergence was attained in three to four iterations. It was stated that the major computational requirements were in the calculation of the $[B']$ matrix, involving partial derivatives of the K_i values, and in the solution of the linearized equations.

Consideration must also be given to the values of the mass transfer rate parameters, $N^1_{ki,j}$, if the physical properties vary within the extractor. Thus, changes in density difference and interfacial tension are likely to lead to local changes in dispersed phase hold-up and droplet diameter, and hence in mass transfer coefficient and interfacial area (Wilburn and Nicholson 1965). Equation (7.113) would therefore be inadequate under these circumstances.

With multicomponent systems involving two or more solutes, account may also need to be taken of possible solute interactions as these can lead in the limit to osmotic and reverse diffusion, with positive flux under zero or negative concentration gradients respectively, and diffusion barrier with zero flux under positive gradient (Toor 1957). A method for the calculation of such mass transfer coefficients, based on an exact solution of the Maxwell–Stefan equations obtained by Krishna and Standart (1976) has been given by Krishna (1976); see also Pratt (1983).

The method has considerable flexibility in that it can readily be extended to take account of more complex situations, as follows:

1. Variations in the phase flow rates, F_x and F_y, resulting from solute transfer between phases and changes in solvent miscibility.
2. Multiple feeds and/or product withdrawal.
3. Single solvent fractional extraction, with reflux.
4. Dual-solvent extraction, with or without reflux (Section 7.4.4, below).
5. Dissociation extraction (example No. 3 of Ricker *et al.* 1981).

Matrix methods of the present type are directly applicable to the calculation of operational performance, that is, the exit compositions for a given extractor given the mass transfer and backmixing parameters and the inlet compositions. They can also be used for parameter estimation from measured concentration profiles (Section 7.6.1.2, p. 475). For design problems, however, it is necessary to assume the number of stages or height, preferably starting from a value obtained by a short cut method, and to iterate on this to obtain the desired performance.

7.4.4 Dual-solvent extraction

The important case of backmixed dual-solvent extractors has been studied by Dongaonkar *et al.* (1987), who proposed two alternative methods of solution. The first method, which is restricted to two solutes, follows similar lines to that of Prochazka and Jiricny (1976) described in Section 6.7.5 (p. 394) of the previous chapter, with the stage-to-stage computations replaced by the generalized analytical solutions given in Table 7.2. The second method is based on the matrix formulation given in Section 7.4.4 (p. 468) and requires considerably more computational effort.

Using either method, it is first necessary to compute the solute mass balances and to select values of the flow ratio parameters P and R, as described in Section 6.7.3 (p. 389). With the first method either the backflow or the diffusion model can be used, and values of the continuous and dispersed phase backmixing parameters, Pe_k^{-1} or α_k as appropriate, are input together with the mass transfer parameters H_{ox} or N_{ox}^1 for each solvent in each column section, together with the phase to be dispersed. Values of E are then computed for each solute in each section and stored, after which the procedure is as follows:

1. A value is assumed for the concentration of component B in the extract phase leaving the extraction section. The B concentration in the raffinate entering this section is then computed by mass balance, followed by $H_{E,b}$ or $N_{E,b}$ using eqn I of Table 7.2, treating the raffinate as 'feed' phase.
2. Noting that the B concentration in the extract phase entering the scrub section is the same as that leaving the extraction section, the concentration in the exit raffinate from the scrub section is calculated by mass balance. Treating the extract phase now as 'feed' phase, $H_{S,b}$ or $N_{S,b}$ is computed using eqn I of Table 7.2.
3. Using eqn II of Table 7.2, the concentration of A in the extract phase entering the scrub section is computed for height $H_{S,b}$ or $N_{S,b}$ stages.
4. Noting that the concentration of A in the extract phase leaving the extraction section is the same as that obtained in step 3, $H_{E,a}$ or $N_{E,a}$ is computed using eqn I of Table 7.2, with the raffinate treated as 'feed' phase.
5. The values of $H_{E,a}$ or $N_{E,a}$ are compared with $H_{E,b}$ or $N_{E,b}$; if these differ by more than, say, 0.01 units, then assume that $H_{E,b} = H_{E,a}$ or $N_{E,b} = N_{E,a}$, compute the B concentration in the extract phase leaving the extraction section using eqn II of Table 7.2 and return to step 2.

A program written on these lines for an 8-bit microcomputer occupied about 10k bytes of memory, including subroutines. Typical problems were found to converge in three iterations. In a further development the method has been extended to systems with non-linear equilibria by approximating the equilibrium line by straight line segments (Dongaonkar *et al.* 1988; Pratt and Stevens 1991).

The second method employs the matrix method described in Section 7.4.3 (p. 460) and gives the component split between the two ends for an assumed number of stages. The equations as given must be modified for the feed stage and for the differing raffinate flows in the two sections; the entering solvents will be solute-free unless reflux is used. The direct method of solution, using eqn (7.127) with iteration if necessary, is applicable provided that the distribution coefficients do not vary greatly with composition.

Results obtained by the two methods were shown by Dongaonkar *et al.* (1987, 1988) to agree closely, both without and with backmixing. The comparison was made for two typical systems by first computing the number of stages required in the extraction and scrub sections to achieve a given separation. These were then used in the matrix method, after rounding off to whole stages, to compute the product split.

Although advantageous in requiring minimal computational effort, the analytical method has the disadvantage of being applicable only to systems with two solutes and constant distribution coefficients. However, the latter disadvantage could be overcome by approximating the equilibrium curves by straight line segments, as described in Section 7.2.5.2 (p. 439). Also, systems with more than two solutes could be handled by applying the method to a pair of solutes with adjacent distribution coefficients, provided that component interactions can be disregarded. This method is convenient, even if only approximate in the latter case, in providing a 'short-cut' estimate of the number of stages required for the matrix method, to avoid excessive trial-and-error in design problems.

7.5 Axial dispersion and polydispersivity

7.5.1 General

The foregoing treatment relates to true backmixing resulting from the transport of a phase in a direction opposite to that of its net flow. As will be shown later (Section 7.6.2, p. 477), the extent of this can be determined by steady state injection of a non-transferring tracer, the concentration of which is measured upstream of the injection point. Alternatively, unsteady state tracer injection (e.g. as a pulse) can be used, with downstream monitoring of the response. This gives a measure of the residence time distribution, which takes into account both backmixing and forward mixing.

In the case of the continuous phase it appears from the available data that results obtained by the two methods do not differ greatly, so that forward mixing is generally small or non-existent. For the dispersed phase, however, it has been shown by Gayler and Pratt (1957) from mass transfer studies that backmixing does not occur in the packed column, and this is probably true for most classes of extractor apart possibly for some rotary agitated types. Despite this, it is found on introduction of a pulse of colour into the dispersed phase distributor that in most cases the resulting band of colour broadens on passage through the column. This indicates that forward dispersion of this phase, arising from the differing velocities of the various droplet sizes within the swarm, may well have an appreciable influence on performance.

7.5.2 Polydispersivity and forward mixing

The dispersed phase in extraction equipment is present in the form of droplet 'swarms' in which the individual droplets cover a wide range of sizes. Rod (1966) showed by means of a simplified model with only two droplet sizes present that this leads to a reduced performance as compared with that of a monodispersion with droplets of the same Sauter mean diameter, d_{32}. This is illustrated on the operating diagram in Fig. 7.11, where the two size fractions enter the column with the same concentration, c_y^I. The smaller size fraction, having the greater specific surface, approaches equilibrium more rapidly than the larger size; however, the gradients become similar further along the column and the exit dispersed phase, after coalescence and mixing, has the same composition although a greater height is required. This effect is enhanced if the velocity of the smaller droplets is less than that of the larger ones.

This example illustrates an important distinction between the effects of polydispersivity and backmixing, namely that the former does not lead to an inlet concentration jump, as does the latter. Further, its magnitude does not increase with column diameter, as sometimes does backmixing, since this is determined entirely by the characteristics of the dispersion.

This form of axial dispersion was first described by Olney (1964), who formulated the equations in terms of the diffusion model without solving

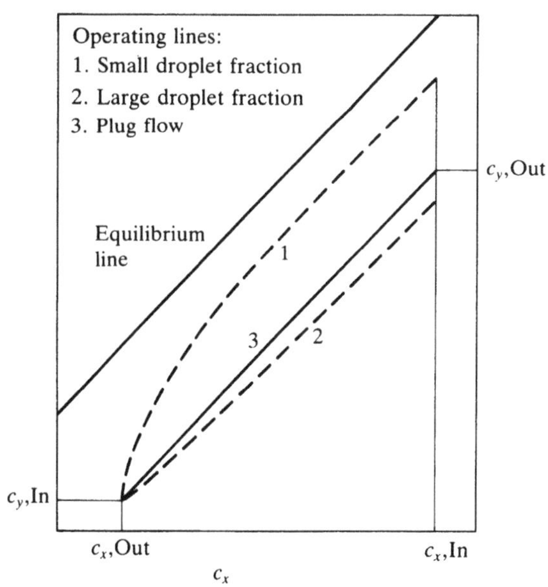

Fig. 7.11. Operating diagram for polydisperse system with two droplet size fractions and no backmixing (based on Rod 1966).

them. The fact that a loss in performance may be expected was confirmed by Charters and Korchinsky (1975) in an extensive computer study using a model similar to that described below.

7.5.2.1 *Theoretical treatment*

The effect of polydispersivity can be introduced into the diffusion model by taking solute balances for each size fraction *i** of the dispersed phase (Olney 1964; Rod 1966). Assuming the *Y*-phase to be dispersed and not backmixed, and expressing the mass transfer coefficients in terms of k_{oy} gives

$$\frac{\mathrm{d}Y_i}{\mathrm{d}Z} = -N_{oy,i}(X - Y_i) \tag{7.131}$$

The specific surface of a droplet of diameter d_i is $6/d_i$, so that $N_{oy,i}$ is given by

$$N_{oy,i} = \frac{6k_{oy,i}\phi f_i H}{U_{y,i} d_i} = \frac{6k_{oy,i} H}{v_i d_i} \tag{7.132}$$

where v_i is the velocity of droplets of size i.

The corresponding expression for the *X*-phase, assuming it to be backmixed, is

$$\frac{\mathrm{d}^2X}{\mathrm{d}Z^2} - P_xB\frac{\mathrm{d}X}{\mathrm{d}Z} = \frac{P_xBH}{EU_y}\sum_{i=1}^{n}\frac{6k_{oy,i}\phi f_i}{d_i}(X - Y_i) \tag{7.133}$$

By introducing a new variable $\mathbf{X}$ defined as follows [cf. eqn (7.73)],

$$\mathbf{X} = X - \frac{1}{P_xB}\frac{\mathrm{d}X}{\mathrm{d}Z}, \tag{7.134}$$

eqn (7.133) can be separated into two first-order differential equations, viz.

$$\frac{\mathrm{d}X}{\mathrm{d}Z} = P_xB(X - \mathbf{X}) \tag{7.135}$$

and

$$\frac{\mathrm{d}\mathbf{X}}{\mathrm{d}Z} = -\frac{L}{EU_y}\sum_{i=1}^{n}\frac{6k_{oy,i}\phi f_i}{d_i}(X - Y_i) \tag{7.136}$$

The boundary conditions for the *X*-phase are given by eqn (7.11 and 7.12) for the inlet and exit respectively; those for the *Y*-phase, in the absence of backmixing, are as follows:

$$Z = 1.0: Y_I = Y^I = 0 \tag{7.137a}$$

* Note that subscript *i* is defined differently from that used in Section 7.4.4 (p. 468).

$$Z = 0: c_y^I = c_{y,I} = \sum_{i=1}^{n} \frac{\phi f_i v_i c_{y,i}^I}{U_y} \tag{7.137b}$$

These equations were solved numerically by Chartres and Korchinsky (1975). A similar study was also carried out in terms of the backflow model (Korchinsky and Azimzadeh-Katyloo 1976).

7.5.2.2 *Droplet size distribution and coalescence*

The droplet size distribution in extraction equipment is determined by a balance between the forces causing droplet breakage and those causing coalescence. In packed columns, droplet breakage results almost entirely from collision with packing elements, whereas in agitated contactors shear forces are the predominant cause. Droplet coalescence on the other hand is controlled primarily by interfacial tension forces, and is influenced by flow pattern and internal geometry. Data on the volume–surface (Sauter) mean diameter, d_{32}, of droplets have been obtained and correlated for a number of extractor types, and are given in the next chapter; in many cases, however, these relate to specific systems and do not take into account differences in physical properties.

Some data are also available on the size distribution of droplet swarms in extractors, and these have been fitted to statistical distributions, for example, of the Mugele–Evans type (1951). However, a more satisfactory procedure would be to calculate the size distribution from the rates of droplet breakage and coalescence, if these were known. Some progress has been made in determining these for packed and pulsed plate columns using a colorimetric method for the measurement of coalescence rate (Hamilton and Pratt 1980). This consisted in first bringing to a steady state size distribution two equal streams of methyl isobutyl ketone (MIBK) containing respectively dithizone (green) and nickel di(ethyl xanthate) (yellow), by passing them through a longitudinally partitioned section of column. The droplets were then allowed to mingle and passed into a second, unpartitioned column section. Coalescence of dissimilar coloured droplets gave red droplets, the proportions and sizes of which were determined by colour photography after leaving the column.

To interpret the results, the droplets were grouped into size ranges with mean diameters based on intervals of $2^{1/3}$, that is, so that the mean droplet volume of any size is twice that of the size below; on this basis two droplets of equal size, on coalescence, give one droplet of the size above, and breakage of a droplet gives two smaller ones of the size below. By writing two sets of population number balances for each size, one on total and the other on red droplets, values of the first-order breakage and second-order coalescence rate constants were derived and correlated empirically by expressions of the form:

$$\text{Coalescence rate}: K_i^c = C_1 d_i^m \phi^n \tag{7.138a}$$

$$\text{Breakage rate}: K_i^b = C_2 d_i^p \phi^q \tag{7.138b}$$

where ϕ represents total dispersed phase hold-up and C_1, C_2 are constants.

Data of this type have been obtained for a packed column (Hamilton and Pratt 1984; Stevens and Pratt 1986) and for a pulsed perforated plate column (Garg and Pratt 1984). Steady state values of d_{32} for the former, calculated by a stepwise procedure from an assumed inlet size using values of K_i^b and K_i^c given by expressions based on eqns (7.138), were in good agreement with the experimental values. Similar calculations for the pulsed plated column, starting from an arbitary monosized droplet dispersion, gave steady state size distributions and d_{32} values in good agreement with experiment over a wide range of pulse frequencies and flow rates.

Studies were also made of the effect of polydispersivity on mass transfer for the extraction of acetic acid with methyl isobutyl ketone in a packed and a pulsed plate column, using a numerical Monte Carlo procedure. For details of the method, with results, the original references should be consulted (Hamilton and Pratt 1984; Garg and Pratt 1984).

7.6 Measurement of axial dispersion and mass transfer parameters

A knowledge of values of the axial dispersion and mass transfer parameters for the various contactor types is essential to rational extractor design and scale-up. In the absence of backmixing the mass transfer parameter is obtainable directly from overall extractor performance data, that is, from the terminal compositions. When backmixing occurs this is not sufficient, however, and it is necessary also to determine the solute concentration profiles of the two phases along the extractor. Alternatively, tracer injection methods can be used to determine the backmixing parameters and knowing these, the mass transfer parameter can be obtained from overall performance data, as described in Section 7.2.5.3 (p. 443); see also Example 5.

Details of the various methods are described below. Values of the backmixing parameters, where available, are given in the next chapter for the various extractor types.

7.6.1 Determination from measured concentration profiles

7.6.1.1 *Graphical method*

Close agreement between experimentally determined solute concentration profiles and those computed from measured values of the backmixing and mass transfer parameters was first demonstrated by Moon *et al.* (1963) for the case of a packed column. The inverse procedure, that is, the determination of the parameters from measured concentration profiles, can be carried out graphically (Rod 1965; Mecklenburgh and Hartland 1969). For this purpose eqn (7.1) is rearranged as follows:

$$\frac{\mathrm{d}}{\mathrm{d}Z}\left(E_x \frac{\mathrm{d}c_x}{\mathrm{d}Z} - U_x c_x\right) = k_{ox} a(c_x - c_x^*) \qquad (7.139)$$

The method is not restricted to linear equilibria, so that actual concentrations can be used and eqn (7.139) becomes, on putting the remaining variables in dimensionless from [cf. eqn (7.4)].

$$\frac{\mathrm{d}}{\mathrm{d}Z}\left(\frac{1}{Pe_x B}\frac{\mathrm{d}c_x}{\mathrm{d}Z} - c_x\right) = N_{ox}(c_x - c_x^*) \qquad (7.140)$$

Integration between $Z = 0$ and $Z = Z$ gives

$$\frac{1}{Pe_x B}\frac{\mathrm{d}c_x}{\mathrm{d}Z} - c_x - \left[\frac{1}{Pe_x B}\left(\frac{\mathrm{d}c_x}{\mathrm{d}Z}\right)_0 - c_{x,0}\right] = N_{ox}\int_o^Z (c_x - c_x^*)\mathrm{d}Z \qquad (7.141)$$

The term in brackets on the left of eqn (7.141) is seen from eqn (7.10), the boundary condition for $Z = 0$, to be equal to $-c_x^0$; inserting this and rearranging gives:

$$-\frac{\mathrm{d}c_x/\mathrm{d}Z}{(c_x^0 - c_x)} = Pe_x B - Pe_x B\, N_{ox}\frac{\int_0^Z (c_x - c_x^*)\mathrm{d}Z}{(c_x^0 - c_x)} \qquad (7.142)$$

A plot of the term on the left of eqn (7.142) against $\int_0^Z (c_x - c_x^*)\,\mathrm{d}Z/(c_x^o - c_x)$ therefore gives a straight line of slope $-Pe_x B\, N_{ox}$ with intercepts of $Pe_x B$ on the Y-axis and $1/N_{ox}$ on the X-axis.

A similar derivation for the Y-phase, starting from eqn (7.2), gives

$$-\frac{\mathrm{d}c_y/\mathrm{d}Z}{(c_y - c_y^1)} = Pe_y B - F_R Pe_y B\, N_{ox}\frac{\int_Z^1 (c_x - c_x^*)\mathrm{d}Z}{(c_y - c_y^1)} \qquad (7.143)$$

so that a plot of the term on the left against the integral term on the right gives a straight line of slope $-F_R Pe_y B\, N_{ox}$ with $Y-$ and X-axis intercepts of $Pe_y B$ and $1/F_R N_{ox}$ respectively. It is to be noted that the value of the integral from Z to 1 in eqn (7.143) is equal to that from 0 to 1 minus the integral from 0 to Z in eqn (7.142).

7.6.1.2 *Numerical method*

The graphical method requires precise evaluation of the gradients and suffers from the inaccuracies normal to such procedures. As an alternative, a statistical approach based on a matrix formulation can be used (Spencer *et al.* 1981; Rod *et al.* 1983). For this purpose, values are assumed for the parameters, P_n, and substituted into the matrix equation [e.g. eqn (7.124)], which is solved for the concentration profile vector (**X**). The Jacobian matrix $[\delta\mathbf{X}/\delta\mathbf{P}]$ is required for the iterative process, and to obtain this each parameter is varied in turn by a small amount $\Delta\mathbf{P}_n$ (typically 1 per cent of $\mathbf{P}_n$) and the matrix equation solved to determine the new concentration vector $(\mathrm{X} + \Delta\mathbf{X})$. The value of $[\delta\mathbf{X}/\delta\mathbf{P}_n]$ is then approximated as $[\Delta\mathbf{X}/\Delta\mathbf{P}_n]$.

The parameters are estimated by minimization of a specified objective function, and for this purpose the non-linear algorithm of Marquardt (1963) is suitable. The iterative procedure consists in substitution of the Jacobian matrix into this algorithm, solving for the parameters (**P**) and using these in the matrix equation to determine new concentration profiles. The objective function is then computed and the procedure repeated until this is minimized.

Rod *et al.* (1983) used the maximum likelihood method for the parameter estimation, requiring the errors in measured concentrations at the various sample points to be mutually independent with zero means and variances σ_x^2 and σ_y^2. The objective function is then

$$\phi(\boldsymbol{\theta}) = \sum_l \frac{1}{\sigma_x^2}[\bar{X}_l - \bar{X}_l^c(\theta)]^2 + \sum_m \frac{1}{\sigma_y^2}[\bar{Y}_m - \bar{Y}_m^c(\theta)]^2 \quad (7.144)$$

where $\bar{X}_l$, $\bar{Y}_m$ are the measured concentrations of the two phases at sample points l and m, and $\bar{X}_l^c$ $(\theta$, $\bar{Y}_{\mathrm{m}}^c$ (θ) are the calculated concentrations for the current values of the parameters. An alternative objective function was used in the study by Spencer *et al.* (1981).

Using Ricker *et al.*'s formulation (Section 7.4.4, see p. 00), two back-mixing parameters are involved, viz. α_x and α_y, together with two mass transfer parameters per solute, N_{xi}^1 and N_{yi}^1. However, the two latter tend to be strongly correlated, and Rod *et al.* (1983) combined these by assuming that $k_{xi}a/k_{yi}a$ is constant. Alternatively, the individual coefficients can be replaced by an overall coefficient, e.g. $k_{ox,i}a$.

An extensive Monte Carlo study of the method was carried out by Rod *et al.* (1983) for a postulated 17 stage extractor with sampling at stages 1, 5, 9, 13 and 17, in which pseudo-random errors drawn from a normal distribution were added to calculated profile compositions. This indicated that the bias was small compared with the standard deviations, and that the method is therefore suitable for use in practice.

7.6.1.3 *Practical considerations*

The major problem with this method is that of obtaining samples of each phase from the dispersion, uncontaminated by the other phase; if any contamination occurs, mass transfer between the phases in the sample will invalidate the results. For this reason, for aqueous systems pairs of sample probes constructed from hydrophilic materials (e.g. stainless steel) and hydrophobic materials (e.g. Teflon) are required for the aqueous and solvent phases respectively. Probes suitable for this purpose have been described in the literature (Lim *et al.* 1979).

The other requirement is that of sample point locations, and in this regard the following should be noted (Rod *et al.* 1983):

1. More information on all parameters is obtained by locating sample points near a phase inlet (in the region of the 'jump') than near the outlet. (See however Section 7.6.3, see p. 00.)
2. Sampling of a given phase provides more information on the backmixing parameter for that phase than for the other.
3. Sampling of a highly backmixed phase provides a fair estimate of all parameters if backmixing is small in the other phase.
4. Information on the mass transfer parameters is at a maximum for the region near the middle of the column.
5. Under normal conditions, the accuracy with which the mass transfer parameters can be estimated is better than that of the backmixing coefficients.

7.6.2 Tracer injection methods

7.6.2.1 *Continuous tracer injection*

By introducing a non-transferring tracer at a steady rate into an extraction column, a decreasing upstream concentration gradient of tracer is established due to backmixing in the phase in which the tracer is soluble (Fig. 7.12(a)). The governing equation for the diffusion model is given by eqn (7.1) with the mass transfer term omitted, as follows

$$E_k \frac{\mathrm{d}^2 c_k}{\mathrm{d}z^2} - U_k \frac{\mathrm{d}c_k}{\mathrm{d}z} = 0 \qquad (7.145)$$

where subscript k refers to the phase containing the tracer. By defining $z' = 0$ as the plane at which the tracer is injected with z' positive in the direction of flow, the boundary conditions are $z' = 0$, $c_k = c_{k,0}$ and $z' \to -\infty$, $c_k = 0$ so that eqn (7.145) gives

$$c_k = c_{k,0} \exp\left[\frac{U_k z'}{E_k}\right] \qquad (7.146)$$

The value of E_k is therefore obtained by measuring the tracer concentration at a number of upstream locations and substituting in eqn (7.146), or alternatively by plotting $\log(c_k/c_{k,0})$ against z' and measuring the slope (Fig. 7.12b).

This solution is applicable to differential columns; for stagewise columns the backflow model should be used, in which case the governing equation is given by eqn (7.39) with the mass transfer term omitted. The solution, with boundary conditions $n' = 0$, $c_k = c_{k,0}$ and $n' \to -\infty$, $c_k = 0$ is then

$$c_{k,n'} = c_{k,0} \exp\left[\frac{\alpha_k}{1+\alpha_k}\right]^{n'} \tag{7.147}$$

where n' refers to the stage number counted upstream from that at which the tracer is introduced. Values of α_k are obtained as before, from eqn (7.147) or from the slope of a plot of $\log c_{k,n'}/c_{k,0}$ against stage number.

The plot of the data in the form shown in Fig. 7.12(b) passes through the point $c_k/c_{k,0} = 1.0$ when z' or $n' = 0$ if mixing of the tracer solution at the inlet is complete. If this is not the case, it is easily shown that the backmixing parameter is still obtainable from the slope of the line through the points for any section of the column above the tracer inlet within which transverse mixing is complete.

It must be emphasized that the values of E_k or α_k obtained by this method are due to backmixing only. A description will be given later of practical considerations associated with the technique.

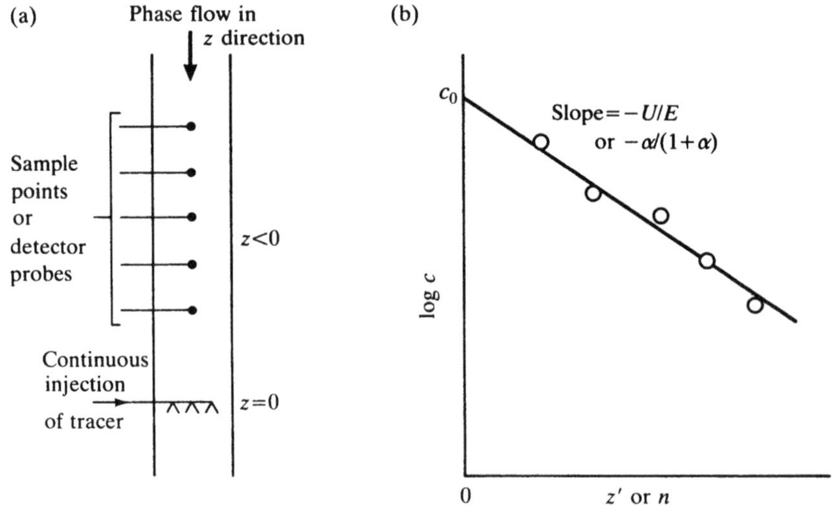

Fig. 7.12. Continuous tracer injection method for measurement of backmixing. (a) Method. (b) Plot of data.

7.6.2.2 *Unsteady tracer injection*

7.6.2.2.1 Diffusion model

In the unsteady state method, shown in Fig. 7.13, a non-transferring tracer is injected intermittently in the form of a pulse, step change or sinusoid into a flowing stream and the response is measured by means of a suitable probe at a point downstream of the injection point. The one-dimensional conservation equation for tracer dispersion, in terms of the diffusion model, is given by eqn (7.145) with an accumulation term added, as follows

$$E_k \frac{\delta^2 c_k}{\delta z^2} - U_k \frac{\delta c_k}{\delta z} = \phi_k \frac{\delta c_k}{\delta t} \tag{7.148}$$

where ϕ_k is the hold-up of phase k.

Mathematically, the simplest case is that of the Dirac delta function, $\delta(t)$, a unit pulse injected in zero time. Levenspiel and Smith (1957) obtained the solution, which for two-phase flow becomes (dropping subscript k)

$$C(\theta) = \left[\frac{Pe_c \phi}{4\theta}\right]^{\frac{1}{2}} \exp\left[-\frac{Pe_c \phi (1 - \theta/\phi)^2}{4\theta}\right] \tag{7.149}$$

where m_T is the mass of tracer injected at time $t = 0$, $C(\theta) = csL/m_T$, the dimensionless concentration of tracer expressed with respect to the mean concentration in the test section, $\theta = tU/(L\phi)$, the dimensionless time since injection, and $Pe_c = UL/E$, the column Peclet number. For the single

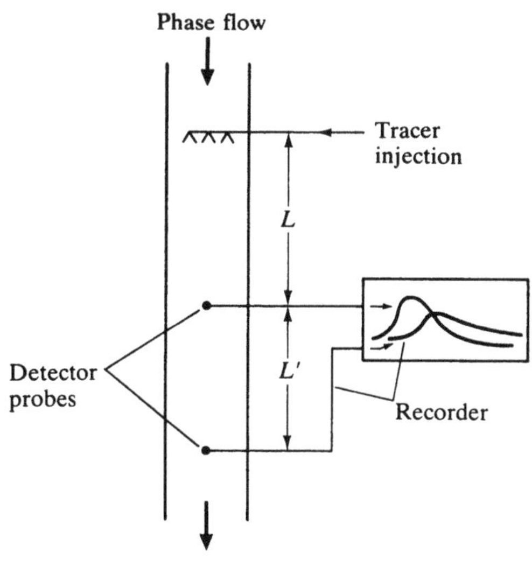

Fig. 7.13. Unsteady state tracer injection method for measurement of axial dispersion.

phase case with no flow (i.e. $U = 0$), eqn (7.149) represents a family of Gaussian error curves symmetric about the origin, with t as parameter. With flow, when plotted in the form $C(\theta)$ vs θ, these become skewed to an extent which becomes significant for $Pe_c \leq 100$ due to the time taken for the tracer to pass the recording point.

It was shown by Levenspiel and Smith (1957) that the value of Pe_c can be obtained from the experimental data, either from the maximum concentration reached or from that at $\theta = 1.0$, using eqn (7.149). An alternative, very simple method for the evaluation of the response curve has been given by Vergnes (1976). This involves measurement of the height of the maximum of the response curve, c_{max}, the time of the maximum since injection, t_{max}, and the area, R, under the response curve. For an 'open' system (see below) and a perfect impulse, the Peclet number is then obtained from the relationship

$$4\pi R^2 = [(1 + Pe_c^2)^{1/2} - 1] \exp [Pe_c - (1 + Pe_c^2)^{1/2}] \quad (7.150)$$

where

$$R = c_{max} t_{max} / \int_0^\infty c(t)\, dt \quad (7.151)$$

Alternatively, as shown by Levenspiel and Smith (1957), Pe_c can be obtained from the mean and variance (1st and 2nd moments) of the recorded curve using the following relationships

$$Pe_c^{-1} = \frac{E}{UL} = \frac{(8\sigma^2 + 1)^{1/2} - 1}{8} \quad (7.152)$$

where

$$\sigma^2 = \frac{\int_0^\infty C(t)\, t^2 dt}{\int_0^\infty C(t)\, dt} - \mu^2 \simeq \frac{\Sigma C(t)\, t^2}{\Sigma C(t)} - \mu^2 \quad (7.153)$$

$$\mu^2 = \frac{\int_0^\infty C(t)\, t dt}{\int_0^\infty C(t)\, dt} \simeq \frac{\Sigma C(t)\, t}{\Sigma C(t)} \quad (7.154)$$

In using this method it is important to record the response with considerable accuracy, especially in the 'tail' of the curve as this has a significant effect on the result. The practical impossibility of producing a perfect delta function pulse and distributing it evenly over the column cross-section also introduces errors. These can be largely overcome by measuring the responses at two downstream locations, as shown in Fig. 7.13, when E is obtained as follows (Aris 1959; Bischoff, 1960)

$$Pe_c^{-1} = \frac{E}{UL'} = \frac{\sigma_2^2 - \sigma_1^2}{2\tau_r^2} \quad (7.155)$$

In this expression, $\tau_r = \mu_2 - \mu_1$ is the residence time, and L' the distance between the two measuring points. As an alternative, the results can be interpreted numerically using the time domain method described below.

The experimental arrangement shown in Fig. 7.13 is doubly 'open', that is, the column goes to infinity in both directions. In practice, an arrangement approximating to this should be used, if possible, to avoid end effects. Solutions are also available for cases in which one or other, or both ends are of finite or zero length (van der Laan 1958; Bischoff and Levenspiel 1962; see also Mecklenburgh and Hartland, 1975); these are obtained by incorporation of the appropriate boundary conditions, eqns (7.11–7.14), and are of greater complexity.

The moments method is also applicable to systems involving more than the two parameters τ_r and Pe, for example, the mass transfer coefficient; however, a higher moment is required for each additional parameter, increasing the sensitivity to measurement errors in the 'tail'. A comparison made by Michelsen (1982) of this method with those of weighted moments, transfer function, Fourier transform and time domain showed that the last gave the greatest freedom from error.

The time domain method involves an iterative least squares fitting of the model parameters to the experimental data to minimize the objective function θ, defined by

$$\theta = \int_0^{t_s} [c_{\mathrm{exp}}(t) - c(t, \mathbf{P})]^2 \, w(t) \, \mathrm{d}t \qquad (7.156)$$

where $c_{\mathrm{exp}}(t)$ is the experimental response, $c(t, \mathbf{P})$ the theoretical response with assumed parameter values $\mathbf{P}$ and $w(t)$ is an optional weighting factor. The value of $c(t, \mathbf{P})$ is expressed by means of the convolution integral

$$c(t, \mathbf{P}) = \int_0^{t_s} f(\tau, \mathbf{P}) \, x(t - \tau) \, \mathrm{d}\tau \qquad (7.157)$$

in which the kernel, $f(\tau, \mathbf{P})$ is the impulse response function containing the system parameters $\mathbf{P}$, $x(t)$ is the input signal and τ is a dummy time variable; the integration is taken to a practical upper limit t_s rather than ∞. The minimum of eqn (7.156) is located by solving the set of equations

$$\frac{\delta \phi}{\delta \mathbf{P}} = 0 \qquad (7.158)$$

The solution of this set of equations involves considerable computational effort. However, Michelsen (1982) described an accurate and rapid procedure for the numerical evaluation of the convolution integral in terms of the Fourier transform of the signal and impulse responses, together with the Fast Fourier Transform algorithm for transformation and inversion. The method is best applied to responses measured at two downstream positions. The sensitivity of the method to measurement and model errors was also discussed, and examples were given of its use; for details the original paper should be consulted.

7.6.2.2.2 Backflow model

The foregoing solutions relate to the diffusion model and are therefore strictly applicable only to differential columns. However, they are frequently used also for compartment-type columns, giving the backmixing in terms of the Peclet number. Although the backmixing ratio is obtainable from this using eqn (7.99), it is preferable to use a solution obtained directly from the corresponding difference equation, that is, from eqn (7.39) with the mass transfer term replaced by an accumulation term; this is as follows [cf. eqn (7.148)]

$$(1 + \alpha_k)\, c_{k,n-1} - (1 + 2\alpha_k)\, c_{k,n} + \alpha_k c_{k,n+1} = \frac{V\phi_k}{F_k}\frac{\mathrm{d}c_k}{\mathrm{d}t} \qquad (7.159)$$

Solutions to eqn (7.159) have been obtained by Haddad and Wolf (1967), Roemer and Durbin (1967) and Sawinsky and Hunek (1981) by Laplace transformation; see also Dongaonkar (1989) for a generalized solution.

The solution obtained by Sawinsky and Hunek (1981), which is particularly convenient in use, was obtained on the assumption of an extractor of N stages with tracer injected into stage p and the response measured in stage n. This gave the following result (dropping subscript k)

$$C_m(\theta) = 2N a^{n-p} \alpha \sum_{i=1}^{N} D_i \exp(-z_i \theta) \qquad (7.160)$$

where

$$\begin{aligned}
c_m(\phi) &= \text{dimensionless concentration on stage } m \text{ at time } \theta \\
\theta &= \text{dimensionless time, } tF/(V_t \phi_k) \\
a &= [(1+\alpha)/\alpha]^{1/2} \\
D_i &= (-1)^{i+1} E_i F_i/(1+z_i) \\
E_i &= a \sin(N+1-n)\Psi_i - \sin(N-n)\Psi_i \\
F_i &= a \sin p\Psi_i - \sin(\mathrm{p}-1)\Psi_i \\
z_i &= N[1 + 2\alpha(1 - a\cos\Psi_i)]
\end{aligned}$$

and Ψ_i is the ith root of the equation

$$(a-1)\tan(i\pi/2 - N\Psi_i/2) = (a+1)\tan(\Psi_i/2) \qquad (7.161)$$

This solution was found to give satisfactory results for a case in which $m - p = 27$, provided that $\alpha \geqslant 0.5$; at lower values of α, however, the calculated response curve oscillated strongly at small values of the dimensionless time, θ (Dongaonkar 1989). The cause of this was identified as due to the influence of the term a^{n-p}, which can attain very large values, requiring high numerical precision in the computation of α. Similar results were observed with the solutions of Haddad and Wolf (1969) and Roemer and Durbin (1967), referred to earlier. Dongaonkar (1989) also obtained a more comprehensive solution which is to be preferred in such circumstances.

7.6.2.3 *Applications of tracer methods*

7.6.2.3.1 Continuous phase

The accurate determination of continuous phase axial dispersion coefficients using tracer methods requires careful attention to be given to experimental detail. Thus, in the first place the column must be of adequate length, with the injection and sampling points sufficiently remote from the two ends for the continuous phase to be evenly distributed and the dispersed phase droplets to have attained a steady state size distribution over the measuring length. When using the unsteady state method it is also desirable to ensure that the column is effectively 'doubly infinite' (Section 7.6.2.2, see p. 00), to simplify the interpretation of the data.

A second requirement is that of ensuring adequate mixing of the tracer at the injection point. With rotary agitated columns it may be sufficient to inject into the eye of the impeller. With other types a multi-orifice injector should be used, supplemented where possible by a simple 2 or 4-blade agitator driven by a shaft passing through a sleeve down the centre of the column. This can be rotated quite slowly as the mixing is increased by the droplet swirl; however, care must be taken to avoid the breakage of droplets when these are large, as in non-mechanical columns.

Finally, care must be taken in selecting a suitable tracer. Frequently salt tracers have been used with aqueous systems, e.g. NaCl or KNO_3, which can be analysed by electrical conductivity, or K_2CrO_4, measured spectrophotometrically at 350 nm in 0.1 M H_2SO_4. However, although satisfactory with non-polar solvents, e.g. kerosene and toluene, the use of salts with polar solvents such as butyl acetate and methyl isobutyl ketone has been found to result in a reduced droplet size, with consequent increased dispersed phase hold-up and decreased flood-point (Fitzpatrick *et al.* 1986a). In such cases a dye tracer should be used; although methylene blue has been used for the purpose, it has been found to absorb on to the stainless steel plates in a pulsed plate column and to modify the wetting properties (Fitzpatrick *et al.* 1986b). However tartrazine (Acid Yellow 22), with an absorption peak at 425 mm, did not present any such problems.

7.6.2.3.2 Dispersed phase

The measurement of axial dispersion in the dispersed phase presents much greater difficulty due to the impossibility of adding tracer to a droplet dispersion. Irradiation of a dispersion to stimulate fluorescence or radioactivity might provide a basis for a method, but no such study of continuous injection has been reported.

Some transient response data have been reported for which a pulse of colour tracer was introduced into the dispersed phase distributor and the response measured within the column, or above the interface after coalescence. However, apart from inherent measurement inaccuracies, such methods involve an additional error due to the appreciable column length required for the droplets to achieve a steady-state size and velocity distribution. Further, Levenspiel and Fitzgerald (1983) issued a warning on the application of the 'dispersion' (i.e. backmixing) model to the dispersed phase, and showed that a similar Gaussian spread of tracer is predicted by a 'convective' (i.e. forward dispersion) model which takes into account differences in velocity of the various droplet sizes. It was also shown that the variance, σ, of the residence time distribution, which indicates the spread of the response curve, is proportional to column length, L, for the convective model, as compared with $L^{1/2}$ for the dispersion model.

Evidence that this interpretation is correct in the case of a pulsed plate column was obtained by Aufderheide and Pilhofer (1986) from their data on the measured responses at two downstream locations to a pulse injection of luminescent tracer injected into the dispersed phase distributor. Although further study is required, this result is in accordance with the opinion expressed in Sections 7.1.1 and 7.5.1 (pp. 416, 470) that true backmixing of dispersed phase is absent or negligible in many extractor types.

7.6.3 Discussion

The use of axial dispersion models is open to criticism in that they are too simple to account satisfactorily for the complex flow behaviour which occurs in extraction columns, and that tracer methods may not give valid results in the presence of mass transfer. In this regard Dongaonkar (1989; see also Dongaonkar *et al.* 1991) in a comprehensive study of a Kühni column found that backmixing ratios for the continuous phase obtained from concentration profile measurements under conditions where interfacial effects were absent had a mean value 55 per cent below that given by the unsteady state tracer method with mass transfer taking place. However, agreement was closer, to a mean of 34 per cent, on disregarding stages at the two ends of the column in assessing the results given by the concentration profile method; this was ascribed to conditions in the interface and dispersed phase inlet regions differing from those in the remainder of the column where the hydrodynamical conditions were more uniform.

In the same study it was found that the unsteady state and the continuous tracer injection methods both gave similar values for the continuous phase backmixing ratios in the absence of mass transfer. Surprisingly, however, these were some 30 per cent higher than those obtained by the unsteady state method with mass transfer taking place, which as stated above were themselves higher than those obtained from concentration profiles. A possible explaination of this has been given by Korchinsky (1988) it may be concluded that the use of either tracer method in the absence of mass transfer should give 'safe' values of the backmixing ratio for design purposes, at least for the Kühni column.

Axial mixing data obtained by tracer injection are available in the literature for a number of extractor types. However, most of these have been obtained for relatively small diameter columns and, in some cases, for single phase flow. Data of the latter type must be treated with particular caution, as the flow patterns would normally differ considerably from those obtained with both phases present. Thus, with single phase flow, the fourth of the factors listed in Section 7.1.1 (p. 416) as contributing to axial dispersion, that is, non-uniform velocity profiles, would predominate, together with the third, if mechanical agitation were used. However, with dispersed phase droplets present the first two factors, viz. droplet wake shedding and induced circulatory flow, would have the major influence.

This is illustrated by data obtained by Fitzpatrick *et al.* (1986b) for 73, 151, and 350 mm diameter pulsed plate columns using continuous tracer injection. Under single-phase conditions, values of the backmixing ratio, α_c, decreased with flow rate and increased with column diameter, whereas for two-phase flow under the same pulsation conditions they were independent of column diameter; moreover, they increased sharply at high values of the flow ratio, U_d/U_c.

In the case of the dispersed phase it has been shown that backmixing can be disregarded for most extractor types (Section 7.5.1, p. 470). Forward mixing is likely to prove important, however, but in the absence of data on the velocities of individual droplets in swarms it is not at present possible to make full use of the theory given in Section 7.5.2 (p. 471).

Nomenclature

A_i	constants (i = 1 to 4) in eqns (7.15–7.16) and (7.51–7.52).
a	superficial area of contact of phases; a_j refers to the value for stage j, and $\bar{a}$ to the mean value over the whole extractor, $m^2m^{-3} = m^{-1}$
a_i	defined by eqns (7.18, 7.19, 7.38, 7.54, and 7.55); see also Table 7.3
B	H/d_c
(**C**)	vector of external feed rates to stages [eqn (7.125)]
C_k	solute concentration in phase k, defined by eqns (7.73–7.74 and 7.88–7.89), kg/m^3 or kmol/m^3
c_k	solute concentration in phase k, kg/m^3 or kmol/m^3

c_y^*	value of c_y at equilibrium with c_x, kg/m^3 or kmol/m^3.
c_{ki}^0	concentration of component i in feed of k-phase, kg/m^3 or kmol/m^3
c_{yk}	value of c_y at intersection of straight equilibrium lines (Fig. 7.5), kg/m^3 or kmol/m^3.
d	column diameter, m
d_c	characteristic dimension, e.g. packing size, compartment height, etc., m
d_i	mean diameter of droplets of size range i, m
E_k	effective backmixing diffusion coefficient in phase k, m^2/s
E	extraction factor, $mU_x/U_y = mF_R$
E_{Mk}	Murphree efficiency based on phase k
(**F**)	discrepancy vector with elements $F_{k'i,j}$ (replaces (**0**) on right-hand side of eqn (7.125) when not balanced).
F_k	flow rate of phase k, m^3/s
$F_{k'i,j}$	elements of discrepancy vector (**F**) ($k' = 1$–4), defined by eqns (7.107–7.110).
F_R	flow ratio, F_x/F_y
f_i	volume fraction of droplets of size range i in the hold-up (Section 7.5.2.1, p. 472)
f_k	backflow rate of phase k, m^3/s
H	height of column, m
H_{ox}	H/N_{ox}, height of a 'true' overall transfer unit based on the X-phase, m
h_c	height of compartment or stage, m
J, J'	net flow of solute [eqns (7.72, 7.76, 7.91)]
$K_{i,j}$	distribution coefficient for component i in stage j [eqn (7.106)]
K_i^b	first order breakage rate constants for droplets of size range i, s^{-1}
K_i^c	second order coalescence rate constants for droplets of size range i, m^3/s
k_k	mass transfer coefficient for phase k, m/s
k_{ok}	overall mass transfer coefficient based on phase k, m/s
L, L'	distance between injection and measuring points, or between two measuring points respectively, m
[**M**], [**M**′]	square matrices defined by eqns (7.125) and (7.129) respectively.
m	slope of equilibrium line on (c_y, c_x) coordinates [eqn (7.3)]
N_{ox}, N_{oy}	number of overall transfer units based on X- and Y-phase respectively
N_{ox}^1, N_{oy}^1	number of overall transfer units per stage or compartment based on X- and Y-phase respectively.
N	total number of stages
n	number of stage counted from feed (X-) phase inlet
Pe_k	d_cU_k/E_k, local Peclet number for phase k
Pe_c	LU/E or $L'U/E$, column Peclet No. (Section 7.6.2.2, p. 479)
P	vector of parameters to be evaluated (Sections 7.6.1.2, p. 475, and 7.6.2.2, p. 479)
q	intercept of equilibrium line on (c_y, c_x) coordinates
$R_{ki,j}$	dimensionless mass transfer rate of solute i in phase k and stage j [defined by eqns (7.111) and (7.112)]

Re_d	droplet Reynolds number, $d_i v_i / \nu_c$
$r_{i,j}$	rate of interphase mass transfer of solute i in stage j, defined by eqn (7.105)
S	cross-sectional area of extractor, m^2
t	time since injection of pulse, s
U_k	superficial velocity of phase k, m/s
V	volume of stage, m^3
V_j	volume of stage j [eqn (7.113)], m^3
V_t	total volume of extractor [eqn (7.113)], m^3
v_i	$U_{y,i}/\phi f_i$, velocity of droplets of size fraction i (Section 7.5.2.1, p. 472), m/s
X	$\frac{[c_x - (c_y^I - q)/m]}{[c_x^0 - (c_y^I - q)/m]}$, dimensionless concentration of feed phase (for backflow model, c_y^I is replaced by c_y^{N+1})
$\bar{X}$	c_x / c_x^0 (Section 7.4.4, p. 468)
$(\mathbf{X})$	vector defined by eqns (7.126a and 7.126b)
Y	$(c_y - c_y^I)/[mc_x^0 - (c_y^I - q)]$, dimensionless concentration of extractant phase (for backflow model, c_y^I is replaced by c_y^{N+1})
$\bar{Y}$	c_y / c_x^0 (Section 7.4.4, p. 468)
Y_k	value of Y corresponding to c_{yk}
Z	z/H, dimensionless height
z	height within column from feed phase inlet, m

Greek letters

α, β, γ	defined by eqns (7.7–7.9) and (7.44–7.46); also in Table 7.4.
α', β', γ'	defined by eqns (7.31–7.33)
α_k	backmixing ratio for phase k, i.e. ratio of backflow to net forward flow, f_k/F_k.
θ	variable proportional to time [eqn (7.102)]
θ	tU/L, dimensionless time
(θ)	vector of parameter values.
$\kappa_i\ (i = 2\text{–}4)$	roots of generalized characteristic equation [eqn (7.68)].
$\lambda_i\ (i = 2\text{–}4)$	roots of characteristic equation for diffusion model [eqn (7.17)]
$\lambda_i'\ (i = 2\text{–}4)$	roots of modified characteristic equation for diffusion model [eqn (7.30)]
μ	first moment of tracer response curve
$\mu_i\ (i = 2\text{–}4)$	roots of characteristic equation for backflow model (eqn (7.53)]
ν	kinematic viscosity, m^2/s
ρ_s	solute density, kg/m^3
σ	second moment of tracer response curve.
τ	dummy time variable, s.
τ_r	residence time.
ϕ	fractional volumetric hold-up of dispersed phase.

Subscripts

adj	adjusted value obtained from mass balance
c	continuous phase

d	dispersed phase
e	based on equilibrium line for extractant inlet section of column (Section 7.2.5.2, p. 439)
f	based on equilibrium line for feed inlet section of column (Section 7.2.5.2, p. 439)
f	feed
I	extractant phase inlet end, within extractor
i	number of root (i = 2–4)
i	number of solute (Section 7.4.3, p. 460)
i	droplet size fraction (Section 7.5, p. 470)
j	stage or step number
k	X- or Y-phase
k'	components of discrepancy vector (**F**) [k' = 1 and 2 refer to the X- and Y-phase balances and k' = 3 and 4 to the mass transfer rates and equilibria respectively; see eqns (7.107–7.110)]
M	number of solutes present
m	X-phase inlet stage (Section 7.4.3, p. 460); tracer injection stage (Section 7.6.2.2.2, p. 482)
N	total number of stages in extractor
n	number of stage counted from the X-phase inlet (except in Section 7.4.4, p. 468)
n	Y-phase inlet stage (Section 7.4.4, p. 468)
0	feed phase inlet end, within extractor
x	feed (i.e. X-) phase
y	extractant (i.e. Y-)phase

Superscripts

I	extractant phase inlet, external to column (diffusion model)
i	value at interface
$N+1$	extractant phase inlet, external to extractor (backflow model)
0	feed phase inlet, external to extractor (both models)
o	overall height or number of stages (Section 7.2.5.2, p. 439).
T	transpose of vector

References

ANDERSON, W.J. and PRATT, H.R.C. (1978). *Chemical Engineering Science*, **33**, 995.

ARIS, R. (1959). *Chemical Engineering Science*, **9**, 266.

ARIS, R. and AMUNDSEN, N.R. (1957). *American Institute of Chemical Engineers Journal*, **3**, 280.

AUFDERHEIDE, E. and VOGELPOHL, A. (1986). *Chemical Engineering Science*, **41**, 1747.

BISCHOFF, K.B. (1960). *Chemical Engineering Science*, **12**, 69.

BISCHOFF, K.B. and LEVENSPIEL, O. (1962). *Chemical Engineering Science*, **17**, 245, 257.

CHARTRES, R.H. and KORCHINSKY, W.J. (1975). *Transactions of the Institution of Chemical Engineers*, **53**, 247.

DONGAONKAR, K.R. (1989). PhD thesis. University of Melbourne.

DONGAONKAR, K.R., STEVENS, G.W., and PRATT, H.R.C. (1987). *Institution of Chemical Engineers Symposium Series*, No. 103, 279.

DONGAONKAR, K.R., STEVENS, G.W., and PRATT, H.R.C. (1988). *Proceedings of the International Solvent Extraction Conference*, Moscow (ISEC '88), **II**, 343.

DONGAONKAR, K.R., PRATT, H.R.C., and STEVENS, G.W., (1991). *American Institute of Chemical Engineers Journal*, **37**, 694.

FITZPATRICK, L.M., PRATT, H.R.C., and STEVENS, G.W. (1986a). *Proceedings of the International Solvent Extraction Conference*, Munich (ISEC '86), **III**, 325.

FITZPATRICK, L.M., PRATT, H.R.C., and STEVENS, G.W. (1986b). *Proceedings of the International Solvent Extraction Conference*, Munich (ISEC '86), **III**, 175.

GARG, M.O. and PRATT, H.R.C. (1984). *American Institute of Chemical Engineers Journal*, **30**, 432.

GAYLER, R. and PRATT, H.R.C. (1957). *Transactions of the Institution of Chemical Engineers*, **35**, 267.

GEANKOPLIS, C.F. and HIXSON, A.N. (1950). *Industrial and Engineering Chemistry*, **42**, 1,141.

GEANKOPLIS, C.F., WELL, P.J., and HAWK, E.L. (1951). *Industrial and Enginnering Chemistry*, **43**, 1,848.

GIER, T.E. and HOUGEN, J.D. (1953). *Industrial and Engineering Chemistry*, **45**, 1,362.

HADDAD, A.H. and WOLF, D. (1967). *Canadian Journal of Chemical Engineering*, **45**, 100.

HAMILTON, J.A. and PRATT, H.R.C. (1980). *Proceedings of the International Solvent Extraction Conference*, Liege (ISEC '80), Vol. 1, Session 3, Paper 80-19.

HAMILTON, J.A. and PRATT, H.R.C. (1984). *American Institute of Chemical Engineers Journal*, **30**, 442.

HARTLAND, S. and MECKLENBURGH, S.C. (1966). *Chemical Engineering Science*, **21**, 1,209.

KORCHINSKY, W.J. (1988). *Chemical Engineering Science*, **3**, 349.

KORCHINSKY, W.J. and AZIMZADEH-KATYLOO, S. (1976). *Chemical Engineering Science*, **31**, 876.

KRISHNA, R.(1976). *Letters in Heat and Mass Transfer*, **3**, 153.

KRISHNA, R. and STANDART, G.L. (1976). *American Institute of Chemical Engineers Journal*, **22**, 383.

LETAN, R. and KEHAT, E. (1968). *American Institute of Chemical Engineers Journal*, **14**, 398.

LEVENSPIEL, O. and FITZGERALD, T.J. (1983). *Chemical Engineering Science*, **38**, 489.

LEVENSPIEL, O. and SMITH, W.K. (1957). *Chemical Engineering Science*, **6**, 227.

LIM, C.J., HENTON, J.E., BERGERON, G., and CAVERS, S.D. (1979). *Proceedings of the International Solvent Extraction Conference*, Toronto (ISEC '77), Vol. 1, 248. Canadian Institute of Mining and Metallurgy.

MCSWAIN, C.V. and DURBIN, L.D. (1966). *Separation Science*, **1** (6), 677.

MARQUARDT, D.W. (1963). *Journal of the Society of Industrial Applied Mathematics*, **II**, 431 (1963).

MECKLENBURGH, J.C. and HARTLAND, S. (1967). *Institution of Chemical Engineers Symposium Series* No. 26, (a) p. 115, (b) p. 121, (c) p. 130. Institution of Chemical Engineers, London.

MECKLENBURGH, J.C. and HARTLAND, S. (1969). *Canadian Journal of Chemical Engineering*, **47**, 453.

MECKLENBURGH, J.C. and HARTLAND, S. (1975). *The theory of backmixing.* WILEY, London.

MICHELSEN, M.L. (1972). *The Chemical Engineering Journal*, **4**, 171.

MIYAUCHI, T. and VERMEULEN, T. (1963a), *Industrial and Engineering Chemistry Fundamentals*, **2**, 113.

MIYAUCHI, T. and VERMEULEN, T. (1963b), *Industrial and Engineering Chemistry Fundamentals*, **2**, 304.

MOON, J.S., HENNICO, A., and VERMEULEN, T. (1963). University of California Lawrence Radiation Laboratory Report UCRL - 10928.

MUGELE, R.A. and EVANS, H.D. (1951). *Industrial and Engineering Chemistry*, **43**, 1,317.

NAPHTALI, L.M. and SANDHOLM, D.P. (1971). *American Institute of Chemical Engineers Journal*, **17**, 148.

OLNEY, R.B. (1964). *American Institute of Chemical Engineers Journal*, **10**, 827.

PRATT, H.R.C. (1975). *Industrial and Engineering Chemistry Process Design and Development*, **14**, 741.

PRATT, H.R.C. (1976a). *Industrial and Engineering Chemistry Process Design and Development*, **15**, 34.

PRATT, H.R.C. (1976b). *Industrial and Engineering Chemistry Process Design and Development*, **15**, 544.

PRATT, H.R.C. (1983). *Solvent Extraction and Ion Exchange*, **1**, 669; also (1985) **3**, 377.

PRATT, H.R.C. (1983). *Handbook of solvent extraction* (eds T.C. LO, M.H.I. BAIRD, and C. HANSON), Ch. 3, pp. 108–111. WILEY, New York.

PRATT, H.R.C. and BAIRD, M.H.I. (1983). *Handbook of solvent extraction* (eds T.C. LO, M.H.I. BAIRD, and C. HANSON), Ch. 6. WILEY, New York.

PRATT, H.R.C. and STEVENS G.W. (1991). *Industrial and Engineering Chemistry Research*, **30**, 733.

PROCHAZKA, J. and JIRICNY, V. (1976). *Chemical Engineering Science*, **31**, 179.

RICKER, N.L., NAKASHIO, F., and KING, C.J. (1981). *American Institute of Chemical Engineers Journal*, **27**, 277.

ROD, V. (1964). *British Chemical Engineering*, **9**, 300.

ROD, V. (1965). *Collection of Czecho-Slovak Chemical Communications*, **30**, 3,822.

ROD, V. (1966). *British Chemical Engineering*, **11**, 483.

ROD, V. (1971). *Recent advances in chemical engineering* (ed. C. HANSON), Ch. 7. Pergamon Press, Oxford.

ROD, V., FEI, W.Y., and HANSON, C. (1983). *Chemical Engineering Research and Design*, **61**, 290.

ROEMER, M.H. and DURBIN, L.D. (1967). *Industrial and Engineering Chemistry Fundamentals*, **6**, 120.

SAWINSKY, J. and HUNEK, J. (1981). *Transactions of the Institution of Chemical Engineers*, **59**, 64.

SLEICHER, C.A. (1959). *American Institute of Chemical Engineers Journal*, **5**, 145.

SLEICHER, C.A. (1960). *American Institute of Chemical Engineers Journal*, **6**, 529.

SPENCER, J.L. STEINER, L., and HARTLAND, S. (1981). *American Institute of Chemical Engineers Journal*, **27**, 1,008.
STEVENS, G.W. and PRATT, H.R.C. (1986). *American Institute of Chemical Engineers Journal*, **32**, 2,079.
TOOR, H.L. (1957). *American Institute of Chemical Engineers Journal*, **3**, 198.
VAN DER LAAN, E.TH. (1958). *Chemical Engineering Science*, **7**, 187.
VERGNES, F. (1976). *Chemical Engineering Science*, **31**, 88.
WEHNER, J.F. and WILHELM, R.H. (1956). *Chemical Engineering Science*, **6**, 89.
WILBURN, N.P. (1964). *Industrial and Engineering Chemistry Fundamentals*, **3**, 189.
WILBURN, N.P. and NICHOLSON, W.L. (1965). *American Institute of Chemical Engineers/Institution of Chemical Engineers Symposium Series* No. 1. Institution of Chemical Engineers, London.
YEHESKEL, J. and KEHAT, H. (1971). *Chemical Engineering Science*, **26**, 1, 223; 2,037.

8

SELECTION, DESIGN, PILOT-TESTING, AND SCALE-UP OF EXTRACTION EQUIPMENT

H.R.C. PRATT and G.W. STEVENS

8.1. Selection of extraction equipment

8.1.1 Extractor types

The range of extractor types available for consideration for a given duty is considerable, as is the range of design variables involved. The selection of a suitable extractor therefore involves complex decisions which, in the absence of prior experience with the same system on at least pilot scale, must at best be subjective.

An early, but useful survey of extractor types was given by Morello and Poffenburger (1950). This was extended by Pratt (1954) in the form of an extractor selection chart giving numerical ratings for each type against each of a range of design requirements. An updated version of this chart was given by Pratt and Hanson (1983), who also reproduced in their Fig. 1 a classification chart presented earlier by Lo (1975).

A classification of the more important extractor types, which is further developed below, is given in Table 8.1 together with general indications of the range of industrial applications of each type. Sections 8.3 (p. 514) and 8.4 (p. 573) should be consulted for detailed information on specific extractor types.

It is to be noted that various other types of extractor, apart from those included in Table 8.1, have been described in the literature from time to time. However, these are not included in the present classification as they do not yet appear to have industrial application.

8.1.2 Design requirements

The industrial applications listed in Table 8.1 for the various extractors must be considered as indicative only. A realistic choice of extractor for a given duty must therefore be based on a detailed consideration of a range of design requirements. These are discussed in some detail below, and ratings are given in the extractor selection chart, shown in Table 8.2, to each extractor type against each design requirement.

TABLE 8.1 *Classification and range of application of extraction equipment*

Classification and type	Industrial applications*
A. Gravity-separated extractors	
a. Continuous-contact (i.e. differential)	
1. Non-mechanical columns	
(i) Spray columns	CR
(ii) Baffle-plate columns	C, P
(iii) Packed columns	C, P, N
2. Mechanically-agitated columns	
(i) Pulsed packed columns	C, P
(ii) Raining bucket (Graesser RTL) contactor	C
b. Discontinuous (i.e. stagewise) contact, no interstage settling	
1. Rotary agitated columns	
(i) Rotary disc column (RDC)	C, P
(ii) Multi-impeller columns	C, P, M
2. Reciprocating types	
(i) Pulsed perforated plate column	C, P, Ph, M, N
(ii) Oscillating plate (Karr) column	C, P, Ph
c. Discontinuous (stagewise) contact, with interstage settling	
1. Partial Settling	
(i) Scheibel column	C, Ph
(ii) Asymmetric rotary disc column (ARDC)	C, Ph
2. Total Settling	
(i) Perforated plate column (with downcomers or risers)	P, C
(ii) Rotary film extractor	M
(iii) Vertical mixer–settlers	P, C, M
(iv) Horizontal mixer–settlers	P, C, M, N
B. Centrifugally-separated extractors	
a. Continuous contact	
1. Perforated plate (Podbialniak)	C, Ph, (N)
2. Film-flow (de Laval)	C, Ph, (N)
b. Mixer–settler	
1. Luwesta	C, Ph
2. Robatel	C, Ph

*Key to applications: C = chemical manufacture; CR = fast homogeneous chemical reaction only; P = petroleum refining, petrochemicals; Ph = pharmaceutical; M = hydrometallurgy; N = Nuclear

TABLE 8.2 *Extractor selection chart*

Ref. No.	Design requirements	A. Gravity-separated extractors																		B Centrifugally-separated extractors			
		a. Continuous contact					b. Discontinuous contact no interstage settling					c. Discontinuous contact, with interstage settling								a. Continuous contact		b. Mixer-settler	
		1. Non-mechanical			2. Mechanical		1. Rotary agitated			2. Reciprocating		1. Partial		2. Total		3. Mixer-settlers: Horizontal		3. Mixer-settlers: Vertical					
	Description	Aa1(i) Spray	Aa1(ii) Baffle-plate	Aa1(iii) Packed	Aa2(i) Pulsed packed	Aa2(ii) Raining bucket (RTL)	Ab1(i) Rotary disc	Ab1(ii) Multi-impeller	Ab1(iii) Kühni	Ab2(i) Pulsed plate	Ab2(ii) Reciprocating plate	Ac1(i) Scheibel	Ac1(ii) Asymmetric RDC	Ac2(i) Perforated plate	Ac2(ii) Rotary film	Ac3(i) Pump-settler	Ac3(ii) Mixer-settler	Ac3(iii) Pump-settler	Ac3(iv) Mixer-settler	Ba(i) Perforated plate	Ba(ii) Film flow	Bb(i) Multi-stage	Bb(ii) Single-stage
1	Total throughput:																						
	$\geqq 0.25$ m^3/h	S	S	S	S	S	S	S	S	S	S	S	P	S	D	U	S	U	S	U	U	U	U
	0.25–2.5 "	S	S	S	S	S	S	S	S	S	S	S	S	S	S	P	S	P	S	P	P	P	P
	2.5–25 "	S	P	S	S	P/D	S	S	S	S	S	S	S	S	S	S	S	S	S	S	S	S	S
	25–250 "	S	D	S	S	U	S	S	S	D	P	D	S	S	P	S	S	P	P	P	D[1]	P[1]	P[1]
	>250 "	S	U	S	P	U	P	P	P	U	D	U	P	S	D	E	E	D	D	U[1]	U[1]	U[1]	U[1]
2	NTS:																						
	$\leqq 1.0$	E[3]	S	S	S	S	S	S	S	S	S	S	S	S	S	S	S	S	S	—	—	—	S
	1–5	D[4]	S	S	S	S	S	S	S	S	S	S	S	S	S	S	S	S	S	S	S	S	S
	5–10	U	P	S	S	S	S	S	S	S	S	S	S	S	S	S	S	S	S	P[1]	P[1]	P[1, 2]	S
	10–15	U	D	P	P	P	P	P	P	P	P	P	P	P	P	S	S	P	P	U[1]	U[1]	U[1, 2]	S
	>15	U	D	P	P	D	D	D	D	D	D	D	D	D	D	S	S	D	D	U[1]	U[1]	U[1, 2]	S
3	Physical Properties:																						
	(a) $(\gamma/\Delta\rho g)^{\frac{1}{2}} > 0.60$	P	P	P	S	P	S	S	S	S	S	S	S	P	P	S	S	S	S	E	E	E	E
	(b) $50 \geqq \Delta\rho \geqq 30$ kg/m^3	S	P	S	D	P	U	U	U	U	U	U	U	D	D	D	D	D	D	E	E	E	E
	(c) μ_c and/or $\mu_d > 0.02$ Pa s	D	D	D	D	D	D	D	D	D	D	D	D	D	D	D	D	D	D	D	D	D	D
4	Slow heterogeneous reaction: $k \leq 4 \times 10^{-5}$ m/s	U	D	P	S	D	P	S	S	S	S	S	S	D	D	U	S	U	S	S	S	S	S

5 Slow homogeneous reaction:																						
$t_{1/2} = 30 - 300$ s	D	P	P	P	P	P	P	P	P	P	P	P	P	P	S	E	S	S	U	U	P	P
> 300 s	U	U	U	U	D	U	U	U	U	U	U	U	D	D	P	E	S	S	U	U	U	U
6 Extreme phase ratio:																						
$F_d/F_c < 0.2$ or > 5	S	P	P	P	E	P	P	P	P	P	P	P	S	S	D	E[5]	D	E[5]	S	S	S	S
7 Short residence time	U	U	U	D	U	D	D	D	D	D	U	U	U	U	U	U	U	U	E	E	S	S
8 Ability to handle solids:																						
Trace	S	S	P	P	E	S	S	S	S	S	P	P	P	E	S	S	S	S	D	D	P	P
Appreciable	P	P	D	D	P	P	P	P	P	P	U	D	U	E	D[6]	D[6]	D[6]	D[6]	D	D	P	P
Heavy	D	D	U	U	D	D	D	D	D	D	U	U	U	E	U	U	U	U	U	U	D	D
9 Tendency to emulsify:																						
Slight	S	S	P	D	S	P	D	D	P	P	D	P	P	E	D	D	D	D	E	E	E	E
Marked	P	P	D	U	P	D	U	U	U	U	U	U	D	P	U	U	U	U	S	S	S	S
10 Space limitation:																						
Height	U	D	D	D	E	D	D	D	D	D	D	D	D	E	E	E	U	U	E	E	E	E
Floor area	E	E	E	E	U	E	E	E	E	E	E	E	E	U	U	U	E	E	E	E	E	E
11 Special materials required:																						
Metals (SS, Ti, etc)	E	S	E	E	P	S	S	S	S	S	S	S	S	S	S	S	S	S	S	S	S	S
Non-metals	E	S	E	P	D	D	D	D	D	D	D	D	P	P	P	S	P	D	U	U	U	U
12 Radioactivity present:																						
Weak α, β	E	E	E	S	D	D	D	D	E	D	D	D	S	P	D	E	D	S	D	D	D	S
Strong α, β and/or γ[8]	E	E	E	S	U	U	U	U	E	U	U	U	D	U	U	E[8]	U	D	U	U	D[8]	S[8]
13 Ease of cleaning	E	S	P	P	P	S	S	P	S	S	D	P	P	S	S	S	S	S	D	P	P	P
14 Low maintenance	E	E	E	S	P	S	S	S	S	S	P	P	S	S	P	P	P	P	D	D	D	D
15 Solvent properties																						
Volatile	E	E	E	S	P	P	P	P	S	S	P	P	E	P	P	P	P	P	S	S	S	S
Flammable	E	E	E	S[7]	P[7]	P[7]	P[7]	P[7]	S[7]	S[7]	P[7]	P[7]	E	D[7]	D[7]	D[7]	D[7]	D[7]	P[7]	P[7]	P[7]	P

Key to ratings:

U = unsuitable
D = doubtful; requires detailed study
P = possibly suitable, subject to careful consideration
S = suitable
E = excellent

Notes:

1. Multiple series and/or parallel units can be used.
2. Maximum of three stages per machine (Luwesta).
3. For immeasurably fast homogeneous reactions.
4. For diameters ≤ 150 mm.
5. With recirculation of separated phases.
6. Requires provision for solids removal from settlers.
7. Flameproof drives required.
8. With appropriate shielding and (if required) motor drives external to this.

1. Total throughput. The smallest total throughput range considered is $\leqq 0.25\,m^3/h$ of feed plus extractant phases, corresponding to columns of about 10 cm diameter or less for typical systems in large laboratory or small pilot scale units. Capacity ranges in increments of an order of magnitude each are considered up to a maximum of $250\,m^3/h$; these embrace the majority of industrial processes used in the chemical and petroleum industries, although still larger units have been used in some metallurgical operations, especially in the extraction of copper from dilute mine liquors.

2. Number of theoretical stages (NTS). For some extractor types the height required for a given duty increases with increasing diameter, for example, with rotary agitated columns, due to the need to increase the compartment height with increase in diameter, and with pulsed and reciprocating plate columns due to the reduced performance obtained as the diameter is increased. This is taken into account under the previous heading and is therefore not included under the present one. However, the height of most vertical extractors tends to become excessive if more than 15–20 actual stages are required, necessitating 'doubtful' ratings. This limitation does not apply to horizontal extractors, especially mixer–settlers, although the increased floor area required is taken into account under item 10.

3. Physical properties of system.

(a) General. For packed columns the Sauter mean droplet diameter, d_{32}, at low dispersed phase hold-up is given as follows (Gayler and Pratt 1953)

$$d_{32} = 0.92(\gamma/\Delta\rho g)^{1/2} \tag{8.1}$$

The parameter $(\gamma/\Delta\rho g)^{1/2}$ appears to have relevance for all types of extractor. Thus, for non-mechanical types a value above about 0.6 results in large droplets and a low performance; for mechanical types the effect of a large value of this parameter can generally be overcome by an increase in power input.

(b) Density difference. Phase separation can be slow and interface level control difficult in the density difference range $0.05 > \Delta\rho > 0.03$. With values below 0.03 it would generally be necessary to consider the use of a centrifugal extractor.

(c) Viscosity. A high viscosity of one or other phase (≥ 20 cP) is likely to have a marked effect on performance. However, no simple rule is available for predicting this, and each case must be investigated experimentally.

4. Slow heterogeneous reaction. A slow reaction at the interface will always reduce the mass transfer rate and hence increase the height of column required, or reduce the stage efficiency (Section 7.8.2, see p. 00). This can be minimized by increasing the interfacial area by an increase in agitation

rate, although this would be at the expense of a reduced throughout. The ratings given in Table 8.2 relate to an effective first-order reaction rate constant k_I or $k_{II}c_r$ (where c_r is the reagent concentration) of about 4×10^{-5} m s^{-1}, which would reduce the overall mass transfer rate by about 50 percent.

5. Slow homogeneous reaction. Homogeneous reactions normally result in an increased mass transfer rate, provided that the reaction occurs within the boundary layer or 'film'. However if the reaction is slow, part will occur in the bulk extract phase, outside the 'film', and the residence time must then be large enough for the reaction to go to completion within each stage of a multi-stage extractor. Such slow reaction are rare in liquid extraction operations. Allowance is made in the selection chart, Table 8.2, by relating the residence time to the half-time, $t_{1/2}$, of the reaction. As reactions are normally of higher order than first, a pseudo-half time, suitably averaged to allow for changes in reagent concentration, must generally be used. Thus, for an nth order reaction in which one solute molecule reacts with $n - 1$ reagent molecules, this is given by

$$t_{1/2} = 0.639/(k_n c_r^{n-1}) \tag{8.2}$$

6. Extreme phase ratios. A low value of F_d/F_c, the ratio of dispersed to continuous phase flow, has the effect of reducing the dispersed phase hold-up and hence the interfacial area; a high value, on the other hand, may cause increased continuous phase backmixing, so that in both cases the performance suffers. An important exeption is that of mixer–settlers, in which provision can be made for recirculation of the separated phases back to the mixer.

7. Short residence time. It is sometimes necessary, as when handling labile materials such as penicillin, to minimize the contact time to avoid loss of valuable product. Centrifugal extractors are the most favorable in such cases, and other types are assessed according to the approximate residence time per stage.

8. Ability to handle solids. Some extractor types are liable to blockage by solids and require dismantling for cleaning. Exceptions are the rotary agitated, pulsed and reciprocating plate columns (types Ab1 and 2), which can usually accept appreciable amounts of solids. The only type which can handle large amounts satisfactorily is the rotary film [type Ac1(iii)]; to a lesser extent, the raining bucket [type Aa2(ii)] extractor, and the Luwesta centrifugal extractor, in which special provision is made for solids removal, can be used.

9. Tendency to emulsify. Non-mechanical columns and the raining bucket and rotary film extractors will generally handle satisfactorily

materials with a small tendency to emulsify, although pilot-scale tests should always be considered for materials of biological origin. Mechanical extractors, on the other hand, with the notable exception of the centrifugal types, produce such a fine dispersion that phase separation may be impossible. An attempt is made in the selection chart (Table 8.2) to differentiate between feeds with a slight foaming tendency and those with a marked tendency to emulsify.

10. Limitations of space. Separate entries are given in Table 8.2 for limitations on height and on floor area required. A high rating under both headings infers overall compactness, as with the centrifugal extractors. It was assumed in making the entries that at least five theoretical stages are required.

11. Special materials of construction required. Special materials are likely to be required when handling corrosive materials, and two cases are considered in Table 8.2, viz. special metals or alloys such as titanium or stainless steel, and non-metals, e.g. rubber or enamel lining, or plastics such as polypropylene. In general, the former are applicable to all extractor types, although likely to be expensive. The latter, on the other hand, are unsuitable for mechanically agitated extractors unless the shafts can be carried in external bearings, as with horizontal mixer–settlers. Pulsed columns can also be used with an external pulsing mechanism, e.g. air pulsing.

12. Radioactivity present. Most extractor types can be used with predominantly weak α or β radioactivity present, although the more complex mechanical types are less suitable if frequent cleaning is required. With strong γ radiation present, as in nuclear fuel reprocessing, the requirement is for long maintenance-free operation behind heavy shielding. In this case only non-mechanical columns and mechanically agitated types with external drives, i.e. pulsed columns and mixer–settlers, can be used, although disposable centrifugal extractors have been considered.

13. Ease of cleaning. It is assumed in making entries under this heading that cleaning is required for the removal of solids and contaminant build-up, rather than for the changing of feedstock, which requires only simple washing-out.

14. Low maintenance. This item refers to mechanical maintenance associated with shaft drives, bearings and seals, pulsing mechanisms, etc. and not to the effects of corrosion, which is covered by requirement 11.

15. Solvent properties. Most applications of liquid extraction involve the use of at least one organic solvent, which may have appreciable volatility. In such cases the resulting solvent loss may involve an appreciable replacement cost, and also a hazard if flammable. In general, columns are

easier to contain as compared with mixer–settlers, as a single vent only is required, above the upper interface. However, flameproof drives must be provided for the mechanically-agitated types.

No account is taken under this heading of solvent losses due to entrainment or solubility in the other phase.

8.1.3 Extractor solection chart

In the extractor selection chart shown in Table 8.2 the various extractor types given in Table 8.1 are listed across the top, and the various design requirements, described above, down the left-hand side. Alphabetical ratings are then allocated to each extractor under each design requirement as follows:

U = Unsuitable
D = Doubtful; requires experimental study
P = Possibly suitable, subject to detailed consideration
S = Suitable
E = Excellent (but does not cancel a 'U' rating).

To use the selection chart, the equilibrium data must first be obtained for the given system, and a suitable solvent ratio selected; the total throughput of the two phases and the number of plug flow theoretical stages are then calculated. Any extractors for which a rating of 'U' appears against any of the design requirements are next rejected. The remaining extractors are then listed together with their ratings and the most appropriate types selected for further consideration, giving particular weight to ratings of 'E'.

Stichelmair (1980) has given a comparison of a number of commonly used extractors in the form of a plot of the number of theoretical stages per unit of length against total volumetric throughput of both phases. This is of value in providing a rapid comparison of the relative areas of application of the various extractor types, despite being based on data for a single system, viz. toluene–acetone–water at a phase ratio of 1.5, and it is therefore reproduced in a somewhat modified form in Fig. 8.1. This plot is of particular interest in showing the extreme opposite position occupied by the rotating bucket (RTL) contactor, which has a high performance with low throughput, and the perforated plate column with high throughput and relatively low performance.

Also reproduced, in Fig. 8.2, is a further plot given by Stichelmair (1980) of a cost factor *K* against total throughput, where *K* is defined as the capital cost divided by the product of the total throughput in m^3/h and the NTS. This shows that *K* is 3–5 times greater for the RTL extractor, mixer–settler and RDC than for the other extractor types.

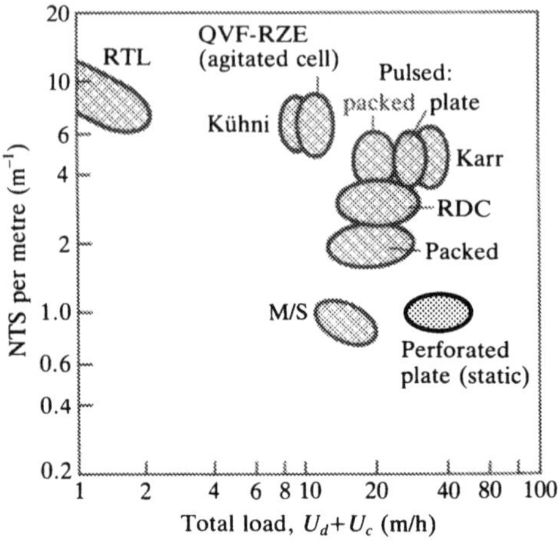

Fig. 8.1. Comparison of extractor performance. System toluene–acetone–water, $U_d/U_c = 1.5$ (Based on Stichlmair 1980).

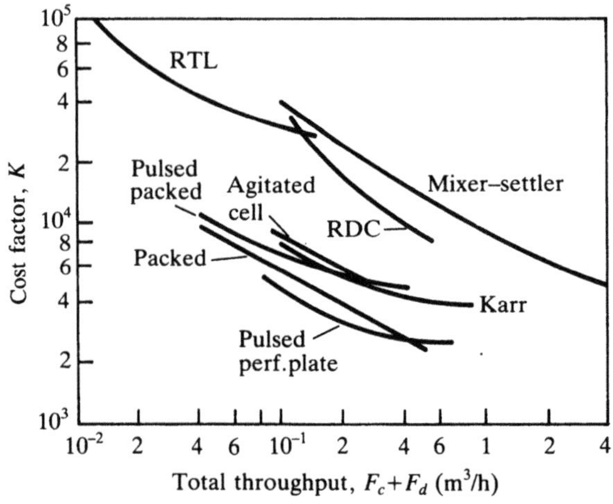

Fig. 8.2. Plot of cost factor K against total throughput. (Based on Stichlmair 1980).

8.1.4 Examples of use of selection chart

Example 1. It is desired to separate the aromatics (BTX) from a 3000 bbl/day petroleum reformate stream using Sulfolane (tetrahydrothiophene-1, 1-dioxide) as solvent in a centre-fed extractor provided with benzene-rich reflux from a subsequent stage of the process.

The design requirements are as follows:

1. Total throughput: $120 m^3/h$ (in enriching section)
2. NTS: 15
3. Physical properties*: $\gamma/(\Delta\rho g)^{1/2} = 0.26$
   $\Delta\rho < 0.3$
   $\mu_c, \mu_d < 20$ cP
6. Phase ratio: 5 : 1
10. Floor area limited (to be in open air)
14. Low maintenance
15. Flammable system.

*Not restrictive (can be disregarded)

Referring to Table 8.2, the following types are considered unsuitable due to entries of 'U', or of 'D' (bracketed)

Ref	Type	Unsuitable due to requirement no(s)
Aa1(i)	Spray	2
Aa1(ii)	Baffle-plate	(1)
Aa2(ii)	RTL	(1)
Ac1(i)	Scheibel	(1)
Ac1(iii)	Rotary film	10, (15)
Ac3(iii)–(iv)	Mixer–settlers (horizontal)	10, 15
Ba, Bb	Centrifugal types	1, 2, (14)

The ratings of the remaining types are listed below

		Design requirement					
Ref.	Type	1	2	6	10	14	15
Aa1(iii)	Packed	S	P	P	E	E	E
Aa2(i)	Pulsed packed	S	P	P	E	S	S
Ab1(i)–(iii)	Rotary agitated	S	P	P	E	S	P
Ab2(i)	Pulsed plate	D	P	P	E	S	S
Ac1(ii)	Reciprocating	D	P	P	E	S	P
Ac1(ii)	ARDC	S	P	P	E	P	P
Ac2(i)	Perforated plate	S	P	S	E	S	E
Ac3(i)–(ii)	Mixer–settler (vertical)	P	D	E	E	P	D

A consideration of these entries suggests that the pulsed and reciprocating plate columns are unsuitable for the large throughput required, and vertical mixer-settlers on grounds of height and safety. The remaining types should all be acceptable; of these, the rotary agitated columns (especially the RDC and Kühni) and the perforated plate column are the most favoured in the industry.

Example 2. An extractor is required for the purification of 30 te/day of uranium, in the form of yellow cake, using 25 vol. per cent tributyl phosphate (TBP) in odourless kerosene (OK) as solvent. The feed solution, obtained by leaching the yellow cake with nitric acid, will contain $100\,kg/m^3$ of uranium in 1.5M HNO_3 and will not be filtered.

The design requirements for the extraction column are as follows (a separate scrub column will be used):

1. Total throughput: $17\,m^3/h$
2. NTS: 10
3. Physical properties*: $(\gamma/\Delta\rho g)^{1/2} = 0.27$
   $\Delta\rho < 0.21$
   $\mu_c, \mu_d < 20$ cP
4. Heterogeneous reaction: $k_1 \simeq 10^{-5}$ m/s
5. Phase ratio: 3*
8. Suspended solids in feed: appreciable
11. Materials of construction: SS*
12. Radioactivity present: weak α, β, (γ)
14. Low maintenance required
15. Solvent: minor hazard only*

*Not restrictive—can be disregarded.

On reference to Table 8.2, the following types are unsuitable due to entries of U, or of D (bracketed).

The remaining extractors comprise the rotary agitated, pulsed plate and reciprocating plate types, all of which should receive further consideration.

Ref	Type	Unsuitable due to requirement no(s)
Aa1(i)	Spray	2
Aa1(ii)	Baffle plate	(2), (4)
Aa1(iii)	Packed	(8)
Aa2(i)	Pulsed packed	(8)
Aa2(ii)	RTL	(1)
Ac1(i)	Scheibel	8
Ac1(ii)	ARDC	8
Ac2(i)	Perforated plate	8
Ac2(ii)	Rotary film	(4)
Ac3	Mixer-settlers	(8)
B(a), B(b)	Centrifugal types	(14)

8.2 Basis of extractor design

8.2.1 Introduction

The design of liquid extraction equipment involves the prediction of two quite separate parameters, namely the cross-section and geometry required to accommodate the specified flows, and the height or number of real stages required to achieve the desired separation. Of these, the former is determined by the hydrodynamical characteristics of the two-phase system, which generally involves a dispersion of droplets of one of the phases in the second, continuous phase. The latter, on the other hand, requires a knowledge of the equilibrium relationship and the rate of mass transfer between the phases.

In the following treatment, an account will be given of appropriate hydrodynamical aspects of two-phase liquid dispersions, and of the rates of mass transfer in these. This will be followed by a description of the fundamental principles involved in extractor design.

8.2.2 Hydrodynamics of liquid extraction equipment

8.2.2.1 *Dispersed-phase hold-up*

On admission of dispersed phase solvent into an extraction column via a distributor, the droplets initially formed, whether or not uniform in size, undergo repeated coalescence and breakage until a final equilibrium size distribution is reached which generally remains constant, provided that the physical properties are not changed by solute mass transfer. As the dispersed phase flow is increased while that of the continuous phase is maintained constant, it is found for most extractor types that the fractional volumetric hold-up of dispersed phase increases approximately linearly with flow rate up to a transition value of 0.10–0.12, after which it increases as U_d to the power of 2–3. This region of rapidly increasing hold-up persists up to an upper transition point, when a second interface forms at the dispersed phase inlet. Further increase in dispersed phase flow is rejected from the column, and the column is said to flood. (In the case of the packed column, a third region may occur above the upper transition point, as will be described later.)

The hold-up, which can be determined by a drainage method involving sudden stoppage of the flows and measurement of the change in interface position (Gayler and Pratt 1951), or by optical density, is found to have a maximum of around 0.50 at zero continuous phase flow which decreases with decrease in solvent ratio, F_d/F_c; the flood point also decreases correspondingly. It is therefore apparent that either hold-up or flood-point can be used as a basis for determining column diameter; as will be seen later, hold-up also has a major influence on the interfacial area, and hence on the mass transfer rate.

The concept of a 'slip', or relative velocity of dispersed phase droplets has proved useful in relating hold-up to the phase flow rates (Gayler and Pratt 1951; Gayler *et al.* 1953). This is defined as follows

$$\bar{v}_s = \frac{U_d}{\epsilon\phi} + \frac{U_c}{\epsilon(1-\phi)} \tag{8.3}$$

where ϕ is the fractional volumetric hold-up of dispersed phase, and ϵ the fractional void volume in the column; the latter is especially relevant to packed columns. By analogy with the work of Steinour (1944) on the rate of settling of solid suspensions, the value of $\bar{v}_0$ can be expressed in the form of the Stokes law (or other drag coefficient) relationship if the buoyancy force is assumed proportional to the difference in densities of the mixture and of the dispersed phase, $(\rho_m - \rho_d)$, rather than of continuous and dispersed phases, $(\rho_c - \rho_d)$. On further expressing ρ_m in terms of ρ_c and ρ_d by the mixture law, eqn (8.3) becomes

$$U_d + U_c \frac{\phi}{1-\phi} = \epsilon\bar{v}_0\phi(1-\phi) \tag{8.4}$$

where $\bar{v}_0$, termed the 'characteristic velocity', is defined as the mean droplet velocity in the extractor when $U_c = 0$ and $U_d \rightarrow 0$. This can be obtained by plotting the left-hand side of eqn (8.4) against $\phi(1-\phi)$, giving a line of slope ϵv_0 through the origin.

Equation (8.4) has been found to correlate hold-up data successfully for a number of extractor types, especially the packed column; further $\bar{v}_0$ in turn can generally be correlated with reasonable accuracy in terms of physical properties and column variables. However, discrepancies have been reported for a number of types of mechanically-agitated extractors, and have led to proposals for modifications to the right-hand side of eqn (8.4). These will be discussed later, in Section 8.3.4 (p. 517), in relation to the various extractor types.

8.2.2.2 *Flood-point*

As described earlier, a continuous increase in flow rate of dispersed (or continuous) phase results finally in flooding of the extractor, which is manifested by the appearance of a second interface at the opposite end to the main interface. This condition corresponds to a maximum in the value of the hold-up, or alternatively, to maximum values of the flows beyond which the hold-up remains constant. This condition can be found by maximising U_c and U_d in eqn (8.4), as follows (Thornton and Pratt 1953)

$$\frac{\mathrm{d}U_d}{d\phi} = 0 : U_c = \epsilon\bar{v}_0(1 - 2\phi_f)(1 - \phi_f)^2 \tag{8.5}$$

$$\frac{\mathrm{d}U_c}{\mathrm{d}\phi} = 0 : U_d = 2\epsilon\bar{v}_0\phi_f^2(1 - \phi_f) \tag{8.6}$$

where ϕ_f, the limiting value of ϕ at the flood-point, is obtained by solving eqns (8.5) and (8.6) for ϕ_f and putting $U_r = U_d/U_c$, thus

$$\phi_f = \frac{(U_r^2 + 8U_r)^{1/2} - 3U_r}{4(1 - U_r)} \tag{8.7}$$

It is to be noted that eqns (8.5)–(8.7) do not apply to the packed column, due to the existence of a third region of operation above the upper transition point. This is also true of the spray column, for which the droplet size is determined at the distributor and is dependent upon the flow rate through the nozzles. Flood-point data for a given extractor (other than the packed and spray columns) can be represented by calculating the value of ϕ_f for each run from eqn (8.7) and plotting U_d against $\phi_f^2(1 - \phi_f)$, when a straight line through the origin of slope $2\epsilon\bar{v}_0$ should be obtained. The value of $\bar{v}_0$ in eqns (8.5) and (8.6) should also be the same as that in eqn (8.4), so that flood-point values can be obtained from hold-up data and vice versa. However, this statment may require revision in the case of extractors for which the right hand side of eqn (8.4) requires modification (Section 8.2.2.1, p. 503).

8.2.2.3 *Droplet size distribution*

The size of droplets formed at a distributor nozzle is generally uniform at lower nozzle velocities, below the jetting range. However, on passage through a column (with the exception of the spray column at low loadings), the size distribution changes as a result of random droplet breakage and coalescence until it finally remains constant, in the absence of changes in physical properties due to mass transfer. The establishment of such a steady state size distribution, which closely matched the experimentally determined distribution, was demonstrated convincingly by Garg and Pratt (1984) for a pulsed plate column by stagewise computation using measured breakage and coalescence constants.

It is convenient in practice to define the droplet size in extraction equipment in terms of a suitable statistical mean diameter. For this purpose the volume–surface (Sauter) mean diameter, d_{32}, is always used, as this defines the diameter of droplet which has the same surface area per unit volume as that of the dispersion as a whole.

The droplet size distribution can be measured photographically in a flat sided cell with transparent windows located above or below the column, as described by Lewis *et al.* (1951). Alternatively the column itself, if transparent and free from packing internals close to the wall, can be photographed directly if fitted externally with a transparent-sided 'box' containing continuous phase, to eliminate distortion; in this case the camera must be focused on to a shallow plane of the dispersion not immediately adjacent to the wall. A millimeter scale located close to the plane of the droplets should also be included in the photographs so that on projecting an enlarged image on to a screen, all droplets within selected areas of the field can be recorded and measured. In the case of non-spherical droplets, the major and minor axes, d_1 and d_2, should be measured and the equivalent diameters, d_e, calculated from the relation

$$d_e = (d_1^2 d_2)^{1/3} \tag{8.8}$$

The Sauter mean diameter is then calculated from the relation

$$d_{32} = \sum_{i=1}^{N} n_i d_i^3 \Big/ \sum_{i=1}^{N} n_i d_i^2 \tag{8.9}$$

where n_i is the number of droplets of mean diameter d_i within a narrow size range i. The effective interfacial area of contact for mass transfer is given by

$$a = 6\epsilon\phi/d_{32} \tag{8.10}$$

where ϕ is the fractional hold-up of dispersed phase.

The interfacial area can also be determined directly by a method involving measurement of the rate of a liquid–liquid chemical reaction such as the hydrolysis of an organic phase consisting of an ester, e.g. *n*-butyl formate, by an aqueous phase containing sodium hydroxide. The rate of such a pseudo-first order reaction is dependent only upon the interfacial area, so that this can be estimated from the experimental results, knowing the reaction rate constant. This method was used by Puranik and Sharma (1970) for a packed column, and gave results in good agreement with those obtained photographically over the range of flow rates covered.

Values of d_{32} for a packed column for packings above a 'critical size' have been correlated with acceptable accuracy in terms of physical properties by Lewis *et al.* (1951) and Gayler and Pratt (1953). For mechanically agitated extractors the droplet size is strongly dependent upon the energy input and column geometry, so that predictions of d_{32} are much less precise. A knowledge of the actual droplet size distribution in extraction columns is also of importance in relation to the interpretation of forward mixing effects (Section 7.5.2, p. 471). These subjects will be discussed in Section 8.3.4 (p. 517), in connection with the various extractor types.

8.2.2.4 *Droplet coalescence*

For successful operation of liquid extraction equipment it is of vital importance that effectively complete coalescence of dispersed phase droplets be obtained at the liquid–liquid interface at one end or other of a column, or within individual settlers in a mixer–settler cascade. It is true that liquid–liquid dispersions are thermodynamically unstable because the free energy associated with the large interfacial area between dispersed and continuous phases can be decreased by coalescence. However, coalescence is hindered by the presence of surface active agents, as, for example, in extracts of natural products and in fermentation liquors; indeed, with such materials, liquid extraction may well be precluded, except possibly by the use of centrifugally separated extractors.

a. *Single droplet coalescence at interfaces*. Coalescence at interfaces takes place by means of two mechanisms, viz. droplet–droplet interaction as the interface is approached, and droplet–interface coalescence at the interface itself. The latter stage has received considerable attention by many workers, who all accept that the coalescence process takes place in five stages, as follows:

(1) approach of droplet to interface, with deformation of both;

(2) damped oscillation of droplet at the interface;

(3) formation of a film of continuous phase trapped between droplet and interface;

(4) drainage of the film against viscous forces until rupture occurs by a small disturbance causing rippling; and

(5) transfer of the droplet contents into the bulk phase.

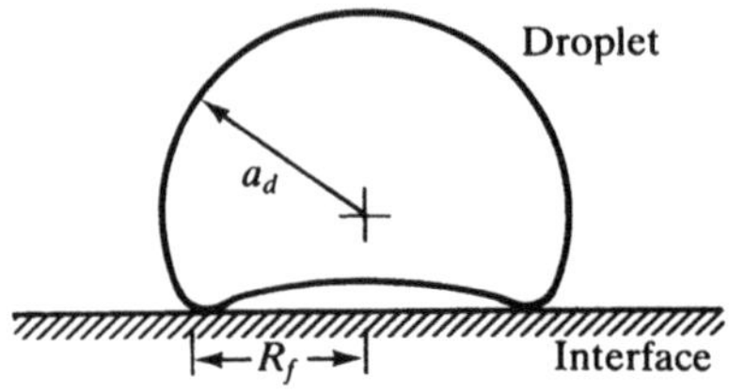

Fig. 8.3. Film drainage area for droplet resting on interface.

Gillespie and Rideal (1956) proposed that the area of the drainage film in stage (4) is that of radius R_f shown in Fig. 8.3, which can be expressed as follows in terms of physical properties

$$R_f = a_d^2\left(\frac{2\Delta\rho g^{1/2}}{\gamma}\right) \tag{8.11}$$

They then incorporated the area of this ring into an expression obtained by Reynolds (1886) to predict the rate of drainage of fluid from between two parallel plates approaching one another, obtaining an expression for the elapsed time from the instant when the droplet comes to rest to that when the film ruptures. They further proposed that rupture was initiated by small disturbances causing ripples to propagate through the film. By thus establishing the probability of rupture they finally obtained the following expression for the fraction of the droplets coalesced

$$\log(N_0/N_{c,0}) = -K(t - t_0)^{3/2} \tag{8.12}$$

where t_0 is the time at which drainage starts, N_c is the number of uncoalesced droplets and $N_{c,0}$ the original number, with K, the coalescence constant given by

$$K = fC_oA_o(6\gamma/a_d\mu)^{1/2} \tag{8.13}$$

where fC_oA_o characterises the disturbance. Gillespie and Rideal (1956) reported data for benzene–water and liquid paraffin–water which showed K and t_0 to increase with temperature.

Elton and Picknett (1957), in an experimental study, found no evidence that the probability of coalescence was zero for a finite film thickness at time t_0, and proposed the alternative relation

$$\log(N_c/N_{c,0}) = -ct^n \tag{8.14}$$

where n had values of 2 for electrolyte solutions and 3 for other systems. Results obtained by other workers tend to favour eqn (8.12), although the value of t_0 is sometimes very small. However, the different exponent obtained in eqn (8.14) for electrolytes is significant, and will be discussed later. An alternative model, based on a concept originally proposed by Bashforth and Adams (1883) and later developed by Princen and Mason (1965) and Hartland (1969), takes account of the deformities of droplet and interface when in close proximity.

b. *Inter-droplet coalescence.* Relatively little experimental work has been carried out on coalescence between droplets because of the difficulty of keeping two droplets together in close contact for long enough to permit film drainage to occur. A theoretical model has been proposed by McAvoy and Kintner (1965) for spheres of equal size, and Princen (1963) considered the case of larger droplets which deform and give a non-uniform film.

Measurements of inter-droplet coalescence and droplet breakage have been reported for a packed column (Hamilton and Pratt 1984) and a pulsed plate column (Garg and Pratt 1984), using a colorimetric method, as described in Section 7.5.2.2, p. 473. These have not, however, been interpreted in terms of a film drainage model.

Fitzpatrick *et al.* (1986) reported a reduced coalescence rate in a pulsed plate column on addition of a salt tracer to systems consisting of water and a polar solvent, e.g. butyl acetate or methyl isobutyl ketone. This manifested itself by an increase in dispersed phase hold-up and a reduction in flood-point, suggesting that only the droplet coalescence rate and not the breakage rate was affected. In this regard it appears that the reduced value of exponent n in eqn (8.14) for electrolyte solutions may be relevant.

c. *Coalescence in settlers.* In general, settling of dispersed phase at the top or bottom of an extraction column is satisfactory for 'clean' systems, without the need to increase the diameter of the settling section for most extractor types. Some work on droplet mechanics in a coalescing dispersion at an interface has been reported by Smith and Davies (1970) and by Lee and Lewis (1967). The latter found that the presence of small amounts of contaminants changed the mechanism from one in which inter-droplet coalescence was frequent, to one in which only droplet-to-interface coalescence took place.

The presence of contaminants such as solids and surface-active materials, such as may occur in plant extracts and other natural products, may necessitate an increase in settler diameter of about 50 per cent or, in extreme cases, the use of an external settler. It must be remembered, however, that an increased settler cross-section reduces the sensitivity of the interface level control system.

The requirements in the case of settlers for horizontal mixer–settlers differ from the above, as they are relatively shallow and involve longitudinal flow of dispersion from inlet, adjacent to the mixer, to the exit. Such a situation was studied by Jeffries *et al.* (1970), who concluded that the dispersion was wedge-shaped due to the progressive coalescence, as shown in Fig. 8.4. Attempts have been made to model this process, e.g. by Jeffries *et al.* (1967),

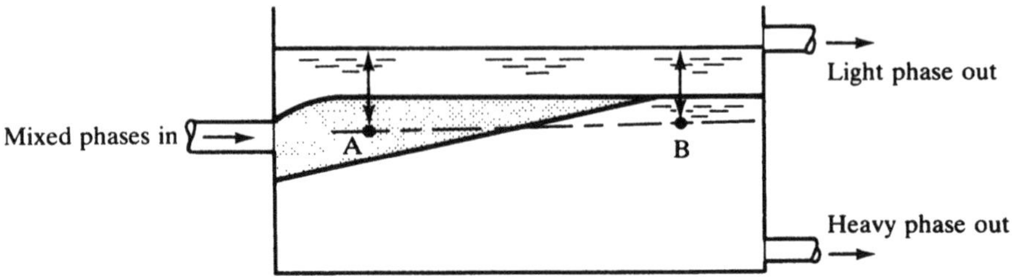

Fig. 8.4. Coalescence 'wedge' in settler.

and Jeffries *et al.* (1970); see also Jeffries and Davies (1971) and Jeelani and Hartland (1988). However, the presence of the mixed phases in the wedge results in longitudinal pressure gradients, e.g. between *A* and *B* (Fig. 8.4), which in turn can lead to non-uniform flow and instability in large units. This can be overcome by the provision of two or more rows of staggered vertical picket-fence baffles near the mixed-phase inlet, followed by horizontal baffle sets (Lewis 1977).

8.2.2.5 *Interfacial tension gradient (Marangoni) effects*

The dramatic effect of an interfacial tension gradient causing motion of liquid surfaces was first reported by Thomson (1855), who observed on placing a small quantity of ethanol on a water surface that 'a rapid rushing of the surface is found to occur outwards from the place where the spirit is introduced'. Marangoni (1865) described a spreading of drops of one liquid on the surface of another liquid, and similar observations were reported by other workers shortly afterwards; a review of this early work has been given by Sawistowski (1971).

The spreading and foam-breaking effect of capryl and other higher alcohols has been known for a long time and has been used to advantage in laboratory and industrial practice. The spreading effect results from a gradient of surface pressure in the opposite direction to that of interfacial tension produced by the additive, as shown in Fig. 8.5. Due to viscous drag, a thin layer of underlaying liquid up to 10 μm thick is carried with the surface, thereby destroying the liquid laminae between adjacent gas bubbles or liquid droplets and breaking the foam or emulsion (Ewers and Sutherland 1952; Davies and Rideal 1963).

Mass transfer from single droplets suspended in a second liquid phase was found by Lewis and Pratt (1953) to give rise to violent kicks and oscillations with many solutes, especially organic ones. This is a result of random fluctuations of solute concentration around the droplet, produced by eddies; thus a high concentration at some point on the surface resulting in a local

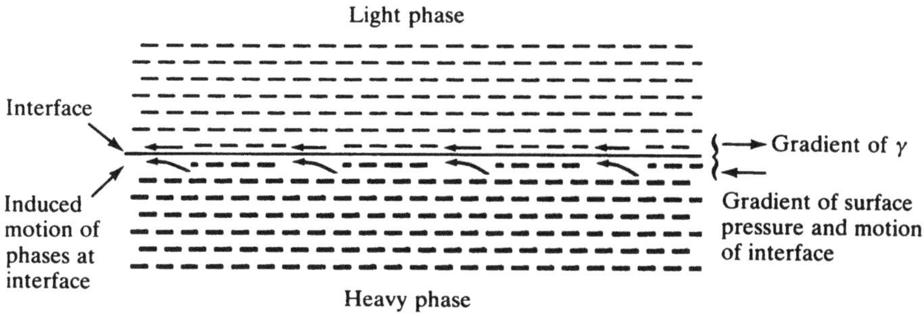

Fig. 8.5. Motion of liquid–liquid interface due to interfacial tension gradient.

reduction in interfacial tension leads to a consequent increase in surface pressure, and hence to rapid motion of the surface to restore stability. This effect can also result in either enhancement or inhibition of coalescence of two adjacent droplets, according to the direction of transfer, as was demonstrated by Groothuis and Zuiderweg (1960). In their experiments, two droplets were held together on nozzle tips in a cell containing continuous phase, as shown in Fig. 8.6. When solute transfer occurred from droplet into continuous phase, the resulting increase in concentration in the film between

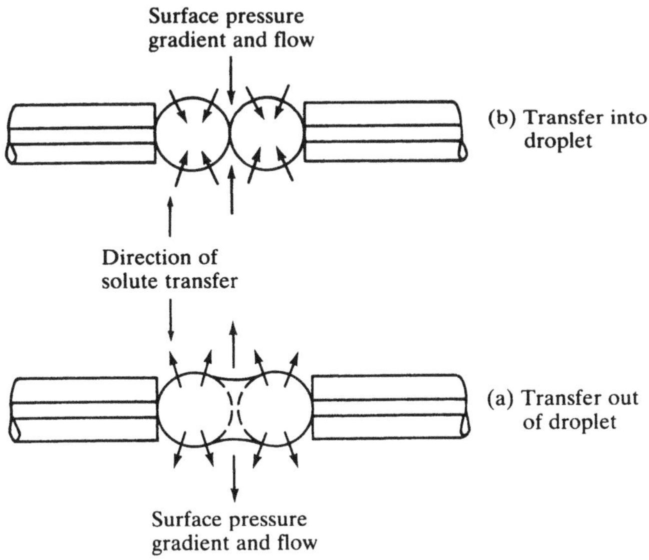

Fig. 8.6. Effect of solute transfer on inter-droplet coalescence: (a) transfer from $d \to c$; (b) transfer from $c \to d$.

the droplets led to a reduction there in interfacial tension; the consequent increase in surface pressure caused the film to be expelled so that coalescence was enhanced. With solute transfer in the opposite direction the surface tension increased locally and the surface pressure decreased, so that continuous phase was drawn in between the droplets and coalescence was inhibited.

The performance of liquid extraction equipment is strongly influenced by such interfacial effects. Thus in columns, mass transfer from dispersed into continuous phase generally results in droplet coalescence, but the resulting larger droplets distort more readily and pass through the column internals more rapidly than they would in the absence of mass transfer, or in transfer from continuous into dispersed phase. As a consequence, the dispersed phase hold-up is reduced, and the flood point increased so that the capacity of the column is increased. However, this effect decreases with depletion of the solute in the dispersed phase, and the column cross-section should be specified to accommodate this. In mixer–settlers, an increase in droplet size and a reduced mass transfer rate is normally observed for transfer from dispersed into continuous phase. However, this can usually be compensated by an increase in agitator speed.

A further result of interfacial instability is an enhancement of mass transfer due to the formation and propagation of roll cells (Sternling and Scriven 1959; Scriven and Sternling 1964). The onset of instability is sometimes related to a critical value of the Marangoni number, *Ma*, defined as follows

$$Ma = \frac{d\gamma}{dc} \frac{(c_i - c)\delta}{\mu D} \tag{8.15}$$

where δ is the boundary layer thickness or penetration depth. However, despite the increase in mass transfer rate, it is generally found that the reduced hold-up and increased droplet size, leading to a reduced interfacial area, more than compensates for this so that the performance of columns often decreases for transfer from dispersed to continuous phase.

An extensive review of this subject has been given by Sawistowski (1971), covering both theoretical and practical aspects of interfacial phenomena.

8.2.3 Mass transfer coefficients

The prediction of the overall mass transfer coefficient on an area basis for dispersed systems requires a knowledge of the external coefficient for transfer between bulk continuous phase and droplet interface, and the internal coefficient for transfer between interface and the inside of the droplets. Much fundamental work on these subjects, both theoretical and experimental, has been reported in the literature for mass transfer to single droplets, and is reviewed in Chapter 2 (p. 40). However, there is much inconsistency in the data, due *inter alia* to droplet distortion, oscillation and degree of internal circulation, which in turn result in the droplets following non-linear flow paths. A further complication results from entrainment and circulation (i.e. backmixing) of continuous phase, which is dependent to some degree on the height of rise or fall of the droplets.

The situation is considerably more complex in the case of the droplet 'swarms' present in practical extraction columns. Thus, interaction effects involving, e.g. changes in flow pattern, in droplet velocity relative to the continuous phase, and in effective interfacial area obviously occur, and become greater with increasing hold-up. However, virtually no work of a fundamental nature has been published on this subject, although some data on apparent overall mass transfer coefficients, generally on a volume basis, have been reported for specific extractor types.

8.2.4 Fundamental design procedures

8.2.4.1 *Columns*

The cross-section of a column required to accommodate given flows can be estimated if suitable correlations are available for flood-point or dispersed phase hold-up in terms of physical properties and geometric variables. If this is the case, the diameter should be selected so that the flows correspond to 50–70 percent of those at the flood-point. Alternatively, a maximum hold-up in the range 15–20 percent can be assumed, with the lower values for flow ratios, U_d/U_c, below 1.0. If such correlations are not available it is necessary to determine the flood-point as described in Section 8.3.2 (p. 514).

Calculation of the length of contacting section required to accomplish a specified duty is more demanding, requiring a knowledge of (i), the interfacial area, as obtained from the dispersed phase hold-up and Sauter mean droplet diameter; (ii), the individual or overall mass transfer coefficients, incorporating if necessary the effects of homogeneous or heterogeneous chemical reaction; and (iii), values of the backmixing coefficients for both phases, and forward dispersion parameters for the dispersed phase. In practice, such information is not yet available for any column type, and design therefore involves scaling up from pilot plant data, as described in Section 8.3 (p. 514).

8.2.4.2 *Mixer-settlers*

The design of mixer-settlers on a fundamental basis involves similar problems. In this case the design of the mixer requires a knowledge of the interfacial area of contact of the phases, the individual or overall mass transfer coefficients, and the flow pattern in the mixing vessel. Similarly, information is required on the coalescence rate in the settler, which cannot be predicted with sufficient certainty. Again therefore, scale-up from pilot-scale equipment is required, as described in Section 8.4; see also Chapter 2 for a more detailed account of mass transfer and Chapter 5 for settler design.

8.3 Pilot testing and scale-up: columns and related types

8.3.1 General

This section considers all extractors in Class A of the selection chart (Table 8.1), with the exception of the mixer–settlers. Where possible, the scale-up procedures are based on considerations of the fundamental hydrodynamic and mass transfer mechanisms. For some extractor types, however, for example the baffle-plate column [type Aa1(ii)] and the rotary film extractor [type Ac2(ii)], this is not possible due to lack of relevant data and in such cases an account is given of the available information on design.

8.3.2 Pilot test procedure

It is assumed in the following that the aim of the pilot testing is to obtain data for scale-up to a commercial-sized unit. The experimental programme must therefore be designed to yield data of a fundamental nature, including that of the effect of solute concentration and direction of transfer. For this purpose the following requirements should be met:

1. The pilot unit should have a throughput of preferably not less than 10 percent of that of full scale, subject to a minimum diameter in the case of columns of not less than 7–10 cm.

2. The length of the column should correspond to that of not less than 5–6 true transfer units (about 3–4 plug-flow transfer units), or, for stagewise extractors, a minimum of 8–10 compartments.

3. The column internals should be designed to maintain the same dispersion characteristics in pilot and full scale units. Rules for three common column types are as follows:

(i) Packed column. The packing size should be of the same type, and preferably the same size as in the full scale unit. However, the latter may not be possible, in which case a smaller size, provided it is above the 'critical size' defined by Gayler *et al.* (1953), can be used subject to suitable allowance being made as described later.

(ii) Reciprocating columns. For both reciprocating plate and pulsed plate columns the plate design (i.e. the hole diameter and free area) and the plate spacing must be kept constant.

(iii) Rotary agitated columns. For rotary disc and multi-impeller types, the column internals should be geometrically similar except for the compartment height, which should be varied as approximately $d_c^{2/3}$ in the absence of specific recommendations for the particular column type.

4. For a non-linear equilibrium relationship the tests should preferably be conducted over two or more concentration ranges, for each of which the equilibrium is approximately linear.

5. If Marangoni effects are likely, e.g. in the transfer of solute from dispersed phase droplets into continuous phase, tests should be conducted over the smallest practicable concentration ranges to ascertain the effect of concentration on performance.

Occasions may arise when some of the foregoing requirements cannot be met simultaneously, necessitating compromise. This is particularly so with non-linear equilibria and with Marangoni effects, since then there would be conflict between requirements 4 or 5 (or both) and 2.

The test procedure using the actual system of interest is as follows:

1. Determine the flooding rate and/or the dispersed phase hold-up as a function of the phase flow rates and, if relevant, the agitation rate. These are then interpreted in terms of the characteristic droplet velocity (Sections 8.2.2.1, p. 503, and 8.2.2.2, p. 505). Alternatively, the flood-points can generally be smoothed by plotting $U_{d(f)}^{1/2}$ against $U_{c(f)}^{1/2}$.
2. Mass transfer runs are then carried out at the same flow ratio and column loading as for the full-scale unit (Section 8.2.4.1, p. 513). The agitation rate, if relevant, should be varied to determine the optimum combination of throughput and performance, noting that increased agitation decreases the droplet size, reducing the height required at the expense of an increased diameter.
3. After determining the best conditions, repeated runs are carried out under these conditions, over different concentration ranges if appropriate. Only runs for which the overall mass balances agree to within 2–3 percent are accepted and used to calculate the mass transfer parameters.

8.3.3 Scale-up Procedures

8.3.3.1 *Allowing for axial dispersion*

The scale-up procedure, assuming that values of the backmixing parameters are known as a function of column diameter, is as follows:

1. Obtain values of the backmixing parameters Pe_k or α_k for pilot and full-scale columns from the appropriate correlations or other source.

2. Calculate the value of the exit extractant phase concentration for the pilot column in dimensionless units, i.e. Y^0 (diffusion model) or Y^1 (backflow model) from the measured concentration.

3. Substitute Y^0 or Y^1 together with the backmixing parameters, column height or number of actual stages, and extraction factor **E** into the simplified generalized expressions given in Section 7.2.5 (p. 437) and Tables 7.2–7.4 (pp. 436–41) and solve iteratively for H_{ox} (diffusion model) or N^1_{ox} (backflow model). The resulting values will include the effect of forward dispersion of the droplet phase, which will be unchanged on scale-up.

4(a). Diffusion model. The value of H_{ox} for the diffusion model can generally be assumed to apply on the full scale. This value is therefore substituted, together with the backmixing parameters and desired Y-phase exit composition for the full scale column, into the simplified model equations of Section 7.2.5 (p. 437), which are then solved for height or length of contacting section.

4(b). Backflow model. The agitation rate for the full-scale unit is scaled in such a way, e.g. on a basis of equal power input per unit volume, that the characteristic droplet velocity, and hence the Sauter mean droplet diameter, are the same as for the pilot-scale unit. The value of N^1_{ox} for the latter is therefore increased in proportion to the increase in compartment height, if appropriate, and substituted together with the values of the backmixing parameters and desired Y-phase exit composition for the full-scale column, into the simplified model equations of Section 7.2.5 (p. 437). These are then solved for the number of stages required.

Further details of the scale-up procedure are given in Section 8.3.4 (p. 517) for each extractor type.

8.3.3.2 *Plug flow basis*

In many cases the appropriate data on backmixing are not available so that it is necessary to scale-up empirically on a plug flow basis. For this purpose, using the diffusion model the value of H_{oxP}, the height of a plug flow transfer unit is calculated from the data for the pilot column, and is assumed to apply to the full scale. For the backflow model the number of theoretical stages, or preferably, the plug flow value of N_{ox}^{1}, is calculated and used as the basis for scale-up.

Equations for the calculation of the plug-flow values of H_{ox} and of the numbers of both theoretical and non-ideal plug flow stages are given in Chapter 6 (p. 345) and are summarized in Table 6.2 (p. 348). However, the generalized equations in Section 7.2.5 (p. 437) are also applicable, so that the computer program described there can also be used for the plug flow case.

Scale-up on a plug flow basis involves greater uncertainty than the method which takes backmixing into account. For this reason it is desirable not to extrapolate appreciably outside the concentration range used in the pilot scale experiments. For some extractor types, specialized plug flow scale-up methods have been developed. These are described in the following section, and should be used in the absence of appropriate backmixing data.

8.3.4 Specific procedures: gravity-separated extractors

8.3.4.1 *Continuous contact, non-mechanical (class Aa1)*

8.3.4.1.1 Spray column

The spray column is the simplest of all the extractor types, comprising an empty shell with the dispersed phase distributor at one end and the interface at the other. The arrangement shown in Fig. 8.7, for light phase dispersed, is due to Blanding and Elgin (1942); this employs expanded end sections which permit higher throughputs without premature flooding. These ends are inverted when the heavy phase is dispersed. The droplets are produced by means of a perforated distributor plate, preferably provided with sharp-edged nozzles of 1.5 to 3 mm inside diameter.

Spray columns have been little used in industry despite their simplicity, high throughput and low cost. This is because the lack of internals results in a strong circulatory flow of continuous phase in the same direction as the droplets, with a return flow around the column wall, as shown in Fig. 8.7(b), (Anderson and Pratt 1978). Consequently, in all but the smallest diameters such as are used in the laboratory, these columns approach the fully-backmixed condition with a performance equivalent to only 1–2 stages. The only situation in which they would find application, therefore, is for mass transfer accompanied by a fast homogeneous reaction, as, for example, in the neutralization of acid or alkali in a solvent by treatment with an aqueous phase containing respectively a strong alkali or acid, for which only a single theoretical stage is required.

Cross-section. In the absence of Marangoni effects, droplet coalescence is

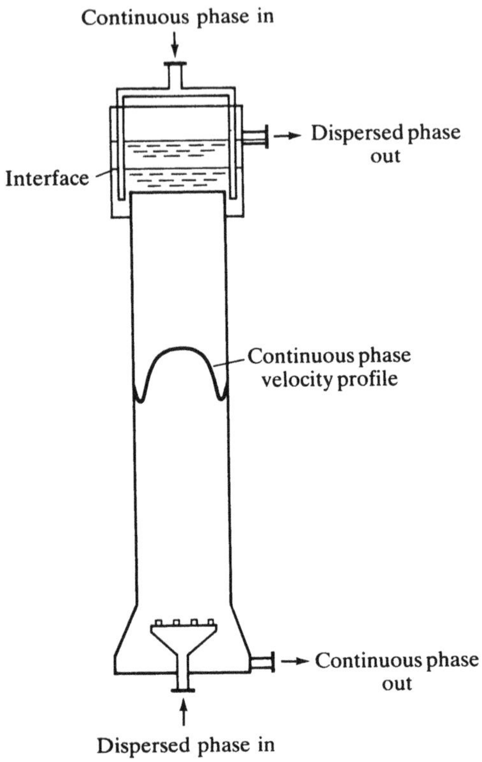

Fig. 8.7. Spray column, showing fully developed continuous phase profile.

usually minimal in spray columns except possibly at high hold-up, so that the droplet size is determined by the nozzle diameter and velocity. A number of relationships have been obtained for the size of droplets formed individually at a nozzle tip from a force balance involving buoyancy, momentum, interfacial tension and continuous phase drag (de Chazal and Ryan 1971; Izard 1972; see also Horvath *et al.* 1978 and Bühler *et al.* 1977). At higher nozzle velocities, droplets are formed above the nozzle tip by break-up of a jet; the velocity at which jet formation starts has been given by Scheele and Meister (1968) as follows

$$v_j = \left[\frac{3\gamma}{\rho_d d_0}\left(1 - \frac{d_N}{d_f}\right)\right]^{0.5} \tag{8.16}$$

where d_f is the diameter of droplet which would form at the nozzle tip in the absence of jetting; this is obtainable from eqn (30) of Part 1 of Scheele and Meister's paper, or eqn (6) of that of Horvath *et al.* (1978). Mayfield and

Church (1952) have recommended the use of jet velocities of 10–25 cm/s; Hughmark (1967) gives the mean droplet diameter as twice that of the jet, d_j, which can be estimated from the following relation (Christiansen and Hixson 1957)

$$d_N/d_j = p + q(\theta)^n \tag{8.17}$$

where $\theta = d_N/(\gamma/\Delta\rho g)^{0.5}$ and the values of p, q, and n are given as follows

$$\theta < 0.785 : p = 1.0; q = 0.485; n = 2$$

$$\theta > 0.785 : p = 0.12; q = 1.51; n = 1$$

Within the column, the droplet velocity relative to an observer is greatest at the centre and diminishes towards the wall due to the continuous phase velocity gradient (Fig. 8.7). Overall, the mean apparent droplet velocity is greater than the droplet terminal velocity, but this decreases at higher hold-up due to suppression of the continuous phase circulation. There is, further, a relatively small decrease in droplet size with increasing nozzle velocity. As may be expected, therefore, the plot of the characteristic velocity expression, eqn (8.4) departs from linearity and is convex upwards. Despite this, Thornton (1956) showed that the flood-point for spray columns can be predicted by substituting for $\bar{v}_0$ in eqn (8.5 and 8.6) the terminal velocity of droplets of size d_N produced at the nozzle at the flood-point. For this purpose he used the expression of Hayworth and Treybal (1950) for droplet diameter, but the more recent relations of de Chazel and Ryan (1971), Izard (1972), and Kumar and Hartland (1984) could be used in a similar manner.

Equation (8.7) indicates that the hold-up, ϕ, attains a maximum value of 0.50 when $U_d/U_c \rightarrow \infty$, i.e. when $U_c = 0$. However, Woodward (1961) found that spray columns used for heat transfer could be operated with dense-packed droplet beds with a hold-up similar to that for a close packing of spheres, i.e. 74 percent. This was confirmed by Loutaty *et al.* (1969), who obtained hold-up values up to 90 percent and found that axial dispersion was much reduced in this range. Perrut *et al.* (1973) found that 50 mm diameter columns could be operated successfully with dense-packed droplets for mass transfer from continuous to dispersed phase, but not in the opposite direc-ion. However, an attempt to scale up a heat exchanger of this type was unsuccessful (Thompson *et al.* 1962).

Height. Mass transfer coefficient data for droplet formation, rise or fall, and coalescence are summarized in Chapter 2, together with recommended correlations. The interfacial area is obtainable from the hold-up and droplet diameter using eqn (8.10), so that the overall volume coefficient can be calculated. For design purposes it would be necessary in principle to use this, together with the axial dispersion coefficients in the simplified diffusion model equations given in Section 7.2.5 (p. 437).

Backmixing of dispersed phase droplets does not occur in spray columns, although there is probably some forward dispersion as a result of the velocity gradients described above. For the continuous phase, Zheleznyak and Landau (1973) have given the following expression for $d_c > 10\,\text{cm}$ and no mass transfer, or transfer from continuous to dispersed phase

$$E_c/\nu_c = 6.5\,Re^{0.987}\,\phi^{0.814}\,\bar{\mu}^{3.89} \qquad (8.18)$$

where ${}_e\bar{\mu} = (\mu_c + \mu_d)/(0.67\mu_c + \mu_d)$, $Re = d_e\bar{v}_s/\nu_c$ and $d_e = d_c(1 - \phi)/(1 + 1.5d_c\phi/d_{32})$.

In practice, these methods are insufficiently accurate for design purposes. Further, extreme care is needed in scaling up from laboratory-sized units due to the over-riding effect of the increasing continuous phase backmixing resulting from the circulating flow, described earlier. The only situation in which scale-up could be reliably carried out would be for fast homogeneous reaction, where it could be assumed that the volume mass transfer coefficient would be unchanged if the column loadings and droplet diameter were kept constant on pilot and full-scale.

8.3.4.1.2 Baffle-plate column

This can be considered as a spray column in which side-to-side or 'disc and doughnut' baffles are used, as shown in Fig. 8.8, to reduce continuous phase backmixing and to induce coalescence and redispersion of dispersed phase. It was once used extensively for the extraction of acetic acid from demethanolized pyroligneous acid from wood carbonization, and has since been used for acetic acid recovery from the dilute liquors obtained, for example, in cellulose acetate manufacture. These columns give a moderate performance, and should be considered for use with 'dirty' systems, i.e. those containing solid matter, or with a tendency to emulsify, as for example when handling biological materials.

In the designs shown in Fig. 8.8, the baffles overlap so that the dispersed phase repeatedly impinges on, and flows across the next plate. A design has been described for the disc and doughnut type in which the discs are rotated at slow speeds against fixed scrapers to remove solids, but this detracts from the essentially simple design of this type of extractor.

In many ways the baffle plate column is a useful type of extractor, but unfortunately very little information is available on its design. Baffle spacings can be of the order of 10–15 cm, with a free space in the baffles of 10–30 percent. It is apparent that the factor which limits the throughput is the continuous phase velocity in the openings in the baffles, in relation to the terminal velocity of the dispersed phase droplets, but unfortunately the size of the latter cannot be predicted at present. Morello and Poffenburger (1950) have given an empirical design chart for this purpose in which the maximum velocity through the opening is plotted against the density difference of the

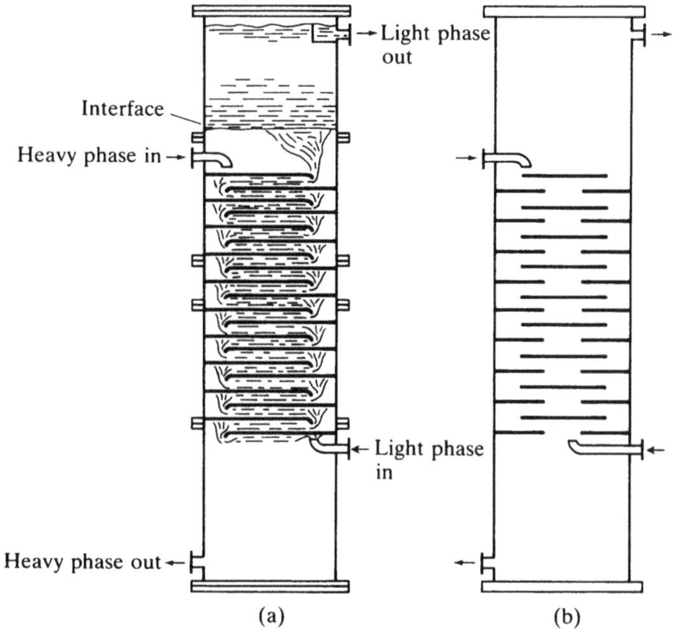

Fig. 8.8. Baffle-plate column: (a) side-to-side type; (b) disc-and-doughnut type.

phases, with the continuous phase density as a parameter. They also state that the contact efficiency is of the order of 0.05–0.10 theoretical stages per plate, which although low, leads to a reasonable column height for many purposes with the plate spacings normally used.

8.3.4.1.3 Packed column

These can be considered as the ultimate development of the previous type in which continuous phase circulation is further reduced by the use of a relatively small size of packing material. They are simple, relatively low in cost and can be used in the largest diameters without loss in performance.

The general arrangement of such columns is shown in Fig. 8.9(a). The expanded ends used for spray columns (Fig. 8.7) can generally be omitted with careful design. Thus, the packing support plate should have a free area at least as large as that of the packing, and should preferably take the form of a 'bar grid', as shown in Fig. 8.9(b). The dispersed phase can be fed through relatively large tubes of 5–7 mm diameter, as the droplets break down quite rapidly to a steady-state size distribution (Hamilton and Pratt 1984; Stevens and Pratt 1986); for a light dispersed phase, these should protrude through the packing support plate, as shown. Similarly, the

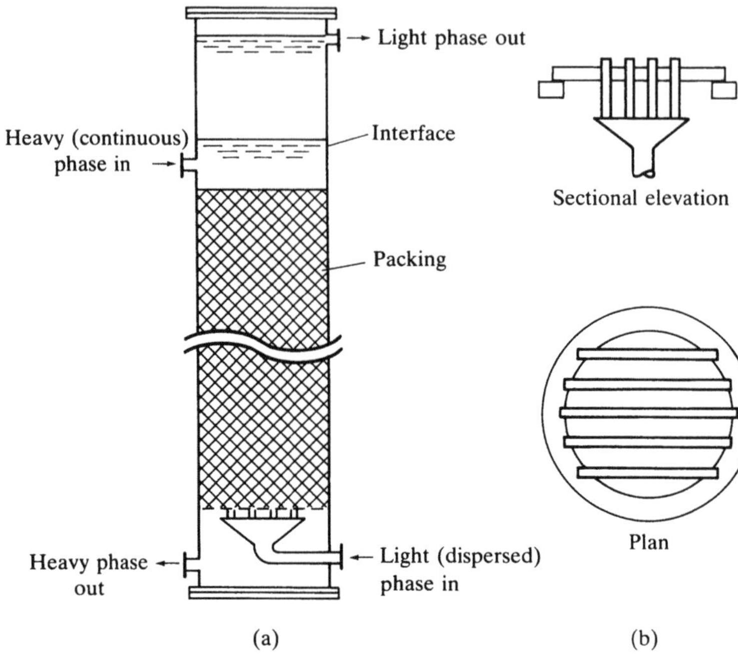

Fig. 8.9. Packed column: (a) general arrangement; (b) packing support grid and distributor for light dispersed phase.

coalescence section can be of the same diameter as the column for 'clean' systems which coalesce easily. If contaminants are present which hinder coalescence, this section may need to be expanded, or even an external coalescer provided. The continuous phase can be distributed by means of a relatively simple sparge pipe arrangement.

For most purposes, with 'clean' systems, random-packed Raschig rings of size greater than the 'critical', as given by eqn (8.21) are suitable, and for aqueous systems can be of ceramic or metal if it is desired to disperse an organic phase as droplets; alternatively, plastics such as polyethylene or polypropylene must be used if an aqueous phase is to be dispersed as droplets. If these materials are interchanged, e.g. using ceramic or metal packing with a dispersed aqueous phase, the dispersed phase then flows over the packing in thick films with few, if any droplets. Under these conditions the column throughput and performance are much less predictable; however, this arrangement is sometimes advantageous for handling biological materials such as plant extracts, which generally contain resinous and surface active substances. In such cases it is also sometimes better to use a more open packing such as Tellerettes, made from carefully selected plastic materials. Similarly, a dispersed organic phase on a hydrophobic plastic packing will also give rise to film flow.

It is sometimes desirable to use packed columns for non-aqueous systems, as in aromatics extraction from paraffins. In such cases droplet dispersions can usually be obtained with the more polar phase continuous, as shown by Moore (1987) for the Sulfolane–hexane system using ceramic rings, which gave hexane droplets with the Sulfolane phase continuous.

Cross-section. The dispersed phase hold-up in packed columns, for droplet flow, can be predicted in the absence of Marangoni effect (Section 8.2.2.5, p. 510) to within about ±15 percent by means of eqn (8.4), using values of $\bar{v}_o$ given by the expression

$$y = 1 - y/z\,[1 - \exp(-z/y)] \tag{8.19}$$

where

$$y = \frac{\bar{v}_o}{V_t}\,[1 - \exp(-23.6\,d_c)]\,\exp\,[(2.12\phi - 0.45)\,M \times 10^5] \tag{8.19a}$$

and $z = \Delta\rho g s/\rho_d V_t$, with d_c in metres. Equation (8.19) was derived by Gayler *et al.* (1953) on the basis of a 'collision theory', which assumes that the droplets are repeatedly stopped by collision with a packing element, after which they accelerate from rest to a velocity below the droplet terminal velocity, V_t; the exponential term in d_c corrects for droplet bypassing at the wall and is negligible for columns of 0.2 m diameter and above. The term containing M, the Morton No., was added by Moore (1987) to account for a progressive increase in $\bar{v}_0$ above 10 percent hold-up which was observed with the non-aqueous system Sulfolane–hexane (see also Moore *et al.* 1990); this was considered to be due to the onset of 'chain flow' of the droplets, with a reduction in drag coefficient and consequent increase in velocity. For normal aqueous systems this term has a value close to 10^{-5} and can be disregarded.

In eqn (8.19), $\bar{s}$ represents the mean path length traversed by the droplets, given by

$$\bar{s} = 0.38d_p - 0.92(\gamma/\Delta\rho g)^{1/2} \tag{8.20}$$

By setting $\bar{s} = 0$ and solving for d_p, an expression is obtained for the 'critical' packing size, as follows

$$d_{p(\text{crit})} = 2.42\,(\gamma/\Delta\rho g)^{1/2} \tag{8.21}$$

This corresponds to the condition for which the mean droplet diameter is equal to the size of void in the packing. Although flow of dispersed phase can still occur, with deformation of the droplets, the hold-up is much

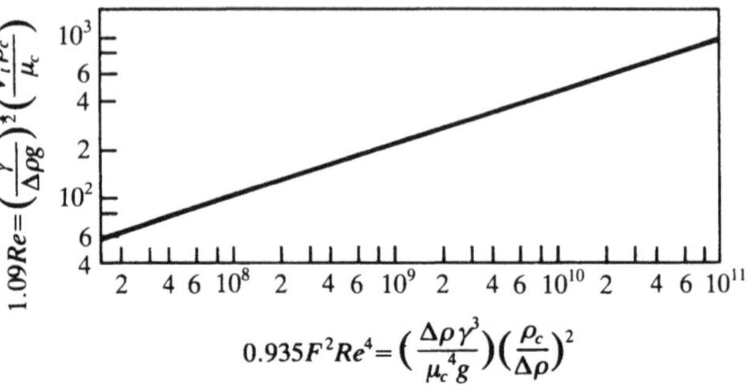

Fig. 8.10. Plot to determine droplet terminal velocity.

increased so that premature flooding occurs and the next larger size of packing should be used.

It has been found that the appropriate value of the terminal velocity, V_t, to be used in eqn (8.19) is that of solid spheres rather than droplets, presumably because the droplets accelerate to a maximum velocity below that at which deformation sets in; this can be obtained in terms of physical properties from Fig. 8.10. Figure 8.11 shows a plot of eqn (8.19) in the form y vs z, from which $\bar{v}_0$ may be determined; alternatively, eqn (8.19) can be solved iteratively for $\bar{v}_0$. The hold-up is then obtained by substitution in eqn (8.4) and solving for the root of the resulting cubic equation which lies between 0 and 0.50, or alternatively by means of the nomogram shown in Fig. 8.12.

As stated earlier (Section 8.2.2.2, p. 505), packed columns generally flood at flow rates above those corresponding to maximum hold-up.

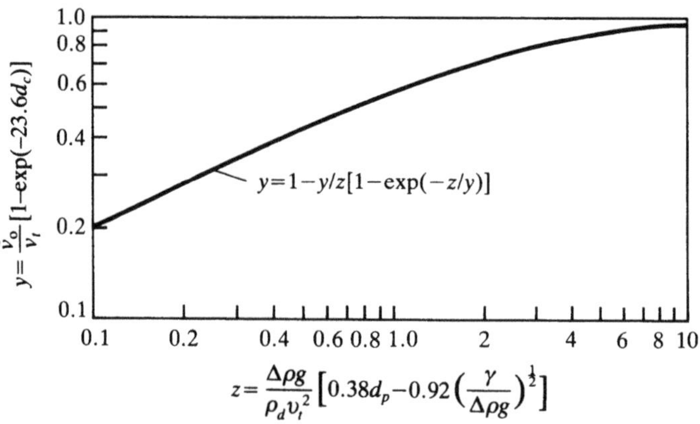

Fig. 8.11. Plot of eqn (8.19) in form y vs z.

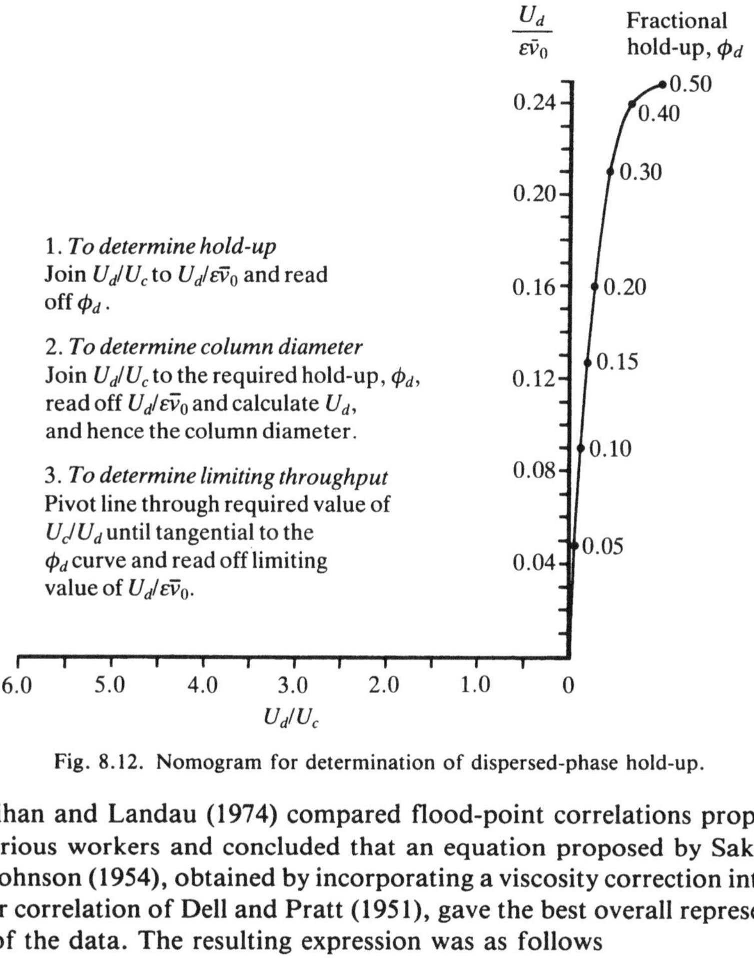

Fig. 8.12. Nomogram for determination of dispersed-phase hold-up.

Houlihan and Landau (1974) compared flood-point correlations proposed by various workers and concluded that an equation proposed by Sakiadis and Johnson (1954), obtained by incorporating a viscosity correction into an earlier correlation of Dell and Pratt (1951), gave the best overall representation of the data. The resulting expression was as follows

$$\left[1 + 0.835\left(\frac{\rho_d}{\rho_c}\right)^{-0.25}\left(\frac{U_{d,f}}{U_{c,f}}\right)^{0.50}\right] = C\left[\left(\frac{U_{c,f}^2 a_p}{g\phi^3}\right)\left(\frac{\rho_c}{\Delta\rho}\right)\gamma^{0.25}\right]^{-0.25} \tag{8.22}$$

where for Raschig rings $C = 0.92\,\epsilon^{0.0068}/a^{0.043}$, with γ in dynes/cm and μ_c in centipoises.

For design purposes, the column diameter can be predicted with reasonable assurance in the absence of Marangoni effects either in terms of hold-up or flood-point. For the former, using eqn (8.19) in conjunction with eqn (8.4) or Fig. 8.12, a working value of the hold-up of 0.20 can be assumed for values of the flow ratio, $U_d/U_c \geqq 1.0$, decreasing progressively to about 0.12 for flow ratios down to 0.2. In terms of flood-point, an operating throughput of 50–70 per cent of the flooding rate should be assumed.

Mass transfer. The rates of mass transfer to or from droplets within packings differ from those for spray columns due to the tortuous paths followed, with repeated deceleration and acceleration, distortion, and coalescence and breakage. The Sauter mean droplet diameter in the absence of Marangoni effects, determined photographically, has been correlated to within ±20 percent as follows for packings above the critical size (Lewis *et al.* 1951; Gayler and Pratt 1953)

$$d_{32} = 0.92(\gamma/\Delta\rho g)^{0.50}(\bar{v}_0\epsilon\phi/U_d) \tag{8.23}$$

$$= d_{32}^0(\bar{v}_0\epsilon\phi/U_d) \tag{8.23a}$$

where d_{32}^0 is the 'characteristic droplet diameter', i.e. the value of d_{32} when $U_c = 0$ and $U_d \rightarrow 0$. The interfacial area is then given by

$$a = 6\epsilon\phi/d_{32} \tag{8.24}$$

$$= 6U_d/(d_{32}^0\bar{v}_0)\ [\text{from eqn}(8.23\text{a})] \tag{8.24a}$$

Axial dispersion of the phases must also be taken into account in assessing mass transfer performance. Gayler and Pratt (1957a) showed that backmixing of dispersed phase droplets does not occur; however, forward dispersion due to differing droplet sizes and velocities does take place, as can be shown by the spread of the colour band obtained on injecting a pulse of coloured solvent into the dispersed phase distributor. Backmixing of continuous phase is appreciable, however, and Pratt and Anderson (1977) recorrelated data obtained by Vermeulen *et al.* (1966) for 12.5 and 19 mm Raschig rings as follows (see also Watson and McNeese 1972)

$$\epsilon Pe_c = 0.00786\left(\frac{d_eU_c}{\nu_c}\right)^{0.47}\left(\frac{U_c}{U_d}\right)^{0.43}\left(\frac{d_e}{d_{32}}\right) \tag{8.25}$$

where $d_e = 4\epsilon/a$, the equivalent diameter of the packing voids.

An attempt was made by Pratt and Anderson (1977) to re-interpret the overall mass transfer coefficient data of Gayler and Pratt (1957b) for the transfer of acetone from water into four different solvents in a 10.2 cm diameter column packed with 12.5 mm Raschig rings. For this purpose, individual 'film' coefficient data, obtained by Gayler and Pratt (1957a) for the system ethyl acetate–water by the mutual saturation technique of Colburn and Welsh (1942), were expressed on a corrected area basis with allowance made for backmixing in the case of the continuous phase coefficients. They were then fitted to the expression,

$$k_k/\bar{v}_0 = C_k(d_{32}^0\bar{v}_0/\nu_k)^{0.5}(\nu_k/D_k)^{0.5} \tag{8.26}$$

giving values of the constant C_k of $C_c = 0.993$ for the continuous phase and $C_d = 0.870$ for the dispersed phase; the exponent of 0.5 on the Schmidt group, (ν_k/D_k), assumes the applicability of the penetration theory to both phases.

Equation (8.26) was then used to estimate the individual coefficients for the acetone systems, and these were combined to give the 'true' values for the overall coefficients k_{oc}, and of H_{oc}, as follows

$$k_{oc} = [1/k_c + m/k_d]^{-1} \tag{8.27}$$

$$H_{oc} = U_c/k_{oc}a \tag{8.28}$$

Finally, predicted values of H_{oc} were compared with the experimental valued obtained from the acetone data of Gayler and Pratt (1957b). These were found to be generally low, ranging from −60 percent to −13 percent for Chlorex ($\beta\beta'$ dichlorodiethyl ether), −36 percent to +22 percent for ethyl acetate, and −10 percent to +8 percent for butyl acetate; the results for toluene were unsatisfactory, probably because the packing was too close to the critical size for this solvent.

It is apparent that reliable prediction of the performance of packed columns is not at present possible, even for simple systems. Pilot scale tests should therefore always be considered, and are mandatory for 'dirty' systems.

Design and Scale-up. Using Raschig ring packing, a size greater than the critical, as given by eqn (8.21), and preferably not smaller than 12.5 mm should be used. To minimize excessive voidage at the wall, the column should have a diameter of not less than 8 times the packing size, i.e. at least 10 cm, and a packed length of 1.5–2 mm.

Initially it is desirable, and essential if Marangoni effects are likely, to determine the flood-point and, preferably, the hold-up for flow ratios close to the desired operating value and to compare these with the appropriate correlations. Mass transfer runs are then carried out at the desired column loading (Section 8.2.4.1, p. 513) and flow ratio and the value of H_{ox} calculated, as described in Sections 8.3.2 (p. 514) and 8.3.3 (p. 516), using the diffusion model. This value can be assumed to apply on the full scale if the same packing size is used, but may require to be adjusted if a larger size is to be used. No data are available on 'true' H_{ox} values for Raschig ring packings above 19 mm in size, but Gayler and Pratt (1957b) reported plug flow values for the butyl acetate–acetone–water system which showed increases above those for 12.5 mm rings of up to 12 percent, 25 percent, and 50 percent respectively for 19.0, 25.4, and 38.0 mm rings, according to the throughput and flow rates. However, these almost certainly reflect the effect of increasing backmixing with the larger sizes, and it may be expected that there would be much less change in the 'true' H_{ox} values.

Particular care must be taken if Marangoni effects occur, as these will diminish towards the low concentration (i.e. extractant inlet) end of the column. The throughput and column diameter must therefore be based on conditions at the latter end, where the flood point would be lower. Similarly, the value of H_{ox} is normally higher, by a factor as high as two, when interfacial effects are strong, but will diminish to normal values at low solute concentration. It must be remembered that eqn (8.25), for continuous phase backmixing, may not be reliable when these effects occur, and it may then be necessary to scale-up on a plug flow basis.

The scale-up procedure is much more subjective in cases where special packing materials are required. This is especially so if film rather than droplet flow has to be used, as in the extraction of biological materials.

8.3.4.2 *Continuous contact, mechanically agitated (Class Aa2)*

8.3.4.2.1 Pulsed Packed Column

The performance of Raschig ring packed columns can be enhanced, at the expense of a reduced throughput, by mechanical pulsation of the continuous phase. This is a result of an increase in shear forces and consequent reduction in size of the dispersed phase droplets so that the interfacial area, and hence the mass transfer rate, is increased. Columns of this type have been built in sizes up to 3.0 m with throughputs in excess of 200 m^3/h (Simons 1978, 1983), and can be scaled up in diameter without loss in performance.

The construction of the column and internals is similar to that of the normal packed column [Section 8.3.4.1(iii)], with the addition of a pulse generator. A packing size of 25 mm is generally used, and this can be made of any suitable material, e.g. ceramic, metal, or plastic; however, except possibly when using heavy metallic packings, a perforated packing restraint plate should be fitted at the top to prevent movement and consequent alignment of the packing elements due to the pulsation, forming 'chimneys' which reduce performance.

The pulse generator must be robust, as very large quantities of liquid must be moved. Mechanically driven units can take the form of a valveless plunger pump, either separate or mounted directly on the base of the column, or of a cam-driven diaphragm or bellows; typical arrangements are shown in Figs 8.13(a)–(c). Alternative methods comprise the hydraulic pulsator, driven by a pump and rotary valve (Simons 1978, 1983), and the pneumatic pulse generator, shown in Figs 8.13(d) and (e). The power requirements for pulsation are reportedly of the same order as for the pulsed plate column [Section 8.3.4.4.(i)].

Hold-up and flood-point. Hold-up data for pulsed packed columns can be correlated in terms of the characteristic velocity by means of eqn (8.4); this is expressed in terms of the pulsation velocity, fA, as $\bar{v}_{fA}$. Two distinct regions of operation have been observed, as follows (Simons 1983).

Region 1. High pulse velocity, small droplet size, $\bar{v}_{fA}$ strongly dependent upon d_{32} and almost independent of packing size.

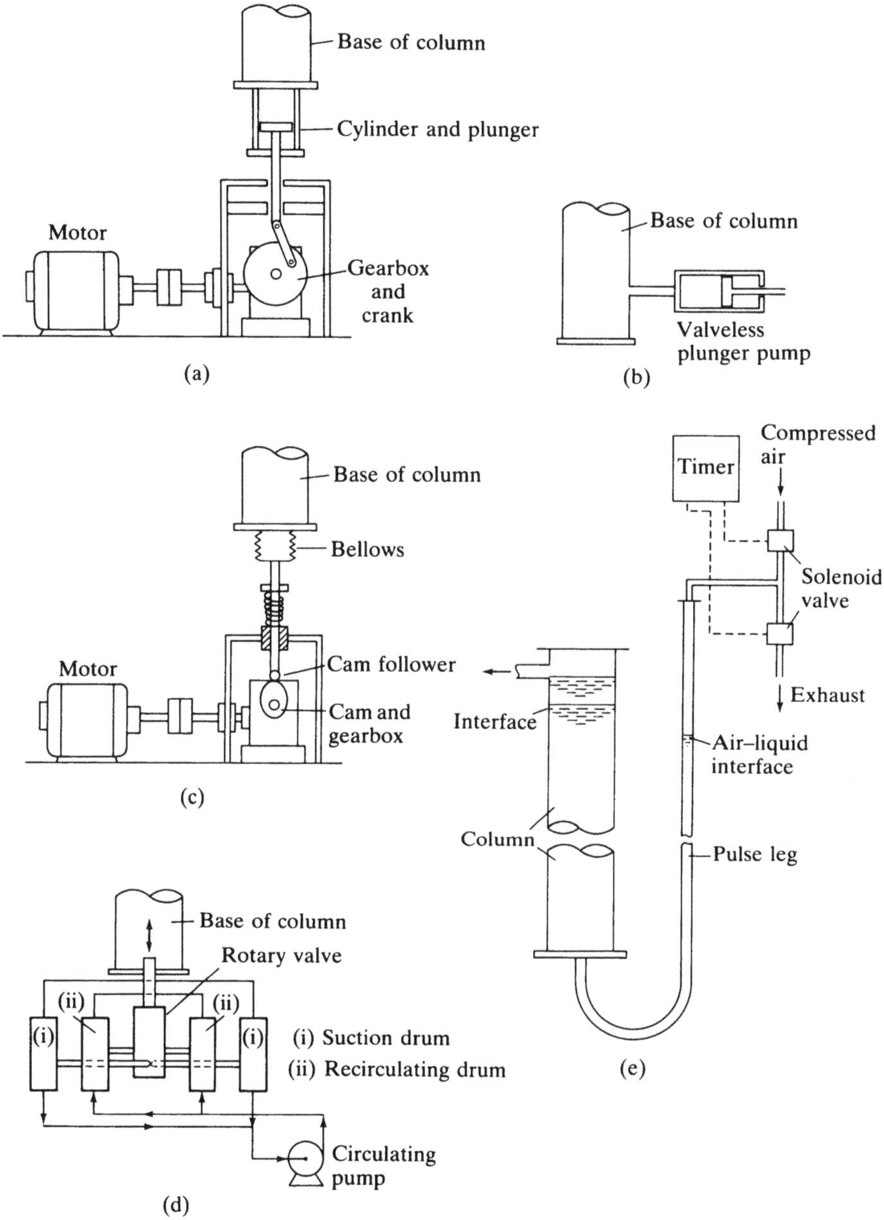

Fig. 8.13. Pulsation generators: (a) and (b) valveless plunger pump; (c) cam-operated bellows; (d) hydraulic pulsator; (e) air-operated pulse generator.

Region 2. Low pulse velocity, large droplet size, $\bar{v}_{fA}$ less dependent upon d_{32} but more dependent upon packing size.

A satisfactory theoretical model for the prediction of $\bar{v}_{fA}$ has not been obtained, but the following dimensional correlations, in SI units, have been proposed (Spaay *et al.* 1971; Simons 1983):

Region 1, $L \leqq 0.406$

$$\bar{v}_{fA} = 6.32 \times 10^{-3}\, d_{32}^{0.73}\, \Delta\rho^{0.82}\, (f^2 A)^{-0.25}\, [a_p(1-\phi)]^{-0.18}\, \bar{\mu}_c^{-0.35} \quad (8.29)$$

Region 2, $L > 0.406$

$$\bar{v}_{fA} = 2.57 \times 10^{-3}\, d_{32}^{-0.06}\, \Delta\rho^{0.56}\, (f^2 A)^{-0.11}\, [a_p(1-\phi)]^{-0.61}\, \mu_c^{-0.35} \quad (8.30)$$

where

$$L = d_{32}^{0.79}\, \Delta\rho^{0.26}\, (f^2 A)^{-0.14}\, [a_p(1-\phi)]^{0.43} \quad (8.31)$$

and

$$d_{32}/d_p \leqq 1.37\, \Delta\rho^{-0.5} \quad (8.32)$$

The flood-point for these columns is also dependent upon the pulse velocity. Few data are available on this subject, but an approximate estimate of the flooding rate can be obtained from eqns (8.5 or 8.6 and 8.7), using the value of v_{fA} given by eqn (8.29 or 8.30) Simons 1983).

Mass transfer. The prediction of packing height requires a knowledge of the interfacial area of contact, the backmixing parameters, and the individual mass transfer coefficients as for the unpulsed column [Section 8.3.4.1 (iii)]. The interfacial area, given by eqn (8.24), requires in turn a knowledge of the hold-up, obtained from eqn (8.4) together with the value of $\bar{v}_0$, and the mean droplet diameter, for which Spaay *et al.* (1971) have obtained the following semi-empirical correlation (see also Simons 1983)

$$\frac{1}{d_{32}} - 0.72\left(\frac{\Delta\rho g}{\gamma}\right)^{0.50} = 6700\left(\frac{1-\phi}{\phi}\right)^{0.95}\left(\frac{fA\rho_c}{\mu_c a_p}\right)\left(\frac{\mu_c^2 a_p}{\gamma\rho_c}\right)^{0.5}\left(\frac{\gamma a_p^2}{\Delta\rho g}\right)^{0.23} \quad (8.33)$$

It is probable that backmixing (but not forward dispersion) of the dispersed phase is absent in these columns, as for the unpulsed type. For the continuous phase, Spaay *et al.* (1971) showed that the backmixing could be expressed in terms of contributions due to the pulse rate and to the phase flow rates respectively, as follows

$$Pe_c^{-1} = E_c/d_p U_c \simeq 1.2[(\Psi_1 fA/U_c) + \Psi_2 B] \quad (8.34)$$

In this expression Ψ_1 has a value of 0.18 for 25 mm Raschig rings at a pulse amplitude A within the packing of 1.02 cm, increasing linearly to

$\Psi_{1,\max} = 0.51$ for $A = 3.7$ cm, above which it remains constant. The constant B has a value of 0.31 for the same packing, and Ψ_2 decreases linearly from 0.6 to 0.2 with increase in $(\Psi_{1,\max} fA/\Psi_1) \times 10^2$ from 3.0 to 10.0. The constant 1.2 in eqn (8.34) represents the influence of the dispersed phase flow on Pe_c.

Design and Scale-up. Mass transfer coefficient data for pulsed packed columns are not available, and design must be effected by scale-up from pilot plant results. The procedure for this follows that described in Sections 8.3.2 (see p. 00) and 8.3.3 (see p. 00). Thus, flood-point data are obtained as a function of flow rates and pulse velocity, fA, and interpreted in terms of the characteristic velocity, $\bar{v}_{fA}$, using eqns (8.5 and 8.7); these are compared with the values predicted by eqns (8.29–8.31). Mass transfer runs are then carried out at selected pulse velocities within the range 50–70 percent of flooding, and values of H_{ox} calculated using values of the backmixing parameter given by eqn (8.34). Finally, scale-up is effected on the basis of the resulting H_{ox} values, using the pulse velocity giving the optimum combination of column diameter and height.

8.3.4.2.2 Raining bucket (Graesser RTL) extractor

This extractor, now termed the RTL, was developed to handle systems of low density difference and low interfacial tension with marked emulsification tendencies. It consists of a rotor mounted on a horizontal shaft within a cylindrical shell, as shown in Fig. 8.14. The rotor carries a series of circular baffles on which are mounted cylindrical buckets, open in the direction of rotation. In operation, the extractor is filled with the two immiscible liquids which pass through countercurrently with the interface controlled at the mid-plane. Slow rotation of the shaft therefore causes each phase to be picked up in turn and cascaded through the other phase. Annular gaps are provided between the rotor baffles and outer shell to permit longitudinal flow of the phases. The available sizes are stated by Coleby (1983) to range from pilot units 0.1 m diameter × 1 m long in glass or stainless steel, to full scale units up to 2 m diameter × 8 m long in mild or stainless steel. Phase ratios up to 6 : 1 can be used, and with special design modifications, up to 20 : 1.

For a given extractor, the two major design variables are the rotor speed and the width of the annular gap between rotor and outer shell. The rotor speed must be adjusted according to the bucket capacity and the settling properties of the dispersion. The droplets are relatively large in comparison to those in other extractor types, but despite this the performance, expressed as number of theoretical stages per metre length, is high as shown by the data of Stichlmair (1980) in Fig. 8.15 for a pilot unit. On the other hand, the throughput is relatively small, as shown by Fig. 8.1.

The annular gap has little effect on droplet size and interfacial area, but has a major influence on axial dispersion. Thus, a narrow gap gives reduced

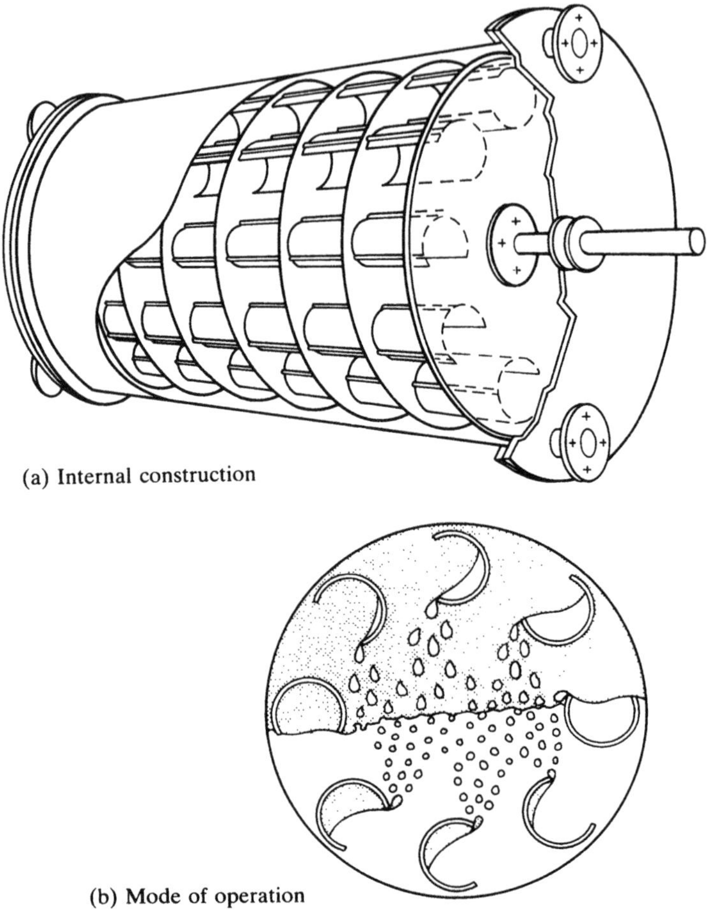

(a) Internal construction

(b) Mode of operation

Fig. 8.14. Rotating bucket (Graesser, RTL) extractor.

backmixing, but the gap must not be so narrow that the countercurrent flow is hindered. Sheik *et al.* (1972) obtained data on both axial dispersion and backmixing, using step and continuous tracer injection respectively, for a unit 150 mm diameter × 450 mm long; surprisingly, the diffusion coefficients for backmixing were higher than those for total axial dispersion. Mass transfer data for the same unit has been described by Wang *et al.* (1976). It is apparent that the design and scale-up procedures for the column extractors, described earlier, are not directly applicable to the RTL contactor. Scale-up from pilot plant tests must therefore be effected on an empirical plug flow basis, noting that efficiencies of up to 0.3 theoretical stages per compartment can be achieved with extraction factors, ***E***, of around unity (Coleby 1983).

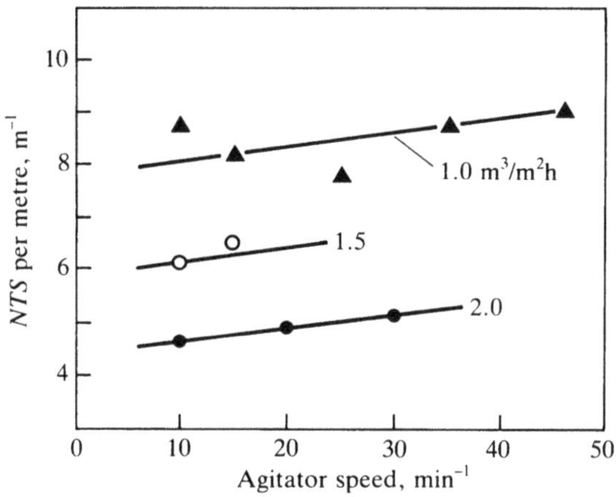

Fig. 8.15. Performance of RTL extractor. System toluene–acetone–water, $U_d/U_c = 1.5$, transfer from $c \rightarrow d$ (Stichlmair 1980).

8.3.4.3 *Discontinuous contact, rotary agitated, without interstage settling (Class Ab1)*

8.3.4.3.1 Rotary disc column

This extractor, known as the RDC, was developed in the Royal Dutch/Shell laboratories in Amsterdam in 1948–52. As shown in Fig. 8.16(a), it comprises a vertical shell divided into compartments by means of annular stator rings, with a vertical shaft, driven by an external electric motor, carrying a series of rotor discs located in the middle of each compartment. The diameter of the rotor discs is smaller than the openings in the stator baffles so that the entire rotor assembly can be installed or withdrawn for maintenance. In operation, the entire contents of the column rotate and horizontal vortex pairs are superimposed on this motion, between each rotor and the two adjacent stator discs, as shown in Fig. 8.16(b). This vortex motion serves to break down the dispersed phase into fine droplets with a net vertical flow component which is countercurrent to that of the continuous phase.

Some hundreds of RDC's are in use worldwide, with diameters ranging from below 1 m to 4.5 m (Kosters 1983). Many of these are employed in the petroleum industry, for example, in the refining of lubricating oil, in aromatics separation with Sulfolane, and in propane–butane deasphalting of crude oil residues. They have also found extensive application in the chemical industry, and uses have been reported, *inter alia*, in edible oil refining and metals extraction. As the RDC has a relatively open structure, it is less prone to fouling than most other types of extractor and can therefore

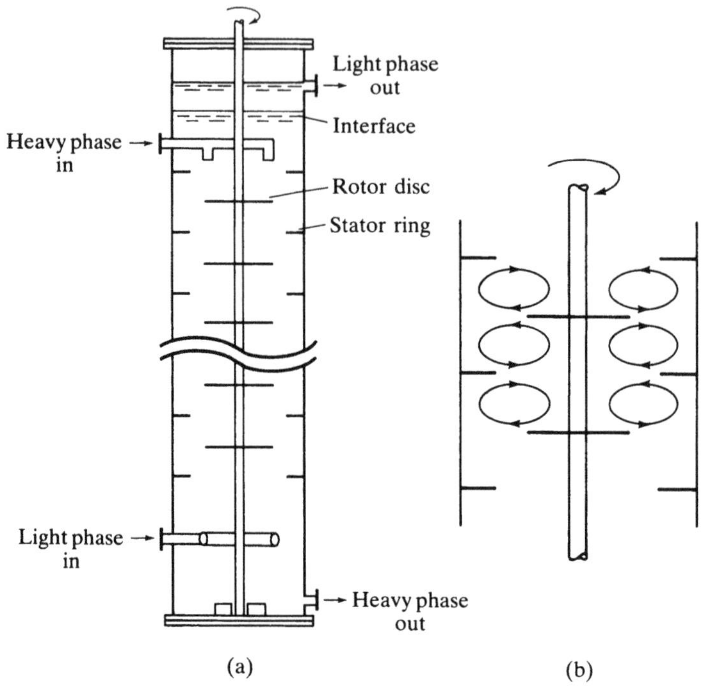

Fig. 8.16. Rotary disc column: (a) general arrangement; (b) flow pattern in stage.

handle systems containing solids, as in the purification of crystals, the washing of solids, and the extraction of glycerine from soap curd.

Geometry. Stabilization of the vortices requires an adequate baffling effect of the rotor discs and stator rings. In addition, the free area for vertical flow of the phases must be such that a reasonable capacity is obtainable; this requires that $(1 - d_r^2/d_c^2)$ and d_s^2/d_c^2 should be approximately equal.

The compartment (i.e. stage) height must also be chosen with care, as excessive height may lead to vortex instability, while an inadequate height can result in excessive damping of the vortices. Scaling-up on a geometric basis results in increased mass transfer per stage due to the increased residence time, but despite this the overall extractor height will be greater. In practice therefore the stage height is increased to a lesser extent than in proportion to the diameter.

On the basis of these considerations, the recommended dimensions for these extractors is as follows:

$$d_r/d_c = 0.6$$

$$d_s/d_c = 0.7$$

$$h_c(\text{approx}) = 0.13\, d_c^{0.67} \text{ (in m)}$$

Power requirement. The power input per rotor disc is given by the following relationship between the power number, *Po* and the Reynolds number

$$Po = P/(\rho N^3 d_r^5) = f(d_r^2 N\rho/\mu) \tag{8.35}$$

The form of the functional relationship is shown in Fig. 8.17. In most cases RDC's are operated in the region $Re > 10^5$, for which the power number is constant.

In scaling-up it is normal to maintain the power input per unit mass constant. For the constant power number region the power input to a single rotor disc is proportional to $\rho N^3 d_r^5$. The power input per unit mass in a compartment is therefore proportional to $N^3 d_r^5/d_c^2 h_c$, termed the *specific power input group*, which is used as the basis for scale-up of the RDC.

Cross-section. Extensive hold-up and flood-point data obtained for a 7.5 cm diameter RDC with a range of internal geometries were found to be well represented by eqns (8.4 and 8.6) with $\bar{v}_0$ replaced by $\bar{v}_N$, the characteristic velocity at rotor speed N (Logsdail *et al.* 1957). The resulting values of $\bar{v}_N$, obtained for a number of solvent–water systems, were found to be correlated to within ± 15 percent in terms of column variables and physical properties as follows

$$\frac{\bar{v}_N \mu_c}{\gamma} = 0.012\left(\frac{\Delta\rho}{\rho_c}\right)^{0.90}\left(\frac{g}{d_r N^2}\right)^{1.0}\left(\frac{d_s}{d_r}\right)^{2.3}\left(\frac{h_c}{d_r}\right)^{0.90}\left(\frac{d_r}{d_c}\right)^{2.7} \tag{8.36}$$

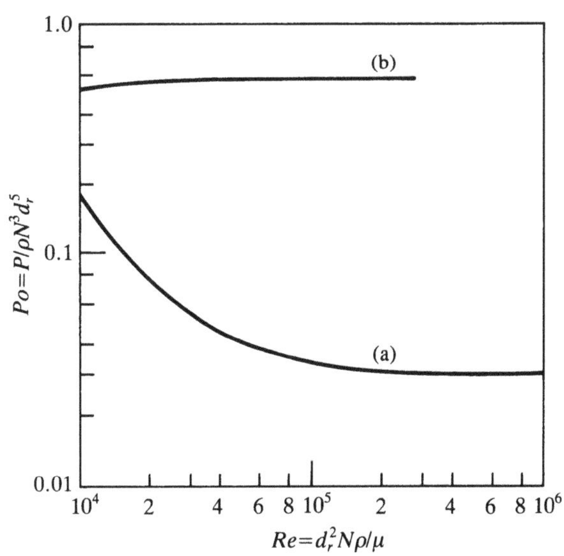

Fig. 8.17. Plot of power number vs Reynolds number (per stage) for rotary agitated columns.

This expression was obtained for a single column diameter of 0.076 m; however, a comparison with data given by Reman and Olney (1955) for three columns with diameters of 0.079, 0.63, and 1.86 m treating a lubricating oil distillate with furfural gave an exponent on the group (d_r/d_c) of 2.6, in close agreement with that in eqn (8.36).

It may be concluded that eqn (8.36), in conjunction with eqns (8.4 and 8.6), can be used with reasonable confidence for the prediction of hold-up and flood-point. A somewhat similar method has been described by Kosters (1983). However, it should be noted that measurements of the hold-up profile along the column show that the local hold-up increases from the base to a maximum at a point midway up the column and subsequently decreases (Strand *et al.* 1962; Sarkar *et al.* 1985). In the absence of data for a given system it is therefore always desirable to determine the column diameter on the basis of flood-point measurements in a pilot-scale column.

Mass transfer. As with other column types, the prediction of mass transfer performance and height requires a knowledge of the interfacial area and of the interphase mass transfer coefficients. The former can be expressed in terms of hold-up and d_{32} by an expression similar to eqn (8.10), viz.

$$a = 6\phi / d_{32} \tag{8.37}$$

Data on droplet size in an RDC, obtained by Mumford and Al-Hemeri (1974) and by Sarkar *et al.* (1985) showed that d_{32} increases with hold-up and decreases with column length. The former workers also gave a correlation of d_{32} in terms of column variables, including length, and physical properties. However, the reported variations in hold-up and d_{32} with position in the column would result in inaccuracies in the interfacial area as given by eqn (8.37), and these are likely to be accentuated under mass transfer conditions.

Mass transfer coefficient data for a 7.5 cm diameter column, interpreted on a plug flow basis, have been reported by Logsdail *et al.* (1957). Strand *et al.* (1962) have obtained similar data for 15 and 107 cm diameter columns, and presented the results as 'true' overall coefficients, i.e. corrected for axial dispersion. However, reliable methods for the prediction of mass transfer coefficients are not available, so that it is necessary for design purposes to scale-up from pilot plant results.

Axial dispersion data for 15 and 107 cm diameter columns have been obtained by Strand *et al.* (1962), using the pulse injection method. Their results, expressed in terms of backmixing ratios, were correlated as follows:

Continuous phase (for $d_r N/U_c > 30$):

$$\alpha_c = 0.09\,(1 - \phi)\left(\frac{d_r N}{U_c}\right)\left(\frac{d_r}{d_c}\right)^2\left[\left(\frac{d_s}{d_c}\right)^2 - \left(\frac{d_r}{d_c}\right)^2\right] \tag{8.38}$$

Dispersed phase (for $d_r N/U_c > 30$):

$$\alpha_d = 0.09\phi\left(\frac{d_r N}{U_d}\right)\left(\frac{d_r}{d_c}\right)^2\left[\left(\frac{d_s}{d_c}\right)^2 - \left(\frac{d_r}{d_c}\right)^2\right] \tag{8.39}$$

Stainthorp and Sudall (1964) obtained similar data for a 3.5 cm diameter column and showed that the results for the continuous phase agreed fairly well with eqn (8.38). Those for the dispersed phase were considerably higher than the values given by eqn (8.39); however, it should be noted that, as the pulse injection method was necessarily used for this phase, much of the response must have been due to forward dispersion of the droplets.

Scale-up procedure. A pilot column with a diameter of at least 7.5 cm, and preferably 10–15 cm, and with not less than 15 compartments (i.e. stages) should be used. This should be provided with a continuously variable speed drive to the rotor.

Measurements of the flood-point should first be made over a range of flow ratios, U_d/U_c, from one-half to twice that which is to be used in practice, and with a range of rotor speeds. Four sets of conditions should be covered, viz. without solute present and with each phase dispersed in turn (i.e. with the interface first at one end, and then at the other), and similarly with solute present. The results are then plotted according to eqns (8.5 and 8.7), that is, as U_d vs ϕ^2 $(1 - \phi)$ for each rotor speed, and the characteristic velocities obtained from the slopes of the lines through the origin; these should be compared with the values given by eqn (8.36).

The mass transfer runs should be conducted over a suitable range of rotor speeds, with the desired flow ratio and flow rates in the range 50–75 per cent of the flood-point values. Again, results should be obtained first with one phase dispersed, and then with the other, accepting only those runs for which the overall mass balance closes to within 2–3 per cent. The results are interpreted in terms of N_{ox}^{l}, that is, using the backflow model as described in Section 8.3.3 (p. 516).

To scale up, conditions are chosen from the pilot results which appear to offer a favourable combination of throughput and mass transfer performance and the diameter of the full scale column is calculated, together with the corresponding dimensions of the rotor, stator ring and compartment height as described earlier. The corresponding rotor speed is then that for which the specific power input group, $N^3 d_r^5 / d_c^2 h_c$ is the same as for the pilot column. On this basis, the value of N_{ox}^1 is obtained by scaling up the pilot plant value in terms of the residence time per compartment, requiring N_{ox}^1 to be multiplied by the ratio h_c (full-scale)/h_c (pilot scale). Finally, this value of N_{ox}^1, together with the appropriate values of the backmixing ratios and inlet and exit concentrations for the full-scale unit are substituted into the backflow model equations given in Section 7.2.5 (p. 437) and these are solved to obtain the number of compartments required.

The calculations are repeated for a number of values of the rotor speed for each phase dispersed in turn. By this means the most favourable design, that is, the one offering minimum capital cost, can be selected.

8.3.4.3.2 Multi-impeller columns

This classification comprises multi-compartment columns in which agitation is effected by turbines, rather than by discs as in the RDC. Two major types have evolved, the earlier of which was described by Oldshue and Rushton (1952), and was followed by the Kühni column, developed some 10–15 years later. Both types have been used extensively in industry and are described in detail below; due to small differences in certain characteristics, they receive separate listing in the selection chart (Table 8.2). Brief mention is also made below of certain modified forms of these extractors which have been proposed, but not yet used outside the laboratory. As with the RDC, these extractors can all be operated with either phase dispersed.

Oldshue–Rushton column. In this extractor, shown in Fig. 8.18, the compartments are separated by stator rings, as in the RDC, and agitation is provided by means of open 4- or 6-bladed flat turbines; four vertical wall

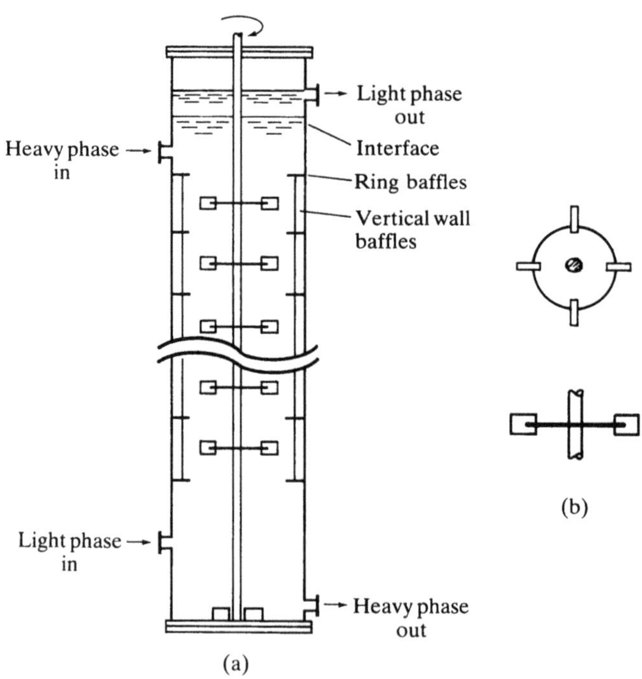

Fig. 8.18. Oldshue–Rushton column: (a) general arrangement; (b) impeller.

baffles are provided to increase the power absorption, and consequently the turbulence, to give a smaller droplet size distribution. Columns of this type have been built in sizes up to 4.57 m diameter × 30 m high.

Many combinations of compartment geometry are possible, but typical values are as follows

$$d_r/d_c = 0.33\text{–}0.50$$

$$d_s/d_c = 0.35\text{–}0.55$$

$$w_b/d_c = 0.08$$

$$h_c/d_c = 0.4\text{–}0.6$$

It is advantageous for the stator opening to be of somewhat larger diameter than the turbines so that the rotor assembly can be removed for maintenance and cleaning without disassembly. Columns of this type with diameters ranging up to 3 m have been operated successfully in a number of industries.

Hold-up data for these columns can be correlated by means of eqn (8.4) with $\epsilon \simeq 1.0$ and the characteristic velocity, $\bar{v}_0$, replaced by $\bar{v}_N$, the value for rotor speed N. Similarly, flood-point data can be represented by means of eqns (8.5 and 8.7); however, a generalized expression for the prediction of v_N for these columns is not available, and values must be obtained from pilot plant data.

Mass transfer data have been obtained by Oldshue and Rushton (1952) for a 15 cm diameter column of this type using the system acetic acid–water–methyl isobutyl ketone. These were interpreted in terms of the HETS, which was found to increase by a factor of 1.8 with increase in compartment height from 7.6 to 15.2 cm despite an increase in stage efficiency from 74 per cent to 86 per cent. A reliable method for the prediction of mass transfer rate is not available, however, and design must be effected by scale-up from pilot plant results as for the RDC.

Ingham (1972) obtained continuous phase backmixing data for two-phase flow in a 15 cm diameter column of this type, using the continuous tracer injection method. The column had a total height of 91 cm and 6-bladed turbines of 5.08 cm diameter; the compartment height was varied from 7.4 to 10.2 cm, and stator plates with openings of 5.6 and 9.25 cm were used. The results, expressed in terms of the backflow model, are as follows:

$$\alpha_c + \frac{1}{2} = 0.375\,\frac{d_r N(1-\phi)}{U_c}\left(\frac{d_s}{d_c}\right)^2 \tag{8.40}$$

It is not known whether this expression is applicable to larger diameter columns.

No data are available on backmixing in the dispersed phase. Visual evidence indicates that this does occur with the smaller drop sizes, which however make a relatively small contribution to mass transfer. Forward dispersion of droplet phase is almost certainly appreciable.

Kühni column. This column, shown in Fig. 8.19(a), differs from the Oldshue–Rushton type in that shrouded double-entry radial flow turbines are used, in conjunction with stator plates provided with a number of relatively small openings. The turbines produce a vortex-type flow pattern, shown in Fig. 8.19(c), which somewhat resembles that obtained in the RDC. The stator plates permit axial flow through the column, but isolate the compartments more thoroughly than is the case with the RDC and Oldshue–

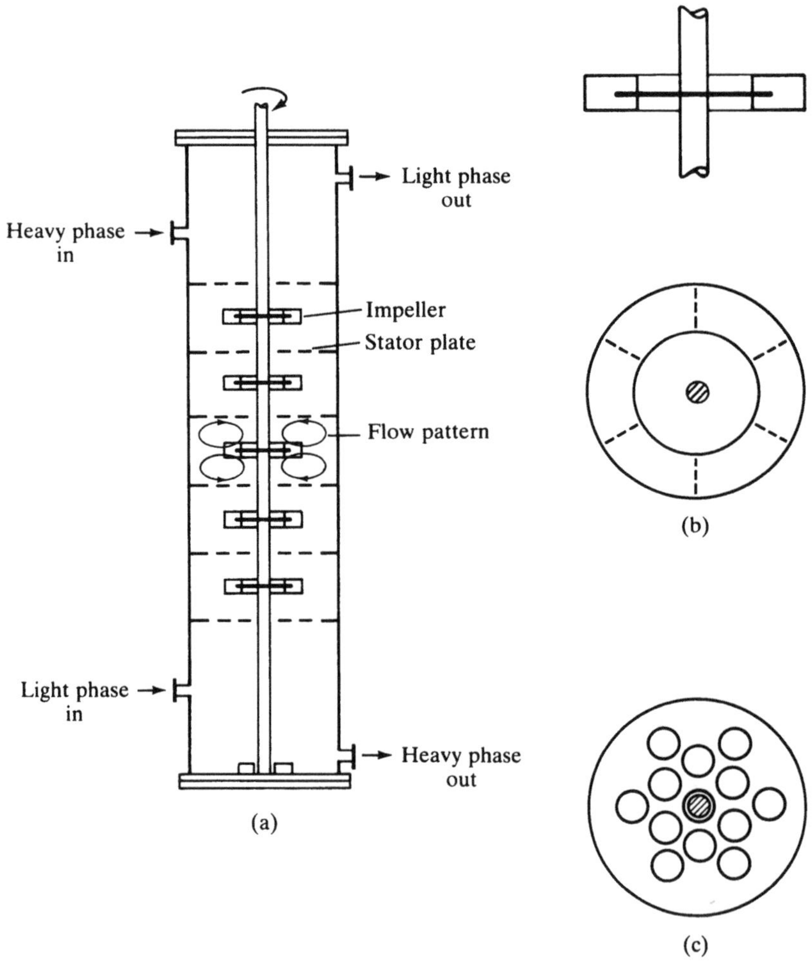

Fig. 8.19. Kühni column: (a) general arrangement; (b) impeller; (c) typical stator plate.

Rushton types; this reduces backmixing, especially of dispersed phase, but makes assembly, and disassembly for cleaning and maintenance, more difficult.

Kühni columns of this type range in size from laboratory units of 60 mm diameter up to 2.5 m diameter industrial units, and are used in a similar wide range of industries to the RDC. For sizes up to 1.5 m diameter the rotor assemblies and stator plates, with bearings, are assembled in units of up to 5 m in length and inserted in the column shell. Larger units, which in sizes greater than 3 m in diameter are provided with three sets of turbine agitators on parallel shafts (the 'Tri-Extractor') are assembled on site through manholes on the side of the column (Mögli and Bühlmann 1983).

Information on compartment geometry for scale-up purposes has not been documented, but the following can be taken as a guide:

$$d_r/d_c = 0.33\text{–}0.5$$

$$h_c = 0.2\text{–}0.3\,d_c^{0.6}$$

$$\text{Stator free area} = 20\text{–}40 \text{ per cent}$$

The arrangement of a typical 6-bladed turbine impeller is shown in Fig. 8.19(b). The free area of the stator plate comprises the annular space around the impeller shaft together with that of one or more concentric rings of staggered holes, as shown typically in Fig. 8.19(d). The effect of increasing the stator free area is to increase both the throughput at flooding and the continuous phase axial dispersion.

Kumar *et al.* (1986) showed that hold-up and flood-point data can best be correlated by means of eqns (8.4, 8.5, and 8.7) with $\epsilon \approx 1.0$ and $\bar{v}_0$ replaced by $\bar{v}_N$. A generalized correlation for $\bar{v}_N$ is not yet available, however, and values must be obtained from pilot scale data, together with the necessary mass transfer data.

Single-phase backmixing data for the continuous phase have been reported by several investigators, and the data have been correlated in terms of the diffusion model (Mögli and Bühlmann 1975). Hody (1975) found no influence of the dispersed phase on continuous phase backmixing for hold-up values up to 7 per cent. Ingham *et al.* (1974), on the other hand, using continuous tracer injection with a 15 cm diameter column, observed a small reduction in continuous phase backmixing due to the presence of dispersed phase. This was explained by the resulting increase in continuous phase velocity in the free area of the partition plates, and was allowed for by the incorporation of the term $(1 - x)$ into the correlation of single-phase data.

Dongaonkar (1989, see also Dongaonkar *et al.* 1991), in a study of the performance of a 7.5 cm diameter Kühni column, obtained continuous phase backmixing coefficient data using both unsteady state and continuous tracer injection. The results, together with the foregoing single- and two-phase data were correlated in the form proposed by Ingham *et al.* (loc. cit.). The result, expressed in terms of the backflow model, was as follows

$$\alpha_c = -0.346 + 0.063\left[\left(\frac{Nd_r(1-x)}{U_c}\right)\left(\frac{d_r}{d_c}\right)\left(\frac{h_r}{d_r}\right)^{0.067}\epsilon_p^{0.587}\right] \quad (8.41)$$

However, it was observed that the value of α_c given by this expression was 10–20 per cent higher than the experimental values, indicating that the effect of the dispersed phase is not fully accounted for by the term $(1 - x)$. Other differences were also observed between the results obtained under various conditions (Section 7.6.3, p. 484); for the present, however, eqn (8.41) appears to provide the best available means for the prediction of α_c values for the Kühni column.

Backmixing of dispersed phase, as determined by solute concentration-profile measurements, was shown by Dongaonkar *et al.* (1991) to be negligible for his 75 mm diameter column. Forward dispersion of dispersed phase droplets was however appreciable.

Other rotary agitated columns. Rotary annular columns, comprising a cylindrical rotor within a stationary outer shell, have been used over many years in the laboratory. In these columns the droplets are broken down and dispersed by the Taylor vortices set up between rotor and column wall; they are not suitable for large throughputs, however, as the annular gap cannot be scaled up in proportion to the diameter. Extensive data on the throughput and performance of these columns have been reported by Thornton and Pratt (1953).

A modified form of the Oldshue–Rushton column, with draft tubes on the stator plates, has been described by Ingham (1972); this was stated to reduce single-phase backmixing by over 90 per cent. A so-called SHE extractor has been described by Gaubinger *et al.* (1983), which uses conventional agitators with funnel-shaped stators which become narrower in the flow direction of each phase; it is stated that the high exit velocities prevent penetration by the other phase.

Steiner and Hartland (1980) described an 'enhanced coalescence' stator plate for a multi-impeller column consisting of a box type or similar grid of material which is preferentially wetted by dispersed phase. Designs such as this in which partial or complete coalescence occurs between stages, are more correctly classified under the next heading, i.e. class Ac2.

Scale-up procedure. Pilot-scale tests with the required system are conducted as described for the RDC, using preferably a column at least 15 cm in diameter. These involve firstly flood-point runs which are interpreted in terms of the characteristic velocity, followed by mass transfer runs.

Having selected favorable pilot plant conditions, the diameter of the full scale column is determined, assuming the same loadings, together with the dimensions of the column internals as given above for the appropriate column type. The rotor speed required to maintain the same power input per unit mass is then calculated, assuming the specific power input group, $N^3d_r^5/d_c^2 h_c$ to be the same in pilot and full scale columns. The actual power required can be obtained from the values of the power number given in Fig. 8.17.

In principle, the number of stages (i.e. compartments) required in the full-scale column should be determined by scale-up on the basis of the value of N_{ox}^I obtained from the pilot plant data, as described for the RDC. However, this requires a knowledge of the backmixing ratios for the pilot and full scale columns, which may not be available. In the absence of such data, scale-up must therefore be effected on a plug flow basis as described in Section 8.3.3.2 (p. 516).

8.3.4.4 *Discontinuous contact, reciprocating, without interstage settling (class Ab2)*

8.3.4.4.1 *Pulsed perforated plate column*

The use of a reciprocating motion to assist droplet breakdown in perforated plate columns was first proposed by van Dijck (1935), whose patent covered both pulsation of the continuous phase, and oscillation of the plates. The pulsed plate column comprises an outer shell provided with stationary perforated plates at regular intervals and a pulse generator which imparts a reciprocating motion to the continuous phase at the base. Downcomers or risers for continuous phase are not provided, as the reciprocating motion causes both phases to pass alternately through the same small perforations. Figure 8.20 shows the general arrangement of such columns; the pulse, generally sinusoidal, is produced by a pulse generator, several forms of which are shown in Fig. 8.13.

The pulsed plate column has found application in diameters up to 2.5 m in a number of industries. It is of particular importance in the nuclear power industry for the processing of highly radioactive materials, especially spent reactor fuel, as it is the only mechanically-agitated column extractor without internal moving parts requiring maintenance; thus, using the air-pulsing system shown in Fig. 8.13(e), the pulse mechanism can be located outside the highly active area. It seems probable that the intensive development that this extractor has received for nuclear purposes will lead to an extension of its use to other industries.

The range of geometric and operational variables for these columns is given below, with the most common values in brackets

Compartment height, h_c = 2.5–10.0 cm (5.0 cm)

Diameter of holes, d_0 = 1.6–6.5 mm (3.2 mm)

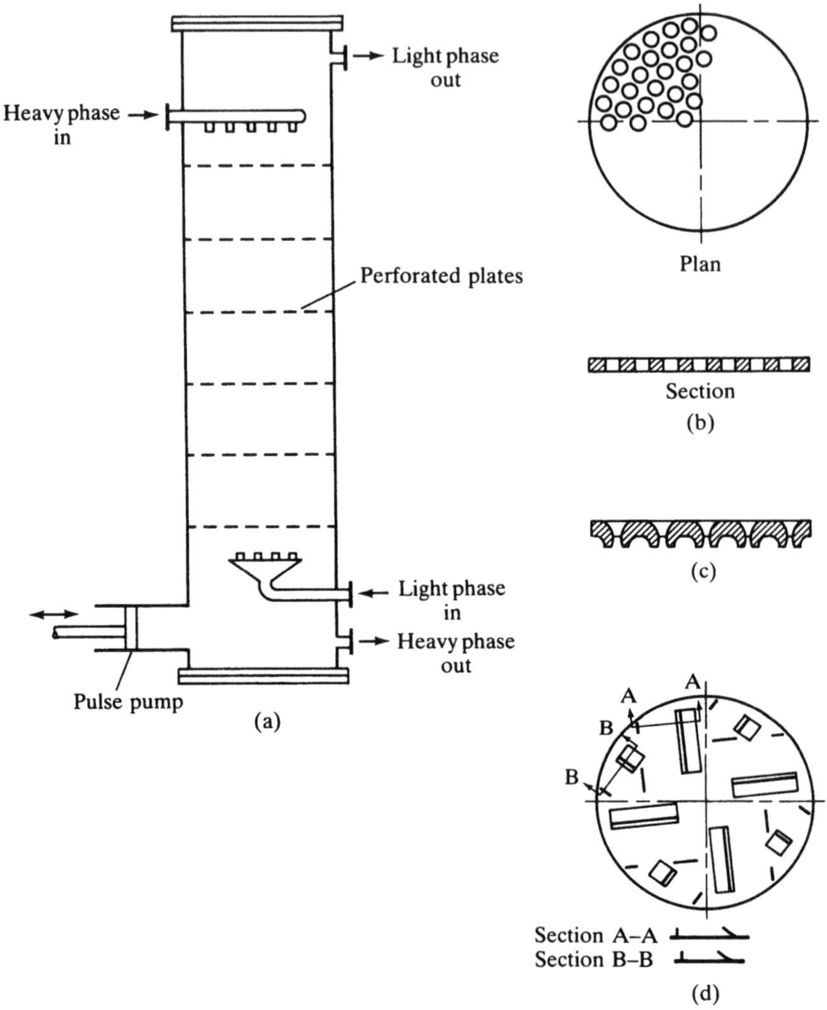

Fig. 8.20. Pulsed perforated plate column: (a) arrangement; (b) flat plate; (c) plate with protruded holes; (d) swirl plate.

Free area of plates, ϵ_p = 15–30 per cent (22 per cent)

Amplitude of pulsation, A = 0.5–2.0 cm

Frequency of pulsation, f = 0.5–2 Hz

Unlike rotary-agitated columns, the compartment height is maintained constant on scale-up. The plates, shown in Fig. 8.20(b), are normally made

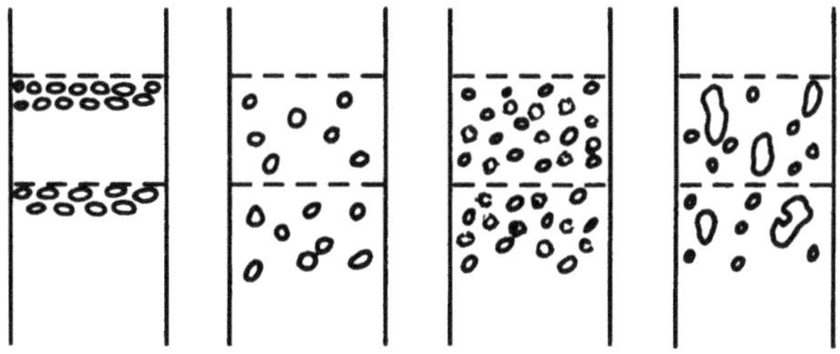

Fig. 8.21. Operational regimes for pulsed plate column: (a) and (b) mixer–settler, during downstroke and upstroke respectively; (c) dispersion (emulsion); (d) unstable.

from metal, usually stainless steel, with drilled or punched holes. However, if operation is required with an aqueous or other highly polar phase dispersed, wetting of the plates results in the formation of excessively large droplets and it is desirable to use nozzle plates, shown in Fig. 8.20(c), with the nozzle outlet directed towards the dispersed phase exit; alternatively, normal plates made from a suitable hydrophobic plastic material such as Teflon can be used. The plates are assembled as cartridges, with tie rods and spacers, before insertion in the column. Clearances between plates and column wall should be minimal, certainly no greater than 1–2 mm, to avoid by-passing of continuous phase.

Hydrodynamics. Three different types of behaviour are observed with pulsed plate columns, according to the value of the pulse velocity, fA, as shown in Fig. 8.21. At the lowest pulse velocities mixer–settler operation occurs in which (assuming the dispersed phase to be the lighter) the droplets repeatedly pass through the plates and rise freely on the upward stroke of the pulse, come to rest forming a droplet 'raft' below the next higher plates, and finally remain there while the continuous phase passes between them on the return stroke.

At higher pulse velocities the dispersion (also termed 'emulsion') region sets in. Here the droplets are distributed evenly throughout each compartment, and become smaller as the pulse velocity is increased. In this region fully-mixed zones appear on each side of each plate; the significance of these will be discussed later, in relation to backmixing. In both mixer–settler and dispersion regimes droplet coalescence and breakage occur in the immediate vicinity of, and within the holes in the plates (Garg and Pratt 1984). Finally, with some systems an unstable region, in which irregular coalescence occurs between the plates, appears at still higher pulse velocities, shortly before flooding occurs. The relationship between the various regions is shown in Fig. 8.22.

Hold-up and flood-point. In the mixer–settler regime the droplets rest

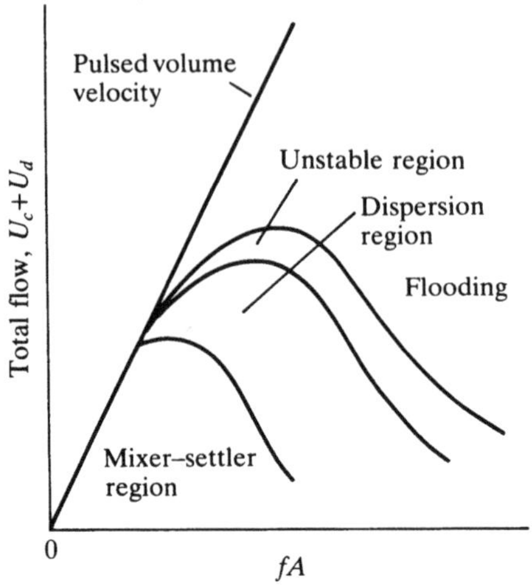

Fig. 8.22. Limiting flows as a function of pulse velocity.

beneath the plates on the downstroke of the pulse, and are forced through on the upstroke. Flooding in this region is therefore due to insufficient pulsing, and is given by

$$U_{d,f} = fA \qquad (8.42)$$

Fitzpatrick *et al.* (1986) obtained data for the kerosene–water system which confirmed this relation; however, if the continuous phase is unable to pass through the accumulated dispersed phase beneath the plates, unaided by the pulse, the flood-point is defined as follows

$$U_{d,f} + U_{c,f} = 2fA \qquad (8.43)$$

The latter expression is usually assumed to apply in this region although for low interfacial tension systems and large holes, the dispersed phase may pass through the holes, independently of the pulsing, resulting in higher throughputs than those predicted by eqn (8.43).

The hold-up in the mixer–settler region is determined by the amount of dispersed phase passing through a single compartment in the time of one pulse (Griffith *et al.* 1952); hence

$$\phi = U_d/fh_c \qquad (8.44)$$

provided the pulse rate is such that the droplets do not pass through more than one plate per stroke.

The flow behaviour is more complex in the dispersion and the unstable region. Thornton (1957) obtained flood-point data for six solvent–water systems of both high and low interfacial tension in a 75 mm diameter column and found that the results were well correlated by eqns (8.5 and 8.7) with $\bar{v}_0$ replaced by $\bar{v}_{fA}$. He also obtained a correlation for $\bar{v}_{fA}$ in terms of column variables and physical properties which was later amended by Gronier *et al.* (1966), taking into account all available data up to that time, as follows

$$\frac{\bar{v}_{fA}\mu_c}{\gamma} = 0.0203\left(\frac{\rho_c\gamma^4}{\Psi_f\mu_c^5}\right)^{0.32}\left(\frac{d_c\rho_c\gamma}{\mu_c^2}\right)^{0.47}\left(\frac{\mu_c^4 g}{\Delta\rho\gamma^3}\right)^{0.81}\left(\frac{\Delta\rho}{\rho_c}\right)^{0.98}\left(\frac{\mu_d}{\mu_c}\right)^{0.12} \tag{8.45}$$

In eqn (8.45) Ψ, the power dissipated per unit mass of fluid by fluid friction in the plate perforations, was obtained by Thornton (1957) as follows for a sinusoidal pulse

$$\Psi = \frac{\pi^2(1-\epsilon_p^2)\,(fA)^3}{2C_o^2\epsilon_p^2 h_c} \tag{8.46}$$

where C_o, the orifice coefficient, has a value of 0.6. A more complex correlation, covering both dispersion and mixer–settler regimes, was subsequently obtained by McAllister *et al.* (1967).

Logsdail and Thornton (1957) obtained flood-point data for two high interfacial tension systems (toluene–water and white spirit–water) in columns of 150 and 300 mm diameter and found that the resulting values of $\bar{v}_{fA}$ agreed well with those for the 75 mm column, so that the flood point was unaffected by column diameter. They also found that the flood-point and $\bar{v}_{fA}$ values were increased in the presence of mass transfer from dispersed into continuous phase; this is a result of Marangoni effects, and is in agreement with results obtained for packed and other column types.

It might be expected that the hold-up in the dispersion regime would be given by eqn (8.4) using values of $\bar{v}_{fA}$, replacing $\bar{v}_0$, obtained from eqn (8.45). However, this is not necessarily the case,; thus Niebuhr and Vogelpohl (1981) proposed the following alternative form in place of eqn (8.4)

$$U_d + U_c\phi/(1-\phi) = \bar{v}_{fA}\phi(1-\phi)^m \tag{8.47}$$

where m is expressed as follows

$$m = 4.6 - 0.13\ln(Ar) \tag{8.47a}$$

Vassello (1983) obtained data for the 3N nitric acid–20 per cent tributyl phosphate/kerosene system which was later correlated by Batey *et al.* (1987) using a form of eqn (8.4) with $\bar{v}_0$ replaced by v_{fA}, given by

$$\bar{v}_{fA} = C_1\exp[-C_2 fA] \tag{8.48}$$

where C_1 and C_2 are constants, the values of which are obtainable from a plot of log $\bar{v}_{fA}$ against fA.

An alternative concept suggested by Batey *et al.* (1987) involved relating the value of $\bar{v}_{fA}$ to the terminal velocity, V_t, of droplets of mean size d_{32} in a spray column. Thus, the time taken for a droplet to rise a height H in a spray column is H/V_t; if plates are present, this is increased by a delay time, Δt, per plate, i.e. by $n\Delta t$ for n plates, so that the droplet velocity is $H/(t + n\Delta t)$. In the limit, as $\phi \rightarrow 0$, this velocity approaches $\bar{v}_{fA}$ so that, taking the ratio of spray to plate column velocities

$$\begin{aligned} V_t/\bar{v}_{fA} &= 1 + N_s\Delta t/t \\ &= 1 + I \end{aligned} \tag{8.49}$$

where I is the plate impedence, which ranges from zero for a spray column to some finite value for a pulsed or other column. Hence if d_{32} is known, V_t is obtainable from a plot of drag coefficient against Re_d as given, e.g. by Hu and Kintner's correlation (1955) and the impedence, I, can be calculated from eqn (8.49).

An entirely different form of correlation for $\bar{v}_0$, obtained by the direct application of dimensional analysis to the hold-up data, was proposed by Kumar and Hartland (1983). This took the form of different expressions for the mixer–settler, 'transition' and dispersion regions. (The two latter were termed 'dispersion' and 'emulsion' regions by these authors).

Prvcic (1987) obtained extensive hold-up data for five solvent–aqueous systems in columns of 75 mm (mainly), 150, and 300 mm diameter. No effect of column diameter was observed, and it was found that Thornton's method (1957), i.e. eqn (8.4) together with eqn (8.44), represented the data well for toluene and heptane; those for kerosene were somewhat higher than predicted, and those for the low interfacial tension systems butyl acetate–water and methyl isobutyl ketone–water were much higher. A comparison with the correlations of Kumar and Hartland (1983) also showed considerable discrepancies in most cases. For the mixer–settler region, the results were in reasonable agreement with eqn (8.43) with the exception of those for butyl acetate.

It is apparent that none of the proposed correlations can be relied upon for the prediction of hold-up, and hence of flood-point, especially in the more important dispersion region. The reason for this probably lies in differences in contact angle of the dispersed phase upon the plates, as a smaller contact angle, corresponding to greater 'wettability', gives larger droplets than for a system of greater contact angle but otherwise similar physical properties. This effect is revealed in data obtained by Logsdail and Larner (1964), which showed that flood-points for the system hexane–water were considerably higher with Teflon than with stainless steel plates.

A further complication arises from a so-called 'salt effect' observed by Prvcic (1987) when a salt tracer was introduced into a column for the measurement of backmixing by the continuous tracer injection method (Section 7.6.2.1, p. 477). Using the polar solvents butyl acetate and methyl isobutyl ketone, it was found on introducing sodium chloride or potassium chromate tracer that there was an immediate increase in both droplet size and hold-up and, in some cases, premature flooding occurred. This behaviour appears to be related to the polar nature of the solvent although the exact mechanism is not clear (Fitzpatrick *et al.* 1986).

As none of the available correlations of hold-up are reliable, it is desirable to determine this experimentally whenever possible. In such cases it has been shown by Prvcic (1987) that the data can be correlated empirically by fitting to the following expression, proposed by Bell and Babb (1969)

$$\phi = U_d[C_1 + (C_2 + C_3 U_c)](fA - C_4)^2 \tag{8.50}$$

where C_1–C_4 are dimensional constants, with C_1 and C_4 dependent on physical properties and C_2, C_3 dependent on column geometry; C_4 represents the pulse velocity at which the transition from mixer-settler to dispersion regime occurs.

Droplet size. Data on droplet size distribution and Sauter mean diameter, d_{32}, have been obtained by a number of authors, although generally only for a single or, at most, two aqueous systems. Some workers have attempted to correlate their data on the basis of Kolmogarov's theory of isotropic turbulence (1949), which leads finally to expressions of the form

$$d_{32} \propto (fA)^{-1.2} \tag{8.51}$$

Pietzsch and Blass (1987), in a review of the work of 16 authors, showed that in fact the exponents on (fA) varied from -0.4 to -1.2, and concluded that models of this type are not valid.

An alternative model for d_{32} has been proposed by Pietzsch and Pilhofer (1984) in which the forces on the droplets in the jets of continuous phase passing through the holes in the plates is taken into account. A balance of the forces due to buoyancy, inertia, drag and interfacial tension led to the expression

$$d_{32} = [(6\gamma/\lambda_1) + (9\lambda_2/64)]^{0.5} - 3\lambda_2/8 \tag{8.52}$$

where

$$\lambda_1 = \Delta\rho g + b\rho_d \tag{8.53a}$$

and

$$\lambda_2 = C_D V_l^2 \rho_c/\lambda_1 \tag{8.53b}$$

Equation (8.53b) is based on a modified form of the drag force equation used by Pietzsch and Pilhofer, in which v_{dr} (see below) is replaced by the droplet terminal velocity, V_t, as given by the Hu and Kintner correlation (1955); this was found by Prvcic (1987) to fit the data better. The value of the drag coefficient, C_D, is given by

$$C_D = 4Ar/(3Re_d^2) \tag{8.53c}$$

and the droplet deceleration b from velocity v_{dr}, the mean velocity of the continuous phase through the holes in the plates during a half-cycle, to zero, by

$$b = v_{dr}^2/z_p \tag{8.53d}$$

where

$$v_{dr} = [(2fA)/(0.6\epsilon_p)] + V_t \tag{8.53e}$$

$$z_p = [0.9065(d_h^2/\epsilon_p)]^{0.5} \tag{8.53f}$$

Equation (8.53d) is based on the assumption that the deceleration occurs over a distance z_p equal to one-half the spacing between adjacent holes.

Equation (8.52) is cubic, and at low values of the pulse velocity, fA, it has three real roots, of which the first lies outside the Hu and Kintner correlation range for V_t and predicts extremely small droplets. The second root increases with increase in fA and has no physical significance so that the third root, which decreases with fA as would be expected, is the required solution. Solutions to the model for specific cases shows that d_{32} varies as fA to an exponent of -0.2 to -0.4 at low fA values, changing gradually with increasing fA to around -1.2 as required by eqn (8.51). Prvcic (1987) showed that the model gave reasonable agreement with published data in a number of cases, although errors of around $\pm$ 40 per cent may be expected.

An alternative form of correlation covering all regimes of operation was proposed by Kumar and Hartland (1986). This employed an exponential function of f^2A, and was claimed to predict droplet diameter with an average deviation of 11.3 percent.

Backmixing. A number of studies of continuous phase axial dispersion and backmixing have been reported for two-phase operation of pulsed columns. These were mostly carried out by pulse or transient injection with the results reported in terms of the axial diffusivity, E_c. Mar and Babb (1959) in an early study used a 5.08 cm diameter column with three different solvent–water systems and correlated the results by dimensional analysis. In a later study by Sehmel and Babb (1964), E_c was observed to pass through a maximum at the transition between mixer–settler and dispersion operation, and separate correlations were given for each region.

The influence of column diameter on backmixing is not clear, as Rouyer *et al.* (1974) observed no difference in E_c for 45 and 600 mm diameter columns, whereas Aufderheide and Vogelpohl (1986), using 80 and 225 mm diameter columns, observed a large effect of diameter for the toluene–water system and only a small one for butyl acetate–water. Tung and Luecke (1986) re-correlated published data for E_c for two-phase emulsion operation as follows:

$$E_c = 0.250 \left(\frac{h_c U_c}{\epsilon_p^{1.30}}\right) \left(\frac{d_o}{h_c}\right)^{0.565} \left(\frac{fA}{U_c}\right)^{0.606} \tag{8.54}$$

Prvcic (1987) conducted an extensive study of continuous phase backmixing in pulsed columns with a view to clarifying the effects of column diameter and system physical properties (see also Prvcic *et al.* 1989). For this purpose columns of 72.5, 150, and 300 mm diameter with standard plate geometry, i.e. $h_c = 50$ mm, $d_o = 3.2$ mm, and $\epsilon_p = 0.213$–0.24, were used. Data were obtained for five solvent–water systems, viz. *n*-butyl acetate, heptane, kerosene, methyl isobutyl ketone and toluene. As tracers, sodium chloride and potassium chromate were used initially, but after the discovery of the 'salt effect' referred to above, Tartrazine dye (Acid yellow 23) was used. The results were expressed in terms of the backmixing ratio, α_c, rather than E_c as this is more appropriate to stagewise columns.

The results for single-phase operation with water, which is not relevant to liquid extraction, revealed an increase in α_c values with increasing diameter. This was not the case for two-phase operation, however, and the results were correlated initially by dimensional analysis, as follows:

Mixer–settler operation:

$$\alpha_c = 5.644 \left(\frac{U_c}{fA}\right)^{-1.45} \left(\frac{h_c}{A}\right)^{-1.42} \left(\frac{d_c}{A}\right)^{0.40} \left(\frac{\mu_c}{\rho_c f A^2}\right)^{0.74} \tag{8.55}$$

$$= 5.644\, f^{0.71} A^{0.99} d_c^{0.40} U_c^{-1.45} h_c^{-1.42} (\mu_c/\rho_c)^{0.74} \tag{8.55a}$$

Dispersion operation:

$$\alpha_c = 0.471 \left(\frac{U_c}{fA}\right)^{-1.34} \left(\frac{h_c}{A}\right)^{-1.03} \left(\frac{d_c}{A}\right)^{0.37} \left(\frac{\mu_c}{\rho_c f A^2}\right)^{0.35} \tag{8.56}$$

$$= 0.471\, f^{0.99} A^{1.30} d_c^{0.37} u_c^{-1.34} h_c^{-1.03} (\mu_c/\rho_c)^{0.35} \tag{8.56a}$$

Although these correlations represented the data satisfactorily, extrapolation is not to be recommended due to their empirical nature and a physical model was therefore developed (Prvcic 1987; Prvcic *et al.* 1989). This was based on a proposal of Novotny *et al.* (1970) for single-phase operation, extended later by Nemecek and Prochazka (1974) to two-phase operation, primarily for reciprocating plate columns. This model considers that backmixing in such columns is due to

(1) backflow of liquid through the plate perforations when it moves against the net flow; and

(2) axial diffusion between adjacent pairs of plates.

As a consequence it is assumed that well-mixed regions, in which the backflow model applies, exist on each side of each plate, connected to similar regions on adjacent plates by diffusion regions in which the axial dispersion model applies, as shown in Fig. 8.23. The realistic nature of this proposal became apparent in Prvcic's experiments (1987) in which a coloured tracer was injected continuously into the continuous phase, when it was observed that the decrease in colour above the injection point occurred in steps, with an oscillating colour 'interface' between each pair of plates.

The relationship between the overall backmixing ratio, α_c, for the column and the backmixing coefficient, q_c, between adjacent stages is given by

$$\alpha_c = \{ (1 + 1/q_c) \exp [U_c (h_c - h_m/E_c)] - 1 \}^{-1} \qquad (8.57)$$

where q_c is the ratio of backflow to net forward flow across a plate, h_m is the total height of the well-mixed regions on the two sides of a plate and E_c is the axial diffusion coefficient for the diffusion region. The value of q_c was obtained as follows by Nemecek and Prochazka (1974) (see also Prvcic 1987):

Mixer–settler operation:

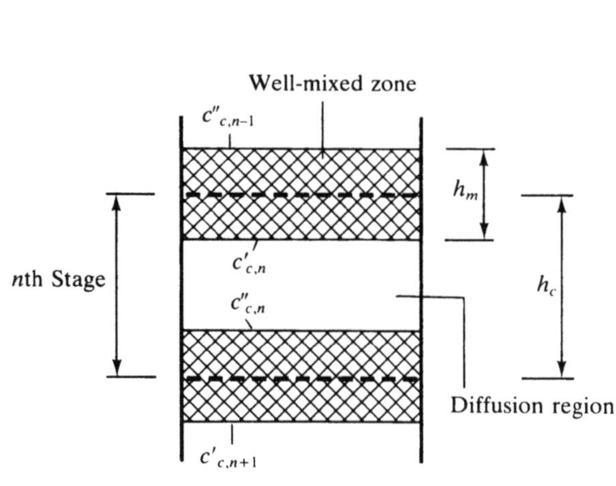

Note. c_c=Steady-state tracer concentration
$c''_{c,n-1}/c'_{c,n}=q_c/(1+q_c)$

Fig. 8.23. Fully-mixed and diffusion regions in pulsed plate column.

$$q_c = (1 - U_d/U_c)[\theta + \cot\theta - \pi/2]/\pi - U_d/U_c, \quad U_d/U_c \neq 1 \tag{8.58a}$$

$$q_c = fA/U_c - 1, \quad U_d/U_c = 1 \tag{8.58b}$$

Dispersion operation:

$$q_c = [\theta + (1 - \phi)(1 - U_d/U_c)\cot\theta - \pi/2]/\pi, \quad U_d/U_c \neq 1 \tag{8.59a}$$

$$q_c = [fA(1 - \phi)/U_c] - 0.5, \quad U_d/U_c = 1 \tag{8.59b}$$

where

$$\theta = a \sin [(U_c - U_d)/\pi fA] \tag{8.60}$$

The value of E_c in eqn (8.57) was expressed by Prvcic (1987) in terms of three components, viz.(1) the turbulence due to pulsing of the liquid through the plates, (2) circulation of continuous phase due to drop motion, and (3) entrainment of the continuous phase in the droplet wakes. These were then expressed in terms of the energy dissipation, giving finally

$$E_c = \frac{2K_2 fA\, d_o}{\epsilon_p^{1.5}} + K_3 d_{32}\left[\frac{4g\Delta\rho(V_t - U_c)\phi(h_c - h_m)}{\rho_c(1 - \phi)}\right)^{1/3} \tag{8.61}$$

where

$$h_m = K_4(2A/\pi) \tag{8.62}$$

Values of the constants K_2, K_3 and K_4 obtained from the data of Prvcic (1987) were as follows

Region:	Mixer–settler	Dispersion
K_2	0.080	0.029
K_3	0.082	0.274
K_4	2.28	1.40

The value of the backmixing coefficient is obtained by substituting eqn (8.61) into eqn (8.58) or (8.59), and then into eqn (8.57). The accuracies of this model and of that based on dimensional analysis, as measured by the correlation coefficients and residual sum of squares, were very similar; however, the physical model is to be preferred, although it is desirable to use experimentally determined values of the hold-up, and preferably of d_{32}, in view of the inadequacy of the available methods for their prediction.

In the case of the dispersed phase, experiments involving the injection of a pulse of colour into the dispersed phase distributor gave no evidence of backmixing, such as would be indicated by an inlet concentration 'jump' (Section 7.2.1.3, see p. 00). They also showed that forward dispersion of that phase was very small in mixer–settler operation, but large in the dispersion region.

Mass transfer. Many workers have reported mass transfer data for pulsed columns in the form of overall coefficients or transfer units, usually uncorrected for backmixing. In most cases small diameter columns (i.e. of 50–80 mm) were used, although Woodfield and Sege (1954) and Sege and Woodfield (1954) obtained data for the system uranyl nitrate–3N HNO_3–30 per cent tributyl phosphate/kerosene in columns of 75, 200, and 600 mm diameter. These showed a mean increase in plug flow H_{oc} of 30 per cent for the 200 mm, as compared with the 75 mm diameter column, and a fourfold increase for that of 600 mm. The latter increase was largely eliminated by the provision of 'swirl-type' mixing plates [Fig 8.20(d)] with 15 per cent free space at intervals of about 20 plates, and with two perforated plates omitted on each side of each mixing plate; however, this was at the expense of a somewhat reduced throughput. Rouyer *et al.* (1974), on the other hand, found little difference in the performance of 50 and 300 mm columns with a similar system.

Thornton (1957) obtained extensive data for the systems toluene–water and butyl acetate–water, with acetone as solute, in a 74 mm diameter column and interpreted the results in terms of the plug flow H_{oc}. Logsdail and Thornton (1957) extended this work to columns of 152 and 305 mm diameter and found a decrease in performance corresponding to a 45 per cent increase in H_{oc} between the 74 and 305 mm diameter columns. This was expressed in the form

$$H_{oc,2}/H_{oc,1} = \exp\left[1.64(d_{c,2} - d_{c,1})\right] \qquad (8.63)$$

where subscripts 1 and 2 refer to the pilot and large diameter columns and d_c is in metres; however, this may overcorrect for diameters above 1.5–2 m.

It is apparent from these results that the effect of column diameter on performance is unresolved. However, Woodfield and Sege (1954) obtained visual evidence of channelling in a 600 mm diameter column which was not apparent in one of 75 mm diameter. Prvcic (1987), in the study of backmixing described earlier, observed considerable maldistribution in single-phase operation of the 300 mm diameter column which was not entirely eliminated in two-phase operation. This was relatively stable, so that samples for analysis of tracer concentration could be withdrawn vertically above one another at a given circumferential position to obtain reliable backmixing ratios. It must be concluded, therefore, that maldistribution is inevitable in large-diameter pulsed columns unless specific precautions are taken to minimize it, for example, by the provision of mixing devices. This results in differing values of the flow ratios, U_d/U_c, over the cross-section, in some parts below and in others above the externally-imposed value; the condition for stability is equality of pressure drop over the cross-section. For modelling purposes it could be assumed that the cross-section is subdivided into two areas, one with a flow ratio above, and the other below the mean value; this would result in a reduction in driving force for mass transfer which could be expressed in terms of a maldistribution factor, F_m, for design purposes.

Design and scale-up. The design of pulsed columns from first principles is not to be recommended on the basis of present knowledge, as for other column types, and scale-up from pilot plant data is therefore required. For this purpose the pilot column should have a diameter of at least 75 mm, and preferably of 100–150 mm.

Flood point measurements should first be made, and interpreted as described in Section 8.3.2; it is also desirable to determine hold-up and, if possible, d_{32} over a range of flow rates for use in the prediction of the continuous phase backmixing ratio using eqns (8.57–8.62). These are then followed by mass transfer runs, which are interpreted in terms of the backflow model, using the predicted values of α_c, to obtain N^1_{ox}.

For relatively limited scale-up the value of N^1_{ox} should be reduced in accordance with the following expression, derived from eqn (8.63)

$$N^1_{ox,2}/N^1_{ox,1} = \exp\left[-1.64\ (d_{c,2} - d_{c,1})\right] \tag{8.64}$$

The number of stages (i.e. compartments) required in the full scale column is then calculated as described in Section 8.3.2 (see p. 00). For scale-up to diameters in excess of 200–500 mm the provision of mixing devices or partitioning baffles should be considered.

The power requirement for pulsing is given as follows (Jealous and Johnson 1955)

$$P = S_1 \frac{\mathrm{d}y}{\mathrm{d}t}\left\{(\rho_1 h_1 - \rho_2 h_3)\, g + \frac{\mathrm{d}^2 y}{dt^2}\,(\rho_1 h_1 + \rho_2 h_2 S_1/S_2) + \tfrac{1}{2}\left[\frac{N\rho_1(1-\epsilon_p^2)}{0.36\epsilon_p^2} + \rho_2 (S_1/S_2 - 1)^2\,(\mathrm{d}y/\mathrm{d}t)^2\right]\right\} \tag{8.65}$$

where S_1 and S_2 are the cross-sectional areas of column and pulse line, y is the linear displacement of liquid in the column, and h_1, h_2, and h_3 are the effective column height, the total length of pulse line to point P (Fig. 8.20), and the height of P above the bottom of the column respectively. Experimentally determined power requirements agree with the calculated value at moderate frequencies, but significant errors occur at higher frequencies.

Weech and Knight (1967) derived three equations characterizing air pulsing which permit calculation of the pulse amplitude in terms of frequency and air pressure. Baird and Ritcey (1974) showed that the energy and air consumption are minimized by pulsing at the natural frequency, which gives a sinusoidal wave form; see also Baird (1965) and Vermijs (1965).

8.3.4.4.2 Reciprocating plate column

Unlike the pulsed column, van Dijck's alternative proposal (1935) to reciprocate the plates to enhance mass transfer received little attention until Karr (1959) reported data for a 75 mm diameter 'open-type' reciprocating type column. This type, which employs plates with large holes and large free area (~ 58 per cent) was further developed by Karr and Lo (1971), and has been widely used in the pharmaceutical, petrochemical, hydrometallurgical and other industries in diameters up to 1.5 m, mainly in North America.

An alternative type of column with small holes and small free area, with or without large openings or downcomers for the continuous phase, has come into use more recently in Eastern Europe and the USSR (Prochazka 1970; Prochazka *et al.* 1971); this is known commercially as the vibrating plate extractor. In an alternative form with countermotion of plates, the plates are arranged in two stacks, each with separate shafts passing through sleeves in alternate plates, the motions of which are 180° out of phase. Some performance data for these columns, and for further modified forms which have been proposed, have been summarized by Lo and Prochazka (1983).

In the account which follows, only the open mesh type of column will be considered.

Geometry and materials. The arrangement of these columns, with internals, and a typical plate stack are shown in Figs 8.24 and 8.25. In all but the smaller laboratory sizes baffles plates are provided at intervals to reduce bypassing at the wall and axial dispersion. Typical values of the geometric and operating variables are as follows, with the most common values in brackets

Plate spacing, h_c	= 2.5 − 5.0 cm (5.0)
Hole diameter, d_o	= 1.3 − 1.6 cm (1.4 cm)
Hole pitch (triangular)	= 1.6–1.9 cm (1.75 cm)
Free area: plates	= 50–60 per cent (~ 58 per cent)
baffles	= 50–60 per cent (58 per cent)
Wall clearance: plates	= 3.2–6.4 mm (4.8 mm)
baffles	= 1.6–4.8 mm (3.2 mm)
Amplitude of reciprocation	= 0.65–2.5 cm
Frequency of reciprocation	= 2–7 Hz

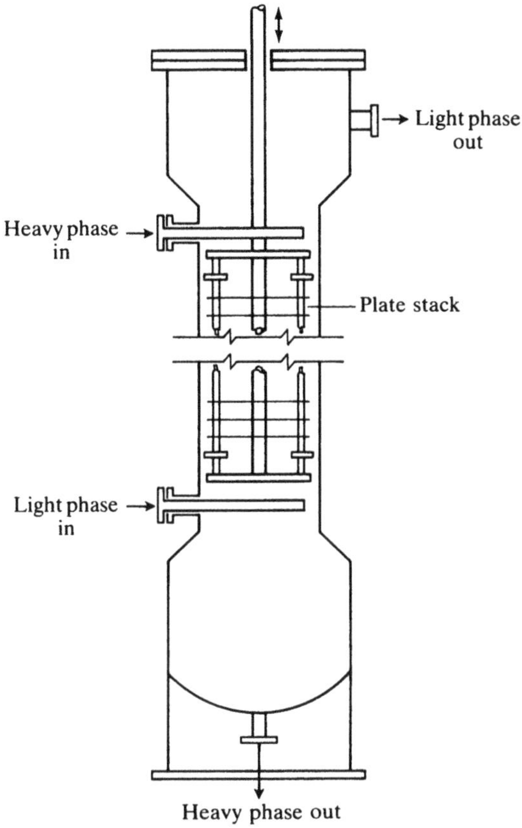

Fig. 8.24. Reciprocating plate column (courtesy of Chem-Pro Equipment Corporation).

For aqueous–continuous systems, the perforated plates are generally made from stainless steel, either by punching or from the expanded metal. The baffle plates are normally located after every fifth perforated plate and are made from stainless steel or a plastic material such as Teflon or polypropylene; using the latter, the external diameter is made slightly greater than that of the perforated plates to avoid metal-to-metal contact of plates and shell. Operation with a dispersed aqueous phase is also possible, using Teflon, polypropylene or other hydrophobic material for the plates.

Hydrodynamics. The purpose of the open-type perforated plates is to provide mild agitation to promote droplet breakage in order to maintain a uniform droplet dispersion, while at the same time greatly reducing the massive circulation of continuous phase which occurs in spray columns. As the droplets can pass freely through the holes in the plates, the mixer–settler

Fig. 8.25. Perforated plate and baffle plate assembly, 1.02 m diameter × 12.2 m long, for open-type reciprocating plate column, type 316 SS with periodic Teflon baffles to prevent metal-to-metal contact (courtesy of Chem-Pro Equipment Corporation).

regime observed with pulsed columns at low pulse rates cannot occur, and operation occurs solely in the dispersion (i.e. emulsion) region. This differs from the situation with the vibrating plate column mentioned earlier, in which the hole size and free area are comparable with those for pulsed plate columns and both operating regions are observed.

Baird and Shen (1984) have shown that, while the characteristic velocity concept is applicable to this type of column, some modification of eqn (8.4) is desirable. For this purpose a form of the Ergun equation (1952) was used to express the drag force on the droplets, leading to the following relationship for the slip velocity in eqn (8.3)

$$\bar{v}_s = u^* (1 - \phi) / \phi^{1/3} \tag{8.66}$$

The velocity u^*, which is analogous to, but not the same as the characteristic velocity $\bar{v}_0$ in eqn (8.4), embodies the Sauter mean droplet size d_{32}, found by Baird and Lane (1973) to be given by

$$d_{32} = 0.36\gamma^{0.6} / (\psi^{0.4} \rho_m^{0.2}) \tag{8.67}$$

In this expression ψ, the power dissipated per unit volume, is given as follows at high agitation levels (Jealous and Johnson 1955)

$$\psi = \frac{2\pi^2 (1 - \epsilon_p^2)\ (fA)^3}{3C_o^2 h_c \epsilon_p^2} \tag{8.68}$$

where C_o, the orifice coefficient, is about 0.7.

The hold-up is obtained by combining eqns (8.3) and (8.67); putting $U_d/U_c = U_r$, the flow ratio, this gives

$$U_c = \frac{u^*(1 - \phi)^2}{\phi^{1/3}(1 - U_r) + U_r \phi^{-2/3}} \tag{8.69}$$

Equation (8.69) cannot be solved explicitly for the hold-up, so that this must be computed by trial-and-error for given values of U_r, U_c, and u^*.

The flooding condition is obtained in the same way as eqns (8.5)–(8.7), after substitution of eqn (8.66) for the slip velocity, giving

$$\phi_f = \frac{(9U_r^2 + 54U_r + 1)^{1/2} - 7U_r - 1}{10(1 - \mathrm{U_r})} \tag{8.70}$$

The hold-up at flooding reaches a limit of 0.40 as $U_r \rightarrow \infty$, as compared with the limit of 0.50 when $U_r \rightarrow \infty$, given by eqn (8.7).

Mass transfer. Considerable mass transfer data have been reported for columns of this type in diameters of 2.5 cm (Lo and Karr 1972), 7.5 cm (Karr 1959), 0.3 m (Karr and Lo 1971), and 0.91 m (Karr and Lo 1976, 1977). The results were expressed in terms of the HETS using the plug flow model and a scale-up procedure was devised on this basis, as described below. Although nominally of stagewise type, the large plate free area and close plate spacing apparently result in an almost continuous concentration gradient, so that the use of either diffusion or backflow model is justifiable.

Continuous phase axial dispersion data for two-phase operation have been obtained by Karr *et al.* (1986) for a 25 mm diameter column, by Bensalem *et al.* (1983) and Bensalem (1985) for one of 76 mm, and by Hafez *et al.* (1979) for one of 150 mm diameter. These indicate that axial dispersion is relatively high at low agitation levels, decreases to a minimum at a pulse velocity of 2–4 cm/s and then increases with increasing pulse velocity. Karr *et al.* (1986) proposed the following expression for the Peclet No.

$$Pe_c = 0.152\, d_c^{0.33}/fA \tag{8.71}$$

However, further data are required, especially for large diameter columns, before they can be relied upon for scale-up purposes.

Some data have also been reported for the dispersed phase (Karr *et al.* 1986). However, these are very scattered, and it is probable that they relate to forward dispersion rather than backmixing.

Design and scale-up. As with other extractor types, design from first principles is unreliable, and scale-up from pilot plant data is therefore necessary. For this purpose, Karr and Lo (1971) developed a procedure using a plug flow model which has proved satisfactory in practice (see also Karr 1980).

For this purpose pilot columns of diameter as small as 2.5–5.0 cm have been used, although ones of 75 mm diameter, none of which require baffle plates, are to be preferred. In general, a plate spacing of 5.0 cm should be used; however, if the physical properties $\Delta\rho$ and γ are likely to vary appreciably throughout the column a minimum of 2.5 cm should be used, with a variation given by

$$h_c \propto [\Delta\rho^{5/3}\gamma^{3/2}]^{-1} \tag{8.72}$$

The procedure is then as follows:

1. Flood-point measurements are made over a range of reciprocation rates and a small range of flow ratios close to the desired value. These are smoothed, if necessary, using eqns (8.69) and (8.70).

2. Mass transfer runs are carried out using the desired flow ratio and at least three different throughputs. The results are interpreted in terms of the HETS (Table 6.2, pp. 368–9), which is plotted against frequency for each throughput. In general, the HETS will exhibit a minimum for each throughput when plotted against frequency; however, in very small diameter columns the minimum may occur very close to the flood-point, in which case the minimum can be taken as the value of the HETS at a frequency of 5–10 per cent below that at flooding.

3. The minimum values of the HETS, in m, are plotted against throughput in $m^3/m^2 s$; normally, this will increase with throughput. The maximum volumetric efficiency, η, for each throughput is then calculated as follows

$$\eta(\mathrm{h}^{-1}) = \frac{\text{total throughput}}{(\mathrm{HETS})_{\min}} \tag{8.73}$$

and plotted against throughput. If sufficient data have been obtained, η will pass through a maximum which is usually close to the economic optimum throughput for scale-up purposes.

4. To scale up, the expected minimum HETS in the large column is calculated from the following

$$\frac{(\mathrm{HETS})_2}{(\mathrm{HETS})_1} = \left(\frac{d_{c,2}}{d_{c,1}}\right)^{0.38} \tag{8.74}$$

where subscripts 1 and 2 refer to the pilot and full scale columns respectively. The exponent of 0.38 in eq (8.74) relates to a 'difficult' extraction system, viz-

o-xylene–acetic acid–water; exponents of 0.19–0.36 have been observed for 'easy' systems, e.g. with methyl isobutyl ketone solvent, depending on the direction of transfer and which phase is dispersed. For safety, however, it is generally desirable to use the value in eqn (8.74).

5. The corresponding reciprocating speed for the large scale column is given by

$$f_2/f_1 = (d_{c,1}/d_{c,2})^{0.14} \tag{8.75}$$

Baffle plates with the dimensions given earlier must be installed in the full-scale column.

This procedure has been reported to have been successful in the scaling up of a number of columns with diameters ranging up to 1.5 m (Karr and Ramanujam 1986). An alternative form of correlation has since been proposed by Tawfik *et al.* (1988) which permits the prediction of both the minimum HETS and the optimum reciprocating speed. However, a method based on one or other of the backmixing models would be preferable, and should be tested against the above method if reliable backmixing data for large diameter columns become available.

8.3.4.5 *Discontinuous contact with partial interstage settling (Class Ac1)*

The extractors in this group comprise rotary-agitated columns in which provision is made to induce partial coalescence between stages. This is claimed to improve performance due to the very large mass transfer rates which are obtained between newly formed droplets and the continuous phase. The modified form of Kühni column with enhanced coalescence plates described by Steiner and Hartland (1980), mentioned in Section 8.3.4.3.2, also belongs to this classification, but will not be discussed further.

8.3.4.5.1 Scheibel column

This extractor, originally intended as an alternative to provide a large number of stages in a single unit, was first described by Scheibel in 1948. As shown in Fig. 8.26(a), it comprises a vertical column with a central shaft carrying a number of flat-bladed impellers, between each of which calming sections filled with knitted wire mesh are provided. The function of the packing is twofold; firstly, it isolates the agitator flow patterns between adjacent stages, as shown in Fig. 8.26(a), thus reducing backmixing; and secondly, it induces coalescence of the droplets and separation of the phases.

Columns of this type, in diameters down to 2.5 cm, are very suitable for use in the laboratory or pilot plant, and can be used in sizes up to about 0.3 m diameter. Beyond this, the need to retain geometric similarity makes the

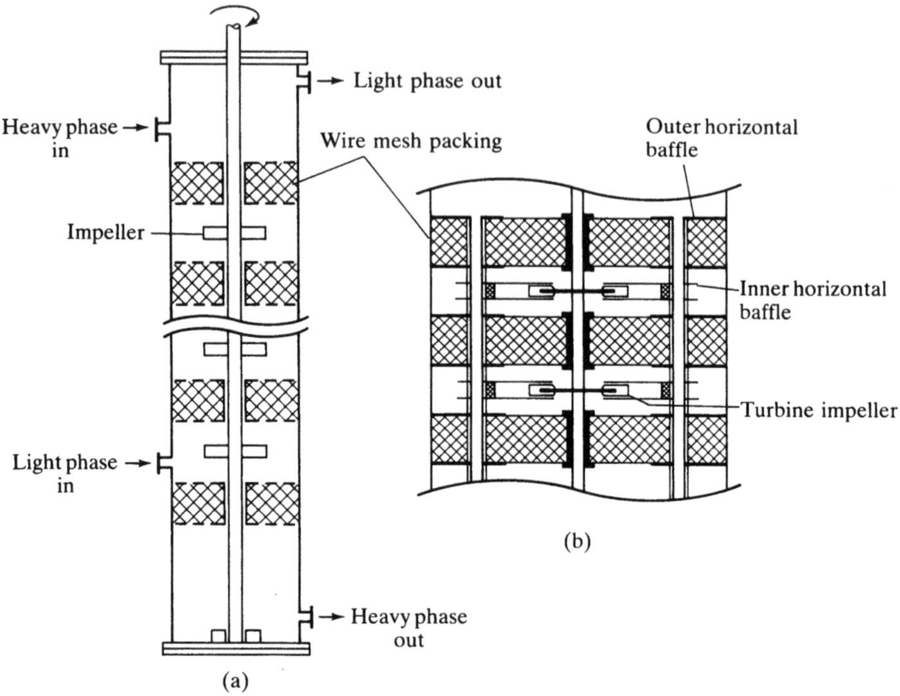

Fig. 8.26. Scheibel column: (a) first type; (b) second type.

height and cost of packing excessive. A second design was therefore developed, as shown in Fig. 8.26(b); this employs a pair of annular baffles around each impeller, with wire mesh between them at the impeller discharge to remove the swirl and break up the droplets further. In addition, annular baffles are provided at the column wall to divert the vertical flow of the phases and reduce mixing. By this means the height of packing between stages can be reduced, making this design economic up to about 1 m in diameter (Scheibel 1956). Performance data on the second type later indicated that removal of the mesh from the packed calming sections did not always reduce efficiency. Further, installation of additional mixing stages in place of the mesh resulted in a greater number of theoretical stages in a given height with systems of high interfacial tension, which require high agitator speeds to break down the droplets. This led to a third design, without packing between stages and with agitator baffles modified for ease of assembly and disassembly; the mesh at the agitator discharge was however retained (Scheibel 1968, 1983). In this form, which is suitable for use in larger diameters, coalescence between stages is eliminated and the design approaches that of the Kühni column (Section 8.3.4.3.2).

The hydrodynamical behaviour of these columns is more complex than for other column types. Performance evaluation and scale-up is therefore best effected on a plug flow basis in terms of stage efficiency, as described in Section 8.3.3.2 (p. 516).

8.3.4.5.2 Asymmetric rotary disc column (ARDC)

This extractor is a rather complex form of rotary disc column in which the shaft carrying the agitator discs is mounted off-centre, and two sets of staggered stator plates are provided, connected by a vertical segmental baffle (see Figs 8.27 and 8.28). By this means the mixing chambers are enclosed, and are connected to one another via openings each side of the vertical baffle leading to chambers in which partial coalescence occurs. Details of the column internals are shown in Fig. 8.28(a); a modified form with additional rotor discs in openings in the stator baffles, developed for ease of dismantling in the larger sizes, is shown at (b) in Fig. 8.28. A large capacity design with three rotors suitable for use in sizes up to 6 m, has also been developed (Misek and Marek 1983).

Columns of this type have been in industrial use since 1965 in sizes of 0.072–3.0 m and more. A standard geometry has been developed, as follows

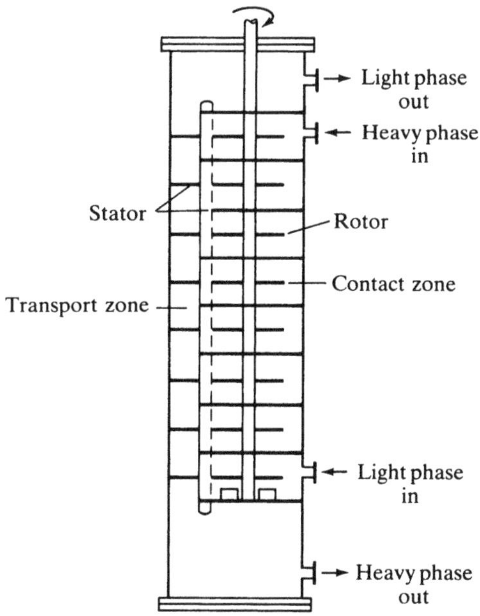

Fig. 8.27. Asymmetric rotary disc column (ARDC).

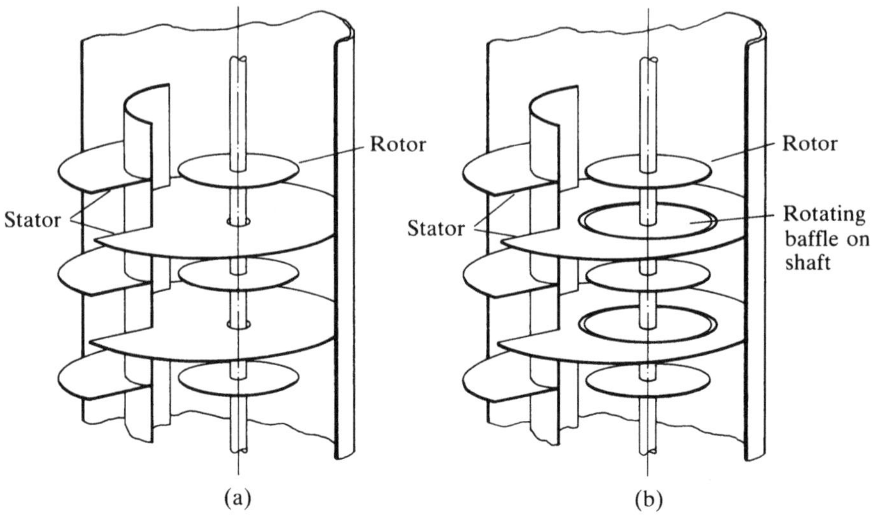

Fig. 8.28. ARDC extraction compartments: (a) standard; (b) with demountable agitators.

d_r/d_c	$= 0.49$
h_c	$= 1.3\, d_c^{0.67}$ (in mm)
Free cross-section* ϵ_p	$= 25$ percent
Nominal r.p.m.	$= 15{,}000/d_c$

*between stages (for phase transport)

If necessary, the disc diameter can be varied along the length to compensate for variations in physical properties. Similarly, the free cross-section and stage height can be increased for higher specific throughput.

Hydrodynamics. Misek (1967) has shown that the dispersed phase hold-up can be expressed in terms of a modified form of the characteristic velocity relation, eqn (8.3), as follows

$$U_d + U_c\phi/(1-\phi) = \epsilon \bar{v}_N \phi(1-\phi)\exp(b\phi) \tag{8.76}$$

in which

$$\bar{v}_N = 0.249 d_{43}(\Delta\rho^2 g^2/\rho_c \mu_c)^{1/3} \tag{8.77}$$

and

$$d_{43} = \sum_i n_i d_i^4 / \sum_i n_i d_i^3 \tag{8.78}$$

Misek and Marek (1970, 1983) have shown that the value of the coalescence constant, b, in eqn (76) is given approximately by

$$b = z/d - 4.1 \tag{8.79}$$

where d, an exponent in the relation between droplet size ratio and velocity ratio which varies with Re, is close to unity in the present case, and z is given by

$$z = 0.0159\ [d_c^2 \gamma / (v_c^2 d_{43} \rho_c)]^{0.25} \tag{8.80}$$

The value of z varies between 2.5 and 6.0 in the absence of mass transfer. For mass transfer from continuous into dispersed phase, coalescence is often negligible, i.e. $z = 0$.

On differentiating eqn (8.76) and setting $dU_c/d\phi = dU_d/d\phi = 0$, as in deriving eqn (8.5–8.7), the following condition is obtained for the flood-point

$$\frac{U_{c,f}}{U_{d,f}} = \frac{2\phi_f^2[(1-\phi_f) - b(1-\phi_f)^2/2]}{(1-\phi_f)^2[(1-2\phi_f) + b\phi_f(1-\phi_f)]} \tag{8.81}$$

This reduces to the ratio of eqs (8.5) and (8.6) when $b = 0$. Misek and Marek (1970, 1983) have found that flooding can also occur by two other mechanisms, viz. by phase inversion at values of $\phi > 0.42$, and by formation of a stable emulsion of dispersed in continuous phase; attempts to predict the onset of the latter were unsuccessful.

Misek and Marek (1970, 1983) have shown that the maximum droplet diameter under mild agitation is independent of rotor speed and is given by [cf. eqn (8.23)]

$$d_{max} = 2.03(\gamma/\Delta\rho g)^{1/2} \tag{8.82}$$

For intensive agitation they obtained expressions for d_{max} in terms of column variables and physical properties, the transition rotor speed and the droplet size distribution; these may be used to calculate approximate values of d_{32} and d_{43} for a given system.

Mass transfer. The rate of mass transfer in the ARDC is considerably influenced by the prevailing hydrodynamic conditions, especially in regard to the degree of coalescence which occurs. The height (i.e. the number of stages) required to achieve the desired performance must therefore be determined by scale-up from pilot plant tests, preferably taking axial dispersion into account.

Backmixing data for the continuous phase have been obtained by Misek (1975; see also Misek and Marek 1983). The results, expressed in terms of the backflow model, were correlated by the following *dimensional* equation (in cgs units)

$$\alpha_c = \frac{1.82 \times 10^{-3}\epsilon_p d_r N}{U_c h_c^{1/3}}\left(\frac{d_r}{d_c}\right)^{2/3}(1-\phi) + 2.71 \times 10^{-4} N^2 d_r^2\ (U_d/U_c) \tag{8.83}$$

Data were also obtained for the dispersed phase, which were correlated in a similar manner. However, it is possible that these related largely, if not entirely to forward dispersion rather than backmixing [Sections 7.5.2, p. 471, and 7.6.2.3.2, p. 483].

Scale-up. Pilot testing and scale-up should follow on similar lines to that described for the RDC (Section 8.3.4.3.1, p. 533). However, particular care must be taken in the determination of the flood-point, in view of the three different types of flooding which can occur. The power input per disc can be obtained as a function of *Re* from the appropriate line on Fig. 8.17.

8.3.4.6 *Discontinuous contact with complete interstage settling (Class Ac2)*

This classification comprises two entirely different contactor types, viz. the non-mechanical perforated (i.e. sieve) plate column, and the mechanical discrete-stage rotary film extractor. Both could in fact be treated as varieties of mixer–settler; however, it is preferable to consider the latter under a separate classification, due to the specific nature of the pilot testing and scale-up methods used.

8.3.4.6.1 Perforated plate column

These columns, also known as sieve tray columns, contain a number of perforated plates or trays, with segmental downcomers or risers for continuous phase (depending on whether ρ_c is greater or less than ρ_d), arranged one above the other in a vertical shell, usually of cylindrical cross-section. In operation they behave in a similar manner to the corresponding type of distillation column; thus, assuming that $\rho_c > \rho_d$, the continuous phase passes across the plates from side to side and down, while the dispersed phase accumulates beneath each plate in turn, coalesces and redisperses as droplets into the stage above, where it rises freely until coalescing beneath the next plate. The general arrangement of the column in its simplest form is shown in Fig. 8.29. Bypasses for 'rag' ('crud'), that is solid and emulsion-like material which accumulates at the interfaces, are often provided; these allow the rag to pass up the column to the main interface, where it can be removed periodically (Mayfield and Church 1952).

Columns of this type are widely used in the petroleum industry, for example, in the refining of lubricating oil, and in mercaptan removal from naphtha by scrubbing with aqueous sodium hydroxide. They can be constructed in very large diameters provided that care is taken to minimize hydraulic gradients and entrainment in the dispersed phase due to the cross-flow of continuous phase.

Many variants of the basic design have been described. These include split flow arrangements to give centre-to-side and side-to-centre flow of

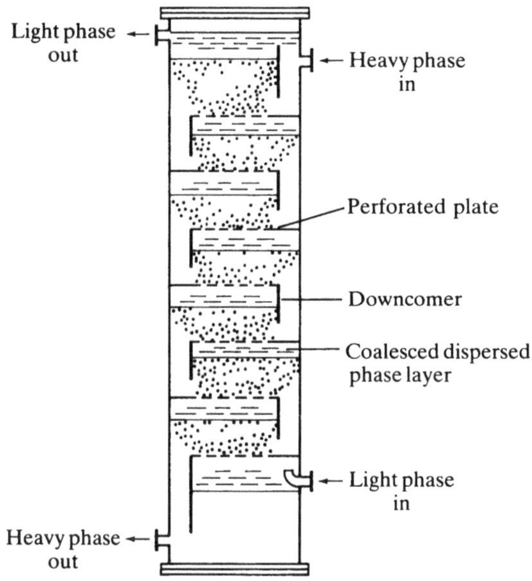

Fig. 8.29. Perforated plate column (light phase dispersed).

continuous phase, and alternatively, a multiplicity of pipes in place of single downcomers or risers. Other designs use trough-type trays with vertically disposed holes, and variable dispersion area trays with second perforated plates over the main ones, adjustable from outside. A more recent development is a cascade weir tray described by Fiocco (1983), in which weirs separate rows of holes so that the optimum number of holes is used for each flow rate and the head of dispersed phase under the trays is relatively constant. These were reported to give capacity increases up to 50 percent and yield increases up to 33 percent, and to have been built in sizes up to 7.6 m diameter.

Geometry. Typical dimensions of the column internals are given below for the basic design shown in Fig 8.29.

Hole diameter	= 3–8 mm
Hole pitch (square or triangular)	= 12–20 mm
Plate spacing	= 150–600 mm
Height of dispersed phase layer below plates	= 0.08–0.10 h_c

The smaller hole sizes are used for systems of high interfacial tension. The velocity through the holes should be such that liquid jets, rather than individual droplets, are formed at individual perforations and break up into droplets a short distance above the plate. This requires linear velocities through the holes of 150–300 mm/s. The plate area opposite the downcomer from the adjacent plate, and within 25–50 mm of the downcomer leading from the plate, should be free of perforations.

The downcomers (or risers) are set flush with the plate from which they lead, i.e. without overflow weirs. Their cross-sectional area is selected such that the linear velocity of continuous phase within them does not exceed the terminal velocity of an arbitarily assumed small droplet of dispersed phase of, say, 1 mm diameter, which otherwise would recycle around the plate. They should extend below the maximum expected depth of coalesced dispersed phase below (or above) the plate, leaving free areas for flow of continuous phase between their ends and the adjacent plates which are larger than their own cross-sections.

Hydrodynamics. The hydrodynamical behaviour of these columns differs from that of the previous types, as the hold-up includes that of the coalesced dispersed phase layers, and flooding occurs when these reach the ends of the downcomers or risers. The characteristic velocity relationship is therefore not applicable to these columns.

The depth of the coalesced layers is made up of contributions due to the head required to maintain the flow in the perforations, h_o; that required for droplet formation, h_γ; and that to overcome the resistance to flow of the continuous phase in the downcomers, h_{cp}; thus

$$h_d = h_o + h_\gamma + h_{cp} \tag{8.84}$$

Bussolari *et al.* (1953) and Major and Hertzog (1955) showed that h_o can be expressed in terms of the orifice equation, viz.

$$h_o = \frac{U_o^2 \rho_d [1 - (S_o/S)^2]}{C_o^2 \Delta p} \tag{8.85}$$

where C_o, the orifice coefficient, is given as follows in terms of the orifice Reynolds No., $Re_o = d_o U_o \rho_d / \mu_d$

$$C_o = 1.0 - 0.71/\log Re_o \tag{8.86}$$

In practice, the value of C_o is generally close to 0.67. Bussolari *et al.* (1953) proposed the following expression for h_γ

$$h_\gamma = 6\gamma/[(d_o)_{0.1} \Delta\rho] \tag{8.87}$$

where $(d_o)_{0.1}$ is the droplet diameter produced at the orifice itself at an orifice velocity of 0.03 m/s; at orifice velocities of 0.3 m/s and above where jets, and not droplets, form at the orifices, h_γ is omitted from eqn (8.84).

Finally, the head loss due to continuous phase is estimated to total 4.5 velocity heads, giving

$$h_{cp} = 4.5\ U_{cd}^2 \rho_c / 2\Delta \rho g \tag{8.88}$$

where U_{cd} is the velocity in the downcomer or riser.

The optimum orifice velocity is that at which the interfacial area obtained by jet break-up is a maximum. This has been shown by Christiansen and Hixson (1957) to occur for orifice/jet diameter ratios given as follows (see also Treybal 1963):

$d_o / (\gamma / \Delta \rho g)^{0.5} < 0.785$

$$\frac{d_o}{d_j} = \frac{0.485 d_o^2}{(\gamma / \Delta \rho g)} + 1 \tag{8.89}$$

$d_o / (\gamma / \Delta \rho g)^{0.5} > 0.785$

$$\frac{d_o}{d_j} = \frac{1.51 d_o}{(\gamma / \Delta \rho g)^{0.5}} + 0.12 \tag{8.90}$$

The corresponding average droplet size produced, which gives maximum interfacial area, is then $d_{av} \simeq 2d_j$, and the corresponding orifice velocity is

$$U_o = 2.69 \left(\frac{d_j}{d_o}\right)^2 \left[\frac{\gamma / d_j}{0.514 \rho_d + 0.472 \rho_c}\right]^{0.5} \tag{8.91}$$

Hughmark (1967) has proposed the following empirical relation as an alternative for the estimation of the orifice velocity giving maximum interfacial area

$$U_o = 2.94 (\gamma / d_o \rho_d)^{0.5} \tag{8.92}$$

Substitution in eqn (8.91) gives the jet diameter, which is doubled to obtain the optimum droplet diameter.

Having thus determined the optimum orifice velocity, the number of perforations required to handle the desired flow of dispersed phase is obtained. Selection of the pitch of the holes then gives the perforated area of the plates; to this must be added the cross-section of the downcomers or risers, together with the areas of a 5 cm wide annular ring around the plate, and a 2.5 cm wide strip alongside the downcomer, both unperforated, to give the total plate area and hence the column diameter.

Mass transfer. The average surface of a spherical droplet during growth is as follows (Coulson and Skinner 1952)

$$a_{d,f} = 3\pi d^{2/5} \tag{8.93}$$

so that the interfacial area per unit column volume for all forming droplets is given by

$$a_f = a_{d,f} N_o / [S(h_c - h_d)] \tag{8.94}$$

It is generally assumed that a_c, the interfacial area for coalescence, is the same as that for formation.

To obtain a_r, the interfacial area during droplet rise (or fall), the droplet diameter is first calculated, using eqn (8.89 or 8.90) together with the relation $d_{av} = 2d_j$. As the continuous phase is in crossflow between the plates, the hold-up is obtainable from the characteristic velocity relation, eqn (8.4), with $U_c = 0$ and $\epsilon = 1.0$; this gives, on rearrangement,

$$\phi(1 - \phi) = U_d/\bar{v}_o \tag{8.95}$$

The value of $\bar{v}_o$ can be estimated from the relation given by Vedaiyan for spray columns (see Laddha and Degaleesan 1976), viz.

$$\bar{v}_o = 1.088(\gamma\Delta\rho g/\rho_c^2)^{0.25}(U_o^2/2gd_N) \tag{8.96}$$

with U_o given by eqn (8.91 or 8.92). Hence a_r is obtainable from eqn (8.10) with d_{32} replaced by d_{av}, using the value of ϕ given by eqn (8.95).

Expressions for the prediction of the overall mass transfer coefficients, k_{od}, for droplet formation, rise (or fall) and coalescence have been summarized by Skelland and Cornish (1965) and Skelland and Conger (1973); see also Chapter 2. The mass transfer rate within a stage n can then be expressed as follows, noting that the continuous phase composition can be considered constant as it is fully backmixed due to the droplet motion:

$$N_a = k_{od,f}a_f(c_{y,n}^* - c_y) + k_{od,r}a_r(c_{y,n}^* - c_y)_{lm} + k_{od,c}a_c(c_y^* - c_{y,n+1}) \tag{8.97}$$

where $(c_{y,n}^* - c_y)_{lm}$ is the logarithmic mean of $(c_{y,n}^* - c_{y,n})$ and $(c_{y,n}^* - c_{y,n+1})$ and subscripts f, r, and c refer to droplet formation, rise, and coalescence respectively. The rate of mass transfer is also given by a mass balance over the stage, viz.

$$N_a = F_d(c_{y,n+1} - c_{y,n}) \tag{8.98}$$

so that eqns (8.97 and 8.98) can be solved by trial-and-error for $c_{y,n+1}$ on each stage.

Conditions within the column can be represented on a plug flow basis on the McCabe–Thiele diagram, as backmixing cannot occur between stages. Skelland and Cornish (1965) therefore determined the number of stages by computing the exit compositions from each stage, starting from the bottom and proceeding upwards to locate the 'pseudo-equilibrium line', as shown in Fig. 8.30. This procedure was later refined by Skelland and Conger (1973) and incorporated into a computer programme. Using this, comparisons of predicted and actual performance for a number of systems reported in the published literature showed reasonable agreement in many cases, although it was emphasized that the method is capable of considerable refinement.

An alternative procedure was described by Treybal (1963) and Laddha and Degaleesan (1976), based on an approximate expression for the overall mass transfer coefficient derived on the assumption that mass transfer is the same during droplet formation and coalescence, viz.

$$k_{od}a = \frac{2k_{od,f}a_f N_o}{S(h_c - h_d)} + k_{od,r}a_r \tag{8.99}$$

The Murphree point efficiency is given by the following expression, derived from Case 2 in Table 6.3 (see p. 00) noting that the Y-phase is dispersed and the X-phase fully backmixed

$$E'_{My} = 1 - \exp(-N_{od}) \tag{8.100}$$

where $N_{od} = k_{od}a\ (h_c - h_d)/U_d$. The overall stage efficiency, which differs appreciably from the point efficiency in all but small diameter columns, is obtained from eqn (6.41), p. 362. The total number of non-ideal stages is then obtained graphically as shown in Fig. 8.30, or analytically as described in Section 6.5.1 (p. 365); the generalized solution in Section 7.2.5 (p. 435) is also applicable. A worked example illustrating this method is given by Laddha and Degaleesan (1976).

Design and Scale-up. It is apparent from the foregoing that design from first principles is better established for the perforated plate column than for other column types. Despite this, pilot scale tests should be considered for any system for which prior information on performance is not available. For such purposes a very short column could be used, even a single stage, provided that tests are conducted at various concentrations within the range of interest. The pilot column should preferably have a diameter of at least 150–200 mm; alternatively, if it is desired to study hydrodynamical

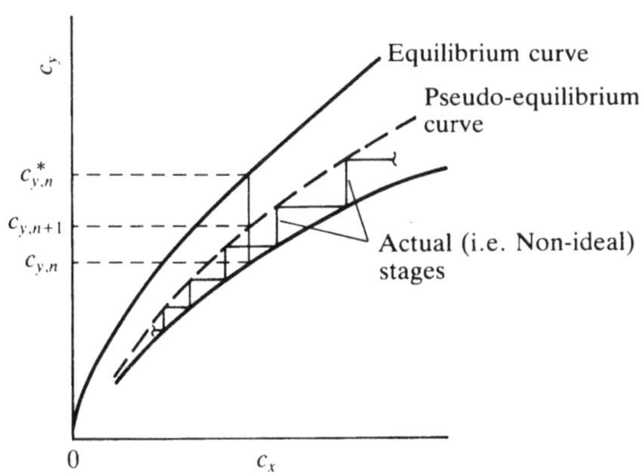

Fig. 8.30. McCabe–Thiele diagram for perforated plate column.

behaviour, the construction of a two-dimensional slice of a very large diameter column should be considered (Fiocco 1983).

8.3.4.6.2 Rotary film extractor

This extractor, of the discrete stage type, was developed specifically for the solvent-in-pulp treatment of ore slurries, that is, for the direct extraction of unfiltered ore leach liquors (North and Wells 1964–65). As shown in Fig. 8.31, it comprises a series of extraction stages arranged horizontally, with a transverse shaft carrying a number of discs within each stage. The static interface is maintained below the shaft in each stage, and in operation the discs rotate typically at 5–15 r.p.m. to keep the slurry in continuous suspension without undue entrainment or emulsification. As the discs rotate they carry a film of one of the phases into contact with the other phase, exposing an area of fresh interface through which mass transfer occurs.

Design and scale-up methods have not yet become available for these extractors. However, much information can be obtained from a single stage pilot unit, which should be used first to determine the maximum rotational speed permissible, without excessive entrainment, for a range of throughputs at the flow ratio dictated by the mass transfer requirements. Having selected the optimal conditions, mass transfer runs are carried out, using real process liquors if possible, and overall mass balances on solute calculated; these should be close to within 2–3 per cent, in which case the overall Murphree efficiency of the stage is calculated (Section 6.4.3, p. 362). These runs should be repeated for a number of concentration levels between those of the initial feedstock and of the specified raffinate. If satisfactory, it would be desirable to add a further one or two stages for testing, to ensure that multistage operation would be satisfactory; by measuring concentrations between stages, the individual stage efficiences could be determined.

In operation, the tangential velocity of the film at a given rotational speed will vary radially across the discs. North and Wells (1964–65) described a theoretical treatment for the case of a homogeneous first-order reaction in

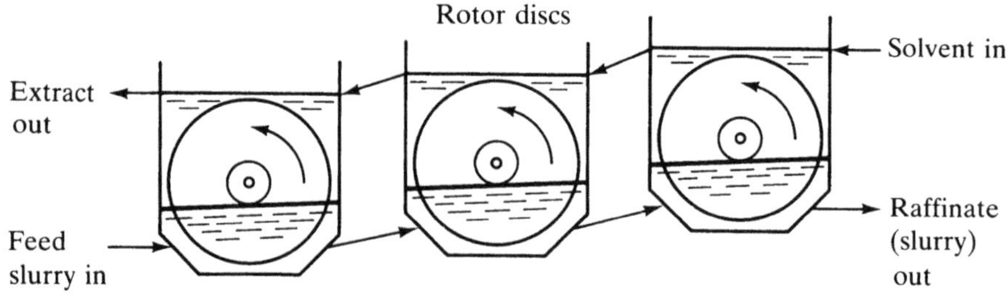

Fig. 8.31. Rotary film extractor.

the film, in relation to the extraction of uranyl salts. For practical design purposes, however, it should be sufficient to scale up the rotor disc diameter in proportion to the square root of the throughput, using the same number of discs, and to operate at a rotor speed giving the same peripheral velocity. The resulting performance, in terms of Murphree efficiency, should be similar to that of the pilot stage; however, it would be desirable to conduct further tests on a single, full-scale stage with a variable speed drive. The total number of stages required is then obtained as follows

$$N_s = NTS/E_{My}(_{\mathrm{av}}) \tag{8.101}$$

8.4 Pilot testing and scale-up: mixer–settlers (Class Ac3)
[See also Davies, Ch. 5]

8.4.1 *General*

The extractors to be considered here relate to the horizontal and vertical pump–settler and mixer–settler cascades of class Ac3 in the selection chart (Table 8.2). In the pump–settler, centrifugal pumps are used to effect both the mixing and the transfer between stages. Typical arrangements of the vertical types are shown in Fig. 8.32, which shows the manner in which individual stages are 'stacked' on one another, with external pumps in the case of the pump–settler. The horizontal types are shown in Fig. 8.33 with the pump–settler, which is now rarely used, at (a). True mixer–settlers in the

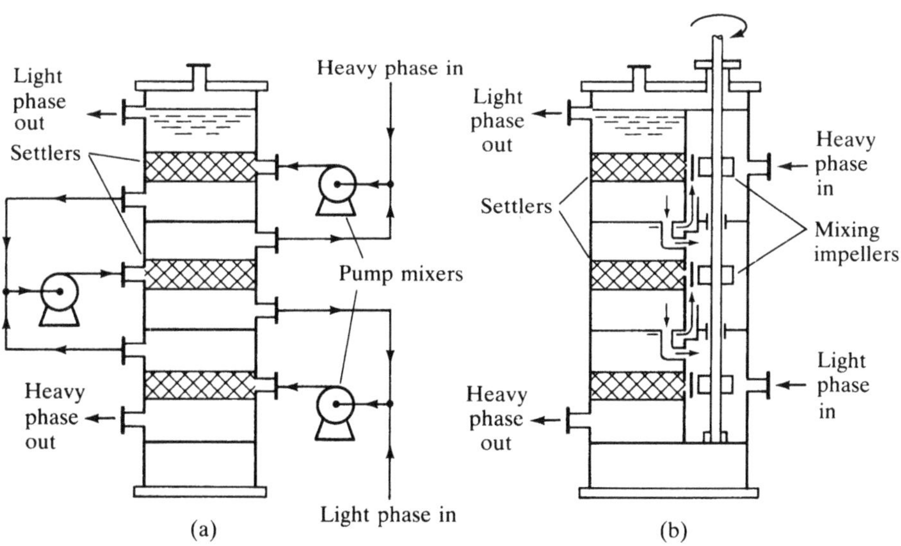

Fig. 8.32. Vertical mixer–settlers: (a) pump–settler; (b) agitated type.

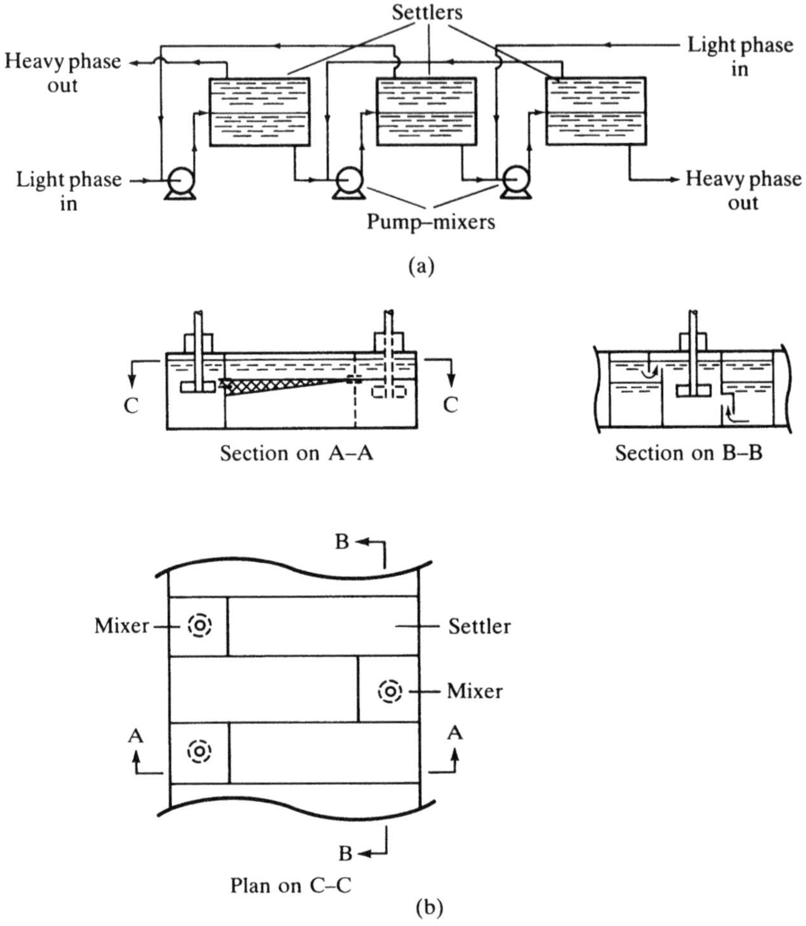

Fig. 8.33. Horizontal mixer–settlers: (a) pump-settler; (b) agitated box-type.

smaller sizes are often of the 'box-type', shown at (b); these have several stages within a single 'box', separated from one another by partitions, and with the mixers at alternate ends. In the larger sizes discrete stages are used, with interconnecting pipework, as shown in Fig. 8.34.

Many mechanical variants in the design of individual stages have been described over the years. An early account of these was given by Morello and Poffenburger (1950). For a description of more recent practice reference should be made to Chapter 5.

Ideally, such extractors would be designed from first principles, involving the determination of (1) the throughput, from the coalescence rate in the

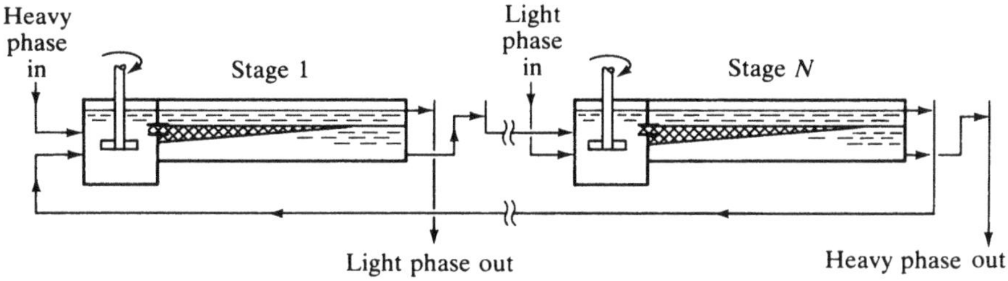

Fig. 8.34. Horizontal mixer–settler of discrete-stage type.

settlers, and (2) the stage efficiency, from the mass transfer rate in the mixers. In practice, however, this is not yet possible, and it is necessary to scale up from pilot plant tests, preferably supplemented by tests on a single full-scale stage.

The scale-up of mixer–settlers is simpler than that of columns in that backmixing and forward dispersion are negligible. The required performance is therefore expressed in terms of the number of ideal plug flow stages required using the graphical or analytical methods described in Section 6.3.1 and 6.3.2 (pp. 354, 358); these are then related to the number of actual stages by means of the Murphree stage efficiency (Section 6.4.3, p. 362). Alternatively, the number of actual stages required can be calculated directly for a linear equilibrium relationship by the method described in Section 6.5.1 (p. 365), or for the more general case, by the method in Section 7.2.5 (p. 437).

8.4.2 *Pilot-scale testing*

The designs of the mixer and the settler are approached separately, while recognizing their interdependence. Thus (1) the mixer must interdisperse the two phases and create sufficient interfacial area for an acceptable stage efficiency to be achieved; if a pump is used as a mixer, the residence time is very short and only physical extraction or extremely fast chemical reaction systems can be handled satisfactorily; and (2) the settler must contain a dispersion band, providing sufficient residence time for coalescence to take place without undue entrainment.

Pilot-scale tests can be either batchwise or continuous, although the former are necessarily limited in scope and only suitable for providing preliminary data. Continuous testing can be conducted with a single stage by varying the conditions in turn to simulate the various stages of a multistage plant. It is, however, desirable to carry out runs with a multistage unit to ensure that hydraulic operation will be satisfactory.

In conducting pilot tests it must be borne in mind that extraction plant performance may be temperature-dependent. To ensure that the final production plant can always meet specification it is therefore necessary to obtain pilot plant data for the least favorable condition likely, which apart from the possible effect on the equilibria will always be the lowest temperature.

8.4.2.1 Batch testing

Batch tests are useful for checking rapidly the effects of various conditions such as temperature, alternative reagants, different impeller types, etc. However, the hydrodynamic conditions in a batch mixer differ significantly from those in a continuous unit so that an accurate simulation of the latter cannot be expected.

The tests should be conducted in a vessel of the same geometry as that of the final unit, that is, of fully baffled cylindrical, or of square section, with a diameter or length of side of at least 15 cm and with the same type of impeller and the same ratio of impeller diameter to vessel size. A turbine with six flat blades and a diameter one-third that of the vessel diameter or side is recommended (Rushton *et al.* 1950), unless a pumping impeller is required.

To conduct the tests, the two liquids are introduced into the vessel in the flowsheet ratio and the agitator is operated for a set time, after which a sample is withdrawn and the phases separated and analysed quickly. This is then repeated for different mixing times and, if appropriate, agitator speeds. At the end of each test the rate of phase separation is determined by stopping the agitator and recording the levels of both air–liquid and liquid–liquid interfaces as a function of time.

It is usually the case that for a given agitator speed a plot of the logarithm of the fractional approach to equilibrium, E_B, against time is approximately linear. The effective first-order rate coefficient k_1 is therefore obtainable from the gradient using the following relation, applicable to a straight equilibrium line (Slater *et al.* 1974).

$$E_B = 1 - \exp(-k_1 t) \tag{8.102}$$

The mean residence time, τ, required to give the desired fractional approach to equilibrium in a single continuous mixer under the same conditions can then be predicted approximately by substituting this value of k_1 into the following expression (Treybal 1963)

$$E_H = \frac{k_1 \tau}{1 + k_1 \tau} \tag{8.103}$$

where E_H is the Hausen efficiency (Section 6.4.4, p. 364). The desired Murphree efficiency is then calculated from the appropriate relationship in Table 6.1 (see p. 366).

From the settling data, the rate of interface movement is determined for the dispersion band depth of interest. The specific coalescence rate is then obtained by combining this with the vessel cross-section, and used to obtain an approximate measure of the size of continuous settler required (Godfrey *et al.* 1979).

8.4.2.2 *Continuous method*

The pilot unit should consist of one or more stages of geometrically similar design to that anticipated for the full-scale plant, with a mixing compartment diameter or length of side of 15 cm or 10 per cent of full-scale, whichever is the greater. With horizontal units it is desirable to provide a movable vertical baffle across the settler to enable the dispersion band depth and effective settler area to be varied. Storage tanks of adequate size must be provided to permit steady state operation to be established, usually requiring a minimum of three to four residence times. Real process liquors should be used if possible, to avoid possible errors resulting from the presence of impurities, and a record of temperature should be kept.

The pilot unit is operated with the flowsheet values of the flow ratio and feed concentrations, and samples of the exit extract and raffinate phases are withdrawn for analysis. Complete phase separation is essential, to avoid errors in interpretation. A solute material balance is then computed, and if this agrees to within 2–3 per cent the overall efficiency is calculated as described in Section 6.4.2 (p. 362). If more than one stage is used, the Murphree efficiency of each stage can be calculated if the compositions are determined between stages.

Recirculation ports can be provided between mixer and settler if desired, on the basis of the Holley–Mott principle (1929) shown in Fig. 8.35. These are effective as a result of the differences in density between the emulsion in the mixer and the separated phases in the settler. Such recirculation is of

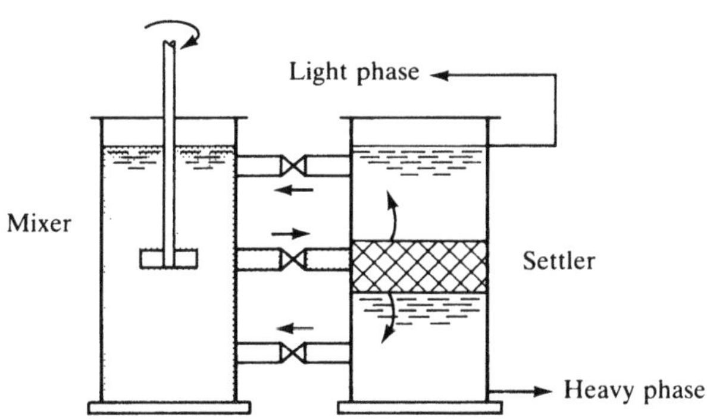

Fig. 8.35. Phase recirculation between settler and mixer (Holley and Mott 1929).

value for increasing the dispersed phase hold-up if the phase ratio, U_d/U_c, is considerably less than unity. It can also be used to determine which phase is to be dispersed in the mixer.

The parameters to be studied should include the total flow rate to show the effect of residence time on stage efficiency. Other parameters, which also influence efficiency, include impeller diameter, type and speed and the effect of making one or other phase continuous. For the settler, measurements should be made of the depth of dispersion band as a function of throughput, which should be varied either by changing the feed rate or by moving the baffle. The rate of collapse of the dispersion in the mixer on stopping the agitator also gives a measure of settling characteristics (Godfrey *et al.* 1977). It should be noted that for geometric reasons the linear velocities of the settled phases will be lower in the production unit than in the full scale unit. As this is an important factor in determining entrainment losses, it may be desirable to provide longitudinal baffles in the pilot settler to obtain velocities comparable with those expected on the full scale (Orjans *et al.* 1977).

8.4.3 Scale-up procedure

8.4.3.1 *Mixer*

The factors contributing to performance, viz. hold-up, droplet size and flow pattern, have different scale-up characteristics so that no simple rule for scale-up can be devised. However, for the pump–settlers [classes Ac3(i) and (iii)], scale-up is effected in terms of pumping capacity, that is, total volume pumped in unit time.

For agitated mixers it is normal to make the residence time equal to that in the pilot unit, thus fixing the mixer volume, and then to scale the agitation rate. Of two criteria generally considered for this purpose, viz. equal tip speed (i.e. constant Nd_r), and equal power per unit volume (equal $N^3d_r^2$), the latter is to be preferred (Orjans *et al.* 1977). However, it is desirable to conduct prior tests on a single full-scale stage in which provision is made for varying the agitator speed and diameter, noting that the power varies as $N^3d_r^5$ in the fully turbulent region, before freezing the design.

8.4.3.2 *Settler*

[See also Davies, Ch. 5]

Experience has suggested that horizontal settlers can be scaled up fairly reliably on a basis of constant specific throughput for a given dispersion band depth, provided that the dispersion characteristics are maintained constant. It is important to ensure, with increase in size, that the dispersion leaving the mixer is distributed across the whole width of the settler without leaving excessive dead space. It has been pointed out in Section 8.4.2.2 (p. 577) that the linear velocities must be kept low to avoid excessive entrainment. On very large units this may necessitate increasing the ratio of width to length in order to reduce the inventory of solvent.

Very little information is available on the coalescence rate and design of settlers for vertical mixer–settlers. However, the mechanism of coalescence differs from that in horizontal settlers in that there is a constant depth of settling emulsion at any given flow rates in the vertical type, as compared with the 'wedge' of emulsion in the horizontal type, so that the coalescing volume should be less.

8.5 Centrifugally-separated extractors

Extractors of this type employ centrifugal force to effect phase separation, and are therefore characterized by extreme compactness, high throughput and low residence time. They are particularly suitable for handling materials of biological origin which tend to form emulsions, and especially those containing labile products such as antibiotic broths.

Two classes of centrifugal extractors are available, viz. those which provide continuous contact, and those of mixer–settler type. These are discussed briefly below in relation to their characterization in the selection chart (Table 8.1).

8.5.1 Continuous-contact units

8.5.1.1 *Perforated plate type [Class Ba(i)]*

The Podbielniak extractor, the earliest centrifugal type available commercially, consists of a series of concentric perforated cylinders contained in a drum rotating about a horizontal shaft. The two liquid phases are introduced into, and removed from the inner- and outermost cylinders via radial pipes connected to passages within the shaft and thence to external seals at each end. The Quadronic extractor is of somewhat similar design, with modifications to improve its versatility.

8.5.1.2 *Film-flow type [Class Ba(ii)]*

In this type the phases contact one another in film flow, rather than in droplet flow as with the preceding type. The Alfa–Laval extractor comprises an underdriven rotating bowl on a vertical axis, with concentric channels within each of which the two phases move countercurrently to one another in film flow; they then pass through ports at top or bottom alternately and thence radially to exits at the inner or outermost ends of the extractor.

8.5.1.3 *Performance*

These machines are mostly available in a number of sizes of varying throughput. Their performance is evaluated in terms of number of theoretical stages, which varies within the range 2–7, depending upon the particular system used and the required flow ratio. It is necessary to operate two or more machines in series (or to repeat the extraction with the same machine on an intermittent basis) if the performance of a single machine is inadequate for the desired separation.

8.5.2 Mixer–settler type

8.5.2.1 *Multi-stage units [Class Bb(i)]*

These extractors comprise vertical arrangements of mixing chambers surmounted by centrifugal separators on a common vertical shaft. The Westfalia 'Luwesta' type, with fixed bowl and rotating shaft, is available with up to three countercurrent stages; for a performance in excess of three near-theoretical stages it would be necessary to employ two or more machines in series.

The Robatel extractor, developed originally for radioactive separations, comprises a rotating bowl divided by baffles into horizontal compartments stacked vertically one above the other. A stationary central shaft carries mixing discs which pass through each compartment and serve both to mix the two phases and to pump them into the corresponding settler.

8.5.2.2 *Single-stage units [Class Bb(ii)]*

Both Westfalia and Robatel also supply single-stage extractors with centrifugal separation. These can be connected together in series to form mixer–settler cascades with any desired number of stages. Specially designed units of this type have been used for the reprocessing of highly radioactive nuclear fuel elements.

Nomenclature

A	pulse amplitude, peak to peak, m
A_i	pulse amplitude, peak to peak within packing, m
A_o	constant [eqn (8.13)]
Ar	Archimedes number, $d_{32}^3 \Delta\rho g \rho_c / \mu_c^2$
a	interfacial mass transfer area per unit volume, $m^2/m^3 = m^{-1}$
a_c	value of a for coalescing droplets, m^{-1}
a_d	radius of droplet [eqns (8.11, 8.13)], m
$a_{d,f}$	average surface of droplet during growth [eqn (8.93)], m^2
a_f	value of a for forming droplets, m^{-1}

a_p	specific surface of packing, i.e. area per unit volume of packing, m^{-1}
a_r	interfacial area of rising (or falling) droplets, m^{-1}
b	coalescence parameter [eqn (8.79)]
b	deceleration of droplet [eqn (8.53d)], m/s^2
C_D	drag coefficient [eqn (8.53c)]
C_o	orifice coefficient
c	solute concentration, kg/m^3
c_i	solute concentration at interface [eqn (8.15)], kg/m^3
c_r	reactant concentration [eqn (8.2)], kg/m^3
D	diffusion coefficient, m^2/s
d	droplet diameter, m
d_{32}	Sauter mean droplet diameter, $\Sigma_i n_i d_i^3 / \Sigma_i n_i d_i^2$, m
d_{43}	mass–volume mean droplet diameter, $\Sigma_i n_i d_i^4 / \Sigma_i n_i d_i^3$, m
d_c	column diameter, m
d_e	$4\varepsilon/a$, equivalent diameter of packing voids, m
d_j	jet diameter, m
d_N	nozzle diameter, m
d_o	orifice or perforation diameter, m
d_p	packing size, m
d_r	rotor diameter, m
d_s	inner diameter of stator ring, m
d_{32}^0	droplet characteristic diameter, m
E_B	efficiency of batch mixer
E_k	effective longitudinal (i.e. backmixing) diffusion coefficient of phase k, m^2/s
E_H	Hausen efficiency (Chapter 6)
E_{My}	Murphree efficiency based on extractant phase (Chapter 6)
F_k	flow rate of phase k, m^3/s
f	frequency, s^{-1}
f	function in eqn (8.13) (Jeffries and Davies 1971)
g	acceleration due to gravity, m/s^2
H	height of rise of droplets, m
H_{ok}	height of an overall transfer unit based on phase k, m
HETS	height equivalent to a theoretical stage, m
h_c	height of compartment (i.e. stage), m
h_d	depth of coalesced dispersed phase below or above plate, m
h_r	height of rotor, m
h_m	height of well-mixed zones across plate, m
I	plate impedence [eqn (8.49)]
K	coalescence constant [eqn (8.12)], $s^{-1.5}$
k_n	nth order rate constant, $m^{3n-3}/s\ (kmol)^{n-1}$
k_k	mass transfer coefficient for phase k, m/s
k_{ok}	overall mass transfer coefficient based on phase k, m/s
L	length, m
Ma	Marangoni number [eqn (8.15)]
M	Morton number, $g\mu_c^4 \Delta\rho/\gamma^3\rho_c^2$
N	revolutions per second, s^{-1}

N_a	flux of solute A across interface, kmol/m^2 s
N_c	number of uncoalesced droplets; $N_{c,0}$ is the initial number
N_0	number of holes in perforated plate [eqn (8.94)]
N_s	number of stages (i.e. compartments)
N^1_{ox}	number of X-phase (i.e. feed phase) transfer units per stage
NTS	number of theoretical stages
n_i	number of droplets in size range i
P	power input, W
Po	Power number, $P/\rho N^3 d_r^5$
Pe_c	Peclet number for continuous phase, $d_p U_c/E_c$
q_c	backmix ratio between well-mixed regions on either side of a plate
R_f	radius of draining film [eqn (8.11)], m
Re_d	droplet Reynolds number, $dV_t\rho_c/\mu_c$
S	cross-section of perforated area of perforated plate [eqns (8.85, 8.94)], m^2
S_o	cross-sectional area of perforations, m^2
S_1, S_2	cross-sectional areas of column and pulse leg respectively [eqn (8.65)], m^2
$\bar{s}$	mean droplet path length, m
t	time, s
$t_{1/2}$	reaction half-time, s
U_{cd}	velocity of continuous phase in downcomer (or riser), m/s
U_k	superficial velocity of phase k in column, m/s
U_o	velocity within orifice, m/s
U_r	flow ratio, $U_d/U_c = F_d/F_c$
u^*	modified characteristic velocity [eqn (8.66)], m/s
V_{dr}	droplet velocity before start of deceleration, m/s
V_t	terminal velocity of droplets, m/s
v_j	nozzle velocity at which jet formation starts, m/s
v_N	nozzle velocity, m/s
$\bar{v}_{fA}$	characteristic droplet velocity for pulse velocity fA, m/s
$\bar{v}_N$	characteristic droplet velocity for rotor speed N, m/s
$\bar{v}_0$	characteristic droplet velocity for non-mechanical column, m/s
$\bar{v}_s$	mean droplet slip velocity [eqns (8.3, 8.66)], m/s
w_b	width of wall baffle, m
Y^0	dimensionless extract phase exit composition (diffusion model)
Y^1	dimensionless extract phase exit composition (backflow model)
y	linear displacement of the liquid [eqn (8.65)], m
z_p	spacing between adjacent holes on perforated plate [eqn (8.53f)], m

Greek Letters

α_k	backmix ratio for phase k, i.e. ratio of backflow to net forward flow between stages, F_k
δ	boundary layer thickness or penetration depth [eqn (8.15)], m
γ	interfacial tension, kg/s^2 = 10^3 dyne/cm
ϵ	fractional void volume in packing or column internals
ϵ_p	fractional free area in perforated plate [within cross-section in Section 8.3.4.5.2, p. 563].
μ	viscosity, Pas = kg/m s^3

ν_k kinematic viscosity of phase k, m^2/s
ρ_k density of phase k, kg/m^3
$\Delta\rho$ absolute density difference, $|\rho_c - \rho_d|$, kg/m^3
τ mean residence time, s
η maximum volumetric efficiency [eqn (8.73)], s^{-1}
ϕ fractional volumetric hold-up of dispersed phase
ϕ_k fractional volumetric hold-up of phase k
ψ power dissipated per unit mass [eqns (8.46 and 8.68)], m^2/s^3

Subscripts
c continuous phase
c during droplet coalescence [eqn (8.97)]
d dispersed phase
f flooding
f during droplet formation [eqns (8.93, 8.97)]
i droplet size range
k *c* or *d* phase; or *X* or *Y* phase
m property of the mixture
n stage number
r during droplet rise or fall
x feed (i.e. *X*−) phase
y extractant (i.e. *Y*−) phase

References

ANDERSON, W. J. and PRATT, H. R. C. (1978). *Chemical Engineering Science*, **33**, 995.
AUFDERHEIDE, E. and VOGELPOHL, A. (1986). *Chemical Engineering Science* **41**, 1747.
BAIRD, M. H. I. (1965). *Proceedings of the joint American Institute of Chemical Engineers/Institution of Chemical Engineers Symposium*, London, **6**, 53.
BAIRD, M. H. I. and LANE, S. J. (1973). *Chemical Engineering Science*, **28**, 947.
BAIRD, M. H. I. and SHEN, Z. J. (1984). *Canadian Journal of Chemical Engineering*, **62**, 218.
BAIRD, M. H. I. and RITCEY, G. M. (1974). *Proceedings of the International Solvent Extraction Conference*, Lyons (ISEC '74), **2**, 1571. Society of Chemical Industry, London.
BASHFORTH, F. and ADAMS, J. C. (1883). *'An attempt to test the theories of capilliary action*. CUP, Cambridge.
BATEY, W., THOMPSON, P. J., and THORNTON, J. D. (1987), *"Extraction '87," Institution of Chemical Engineers Symposium Series* No. 103, 133.
BELL, R. L. and BABB, A. L. (1969). *Industrial and Engineering Chemistry, Process Design and Development*, **8**, 392.
BENSALEM, A. (1985). Ph.D. thesis. Swiss Federal Institute of Technology, Zurich.
BENSALEM, A., STEINER, L., and HARTLAND, S. (1983). *Proceedings of the International Solvent Extraction Conference*, Denver (ISEC '83), 130. American Institute of Chemical Engineers, New York.

BLANDING, E.H. and ELGIN, J.C. (1942). *Transactions of the American Institute of Chemical Engineers*, **38**, 305.

BÜHLER, B., COVELLI, B., and WIDMER, F. (1977). *Chimia*, **31**, 307.

BUSSOLARI, R., SCHIFF, S., and TREYBAL, R.E. (1953). *Industrial and Engineering Chemistry*, **45**, 2413.

CHRISTIANSEN, R.M. and HIXSON, A.N. (1957). *Industrial and Engineering Chemistry*, **49**, 1017.

COLBURN, A.P. and WELSH, D.G. (1942). *Transactions of the American Institute of Chemical Engineers*, **38**, 179.

COLEBY, J. (1983). *Handbook of solvent extraction*, (eds T.C. LO, M.H.I. BAIRD, and C. HANSON), Ch. 13.6. Wiley, New York.

COULSON, J.M. and SKINNER, S.J. (1952). *Chemical Engineering Science*, **1**, 197.

DAVIES, J.T. and RIDEAL, E.K. (1963). *Interfacial phenomena* (2nd edn). Academic Press, New York.

DE CHAZAL, L.E.M. and RYAN, J.T. (1971). *American Institute of Chemical Engineers Journal*, **17**, 1226.

DELL, F.R. and PRATT, H.R.C. (1951). *Transactions of the Institution of Chemical Engineers*, **29**, 89, 270.

DONGAONKAR, K.R. (1989). PhD thesis. University of Melbourne, Australia.

DONGAONKAR, K.R., PRATT, H.R.C. and STEVENS, G.W. (1991). *American Institute of Engineers Journal*, **37**, 694.

ELTON, G.A.H. and PICKNETT, R.G. (1957). *Proceedings of the 2nd International Congress on Surface Activity*, **1**, 288. Butterworths, London.

ERGUN, S. (1952). *Chemical Engineering Progress*, **48**, 89.

EWERS, W.E. and SUTHERLAND, K.L. (1952). *Australian Journal of Scientific Research*, **5**, 697.

FIOCCO, R.J. (1983). *Proceedings of the International Solvent Extraction Conference*, Denver (ISEC '83), 30. American Institute of Chemical Engineers, New York.

FITZPATRICK, L.M., STEVENS, G.W., and PRATT, H.R.C. (1986). *Proceedings of the International Solvent Extraction Conference*, Munich (ISEC '86), **III**, 325. Dechema, Frankfurt-am-Main.

GAUBINGER, W., HUSING, G., and MARR, R. (1983). *German Chemical Engineering*, **6**, 74.

GARG, M.O. and PRATT, H.R.C. (1984). *American Institute of Chemical Engineers Journal*, **30**, 432.

GAYLER, R. and PRATT, H.R.C. (1951). *Transactions of the Institution of Chemical Engineers*, **29**, 110.

GAYLER, R. and PRATT, H.R.C. (1953). *Transactions of the Institution of Chemical Engineers*, **31**, 69.

GAYLER, R. and PRATT, H.R.C. (1957a). *Transactions of the Institution of Chemical Engineers*, **35**, 267.

GAYLER, R. and PRATT, H.R.C. (1957b). *Transactions of the Institution of Chemical Engineers*, **35**, 273.

GAYLER, R., ROBERTS, M.W., and PRATT, H.R.C. (1953). *Transactions of the Institution of Chemical Engineers*, **31**, 57

GILLESPIE, T. and RIDEAL, E.K. (1956). *Transactions of the Faraday Society*, **52**, 173.

GODFREY, J.C., CHANG-KAKOTI, D.K., SLATER, M.J., and THARMALINGAM (1977). *Proceedings of the International Solvent Extraction Conference*, Toronto (ISEC '77), Vol. 1, p. 406. Canadian Institute of Mining and Metallurgy, Montreal (1979).

GRIFFITH, W.L., JASNY, G.R., and TUPPER, H.T. (1952). U.S. Atomic Energy Commission Report AECD-3440.

GRONIER, W.A., MCALLISTER, R.A., and RYON, A.D. (1966). U.S. Atomic Energy Commission Report ORNL-3890.

GROOTHUIS, H. and ZUIDERWEG, F.J. (1960). *Chemical Engineering Science*, **12**, 288.

HAFEZ, M.M., BAIRD, M.H.I., and NIRDOSH, I. (1979). *Canadian Journal of Chemical Engineering*, **57**, 150.

HAMILTON, J.A. and PRATT, H.R.C. (1984). *American Institute of Chemical Engineers Journal*, **30**, 442.

HARTLAND, S. (1969). *Chemical Engineering Science*, **24**, 987.

HAYWORTH, C.B. and TREYBAL, R.E. (1950). *Industrial and Engineering Chemistry*, **42**, 1174.

HODY, D. (1975). Ph.D. thesis. Eidgenössischen Technische Hochschule, Zürich.

HORVATH, M., STEINER, L., and HARTLAND, S. (1978). *Canadian Journal of Chemical Engineering* , **56**, 9.

HOLLEY, A.E. and MOTT, O.E. (1929). British Patent 321,200.

HOULIHAN, R. and LANDAU, J. (1974). *Canadian Journal of Chemical Engineering*, **52**, 758.

HU, S. and KINTNER, R.C. (1955). *American Institute of Chemical Engineers Journal*, **1**, 42.

HUGHMARK, G.A. (1967). *Industrial and Engineering Chemistry, Fundamentals*, **6**, 408.

INGHAM, J. (1972). *Transactions of the Institution of Chemical Engineers*, **50**, 372.

INGHAM, J., BOURNE, J.R., and MÖGLI, A. (1974). *Proceedings of the International Solvent Extraction Conference*, Lyons (ISEC '74), **2**, 1299. Society of Chemical Industry, London.

IZARD, J.A. (1972). *American Institute of Chemical Engineers Journal*, **18**, 634.

JEALOUS, A.C. and JOHNSON, H.F. (1955). *Industrial and Engineering Chemistry*, **47**, 1159.

JEELANI, S.A.R. and HARTLAND, S. (1988). *Proceedings of the International Solvent Extraction Conference*, Moscow (ISEC '88), **II**, 128.

JEFFRIES, G.V. and DAVIES, G.A. (1971). *Recent advances in liquid-liquid extraction* (ed. C. HANSON), Ch. 14. Pergamon Press, Oxford.

JEFFRIES, G.V., DAVIES, G.A., and PITT, K. (1970). *American Institute of Chemical Engineers Journal*, **16**, 823, 827.

JEFFRIES, G.V., SMITH, D.V., and PITT, K. (1967). *Institution of Chemical Engineers Symposium Series* No. 26, 93.

KARR, A.E. (1959). *American Institute of Chemical Engineers Journal*, **5**, 446.

KARR, A.E. (1980). *Proceedings of the International Solvent Extraction Conference*, Liege (ISEC '80), paper 80-28. Association des Ingenieurs sortis de l'Universite de Liege; *Separation Science and Technology*, **15** (4), 877.

KARR, A.E. and LO, T.C. (1971). *Proceedings of the International Solvent Extraction Conference*, The Hague (ISEC '71), **1**, 299. Society of Chemical Industry, London.

KARR, A.E. and LO, T.C. (1976). *Chemical Engineering Progress*, **72**, 68.

KARR, A.E. and LO, TC. (1977). *Proceedings of the International Solvent Extraction Conference*, Toronto (ISEC '77), **1**, 355. Canadian Institute of Mining and Metallurgy, Montreal (1979).

KARR, A.E., and RAMANUJAM, S., LO, T.C., and BAIRD, M.H.I. (1986). *Proceedings of the International Solvent Extraction Conference*, Munich (ISEC '86), **I**, 41. Dechema, Frankfurt-am-Main.

KARR, A.E. and RAMANUJAM, S. (1986). *Proceedings of the International Solvent Extraction Conference*, Munich (ISEC '86), **III**, 943. Dechema, Frankfurtam-Main.

KOLMOGAROV, A.N. (1949). *Doklady Akademii Nauk SSSR*, **66**, 825.

KOSTERS, W.C.G. (1983). *Handbook of solvent extraction*, (eds. T.C. LO, M.C.H. BAIRD, and C. HANSON), Ch. 13.1. Wiley, New York.

KUMAR, A. and HARTLAND, S. (1982). *Chemical Engineering Research and Design*, **60**, 35.

KUMAR, A. and HARTLAND, S. (1983). *Chemical Engineering Research and Design*, **61**, 248.

KUMAR, A. and HARTLAND, S. (1984). *Chemical Engineering Communications*, **31**, 193.

KUMAR, A. and HARTLAND, S. (1986). *Chemical Engineering Communications*, **44**, 163.

KUMAR, A., STEINER, L., and HARTLAND, S. (1986). *Industrial and Engineering Chemistry, Process Design and Development*, **25**, 728.

LADDHA, G.S. and DEGALEESAN, T.E. (1976).*Transport phenomena in liquid extraction*, Ch. 11. Tata McGraw-Hill, New Delhi.

LEE, J.C. and LEWIS, G. (1967). *Institution of Chemical Engineers Symposium Series* No. 26, 13.

LEWIS, I.E. (1977). *Proceedings of the International Solvent Extraction Conference*, Toronto (ISEC '74), **1**, 325. Canadian Institute of Mining and Metallurgy, Montreal (1979).

LEWIS, J.B., JONES, I., and PRATT, H.R.C. (1951). *Transactions of the Institution of Chemical Engineers*, **29**, 126.

LEWIS, J.B. and PRATT, H.R.C. (1953). *Nature*, **171**, 1155.

LOGSDAIL, D.H. and LARNER, G.S. (1964). U.K. Atomic Energy Authority Reports A.E.R.E. 4408, 4409.

LOGSDAIL, D.H. and THORNTON, J.D. (1957). *Transactions of the Institution of Chemical Engineers*, **35**, 331.

LOGSDAIL, D.H., THORNTON, J.D., and PRATT, H.R.C. (1957). *Transactions of the Institution of Chemical Engineers*, **35**, 301.

LO, T.C. (1975), *Recent developments in commercial extractors*, paper presented at Engineering Foundation Conference on Mixing Research at Ridge, New Hampshire; see also PRATT, H.R.C. and HANSON, C. (1983).

LO, T.C. and PROCHAZKA, J. (1983). *Handbook of solvent extraction* (eds T.C. LO, M.H.I. BAIRD, and C. HANSON), Ch. 12. Wiley, New York.

LO, T.C. and KARR, A.E. (1972). *Industrial and Engineering Chemistry, Process Design and Development*, **11**, 495.

LOUTATY, R., VIGNES, A., and LEGOFF, P. (1969). *Chemical Engineering Science*, **24**, 1795.

MCALLISTER, R.A., GRONIER, W.S., and RYON, A.D. (1967). *Chemical Engineering Science*, **22**, 931.

McAvoy, R.M.L and Kintner, R.C. (1965). *Journal of Colloid Science*, **20**, 188.
Major, C.J. and Hertzog, R.R. (1955). *Chemical Engineering Progress*, **51**, 17.
Mar, B.W. and Babb, A.L. (1959). *Industrial and Engineering Chemistry*, **51**, 1011.
Marangoni, C. (1865). *Sull' expansione delle goccie di un liquido galleggiante sulla superficie di altro liquido*. Fusi, Pavia.
Mayfield, F.D. and Church, W.L. (1952). *Industrial and Engineering Chemistry*, **44**, 2253.
Misek, T. (1967). *Collection of Czechoslovak Chemical Communications*, **32**, 4018.
Misek, T. (1975). *Collection of Czechoslovak Chemical Communications*, **40**, 1686.
Misek, T. and Marek, J. (1970). *British Chemical Engineering* , **15** (2), 202.
Misek, T. and Marek, J. (1983). *Handbook of solvent extraction* (eds T.C. Lo, MH.I. Baird, and C. Hanson), Ch. 13.2. Wiley, New York.
Mögli, A. and Bühlmann, U. (1983). *Handbook of solvent extraction* (eds T.C. Lo, M.H.I. Baird, and C. Hanson),Ch. 13.5. Wiley, New York.
Moore, R.J. (1987). M.Eng. thesis. University of Melbourne, Australia.
Moore, R.J., Pratt, H.R.C., and Stevens, G.W. (1989). *Solvent Extraction and Ion Exchange*, **7**, 1043.
Morello, V.S. and Poffenberger, N. (1950). *Industrial and Engineering Chemistry*, **42**, 1021.
Mumford, C.J. and Al-Hemiri, A.A.A. (1974). *Proceedings of the International Solvent Extraction Conference*, Lyons (ISEC '74), **2**, 1591. Society of Chemical Industry, London.
Nemecek, M. and Prochazka, J. (1974). *Canadian Journal of Chemical Engineering*, **52**, 739.
Niebuhr, D. and Vogelpohl, A. (1981). *Proceedings 2nd World Congress of Chemical Engineering*, Montreal, 422.
North, A.A. and Wells, R.A. (1964–65). *Transactions of the Institution of Mining and Metallurgy*, **74**, 463.
Novotny, P., Prochazka, J., and Landau, J. (1970). *Canadian Journal of Chemical Engineering*, **48**, 405.
Oldshue, J.Y. and Rushton, J. (1952). *Chemical Engineering Progress*, **48**, 297.
Orjans, J.R., Notebaart, C.W., Godfrey, J.C., Hanson, C., and Slater, M.J. (1977). *Proceedings of the International Solvent Extraction Conference*, Toronto (ISEC '77), **1**, 340. Canadian Institute of Mining and Metallurgy, Montreal (1979).
Perrut, M., Loutaty, R., and Legoff, P. (1973). *Chemical Engineering Science*, **28**, 1541.
Pietzsch, W. and Blass, E. (1987). *Chemical Engineering Technology*, **10**, 73.
Pietzsch, W. and Pilhofer, T. (1984). *Chemical Engineering Science*, **39**, 961.
Pratt, H.R.C. (1954). *Industrial Chemist*, **30**, 475, 597.
Pratt, H.R.C. and Anderson, W.J., (1977). *Proceedings of the International Solvent Extraction Conference*, Toronto (ISEC '77), **1**, 242. Canadian Institute of Mining and Metallurgy, Montreal (1979).
Pratt, H.R.C. and Hanson, C. (1983). *Handbook of solvent extraction* (eds T.C. Lo, M.H.I. Baird, and C. Hanson), Ch. 16. Wiley, New York.
Princen, H.M. (1963). *Journal of Colloid Science*, **18**, 178.
Princen, H.M. and Mason, S.G. (1965). *Journal of Colloid Science*, **20**, 156.

PROCHAZKA, J. (1970). *Dechema Monographien*, **65**, 325.

PROCHAZKA, J., LANDAU, J., SOUHRADA, F., and HEYBERGER, (1975) A. *British Chemical Engineering*, **16**, 42.

PRVCIC, L.M. (1987). Ph.D. thesis. University of Melbourne, Australia.

PRVCIC, L.M., PRATT, H.R.C., and STEVENS, G.W. (1989). *American Institute of Chemical Engineers Journal*, **35**, 1845.

PURANIK, S.A. and SHARMA, M.M. (1970). *Chemical Engineering Science*, **25**, 257.

REMAN, G.H. and OLNEY, R.B. (1955). *Chemical Engineering Progress*, **51**, 146.

REYNOLDS, O. (1886). *Proceedings of the Royal Society*, **A177**, 157.

ROUYER, H., LEBOUHELLEC, J., HENRY, E., and MICHEL, P. (1974). *Proceedings of the International Solvent Extraction Conference*, Lyons (ISEC '74), **3**, 2339. Society of Chemical Industry, London.

RUSHTON, J.H., COSTICK, E.W., and EVERETT, H.J. (1950). *Chemical Engineering Progress*, **46**, 395, 467.

SAKIADIS, B.C. and JOHNSON, A.I. (1954). *Industrial and Engineering Chemistry*, **46**, 1229.

SARKAR, S., PHILLIPS, C.R., and MUMFORD, C.J. (1985). *Canadian Journal of Chemical Engineering*, **63**, 701.

SAWISTOWSKI, H. (1971). *Recent advances in liquid–liquid extraction* (ed. C. HANSON), Ch. 9, Pergamon Press, Oxford.

SCHEELE, G.F. and MEISTER, B.J. (1968). *American Institute of Chemical Engineers Journal*, **14**, 9.

SCHEIBEL, E.G. (1948). *Chemical Engineering Progress*, **44**, 681.

SCHEIBEL, E.G. (1956). *American Institute of Chemical Engineers Journal*, **2**, 74.

SCHEIBEL, E.G. (1968). U.S. Patent 3,389,970 (June 25, 1968).

SCHEIBEL, E.G. (1983). *Handbook of solvent extraction* (eds T.C. LO, M.H.I. BAIRD, and C. HANSON), Ch. 13.3. Wiley, New York.

SCRIVEN, L.E. and STERNLING, C.V. (1964). *Journal of Fluid Mechanics*, **19**, 321.

SEGE, G. and WOODFIELD, F.W. (1954). *Chemical Engineering Progress*, **50**, 396.

SEHMEL, G.A. and BABB, A.L. (1964). *Industrial and Engineering Chemistry, Process Design and Development*, **3**, 210.

SHEIKH, A.R., INGHAM, J., and HANSON, C. (1972). *Transactions of the Institution of Chemical Engineers*, **50**, 199.

SIMONS, A.J.F. (1978). *Chemistry and Industry*, 748.

SIMONS, A.J.F. (1983). *Handbook of solvent extraction* (eds T.C. LO, M.H.I. BAIRD, and C. HANSON), Ch. 11.1. Wiley, New York.

SKELLAND, A.H.P. and CONGER, W.L. (1973). *Industrial and Engineering Chemistry, Process Design and Development*, **12**, 448.

SKELLAND, A.H.P. and CORNISH, A.R.H. (1965). *Canadian Journal of Chemical Engineering*, **43**, 302.

SLATER, M.J., RITCEY, G.M., and PILGRIM, R.F. (1974). *Proceedings of the International Solvent Extraction Conference*, Lyons (ISEC '74), **1**, 107. Society of Chemical Industry, London.

SMITH, D.V. and DAVIES, G.A. (1970). *Canadian Journal of Chemical Engineering*, **48**, 628.

SPAAY, N.M., SIMONS, A.J.F., and TEN BRINK, G. (1971). *Proceedings of the International Solvent Extraction Conference*, The Hague (ISEC '71), **1**, 281. Society of Chemical Industry, London.

STAINTHORP, F.P. and SUDALL, N. (1964). *Transactions of the Institution of Chemical Engineers*, **42**, T198.
STEINER, L. and HARTLAND, S. (1980). *Separation Science Technology*, **15**, 907.
STEINOUR, H.H. (1944). *Industrial and Engineering Chemistry*, **36**, 618.
STERNLING, C.V. and SCRIVEN, L.E. (1959). *American Institute of Chemical Engineers Journal*, **5**, 514.
STEVENS, G.W. and PRATT, H.R.C. (1986). *American Institute of Chemical Engineers Journal*, **32**, 2079.
STICHLMAIR, J. (1980). *Chemie-Ingenieur-Technik*, **52**, 253.
STRAND, C.P., OLNEY, R.B., and ACKERMAN, G.H. (1962). *American Institute of Chemical Engineers Journal*, **8**, 252.
TAWFIK, W.Y., ECKLES, A.J., and TEDDER, D.W. (1988). *Solvent Extraction and Ion Exchange*, **6**, (4), 563.
THOMSON, J. (1855). *Philosophical Magazine*, **10**, (4), 330.
THOMPSON, W.S., WOODWARD, T., SHRODE, W.A., BAIRD, E.D., and OLIVER, D.A. (1962). U.S. Office of Saline Water Progress Report, 63; see also ibid, 1963, 48.
THORNTON, J.D. (1956). *Chemical Engineering Science*, **5**, 210.
THORNTON, J.D. (1957). *Transactions of the Institution of Chemical Engineers*, **35**, 316.
THORNTON, J.D. and PRATT, H.R.C. (1953). *Transactions of the Institution of Chemical Engineers*, **31**, 289.
TREYBAL, R.E. (1963). *Liquid extraction* (2nd edn). McGraw-Hill, New York.
TUNG, S.L. and LUECKE, R.H. (1986). *Industrial and Engineering Chemistry, Process Design and Development*, **25**, 664.
VAN DIJCK, W. (1935). U.S. Patent 2,011,186.
VASSELLO, G., (1982). Ph.D. Thesis. University of Newcastle-upon-Tyne.
VERMEULEN, T., MOON, J.S., HENNICO, A., and MIYAUCHI, T. (1966). *Chemical Engineering Progress*, **62** (9), 95.
VERMIJS, H.J.A. (1965). *Proceedings of the Joint American Institute of Chemical Engineers/Institution of Chemical Engineers Symposium*, London, **6**, 53.
WANG, P.S.M., INGHAM, J., and HANSON, C. (1977). *Transactions of the Institution of Chemical Engineers*, **35**, 196.
WATSON, J.S. and MCNEEZE, L.E. (1972). *Industrial and Engineering Chemistry, Process Design and Development*, **11**, 1210.
WEECH, M.E. and KNIGHT, B.E. (1967). *Industrial and Engineering Chemistry, Process Design and Development* **6**, 481.
WOODFIELD, F.W. and SEGE, G. (1954). *Chemical Engineering Progress Symposium Series* No. 13, **50**, 14.
WOODWARD, T. (1961). *Chemical Engineering Progress*, **57** (1), 52.
ZHELEZNYAK, A.S. and LANDAU, A.M. (1973). *Theoretical Foundations of Chemical Engineering*, **7**, 525.

INDEX

Abrams–Prausnitz model 9–10
acetic acid, extraction of 179
acetic-acid–tetrachloromethane–triethylamine system, phase diagram for 32
acetylacetone–water system, instability of 173, 176, 178
active separation 303
activity coefficients 6
  prediction of 7–17
    application to ionic systems 32–4
    application to organic systems 32
    correlating equations used 8–10
    correlations for electrolyte solutions 13–17
    Gibbs–Duhem equation used 7–8
    ion interactions and association equilibria 17
    predictive equations used 10–13
    selection of correlation models 18–19
  pressure dependence of 7
  temperature dependence of 6
agitated cell columns
  cost comparison 500
  performance comparison 500
  see also rotary agitated columns
agitated systems
  mass transfer around drops in 120–3
  mass transfer in breaking drops 109–13
Alfa-Laval centrifugal extractors 579–80
ambivalence region 253
  factors affecting 254
  mixer-settlers 314–16
American Petroleum Institute, design standard for oil separators 307
amyl alcohol, water drop coalescence in 285
analytical solution methods (for plug flow)
  countercurrent extraction 358–60
  differential extractors 374–5
  dual-solvent extraction 392–4
anisole–water system, drop coalescence in 193
asymmetric rotary disc columns (ARDCs) 563–6
  backmixing data for 565–6
  characteristics 494–5
  demountable agitators used 564
  hydrodynamics 564–5
  mass transfer in 565
  scale-up for 566
aviation fuel, dewatering of 323
axial dispersion 416–85
  graphical solution methods used 448–57
    differential columns 448–54
    stagewise contactors 454–7
  measurement of parameters 474–85
  nature of 416–17
  nomenclature used 485–8
  numerical solution methods used 457–70
    dual-solvent extraction 468–70
    matrix method 460–8
    stage-to-stage calculation 458–9
    unsteady-state method 459
  and polydispersivity 470–4
  scale-up procedure allowing for 516
axial dispersion coefficients
  continuous-phase parameters, determination of 483
  dispersed-phase parameters, determination of 483–4
  measurement of 474–84
    concentration profiles used 474–7
    tracer injection methods used 477–84
axial dispersion models 417–48
  backflow model 417, 428–33
  comparison of models 433–5
  diffusion model 417, 418–28
  generalized solutions 437–48
  special solutions 435–7

backflow (axial dispersion) model 417, 428–33
  assumptions made 428
  compared with diffusion model 433–5
  derivation of equations 428–30
  generalized solutions 437–48
    linear equilibria 437–9
    non-linear equilibria 439, 442–3
    programming of solutions 443–4
    worked examples 445–6, 447–8
  graphical solution for
    backmixing in both phases 457
    backmixing in one phase 455
  measurement of backmixing parameters in 482–3
  relationship with diffusion model 434–5
  scale-up using 516
  simplified solutions 432–3

backflow (axial dispersion) model—*contd*
solution of equations 431–2
special solutions 435, 437
backmixing
cause of 417
concentration profiles affected by 423
one-phase backmixing
differential extractors 450–2
stagewise contactors 454–6
and transfer units 425–6
two-phase backmixing
differential extractors 452–4
stagewise contactors 456–7
see also axial dispersion
baffle-plate columns 520–1
characteristics 494–5
disc-and-doughnut type 521
models used 417
side-by-side type 521
batch extraction processes 403–11
countercurrent distribution
calculation of distribution 406–8
process described 405–6
countercurrent extraction simulated by 403–4
differential extraction 408–11
process described 408–9
Rayleigh extraction used 409–11
multiple-contact batch extraction 403
single-contact batch extraction 403
batch fractional extraction 411
batch separation time, definition of 308
batch settling, interrelation to continuous behaviour 301–3, 307
Bénard cells 162, 172–3
benzene, miscibility with other solvents 29
binary liquid–liquid systems
heat-produced interfacial instabilities in 172–9
phase diagrams for 2
stability analysis of 174–5
Bond number 267, 280
boundary layer models
laminar flow 66–72
turbulent flow 74
Brownian motion 326
Burkholder–Berg (jet) model 186

capacity mass transfer coefficients 58
carbon tetrachloride, disperse phase with water 253
cell equilibration method, liquid–liquid equilibria measured using 20–2
centrifugal extractors 579–80
characteristics of 494–5
continuous-contact type 579–80
characteristics of 494–5
mixer–settler type 580
characteristics of 494–5
performance of 580
centrifugal pumps, problems in mixer-settlers 316
chemical reactions
interfacial stability affected by 161, 179–80
mass transfer affected by 135–40
heterogeneous reactions 397–8, 496–7
homogeneous reactions 395–7, 497
chlorobenzene–water system, drop size from jet break-up 190–1
circulating drops, mass transfer in 105–6, 224–7
cleaning, ease of, extractor selection affected by 495, 498
coagulated membranes, void structures in 177, 179
coalescence phenomena 255–306
application to design of industrial settlers 306–16
drop-to-drop coalescence 284–5
drop-to-interface coalescence 285–306
film drainage considerations 270–84
nomenclature used 334–7
single drops 257–70
see also drop coalescence
coalescence times
curvature effects 262–3
effect of third component/phase 267
effect of vibration 263
effect of viscosity 267
experimental determination of 262
factors affecting 258, 260–2
coalescer packing
cross-entrainment data for 319
mixer–settler design for 319
non-woven fibre materials used 320–31
cobalt, extraction of 33–4
co-current flow
in coalescers 331–2
mass transfer in 66–72
columns
design of 513
dual-solvent fractional extraction in 394
interfaces in 343
see also baffle-plate . . . ; Kühni . . . ; Oldshue–Rushton . . . ; packed . . . ; perforated plate . . . ; pulsed . . . ; Scheibel . . . ; spray columns
computer programs
axial dispersion models 443–4, 469
dual-solvent extraction 394–5, 469
concave interfaces, drop coalescence at 262–3, 285

concentrated solutions
  electrolytes, diffusivity relationships 52–3
  non-electrolytes, diffusivity relationships 52
concentration units 346
  dimensionless form 346–7
  frames of reference for 345–6
  high solute concentrations 347–8
convex interfaces, drop coalescence at 263, 283, 284
copper, extraction of 33–4, 179–80, 313, 315
countercurrent distribution 405–8
  calculation of distribution 406–8
  operating scheme for 406
  process described 405–6
countercurrent extraction
  batch processes used 403–5
  graphical methods for 354–8
    dual feeds considered 358
    general solution 354–8
  material balance calculations 355, 358–60
  multicomponent separation in 400–2
  non-ideal stages 365, 367
  non-linear equilibria considered 379
  operating diagrams for 347, 356, 360–1
  with reflux 379–81
    graphical solution used 379–81
    process described 379
countercurrent extractors
  design procedures for 343–4
  stage efficiencies for 366
  see also mixer-settlers
Craig extractor tubes 405–6, 407
  process described 405–6
  product distribution curve 408
critical drop size 307, 308, 318
critical film thickness 281
critical sedimentation velocity 307, 308
critical Weber number 248
cross entrainment 314, 315, 316
  techniques to reduce 316–34
crossflow extraction, multicomponent separation in 399–400
crossflow settlers
  design of 308–16
    differential model used 310–12
    experimental data for 312–14
    wedge dispersion band model used 309–10
  inclined plates used 317–18
  packed beds used 318–34
  phase entrainment in 314, 315, 316
    techniques to reduce 316–34
  stage efficiencies for 366
crud, origin of term 267
'cut off' drop size 307, 308
cyclohexane, miscibility with other solvents 29
cyclohexanol–water system, instability of 176
cyclohexanone–water system, drop coalescence in 285, 287

Danckwerts' surface-renewal model 63, 136, 211–12, 215, 220
  interfacial tubulence allowed for in 197, 203
  see also surface-renewal models
D'Arcy's equation 325
daughter (secondary) droplet formation 257, 259–60
Debye–Hückel limiting law 14–15
degrees of freedom 1
density difference ranges 496
density gradients, interfacial convection caused by 162
design (of extraction equipment) 503–14
  requirements listed 494–5
detached drops
  mass transfer in and around 99–123
    around drops in agitated systems 120–3
    around drops in non-agitated systems 113–20
    in breaking drops in agitated systems 109–13
    in circulating drops 105–6
    in oscillating drops 106–9
    in stagnant drops 99–105
diethylene glycol, miscibility with other solvents 29
diethyl ether, miscibility with other solvents 29
di(2-ethylhexyl)phosphoric acid (DEHPA), copper extracted using 179–80
differential columns, graphical solution methods used 448–54
differential extraction 408–11
  process described 408–9
  see also Rayleigh extraction
differential extractors 134, 367–77
  height of extractor 372–7
    analytical solution 374–5
    graphical solution methods 372–4, 451, 453
  material balance for 370
  models used 417
  transfer unit concept 367, 370–2
diffusion 40–55
  Fick's first law 42
  Fick's second law 47–8, 60, 99–100
  multicomponent diffusion 53–5
  steady-state diffusion 44–6
  unsteady-state diffusion 46–9

diffusion (axial dispersion) model 417, 418–28
assumptions made 418
compared with backflow model 433–5
derivation of equations 418–21
effect of entrance sections 424
generalized solutions
linear equilibria 437–9
non-linear equilibria 439, 442–3
programming of solutions 443–4
worked examples 444–5, 446–7
graphical solution for
backmixing in both phases 452
backmixing in one phase 449
material balances 419
measurement of backmixing parameters in 479–82
polydispersivity introduced 472–3
relationship with backflow model 434–5
scale-up using 516–17
simplified solutions 426–8
solutions for concentration profiles 421–4
dilute solutions
electrolytes, diffusivity relationships 52
non-electrolytes, diffusivity relationships 50–2
dimensionless groups, use in instability analysis 168
dispersed phase
determination of 246, 252–3
selection of 125–31
application of criterion 130–1
evaluation of criterion 127–30
formulation of criterion 126
dispersed-phase hold-up
drop size affected by 251
experimental determination of 504
extractor design affected by 503–5
mass transfer affected by 93, 133
dispersion band
collapse with time 293–7
depth varying with flow-rates 286–8, 304
meaning of term 256
modelling of behaviour 289–306
batch settling considered 301–3
collapsing band 293–7
experimental studies 297–8
steady-state band 289–93
stochastic approaches used 300
regions classified 288
structure of 287–8, 289
distillation
countercurrent extraction compared with 381
efficiency terms for 132, 364
drop break-up in agitated systems 247–52
mass transfer in 109–13
drop coalescence
design of industrial extractors affected by 306–16, 507–10
drop-drop coalescence 284–5, 509
drop-interface coalescence 285–306
electrical fields affecting 263–5, 266
experimental factors affecting 260–1
film drainage during 257, 270–84
and Gibbs adsorption 267–8, 269
Marangoni phenomena during 193–4, 268, 270
mass transfer during 124–5, 193–4
physico-chemical properties affecting 265, 267
single drops 257–70, 507–9
size distribution effects 473–4
temperature gradient affecting 268, 270
vibration affecting 263
see also coalescence times
drop drag coefficients, effect of interfacial turbulence 194–5
drop-drop coalescence 284–5, 509
experimental determination of coalescence time 263, 264, 509
drop formation
jet break-up resulting in 80–1, 113
effect of Marangoni phenomena 183–93
laser-tracer technique used to study 96, 108
mass transfer during 95–8
drop-interface coalescence 285–306
experimental determination of coalescence time 262–3
experimental observation of dispersion band 297
droplet (mass transfer) models 223–7
circulating drops 224–7
stagnant drops 224
drop oscillation criteria 93–4
drop phenomena
mass transfer associated with 87–125
measurement of 228–30
drop shape 94–5
drop size 78, 81–2, 87–90
coalescence affected by 261
drops produced from jet break-up 81–2, 190–3
prediction of 249–51
see also Sauter mean drop diameter
drop size distribution, extractor design affected by 505–7
dual feeds (in countercurrent extraction) 358
dual-solvent extraction 343, 381–95
analytical solutions used 392–4
backmixing in 468–70

batch processes used 404, 405
effect of solvent ratio on number of stages 393
feed-point compositions matched 385, 387
graphical solution used 383-9
limiting flows in 390-2
material balances in 384, 389
multicomponent separation in 402
numerical methods used 394-5, 468-70
operating diagrams for 386
operating line intersections 391
process described 381, 383
reflux used 387-9
in staged extractors 384
dual-solvent fractional extraction 381-95

eddy cell models 73, 120, 219
eddy diffusion models 214, 218-19, 221
eddy residence time distribution models 63, 213
listed 215-17
electrical fields, drop coalescence affected by 263-5, 266
electrically enhanced extraction
drop coalescence in 263-5, 266
Marangoni phenomena in 182-3
electrolyte solutions
correlations of activity coefficients for 13-17
diffusivity in concentrated solutions 52-3
diffusivity in dilute solutions 52
electrostatic coalescers 263
emulsification
extractor design affected by 495, 497-8
reactions causing 179
equilibrium compositions, computation of 19-20
equilibrium data
computation of 18-20
graphical representation of 24-6, 348-50
measurement of 20-7
cell equilibration method used 20-2
continuous measurements used 22
titration method used 22-4
equilibrium distribution diagrams 56, 349
countercurrent extraction processes 356
single-extraction processes 351
equilibrium thermodynamics 5-7
equimolal counterdiffusion 44, 45
equipment selection chart 494-5, 499
examples of use 500-3
error function 61
ethyl-acetate-water system, instability of 173, 176, 177
extraction equipment
classification of 493
cost comparison of 499, 500
design of 503-14
hydrodynamics 503-12
mass transfer prediction 512-13
procedures used 513-14
requirements 493, 496-9
performance compared 500
selection of 492-503
chart used 494-5, 499
examples 500-3
Eyring's theory of absolute reaction rates 50

fan model (for non-woven fibre mat) 328
Fenske-Underwood equation 388
fibre bed coalescers 320-34
characteristics 331-2
droplet interception by 326
fibre mat used 320-2
practical unit described 332-4
fibre mat (non-woven)
effect of needling 320
fan model for 328
flow analysis for 323-7
random network models for 330, 331
SEM micrographs 321, 330
trajectory models for 329, 331
use in secondary coalescers 320-34
mechanism of operation 320, 322-3
fibres
droplet interception by 326-8
flow around 324-5
Fick's first law of diffusion 42
Fick's second law of diffusion 47-8, 60, 99-100
film drainage 257, 275-80
approach of drop prior to film drainage 270-5
streamlines due to motion 273, 274
critical film thickness for rupture 281-4
experimental observations 275, 276
pressure considerations 278, 279
simplified models 275-7, 508
symmetrical drainage assumed 280
film-flow centrifugal extractors 579-80
characteristics 494-5
film (mass transfer) model 59-60, 120, 211-12
chemical reactions modelled by 136, 138-40, 396-7
film-penetration model 72-3, 120, 212-13
film rupture 281-4, 508
flood-point 505
flux relations 40-6
foam, drainage from 298-9
Fortescue-Pearson eddy cell model 120, 219
fractional extraction

fractional extraction—*contd*
batch extraction 410, 411
breaking drops in agitated systems 111–12
dual-solvent extraction 381–95
single-solvent extraction 379, 410
stagnant drops 103
Francis miscibility-immiscibility diagram 29
free energy, criteria for equilibrium involving 5
free energy minization method, equilibrium compositions determined by 20
froth-foam transition 288, 297
Fuchs-Stechkina fan model (for non-woven fibre mat) 328
furfural, miscibility with other solvents 29

gamma functions 66, 97, 214, 216
Gibbs adsorption, drop coalescence and 267–8, 269
Gibbs-Duhem equation 7–8
Gibbs free energy, *see* free energy
Godfrey miscibility-immiscibility diagram 28
Gordon's (diffusivity) equation 52
Graesser extractors, *see* RTL extractors
graphical solution methods
axial dispersion 448–57, 474–5
plug flow
countercurrent extraction 354–8
countercurrent extraction with reflux 379–81
crossflow extraction 352–3
differential extractors 372–4
dual-solvent extraction 383–8
graticule-displacement technique
interfacial effects studied by 236, 237
solute concentrations in drops measured by 228–30
gravitational instabilities 176–7
Grober's equation (for mass transfer) 104

Hadamand-Rybczynski equation 305, 307, 308
Handlos-Baron (oscillating-drop mass-transfer) model 107, 226–7
Harkins-Brown correction factor 89
Harriott's surface-renewal model 73, 214, 215
Hausen efficiency 132, 364
countercurrent extractors 132, 364, 365, 366
crossflow extractors 366
mixer-settlers 576
relationship to other efficiencies 366
Haydon's ('kicking drop') model 196
Hayduk-Laudie relationship 51
Henstock-Hanratty eddy diffusion model 218, 221
heptane, miscibility with other solvents 29
heterogeneous reactions
mass transfer rates affected by 397–8
operating diagram compared with that for simple extraction 373
hexamethyldisiloxane jets, drop size from jet break-up 192–3
hexane-aniline system, instability of 176
Higbie's penetration model 60, 203, 211, 215, 220, 222
see also penetration (mass transfer) models
Holley-Mott principle (for phase circulation in mixer-settlers) 577–8
homogeneous reactions, mass transfer rates affected by 395–7
L'Hôpital's rule 45, 104
horizontal settlers, design of 308–16, 574
HTU, *see* transfer units, height of
hydrodynamics
drop phenomena 87–95
extractor design affected by 503–12
jet phenomena 75–82
hydrometallurgy, Marangoni phenomena in 179–80

inclined-plate crossflow settlers 317–18
inter-droplet coalescence, *see* drop-drop coalescence
interfacial area
computation of 110–11, 244, 506
effect of Marangoni phenomena 183–95
experimental determination of 506–7
interfacial boundary layer theory 66–72
interfacial convection
factors affecting 158, 161–2, 172
macroscale turbulence 195–6
mass transfer affected by 195–8, 230–9
mechanisms of 158–62
microscale turbulence 196–7
see also Marangoni phenomena
interfacial tension, phase separation affected by 245
interfacial tension gradient, perturbation caused by 158, 160, 510–12
interphase mass transfer 40–140
nomenclature used 140–7
ion interaction models 14–15
iso-activity method, equilibrium compositions determined by 19–20
isobutanol-water system, instability of 176, 177, 182–3

Janecke diagrams
basis of 350

countercurrent extraction 356
multistage crossflow extraction 353
single-solvent fractional extraction 382
single-stage extraction 351
jet break-up, drop formation from 80–1, 113
drop coalescence 193–4
drop drag coefficients 194–5
drop size calculations 190–3
effect of Marangoni phenomena 183–93
stability analysis 186–90
jet contraction 79–80
jet lengths, mass transfer affected by 76, 77–9, 184–6
jet phenomena
hydrodynamics 75–82
mass transfer associated with 75–87
mass transfer in and around jets 84–7
mass transfer inside jets 82–7
parabolic flow 83
plug flow 82–3
jet stability, mass transfer affecting 186–90
jetting velocity 77

Karr reciprocating plate extractors 556–61
cost comparison 500
performance comparison 500
see also reciprocating plate columns
'kicking' drops 195–6, 510
kinematic viscosity, interfacial convection affected by 161
King's boundary layer model 74, 218, 221
Kishinevsky's eddy diffusion model 73, 214, 222
Kolmogoroff's theory of isotropic turbulence 247, 549
Koppel's surface-renewal model 213, 216, 221
Kronig–Brink (circulating-drop mass-transfer) model 105, 224–6
evalualtion of parameters for 106, 225–6
Kühni columns 540–2
backmixing in 484, 485, 541–2
characteristics 494–5
enhanced coalescence plates used 542, 561
geometry 541
performance comparison 500
sizes available 541
Kuwabara–Happel (fibre) cell model 325, 327

laminar flow boundary layer model 66–72
Lamont–Scott eddy cell model 73, 120, 219
Lamourelle–Sandall eddy diffusion model 218, 221
Laplace's equation 278
laser Doppler anemometry (LDA) techniques 219, 223
laser–dye technique
applications 96, 108, 198–201, 222, 223, 231–5, 275, 276
experimental details 232–3
tracer dye used 232
Leffler–Cullinan relationships 52, 54
lever arm rule 352
Levich's eddy diffusion model 214, 221
Lewis–Whitman two-film model 59–60, 210–11, 220
see also film (mass transfer) model
liquid–liquid equilibria 1–35
computation of 18–20
graphical representation of data 24–6, 348–50
measurement of 20–7
cell equilibration method used 20–2
continuous measurements used 22
titration method used 22–4
nomenclature used 35–8
liquid–liquid extraction, meaning of term 244, 343
Luwesta (Westfalia) extractor 580

McCabe–Thiele (stage calculation) method 357–8, 381
examples 356, 382, 570, 571
maintenance requirements 495, 498
Marangoni disturbances, meaning of term 162
Marangoni instabilities, meaning of term 162
Marangoni number 168, 512
jet stability analysis using 187–8
Marangoni phenomena 157–204
applications of stability analysis 167–83
dimensionless groups used 168
mass transfer with chemical reaction 179–80, 181
mass transfer in electric fields 182–3
mass transfer with heat effects 172–9
solute transfer across flat interface 167–72
theoretical studies 180, 182
and drop coalescence 193–4, 268, 270
experimental observation of 172, 176, 178, 181, 198–201, 203, 231–5
heat-induced instabilities 172–9
interactions between surface-renewal rates and mass-transfer rates 198–204
interfacial area affected by 183–95
mass transfer affected by 163, 195–8, 510–12
mathematical analysis of instability 163–6
general boundary conditions 166

mathematical analysis of instability—*contd*
model analysed 164
meaning of term 157
mechanisms of interfacial convection 158–62
nomenclature used 204–5
and surface renewal 198–204
types of 162–3
Marangoni thin-film phenomena 162–3
mass transfer 40–140
with chemical reaction 135–40
and drop phenomena 87–125, 223–30
in and around detached drops 99–123
during drop coalescence 124–5
during drop formation 95–8
hydrodynamics 87–95
zero-resistance case 104
interfacial effects on 195–8, 230–9
and jet phenomena 75–87
Marangoni phenomena affecting 195–8, 510–12
models for 59–74, 210–23
experimental verification 223
listed 220–1
nomenclature used 140–7, 239–40
mass transfer coefficients 55–8
determination from measured concentration profiles 474–7
graphical method 474–5
numerical method 475–6
practical considerations 476–7
determination using tracer injection methods 477–84
applications 483–4
continuous tracer injection used 477–8
unsteady tracer injection used 479–83
partial mass transfer coefficients 227–30
prediction for design use 512–13
see also partial mass transfer . . .
materials of construction 495, 498
matrix method (for axial dispersion)
applications of method 467–8
derivation of equations 460–3
matrix formulation 463–5
method of solution 465–7
methyl carbonate, water drops in, coalescence of 285
methyl ethyl ketone (MEK)–water system, instability of 174, 176, 177
methyl isobutyl ketone (MIBK), dispersed phase with water 253
microgravity experiments, interfacial turbulence studied during 177, 178
miscibility–immiscibility diagram 29
mixed-solvent systems, solvent selection for 30–1
mixer–settlers 573–9
centrifugal extractors using 494–5, 580
characteristics 494–5
cost comparison 500
design of 245–6, 513–14, 574–5
dual-solvent extraction in 384
horizontal mixer–settlers 574
characteristics 494–5
interfaces in 343
mass transfer in 124
mixing equipment used 247
models used 417
performance comparison 500
phase entrainment in 314, 315, 316–34
techniques to reduce 316–34
pilot-testing of 575–8
batch tests 576–7
continuous method 577–8
pumps used 247, 316, 573, 574
scale-up for 575, 578–9
vertical mixer–settlers 573
characteristics 494–5
see also pump-mix mixer–settlers
mixing equipment 247
mixing phenomena 247–52
nomenclature used 334–7
molality, meaning of term 13
molecular diffusion 42
molecular diffusivities 49–55
multicomponent diffusion 53–5
multicomponent separations
countercurrent extraction 400–2
crossflow extraction 399–400
dual-solvent systems 402
operations used 398–9
multi-impeller columns 538–43
characteristics 494–5
pilot-testing of 515
scale-up for 542–3
see also Kühni . . . ; Oldshue–Rushton columns
multinozzle distributors, drop size correlations for 89–90
multiple-contact batch extraction 403
multistage crossflow extraction
analytical method 353–4
graphical methods 352–3
Murphree efficiencies 132, 357, 362–4
local/point efficiencies 134, 363
relationship to other efficiencies 366

Navier–Stokes equations 165
Nernst's (diffusivity) equation 52
Newman's equation (for mass transfer) 104
new-surface-elements model
mass transfer during drop formation 95, 96
mass transfer in oscillating drops 107, 108, 119
nitrobenzene

water drops in
drag coefficent 194–5
terminal velocity 90–1
non-agitated systems
mass transfer around drops in 113–20
internally circulating spheres 115, 116, 117–19
internally stagnant spheres 114–15, 116
oscillating drops 116, 119–20
non-electrolytes
diffusivity in concentrated solutions 52
diffusivity in dilute solutions 50–2
non-woven fibre materials, coalescers using 320–31
NRTL equation 8–9
applications 32
data sources 34
determination of binary parameters for 9
NTS, *see* theoretical stages, stages of
NTU, *see* transfer units, number of
nuclear fuel reprocessing, Marangoni phenomena in 180
numerical solution methods
axial dispersion 457–70, 475–6
plug flow in dual-solvent extraction 394–5

oil–water separators 307, 317
Oldshue–Rushton columns 538–40
backmixing in 539
geometry 539
hold-up data for 539
mass transfer data for 539
operating diagrams 361
axial dispersion
backmixing in both phases 452, 457
backmixing in one phase 449, 455
balance line shown 424
non-linear equilibria considered 439
countercurrent extraction 347, 356, 360–1
with reflux 382
dual-solvent extraction 386–7, 391
with reflux 389
multicomponent countercurrent extraction 401
polydisperse system 371
Orthogonality, Principle of 102
oscillating drops, mass transfer in 106–9, 226–7
Oswald-fraction-based theories, phase inversion predicted by 253

packed columns 521–8
characteristics 494–5
cost comparison 500
dispersed-phase hold-up determined for 525
drop break-up/coalescence in 228, 473
droplet terminal velocity determined for 524
mass-transfer coefficients determined for 526–7
models used 417
performance comparison 500
pilot-testing of 514, 521–3
scale-up for 527–8
parabolic flow, mass transfer in jets with 83
partial mass transfer coefficients 227–30
meaning of term 228
passive separation 303
penetration (mass transfer) models 60–2, 73, 211
breaking drops in agitated systems 112, 113
during drop coalescence 124
in and around jets 84–7
with varying surface renewal 62–6, 74
perforated-plate centrifugal extractors 579
characteristics 494–5
perforated-plate columns 566–72
applications 566
characteristics 494–5
geometry 567–8
hydrodynamics 568–9
McCabe–Theile diagram for 570, 571
mass transfer in 75, 93, 124, 569–70
mixing models for 134
performance comparison 500
scale-up for 571–2
Perlmutter's surface-renewal models 73, 213, 215, 220
petroleum industry applications 307, 317
phase continuity
determination of 246, 252–3
importance first recognised 313
phase diagrams 1–5
see also Janecke diagrams
phase entrainment 314, 315, 316
techniques to reduce 316–34
phase inversion 246, 254–5
phase ratios, extractor performance affected by 497
phase rule 1
phase separation rates, factors affecting 245
physical properties, extractor selection affected by 494, 496
picric acid, emulsification of 179
pilot test procedures 514–15
Pinczewski–Sideman model 74
Pitzer's ion interaction model 15–17
plait point, meaning of term 2, 350
plate columns
dispersed phase selection for 125–31
mass transfer in 75, 93, 124

plate columns—*contd*
  see also perforated . . . ; pulsed plate columns
plug flow
  mass transfer in jets with 82–3
  scale-up procedure allowing for 516–17
plug flow analysis 345–411
  batch extraction 403–11
  countercurrent extraction 354–61
    multicomponent separations 400–2
  countercurrent extraction with reflux 379–81
  crossflow extraction 352–4
    multicomponent separations 399–400
  dual-solvent extraction 381–95
    multicomponent separations 402
  effect of chemical reaction 395–8
  multicomponent separations 398–402
  multistage crossflow extraction 352–4
    multicomponent separations 399–400
  nomenclature used 411–14
  single-stage extraction 350–2
Podbielniak extractor 579
polydispersivity
  and axial dispersion 470–4
  droplet size distribution effects 473–4
power requirements
  pulsed columns 528, 555
  rotary disc contactors 535
Prasher–Fricke eddy diffusion model 218, 221
pressure mixers 247
primary break time 295
Prochazka–Jiricny numerical method 394, 468
pulsed columns, drop behaviour in 194
pulsed packed columns 528–31
  characteristics 494–5
  cost comparison 500
  hold-up data determined for 528, 530
  mass transfer coefficients predicted for 530–1
  performance comparison 500
  pulsation generators used 528, 529
  scale-up for 531
pulsed plate columns 543–56
  backmixing in 485, 550–4
  characteristics 494–5
  cost comparison 500
  droplet size distribution for 549–50
  effect of column diameter 554–5
  flood-point data for 546–7
  geometry 543–4
  hold-up data for 546, 548
  hydrodynamics 545
  mass transfer data for 554
  perforated plates used 544–5
  performance comparison 500
  pilot-testing of 515
  power requirements 555
  scale-up for 555
pump-mix mixer-settlers 247, 573, 574
  characteristics 494–5
  drop size computation 252
  scale-up for 578

Quadronic extractor 579
quasi-steady-state droplet technique 201, 203, 235–7
quaternary systems, data representation for 26

radioactivity, extractor selection affected by 495, 498
raining-bucket extractors, *see* RTL extractors
Ramababu–Narsimhan surface-renewal model 213–14, 217
random network models (for non-woven fibre mat) 330, 331
Raschig rings 522
Rayleigh extraction 409–11
Rayleigh–Taylor instability 171
RDC, *see* rotary disc contactors
reactions, *see* chemical reactions
reciprocating plate columns 556–61
  characteristics 494–5
  geometry 556
  hydrodynamics 557–9
  materials of construction 557
  pilot-testing of 515
  plate assembly 558
  scale-up for 559–61
reflux, countercurrent extraction with 379–81
regular solution model
  liquid–liquid equilibria predicted using 11
  solvents selected using 28–31
rejection ratios, meaning of term 388
Renon–Prausnitz model 8–9
residence time
  distribution models 63, 213
  extractor performance affected by 495, 497
Robatel centrifugal extractors 580
Rod's graphical solution method 448–9
  worked examples 449, 452
rotary agitated columns 538–43
  pilot-testing of 515
  see also multi-impeller columns
rotary annular columns 542
rotary disc contactors (RDCs) 533–8
  applications 533
  axial dispersion data 536–7

characteristics 494–5
cost comparison 500
geometry 534
mass transfer performance 536
performance comparison 500
pilot-testing of 515
power requirements 535
scale-up for 535, 537–8
rotary film extractors 572–3
characteristics 494–5
number of stages required 573
pilot-testing of 572
rotating bowl centrifugal extractor 580
rotating bucket extractors, *see* RTL extractors
RTL extractors 531–2
characteristics 494–5
cost comparison 500
internal construction 532
mode of operation 532
performance 533
performance comparison 500

salt effect 548–9
Sauter mean drop diameter 110–11, 244, 496
computation of 248–9, 250, 506
Sawistowski-James ('kicking drop') model 196
scale-displacement technique
interfacial effects studied by 236, 237
solute concentrations in drops measured by 228–30
scale-up procedures 516–17
asymmetric rotary disc columns 566
axial dispersion allowed for 516
mixer-settlers 575, 578–9
multi-impeller columns 542–3
packed columns 527–8
perforated-plate columns 571–2
plug flow basis 516–17
pulsed packed columns 531
pulsed plate columns 555
reciprocating plate columns 559–61
rotary disc contactors 535, 537–8
RTL extractors 532
spray columns 520
Scheele-Meister correlation (for drop size) 88, 89
Scheibel columns 561–3
characteristics 494–5
schlieren photographs, interfacial instabilities studied by 159, 173, 182, 197
Schmidt number 168
secondary coalescers, fibre mat used in 320–34
sedimentation, settler design using 307
selection (of extraction equipment) 492–503
chart 494–5, 499
examples of use 500–3
separation of dispersions 255–334
application to design of industrial settlers 306–16
drop coalescence in 257–70, 284–306
film drainage in 270–84
phase entrainment used 316–34
processes occurring in 255–6
SHE extractors 542
sieve tray columns, *see* perforated-plate columns
single-contact batch extraction 403
single drops, coalescence of 257–70
single-solvent fractional extraction 379–81
batch processes used 405
single-stage extraction, mass balance for 350–2
sinuous region (in jets) 76
slip velocity 93, 504
solids-handling capability 497
solute diffusivity, interfacial convection affected by 161
solvents
extractor selection affected by properties 495–6, 498
selection of 27–31
empirical methods used 28
factors affecting 27–8
regular solution model methods used 28–31
UNIFAC model methods used 31
space limitations (in extractor selection) 495, 498
spherical droplets, steady-state diffusion from 44–5
spray columns 517–20
characteristics 494–5
dispersed phase selection for 125–31
drop coalescence in 194
geometry 517–18
hold-up in 93, 519
mass transfer in 227–8
scale-up for 520
spun fibres, void structures in 177, 179
stage efficiencies 131–5, 362–5
Haussen efficiency 132, 364–5
Murphree efficiencies 132, 362–4
overall stage efficiency 131–2, 362
relationship between efficiencies 366
stages, relationships with transfer units 369, 375
stagewise contactors
graphical solution methods used 454–7
backmixing in both phases 456–7

stagewise contactors—*contd*
backmixing in one phase 454–6
stagnant drops, mass transfer in 99–105, 224
steady-state diffusion 44–6
equimolal counterdiffusion 45
from sphere 44–5
unimolal unidirectional diffusion 45–6
stepwise coalescence 260
Sternling–Scriven model 164, 167
experimental results compared with predictions 171
stochastic trajectory models (for non-woven fibre mat) 329, 331
Stokes' law, drag evaluated by 49–50
Stokes' stream function, definition of 271
Stönner–Wöhler equation 303
strong electrolytes, activity coefficients of 14–17
surface-active agents
coalescence affected by 261, 268, 507, 509
disperse-phase mass transfer affected by 131
droplet oscillation criteria affected by 94
jet hydrodynamics affected by 78, 79, 81
Marangoni phenomena affected by 161–2
surface-active contamination
drop size affected by 88
drop terminal velocity affected by 92
mass transfer affected by 128
surface age distribution functions 64–6, 215–17
surface extension by new-element addition 95, 96, 107
see also new-surface-elements model
surface-renewal models 62–6, 73–4, 211–12, 213–14
age/residence time distribution models used 213–14, 215–17
combined with film models 72–3, 212–13
compared with turbulent diffusion models 222
surface-renewal rates, interactions with mass-transfer rates, effect of Marangoni phenomena 198–204
surface-stretch model
mass transfer during drop formation 95, 96
mass transfer in oscillating drops 107–8, 119
surface tension driven flows, interfacial convection caused by 158–62

Tellerettes (column packing) 522–3
terminal velocity (of drops) 90–3
ternary liquid–liquid systems
interfacial instability in 169, 170
molecular diffusion in 53–4
temperature dependence of equilibria in 3, 4
Type I 2, 3, 348
Type II 2, 3, 348
Type III 2, 3, 348
Type IV 5
ternary phase diagrams 2–3, 24–6, 348–50
interpolation of tie lines for 27
rectangular coordinates used 24–5, 349
solutes-free basis 25–6
solvent-free basis 25, 26, 349, 350
see also Janecke diagram
theoretical stages, number of
calculation for 133–4, 360
extractor selection affected by 494, 496
thermally induced interfacial instabilities 172–9
thin-film Marangoni phenomena 162–3
drop coalescence affected by 194
jet-break-up predictions based on 184
throughput ranges, extractor selection affected by 494, 496
titration method, liquid–liquid equilibria measured using 22–4
toluene
water drops in
coalescence of 285
graticule-displacement technique used to study 230
laser–tracer technique used to study 198, 199, 200, 202, 203, 204, 233–5
Toor–Marchello film-penetration model 72–3, 212–13, 220
tracer injection methods (for backmixing parameters) 477–84
applications 483–4, 549, 551
continuous tracer injection 477–8
probes for 477
unsteady tracer injection 479–83
trajectory computations (for crossflow settlers) 308
trajectory models (for non-woven fibre mat) 329, 331
transfer unit concept 367, 370–2
transfer units
and backmixing 425–6
graphical representation of 374
height of 372
and mixing in a stage 375–7
number of 133, 370
analytical determination of 374–5
graphical determination of 373–4
overall transfer units 371–2
relationships with stages 369, 375

tributylphosphate (TBP), uranium extraction using 180, 204, 236, 237, 238
triethylene glycol, miscibility with other solvents 29
TriExtractor 541
turbine contactors 538–43
  phase continuity in 254
  see also multi-impeller columns
turbulence intensity, drop size affected by 249
turbulent diffusion models 74, 214, 218–19, 221
  compared with surface-renewal models 222
two-film (mass transfer) model 59–60, 210–11
  chemical reactions modelled by 136, 138–40, 396–7
  combined with surface-renewal models 72–3, 212–13
  see also film model
two-solvent systems, solvent selection for 30
Type I systems 2, 3, 348
Type II systems 2, 3, 348
Type III systems 2, 3, 348
Type IV systems 5

UNIFAC (LLE) model 18
UNIFAC model
  determination of parameters for 12
  liquid–liquid equilibria predicted using 12–13
  solvents selected using 31
UNIFAC (VLE) model 19
unimolal unidirectional diffusion 44, 45–6
UNIQUAC (COMMON) model 18
  applications 32
UNIQUAC model 9–10
  data sources for 34
  determination of parameters for 10
UNIQUAC (SPECIFIC) model 18
  applications 32
unsteady-state diffusion 46–9
  with constant-time surface renewal 60–2
  with varying surface renewal 62–6
uranium, extraction of 180, 204, 236, 237, 238

varicose region (in jets) 76
vertical settlers, design of 306–8, 573
vibrating plate extractor 556
  see also reciprocating plate columns
vibration, drop coalescence affected by 263
viscosity
  drop coalescence affected by 267
  extractor performance affected by 496
volumetric mass transfer coefficients 58

water, miscibility with other solvents 29
weak electrolytes, activity coefficients of 17
Weber number 248
wedge-shaped dispersion band model 309–10, 509–10
Westfalia centrifugal extractors 580

The manufacturer's authorised representative in the EU for product safety is Oxford University Press España S.A. of el Parque Empresarial San Fernando de Henares, Avenida de Castilla, 2 – 28830 Madrid (www.oup.es/en or product.safety@oup.com). OUP España S.A. also acts as importer into Spain of products made by the manufacturer.

www.ingramcontent.com/pod-product-compliance
Ingram Content Group UK Ltd.
Pitfield, Milton Keynes, MK11 3LW, UK
UKHW020851150926
475595UK00003B/117
* 9 7 8 0 1 9 8 5 6 2 3 6 8 *